ENERGY METHODS AND FINITE ELEMENT TECHNIQUES

ENERGY METHODS AND FINITE ELEMENT TECHNIQUES

Stress and Vibration Applications

MUHSIN J. JWEEG
Al-Farahidi University, Iraq

MUHANNAD AL-WAILY
University of Kufa, Iraq

KADHIM KAMIL RESAN
Mustansiriyah University, Iraq

ELSEVIER

Elsevier
Radarweg 29, PO Box 211, 1000 AE Amsterdam, Netherlands
The Boulevard, Langford Lane, Kidlington, Oxford OX5 1GB, United Kingdom
50 Hampshire Street, 5th Floor, Cambridge, MA 02139, United States

British Library Cataloguing-in-Publication Data
A catalogue record for this book is available from the British Library

Library of Congress Cataloging-in-Publication Data
A catalog record for this book is available from the Library of Congress

ISBN: 978-0-323-88666-6

For Information on all Elsevier publications
visit our website at https://www.elsevier.com/books-and-journals

Publisher: Matthew Deans
Acquisitions Editor: Dennis McGonagle
Editorial Project Manager: Fernanda A. Oliveira
Production Project Manager: Sojan P. Pazhayattil
Cover Designer: Matthew Limbert

Typeset by MPS Limited, Chennai, India

Contents

II

Finite element method

About the authors

Dr. Muhsin J. Jweeg is a professor of applied mechanics. As a chartered engineer, he is a fellow of the Institution of Mechanical Engineers (UK) and currently the Vice Chancellor of Scientific Affairs, Al-Farahidi University, Iraq. Before this, he was a Dean of the College of Engineering—Al-Nahrain University for Oct 2006–Mar 2013. Prof. Jweeg published many journal papers and three books. He attended many international and national conferences and gained an international prize in India for the best-published paper in IMECH 1995 and has received many national awards. Several PhD and MSc students graduated under his supervision. Most of them are working in the field of energy and finite element applications.

Dr. Muhannad Al-Waily is a professor in the Mechanical Engineering Department, Faculty of Engineering, University of Kufa. His areas of specialization include applied mechanics–vibration analysis, composite material, crack analysis, health monitoring of structures, and prostheses and ortheses engineering applications. He supervised many PhD and MSc students, and most of them are working in stress analysis, vibrations, and control in different mechanical applications. Prof. Al-Waily published many papers in international and national journals and attended many national conferences. He was also an Associate Editor of the Applied Mechanics Research Center, International Energy and Environment Foundation (IEEF), Najaf.

Dr. Kadhim Kamil Resan is a professor at the Materials Engineering Department at the Mustansiriyah University. He received a PhD in applied mechanical engineering in 2007 from the University of Technology. His primary research interests are prostheses and ortheses, stress analysis, and biomaterials. Prof. Resan published many journal papers and seven patents. He attended international and national conferences and gained international prizes and many national prizes on different occasions. He is a member of international societies, including The American Society of Mechanical Engineers (ASME) and the International Society for Prosthetics and Orthotics (ISPO). He is a member of national scientific associations.

The authors regard the book as being of interest to the engineers in industry sector, senior undergraduates, and the PhD and MSc students working in finite element simulation for stress and vibration problems. This book also aims to have additional benefits to the researchers wishing to study the application of energy and finite element princilpes to a wide range of significant engineering problems.

Preface

This is the first edition of the *Fundamentals of Energy and Finite Element Methods* as numerical tools for the evaluation of displacements, strains, stresses, and natural frequencies. These techniques are still valid and increasing in use and developed very fast with the advances in computer capabilities. Many packages have been built since the invention of the Finite Element Technique in 1950 and simply the solution needs to change the input data to accommodate the problem in addition to obtaining a very clear picture of the stress distribution. The book is divided into two parts. Part 1 is devoted to present the energy methods in applied mechanics, and Part 2 is concerned with the finite element section.

Part 1 is organized into four chapters. Chapter 1, Fundamentals of Energy Methods, gives the fundamentals of total potential energy and complementary energy principles and their applications in solving simple problems including the bars, beams, and torsion of rectangular sections. The basic equations of the total potential energy are developed. The concept of work, energy, and the minimization principles and principle of virtual work was demonstrated for simple structural analysis. Chapter 2, Direct Methods, illustrates the basic classical approximate direct methods, such as Galerkin, and Rayleigh–Ritz method (R.R.M.). Examples cover the application of total potential energy and the formulation of bar under extension, frame using different models, and approximate functions for static analysis only. Chapter 3, Application of Energy Methods to Plate Problems, presents the application of total potential energy in static analysis of plates under bending, extensional, and bucking. The total potential energy of the plate equations is formulated by presenting different applications. Finally, Chapter 4, Energy Methods in Vibrations, presents the dynamic analysis of the previous cases presented in Chapters 1–3. The Eigenvalue problem is formulated and the relevant differential equations are set using Rayleigh's, R.R.M., Galerkin methods.

Part 2 is devoted to the finite element part. This part contains 10 chapters, and they are arranged in a logical order. Chapter 5, Introduction to Finite Element Method (F.E.M.): Bar and Beam Applications, presents the concept of the finite element technique by deriving the relevant equilibrium equations depending upon the total potential energy principles for one-dimensional problems bar and beam cases. The shape functions, element stiffness matrices, element force vector. The system stiffness matrix, system load vector, application of boundary conditions, and the standard solution required to obtain, displacement, strain, and stresses. Different cases were presented starting from bar case, plane frames, and bending example. Chapter 6, Two-Dimensional Problems: Application of Plane Strain and Stress, is concerned with the finite element modeling of plane stress/ strain problems. The examples are explained in detail starting from the assumption of displacement models, formulation of stiffness and load vector elements and the formulation of system stiffness and load vectors and the application of boundary conditions. The examples are also including the stiffeners' effects on the problem and their contribution on the whole equilibrium equations. Chapter 7, Torsion problem, is devoted to the derivative of torsion problems modeling; stiffness and load vector for both solid bars and plan cases were derived. The formulation of axisymmetric problems by finite element technique is presented in Chapter 8, Axisymmetric Elasticity Problems. Chapter 9, Application to Three-Dimensional Elasticity Problems, deals with the formulation of solid bodies and their modeling in the finite element method. The hexahedral and tetrahedral elements were presented. The application of finite elements in vibration analysis was presented in Chapter 10, Application of F.E. to the Vibration Problems. This covers the formulation of the Eigenvalue problem for bar, beam, axisymmetric triangular, and the 8-node solid elements. The concentration was on the formulation of the mass matrix and get used to the previously derived stiffness matrices. The basic equations were based upon the total energy, strain, and kinetic energy principle. The heat transfer problem is demonstrated in Chapter 11, Steady State Heat Conduction. The basic equations of conduction, convection, and radiations were presented. One-dimensional and two-dimensional heat transfer examples were given in this chapter. The derivations of the shape functions depending upon Lagrange interpolation for different types of applications were derived in Chapter 12, Shape Functions Determinations and Numerical Integration, using the intrinsic coordinates system. The numerical integration scheme was also given in this chapter for one-, two-, and three-dimensional applications using the Gaussian scheme of integration. Chapter 13, Higher-Order Isoparametric Formulation, presents the formulation of higher-order elements to be used in the analysis of plate and shell modeling. The introduction to the analysis of orthotropic is also given in this chapter. For the reader to be familiar with the steps of programming and to be aware of what is going in the available software, Chapter 14, Finite Element Programs Structures, presents the finite element program structure. It is restricted to the flow charts construction to give the basic idea of building the finite element computer programs. It is concentrated to plane stress/strain problems static analysis, free vibration, and dynamic analysis. This is useful to the student to enable him to write his own finite element computer program and to the researcher to know what is going on inside the available built-in packages.

This book is a production of the experience of authors in different Iraqi universities in Iraq. Baghdad University, AL-Nahrain University, University of AL-Mustansiriya, University of Technology, Kufa, and lastly Karbala university. The material written in this book was based mainly upon the given lectures and the examples given to the students in addition to some basic textbooks. Most of the problems were exam papers addition to some of them were borrowed from some textbooks for clarification purposes. This book is mainly useful for researchers, postgraduate students, mechanical, civil, engineering, and applied mechanics students, and people who are interested in energy and finite element applications. Some chapters may be useful to be set as a basic course for the undergraduate students.

We express our gratitude to our students who graduated under the author's supervision and who took the courses of energy and finite element courses. Authors are sure that their contribution in giving us a chance to model the complicated structures in a simple way and set in this book. Our colleagues' encouragement, advice, and support Prof. Dr. Mooner H. Tleph, President of Karbala University, Ass. Prof. Dr. Ayad. M. Takak, Dr. Ali M. AL-Hilli, Prof. Dr. Jumaa Ch. Salman, and Prof. Dr. Mohsin A. AL-Shammari, Dr. Wedad I. Majeed, and Dr. Ebtihal S. Abbas University of Baghdad are never forgotten. Finally, special thanks to engineer Diyaa H. J. Al-Zubaidi who revised and achieved many calculations of some examples presented in the book chapters.

Any comments, criticism in any way concerning the presentation, examples, philosophy of the book are appreciated and will be taken into consideration in the second edition. Deep gratitude to every reader who pays attention and wishes to draw authors' attention to correct the errors in the next edition using the email muhanedl.alwaeli@uokufa.edu.iq

PART I

Energy Method

CHAPTER

1

Fundamentals of energy methods

1.1 Principles of virtual work (P.V.W.)

We suppose a particle is moving from A to B as shown in Fig. 1.1,
The work done is given by:

$$\Delta w = \overline{F}.\Delta\overline{r}$$

Where $\overline{F}$ is the force vector, and $\Delta\overline{r}$ is the position vector.
or,

$$\Delta w = F_x.\Delta u + F_y.\Delta v + F_z.\Delta w$$

Or,

$$w|_A^B = \int_A^B \left(F_x.du + F_y.dv + F_z.dw\right)$$

or,

$$w|_A^B = \int_A^B \overline{F}.d\overline{r} \tag{1.1}$$

Where F_x, F_y, and F_z are the forces in the x, y, and z directions, respectively, and the corresponding displacements are u, v, and w.

We suppose that we have a system of N particles, then,

$$\Delta w = \sum_S^N \overline{F}_s.\Delta\overline{u}_s = \sum_{s=1}^N \left(F_{x_s}.\Delta u_s + F_{y_s}.\Delta v_s + F_{z_s}.\Delta w_s\right) \tag{1.2}$$

We suppose the 3N coordinate system is required to define the configuration of the particle system related by m equations of constraint, then the number of degrees of freedom (d.o.f.) is reduced from 3N to n,

$$n = \underbrace{3N}_{\text{coord.}} - \underbrace{m}_{\text{constraints}}$$

Energy Methods and Finite Element Techniques
DOI: https://doi.org/10.1016/B978-0-323-88666-6.00016-2

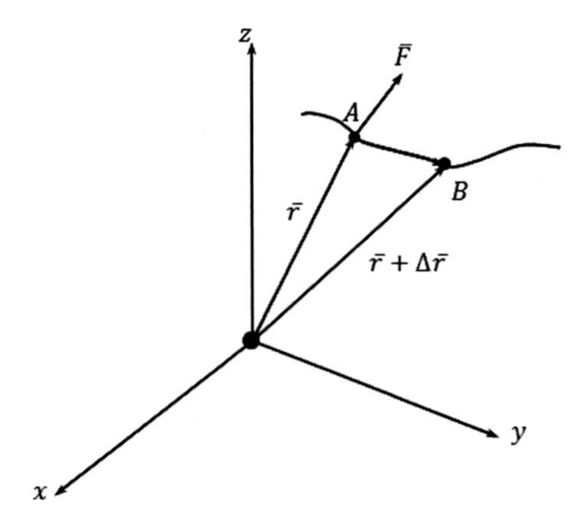

FIGURE 1.1 A particle moving from position A to position B.

Where n is the number of constrained d.o.f.

Denote the n generalized coordinate by $q_1, q_2, \ldots, q_n$, then,

$$u_s = f_s(q_1, q_2, \ldots, q_n)$$
$$v_s = g_s(q_1, q_2, \ldots, q_n)$$
$$w_s = h_s(q_1, q_2, \ldots, q_n)$$

Then,

$$\Delta u_s = \frac{\partial f_s}{\partial q_1}\Delta q_1 + \frac{\partial f_s}{\partial q_2}\Delta q_2 + \ldots = \sum_{i=1}^{n}\frac{\partial f_s}{\partial q_i}\Delta q_i$$

Similarly,

$$\Delta v_s = \sum_{i=1}^{n}\frac{\partial g_s}{\partial q_i}\Delta q_i$$

And,

$$\Delta w_s = \sum_{i=1}^{n}\frac{\partial h_s}{\partial q_i}\Delta q_i$$

Therefore

$$\Delta w = \sum_{i=1}^{N}\left[F_{x_s}\left(\sum_{i=1}^{n}\frac{\partial f_s}{\partial q_i}\Delta q_i\right) + F_{g_s}\left(\sum_{i=1}^{n}\frac{\partial g_s}{\partial q_i}\Delta q_i\right) + F_{z_s}\left(\sum_{i=1}^{n}\frac{\partial h_s}{\partial q_i}\Delta q_i\right)\right]$$

Since it does not matter when we sum on i or s first, so we can write,

$$\Delta w = \sum_{i=1}^{n}\left\{\sum_{s=1}^{N}\left(F_{x_s}\frac{\partial f_s}{\partial q_i} + F_{g_s}\frac{\partial g_s}{\partial q_i} + F_{z_s}\frac{\partial h_s}{\partial q_i}\right)\Delta q_i\right\} \tag{1.3}$$

Or,

$$\underset{\text{scalar}}{\underset{\uparrow}{\Delta w}} = \sum \underset{\text{generalized forces corresponding to G.C.}}{\underset{\uparrow}{Q_i}} \cdot \underset{\text{generalized coordinates}}{\underset{\uparrow}{\Delta q_i}}$$

1.2 Work function and potential energy

In many practical situations,
$\Delta w = d\varnothing =$ differential of a scalar function called the work function
Where,
$\varnothing = \varnothing(q_1, q_2 \ldots\ldots\ldots, q_n)$ then,

$$d\varnothing = \frac{\partial\varnothing}{\partial q_1}dq_1 + \frac{\partial\varnothing}{\partial q_2}dq_2 + \ldots\frac{\partial\varnothing}{\partial q_n}dq_n$$

So that with,
$dw = \sum Q_i.dq_i$, and

$$Q_i = \frac{\partial\varnothing}{\partial q_i}$$

Also,

$$w|_A^B = \int_A^B dw = \int_A^B d\varnothing = \left(\underset{\text{Position2}}{\underset{\nearrow}{\varnothing_B}} - \underset{\text{Position1}}{\underset{\nearrow}{\varnothing_A}}\right) \tag{1.4}$$

In engineering, we use $V = -\varnothing = \text{PE} =$ potential energy
Then,

$$Q_i = \frac{\partial V}{\partial q_i}$$

1.3 Total potential energy

For the forces to perform a virtual work, the system should execute a virtual displacement from its position of equilibrium in order that all the forces will do work. The true equilibrium configuration of the system subjected to forces is given by the principles of virtual work (P.V.W.) which states,

If a system executes a virtual displacement from equilibrium position, the V.W. of all the forces is zero that is,

$$\delta w = 0$$

If internal as well as external forces do work, the P.V.W. becomes,

$$\delta w_i + \delta w_e = 0 \tag{1.5}$$

Where δw_i is the internal work done caused by stresses and δw_e is the external work done caused by the external forces.

This holds true whether we have elasticity or not if in addition we do have elasticity, then,

$$\delta w_i = -\delta U_{\nwarrow \text{potential energy of strain (strain energy)}}$$

Then the P.V.W. becomes,

$$\delta w_e = \delta U$$

Suppose, in addition, the external forces have a potential Ω, then,

$$\delta w_e = -\delta\Omega$$

And the P.V.W. becomes,

$$\delta(U + \Omega) = \delta V_{\nwarrow \text{total P.E. of the whole system}} = 0 \tag{1.6}$$

Example 1.1: A rigid bar is suspended with three springs as shown in Fig. 1.2. Formulate the system equilibrium equations in a form of,

$$[k]\{q\} = \{Q\}$$

Where $[k]$ is the system stiffness matrix, $\{q\}$ is the generalized coordinate vector, and $\{Q\}$ is the generalized force vector.

Solution, Suppose we choose coordinates e_1, e_2, e_3 to define the configuration, they are not independent, but they are related with the equation of constraints,

$$e_2 = e_1 + (e_3 - e_1)\left(\frac{a}{a+b}\right) \tag{E1.1}$$

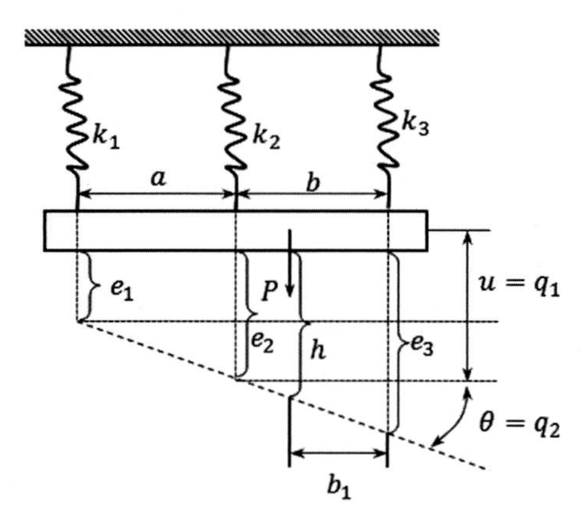

FIGURE 1.2 Beam suspended by three springs and subjected to a concentrated load P.

So that the system has two d.o.f. and we require two generalized coordinates. Since we are free to choose the generalized coordinates, we use u and θ, which are related with the three equations of constraints as follows,

$$e_1 = u - a\theta = q_1 - aq_2$$

$$e_2 = u = q_1$$

$$e_3 = u + b1\theta = q_1 + b1.q_2$$

If we suppose that $b1 = b/2$

$$h = u + \frac{b\theta}{2} = q_1 + b.\frac{q_2}{2} \tag{E1.2}$$

Virtual displacements

$$\delta q_1 \text{ and } \delta q_2$$

Application of P.V.W.

$$\delta w_e = \delta U$$

1. To find the strain energy $U(q_1, q_2)$
   The total strain energy of the system is given by,
   Now,
   $U = U_1 + U_2 + U_3$: strain energy of the springs 1, 2, and 3 with the extensions e_1, e_2, and e_3, respectively.

$$= \frac{1}{2}k_1e_1^2 + \frac{1}{2}k_2e_2^2 + \frac{1}{2}k_3e_3^2 = \frac{1}{2}k_1(q_1 - aq_2)^2 + \frac{1}{2}k_2q_1^2 + \frac{1}{2}k_3(q_1 + bq_2)^2$$

2. To find the equilibrium equations, the P.V.W. is used. Since, $\delta w_e = \delta U$, therefore,

$$P.\delta h = \delta U_1 + \delta U_2 + \delta U_3$$

That is,

$$P\left(\delta q_1 + \frac{b}{2}\delta q_2\right) = k_1(q_1 - aq_2)(\delta q_1 - a\delta q_2) + k_2q_1\delta q_1 + k_3(q_1 + bq_2)(\delta q_1 + b\delta q_2)$$

That is,

$$0 = \delta q_1[k_1(q_1 - aq_2) + k_2q_1 + k_3(q_1 + bq_2) - P] \tag{E1.3}$$

But δq_1 and δq_2 in Eq. (E1.3) are completely independents, so that the square brackets on both sides zero. Therefore we can write,

$$(k_1 + k_2 + k_3)q_1 - (ak_1 - bk_3)q_2 = P$$

$$-(ak_1 - bk_3)q_1 + (a^2k_1 + b^2k_3)q_2 = \frac{P.b}{2} \tag{E1.4}$$

That is,

$$k_{11}q_1 + k_{12}q_2 = (Q_1)_{\text{ext.}}$$
$$k_{21}q_1 + k_{22}q_2 = (Q_2)_{\text{ext.}}$$

In matrix form,

$$\begin{bmatrix} k_{11} & k_{12} \\ k_{21} & k_{22} \end{bmatrix} \begin{Bmatrix} q_1 \\ q_2 \end{Bmatrix} = \begin{Bmatrix} Q_1 \\ Q_2 \end{Bmatrix}_{\text{ext.}} \tag{E1.5}$$

Or,

$$[k]\{q\} = \{Q\} \tag{E1.6}$$

1.4 Application of P.V.W. to generate differential equations for axial member

Consider the bar under axial loading (Fig. 1.3), all forces are conservative,
That is, for equilibrium,

$$\delta V = 0 = \delta(U + \Omega)$$

Now, the potential energy for a bar under axial loading is given by,

$$V = \int_0^l \frac{EA}{2}\varepsilon^2.dx - \int_0^l \varnothing(x).u.dx \tag{1.7}$$

Where, $\varnothing(x) = P(x)$

But, the axial strain is given by,

$\varepsilon = \frac{du}{dx}$, and the total potential energy is given by,

$$V = \int_0^1 \left[\frac{EA}{2}\left(\frac{du}{dx}\right)^2 - p.u\right]dx$$

And,

$$\delta V = \int_0^l \left[\overbrace{EA.\frac{du}{dx}}^{u}.\overbrace{\delta\left(\frac{du}{dx}\right)}^{dv} - \overbrace{P.\delta u}^{v.du}\right]dx$$

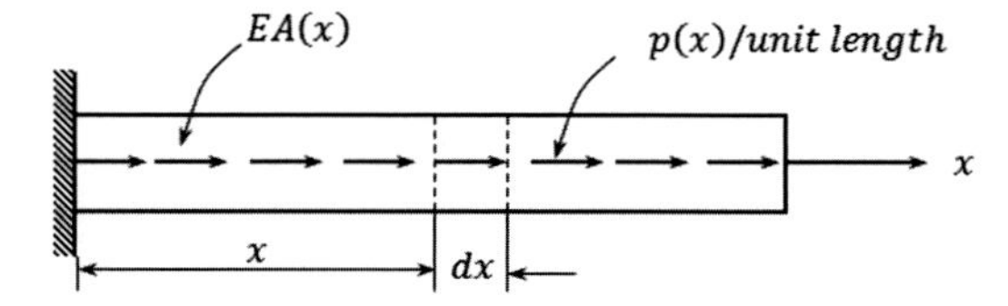

FIGURE 1.3 Bar under axial loading.

Carrying out the integration by parts:

$$\int u.dv = u.v - \int v.du$$

Or,

$$\delta V = EA\frac{du}{dx}\delta u\Big|_0^l - \int_0^l \left[\frac{d}{dx}\left(EA\frac{du}{dx}\right) + P\right]\delta u dx$$

Since $\delta u = 0 \text{at} x = 0$ (u prescribed at $x = 0$), δu arbitrary at $x = l$ and over the entire length of the bar. For equilibrium

$$\delta V = 0$$

Therefore
$\left(EA\frac{du}{dx}\right) = 0$ at $x = l$
And,

$$\frac{d}{dx}\left(EA\frac{du}{dx}\right) + P = 0\text{: Differential equation for bar under axial loading.} \quad (1.8)$$

Example 1.2: XXX.

For an approximate solution for the bar under axial loading, we assume that the displacement is given by:

$$u(x) = a.x \quad (E2.1)$$

The assumed displacement satisfy the prescribed geometric B.Cs on the basis of this assumption,

$$\varepsilon = \frac{du}{dx} = a$$

a = a parameter to be found, it is a generalized coordinate of the system.
The criterion for finding best possible "a" is,

$$\delta V = 0$$

As before and by using Eq. (1.7), the total potential energy is given by,

$$V = \int_0^l \left[\frac{1}{2}EA\left(\frac{du}{dx}\right)^2 - p.u\right]dx = \int_0^l \left[\frac{1}{2}EAa^2 - P.a.x\right]dx$$

Or,

$$= \frac{a^2}{2}EA_0\left(1 - \frac{x}{l}\right) - \frac{P.a.x^2}{2}\Big|_0^l = \frac{8EA_0 la^2}{4} - \frac{Pl^2a}{2} \quad (E2.2)$$

or,

$V = V(a)$, for, $\delta V = 0, \frac{\delta V}{\delta a} = 0$, which gives

$$\frac{3}{2}EA = l.a - \frac{Pl^2}{2}, \ a = \frac{Pl}{3EA_0}$$

Therefore

Approximate solution $u = \frac{Pl}{3EA_0}$, exact $u = \frac{Pl^2}{EA_0}\left(\frac{x}{l} + l\right)$: theory of elasticity

The results of the axial displacements are shown in Table 1.1 for both approximate and exact solutions.

1.5 Principles of stationary total potential energy (P.S.T.P.E.) and trigonometric series for beam bending

Consider a simply supported beam (S.S.B.) for an approximate solution (Fig. 1.4), assume:

$$w(x) = b_1.\sin\frac{\pi x}{l} \tag{1.9}$$

Where w is the displacement in y-direction. The boundary conditions are,

$$w = 0 \text{ at } x = 0 \text{ and } x = l$$

As required by the constraint, then,

$$\frac{dw}{dx} = b_1\frac{\pi}{l}\cos\frac{\pi x}{l}$$

And,

$$\frac{d^2w}{dx^2} = -b_1\left(\frac{\pi}{l}\right)^2\sin\frac{\pi x}{l}, \frac{d^3w}{dx^3} = -b_1\left(\frac{\pi}{l}\right)^3\cos\frac{\pi x}{l} \tag{1.10}$$

TABLE 1.1 Approximate and exact solution of axial displacements of an axial loading bar.

	x = 0	1/4	1/2	31/4	l
Approximate	0	0.1165	0.144	0.25	0.333
Exact	0	0.083	0.212	0.28	0.307

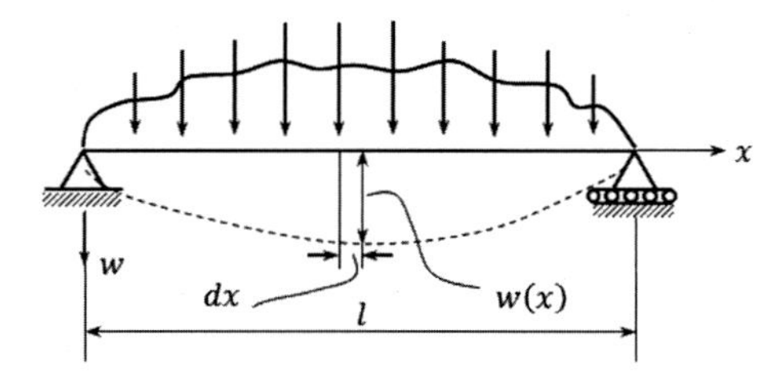

FIGURE 1.4 Simply supported beam under distributed loading.

To find U, strain energy due to bending load is given by,

$$U = \int_0^l EI \frac{d^2w}{dx^2} dvol$$

Using Eq. (1.10),

$$U = \int_0^l \frac{EI}{2} b_1^2 \left(\frac{\pi}{l}\right)^4 \sin\frac{\pi x}{l} dx \tag{1.11}$$

That is,

$$U = \frac{\pi^4 EI b_1^2}{4l^3}$$

To find the potential energy due to applied load Ω, this depends on the details of the applied load.

For this case (Fig. 1.5):

$$\Omega = -Pw_c = -P.b_1 \sin\frac{\pi c}{l} \tag{1.12}$$

To find δV,

The total potential energy is given by,

$$V = U - \Omega$$

Or,

$$V = \frac{\pi^4 EI b_1^2}{4l^3} - P.b_1 \sin\frac{\pi c}{l} = V(b_1)$$

For equilibrium, $\delta V = 0$

$$\frac{\partial V}{\partial b_1} = 0 = \frac{\pi^4 EI b_1}{2l^3} - P\sin\frac{\pi c}{l}$$

This gives,

$$b_1 = \frac{2l^3 P}{\pi^4 EI} \sin\frac{\pi c}{l}$$

Then approximation solution of Eq. (1.9) is given by interchangeability principles,

$$w = \left(\frac{2l^3 P}{\pi^4 EI} \sin\frac{\pi c}{l}\right) \sin\frac{\pi x}{l} \tag{1.13}$$

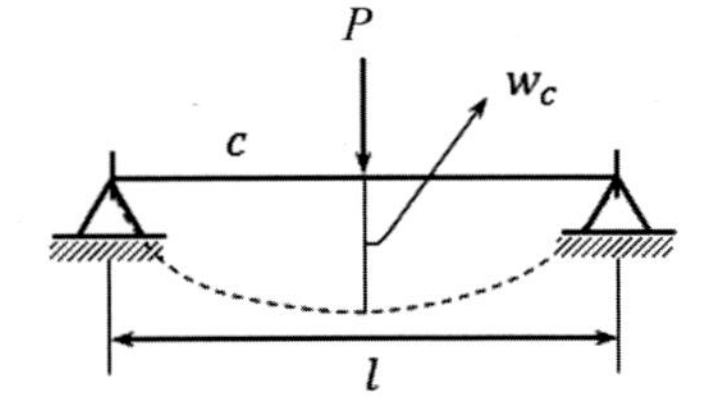

FIGURE 1.5 Simply supported beam under concentrated load.

We notice that interchanging c and x, the value of w is unchanged, so that we can write,

$$w_{xc} = w_{cx} \text{ or } \left(w_{ij} = w_{ji}\right)$$

This is simple illustration of Maxwell reciprocal theorem.

The loading, deflection, shearing, and bending moment diagrams are shown in Fig. 1.6 for both exact and approximate solutions.

To improve the accuracy, we assume,

$$w(x) = \sum_{n=1}^{\infty} b_n \sin\frac{n\pi x}{l} \tag{1.14}$$

Then,

$$w''(x) = -\sum_{n=1}^{\infty} b_n \left(\frac{n\pi}{l}\right)^2 \sin\frac{n\pi x}{l}$$

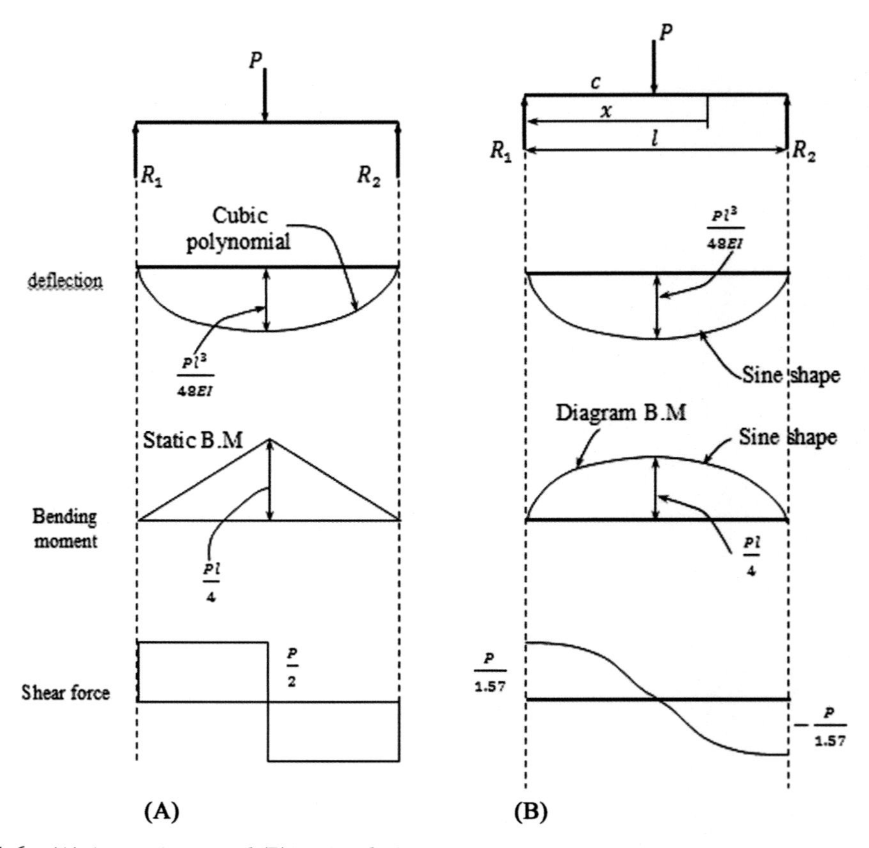

FIGURE 1.6 (A) Approximate and (B) exact solutions.

To find the strain energy U,

$$U = \int_0^l \frac{EI}{2} \sum_{n=1}^{\infty} \left(b_n \left(\frac{n\pi}{l}\right)^2 \sin\frac{n\pi x}{l} \right)^2 dx$$

In view of the orthogonality property of sin w this expression is simplified to,

$$U = \frac{1}{2} \sum \frac{n^4 EI \pi^4}{2l^3} b_n^2 \tag{1.15}$$

To find the total potential energy Ω,

$$\Omega = -\int_0^l P.w dx = -\int_0^l P \sum_{n=1}^{\infty} b_n \sin\frac{n\pi x}{l} dx \tag{1.16}$$

$$= -P \sum b_n \left| \frac{1}{n\pi} \cos\frac{n\pi x}{l} \right|_0^l = -\frac{Pl}{\pi} \sum \frac{b_n}{n} [\cos n\pi - 1] \tag{1.17}$$

To find the coefficient b_n,

The total potential energy $V = U + \Omega$ substitution expressions of $(U + \Omega)$ from Eqs. (1.14) and (1.15) gives,

$$V = \frac{EI\pi^4}{4l^3} \sum_{n=1}^{\infty} n^4 b_n^2 - \frac{pl}{\pi} \sum_{n=1}^{\infty} \frac{b_n}{n} [\cos n\pi - 1] \tag{1.18}$$

Or,

$$= V(b_1, b_2, \ldots, b_n)$$

Therefore

$$b_1 = 2P \frac{l^4}{EI\pi^3} [\cos s\pi - 1] \tag{1.19}$$

Using the vibrational principles we can write,

$$\frac{\partial V}{\partial b_s} = \frac{\pi^4 EI}{4l^3} 2s^4 b_s - \frac{P}{\pi s} (\cos s\pi - 1) = 0 \tag{1.20}$$

Hence,

$$b_s = 2P \frac{l^4}{EI\pi^5 s^5} (\cos s\pi - 1) \tag{1.21}$$

Where $s = 1$ to the number of terms n

For a point load at c position, we have, $b_s = \frac{2Pl^3}{\pi^4 EI} \sin\frac{\frac{s\pi c}{l}}{s^4}$. By using this coefficient in Eq. (1.9), the deflection form can be used in the analysis of beams under bending loading.

Example 1.3: XXX.

Consider the beam shown in Fig. 1.7 which is subjected to both bending load P and axial thrust T, use the energy method employing the trigonometric series to obtain the deflected shape.

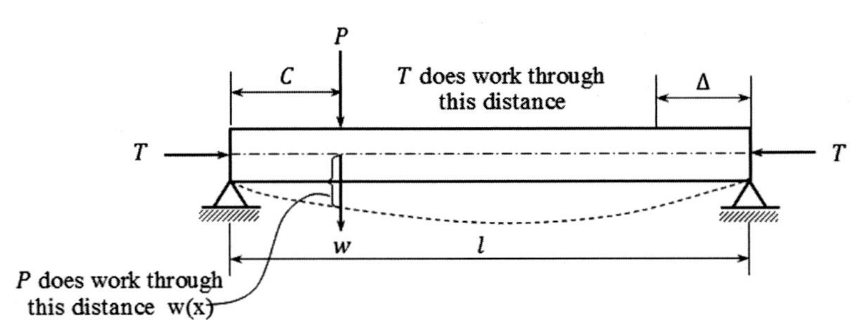

FIGURE 1.7 Simply supported beam subjected to bending and axial loading.

Use principle of stationary total potential energy (P.S.T.P.E.) and assume,

$$w = \sum_{i=1}^{\infty} b_n \sin\frac{n\pi x}{l} \tag{E3.1}$$

1. To find U, as in the previous example since we assume the same deflected shape, therefore,

$$U = \frac{\pi^4 EI}{4l^3} \sum n^4 b_n^2 \tag{E3.2}$$

2. To find Ω_p, Potential energy due to applied load.

$$\Omega_p = -P\sum_{n=1}^{\infty} b_n \sin\frac{n\pi c}{l} \tag{E3.3}$$

3. To find Ω_T
   $\Omega_T = -T.\Delta$: Potential energy due to the application of axial load T, Δ is the corresponding axial displacement. Now,

$$\Delta = \frac{1}{2}\int_0^l \left(\frac{dw}{dx}\right)^2 .dx \tag{E3.4}$$

Therefore

$$\Omega_T = -\frac{T}{2}\int_0^l \left(\sum_{i=1}^{\infty} b_n \frac{n\pi}{l}\cos\frac{n\pi x}{l}\right)^2 .dx \tag{E3.5}$$

In view of orthogonally of the cosine series,

$$\Omega_T = -\frac{\pi T}{4l}\sum_{i=1}^{\infty} b_n^2 n^2 \tag{E3.6}$$

4. To find V, Total potential energy using Eqs. (E3.2) and (E3.3) gives,

$$V = \frac{\pi^4 EI}{4l^3}\sum_{i=1}^{\infty} b_n^2 n^2 - P\sum_{n=1}^{\infty} b_n \sin\frac{\pi s}{l} - \frac{\pi^2 .T}{4.l}\sum_{i=1}^{\infty} b_n^2 n^2 \tag{E3.7}$$

For equilibrium, $\delta V = 0$, so that,

$$\frac{\delta V}{\delta b_n} = \frac{\pi^4 EI}{4l^3} n^4 2b_n - P.\sin\frac{n\pi c}{l} - \frac{\pi^2 T n^2}{4l}.2b_n = 0 \tag{E3.8}$$

Therefore

$$\frac{\pi^4 EI}{2l^3} n^4 b_n \left(1 - T\frac{l^2}{n^2\pi^2 EI}\right) = P\sin\frac{n\pi c}{l}$$

$$b_n = \frac{2l^3 P}{\pi^4 EI}\sin\frac{n\pi c}{l}\frac{\left(\frac{1}{n^4}\right)}{\left(1 - \frac{l^2}{n^2\pi^2 EI}T\right)} \tag{E3.9}$$

Therefore the deflected shape for a beam under bending load may be assumed as follows,

$$w(x) = \sum_{i=1}^{\infty} \frac{2l^3 P}{\pi^4 EI}\sin\frac{n\pi c}{l}\frac{\left(\frac{1}{n^4}\right)}{\left(1 - \frac{l^2}{n^2\pi^2 EI}T\right)}.\sin\frac{n\pi x}{l} \tag{E3.10}$$

For $b_n = \infty$ in Eq. (E3.10) gives,

$$1 - \frac{l^2}{n^2\pi^2 EI}T = 0$$

Then,

$$T = n^2\left(\frac{\pi^2 EI}{l^2}\right), \text{ thrust load.} \tag{E3.11}$$

1.6 Principle of virtual complementary energy (P.V.C.E.)

The load–extension and the stress–strain diagrams are shown in Fig. 1.8A–B for a uniaxial testing specimen. In this case:

$$dW = p.du \tag{1.22}$$

Where dW is the change in virtual work due to virtual displacement.

$$W = \int p.du$$

$$dW^* = u.dp$$

$W^* = \int u.dp$, is the complementary work done due to dp

$$\int p\,\delta u : \int \sigma\,\delta\varepsilon$$

Equation state   Compressible state

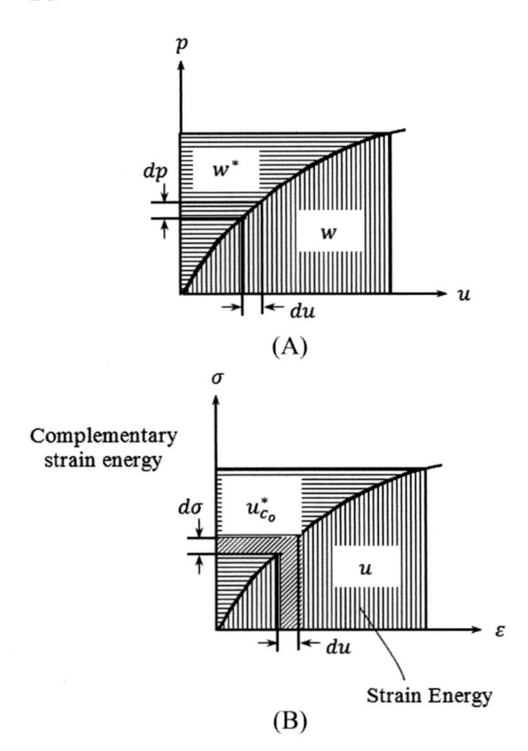

FIGURE 1.8 (A) Load–extension diagram and (B) stress–strain diagram.

In the P.V.W., we are concerned with displacement δu which provides compatible strain $\delta\varepsilon$.

If in the principle, the actual displacement with the corresponding strain, about to satisfy the compatibility conditions if at the same time, we consider changes of variation δp in p with corresponding equilibrium variation $\delta\sigma$ and substitution all this in the P.V.W. (Fig. 1.8B),

$$\int u\delta p = \int \varepsilon\delta\sigma = \text{P.V.C.W.}$$

P.V.W. manipulates compatible displacement to lead to equilibrium, the P.V.C.W. manipulate equilibrium. stresses lead to compatibility conditions.

1.6.1 Castiglianon's theorem of deflections

In the principle of C.V.W., let δp corresponds to actual increments in the statically independent,

$\delta U^* = \Delta U^*$ (actual increment)

$$= \sum_{i=1}^{n} u_i \Delta p_x \quad \uparrow \text{actual increamental displacement}$$

But,

$$\Delta U^* = \sum \frac{\partial U^*}{\partial p_i} \Delta p_n$$

Since $U^* = U^*(p_1, p_2, \ldots\ldots)$,
The P.V.C.W. becomes $\delta U^* = 0$ and,

$$\sum \left(\frac{\partial U^*}{\partial p_i} - u_i \right) \Delta p_i = 0 \tag{1.23}$$

And since the Δp_i are independent, then the terms between brackets = 0 which results in,

$u_i = \frac{\partial U^*}{\partial p_i}$, This is the state of first Castigliano's theorem, the partial derivative of the strain energy with respect to the load gives the displacement in the direction of the load.

Example 1.4: XXX.

For the ring shown in Fig. 1.9, find the B.M. at any section in the ring using the P.V.C.W. reduces to $\delta U^* = 0$

To find B.M.D.,

The imaginary out may be placed any-where in the ring including, P_0, M_0, N_0 Fig. 1.10.

But to reduce calculating efforts, it is desirable to exploit any nature feature in the problem, for example, symmetry since P_0, M_0 = shearing force = 0 as shown in Fig. 1.11.

To find U^*, strain energy density,

$$U^*(\sigma) = \int_{Vol} U_0^* dvol \tag{1.24}$$

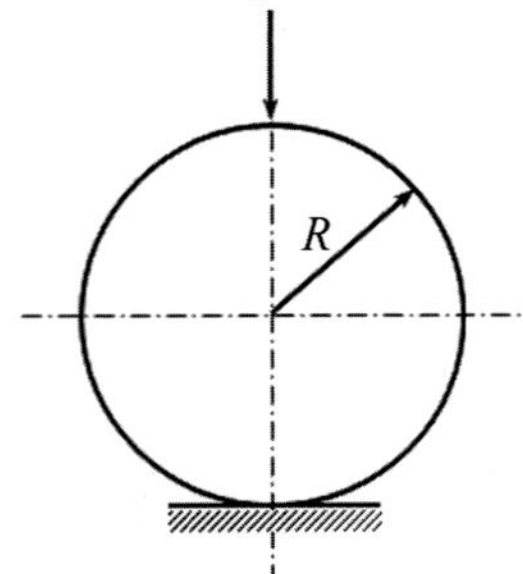

FIGURE 1.9 Full ring under concentrated load.

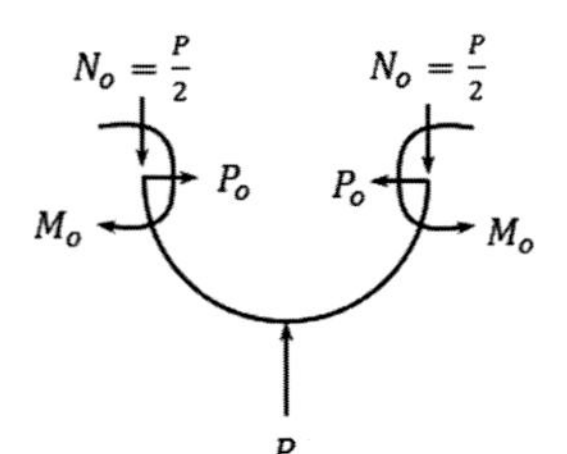

FIGURE 1.10 Half ring under equilibrium.

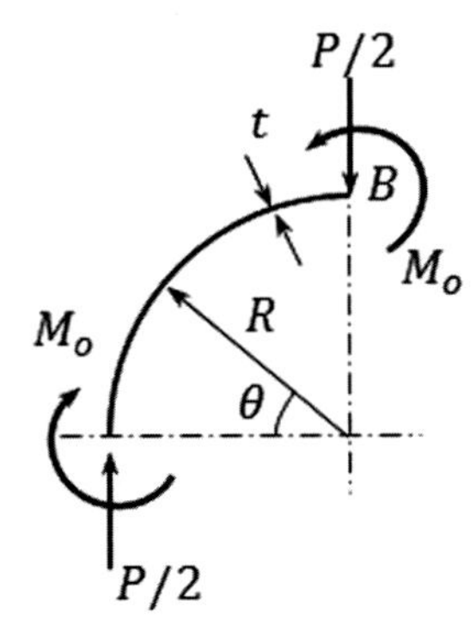

FIGURE 1.11 Quarter ring under equilibrium.

If $t << R$, effect of curvature is negligible, then, the only significant stress is the ordinary bending stress is given by,

$\sigma_b = \frac{M.z}{I}$, Simple bending theory

The direct stress over a typical section will be negligible compared to the bending action, therefore,

$$\int_{\text{intial}}^{\text{final}} dU^* = \int_{\text{intial}}^{\text{final}} \varepsilon d\sigma = \int_{\text{intial}}^{\text{final}} \frac{\sigma}{E} d\sigma$$

That is,

$U_0^* = \frac{\sigma^2}{2E}$, The strain energy per unit volume in an axial member.

The integrating through the thickness of the ring gives the bending stress as follows,

$$\sigma = \frac{Mz}{I}$$

To find M_0,

$$M(\theta) = M_0 + \frac{P}{2} R(1 - \cos\theta) \quad (1.25)$$

And,

$$\frac{\partial M}{\partial M_0} = 1$$

To find M_0, Since,

$$U^* = U^*(M_0, P)$$

But for compatibility $\delta U^* = 0, \frac{\partial U^*}{\partial M_0} = 0$

Instead of computing U^* and then carrying differentiation, we can save some efforts manipulation by differentiation inside the integral sign $\int$, this gives,

$$\frac{\partial U^*}{\partial M} = \int_0^{\pi/2} \frac{M}{EI} \frac{\partial M}{\partial M_0} R.d\theta \quad (1.26a)$$

By using Eq. (1.25) into Eq. (1.26a) gives,

$$= \frac{1}{EI}\int_0^{\pi/2}\left(M_0 + \frac{PR}{2}(1-\cos\theta)\right)R.d\theta = \frac{R}{EI}\left|\left(M_0 + \frac{PR}{2}\right)\theta - \frac{PR}{2}\sin\theta\right|_0^{\pi/2}$$

That is,

$$M_0 = \frac{PR}{2}\left(1 - \frac{\pi}{2}\right)$$

Example 1.5: XXX.

Use complementary energy principle and Castigliano's theorem of deflection to find the redundant forces and deflections in the joints shown in Fig. 1.12.

Solution,

We have from the principles of total complementary potential energy,

$$V^* = U^* - \left(\sum u_i p_i\right)^{\swarrow \text{shown when displacement prescribed}}$$

the part of the structure at which the displacements are prescribed at three points A, B, C and the displacement are prescribed to zero.

$$V^* = U^*$$

The strain energy in the axial-loaded members is given by,

$$U = \frac{T_1^2 l_1}{2(EA_1)_1} + \frac{T_2^2 l_2}{2(EA_2)_2} + \frac{T_3^2 l_3}{2(EA_3)_3}$$

P is the axial load and in our case $P_i = T_i$

That is,

$$V^* = \frac{T_1^2 l_1}{2(EA_1)_1} + \frac{T_2^2 l_2}{2(EA_2)_2} + \frac{T_3^2 l_3}{2(EA_3)_3} \tag{E5.1}$$

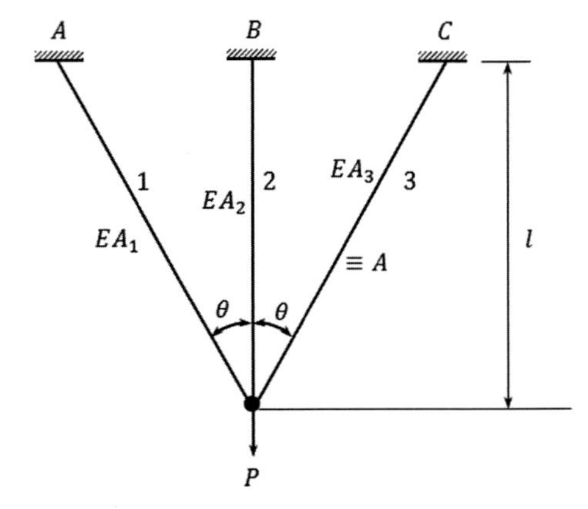

FIGURE 1.12 Three jointed bars.

Where T_1, T_2, and T_3 are the tensions in the cables 1, 2, and 3, respectively.

Before find the variation δV^* we must take Eq. (E5.1).

Taking conditions into account equilibrium conditions,

$$\sum F_y = T_2 + T_1\cos\theta + T_3\cos\theta - P = 0$$

Or,

$$T_2 + 2T_1\cos\theta = P \tag{E5.2}$$

Hence the problem statically indeterminate since there is one equilibrium. Unknown T_1 and T_3 may be one of them regarded as independent or redundant one.

We will choose arbitrary redundant T_1 and $T_1 = T_3$ due to symmetry, Fig. 1.13. Hence, Eq. (E5.3) can be reduced to,

$$V^* = \frac{2T_1^2 l_1}{2(EA_1)_1} + \frac{T_2^2 l_2}{2(EA_2)_2} \tag{E5.3}$$

And taking into account the substitution of T_2 in terms of T_1 in Eq. (E5.1) gives:

$$V^* = \frac{T_1^2 l_1}{EA_1} + \frac{(P-2T_1\cos\theta)^2 l_2}{2EA_2} = V^*(P, T_1) \tag{E5.4}$$

But for compatibility $V^* = 0 = \frac{\partial V^*}{\partial T_1}$, Eq. (E5.4) results,

$$0 = \frac{2T_1 l_1}{(EA)_1} + \frac{(P - 2T_1\cos\theta)(-2\cos\theta)l_2}{(EA)_2} \tag{E5.5}$$

We have,

$(EA)_1 = (EA)_2 = (EA)_3 = EA$ Constant axial stiffness assumption.

Therefore Eq. (E5.3) becomes,

$$0 = \frac{2T_1 l}{EA\cos\theta} - \frac{2Pl\cos\theta}{EA} + \frac{4T_1 l\cos^2\theta}{EA}$$

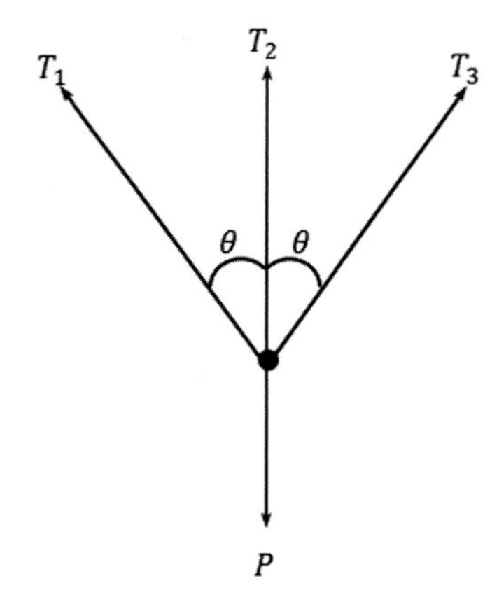

FIGURE 1.13 Equilibrium of the joint of application of load.

Or,

$$\frac{2Pl\cos\theta}{EA} = \frac{2T_1 l}{EA\cos\theta}\left(1 + 2\cos^3\theta\right) \tag{E5.6}$$

From which,

$$T_1 = \frac{P\cos^2\theta}{(1 + 2\cos^3\theta)} \tag{E5.7}$$

Thus the complementary P gives the redundant (statically indeterminate force T_1). Also, by using $\frac{\partial V}{\partial T_2}$, in Eq. (E5.4) gives,

$$T_2 = \frac{P}{(1 + 2\cos^3\theta)} \tag{E5.8}$$

To find the deflection at P by using Castigliagno's theorem and since the forces in the bars are turned as a function of the applied load, and by using Eq. (E5.7), we can write,

$$V = \frac{2l}{2EA\cos\theta}\cdot\left(\frac{P\cos^2\theta}{(1 + 2\cos^3\theta)}\right) + \frac{2}{2EA}\left(\frac{P}{(1+2\cos^3\theta)}\right)^2$$

That is,

$$V = \frac{1}{2}\frac{P^2 l\left(2\cos^3\theta + 1\right)}{(1 + 2\cos^3\theta)} = \frac{P^2 l}{2EA(1+2\cos^3\theta)^2} \tag{E5.8a}$$

Therefore Castigliagno's theorem,
$u_P = \frac{\partial U_P}{\partial P} = \frac{P.l}{EA(1 + 2\cos^3\theta)}$, The deflection at the application of the load P.

1.7 Torsion of a rectangular section bar

Consider the rectangular section shown in Fig. 1.14, the Airy stress function is used to obtain the torsion function ∅ for the shearing stresses calculations using the theory of elasticity concept.

Displacement at two ends are prescribed at clamped end is zero and the other end is θl.

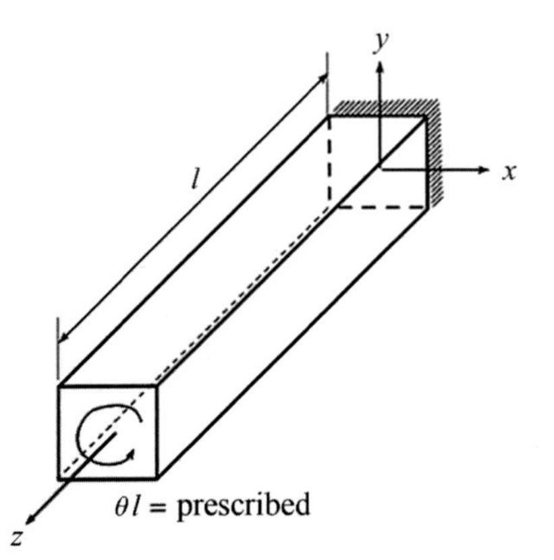

FIGURE 1.14 Rectangular section shaft under torsion.

The total P.E. are given by,

$$V^* = U^* - M_t.l\theta = \int_{vol}\left[\frac{T_{zx}^2}{2G} + \frac{T_{zy}^2}{2G}\right]dvol - M_t.l\theta \tag{1.26}$$

To use Complementary energy principles (C.E.P.), we must work with equilibrium stress systems and the convenient way of insuring this for continuum system is to work of stress function.

$$V^* = \underbrace{\frac{l}{2G}\iint\left[\left(\frac{\partial\varnothing}{\partial x}\right)^2 + \left(\frac{\partial\varnothing}{\partial y}\right)^2\right]dx\ dy}_{\text{Complementary stored in the bar}} - \underbrace{2l\theta\iint\varnothing dx\, dy}_{\text{Due to applied load}}$$

Using the Prandtl stress function $\varnothing(x, y)$,
Or,

$$V^* = \frac{l}{2G}\iint\left[\left(\frac{\partial\varnothing}{\partial x}\right)^2 + \left(\frac{\partial\varnothing}{\partial y}\right)^2\right]dx\ dy - 4G\theta\varnothing dx\ dy \tag{1.27}$$

Example 1.6: For the rectangular section shown in Fig. 1.15, and using the stress torsion function $\varnothing = C_0(a^2 - x^2)(b^2 - y^2)$, derive the formula that given the torsional stiffness using the energy principles.

First approximation,

Assume $\varnothing$ which must satisfy the forced boundary condition, $\varnothing = 0$ (all around the edges).

Therefore we can use,

$$\varnothing(x, y) = C_0(a^2 - x^2)(b^2 - y^2) \tag{E6.1}$$

Then,

$$\frac{\partial\varnothing}{\partial x} = -2C_0x(b^2 - y^2)$$

$$\frac{\partial\varnothing}{\partial y} = -2C_0y(a^2 - x^2)$$

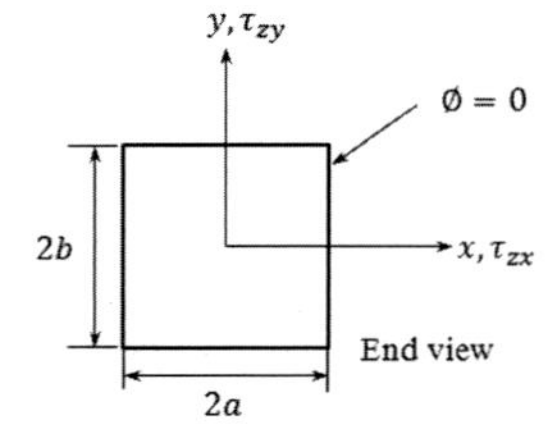

FIGURE 1.15 Rectangular section bar under torsion load.

We find $V^* = V^*(C_0)$, using Eq. (1.20),

$$V^* = \frac{l}{2G}\int_0^b\int_0^a \left[4C_0^2x^2\left(b^2-y^2\right)^2 + 4C_0^2y^2\left(a^2-x^2\right)^2 - 4G\theta C_0\left(a^2-x^2\right)\left(b^2-y^2\right)\right]dxdy$$

We have,

$$\frac{\partial V^*}{\partial C_0} = \frac{l}{2G}\int_0^b\int_0^a \left[8C_0x^2\left(b^2-y^2\right)^2 + 8C_0y^2\left(a^2-x^2\right)^2 - 4G\theta\left(b^2-y^2\right)a^2 + 4G\theta\left(b^2-y^2\right)x^2\right]dx\ dy$$

Or,

$$\frac{\partial V^*}{\partial C_0} = \frac{l}{2G}\int_0^b \left|8C_0\frac{x^3}{3}\left(b^2-y^2\right)^2 + 8C_0y^2\left(a^4x - 2a^2\frac{x^3}{3} + \frac{x^5}{5}\right) - 4G\theta\left(b^2-y^2\right)a^2x + 4G\theta\left(b^2-y^2\right)\frac{x^3}{3}\right|_0^a dy$$

$$\frac{\partial V^*}{\partial C_0} = \frac{l}{2G}\int_0^b \left[8C_0\frac{a^3}{3}\left(b^4 - 2b^2y^2 - y^4\right) + 8C_0y^2\left(a^5 - 2\frac{a^5}{3} + \frac{a^5}{5}\right) - 4G\theta\left(b^2-y^2\right)a^3 + 4G\theta\left(b^2-y^2\right)\frac{a^3}{3}\right]dy$$

Carrying out the integration of terms gives,

$$\frac{\partial V^*}{\partial C_0} = \frac{l}{2G}\left|8C_0\frac{a^3}{3}\left(b^4y - 2b^2\frac{y^3}{3} - \frac{y^5}{5}\right) + 8C_0\frac{y^3}{3}\left(8\frac{a^5}{15}\right) - 4G\theta\left(b^2y - \frac{y^3}{3}\right)a^3 + 4G\theta\frac{a^3}{3}b^2y - 4G\theta\frac{a^3}{3}\frac{y^3}{3}\right|_0^b$$

$$\frac{\partial V^*}{\partial C_0} = \frac{l}{2G}\left[8C_0\frac{a^3}{3}\left(b^5 - 2\frac{b^5}{3} - \frac{b^5}{5}\right) + 8C_0\frac{b^3}{3}\left(8\frac{a^5}{15}\right) - 4G\theta\left(b^3 - \frac{b^3}{3}\right)a^3 + 4G\theta\frac{a^3}{3}b^3 - 4G\theta\frac{a^3}{3}\frac{b^3}{3}\right]$$

For equilibrium, $\frac{\partial V^*}{\partial C_0} = 0$, then,

$$0 = \frac{8C_0a^3}{3}\left(\frac{8b^5}{15}\right) + C_0\frac{64}{45}b^3a^5 - 4G\theta a^3b^3 + 8G\theta\frac{a^3b^3}{3} - \frac{4}{9}G\theta\frac{a^3b^3}{9}$$

Or,

$$0 = \frac{8C_0a^3b^5}{45} + C_0\frac{64C_0b^3a^5}{45} + a^3b^3\left(-4G\theta + \frac{8G\theta}{3} - \frac{4}{9}G\theta\right)$$

And,

$$0 = \frac{64}{45}C_0\left(b^2+a^2\right) - \frac{G\theta 16}{9}$$

From which,

$$C_0 = \frac{5}{4}\frac{G\theta}{\left(b^2+a^2\right)}$$

Therefore Eq. (E6.1) becomes,

$$\varnothing = \frac{5}{4}\frac{G\theta}{\left(b^2+a^2\right)}\left(a^2-x^2\right)\left(b^2-y^2\right) \tag{E6.2}$$

From theory of elasticity, we have,

$$M_t = 2\int\int \varnothing dx\, dy$$

Or,

$$M_t = 2\frac{5}{4}\int\int \frac{G\theta}{(b^2+a^2)}(a^2-x^2)(b^2-y^2)dx\, dy \tag{E6.3}$$

Carrying out the double integration and simplification of Eq. (E6.3) as follows,

$$M_t = 10\int_0^b \left| \frac{G\theta}{(b^2+a^2)}(b^2-y^2)\left(a^2x-\frac{x^3}{3}\right)\right|_0^a dy$$

Or,

$$M_t = 10\frac{G\theta}{(b^2+a^2)}\int_0^b (b^2-y^2)\left(a^3-\frac{a^3}{3}\right)dy$$

Hence,

$$M_t = 10\frac{G\theta}{(b^2+a^2)}\frac{2}{3}a^3\left|a^2y-\frac{y^3}{3}\right|_0^b$$

Thus gives,

$$M_t = \frac{20}{3}\frac{G\theta}{(b^2+a^2)}a^3\frac{2}{3}b^3$$

Or,

$$M_t = \frac{40}{9}\frac{G\theta}{(b^2+a^2)}a^3b^3$$

And the torsion constant of the section is,

$$k_t = \frac{M_t}{G\theta} = \frac{40}{9}\frac{a^3b^3}{(b^2+a^2)} \tag{E6.4}$$

Example 1.7: The framework shown in Fig. 1.16 is made up of a rigid outer member and relatively flexible tie rods. All the joints are pinned and the tension T in any rod is related to the extension e by, $T = k\sqrt{e}, k =$ constant. Use the principle of stationary total potential energy to determine the deflection at the center and the tensions in the bars (Fig. 1.16).

Solution,
Let,

$$V_p = q$$

The extensions of the tie rods are in terms of generalized displacement q(def n̲) at center are illustrated in Fig. 1.17.

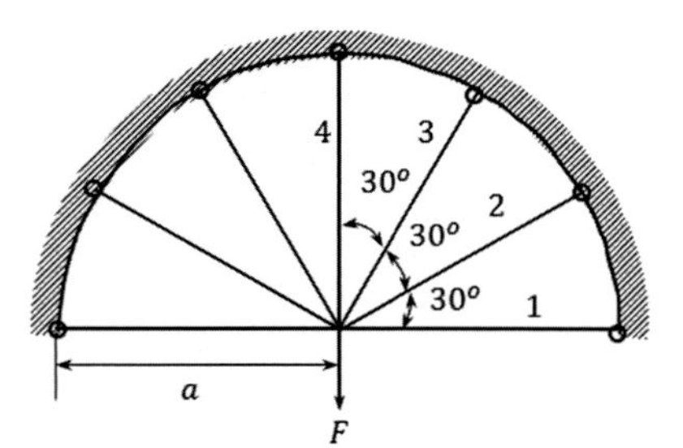

FIGURE 1.16 A framework of seven bars jointed together.

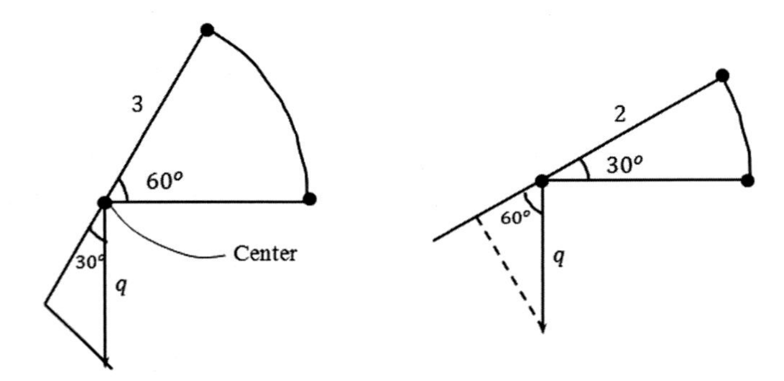

FIGURE 1.17 Free body diagram of bars.

From geometry, the extensions of the rods are as follows,

$$e_2 = q\cos 60$$

$$e_2 = \frac{1}{2}q$$

$$e_3 = q\cos 30 \tag{E7.1a}$$

$$e_3 = \frac{\sqrt{3}}{2}q$$

$$e_4 = q$$

Where $e_1, e_2, \ldots e_4$ are the extensions in the members' 1, 2, 3, and 4, respectively.

We have,

$$T = k\sqrt{e}\{\text{nonlinear relation}\} \tag{E7.1b}$$

The strain energy stored in the cable is given by,

$$U = \int_0^e T\,de$$

Or, using Eq. (E7.1b) gives,

$$= \int_0^e k\sqrt{e}\,de = k.\frac{2}{3}e^{3/2}$$

The total strain energy taking into considerations the symmetry of the members in the structure is given by,

$$U_{\text{total}} = \frac{2}{3}k\left[2e_1^{3/2} + 2e_2^{3/2} + 2e_3^{3/2} + e_4^{3/2}\right]$$
$$= \frac{2}{3}k\left[0 + 2\left(\frac{1}{2}q\right)^{3/2} + 2\left(\frac{\sqrt{3}}{2}q\right)^{3/2} + q^{3/2}\right]$$
$$= \frac{2}{3}kq^{3/2}\{2*0.353 + 2*0.805 + 1\} = 3.317\frac{2}{3}kq^{3/2} = 2.212\, q^{3/2} \quad \text{(E7.1)}$$

Potential energy is given by,

$$\Omega = -P.q \quad \text{(E7.2)}$$

Therefore the total potential energy,

$$V = \Omega + U$$

Using Eqs. (E7.1) and (E7.2) gives,

$$V = -P.q + 2.212q^{3/2} \quad \text{(E7.3)}$$

For equilibrium,

$$\delta V = 0$$

Using Eq. (E7.3),

$$\delta V = -P + 2.212k\frac{3}{2}q^{\left(\frac{3}{2}-1\right)} = 0$$

Which results,

$$P = 2.212*\frac{3}{2}kq^{1/2}$$

Or,

$$q^{1/2} = \frac{P}{k}\frac{2}{3*2.212}$$

And the central deflection becomes,

$$q = \frac{P^2}{k^2}*0.090833$$

The tension in the cables, Fig. 1.18, are,

$$T_1 = 0$$

$$T_2 = k\sqrt{e_2} = k\sqrt{\frac{1}{2}\frac{P^2}{k^2}*0.090833} = P*0.21311\text{ N}$$

$$T_3 = k\sqrt{e_3} = k\sqrt{\frac{\sqrt{3}}{2}\frac{P^2}{k^2}*0.090833} = P*0.2804\text{ N}$$

And,

$$T_4 = k\sqrt{e_4} = k\sqrt{\frac{P^2}{k^2}*0.090833} = P*0.30138 \text{ N}$$

Example 1.8: Using a half rang sine series, and principle of stationary total potential energy, find the deflection at the center of a simply supported beam carrying a uniformly distributed load (Fig. 1.19).

Solution, Assume,

$$w(x) = \sum_{n=1}^{\infty} b_n.\sin\frac{n\pi x}{l} \{\text{Half range sine expansion}\} \quad \text{(E8.1)}$$

The strain energy due to bending is given by,

$$U = \int_0^l \frac{EI}{2}\left(\frac{d^2w}{dx^2}\right)^2 dx \quad \text{(E8.2)}$$

Carrying out the differentiations as required by Eq. (E8.2) using Eq. (E8.1) as follows,

$$w'(x) = \sum_{n=1}^{\infty} b_n.\frac{n\pi}{l}.\cos\frac{n\pi x}{l} \quad \text{(E8.3)}$$

And,

$$w''(x) = -\sum_{n=1}^{\infty} b_n.\left(\frac{n\pi}{l}\right)^2.\sin\frac{n\pi x}{l}$$

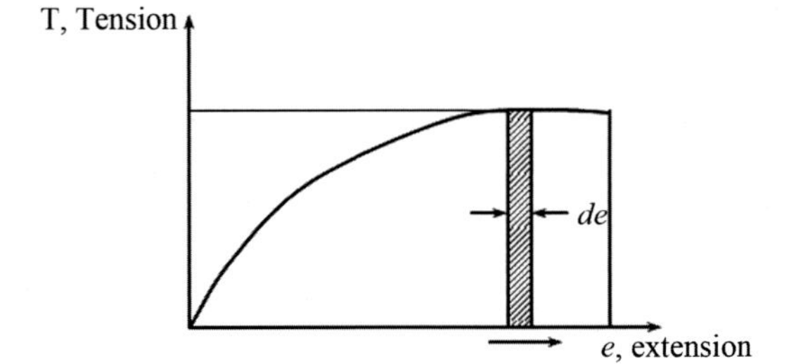

FIGURE 1.18 Load–extension of a typical bar.

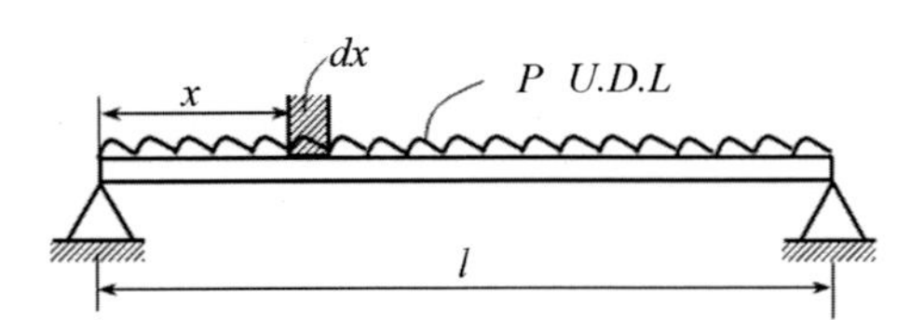

FIGURE 1.19 Simply supported beam under uniform loading.

Therefore

$$U = \int_0^l \frac{EI}{2}\left[-\sum_{n=1}^{\infty} b_n.\left(\frac{n\pi}{l}\right)^2.\sin\frac{n\pi x}{l}\right]^2 dx$$

$$= \int_0^l \frac{EI}{2}\sum_{n=1}^{\infty} b_n^2.\left(\frac{n\pi}{l}\right)^4.\sin^2\frac{n\pi x}{l}dx$$

$$= \frac{\pi^4 EI}{2l^4}\sum_{n=1}^{\infty} n^4 b_n^2 \int_0^l \sin^2\frac{n\pi x}{l}dx$$

$$= \frac{\pi^4 EI}{2l^4}.\frac{l}{2}.\sum_{n=1}^{\infty} n^4 b_n^2$$

$$= \frac{\pi^4 EI}{4l^3}.\sum_{n=1}^{\infty} n^4 b_n^2 \quad \text{(E8.4)}$$

Potential energy due to applied load is given by,

$$\Omega = -\int_0^l P.w.dx$$

Again using Eq. (E8.1) gives,

$$= -P\int_0^l \sum_{n=1}^{\infty} b_n.\sin\frac{n\pi x}{l}dx$$

$$= P\sum_{n=1}^{\infty} b_n.\frac{l}{n\pi}\left|\cos\frac{n\pi x}{l}\right|_0^l$$

$$= \frac{l}{\pi}P\sum_{n=1}^{\infty}\frac{b_n}{n}(\cos n\pi - 1)$$

$$\Omega = \frac{\pi P}{l}\sum_{n=1}^{\infty} b_n(\cos n\pi - 1) \quad \text{(E8.5)}$$

The total potential energy is given by,

$$V = U + \Omega \quad \text{(E8.6a)}$$

Substitute the values of U and Ω from Eqs. (E8.4) and (E8.5), respectively, in Eq. (E8.6a), as follows,

$$= \frac{\pi^4 EI}{4l^3}.\sum_{n=1}^{\infty} n^4 b_n^2 + \frac{\pi P}{l}\sum_{n=1}^{\infty} b_n(\cos n\pi - 1)$$

Where,

$$V = V(b_1, b_2, \ldots b_s)$$

Minimization of the total potential energy is carried out as follows,

$$\frac{\partial V}{\partial b_s} = \frac{\pi^4 EI}{4l^3}.s^4.2.b_s + \frac{\pi P}{l.s}(\cos s\pi - 1)\, for\ s = 1, 2, 3, \ldots \infty$$

Or for equilibrium,

$$0 = \frac{\pi^4 EI}{2l^3} s^4.b_s - \frac{2Pl}{\pi.s} \tag{E8.6}$$

From Eq. (E8.6) can be write,

$$b_s = \frac{2Pl}{\pi.s} * \frac{2.l^3}{\pi^4 EI.s^4}$$

Or,

$$b_s = \frac{4.l^4.P}{\pi^5 EI}.\frac{1}{s^5}$$

Therefore the deflection Eq. (E8.1) becomes,

$$w(x) = \sum_{n=1}^{\infty} \frac{4.l^4.P}{\pi^5 EI}.\frac{1}{n^5}.\sin\frac{n\pi x}{l} \quad \text{for } n = 1, 3, 5, \ldots \tag{E8.7}$$

Let,

$$n = 2m + 1, (m = 0, n = 1), (m = 1, n = 3), (m = 2, n = 5), \ldots$$

Therefore Eq. (E8.7) becomes,

$$w(x) = \frac{4.l^4.P}{\pi^5 EI} \sum_{m=0}^{\infty} \frac{1}{(2m+1)^5}.\sin\frac{(2m+1)\pi x}{l}$$

At, $x = l/2$,

$$w(l/2) = \frac{4.l^4.P}{\pi^5 EI} \sum_{m=0}^{\infty} \frac{1}{(2m+1)^5}.\sin\frac{(2m+1)\pi l}{2l}$$

For $m = 0$, first term of series is,

$$w(l/2) = \frac{4.l^4.P}{\pi^5 EI}.\frac{1}{1}.\sin\frac{\pi}{2}$$

$w(l/2) = \frac{1}{76.5049}.\frac{P.l^4}{EI}$ (approximate solution)
Exact ensure,
$w(x) = \frac{5}{384}.\frac{P.l^4}{EI}$ (for strength of Maxwell)

$$w(\text{center}) = \frac{1}{76.8}.\frac{P.l^4}{EI} \tag{E8.8}$$

A good agreement between the results of approximate and the exact solutions for the central deflection of the beam.

Example 1.9: Denoting $\frac{P}{P_0} = r$ where $P_0 = \frac{\pi^2 EI}{l^2}$ derive a formula for $\frac{R}{F}$ for the beam shown, Fig. 1.20A.

Solution,

Using the superposition principles for the beam shown in Fig. 1.20A, the representation based upon these principles is shown in Fig. 1.20B and the reactions are shown in Fig. 1.20C.

Solution,

Using the superposition principles for the beam shown in Fig. 1.20, the representation based upon these principles is shown in Fig. 1.20A

Assume,

$$w(x) = \sum_{n=1}^{\infty} b_n.\sin\frac{n\pi x}{l}$$

Therefore

$$w'(x) = \sum_{n=1}^{\infty} b_n.\frac{n\pi}{l}.\cos\frac{n\pi x}{l}$$

And,

$$w''(x) = -\sum_{n=1}^{\infty} b_n.\left(\frac{n\pi}{l}\right)^2.\sin\frac{n\pi x}{l}$$

Since the strain energy due to bending loading is given by,

$$U = \int_0^l \frac{EI}{2}(w'')^2 dx \tag{E9.1}$$

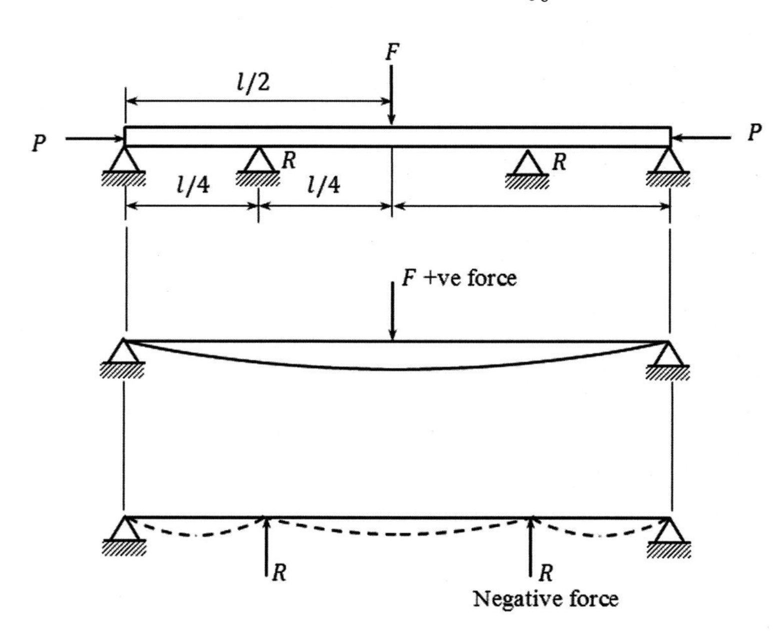

FIGURE 1.20 (A) The given beam. (B) Effect of the concentrated load. (C) Reaction of the supports R.

The total potential energy Ω_F is given by,

$$\Omega_F = -F.w\left(\frac{l}{2}\right) = -F\sum_{n=1}^{\infty} b_n.\sin\frac{n\pi l}{2l} \quad \text{(E9.2)}$$

Or,

$$\Omega_F = -F\sum_{n=1}^{\infty} b_n.\sin\frac{n\pi}{2}$$

The potential energy Ω_p due to the axial load effect in bending direction is given by,

$$\Omega_p = -\frac{P}{2}\int_0^l (w')^2 dx$$

Or,

$$= -\frac{P}{2}\int_0^l \left[\sum_{n=1}^{\infty} b_n.\frac{n\pi}{l}.\cos\frac{n\pi x}{l}\right]^2 dx$$

View of orthogonally of cosine series, gives

$$= -\frac{P}{2}\sum_{n=1}^{\infty} b_n^2.\left(\frac{n\pi}{l}\right)^2.\frac{l}{2}$$

$$= -\frac{\pi^2 P}{4.l}\sum_{n=1}^{\infty} b_n^2.n^2$$

And the strain energy Eq. (E9.1) becomes,

$$U = \int_0^l \frac{EI}{2}\left(\sum_{n=1}^{\infty} b_n.\left(\frac{n\pi}{l}\right)^2.\sin\frac{n\pi x}{l}\right)^2 dx \quad \text{(E9.3)}$$

Ω_R = Potential due to external point load at $x = \frac{l}{4}$,

$$\Omega_R = R.w\left(\frac{l}{4}\right)$$

But they are two forces at the same distances from two sides,

$$\Omega_R = 2R\sum_{n=1}^{\infty} b_n.\sin\frac{n\pi l}{4l}$$

$$= 2R\sum_{n=1}^{\infty} b_n.\sin\frac{n\pi}{4} \quad \text{(E9.4)}$$

Therefore the total potential energy is given by,

$$V = U + \Omega_F + \Omega_P + \Omega_R + \Omega_R \quad \text{(E9.5a)}$$

Substitute their corresponding values in Eqs. (E9.2)–(E9.4) in Eq. (E9.5a) gives,

$$= \int_0^l \frac{EI}{2}\left(\sum_{n=1}^{\infty} b_n.\left(\frac{n\pi}{l}\right)^2.\sin\frac{n\pi x}{l}\right)^2 dx - F\sum_{n=1}^{\infty} b_n.\sin\frac{n\pi}{2} - \frac{P}{2}\int_0^l \left[\sum_{n=1}^{\infty} b_n.\frac{n\pi}{l}.\cos\frac{n\pi x}{l}\right]^2 dx + 2R\sum_{n=1}^{\infty} b_n.\sin\frac{n\pi}{4}$$

For equation $\delta V = 0$

$$\frac{\delta V}{\delta b_n} = EI\int_0^l \sum_{n=1}^{\infty} b_n.\left(\frac{n\pi}{l}\right)^4.\sin^2\frac{n\pi x}{l}dx - F\sum_{n=1}^{\infty}\sin\frac{n\pi}{2} - P\int_0^l \sum_{n=1}^{\infty} b_n.\left(\frac{n\pi}{l}\right)^2.\cos^2\frac{n\pi x}{l}dx + 2R\sum_{n=1}^{\infty}\sin\frac{n\pi}{4} = 0$$

Which gives,

$$EI\frac{n^4\pi^4}{l^4}b_n\int_0^l \sin^2\frac{n\pi x}{l}dx - F\sin\frac{n\pi}{2} - P\frac{n^2\pi^2}{l^2}b_n\int_0^l \cos^2\frac{n\pi x}{l}dx + 2R\sin\frac{n\pi}{4} = 0$$

$$EI\frac{n^4\pi^4}{l^4}b_n.\frac{l}{2} - F\sin\frac{n\pi}{2} - P\frac{n^2\pi^2}{l^2}b_n.\frac{l}{2} + 2R\sin\frac{n\pi}{4} = 0$$

$$\underbrace{\left(\frac{\pi^2 EI}{l^2}\right)}_{P_e}.\frac{n^4\pi^2}{l^2}.b_n.\frac{l}{2} - F\sin\frac{n\pi}{2} - P\frac{n^2\pi^2}{2l}b_n + 2R\sin\frac{n\pi}{4} = 0 \quad \text{(E9.5)}$$

Since $\sin\frac{n\pi}{2}$ and $\sin\frac{n\pi}{4}$ if n is even these terms will be vanished so we can say that, $n = (2m+1)$, replace $n = (2m+1), b_n = b_m$.

Therefore Eq. (E9.4) become,

$$P_e\frac{(2m+1)^4\pi^2}{2l}.b_m - F\sin\frac{(2m+1)\pi}{2} - P\frac{(2m+1)^2\pi^2}{2l}b_m + 2R\sin\frac{(2m+1)\pi}{4} = 0$$

Collect similar terms gives,

$$b_m\left[P_e\frac{(2m+1)^4\pi^2}{2l} - P\frac{(2m+1)^2\pi^2}{2l}\right] = F\sin\frac{(2m+1)\pi}{2} - 2R\sin\frac{(2m+1)\pi}{4}$$

Or,

$$b_m\left[\frac{(2m+1)^2\pi^2}{2l}\left(P_e(2m+1)^2 - P\right)\right] = F\sin\frac{(2m+1)\pi}{2} - 2R\sin\frac{(2m+1)\pi}{4}$$

And,

$$b_m\left[\frac{(2m+1)^2\pi^2}{2l}\left((2m+1)^2 - \underbrace{r}_{P/P_e}\right)\right] = \frac{1}{P_e}F\sin\frac{(2m+1)\pi}{2} - 2R\sin\frac{(2m+1)\pi}{4}$$

Therefore

$$b_m = \frac{1}{P_e}.\frac{\left(F\sin\frac{(2m+1)\pi}{2} - 2R\sin\frac{(2m+1)\pi}{4}\right)}{\left(\frac{(2m+1)^2\pi^2}{2l}\left((2m+1)^2 - r\right)\right)} \quad \text{(E9.6)}$$

Thus the deflected shape becomes,

$$w(x) = \sum_{m=0}^{\infty} b_m.\sin\frac{n\pi x}{l}$$

Substitute the value of b_m, the deflected shape is given by,

$$w(x) = \sum_{m=0}^{\infty}\frac{1}{P_e}.\frac{\left(F\sin\frac{(2m+1)\pi}{2} - 2R\sin\frac{(2m+1)\pi}{4}\right)}{\left(\frac{(2m+1)^2\pi^2}{2l}\left((2m+1)^2 - r\right)\right)}.\sin\frac{(2m+1)\pi x}{l} \tag{E9.7}$$

$w = 0$ at place of acting force R (hinged) that is, $w(x) = 0$ at $x = \frac{l}{4}$

$$0 = \sum_{m=0}^{\infty}\frac{1}{P_e}.\frac{\left(F\sin\frac{(2m+1)\pi}{2} - 2R\sin\frac{(2m+1)\pi}{4}\right)}{\left(\frac{(2m+1)^2\pi^2}{2l}\left((2m+1)^2 - r\right)\right)}.\sin\left[\frac{(2m+1)\pi}{l}.\frac{l}{4}\right]$$

To find the ratio $\frac{R}{F}$, we should separate these terms and then simplified gives,

$$0 = \sum_{m=0}^{\infty}\frac{2l}{P_e\pi^2(2m+1)^2}.\left(\frac{F\sin\frac{(2m+1)\pi}{2}.\sin\frac{(2m+1)\pi}{4}}{(2m+1)^2 - r} - \frac{2R\sin\frac{(2m+1)\pi}{4}.\sin\frac{(2m+1)\pi}{4}}{(2m+1)^2 - r}\right)$$

Or,

$$\sum_{m=0}^{\infty}\frac{F\sin\frac{(2m+1)\pi}{2}.\sin\frac{(2m+1)\pi}{4}}{(2m+1)^2\left((2m+1)^2 - r\right)} = \sum_{m=0}^{\infty}\frac{2R\sin\frac{(2m+1)\pi}{4}.\sin\frac{(2m+1)\pi}{4}}{(2m+1)^2\left((2m+1)^2 - r\right)} \tag{E9.8}$$

Therefore

$$\frac{R}{F} = \frac{\sum_{m=0}^{\infty}\left(\frac{\sin\frac{(2m+1)\pi}{2}.\sin\frac{(2m+1)\pi}{4}}{(2m+1)^2\left((2m+1)^2 - r\right)}\right)}{\sum_{m=0}^{\infty}\left(\frac{2\sin\frac{(2m+1)\pi}{4}.\sin\frac{(2m+1)\pi}{4}}{(2m+1)^2\left((2m+1)^2 - r\right)}\right)} \tag{E9.9}$$

Or,

$$\frac{R}{F} = \frac{\left\{\frac{1}{1-r} - \frac{1}{9(9-r)} - \frac{1}{25(25-r)} + \frac{1}{49(49-r)} + \frac{1}{81(81-r)} - \frac{1}{121(121-r)} + \ldots\right\}}{\sqrt{2}\left\{\frac{1}{1-r} + \frac{1}{9(9-r)} + \frac{1}{25(25-r)} + \frac{1}{49(49-r)} + \frac{1}{81(81-r)} + \frac{1}{121(121-r)} + \ldots\right\}} \tag{E9.10}$$

Example 1.10: The simply supported beam of Fig. 1.21 carries a uniformly distributed load and end thrust. Assuming the deflection curve is approximated by $P(x) = c_1\sin\frac{\pi x}{l}$, use the potential energy principle to find c_1. "Note: to calculate the potential energy of the end load", accept that:

$$\Delta = \frac{1}{2}\int_0^l \left(\frac{dw}{dx}\right)^2 dx), \text{ and } T_e = \pi^2 EI/l^2 = \text{Euler buckling load.})$$

(Note: introduction the end, thrust makes a significant difference to the complexity of the governing differential equation for beams. In contrast, using the energy method merely requires the addition of the potential energy, $-\Delta T$, of the end thrust.)

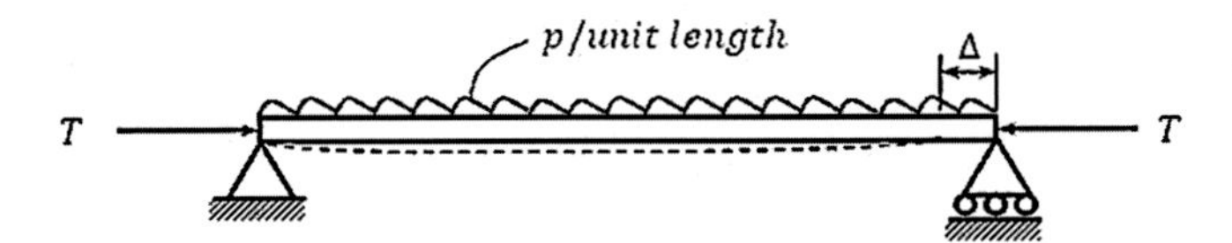

FIGURE 1.21 A simply supported beam under a U.D.L. and thrust T.

Solution,
To find the strain energy U, assume that the deflected shape is given by,

$$w(x) = c_1 \sin\frac{\pi x}{l}$$

Therefore

$$\frac{dw}{dx} = c_1 \frac{\pi}{l} \cos\frac{\pi x}{l}$$

And,

$$\frac{d^2w}{dx^2} = -c_1 \left(\frac{\pi}{l}\right)^2 \sin\frac{\pi x}{l}$$

Then:

$$U = \frac{1}{2}\int_0^l EI(w'')^2 dx = \frac{EI}{2}\int_0^l c_1^2 \left(\frac{\pi}{l}\right)^4 \sin^2\frac{\pi x}{l} dx$$

Or,

$$U = \frac{\pi^4 EI}{4l^3} c_1^2 \tag{E10.1}$$

To find the potential energy, Ω:

1. Due to distributed load:

$$\Omega_d = -\int_0^l Pw dx = -\int_0^l Pc_1 \sin\frac{\pi x}{l} dx$$
$$= Pc_1 \frac{l}{\pi}\left|\cos\frac{\pi x}{l}\right|_0^l = Pc_1 \frac{l}{\pi}(-1-1)$$

Therefore

$$\Omega_d = -2P\frac{l}{\pi}c_1 \tag{E10.2}$$

2. Due to end thrust T:

$$\Omega_T = -\frac{T}{2}\int_0^l c_1^2 \frac{\pi^2}{l} \cos^2\frac{\pi x}{l} dx$$

$$\Omega_T = -\frac{\pi^2 T}{4l} c_1^2 \tag{E10.3}$$

To find c_1:

The total potential energy,

$$V = U + \Omega_d + \Omega_T. \tag{E10.4}$$

Substitute their corresponding values from Eqs. (E10.1)–(E10.3), respectively, into Eq. (E10.4) gives,

$$V = \frac{\pi^4 EI}{4l^3}c_1^2 - \frac{\pi^2 T}{4l}c_1^2 - 2P\frac{l}{\pi}c_1$$

Applying the T.P.E. principles gives,

$$\frac{dV}{dc_1} = \left(\frac{\pi^4 EI}{2l^3} - \frac{\pi^2 T}{2l}\right)c_1 - 2P\frac{l}{\pi} = 0$$

$$= \frac{\pi^4 EI}{2l^3}\left(1 - \frac{Tl^2}{\pi^2 EI}\right)c_1 - 2P\frac{l}{\pi} = 0 \text{ and } T_e = \pi^2 EI/l^2$$

Or,

$$c_1 = \frac{4Pl^4}{\pi^5 EI}\frac{1}{(1 - T/T_e)}$$

Hence, the deflected shape is given by,

$$w(x) = \frac{4Pl^4}{\pi^5 EI} \cdot \frac{\sin(\pi x/l)}{(1 - T/T_e)} \tag{E10.5}$$

Example 1.11: A beam of length l, stiffness EI and simply supported at both ends, carries a uniformly distributed load $\frac{p_0}{\text{unitlength}}$ and an end thrust T as shown in Fig. 1.22.

Additional support is provided at midspan by a nonlinear spring having the characteristic $P = k\delta^2, p = \text{force}$, $\delta = \text{extension}$, $k = \text{constant}$. Use the principle of stationary total potential energy to formulate the equilibrium equations appropriate to a Rayleigh–Ritz treatment if the deflection curve is approximated by,

$$w(x) = \sum_{n=1,3,5,\ldots}^{N}\left(a_n \sin\frac{n\pi x}{l}\right)$$

In a particular design situation, the axial thrust is negligible; reformulate the problem in terms of the stationary complementary energy principle and deduce the value of the coefficient a_n.

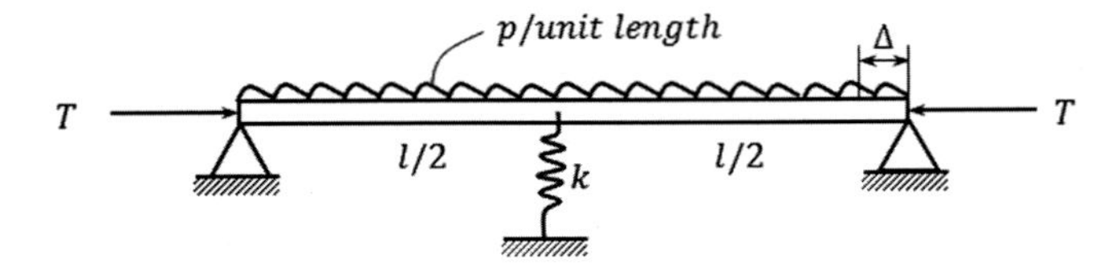

FIGURE 1.22 Simply supported beam and axial and uniformly distributed load.

Solution, Assume,

$$w(x) = \sum_{n=1,...}^{N} a_n.\sin\frac{n\pi x}{l} \tag{E11.1}$$

Then, the strain energy is given by,

$$U = \frac{EI}{l}\int w''^2 dx$$

Where,

$$w'' = \sum_{n=1}^{N} -\left(\frac{n\pi}{l}\right)^2 a_n.\sin\frac{n\pi x}{l}$$

Therefore

$$U = \frac{EI}{2}\int_0^l \sum\left(\frac{n\pi}{l}\right)^4 a_n^2 \sin\frac{n\pi x}{l}dx = \frac{EI}{2}\sum a_n^2\left(\frac{n\pi}{l}\right)^4 \frac{1}{2}\int_0^l \left(1+\cos^2\frac{n\pi x}{l}\right)dx$$

$$= \frac{EI}{2}\sum a_n^2\left(\frac{n\pi}{l}\right)^4 \frac{1}{2}\left[x - \frac{l}{2n\pi}\sin\frac{2n\pi x}{l}\right]_0^l \tag{E11.2}$$

Carrying out the integration of Eq. (E11.1) and simplification gives,

$$U = \frac{EI}{2}\sum a_n^2\left(\frac{n\pi}{l}\right)^4 \frac{l}{2}$$

The potential energy of the applied load is given by,

$$\Omega = -\int_0^l P.wdx = -\int P_0 \sum a_n \sin\frac{n\pi x}{l}dx = -\left[P_0\sum a_n \frac{n\pi}{l}\cos\frac{n\pi x}{l}\right]_0^l \tag{E11.3}$$

Carrying out the integration of Eq. (E11.3) and simplification gives,

$$\Omega = -P_0\sum a_n\left(\frac{l}{n\pi}\right)(\cos n\pi - 1) \tag{E11.4}$$

Potential energy due to thrust is given by,

$$\Omega_{\text{due to thrust}} = -T\Delta$$

$$\Delta = \frac{1}{l}\int_0^l \left(\frac{dw}{dx}\right)^2 dx$$

Or,

$$\Omega = -T\frac{1}{l}\int_0^l \left(\sum\left(\frac{n\pi}{l}\right)a_n\cos\frac{n\pi x}{l}\right)^2 dx = -\frac{T}{2}\int_0^l \sum a_n^2\left(\frac{n\pi}{l}\right)^2 \cos^2\frac{n\pi x}{l}dx \tag{E11.5}$$

Carrying out the integration of Eq. (E11.4) and simplification gives,

$$\Omega_T = -\frac{T}{2}\frac{l}{2}\sum a_n^2\left(\frac{n\pi}{l}\right)^2$$

Strain energy of spring,

$$U = P.w$$

$$P = \int_0^l k\delta^2 dx = k\frac{\delta^3}{3}\Big|_0^l = k\frac{l^3}{3}$$

Therefore

$$U = Pw_{\ x=\frac{l}{2}} = k\frac{l^3}{3}.\sum a_n \sin\frac{n\pi}{2} \tag{E11.6}$$

Total potential energy $V = U + \Omega + \Omega_{\text{due to thrust}}$
Thus

$$V = \frac{EI}{2}\sum a_n^2\left(\frac{n\pi}{l}\right)^4\frac{l}{2} + k\frac{l^3}{3}.\sum a_n \sin\frac{n\pi}{2} - \frac{T}{4}l\sum a_n^2\left(\frac{n\pi}{l}\right)^2 - P_0\sum a_n\left(\frac{l}{n\pi}\right)(\cos n\pi - 1) \tag{E11.7}$$

Minimization principles states that, $\frac{\partial V}{\partial a_n} = 0$ using Eq. (E11.7) gives,

$$\frac{EI}{2}.2a_n\left(\frac{n\pi}{l}\right)^4\frac{l}{2} + k\frac{l^3}{3}\sin\frac{n\pi}{2} - \frac{T}{4}l2a_n\left(\frac{n\pi}{l}\right)^2 - P_0\left(\frac{l}{n\pi}\right)(\cos n\pi - 1) = 0$$

Or,

$$a_n\left\{EI\left(\frac{n\pi}{l}\right)^4\frac{l}{2} - \frac{T}{2}l\left(\frac{n\pi}{l}\right)^2\right\} = P_0\left(\frac{l}{n\pi}\right)(\cos n\pi - 1) - k\frac{l^3}{3}\sin\frac{n\pi}{2} = \frac{P_0\left(\frac{l}{n\pi}\right)(\cos n\pi - 1) - k\frac{l^3}{3}\sin\frac{n\pi}{2}}{\left\{EI\left(\frac{n\pi}{l}\right)^4\frac{l}{2} - \frac{T}{2}l\left(\frac{n\pi}{l}\right)^2\right\}}$$

$$\text{since, } w(x) = a_n \sin\frac{n\pi x}{l}$$

For $n = 1$

$$a_1 = \frac{2P_0\left(\frac{l}{\pi}\right) - k\frac{l^3}{3}}{\left\{\frac{EI}{2}\frac{\pi^4}{l^3} - \frac{T}{2l}\pi^2\right\}}, \text{ For odd numbers.}$$

$a_2 = 0$, For even numbers.

$$a_3 = \frac{-2P_0\left(\frac{l}{3\pi}\right) - k\frac{l^3}{3}}{\left\{\frac{EI}{2}\frac{(3\pi)^4}{l^3} - \frac{T}{2l}(3\pi)^2\right\}}$$

$$\vdots$$

Therefore

$$w(x) = \left(a_1 \sin\frac{n\pi x}{l} + a_3 \sin\frac{n\pi x}{l} + \ldots\right)$$

Example 1.12: A particular machine is modeled by a uniform beam S.S. at each end with the possibility of a simple linear spring, stiffness k provided additional support of mid span (Fig. 1.23). If to provide a basic for CAD package,

$$w(x) = \sum_{n=1}^{N} q_n \sin\frac{n\pi x}{l} \tag{E12.1}$$

Is acceptable as a displacement model. On the basis of total potential energy principles, find the expression of q_n which can be used in Eq. (E12.1) and implemented in CAD package.

Solution,

The deflected shape of the simply supported beam starts from zero value at the first support and increases and reaches a maximum value at the center and then reaches zero value at the other support. This behavior for a beam-loaded vertical, but in this example the case is different, the thrust increases the deflection and the spring acts as an elastic support. Therefore the best assumption of the deflected shape is the sine function, and the validity can be checked by using the boundary conditions.

Using the given assumptions,

$$\text{at,} \quad x = 0, \quad w_0 = q_n \sin(0) = 0$$

$$\text{at,} \quad x = \tfrac{l}{2}, \quad w_{\frac{l}{2}} = q_n \sin\left(\tfrac{n\pi}{l}\tfrac{l}{2}\right) = q_n = \text{Maximum deflection.}$$

$$\text{at,} \quad x = l, \quad w_l = q_n \sin(n\pi) = 0$$

As before, the strain energy U, potential energy due to applied load Ω_p, and the potential energy due to thrust Ω_T, are given as follows,

$$U = \int_0^l \frac{EI}{2} (w'')^2 dx \tag{E12.2}$$

$$\Omega_p = -\int_b^{2b} P.w dx \tag{E12.3}$$

$$\Omega_T = -\frac{T}{2} \int_0^l (w')^2 dx \tag{E12.4}$$

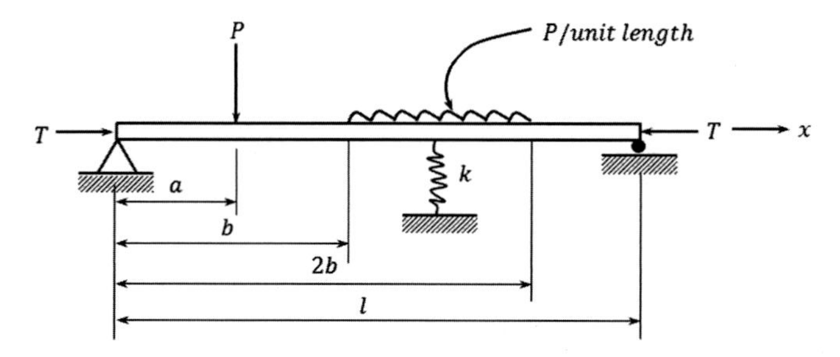

FIGURE 1.23 A simply supported beam under thrust and bending loading with an addition of spring support.

According to the assumption of the deflected shape of the beam, we have,

$$w(x) = \sum q_n \sin\left(\frac{n\pi x}{l}\right)$$

Carrying out the differentiation as required in the calculation of energies as follows,

$$\frac{\partial w}{\partial x} = \sum \frac{n\pi}{l} q_n \cos\left(\frac{n\pi x}{l}\right)$$

Therefore

$$\frac{\partial^2 w}{\partial x^2} = -\sum \left(\frac{n\pi}{l}\right)^2 q_n \sin\left(\frac{n\pi x}{l}\right)$$

And,

$$\left(\frac{\partial w}{\partial x}\right)^2 = \sum \left(\frac{n\pi}{l}\right)^2 q_n^2 \cos^2\left(\frac{n\pi x}{l}\right)$$

$$\left(\frac{\partial^2 w}{\partial x^2}\right)^2 = \sum \left(\frac{n\pi}{l}\right)^4 q_n^2 \sin^2\left(\frac{n\pi x}{l}\right)$$

Substitute to calculate the strain energy in Eq. (E12.2) as follows,

$$U = \int_0^l \frac{EI}{2} \sum \left(\frac{n\pi}{l}\right)^4 q_n^2 \sin^2\left(\frac{n\pi x}{l}\right) dx + \frac{1}{2} k w_{x=\frac{l}{2}}^2$$

Or,

$$= \frac{EI\pi^4}{2l^4} \sum n^4 q_n^2 \int_0^l \sin^2\left(\frac{n\pi x}{l}\right) dx + \frac{1}{2} k w_{x=\frac{l}{2}}^2$$

Simplification gives,

$$= \frac{EI\pi^4}{2l^4} \sum n^4 q_n^2 \left[\int_0^l \left(\frac{1}{2} - \frac{1}{2}\cos\left(\frac{2n\pi x}{l}\right)\right) dx\right] + \frac{1}{2} k \sum q_n^2 \sin^2\left(\frac{n\pi}{2}\right)$$

$$= \frac{EI\pi^4}{2l^4} \sum n^4 q_n^2 \left[\int_0^l dx - \frac{1}{2}\int_0^l \cos\left(\frac{2n\pi x}{l}\right) dx\right] + \frac{1}{2} k \sum q_n^2 \sin^2\left(\frac{n\pi}{2}\right)$$

$$= \frac{EI\pi^4}{2l^4} \sum n^4 q_n^2 \left[\frac{l}{2} - \frac{l}{4n\pi}\int_0^l \frac{2n\pi}{l} \cos\left(\frac{2n\pi x}{l}\right) dx\right] + \frac{1}{2} k \sum q_n^2 \sin^2\left(\frac{n\pi}{2}\right)$$

$$= \frac{EI\pi^4}{2l^4} \sum n^4 q_n^2 \left[\frac{l}{2} - \overbrace{\frac{l}{4n\pi} \sin\left(\frac{2n\pi x}{l}\right)\bigg|_0^l}^{0}\right] + \frac{1}{2} k \sum q_n^2 \sin^2\left(\frac{n\pi}{2}\right)$$

$$= \frac{EI\pi^4}{2l^4} \sum n^4 q_n^2 \frac{l}{2} + \frac{1}{2} k \sum q_n^2 \sin^2\left(\frac{n\pi}{2}\right)$$

Or,

$$U = \frac{EI\pi^4}{4l^3}\sum n^4 q_n^2 + \frac{k}{2}\sum q_n^2 \sin^2\left(\frac{n\pi}{2}\right) \tag{E12.5}$$

The potential energy due to applied load Eq. (E12.3) is calculated as follows,

$$\Omega_p = -\int_{b=\frac{l}{3}}^{2b=\frac{3l}{2}} P.w(x)dx$$

Or,

$$= -\int_{b=\frac{l}{3}}^{2b=\frac{3l}{2}} \sum q_n P \sin\frac{n\pi x}{l} dx = -P\sum_{n=1}^{N} q_n \int_0^l \sin\frac{n\pi x}{l} dx$$

$$= \sum q_n \frac{l}{n\pi}\int_0^l \frac{-n\pi}{l}\sin\frac{n\pi x}{l} dx = P\sum q_n \frac{l}{n\pi}\cos\frac{n\pi x}{l}\Big|_{b=\frac{l}{3}}^{2b=\frac{3l}{2}}$$

Carrying out the integration and simplification gives,

$$\Omega_p = P\sum_{n=1}^{N} q_n \frac{l}{n\pi}\left[\cos\frac{2n\pi}{3} - \cos\frac{n\pi}{3}\right] \tag{E12.6}$$

The potential energy given by Eq. (E12.4) is calculated as follows,

$$\Omega_T = -\frac{T}{2}\int_0^l (w')^2 dx$$

Substituting the value of w' and simplification gives,

$$= -\frac{T}{2}\int_0^l \sum \frac{n^2\pi^2}{l^2} q_n^2 \cos^2\frac{n\pi x}{l} dx = -\frac{T}{2}\sum \frac{n^2\pi^2}{l^2} q_n^2 \int_0^l \cos^2\frac{n\pi x}{l} dx$$

Simplification and carrying out the integration gives,

$$= -\frac{T}{2}\sum \frac{n^2\pi^2}{l^2} q_n^2 \int_0^l \left[\frac{1}{2} + \frac{1}{2}\cos\frac{2n\pi x}{l}\right] dx$$

$$= -\frac{T}{2}\sum \frac{n^2\pi^2}{l^2} q_n^2 \left[\frac{1}{2}\int_0^l dx + \frac{1}{2}\int_0^l \cos\frac{2n\pi x}{l} dx\right]$$

$$= -\frac{T}{2}\sum \frac{n^2\pi^2}{l^2} q_n^2 \left[\frac{l}{2} + \frac{l}{4n\pi}\int_0^l \frac{2n\pi}{l}\cos\frac{2n\pi x}{l} dx\right]$$

$$= -\frac{T}{2}\sum \frac{n^2\pi^2}{l^2} q_n^2 \left[\frac{l}{2} + \frac{l}{4n\pi}\sin\frac{2n\pi}{l}\Big|_0^l\right] = -\frac{T}{2}\sum \frac{n^2\pi^2}{l^2} q_n^2 \frac{l}{2}$$

Or,

$$\Omega_T = -\frac{T\pi^2}{4l}\sum n^2 q_n^2 \tag{E12.7}$$

The potential energy at $x = a$ is,

$$\Omega_{p_N} = -P_N w_{x=a}$$

$$\Omega_{p_N} = -P_N \sum q_n \sin\left(\frac{n\pi a}{l}\right) \tag{E12.8}$$

The total potential energy is given by,

$$V = U + \Omega_p + \Omega_{p_N} + \Omega_T \tag{E12.9a}$$

Inserting the values of (E12.5)–(E12.8) gives,

$$V = \frac{EI\pi^4}{4l^3}\sum n^4 q_n^2 + \frac{k}{2}\sum q_n^2 \sin^2\left(\frac{n\pi}{2}\right) + P\sum_{n=1}^{N} q_n \frac{l}{n\pi}\left[\cos\frac{2n\pi}{3} - \cos\frac{n\pi}{3}\right] - \frac{T\pi^2}{4l}\sum n^2 q_n^2 - P_N \sum q_n \sin\left(\frac{n\pi a}{l}\right)$$

Minimization of total potential energy gives,

$$\delta V = 0, \quad \text{or} \quad \frac{\partial V}{\partial q_n} = 0$$

And,

$$\frac{\partial V}{\partial q_n} = \frac{EI\pi^4}{2l^3}\sum n^4 q_n + \frac{K}{2}2\sum q_n \sin^2\left(\frac{n\pi}{2}\right) + P\sum_{n=1}^{N}\frac{l}{n\pi}\left[\cos\frac{2n\pi}{3} - \cos\frac{n\pi}{3}\right] - \frac{T\pi^2}{4l}2\sum n^2 q_n$$
$$- P_N \sum \sin\left(\frac{n\pi a}{l}\right) = 0$$

Simplification gives,

$$q_n = \frac{\left(P_N \sum \sin\left(\frac{n\pi a}{l}\right) - P\sum\limits_{n=1}^{N}\frac{l}{n\pi}\left[\cos\frac{2n\pi}{3} - \cos\frac{n\pi}{3}\right]\right)}{\left(\frac{EI\pi^4}{2l^3}\sum n^4 + \frac{K}{2}2\sum \sin^2\left(\frac{n\pi}{2}\right) - \frac{T\pi^2}{4l}2\sum n^2\right)} \tag{E12.9}$$

So, the deflected shape is as follows,

$$w(x) = \sum_{n=1}^{N} q_n \sin\left(\frac{n\pi x}{l}\right)$$

Or,

$$w(x) = \sum_{n=1}^{N}\left[\frac{\left(P_N \sum \sin\left(\frac{n\pi a}{l}\right) - P\sum\limits_{n=1}^{N}\frac{l}{n\pi}\left[\cos\frac{2n\pi}{3} - \cos\frac{n\pi}{3}\right]\right)}{\left(\frac{EI\pi^4}{2l^3}\sum n^4 + \frac{K}{2}2\sum \sin^2\left(\frac{n\pi}{2}\right) - \frac{T\pi^2}{4l}2\sum n^2\right)}\right]\sin\left(\frac{n\pi x}{l}\right) \tag{E12.10}$$

Example 1.13: The rigid bar shown in Fig. 1.24 is supported by three wire as shown. If $E = 207000\ \mathrm{N/mm^2}$ and cross-sectional area $A = 25\ \mathrm{mm^2}$ for each bar, use the principle of S.T.P.E. to determine the angular deflection θ of the bar.

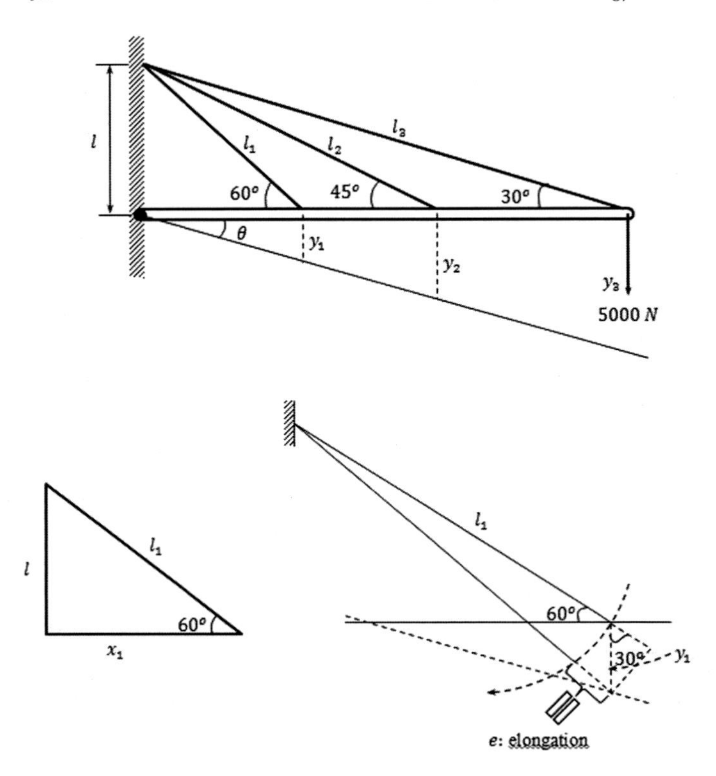

FIGURE 1.24 A rigid bar suspended by three cables and subjected to a concentrated load at the end with a free body diagram for the cable inclined 60 degrees.

From geometry,

$$x_1 = \frac{l}{\sqrt{3}}$$

$$y_1 = x_1.\theta = \frac{l}{\sqrt{3}}.\theta$$

Similarly, for other cables, 2 and 3, respectively,

$$y_2 = l.\theta$$

$$y_3 = \sqrt{3}.l.\theta$$

The extension of the cables, 1, 2, and 3, respectively,

$$e_1 = \cos\theta.y_1 = \frac{l}{2}.\theta$$

$$e_2 = y_2 \cos 45 = \frac{l}{\sqrt{2}}.\theta$$

And,

$$e_3 = y_3 \cos 60 = \frac{\sqrt{3}}{2}.\theta$$

Strain energy in tension is given by,

$$U = \frac{EA}{2l}e^2$$

And,

$$U_{\text{total}} = \frac{EA}{2}\left[\frac{e_1^2}{l_1} + \frac{e_2^2}{l_2} + \frac{e_3^2}{l_3}\right] \tag{E13.1}$$

Potential energy Ω due to applied load P is given by,

$$\Omega = -P.y_3 = -P.\frac{\sqrt{3}}{2}l.\theta \tag{E13.2}$$

Total potential energy V is given by,

$$V = U + \Omega$$

By Substitution their corresponding values from Eqs. (E13.1) and (E13.2) as follows,

$$= \frac{EA}{2}\left[\frac{\sqrt{3}}{8}l.\theta^2 + \frac{1}{2\sqrt{2}}l.\theta^2 + \frac{3}{8}l.\theta^2\right] - P.\frac{\sqrt{3}}{2}l.\theta \tag{E13.3}$$

$\delta V = 0$, minimization principles of Eq. (E13.3), gives,

$$0 = \frac{EA}{2}\left[\frac{\sqrt{3}}{4}.\theta + \frac{1}{\sqrt{2}}\theta + \frac{3}{4}\theta\right] - 5000.\frac{\sqrt{3}}{2}$$

From which,

$$\theta = 0.2626$$

Example 1.14: A spring-loaded pantograph provides electrical contact through a shoe designed to support a load W when the angle between the equal arms is θ, Fig. 1.25. Each spring has stiffness k and is free of load when $oA = a$. Find the necessary value of k if the weights of all the parts are negligible compared with W (Fig. 1.25).

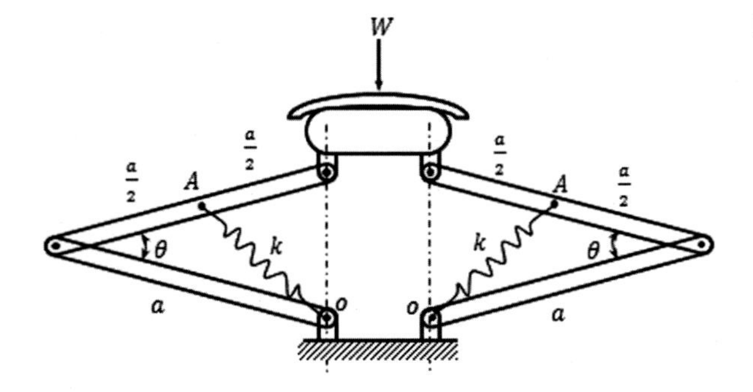

FIGURE 1.25 Spring-loaded pantograph.

Solution,

Fig. 1.26 show the positions of the springs poised at the rods and Fig. 1.27 shows the extension positions of the frame,

Let the case that when there is load applied, (cosine law) Fig. 1.27,

$$l^2 = a^2 + \frac{a^2}{4} - 2.a.\frac{a}{2}\cos\theta \tag{E14.1}$$

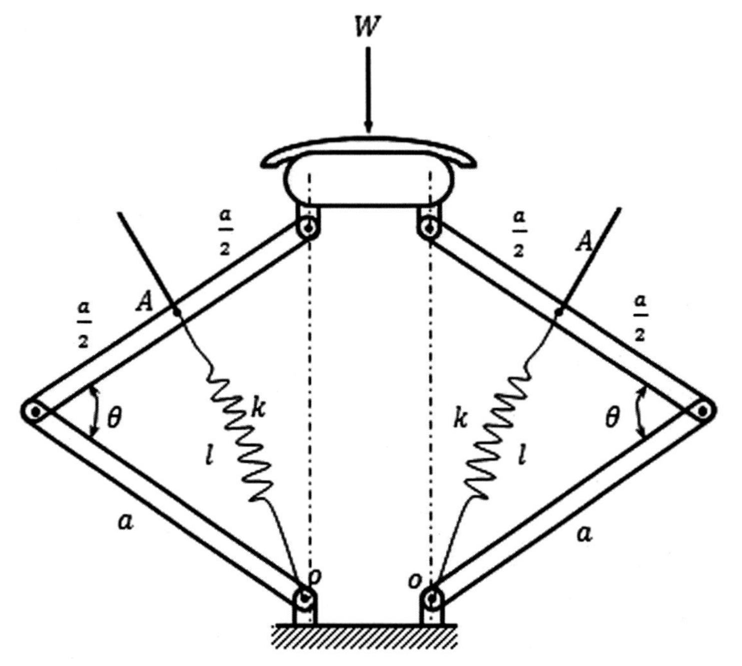

FIGURE 1.26 Extension position of the frame.

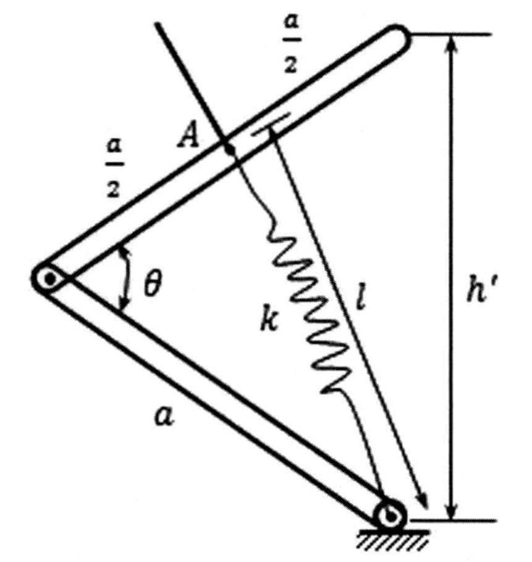

FIGURE 1.27 Extension of the left spring.

Extension of left spring due to the applied load is $= l - a$
And,

$$l - a = \sqrt{a^2 + \frac{a^2}{4} - a^2\cos\theta} - a$$

$$l - a = \sqrt{\frac{5}{4}a^2 - a^2\cos\theta} - a \quad \text{(E14.2)}$$

Or,

$$l - a = a\left(\sqrt{\frac{5}{4} - \cos\theta} - 1\right) \quad \text{(E14.3)}$$

Using cosine law in Fig. 1.28 gives,

$$a^2 = a^2 + \frac{a^2}{4} - 2a.\frac{a}{2}\cos\beta$$

Therefore

$$\cos\beta = \frac{5}{4} - 1 = \frac{1}{4} \quad \text{(E14.4)}$$

Where β is the angle that makes the length of the spring $= a$.
To find h, again from the cosine law (Fig. 1.28),

$$h^2 = a^2 + a^2 - 2a^2\cos\beta$$

Or, using Eq. (E14.2),

$$= 2a^2\left(1 - \frac{1}{4}\right) = 2*\frac{3}{4}a^2$$

Therefore

$$h = \sqrt{\frac{3}{2}}a \quad \text{(E14.5)}$$

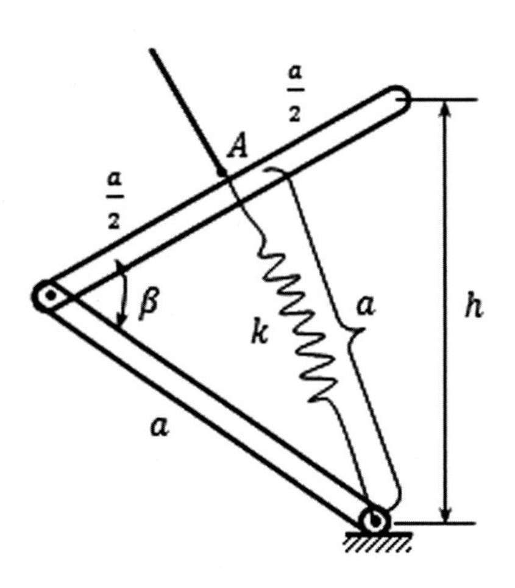

FIGURE 1.28 F.B.D. of the left spring in an angle β.

and,

$$h'^2 = a^2 + a^2 - 2a^2\cos\theta$$
$$= 2a^2 - 2a^2\cos\theta$$
$$= 2a^2(1 - \cos\theta)$$

And,

$$= 2a^2.2\sin\frac{\theta}{2}$$

Or,

$$h' = 2a*\sin\frac{\theta}{2}$$

Then, Displacement of the load W is,

$$h - h' = \sqrt{\frac{3}{2}}a - 2a*\sin\frac{\theta}{2} \tag{E14.6}$$

Strain energy

For the two springs, using the extension given in Eq. (E14.3), the strain energy of the springs is gives by,

$$U = 2*\frac{1}{2}k\left[a\sqrt{\frac{5}{4}-\cos\theta}-1\right]^2$$

$$U = ka^2\left[\sqrt{\frac{5}{4}-\cos\theta}-1\right]^2 \tag{E14.7}$$

Potential energy, using Eq. (E14.5),

$$\Omega = -W\left(\sqrt{\frac{3}{2}}a - 2a*\sin\frac{\theta}{2}\right) \tag{E14.8}$$

The total potential energy is given by,

$$V = U + \Omega$$

Inserting the values of U and Ω from Eqs. (E14.6) and (E14.8), respectively, gives,

$$= ka^2\left[\sqrt{\frac{5}{4}-\cos\theta}-1\right]^2 - Wa\left(\sqrt{\frac{3}{2}} - 2*\sin\frac{\theta}{2}\right) \tag{E14.9}$$

From minimization principles using Eq. (E14.3), $\delta V = 0$

$$\delta V = 2ka^2\left(\sqrt{\frac{5}{4}} - \cos\theta - 1\right).\frac{1}{2}\left(\frac{5}{4}-\cos\theta\right)^{-1/2}(\sin\theta)\delta\theta + 2Wa\cos\frac{\theta}{2}.\frac{1}{2}\delta\theta$$

$$0 = ka^2\left(\sqrt{\frac{5}{4}} - \cos\theta - 1\right)\left(\frac{5}{4}-\cos\theta\right)^{-1/2}\sin\theta.\delta\theta + Wa\delta\theta\cos\frac{\delta\theta}{2}$$

Or,

$$ka^2\left(\sqrt{\frac{5}{4}-\cos\theta}-1\right).\frac{1}{\sqrt{\left(\frac{5}{4}-\cos\theta\right)}}\sin\theta+wWa\ \cos\frac{\delta\theta}{2}=0$$

$$ka^2\left(\sqrt{\frac{5}{4}-1+2\sin^2\frac{\delta\theta}{2}}-1\right)\frac{1}{\sqrt{\left(\frac{5}{4}-1+2\sin^2\frac{\delta\theta}{2}\right)}}.2\sin^2\frac{\delta\theta}{2}\cos\frac{\delta\theta}{2}+Wa\cos\frac{\delta\theta}{2}=0 \quad \text{(E14.10)}$$

We have,

$$\sin^2\frac{\delta\theta}{2}=\frac{1}{2}(1-\cos\theta)$$

Or,

$$\frac{1}{2}\cos\theta=\frac{1}{2}-\sin^2\frac{\delta\theta}{2}$$

Therefore

$$\cos\theta=1-2\sin^2\frac{\delta\theta}{2}$$

And,

$$\sin\theta=2\sin\frac{\delta\theta}{2}\cos\frac{\delta\theta}{2}$$

$$\sin\theta=\sqrt{1-\cos^2\theta}$$

Or,

$$\sin\theta=\sqrt{1-\left(1-2\sin^2\frac{\delta\theta}{2}\right)^2}$$

$$=\sqrt{1-1+4\sin^2\frac{\delta\theta}{2}-4\sin^4\frac{\delta\theta}{2}}$$

$$=2\sin\frac{\delta\theta}{2}\sqrt{1-\sin^2\frac{\delta\theta}{2}}$$

Or,

$$=2\sin\frac{\delta\theta}{2}\cos\frac{\delta\theta}{2}$$

Put into Eq. (E14.9),

$$ka^2\left(\sqrt{\frac{1}{4}+2\sin^2\frac{\delta\theta}{2}}-1\right)\frac{1}{\sqrt{\frac{1}{4}+2\sin^2\frac{\delta\theta}{2}}}2\sin\frac{\delta\theta}{2}=-\ Wa \quad \text{(E14.11)}$$

Simplification of Eq. (E14.11) gives,

$$k = -\frac{Wa}{2a\sin\frac{\delta\theta}{2}} \frac{\sqrt{1 + 8\sin^2\frac{\delta\theta}{2}}}{\left(\sqrt{1 + 8\sin^2\frac{\delta\theta}{2}} - 2\right)}$$

Or,

$$k = \frac{Wa}{2a\sin\frac{\delta\theta}{2}} \frac{\sqrt{1 + 8\sin^2\frac{\delta\theta}{2}}}{\left(\sqrt{1 + 8\sin^2\frac{\delta\theta}{2}} - 2\right)} \quad \text{(E14.12)}$$

Example 1.15: A component in a test rig and the load applied to it are modeled as shown in Fig. 1.29. Assuming that the axially displacement in each element varies linearly with axial position, (so that this is only an approximation for the tapered element), use the P.V.W. to estimate the displacement at the load. How would you improve on the accuracy of your predictions assuming that the conceptual model of Fig. 1.29 is adequate? The material is steel.

Solution,
From Fig. 1.29. we have,

$$A = A_0\left(1 - \frac{x}{2l_1}\right)$$

For elements (1), (2):
Assume that the displacement model is given by,

$$u(x) = cx + b \quad \text{(E15.1)}$$

Boundary conditions are,
$u = u_0$ at $x = l_1$ (Fig. 1.30),

$$u_0 = cl_1 \rightarrow \text{which gives} c = \frac{u_0}{l_1}$$

Therefore

$$u = \frac{u_0}{l_1}x \text{ and } \frac{du}{dx} = \frac{u_0}{l_1} \quad \text{(E15.2)}$$

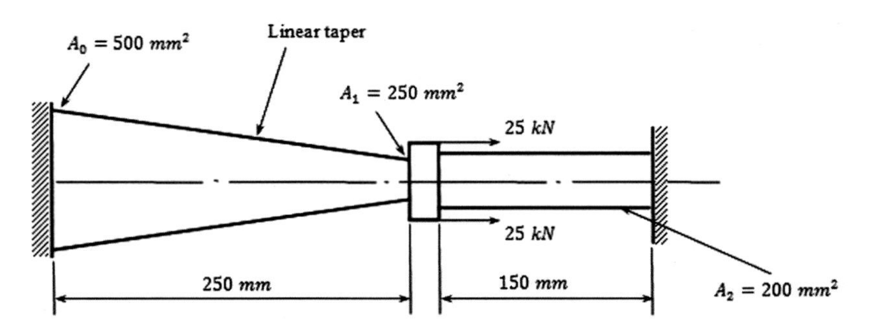

FIGURE 1.29 Model of component rig.

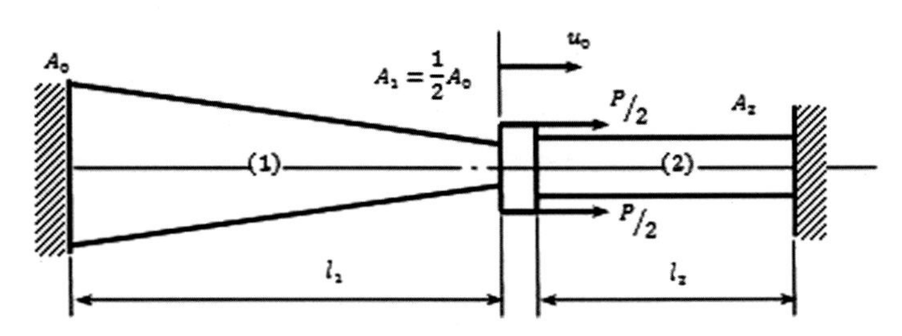

FIGURE 1.30 Model of the component.

Element (1) – strain energy stored:
The strain energy for element No. 1 is given by,

$$U = \frac{1}{2}\int_0^{l_1} EA\left(\frac{du}{dx}\right)^2 dx \tag{E15.3}$$

We have,

$$u = \frac{x}{l_1}u_0 \text{ and the strain } \varepsilon = \frac{du}{dx} = \frac{u_0}{l_1}$$

Hence, Eq. (E15.1) becomes,

$$U = \frac{EA_0}{2}\frac{u_0^2}{l_1^2}\int_0^{l_1}\left(1 - \frac{x}{2l_1}\right)dx = \frac{EA_0u_0^2}{2l_1^2}\left|x - \frac{x^2}{4l_1}\right|_0^{l_1}$$

Carrying out the integration gives,

$$U = \frac{3EA_0}{8l_1}u_0^2 \tag{E15.4}$$

Element (2) – strain energy stored:
The strain energy for element No. 2 is given by,

$$U = \frac{1}{2}\int_0^{l_2} EA\left(\frac{du}{dx}\right)^2 dx \tag{E15.5}$$

The boundary conditions are,
$u = u_0\left(1 - \frac{x}{l_2}\right)$, Hence, the strain becomes, $\varepsilon = \frac{du}{dx} = -\frac{u_0}{l_2}$,
Put into Eq. (E15.2) becomes,

$$U = \frac{1}{2}\int_0^{l_2} EA_2\frac{u_0^2}{l_2^2}dx = \frac{EA_2}{2l_2}u_0^2 \tag{E15.6}$$

P.E. due to the applied load:

$$\text{P.E.} = -Pu_0$$

To find u_0,
The total potential energy $V = U_1 + U_2 + \text{P.E.}$ is as follows,

$$V = \left(\frac{3EA_0}{8l_1} + \frac{1}{2}\frac{EA_2}{l_2}\right)u_0^2 - Pu_0$$

for $\delta V = 0$, total potential energy principles,

$$u_0 = \frac{P}{\left[\frac{3EA_0}{4l_1} + \frac{EA_2}{l_2}\right]} = \frac{\frac{1}{2} \times \frac{100000}{207000}}{\left[\frac{3}{4} \times \frac{500}{250} + \frac{4}{3} \times \frac{200}{150}\right]}$$

$$\text{Hence,} \quad u_0 = 0.085 \text{ mm.}$$

Example 1.16: Accepting that when Hook's law springs experience a temperature rise T, the constitutive equation becomes:

$$F = k(e - \alpha l T)$$

Generate the matrix equilibrium equation for the assemblage of Fig. 1.31, where α = temperature coefficient of expansion, l = length, and$e$ = spring extension. Use q_1 and q_2 as generalized coordinates.

Solution,

Equilibrium conditions, Fig. 1.32:

Equilibrium of forces gives,

$$\sum F = F_1 + F_2 + F_3 - W = 0 \tag{E16.1}$$

Equilibrium of moments,

$$\sum M_0 = F_3 b + F_1 a - \frac{Wb}{2} = 0 \tag{E16.2}$$

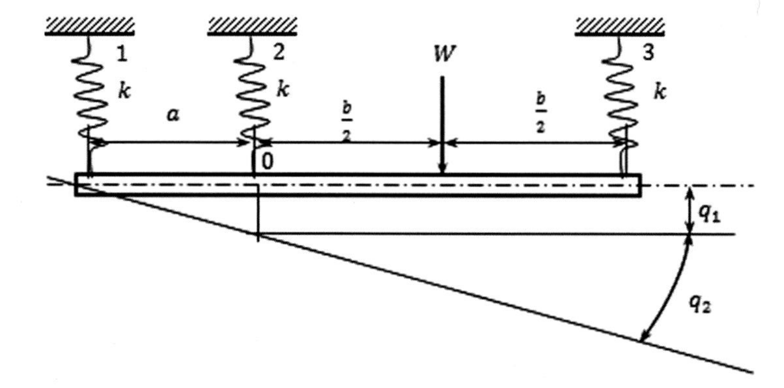

FIGURE 1.31 A beam suspended by three springs.

FIGURE 1.32 F.B.D. of the beam.

Compatibility conditions, the extension in these springs are,

$$e_1 = q_1 - aq_2,\ e_2 = q_1,\ e_3 = q_1 + bq_2 \tag{E16.3}$$

Hooke's law:

$$F_i = k_i(e_i - \alpha_i l_i T_i) \tag{E16.4}$$

Put Eq. (E16.3) into Eq. (E16.4) in Eqs. (E16.1) and (E16.2) gives,

$$k_1(e_1 - \alpha_1 l_1 T_1) + k_2(e_2 - \alpha_2 l_2 T_2) + k_3(e_3 - \alpha_3 l_3 T_3) = W$$

Or,

$$k_1(q_1 - aq_2) + k_2(q_1) + k_3(q_1 + bq_2) - k_1\alpha_1 l_1 T_1 - k_2\alpha_2 l_2 T_2 - k_3\alpha_3 l_3 T_3 = W$$

$$(k_1 + k_2 + k_3)q_1 + (-ak_1 + bk_3)q_2 = \left(W + \sum_{i=1}^{3} k_i\alpha_i l_i T_i\right) \tag{E16.5}$$

$$k_3(q_1 + bq_2 - \alpha_3 l_3 T_3)b - k_1(q_1 - aq_2 - \alpha_1 l_1 T_1)a = \frac{Wb}{2}$$

$$(-ak_1 + bk_3)q_1 + \left(a^2k_1 + b^2k_3\right)q_2 = \frac{Wb}{2} - k_1 a\alpha_1 l_1 T_1 + k_3 b\alpha_3 l_3 T_3 \tag{E16.6}$$

Put Eqs. (E16.5) and (E16.6) in matrix form as follows,

$$\begin{bmatrix} (k_1 + k_2 + k_3) & (-ak_1 + bk_3) \\ (-ak_1 + bk_3) & (a^2k_1 + b^2k_3) \end{bmatrix} \begin{Bmatrix} q_1 \\ q_2 \end{Bmatrix} = \begin{Bmatrix} W + \sum_{i=1}^{3} k_i\alpha_i l_i T_i \\ \frac{Wb}{2} - k_1 a\alpha_1 l_1 T_1 + k_3 b\alpha_3 l_3 T_3 \end{Bmatrix} \tag{E16.7}$$

Example 1.17: A loaded projection on a steel casting is modeled as a tapered cantilever as shown in Fig. 1.33. Assuming the one parameter deflection function:

$$w(x) = c_1 x^2(3 - x/l)$$

Use the total potential energy principles to obtain an approximation to the tip deflection. Comment on this choice of trial function. You may accept that the strain energy due to bending is given by,

$$U = \frac{1}{2}\int_0^l EI(w'')^2 dx$$

Solution,
Properties of the cantilever

$$EI(x) = EI_0\left(1 - \frac{x}{2l}\right) \tag{E17.1}$$

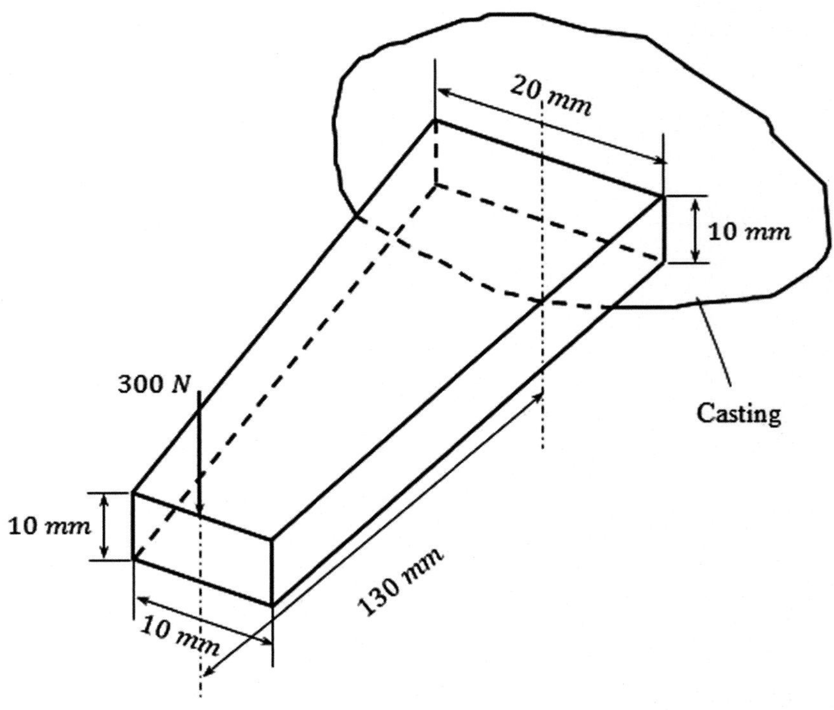

FIGURE 1.33 A tapered cantilever beam.

Where,

$I_0 = \frac{20 \times 10^3}{12}$, calculated at $x = 0$.

Displacement model

The displacement model is given by,

$$w(x) = c_1 x^2 \left(3 - \frac{x}{l}\right) \quad \text{(E17.2)}$$

And,

$$\frac{dw}{dx} = c_1 \left(6x - \frac{3x^2}{l}\right) \quad \text{(E17.3)}$$

$$\frac{d^2w}{dx^2} = 6c_1 \left(1 - \frac{x}{l}\right) \quad \text{(E17.4)}$$

Note that,

$w = \frac{dw}{dx} = 0$ at $x = 0$, as required.

Further, this function has $\left(d^2w/dx^2\right)$ ranging linearly from zero to $x = l$ to maximum at $x = 0$. It represents the deflection case for a uniform cantilever beam and so will be a useful trial function for a tapered cantilever where c_1 must be found by the P.V.W.

To find the strain energy U, we have,

$$U = \frac{1}{2}\int_0^l EI(w'')^2 dx = \frac{EI_0}{2}\int_0^l \left(1 - \frac{x}{2l}\right) 36c_1^2 \left(1 - \frac{x}{l}\right)^2 dx$$

Or,

$$= 18EI_0c_1^2\int_0^l\left[1-\frac{5}{2}\frac{x}{l}+\frac{2x^2}{l^2}-\frac{x^3}{2l^3}\right]dx$$

From which,

$$= 18EI_0c_1^2\left|x-\frac{5x^2}{4l}+\frac{2x^3}{3l^2}-\frac{x^4}{8l^3}\right|_0^l$$

Carrying out the integration and simplification gives,

$$U=\frac{21}{4}EI_0lc_1^2 \quad \text{(E17.5)}$$

To find the potential energy due to the applied load Ω:

$$\Omega=-2Pc_1l^2,\ \text{at } x=l \quad \text{(E17.6)}$$

To find c_1:
Total potential energy $V=U+\Omega$.
Put their corresponding values of U and Ω from Eqs. (E17.5) to (E17.6) as follows,

$$V=\frac{21}{4}EI_0lc_1^2-2Pl^2c_1$$

For equilibrium,

$$\delta V=0$$

$$\frac{dV}{dc_1}=0=\frac{21}{2}EI_0lc_1-2Pl^2$$

Therefore

$$c_1=\frac{4pl^2}{21EI_0l}=\frac{4}{21}\frac{pl}{EI_0}$$

Then, the approximate solution is,

$$w(x)=\frac{4plx^2}{21EI_0}(3-x/l)$$

At the tip, $x=l$,

$$w(l)=\frac{8pl^3}{21EI_0}=\frac{8\times 300\times 130^3\times 12}{21\times 207\times 10^3\times 20\times 10^3}$$

Or,

$$w(l)=0.73\ \text{mm}$$

Problems

P.1.1 A structure member is modeled as shown in Fig. P.1.1. If the material behavior is given by $\sigma=k\varepsilon^2$, where, k is constant, find the strain energy stored in the member. Neglect the dead weight and assume AE is constant.

P.1.2 The system shown in Fig. P.1.2, is used as a vehicle for demonstrating how elements having a suitable mixture of characteristics can be accommodated systematically in using the energy principles, Note that, the generalized coordinates are shown and the strain energy is calculated for the beam and the springs, that is,

$$U = \sum U_{\text{beam}} + \sum U_{\text{springs}}$$

Deduce the system stiffness matrix.

P.1.3 Apply the method of stationary total potential energy to prove that the central deflection for the beam shown in Fig. P.1.3. is given by,

$$w_{\max} = \frac{Wl^3}{106.2\, EI}$$

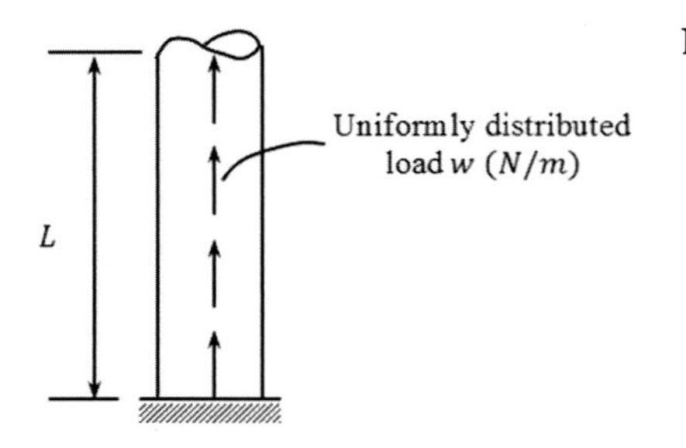

FIGURE P.1.1

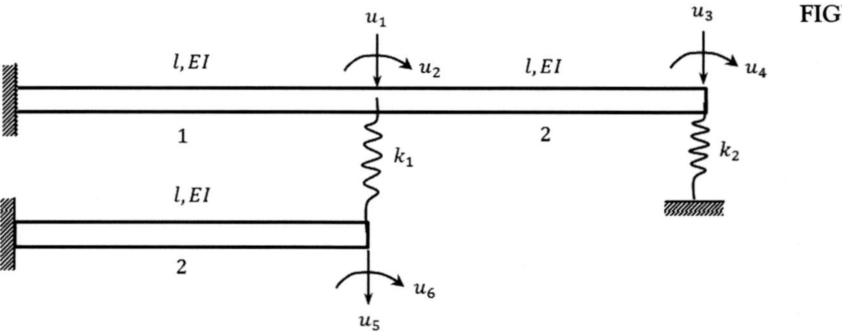

FIGURE P.1.2

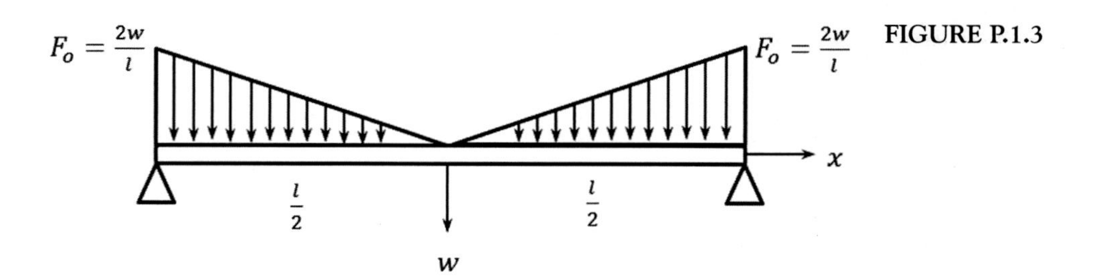

FIGURE P.1.3

The total applied load is W distributed as shown in the Fig. P.1.3. Use the trigonometric series in the assumption of the deflected shape.

P.1.4 For the purposes of static analysis, a mechanical system has been modeled into the assemblage of springs and light rigid bars shown in Fig. P.1.4. Each spring is linear so that Hooke's law $F = ke$ holds. Choosing q_1, q_2, q_3 as generalized coordinates, prove that the synthesis yields the following matrix equilibrium equation for the assemblage,

$$\begin{bmatrix} k_1 + k_2 + k_3 & -ak_1 + bk_3 & -k_3 \\ -ak_1 + bk_3 & (a^2k_1 + b^2k_3) & -bk_3 \\ -k_3 & -bk_3 & (2k_4 + k_5 + k_3) \end{bmatrix} \begin{Bmatrix} q_1 \\ q_2 \\ q_3 \end{Bmatrix} = \begin{Bmatrix} P \\ M \\ 0 \end{Bmatrix}$$

P.1.5 For question P.1.4, explain why the system has three d.o.f.

Denoting the extension of each spring by $e_1.e_2\cdots$, Show that the compatibility matrix relating the e's to generalized coordinates $q_1.q_2.q_3$ is given by,

$$\begin{Bmatrix} e_1 \\ e_2 \\ e_3 \\ e_4 \\ e_5 \end{Bmatrix} = \begin{bmatrix} 1 & -a & 0 \\ 1 & 0 & 0 \\ 1 & b & -1 \\ 0 & 0 & 1 \\ 0 & 0 & 1 \end{bmatrix} \begin{Bmatrix} q_1 \\ q_2 \\ q_3 \end{Bmatrix}$$

P.1.6 From a hand book, the deflection curve for cantilever of Fig. P.1.6A is given by:

$$w(x) = \frac{1}{2EI}\left(ax^2 - \frac{x^3}{3}\right) \text{for } 0 \le x \le a$$

Use this result to prove that the force in the spring of Fig. 1.6B is given by:

$$F = \frac{W}{2}\frac{[3(l_1/l_2) - 1]}{[1 + I_1/I_2][1 + 3EI_1/(kl_2^3(1 + I_1/I_2))]}$$

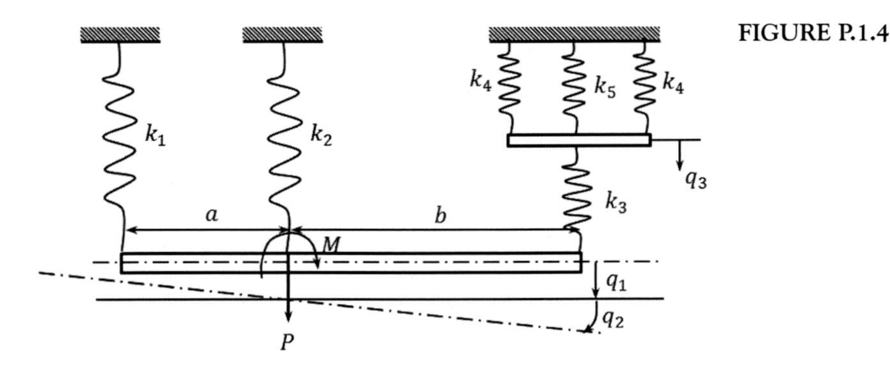

FIGURE P.1.4

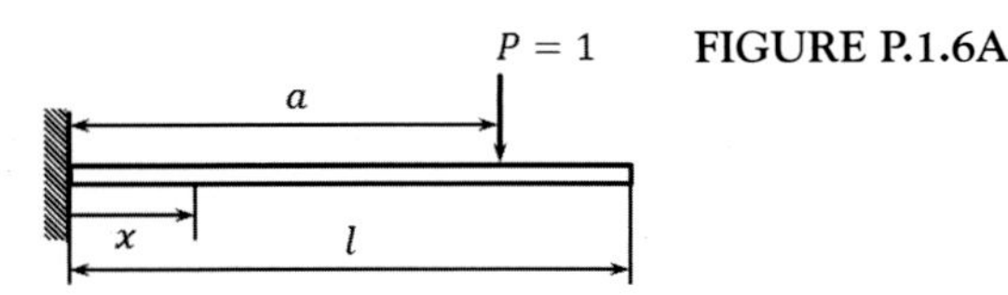

FIGURE P.1.6A

P.1.7: Establish the stiffness equilibrium equations for the systems shown in Figs. P.1.7A and P.1.7B,

P.1.8 A uniform beam of length l, simply supported at both ends, carries a uniformly distributed load $\frac{p_0}{\text{unitlength}}$ over its entire span and an end thrust T. Additional support is provided by a linear spring k, at mid span, with the origin at one support, and assuming a deflection mode,

$$w \sum_{n=1}^{N} b_n \sin \frac{n\pi x}{l}$$

Use the potential energy principle to show that the b_n's are given by simultaneous equations of the type,

$$\left[\frac{\pi^4 n^4 EI}{2l^3} - \frac{\pi^2 n^2 T}{2l}\right] b_n + k\sin\frac{n\pi}{2} \sum_{n=1}^{N} b_n \sin\frac{n\pi}{2} = \frac{p_0.l}{\pi}\left(1 - \frac{\cos n\pi}{n}\right)$$

Find also the spring force with the end thrust T removed, using the principle of complementary energy.

P.1.9 The bent is hinged at its feet, Fig. P.1.9. The diagonal tie has no flexural stiffness. Derive formulas for the rotation of each joint and for the displacement of the top member by the principle of stationary potential energy.

P.1.10 Show by considering the behavior of an element of length dx as shown in Fig. P.1.10, that the following equations describe the behavior of a beam,

$$\frac{dM}{dx} = F, \quad \frac{dF}{dx} = -p, \quad \frac{d^2}{dx^2}\left(EI\frac{d^2}{dx^2}\right) = p$$

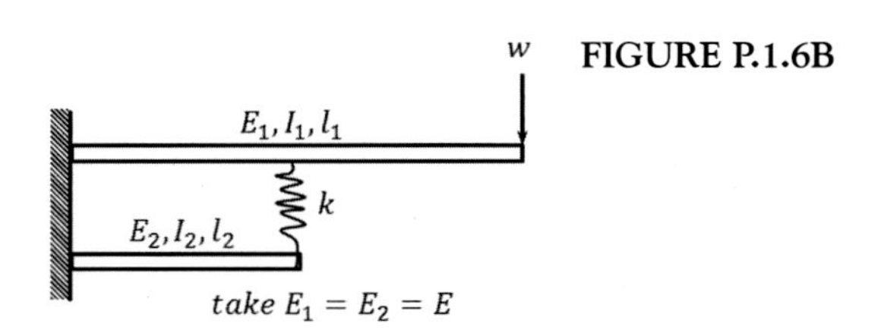

FIGURE P.1.6B

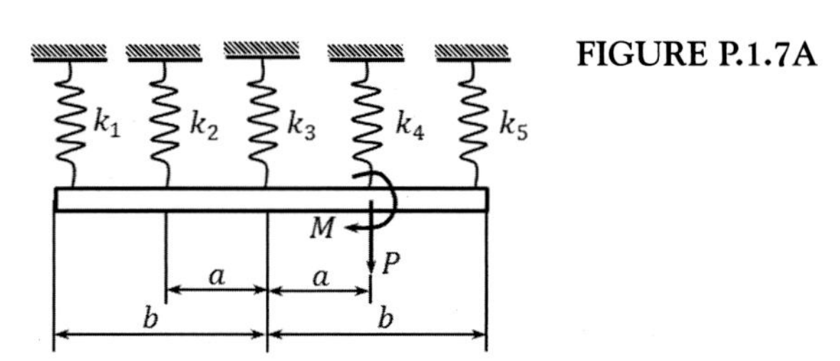

FIGURE P.1.7A

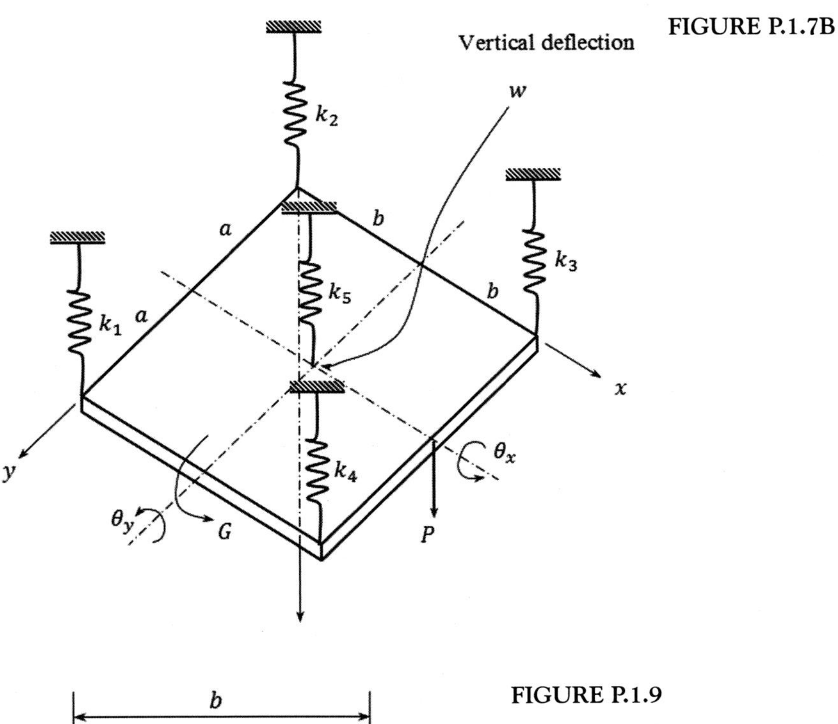

FIGURE P.1.7B

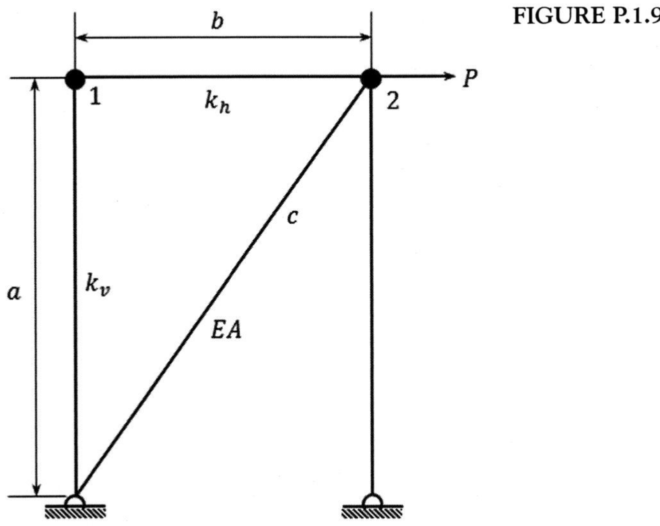

FIGURE P.1.9

Accept that the bending is in the form in the form $\frac{1}{R} = \frac{M}{EI}$ and take positive moment, shear and deflection as shown. Express the boundary conditions for the system in terms of displacements v and its derivatives.

P.1.11 Ignoring the weight of the members, find the deflection at D and the tensions in the members of the pin jointed crane structure shown in Fig. P.1.11, establish the stiffness equation for the structure and then solve for the deflection unknowns. $E = 200,000\ \text{N/mm}^2$, For all members.

P.1.12 For the following problem, shown in Fig. P.1.12 calculate the unknown tensions on the members by P.S.T.P.E. check the results by the condition of static equilibrium of forces and moments. Use

$$(EA)_1 = 3/2,\ \ (EA)_2 = 1/2,\ \ (EA)_3 = 1.$$

P.1.13 By means of P.S.T.P.E., derive tensions in the members of the wall bracket, assuming that the displacement of the joint is small and that the stress–strain relation of the material is unknown. Show that the results agree with those obtained by balancing forces at the right hand joint as shown in Fig. P.1.13. $EA =$ constant.

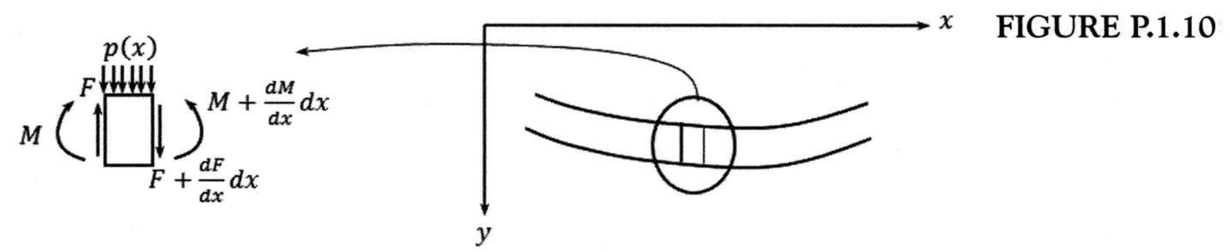

FIGURE P.1.10

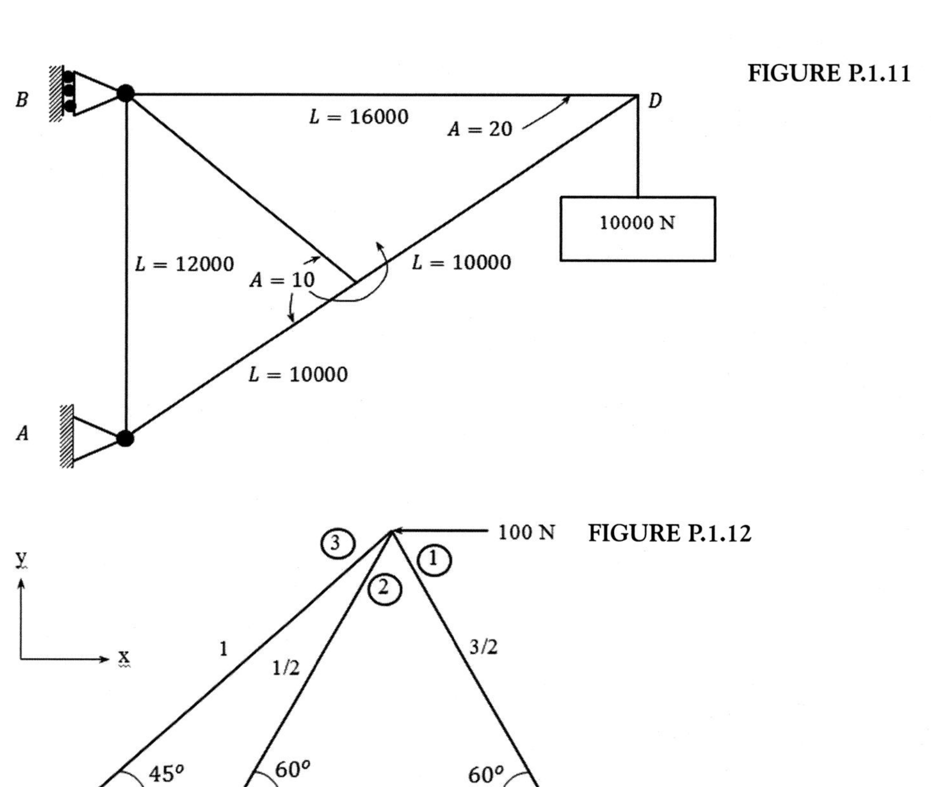

FIGURE P.1.11

FIGURE P.1.12

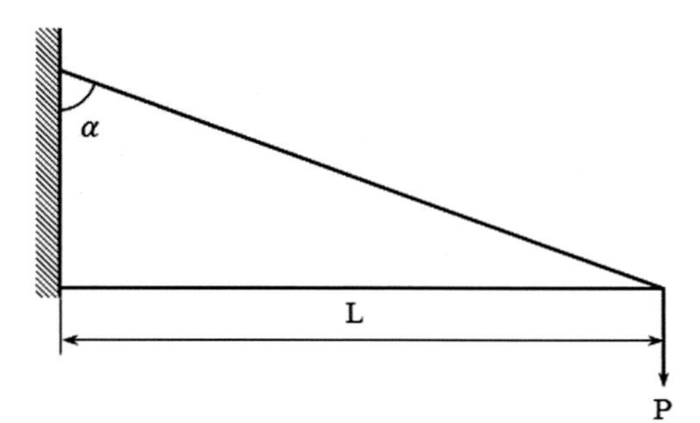

FIGURE P.1.13

Fig. P.1.14 shows the details of the bounded joint. Each cover plate has thickness t_1 and material having young's modulus E_1. The main plate is t_2 thick and its material has young's modulus E_2. The adhesive is t_a thick with a shear modulus G_a. The assembly is b wide. Using a modified complementary energy principle, establish an approximate theory for the shearing stress in the adhesive on the basis of the following assumptions:

1. The adhesive is so thin that its shearing stress τ_a does not change through the thickness although it is a function of x. Assume that, after curing, the adhesive behaves in a linear elastic manner.
2. The main and cover plates are so stiff compare with the adhesive that their plane normal sections remain plane and normal when the load is applied.

Show that for equilibrium in the region $0 < x < 1$

$$\tau_a = -t_1 \frac{d\sigma_1}{dx}, \quad \tau_a = \frac{d\sigma_2}{dx} \tag{P.14.1}$$

Introduction constraints (P.14.1) by means of Lagrange multipliers λ_1, λ_2 show that a suitable modified complementary energy functional is

$$\overline{U}^* = \left(\begin{array}{c} \frac{b}{2}\left\{ \int_0^l \left[\frac{2t_1}{E_1}{\sigma_1}^2 + \frac{t_2}{E_2}{\sigma_2}^2 + \frac{2t_a}{G_a}{\tau_a}^2 \right] dx + \frac{\sigma_2^2 a t_2}{E_2} \right\} - \\ b\left\{ \int_0^l \lambda_1 \left(\tau_a + t_1 \frac{d\sigma_1}{dx} \right) dx + \int_0^l \lambda_2 \left(\tau_a - \frac{t_2}{2}\frac{d\sigma_2}{dx} \right) dx \right\} \end{array} \right)$$

Show that $\delta\overline{U}^* = 0$ leads to,

$$\frac{d^2\tau_a}{dx^2} - \mu^2\tau_a = 0$$

Where μ a constant for the system, and that two boundary conditions is on $\frac{d\tau_a}{dx}$ are available (Fig. P.1.14).

P.1.15 Work in terms of Prandtl's stress functions and apply the complementary energy principle $\delta v^* = 0$, where,

$$\delta v^* = \frac{l}{2G} \iint \left[\left(\frac{\partial \varnothing}{\partial x} \right)^2 + \left(\frac{\partial \varnothing}{\partial x} \right)^2 - 4G \propto \varnothing \right] dxdy$$

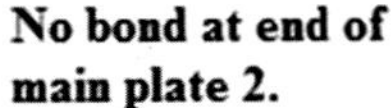

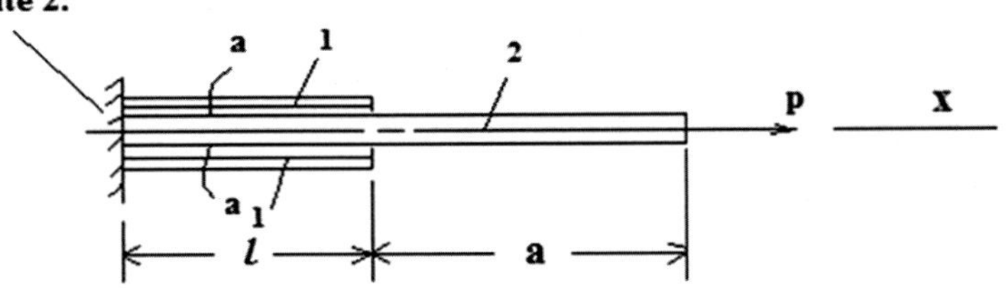

FIGURE P.1.14

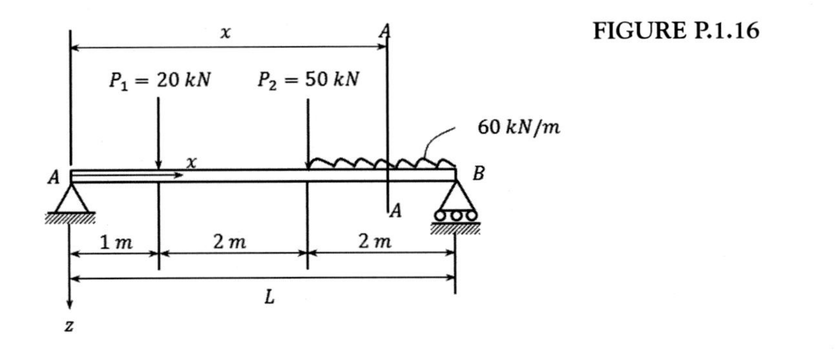

FIGURE P.1.16

For a rectangular section $2a \times 2b$, obtain an approximate value for the torsion constant assuming a one term expansion for $\varnothing(x.y)$ in the Rayleigh–Ritz process. That is, assume:

$$\varnothing(x.y) = q_{11}\cos\frac{m\pi x}{2a}\cos\frac{n\pi y}{2b}$$

(Note: for a square section, side $2a$, Timoshenko and Goodier give the value $k = 2.2496a^4$ for the torsion constant.)

P.1.16 Use the energy principles, find the deflection order the 50 KN load for the S.S. beam loading system shown in Fig. P.1.16. Also find the position and magnitude of the maximum deflection. $E = 200\ \text{GN/m}^2$, $I = 83 * 10^{-6}\ \text{m}^4$.

Bibliography

[1] C. Jordan, H.C. Carver, Calculus of Finite Differences, second ed., Book, Chelsea Publishing Company, New York, 1950.

[2] R. Weinstock, Calculus of Variations, McGraw-Hill, 1952.

[3] Z. Kodal, Numerical Analysis, second ed., Chapman and Hall, 1961.

[4] S.P. Timoshenko, J.M. Gere, Theory of Elastic Stability, second ed., McGraw-Hill, New York, 1961.

[5] H.L. Langhaar, Energy Methods in Applied Mechanics, John Wiley and Sons, Inc, 1962.

[6] R.J. Melosh, Basis for derivation of matrices for the direct stiffness method, Journal of the American Institute of Aeronautics and Astronautics I 7 (1963) 1631–1637.

[7] E. Wilsonural analysis of axisymmetric soli, Structural analysis of axisymmetric solids, Journal of the American Institute of Aeronautical and Astronautics 3 (12) (1965) 2269–2274.

[8] R. Clough, Y. Rashid, Finite element analysis of axisymmetric solids, Journal of the Engineering Mechanics, Division, American Society of Civil Engineers 91 (1965) 71–85.

[9] B.M. Irons, Engineering applications of numerical integrations in stiffness methods, Journal of the American Institute of Aeronautics and Astronautics 4 (11) (1966) 2035–2037.
[10] G.R. Cower, The shear coefficients in Timoshenko's beam theory, Journal of Applied Mechanics 33 (1966) 335–340.
[11] T.H. Richards, Energy Methods in Stress Analysis, Ellis Horwood Series in Engineering Science (1977).
[12] Y.Y. Hsieh, Elementary Theory of Structures, 2nd edition, Prentice-Hall, Englewood, Cliffs., 1982.
[13] W. Weaver Jr., P.R. Johnston, Finite Element for Structural analysis, Prentice-Hall Inc, 1984.
[14] L.J. Segerling, Applied Finite Element Analysis, Book, John Wiley and Sons, Inc., 1984.
[15] H.C. Haung, E. Hinton, A nine-node Lagrangian plate element with enhanced shear interpolation, Engineering Computer (1984) 369–379.
[16] Internal Report, Finite Element Metohd, Granfield Institute of Technology, School of Mechanical Engineering, 1985.
[17] C. Lanczos, The Variational Principles of Mechanica, fourth ed., Dover, New York, 1986.
[18] D.L. Logan, A First Course in the Finite Element Method, PWS Publishers, 1986.
[19] W. Weaver Jr., S.P. Timoshenko, D.H. Young, Vibration Problems in Engineering, fifth ed., John Wiley and Sons, Inc, New York, 1990.
[20] R.D. Cook, Finite Element Modeling for Stress Analysis, Book, John Wiley and Sons, Inc, 1995.
[21] E.J. Hearn, Mechanics of Materials, International Series on Materials Science and Technology I (1997).
[22] E.J. Hearn, Mechanics of Materials, International Series on Materials Science and Technology II (1997).
[23] P. Hunter, A. Pullan, FEM/BEM Notes, Book, Department of Engineering Science, 1997.
[24] M.J. Jweeg, K.K. Al-Kinani, M. Al-Waily, Lecture Notes, Al-Nahrain University, Al-Mustansirya University, and Kufa Unversity.

CHAPTER

2

Direct methods

In this chapter, the approximate direct methods are used to formulate the static and dynamic problems to find the displacements, strains, stresses, and natural frequencies based upon the differential equations prescribed the problem. The static applications only will be presented in this chapter. The direct methods of solving problems using the energy principles involve the Galerkin's and, Rayleigh Ritz methods. The solutions of the presented examples depend upon the formulation of the total potential energy and the minimizing principles from which the undetermined coefficients are obtained and accordingly, the assumed solution will be defined. The presented examples are aimed to give the effectiveness of the direct methods such as the axial loaded bars, subjected to concentrated direct load and the assumed distributed loading in addition to the bending cases of beams. The examples covered straight and tapered loading. The assumed functions will include polynomial with different degrees for modeling the assumed deflection of beams with different type of loading and support conditions. At the end of the chapter, different problems are included together with the references used in this chapter.

2.1 Galerkin method (G.M.)

We assume that the wanted quantity can be approximated by a series of prescribed functions and in terms of parameters to be found. These functions must satisfy all the boundary conditions, but in general, they will not satisfy the differential equation and this is the source of approximation.

Suppose we wish to solve a linear partial differential (P.D.) equation, in general, the disturbing function is given by,

$$L[w(x,y)] = f(x,y) \tag{2.1}$$

For a beam,

$$\frac{d^2}{dx^2}\left(EI\frac{d^2w}{dx^2}\right) = p \tag{2.2}$$

Assume that,

$$L = \frac{d^2}{dx^2}\left(EI\frac{d^2}{dx^2}\right)$$

Energy Methods and Finite Element Techniques
DOI: https://doi.org/10.1016/B978-0-323-88666-6.00001-0

or,

$$L[\varnothing] = g \tag{2.3}$$

And for torsion problem,

$$\nabla^2\varnothing + 2G\theta = 0 \tag{2.4}$$

Representing $w(x, y)$ for a two dimensional plate application,

$$w(x, y) = \sum_{k=1}^{\infty} C_k\varnothing_k(x, y) \tag{2.5}$$

And for a beam,

$$w = \sum_{k=1}^{n} C_k\varnothing_k(x) \tag{2.6}$$

Where, $\varnothing_k(x)$ are prescribed that is, known and the C_k's are to be found.

Since the $\varnothing$'s or $C_k(x)$ don't satisfy the differential equation exactly. Therefore, $L[\sum_{k=1}^{n} C_k\varnothing_k] - f =$ Residual $\neq 0$ in general,

The task is to make this residual as small as possible,

$$\int_0^l \left\{ L\left[\sum_{k=1}^{n} C_k\varnothing_k\right] - f \right\}\varnothing_j.dx = 0,\ j = 1, 2, 3, \ldots$$

Then,

$$\sum_{k=1}^{n} C_k\left(\int_0^l L(\varnothing_k)\varnothing_j dx\right) - \int_0^l f.\varnothing_j.dx = 0$$

That is,

$$\sum_{k=1}^{n} \alpha_{jk}.C_k - \beta_j = 0,\ j = 1, 2, 3, \ldots$$

Where,

$$\alpha_{jk} = \int_0^l L(\varnothing_k)\varnothing_j dx \tag{2.7}$$

$$\beta_j = \int_0^l f.\varnothing_j.dx \tag{2.8}$$

In matrix form, we may write,

$$[\alpha]\{C\} = \{\beta\} \tag{2.9}$$

Eq. (2.9) represents a set of equilibrium equations resulted from the application of Galerkin's method.

2.1.1 Boundary value problems

For bar under axial loading (Fig. 2.1), Euler's formula for equilibrium is given by,

$$\frac{d}{dx}\left(EA\frac{du}{dx}\right)+P=0$$

In the region shown, the boundary conditions are, at $x=0, u=0$ (Fixed end), and

$$\text{at} x=l, EA\frac{du}{dx}=0 \text{ or } \frac{du}{dx}=0 \text{ (Free end)}$$

Selection of a suitable function, assume that the displacement field is given by,

$$u(x)=a+b.x+c.x^2 \tag{2.10a}$$

And the axial strain is given by,

$$\frac{du}{dx}=b+2.c.x$$

But,
at $x=0,\ u=0$, from which, $a=0$, boundary conditions

$$\text{Also, at } x=l, \frac{du}{dx}=0=b+2.c.l,\ c=-\frac{b}{2l}$$

Therefore, Eq. (2.10a) becomes,

$$u(x)=b.x-\frac{b}{2l}x^2=b\left(x-\frac{x^2}{2l}\right)=c_1\varnothing_1(x)$$

This is the starting for applying G.M.

$$\int_0^L\left\{\frac{d}{dx}\left[EA_0\left(1-\frac{x}{l}\right)\right]+P_0\right\}\varnothing_j dx=0$$

That is,

$$EA_0c_1\int_0^l\frac{d}{dx}\left(1-\frac{x}{l}\right).\left(x-\frac{x^2}{2l}\right)dx+\int_0^l P_0.\left(x-\frac{x^2}{2l}\right)dx=0 \tag{2.10b}$$

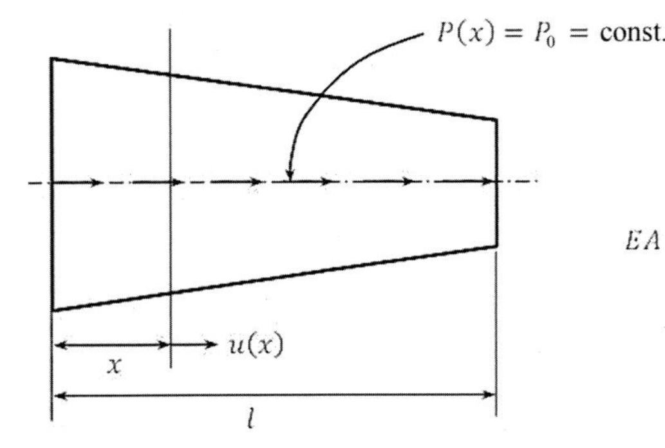

FIGURE 2.1 Bar under axial load P_0.

Integration Eq. (2.10b) gives,

$$c_1 = \frac{4}{7}\frac{P_0.l}{EA_0}$$

Hence, the displacement field in Eq. (2.10a) will be as follows,

$$u(x) = \frac{4}{7}\frac{P_0.l}{EA_0}\left(x - \frac{x^2}{2l}\right)$$

At the free end, the displacement in Eq. (2.10a) becomes,

$$\text{at } x = l, u(l) = \frac{2}{7}\frac{P_0.l^2}{EA_0} = 0.286\frac{P_0.l^2}{EA_0} \tag{2.10c}$$

2.1.2 Assessment of accuracy

The exact solution from strength of materials hand books is given by,

$$u(x) = \frac{P_0.l^2}{EA_0}\left[\frac{x}{l} + \ln\left(1 - \frac{x}{2l}\right)\right]$$

And the displacement at the free end in this is,

$$\text{at} x = l, u(l) = 0.307\frac{P_0.l^2}{EA_0} \tag{2.10d}$$

This shows that a good approximation is obtained between the results of Eqs. (2.10b) and (2.10c) with a percentage of discrepancy is 6.84%.

2.2 Rayleigh ritz method (R.R.M.)

Suppose that we wish to find a function $y(x)$ which is the external to some variational problem. R.R. procedure is to assume,

$$y_n(x) = c_1\varnothing_1(x) + c_2\varnothing_2(x) + \ldots + c_n\varnothing_n(x) \tag{2.10}$$

Where, $\varnothing_1(x), \varnothing_2(x), \ldots, \varnothing_n(x)$ are the linear combinations of suitably chosen coordinate functions.

$c_1, c_2, \ldots, c_n$ are to be determined, and found using the minimization principles as follows,

$$\frac{\partial I}{\partial c_k} = 0, k = 1, 2, \ldots, n$$

The mean square error of this approximation in the region of interest (a, b) is,

$$\lim_{n\to\infty} M_n = \lim_{n\to\infty} \int_a^b \left[y(x) - \sum_{k=1}^{n} c_k\varnothing_k(x)\right]^2 dx = 0$$

Suppose that the $\varnothing_k$ above are chosen to be orthogonal over the region (a,b), we can write,

$$\int_0^b \varnothing_i(x).\varnothing_j(x).dx = \delta_{ij}, \quad \delta_{ij} = \begin{cases} 1 \text{ if } i = j \\ 0 \text{ if } i \neq j \end{cases} \tag{2.11}$$

Where δ_{ij} is the Kronecker delta.

Example 2.1: The simply supported beam-column of span L and uniform cross-section shown in Fig. 2.2 is subjected to an axial load (P) together with a concentrated transverse (W) applied at distance C from the left hand end. By minimizing the total potential energy, determine the deflection under the transverse load.

Assume that the deflected shape is given by,

$$y(x) = \sum_1^\infty a_n \sin\frac{n\pi x}{L} \tag{E1.1}$$

And,

$$y' = \frac{\pi}{L}\sum_1^\infty n a_n \cos\frac{n\pi x}{L}, y'' = -\frac{\pi^2}{L^2}\sum_1^\infty n^2 a_n \sin\frac{n\pi x}{L}$$

The strain energy of the beam due to bending is given by,

$$U = \frac{\pi^4 EI}{2L^4}\int_0^L \left(n^2 a_n \sin\frac{n\pi x}{L}\right)^2 dx$$

Or,

$$U = \frac{\pi^4 EI}{4L^3}\sum_1^\infty n^4 {a_n}^2 \tag{E1.2}$$

The work done due to W and P is given by,

$$W_d = -W\left(\sum_1^\infty a_n \sin\frac{n\pi c}{L}\right) - Ph \tag{E1.3}$$

$$h = \int_0^L (ds - dx)$$

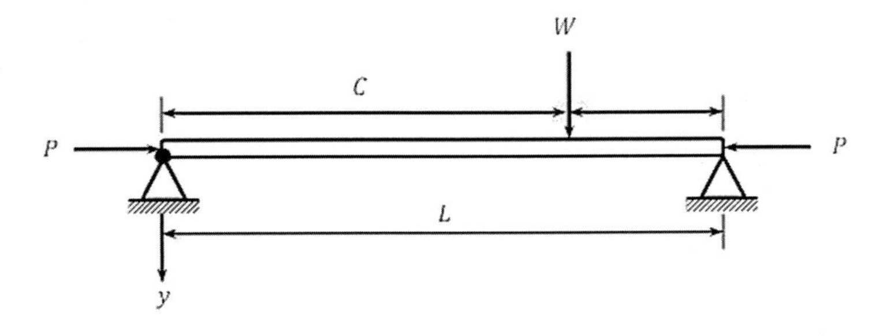

FIGURE 2.2 A Simply supported beam under axial and bending loading.

Therefore,

$$h = \frac{1}{2}\int_0^L (y')^2 dx$$

$$h = \frac{\pi^2}{2L^2}\int_0^L \left(\sum_1^\infty n a_n \cos\frac{n\pi x}{L}\right)^2 dx$$

And since,

$$\int_0^L \cos^2\frac{n\pi x}{L} dx = \frac{L}{2}$$

And,

$$\int_0^L \cos\frac{n\pi x}{L}\cos\frac{m\pi x}{L} dx = 0.\ n \neq m$$

Hence, $h = \frac{\pi^2}{4L}\sum_1^\infty n^2 {a_n}^2$

Therefore, the work done due to W and P of Eq. (E1.3) is given by,

$$W_d = -W\sum_1^\infty a_n \sin\frac{n\pi c}{L} - \frac{\pi^2 P}{4L}\sum_1^\infty n^2 {a_n}^2 \tag{E1.4}$$

The total potential energy $V = U + W_d$ is calculated by substituting the expressions of U and W_d from Eqs. (E1.2) and (E1.4) respectively to give,

$$V = \frac{\pi^4 EI}{4L^3}\sum_1^\infty n^4 {a_n}^2 - W\sum_1^\infty a_n \sin\frac{n\pi c}{L} - \frac{\pi^2 P}{4L}\sum_1^\infty n^2 {a_n}^2 \tag{E1.5}$$

Minimization of total potential energy of Eq. (E1.5) is achieved as follows,
$\frac{\partial V}{\partial a_n} = 0$, or,

$$\frac{\pi^4 EI}{4L^3} n^4 2a_n - W\sin\frac{n\pi c}{L} - \frac{\pi^2 P}{4L}\sum_1^\infty n^2 2{a_n}^2 = 0$$

From which,

$$a_n = \frac{2WL^3}{\pi^4 EIn^4}\frac{\sin(n\pi c/L)}{(1 - (PL^2/\pi^2 n^2 EI))} \tag{E1.6}$$

Using Eq. (E1.1) with the substitution of the expression of a_n in Eq. (E1.6) and at $x = c$, which results in,

$$y(c) = \frac{2WL^3}{\pi^4 EI}\sum_1^\infty \frac{\sin(n\pi c/L)}{(1 - (PL^2/\pi^2 n^2 EI))} \tag{E1.7}$$

2.3 Examples using the P.S.T.P.E. with non-trigonometric coordinate functions

For beams say, we assume,

$$w(x) = \sum_{i=1}^{N} C_i . \emptyset_i(x)$$

Generalized Coordinates ↗ ↖ Coordinate Function

This approach is used in the P.E. method, it is essential that the selected $\emptyset$'s satisfy the prescribed geometric B.Cs.

Example 2.2: For the axial bar shown in Fig. 2.3, find the approximate deflected shape using P.S.T.P.E.

The solution of this problem by Rayleigh-Ritz technique, the displacement medal is given by,

$$u(x) = \sum_{i=1}^{n} c_i . \emptyset_i(x) \tag{E2.1}$$

First approximation
If we take one term series,
$u(x) = c_i . \emptyset(x)$ which satisfies the geometric constraints (compatibility requirements)

$$u = 0 \text{ at } x = 0 \text{ (Fixed end)}$$

Then, the strain is given by,
$\frac{du}{dx} = c_1$ axial strain is constant
To find the strain energy U stored in the bar,

$$U = \int_0^l \frac{E.\varepsilon^2}{2} A.dx$$

Since, $\varepsilon = \frac{du}{dx}$
Therefore,

$$U = \frac{1}{2} \int_0^l EA_0 \left(2 - \frac{x}{l}\right) c_1^2 .dx$$

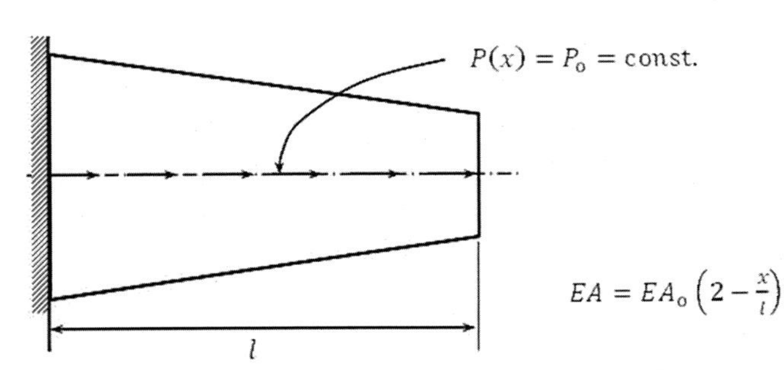

FIGURE 2.3 Tapered bar under uniformly distributed axial load.

Or,

$$= \frac{EA_0c_1^2}{2}\left|2x - \frac{x^2}{2l}\right|_0^l = \frac{1}{2}\left(\frac{3}{2}EA_0.l\right)c_1^2 \tag{E2.2}$$

To find the total potential energy Ω is given by,

$$\Omega = -\int_0^l P_0.u.dx = -\int_0^l P_0.c_1.x.dx = -\frac{P_0l^2}{2}c_1 \tag{E2.3}$$

To find the total potential energy

$$V = U + \Omega \tag{E2.4}$$

Put the values of Eqs. (E2.1) and (E2.3) into Eq. (E2.4) and carrying out the integrations gives,

$$V = \frac{1}{2}\left(\frac{3}{2}EA_0.l\right)c_1^2 - \frac{P_0l^2}{2}c_1$$

Note, it is important to realize that because the chosen displacement pattern is approximate, we impose, $\delta V = 0$, and we can't satisfy equilibrium exactly or point by point basis throughout the solid, but only in some average way, this is the source of approximation. That is, equilibrium is satisfied in some average way which is the nature of approximation, Therefore,

$$\delta V = 0, \frac{\partial V}{\partial c_1} = 0$$

Hence,

$$\frac{3}{2}EA_0.l.c_1 - \frac{P_0l^2}{2} = 0$$

Or, $c_1 = P_0.\frac{l}{3EA_0}$

Hence, the first approximation solution is,

$$u(x) = P_0.\frac{l}{3EA_0}.x = P_0.\frac{l^2}{3EA_0}\left(\frac{x}{l}\right) \tag{E2.5}$$

The boundary conditions are,

At $x = l, u(l) = P_0.\frac{l^2}{3EA_0}$ (Displacement at the free end)

So that the displacement pattern is good compared to the exact solution.

Exact, $u = 0.307P_0.\frac{l^2}{EA_0}$ (strength of materials books).

Using the exact and approximate values of the axial displacement, the stress distribution may be obtained as shown in Fig. 2.4.

Second approximation

Assume that the displacement model is given by,

$$u(x) = c_1.\underbrace{x}_{\varnothing_1} + c_2.\underbrace{x^2}_{\varnothing_2}\quad\left(\sum_{i=1}^{2} c_i\varnothing_i(x)\right) \tag{E2.6}$$

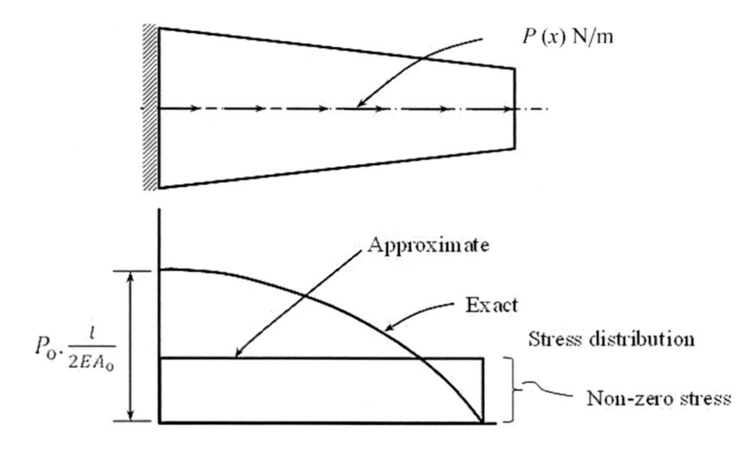

FIGURE 2.4 Bar under axial load with approximate and exact stress distribution.

On the basis of this assumption, the strain is given by,

$$\varepsilon = \frac{du}{dx} = c_1 + 2c_2x \tag{E2.7}$$

It varies linearly, in contrast with the constant model in the first approximation. The strain energy stored in the element U is given by,

$$U = \frac{1}{2}\int_0^l EA_0\left(2 - \frac{x}{l}\right)(c_1 + 2c_2x)^2.dx \tag{E2.8}$$

The total potential energy Ω is given by,

$$\Omega = -\int_0^l P_0\left(c_1.x + c_2.x^2\right)dx \tag{E2.9}$$

And the total potential energy in this case,

$$V = U + \Omega = V(c_1, c_2) \tag{E2.10}$$

Put the values of strain energy Eq. (E2.8) and the potential energy Eq. (E2.9) into the total potential energy Eq. (E2.10) and applying the T.P.E. principles as follows,

$$\frac{\partial V}{\partial c_1} = \frac{1}{2}\int_0^l EA_0\left(2 - \frac{x}{l}\right).2.(c_1 + 2c_2x).dx - \int_0^l P_0.x.dx = 0$$

$$\frac{\partial V}{\partial c_2} = \frac{1}{2}\int_0^l EA_0\left(2 - \frac{x}{l}\right).2.(c_1 + 2c_2x).2x.dx - \int_0^l P_0.x^2.dx = 0$$

Arrange the above two equations in matrix form gives,

$$\begin{bmatrix} \frac{3}{2}EA_0l & \frac{4}{3}EA_0l^2 \\ \frac{4}{3}EA_0l^2 & \frac{5}{3}EA_0l^3 \end{bmatrix} \begin{Bmatrix} c_1 \\ c_2 \end{Bmatrix} = \begin{Bmatrix} \frac{P_0.l^2}{2} \\ \frac{P_0.l^3}{3} \end{Bmatrix} \tag{E2.11}$$

Or,

$$[K]\{u\} = \{P\}$$

Inverting Eq. (E2.11) yields $c_1$andc_2 as follows,

$$c_1 = \frac{7}{13}\frac{P_0l}{EA_0}$$

$$c_2 = -\frac{3}{13}\frac{P_0}{EA_0}$$

Inserting the values of c_1, and c_2 into Eq. (E2.6) gives,

$$u(x) = \frac{7}{13}\frac{P_0l}{EA_0}x - \frac{3}{13}\frac{P_0}{EA_0}x^2$$

Therefore,
$u(l) = 0.308\frac{P_0l^2}{EA_0}$, displacement at the free end.
Exact, $u(l) = 0.307\frac{P_0l^2}{EA_0}$
Comparison of stresses,
$\sigma = E.\frac{du}{dx} = \frac{7}{13}\frac{P_0l}{A_0} - \frac{6}{13}\frac{P_0x}{A_0}$ (Linear variation of axial stress distribution)
However, the stress distribution for both exact and first and second approximations is shown in Fig. 2.5.

Example 2.3: To improve the solution of Example 2.2, assume a displacement model $u(x) = c_1x + c_2x^2$ for the tapered element and a linear model for the uniform bar. Show that the parameters c_1 and c_2 are given by:

$$\begin{bmatrix} 3.6656250 & 8.6249999 \times 10^2 \\ 8.6249999 \times 10^2 & 2.4257812 \times 10^5 \end{bmatrix} \begin{Bmatrix} c_1 \\ c_2 \end{Bmatrix} = \begin{Bmatrix} 1.25 \times 10^{-3} \\ 3.125 \times 10^{-1} \end{Bmatrix}$$

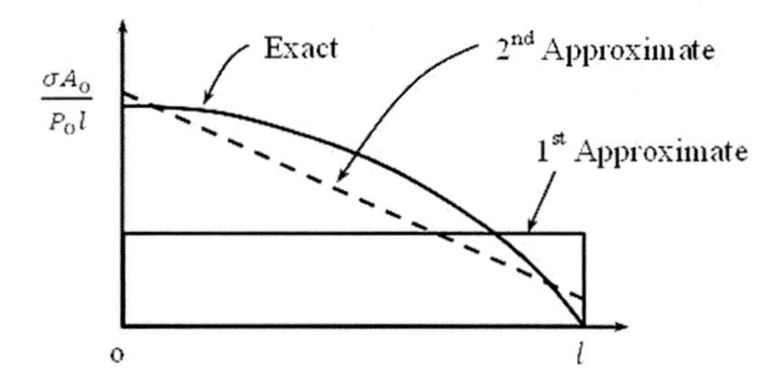

FIGURE 2.5 Exact and 1st and 2nd approximations of stress distribution.

Accepting that the solution is:

$$\begin{Bmatrix} c_1 \\ c_2 \end{Bmatrix} = \begin{Bmatrix} 2.31884 \times 10^{-4} \\ 4.63768 \times 10^{-7} \end{Bmatrix}$$

Show that the displacement at the load is 0.87 mm. Contrast this with the result of the first approximation and give comments. Improving solution of first approximation

Solution

For the tapered bar

Assume $u(x) = c_1 x + c_2 x^2$

Therefore, the strain is given by,

$$\frac{du}{dx} = c_1 + 2c_2 x$$

Then, the strain energy stored in the element is given by,

$$U_1 = \frac{EA_0}{2}\int_0^{l_1}\left(1 - \frac{x}{2l_1}\right)(c_1 + 2c_2 x)^2 dx \tag{E3.1}$$

For the uniform bar

Strain is constant in this member and for compatibility

$(U_{\text{RHS}})_1 = (U_{\text{LHS}})_2 =$ compression in member 2

Therefore, the extension of this member is given by,

$$e_2 = \left(c_1 l_1 + c_2 l_1^2\right)$$

And the strain energy stored in this element is given by,

$$U_2 = \frac{EA_2}{2l_2}\left(c_1 l_1 + c_2 l_1^2\right)^2 \tag{E3.2}$$

The total potential energy is given by V:

$$V = U_1 + U_2 + \Omega = U - P\left(c_1 l_1 + c_2 l_1^2\right) \tag{E3.3}$$

And for equilibrium, $\delta V = 0$, and since V is a function of c_1 and c_2, therefore, $\left(\frac{\partial V}{\partial c_1} = 0, \frac{\partial V}{\partial c_1} = 0\right)$, that is,

$$\frac{\partial V}{\partial c_1} = EA_0\int_0^{l_1}\left(1 - \frac{x}{2l_1}\right)(c_1 + 2c_2 x)dx + \frac{EA_2}{l_2} l_1\left(c_1 l_1 + c_2 l_1^2\right) - Pl_1 = 0$$

$$c_1 EA_0\left|x - \frac{x^2}{4l_1}\right|_0^{l_1} + 2c_2 EA_0\left|\frac{x^2}{2} - \frac{x^3}{6l_1}\right|_0^{l_1} + EA_2\frac{l_1}{l_2}\left(c_1 l_1 + c_2 l_1^2\right) - Pl_1 = 0$$

$$\left(\frac{3}{4}EA_0 l_1 + EA_2\frac{l_1^2}{l_2}\right)c_1 + \left(\frac{2}{3}EA_0 l_1^2 + EA_2\frac{l_1^3}{l_2}\right)c_2 = Pl_1 \tag{E3.4}$$

Writing Eq. (E3.4) in compact form gives,

$$k_{11}c_1 + k_{12}c_2 = Q_1 \tag{E3.5}$$

And,

$$\frac{\partial V}{\partial c_2} = EA_0 \int_0^{l_1} \left(1 - \frac{x}{2l_1}\right)(c_1 + 2c_2x)2xdx + \frac{EA_2}{l_2}(c_1l_1 + c_2l_1^2)l_1^2 - Pl_1^2 = 0$$

$$\frac{\partial V}{\partial c_2} = 2c_1EA_0 \int_0^{l_1} \left(x - \frac{x^2}{2l_1}\right) dx + 4c_2EA_0 \int_0^{l_1} \left(x^2 - \frac{x^3}{2l_1}\right) dx + \frac{EA_2}{l_2}(c_1l_1 + c_2l_1^2)l_1^2 - Pl_1^2 = 0$$

$$2c_1EA_0 \left|\frac{x^2}{2} - \frac{x^3}{6l_1}\right|_0^{l_1} + 4c_2EA_0 \left|\frac{x^3}{3} - \frac{x^4}{8l_1}\right|_0^{l_1} + \frac{EA_2}{l_2}(c_1l_1 + c_2l_1^2)l_1^2 = Pl_1^2$$

$$\left(\frac{2}{3}EA_0l_1^2 + EA_2\frac{l_1^3}{l_2}\right)c_1 + \left(\frac{5}{6}EA_0l_1^3 + EA_2\frac{l_1^4}{l_2}\right)c_2 = Pl_1^2 \tag{E3.6}$$

Writing Eq. (E3.6) in compact form gives,

$$k_{21}c_1 + k_{22}c_2 = Q_2 \tag{E3.7}$$

Evaluating k_{ij} etc.: Substitute the values of geometry and materials properties in the form of the coefficients k_{ij}

$$k_{11} = E\left(\frac{3}{4} \times 500 \times 250 + 200\frac{250^2}{150}\right) = 177083 \times E = 3.666 \times 10^{10}\text{N.mm}$$

$$k_{12} = k_{21} = E\left(\frac{2}{3} \times 500 \times 250^2 + 200\frac{250^3}{150}\right) = 41666666 \times E = 8.625 \times 10^{12}\text{N.mm}^2$$

$$k_{22} = E\left(\frac{5}{6} \times 500 \times 250^3 + 200\frac{250^4}{150}\right) = 1.171875 \times E = 2.426 \times 10^{15}\text{N.mm}^3$$

$$Q_1 = Pl_1 = \frac{1}{2} \times 100000 \times 250 = \frac{1}{2} \times 2.5 \times 10^7\text{N.mm}$$

$$Q_1 = Pl_1^2 = \frac{1}{2} \times 100000 \times 250^2 = \frac{1}{2} \times 6.25 \times 10^9\text{N.mm}^2$$

Then:

$$\begin{bmatrix} 3.6656250 \times 10^{10} & 8.6249999 \times 10^{12} \\ 8.6249999 \times 10^{12} & 2.4257812 \times 10^{15} \end{bmatrix} \begin{Bmatrix} c_1 \\ c_2 \end{Bmatrix} = \begin{Bmatrix} 1.25 \times 10^7 \\ 3.125 \times 10^9 \end{Bmatrix}$$

Or,

$$\begin{bmatrix} 3.6656250 & 8.6249999 \times 10^2 \\ 8.6249999 \times 10^2 & 2.4257812 \times 10^5 \end{bmatrix} \begin{Bmatrix} c_1 \\ c_2 \end{Bmatrix} = \begin{Bmatrix} 1.25 \times 10^{-3} \\ 3.125 \times 10^{-1} \end{Bmatrix}$$

$$\begin{Bmatrix} c_1 \\ c_2 \end{Bmatrix} = \begin{Bmatrix} 2.31884 \times 10^{-4} \\ 4.63768 \times 10^{-7} \end{Bmatrix}$$

Then, solution for tapered bar is:
$u(x) = 2.31884 \times 10^{-4}x + 4.63768 \times 10^{-7}x^2(\text{mm})$
Displacement at load:

$$u_0 = 2.31884 \times 10^{-4} \times 250 + 4.63768 \times 10^{-7} \times 250^2$$

Then,

$$u_0 = 0.087\text{mm}$$

Note: This displacement is slightly greater than the first approximate solution in Example 2.2. We would expect this since we have provided increased flexibility to the tapered bar.

We note that in spite of this considerable extra calculating effort, the improvement in displacement predication is not very much in this case. However, we expect the stresses to be improved of the constant value in first approximation.

Example 2.4: Calculate the bar stresses according to the solution of Example 1.14. Assess how well equilibrium is satisfied by isolating the rigid boss as a free body and considering the forces acting on it.

Solution:

The tapered element
The section variation is given by,

$$A = A_0\left(1 - \frac{x}{2l}\right) = \frac{A_0}{2l}(2l - x)$$

The axial stress is given by,

$$\sigma = \frac{2Pl}{A_0(2l - x)} = E\frac{du}{dx} \tag{E4.1}$$

Where the axial strain,

$$\frac{du}{dx} = \frac{2Pl}{EA_0}\frac{1}{(2l - x)} \tag{E4.2}$$

The displacement field in the tapered element is as follows,

$$u(x) = -\frac{2Pl}{EA_0}\ln(2l - x) + \frac{2Pl}{EA_0}\ln c_1 \tag{E4.3}$$

The boundary conditions are,
At, $x = 0, u = 0$ (Fixed end)

Hence,

$$c_1 = 2l$$

Therefore, Eq. (E4.3) becomes,

$$u(x) = -\frac{2Pl}{EA_0}\ln\left(\frac{1}{1 - x/2l}\right) \tag{E4.4}$$

Check the strain using the displacement field of Eq. (E4.4) with that obtained in (E4.1) is as follows,

$$\frac{du}{dx} = \frac{2Pl}{EA_0}\frac{1}{1/\left(1-\frac{x}{2l}\right)}\frac{1}{\left(1-\frac{x}{2l}\right)^2}\left(\frac{-1}{2l}\right) = \frac{P}{EA_0(1 - x/2l)} \tag{E4.5}$$

They are identical which proves the effectiveness of the displacement field.

Total elongation

For the tapered element, using Eq. (E4.5), we can obtain the extension in the tapered element that is, at $x = l$,

$$e_t = \frac{2P_1l_1}{EA_0}\ln 2 = 1.3863\frac{P_1l_1}{EA_0}$$

And for the uniform element, the extension is as follows,

$$e_u = \frac{P_2l_2}{EA_3}$$

Assemblage

The equilibrium of forces for the element shown in Fig. 2.6 is as follows,

$$\sum F_{\text{forces}} = P_1 + P_2 - W = 0 \tag{E4.6}$$

Then, since:

$$P_1 = \frac{EA_0}{1.3863l_1}e_t; P_2 = \frac{EA_3}{l_2}e_u$$

Where e_t, e_u are the extensions of the taper and uniform segments respectively.

We have $e_t = e_u = e$, therefore, Eq. (E4.6) becomes,

$$\left(\frac{EA_0}{1.3863l_1} + \frac{EA_3}{l_2}\right)e = W = \frac{EA_0}{1.3863l_1}\left(1 + \frac{1.3863A_3l_1}{A_0l_2}\right)$$

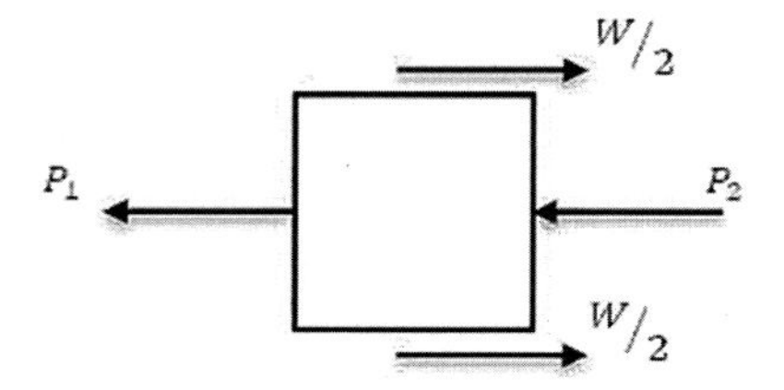

FIGURE 2.6 Equilibrium of forces.

Hence,

$$e = \frac{1.3863 W l_1}{E A_0 \left[1 + \frac{1.3863 A_3 l_1}{A_0 l_2}\right]}$$

$$e = \frac{1.3863 \times 50 \times 250}{207 \times 500 \left[1 + \frac{1.3863 \times 200 \times 250}{500 \times 150}\right]}$$

Or,

$$e = 0.08701 \text{ mm}$$

Calculation of stresses:
Tapered bar segment:
Now,

$$P_1 = \frac{E A_0}{1.3863 l_1} \times 0.08701$$

$$P_1 = \frac{207 \times 500 \times 0.08701}{1.3863 \times 250}$$

Or,

$$P_1 = 25.98 \text{ KN}$$

Then at the end A_0,

$$\sigma = 25.98/500 = 0.052 \text{ KN/mm}^2 \tag{E4.7}$$

And at end A_1,

$$\sigma = 25.98/250 = 0.104 \text{ KN/mm}^2 \tag{E4.8}$$

Uniform bar segment
Using Eq. (E4.1), the stress in the uniform bar is given by,

$$\sigma = \frac{207 \times 0.08701}{150} = 0.120 \text{ KN/mm}^2 \tag{E4.9}$$

2.4 Case studies for bar and beam problems under different loading and support conditions

2.4.1 Bar problem by Galerkin's method using polynomial assumption

To find an approximate value of displacement for the bar shown in Fig. 2.7. Use,

$$u(x) = \alpha_1 + \alpha_2 x + \alpha_1 x^2 + \alpha_4 x^3$$

Assume that the expression for bar under extension is given by,

$$\frac{d}{dx}\left(EA\frac{du}{dx}\right) + P = 0 \tag{2.12a}$$

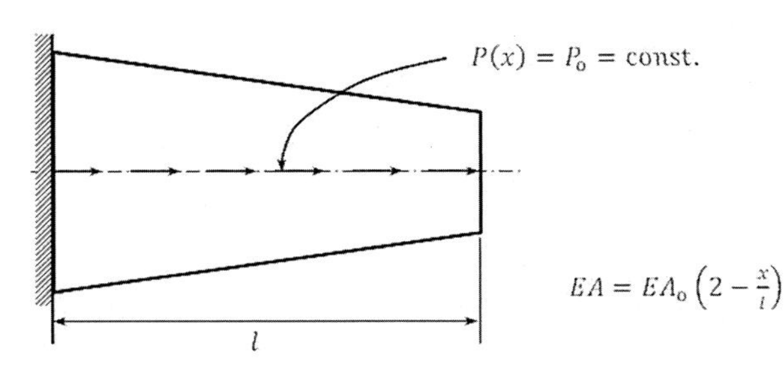

FIGURE 2.7 Tapered Bar under Axial Loading.

Let,

$$u(x) = \alpha_1 + \alpha_2 x + \alpha_3 x^2 + \alpha_4 x^3$$

The boundary conditions
$u = 0$, at, $x = 0$, (fixed end), therefore, $\alpha_1 = 0$, and,
$\frac{du}{dx} = 0$, at, $x = 1$, (free end), therefore,

$$\alpha_2 = -2\alpha_3 l - 3\alpha_4 l^2$$

Hence, the displacement model is given by,

$$u(x) = -\left(2\alpha_3 l + 3\alpha_4 l^2\right)x + \alpha_3 x^2 + \alpha_4 x^3 = -2l\alpha_3\left(x - \frac{x^2}{2l}\right) - 3l^2\alpha_4\left(x - \frac{x^3}{2l^2}\right)$$

Or,

$$u(x) = a_1\left(x - \frac{x^2}{2l}\right) - a_2\left(x - \frac{x^3}{2l^2}\right) \tag{2.12b}$$

From which,

$$\varnothing_1(x) = x - \frac{x^2}{2l}, \text{ and, } \varnothing_2(x) = x - \frac{x^3}{2l^2} \tag{2.12c}$$

Now applying Galerkin's method to find a_1 and a_2

$$\int_0^l \varnothing_1(x).R(x)dx = 0$$

And,

$$\int_0^l \varnothing_2(x).R(x)dx = 0 \tag{2.12d}$$

Where, $R(x)$ = residual weighing function

$$R(x) = EA_0 \frac{d}{dx}\left(2 - \frac{x}{l}\right)\frac{d}{dx}\left[a_1\left(x - \frac{x^2}{2l}\right) + a_2\left(x - \frac{x^3}{3l^2}\right)\right]$$

Using Eq. (2.12d) gives,

$$EA_0\left\{\int_0^l \left(x - \frac{x^2}{2l}\right)\left[\frac{d}{dx}\left(2 - \frac{x}{l}\right)\frac{d}{dx}\left(a_1\left(x - \frac{x^2}{2l}\right) + a_2\left(x - \frac{x^3}{3l^2}\right)\right)\right]dx\right\} + P_0\int_0^l \left(x - \frac{x^2}{2l}\right)\, dx = 0 \tag{2.12e}$$

And,

$$EA_0\left\{\int_0^l \left(x - \frac{x^3}{3l^2}\right)\left[\frac{d}{dx}\left(2 - \frac{x}{l}\right)\frac{d}{dx}\left(a_1\left(x - \frac{x^2}{2l}\right) + a_2\left(x - \frac{x^3}{2l^2}\right)\right)\right]dx\right\} + P_0\int_0^l \left(x - \frac{x^3}{3l^2}\right)\, dx = 0$$

From Eq. (2.12e),

$$EA_0\left\{\int_0^l \left(x - \frac{x^2}{2l}\right)\left[\frac{d}{dx}\left(2 - \frac{x}{l}\right)\left(a_1\left(1 - \frac{x}{l}\right) + a_2\left(1 - \frac{x^2}{l^2}\right)\right)\right]dx\right\} + P_0\left(\frac{l^2}{2} - \frac{l^2}{6}\right) = 0$$

Carrying out the integrations and simplification gives,

$$EA_0\left\{\int_0^l \left(x - \frac{x^2}{2l}\right)\left[a_1\left(-\frac{3}{l} + \frac{2x}{l^2}\right) + a_2\left(-\frac{4x}{l^2} - \frac{1}{l} + \frac{3x^2}{l^3}\right)\right]dx\right\} = -P_0\frac{l^2}{3} \tag{2.12f}$$

Or, $\frac{7l}{12}a_1 + \frac{43l}{60}a_2 = \frac{P_0 l^2}{3EA_0}$
Therefore,

$$a_1 = 0.5714285\frac{P_0 l}{EA_0} - 1.22857a_2 \tag{2.12g}$$

From Eq. (2.12f),

$$EA_0\left\{\int_0^l \left(x - \frac{x^3}{3l^2}\right)\left[a_1\left(-\frac{3}{l} + \frac{2x}{l^2}\right) + a_2\left(-\frac{4x}{l^2} - \frac{1}{l} + \frac{3x^2}{l^3}\right)\right]dx\right\} = -P_0\left(\frac{l^2}{2} - \frac{l^2}{12}\right) = -\frac{5P_0 l^2}{12}$$

Carrying out the integrations and simplifications gives,

$$a_1\left(-\frac{3l}{2} + \frac{2l}{3} + \frac{l}{4} - \frac{2l}{15}\right) + a_2\left(-\frac{4l}{3} - \frac{l}{2} + \frac{3l}{4} + \frac{4l}{15} + \frac{l}{12} - \frac{l}{6}\right) = -\frac{5P_0 l^2}{12EA_0}$$

Hence,

$$a_1 = 0.5813953\frac{P_0 l}{EA_0} - 1.255814a_2 \tag{2.12h}$$

Substitute Eq. (2.12h) into Eq. (2.12g) to obtain a_2

$$a_2 = 0.3658353\frac{P_0 l}{EA_0}$$

And,

$$a_1 = 0.121974 \frac{P_0 l}{EA_0}$$

Therefore, we can write the axial displacement pattern is given by,

$$u(x) = 0.121974 \frac{P_0 l}{EA_0}\left(x - \frac{x^2}{2l}\right) + 0.3658353 \frac{P_0 l}{EA_0}\left(x - \frac{x^3}{2l^2}\right) \quad (2.12i)$$

At, $x = l, u(l) = 0.3048773 \frac{P_0 l^2}{EA_0}$
Where,

$$u(l)_{\text{exact}} = 0.307 \frac{P_0 l^2}{EA_0}$$

$$\sigma = E\varepsilon = E\frac{du}{dx}, \sigma = 0.1219742 \frac{P_0 l}{A_0}\left(1 - \frac{x}{l}\right) + 0.3658353 \frac{P_0 l}{A_0}\left(1 - \frac{x^2}{l^2}\right)$$

Hence,

$$\sigma = 0.4878095 \frac{P_0 l}{A_0}, \text{ stress at the root section.} \quad (2.12j)$$

The exact value of the stress at the root is, $\sigma = 0.5 \frac{P_0 l}{A_0}$ which is very close to the value obtained by using the energy method.

2.4.2 Beam under uniformly distributed load by using a higher order polynomial deflected shape

In this respect, to find the deflection at midpoint of the beam using G.M. for the beam shown in Fig. 2.8.

Assume that the deflected shape is given by,

$$w(x) = \alpha_1 + \alpha_2 x + \alpha_3 x^2 + \alpha_4 x^3 + \alpha_5 x^4 \quad (2.13a)$$

The boundary conditions are,
Deflection $w = 0$ at $x = 0$ and l (Fixed-Fixed ends)
Slope $w' = 0$ at $x = 0$ and$l$ (Fixed-Fixed ends)
Inserting the above boundary conditions in Eq. (2.13a) gives,

$$\alpha_1 = 0, \ \alpha_2 = 0$$

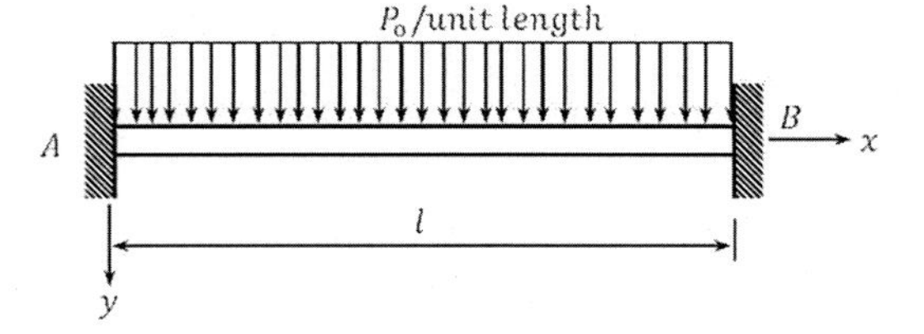

FIGURE 2.8 Fixed-fixed beam under uniformly distributed loading.

Therefore, Eq. (2.13a) becomes,

$$w(x) = \alpha_5\left(x^2l^2 - 2x^3l + x^4\right)$$

Or,

$$= \alpha_5 x^2(x-l)^2$$

And, we can write,

$$= ax^2(x-l)^2 \quad (2.13b)$$

Where,

$$\varnothing_1(x) = x^2(x-l)^2$$

Using G.M. to find the value of a as follows,

$$\int_0^l \left(\frac{d^2}{dx^2}EI\frac{d^2w}{dx^2} - P\right)\varnothing_1(x)dx = 0 \quad (2.13c)$$

We have,

$$\frac{d^2w}{dx^2} = 2a\left(l^2 - 6xl + 6x^2\right)$$

And,

$$\frac{d^4w}{dx^4} = 24a$$

Substitute into Eq. (2.13c) gives,

$$24aEI\int_0^l \varnothing_1(x)dx - P\int_0^l \varnothing_1(x)dx = 0 \quad (2.13d)$$

We have,

$$\int_0^l \varnothing_1(x)dx = \frac{l^5}{30}$$

Hence, Using Eq. (2.13d),

$$\frac{24aEIl^5}{30} - \frac{Pl^5}{30} = 0$$

Therefore,

$$a = \frac{P}{24EI}$$

Hence, Eq. (2.13b) becomes,

$$w(x) = \frac{Px^2}{24EI}(x-l)^2$$

$w = w_{\text{cent.}}$, at, $x = \frac{l}{2}$, central deflection

$$w_{\text{cent.}} = \frac{Pl^2}{24EI}\frac{l^2}{4} = \frac{Pl^4}{384EI} = w_{\text{exact}} \tag{2.13e}$$

2.4.3 Cantilever beam under uniformly varying load

Example 2.5: Find the maximum deflection of the loaded beam shown in Fig. 2.9 using the G.M. method.

Solution

The loading at section xx is,

$$P(x) = EI\frac{d^4w}{dx^4} = P\left(1 + \frac{2x}{l}\right) \tag{E5.1}$$

Let the assumed $w(x)$ is given by,

$$w(x) = \alpha_1 + \alpha_2 x + \alpha_3 x^2 + \alpha_4 x^3 + \alpha_5 x^4$$

Boundary conditions are,
at$x = l, w = 0$ and $w' = 0$, Fixed end.
at$x = 0, w = 0\, w'' = w''' = 0$, Free end—free of shear and moment.
Therefore,

$$\begin{aligned} w(x) &= \alpha_5\left(x^4 - 4xl^3 + 3l^4\right) \\ &= a\left(x^4 - 4xl^3 + 3l^4\right) \\ &= a\varnothing_1(x) \end{aligned} \tag{E5.2}$$

Where,

$$\varnothing_1(x) = \left(x^4 - 4xl^3 + 3l^4\right)$$

Using Galerkin's method to find the value of a as follows,

$$\int_0^l \left[EI\frac{d^4w}{dx^4} - P\left(1 + \frac{2x}{l}\right)\right]\varnothing_1(x).dx = 0 \tag{E5.3}$$

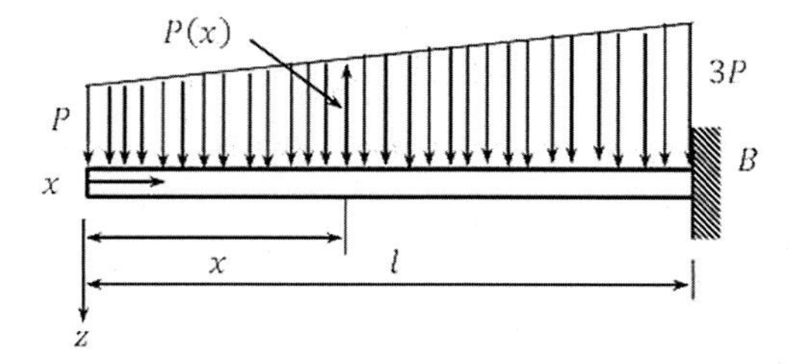

FIGURE 2.9 A cantilever beam under uniform loading.

We have,

$$\frac{d^4w}{dx^4} = 24a$$

Therefore,

$$24aEI\int_0^l \varnothing_1 dx - P\int_0^l \left(1 + \frac{2x}{l}\right)\varnothing_1(x).dx = 0$$

Evaluating the above integral gives,

$$a = \frac{7P}{108EI}$$

Hence, Eq. (E5.2) becomes,

$$w(x) = \frac{7P}{108EI}\left(x^4 - 4xl^3 + 3l^4\right) \tag{E5.4}$$

Therefore, the deflection at the tip of the cantilever where = 0, is,

$$w(0) = \frac{21Pl^4}{108EI} = 0.1944\frac{Pl^4}{EI}$$

While the exact value is,

$$w_{\text{exact}}|_{x=0} = \frac{23Pl^4}{120EI} = 0.1916\frac{Pl^4}{EI}$$

This indicates that the exact value of the maximum deflection is obtained by using G.M. with the assumption of a deflected shape of a polynomial of fourth order.

2.4.4 A cantilever beam under a concentrated load at the free end

Example 2.6: Use G.M. to find the deflection and the slope of point (A) on the Fig. 2.10.

Solution

In this case,

$$EI\frac{d^4w}{dx^4} = 0$$

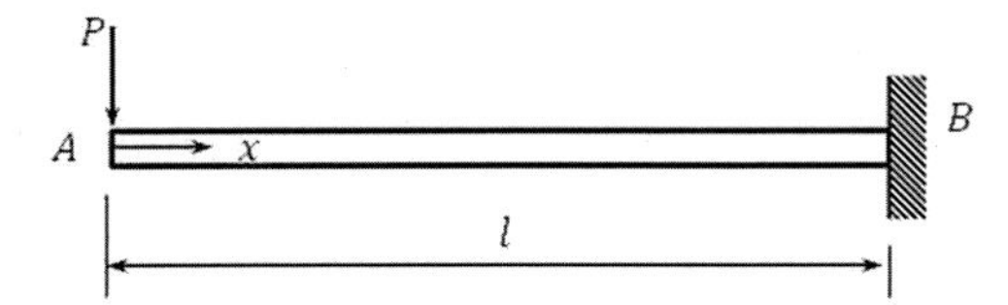

FIGURE 2.10 A cantilever beam under concentrated load.

Carrying out the integration twice gives,

$$EI\frac{d^2w}{dx^2} = c_1x + c_2 \tag{E6.1}$$

Load boundary conditions are,

$$\text{at} x = 0 \; M = 0$$

$$\text{at} x = l \; M = Pl$$

Where, $M(x) = EI\frac{d^2w}{dx^2}$

Substitute the above load boundary conditions in (E6.1) and using the above boundary conditions, gives,

$$c_2 = 0 \text{ and } c_1 = P$$

Therefore,

$$EI\frac{d^2w}{dx^2} - Px = 0$$

Let the deflected shape is given by,

$$w(x) = \alpha_1 + \alpha_2 x + \alpha_3 x^2 \text{ (As a first approximation)} \tag{E6.2}$$

And,

$$w'(x) = \alpha_2 + 2\alpha_3 x \tag{E6.3}$$

Boundary conditions are,
$w = w' = 0$ at $x = l$ (Fixed end)
Apply the above boundary conditions in Eq. (E6.3) and simplifications give,

$$w(x) = \alpha_3(x-l)^2 = a(x-l)^2 = a\varnothing_1(x) \tag{E6.4}$$

Now, using G.M. as follows,

$$\int_0^l \left[EI\frac{d^2\tilde{w}}{dx^2} - Px\right]\varnothing_1(x).dx = 0 \tag{E6.5}$$

From Eq. (E6.4), we have,

$$\frac{d^2w}{dx^2} = 2a$$

And,

$$\int_0^l \varnothing_1(x)dx = \frac{l^3}{3}, \text{and} \int_0^l x\varnothing_1 dx = \frac{l^4}{12}$$

Substitute the above values in Eq. (E6.5) gives,

$$2aEI\frac{l^3}{3} - P\frac{l^4}{12} = 0$$

Hence,

$$a = \frac{Pl}{8EI}$$

Therefore, Eq. (E6.4) becomes,

$$w(x) = \frac{Pl}{8EI}(x-l)^2 \tag{E6.6}$$

At point A, ($x = 0$), the deflection is,

$$w(0) = \frac{Pl^3}{8EI}, \text{ while, } w_{\text{exact}} = \frac{Pl^3}{3EI}$$

And,

$$w' = -\frac{Pl^2}{4EI}, \text{ while, } w'_{\text{exact}} = -\frac{Pl^2}{2EI} \tag{E6.7a}$$

A difference is found between the exact and the approximate solutions and to improve this value, we take four terms of the deflection model $w(x)$ as follows,

$$\text{Let, } w(x) = \alpha_1 + \alpha_2 x + \alpha_3 x^2 + \alpha_4 x^3 \text{ (as a second approximation)} \tag{E6.7}$$

Boundary conditions are,

$$w = w' = 0 \text{ at } x = l$$

$$w'' = 0 \text{ at } x = 0$$

Substitute the above boundary conditions in (E6.7) gives,

$$w(x) = a\left(2l^3 - 3l^2x + x^3\right) = a\varnothing_1(x) \tag{E6.8}$$

Where, $\varnothing_1(x) = \left(2l^3 - 3l^2x + x^3\right)$

We have,

$$\frac{d^2w}{dx^2} = 6ax$$

Then, using G.M. results,

$$6aEI\int_0^l x\varnothing_1(x).dx - P\int_0^l x\varnothing_1(x).dx = 0$$

Carrying out the above integrations and simplifications give,

$$6aEI - P = 0. \text{ Or } a = \frac{P}{6EI}$$

Hence, Eq. (E6.8) becomes,

$$w(x) = \frac{P}{6EI}\left(2l^3 - 3l^2x + x^3\right) \tag{E6.9}$$

And, the slope is as follows,

$$w' = \frac{P}{6EI}\left(-3l^2 + 3x^2\right)$$

At point A ($x = 0$), the deflection is,
$w = \frac{Pl^3}{3EI}$, and the slope is, $w' = -\frac{Pl^2}{2EI}$,
We see that, the value of w is different from its value in the first approximation. Therefore, w and w' are equal to those obtained using the exact methods (strength of materials book).

$$w_{\text{exact}} = \frac{Pl^3}{3EI}$$

$$w'_{\text{exact}} = -\frac{Pl^2}{2EI} \tag{E6.10}$$

Example 2.7: A cantilever beam (Fig. 2.11) is tapered such that,

$$EI = EI_0\left[1 - \frac{1}{2}\left(\frac{x}{l}\right)^2\right]$$

Where x is measured from the clamped end. the beam caries a load having intensity,

$$p = p_0\left(1 - \frac{x}{l}\right)$$

Assuming a deflection beam,

$$w(x) = c.l^2\left[\left(\frac{x}{l}\right)^2 - \frac{2}{3}\left(\frac{x}{l}\right)^3 + \frac{1}{6}\left(\frac{x}{l}\right)^4\right]$$

1. Use Galerkin's method to obtain an expression for the tip deflection, taking that the one term expansion (c) is, indeed permissible as a comparison a function.
2. Determine an approximate value for the first natural frequency of beam over vibration of the cantilever of (a), using the coordinate function of (a).

Solution,

1. For the extension of the given beam,

Euler's Equation,

$$\frac{d^2}{dx^2}\left(EI\frac{d^2w}{dx^2}\right) = P, \text{and} \tag{E7.1}$$

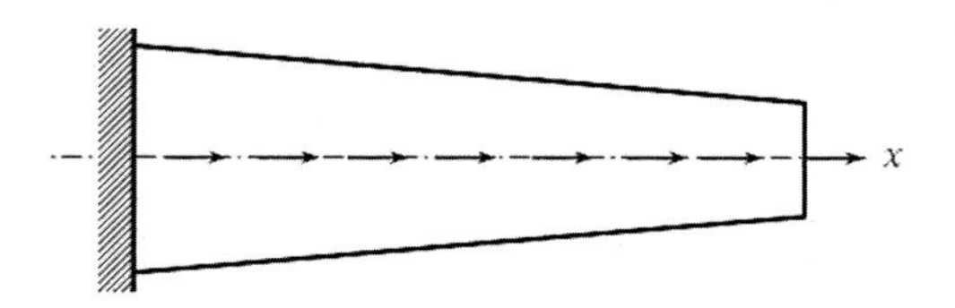

FIGURE 2.11 Tapered beam under variable intensity load.

$L[w] = f$ Disturbing function
Also, we have,

$$EI = EI_0\left[1 - \frac{1}{2}\left(\frac{x}{l}\right)^2\right], P = p_0\left(1 - \frac{x}{l}\right)$$

Let, the deflected shape is given by,

$$w(x) = c.l^2\left[\left(\frac{x}{l}\right)^2 - \frac{2}{3}\left(\frac{x}{l}\right)^3 + \frac{1}{6}\left(\frac{x}{l}\right)^4\right] \tag{E7.2}$$

Or, $w(x) = c.\varnothing_j(x)$
Where,

$$\varnothing_j(x) = l^2\left[\left(\frac{x}{l}\right)^2 - \frac{2}{3}\left(\frac{x}{l}\right)^3 + \frac{1}{6}\left(\frac{x}{l}\right)^4\right] \tag{E7.3}$$

Now, applying the G.M. as follows,

$$\int_0^l \left\{L\left[\sum_{k=1}^{n} c_k\varnothing_k\right] - P\right\}\varnothing_j(x).dx = 0$$

Or,

$$c.l^2\int_0^l \frac{d^2}{dx^2}\left(EI_0\left[1 - \frac{1}{2}\left(\frac{x}{l}\right)^2\right]\frac{d^2}{dx^2}\left[\left(\frac{x}{l}\right)^2 - \frac{2}{3}\left(\frac{x}{l}\right)^3 + \frac{1}{6}\left(\frac{x}{l}\right)^4\right] - p_0\left(1 - \frac{x}{l}\right)\right)\varnothing_j(x).dx = 0 \tag{E7.4}$$

For the term, $\frac{d^2}{dx^2}\left[\left(\frac{x}{l}\right)^2 - \frac{2}{3}\left(\frac{x}{l}\right)^3 + \frac{1}{6}\left(\frac{x}{l}\right)^4\right]$

$$\frac{d}{dx}\left(\frac{2x}{l^2} - 2\frac{x^2}{l^3} + \frac{4}{6}\frac{x^3}{l^4}\right) = \frac{2}{l^2} - 4\frac{x}{l^3} + 2\frac{x^2}{l^4}$$

Therefore, the first part becomes,

$$\frac{d^2}{dx^2}EI_0\left[1 - \frac{1}{2}\left(\frac{x}{l}\right)^2\right]\left(\frac{2}{l^2} - 4\frac{x}{l^3} + 2\frac{x^2}{l^4}\right) \tag{E7.5a}$$

Carrying out the differentiations of the first term in Eq. (E7.5a) as follows,

$$EI_0\frac{d^2}{dx^2}\left(\frac{2}{l^2} - 4\frac{x}{l^3} + 2\frac{x^2}{l^4} - \frac{x^2}{l^4} + 2\frac{x^3}{l^5} - \frac{x^4}{l^6}\right)$$

$$EI_0\left(\frac{4}{l^4} - \frac{2}{l^4} + 12\frac{x}{l^5} - 12\frac{x^2}{l^6}\right)$$

$$= EI_0\left(\frac{2}{l^4} + 12\frac{x}{l^5} - 12\frac{x^2}{l^6}\right) \tag{E7.5b}$$

Hence,

$$EI_0 \varnothing_j(x)\left(\frac{2}{l^4} + 12\frac{x}{l^5} - 12\frac{x^2}{l^6}\right)$$

Insert the value of $\varnothing_j(x)$ from Eq. (E7.3) gives,

$$EI_0\left(c.l^2\right)\left[\left(\frac{x}{l}\right)^2 - \frac{2}{3}\left(\frac{x}{l}\right)^3 + \frac{1}{6}\left(\frac{x}{l}\right)^4\right]\left[\frac{2}{l^4} + 12\frac{x}{l^5} - 12\frac{x^2}{l^6}\right] \quad \text{(E7.5)}$$

First part

$$\int_0^l \underbrace{EI_0 \frac{d^2}{dx^2}\left(\frac{2}{l^2} - 4\frac{x}{l^3} + 2\frac{x^2}{l^4} - \frac{x^2}{l^4} + 2\frac{x^3}{l^5} - \frac{x^4}{l^6}\right)}_{EI_0\left(\frac{4}{l^4} - \frac{2}{l^4} + 12\frac{x}{l^5} - 12\frac{x^2}{l^6}\right)} \varnothing_j\,(x).dx$$

Carrying out the multiplications of the terms of Eq. (E7.5) as follows,

$$\int_0^l EI_0 c.l^2\left\{\frac{2}{l^4}\left(\frac{x}{l}\right)^2 + 12\frac{x^3}{l^7} - 12\frac{x^4}{l^8} - \frac{4}{3}\left(\frac{x}{l}\right)^3\frac{1}{l^4} - 8\frac{x^4}{l^8} + 8\frac{x^5}{l^9} + \frac{1}{3l^4}\left(\frac{x}{l}\right)^4 + 2\frac{x^5}{l^9} - \frac{x^6}{l^{10}}\right\}dx$$

$$= EI_0 c^2 l^4 \left|\left[\frac{2}{l^4}\frac{x^3}{3.l^3} + \frac{12}{4}\frac{x^4}{l^7} - \frac{12}{5}\frac{x^5}{l^8} - \frac{4}{4x^3}\frac{x^4}{l^7} - \frac{8}{5}\frac{x^5}{l^8} + \frac{8}{6}\frac{x^6}{l^9} + \frac{1}{12l^4}\frac{x^5}{l^4} + \frac{2}{6}\frac{x^6}{l^9} - \frac{2}{7}\frac{x^7}{l^{10}}\right]\right|_0^l$$

$$= EI_0 c^2 l^4 \left\{\frac{2}{3}\frac{1}{l^3} + \frac{3}{l^3} - \frac{12}{5}\frac{1}{l^3} - \frac{1}{3}\frac{1}{l^3} - \frac{8}{5}\frac{1}{l^3} + \frac{8}{6}\frac{1}{l^3} + \frac{1}{12}\frac{1}{l^3} + \frac{1}{3}\frac{1}{l^3} - \frac{2}{7}\frac{1}{l^3}\right\}$$

$$= EI_0 c^2 l^4 \frac{1}{l^3}\left\{\frac{2}{3} + 3 - \frac{12}{5} - \frac{8}{5} + \frac{4}{3} + \frac{1}{12} - \frac{2}{7}\right\}$$

$$= EI_0 c^2 l^4 \frac{1}{l^3}\left[\frac{6}{3} - \frac{20}{5} + 3 + \frac{1}{12} - \frac{2}{7}\right]$$

$$= EI_0 c^2 l\left[1 + \frac{1}{12} - \frac{2}{7}\right] = EI_0 c^2 l\left[\frac{12*7 + 7 - 24}{12*7}\right] = EI_0 c^2 l*0.797 \quad \text{(E7.6)}$$

To find the integration of the second term,

$$\int_0^l p_0\left(1 - \frac{x}{l}\right)\varnothing_j(x).dx$$

Substitute the value of $\varnothing_j(x)$ from Eq. (E7.3) and carry out the required multiplications and integrations as follows,

$$c.l^2 p_0 \int_0^l \left(1 - \frac{x}{l}\right)\left(\left(\frac{x}{l}\right)^2 - \frac{2}{3}\left(\frac{x}{l}\right)^3 + \frac{1}{6}\left(\frac{x}{l}\right)^4\right)dx$$

$$c.l^2p_0\int_0^l\left(\left(\frac{x}{l}\right)^2-\frac{2}{3}\left(\frac{x}{l}\right)^3+\frac{1}{6}\left(\frac{x}{l}\right)^4-\left(\frac{x}{l}\right)^3+\frac{2}{3}\left(\frac{x}{l}\right)^4-\frac{1}{6}\left(\frac{x}{l}\right)^5\right)dx$$

$$=c.l^2p_0\left[\frac{x^3}{3.l^2}-\frac{2}{12}\frac{x^4}{l^3}+\frac{1}{30}\frac{x^5}{l^4}-\frac{1}{4}\frac{x^4}{l^3}+\frac{2}{15}\frac{x^5}{l^4}-\frac{1}{36}\frac{x^6}{l^5}\right]_0^l$$

$$=c.l^2p_0\left[\frac{1}{3}l-\frac{2}{12}l+\frac{1}{30}l-\frac{1}{4}l+\frac{2}{15}l-\frac{1}{36}l\right]$$

$$=c.l^3p_0\left[\frac{1}{3}-\frac{2}{12}+\frac{1}{30}-\frac{1}{4}+\frac{2}{15}-\frac{1}{36}\right]$$

$$=c.l^3p_0\frac{10}{180}=0.0555\,c.l^3p_0 \tag{E7.7}$$

$$EI_0c^2l{*}0.797-0.0555c.l^3p_0$$

Collecting the two parts of integrations given in Eqs. (E7.6) and (E7.7), Then solving for c gives,

$$c=\frac{0.0555}{0.797}\frac{p_0l^2}{EI_0} \tag{E7.8a}$$

Hence, the deflected shape given by Eq. (E7.1) becomes,

$$w(x)=0.073\frac{p_0l^4}{EI_0}\left[\left(\frac{x}{l}\right)^2-\frac{2}{3}\left(\frac{x}{l}\right)^3+\frac{1}{6}\left(\frac{x}{l}\right)^4\right] \tag{E7.8}$$

$$w_{\text{tip}}=0.073\frac{p_0l^4}{EI_0}\left(1-\frac{2}{3}+\frac{1}{6}\right)$$

$$=0.0365\frac{p_0l^4}{EI_0}$$

2. Equation of motion for the beam is,

$$\frac{\partial^2}{\partial x^2}\left(EI\frac{\partial^2 w}{\partial x^2}\right)=m\frac{\partial^2 w}{\partial t^2} \tag{E7.9}$$

And the deflected shape is assumed as follows,

$$w(x,t)=c.l^2\left[\left(\frac{x}{l}\right)^2-\frac{2}{3}\left(\frac{x}{l}\right)^3+\frac{1}{6}\left(\frac{x}{l}\right)^4\right]\sin\omega t \tag{E7.10}$$

We have,

$$\frac{\partial^2 w}{\partial t^2}=-\omega^2c.l^2\left[\left(\frac{x}{l}\right)^2-\frac{2}{3}\left(\frac{x}{l}\right)^3+\frac{1}{6}\left(\frac{x}{l}\right)^4\right]\sin\omega t \tag{E7.11}$$

Applying G.M. to find the fundamental natural frequency – free vibration- as follows,

$$\int_0^l \left[\frac{\partial^2}{\partial x^2}\left(EI\frac{\partial^2 w}{\partial x^2}\right) + m\frac{\partial^2 w}{\partial t^2}\right]\varnothing_j(x).(x).dx = 0 \tag{E7.12}$$

The values of $\frac{\partial^2 w}{\partial x^2}$ and $\varnothing_j(x)$ are taken from the part (E7.1) of the problem, while the term $\frac{\partial^2 w}{\partial t^2}$ is calculated in Eq. (E7.11). The expressions of these terms are substituted in Eq. (E7.11) as follows,

$$c.l^2\frac{d^2}{dx^2}\int_0^l \left\{EI_0\left(1-\frac{1}{2}\left(\frac{x}{l}\right)^2\right)\frac{d^2}{dx^2}\left[\left(\frac{x}{l}\right)^2 - \frac{2}{3}\left(\frac{x}{l}\right)^3 + \frac{1}{6}\left(\frac{x}{l}\right)^4\right]\right\}$$
$$+\int_0^l m\omega^2 c.l^2\left[\left(\frac{x}{l}\right)^2 - \frac{2}{3}\left(\frac{x}{l}\right)^3 + \frac{1}{6}\left(\frac{x}{l}\right)^4\right]\varnothing_j(x).dx = 0$$

For the forced vibration of beam, the equation of motion is given by,

$$\frac{\partial^2}{\partial x^2}\left(EI\frac{\partial^2 w}{\partial x^2}\right) = p(x,t) - m\frac{\partial^2 w}{\partial t^2} \tag{E7.13}$$

Again, by using Galerkin's method as follows,

$$L[w] = f$$

And, we have,

$$EI = EI_0\left[1-\frac{1}{2}\left(\frac{x}{l}\right)^2\right],$$

$$p = p_0\left(1-\frac{x}{l}\right).\sin\omega t$$

Again, the deflected shape is given by,

$$w(x,t) = c.l^2\left[\left(\frac{x}{l}\right)^2 - \frac{2}{3}\left(\frac{x}{l}\right)^3 + \frac{1}{6}\left(\frac{x}{l}\right)^4\right]\sin\omega t \tag{E7.14}$$

And,

$$\ddot{w} = -\omega^2 c.l^2\left[\left(\frac{x}{l}\right)^2 - \frac{2}{3}\left(\frac{x}{l}\right)^3 + \frac{1}{6}\left(\frac{x}{l}\right)^4\right]\sin\omega t$$

Applying G.M. in this case as follows,

$$\int_0^l \left\{L\left[\sum_{k=1}^{n} c_k\varnothing_k - p + m\frac{\partial^2 w}{\partial t^2}\right]\right\}\varnothing_j(x,t).dx = 0$$

Where,

$$L = \frac{\partial^2}{\partial x^2}\left(EI\frac{\partial^2}{\partial x^2}\right)$$

$$c.l^2\int_0^l\left[\frac{d^2}{dx^2}\left\{EI_0\left(1-\frac{1}{2}\left(\frac{x}{l}\right)^2\right)\frac{d^2}{dx^2}\left[\left(\frac{x}{l}\right)^2-\frac{2}{3}\left(\frac{x}{l}\right)^3+\frac{1}{6}\left(\frac{x}{l}\right)^4\right]\right\}\sin\omega t - p_0\left(1-\frac{x}{l}\right).\sin\omega t + m\ddot{w}\right]\varnothing_j(x,t).dx = 0 \tag{E7.15}$$

We have,

$$\frac{d^2}{dx^2}\left[\left(\frac{x}{l}\right)^2-\frac{2}{3}\left(\frac{x}{l}\right)^3+\frac{1}{6}\left(\frac{x}{l}\right)^4\right] = \frac{d}{dx}\left(\frac{2x}{l^2}-2\frac{x^2}{l^3}+\frac{4}{6}\frac{x^3}{l^4}\right) = \frac{2}{l^2}-4\frac{x}{l^3}+2\frac{x^2}{l^4}$$

Consider the first part of Eq. (E7.15),

$$c.l^2\int_0^l\left[\frac{d^2}{dx^2}EI_0\left(1-\frac{1}{2}\left(\frac{x}{l}\right)^2\right)\left(\frac{2}{l^2}-4\frac{x}{l^3}+2\frac{x^2}{l^4}\right)\right]\varnothing_j(x,t).dx$$

$$= c.l^2\int_0^l\left[\frac{d^2}{dx^2}EI_0\left(\frac{2}{l^2}-4\frac{x}{l^3}+2\frac{x^2}{l^4}-\frac{x^2}{l^4}+2\frac{x^3}{l^5}-\frac{x^4}{l^6}\right)\right]\varnothing_j(x,t).dx$$

$$= c.l^2\int_0^l EI_0\left[\frac{4}{l^4}-\frac{2}{l^4}+12\frac{x}{l^5}-12\frac{x^2}{l^6}\right]\varnothing_j(x).dx \tag{E7.16}$$

We have,

$$\varnothing_j(x,t) = l^2\left[\left(\frac{x}{l}\right)^2-\frac{2}{3}\left(\frac{x}{l}\right)^3+\frac{1}{6}\left(\frac{x}{l}\right)^4\right]\sin\omega t$$

Therefore, Eq. (E7.15) becomes,

$$= c.l^4\sin\omega t\int_0^l EI_0\left[\frac{2}{l^4}+12\frac{x}{l^5}-12\frac{x^2}{l^6}\right]\left[\left(\frac{x}{l}\right)^2-\frac{2}{3}\left(\frac{x}{l}\right)^3+\frac{1}{6}\left(\frac{x}{l}\right)^4\right]dx$$

$$= c.l^4EI_0\sin\omega t\int_0^l\left[\frac{2}{l^6}x^2-\frac{4}{3l^3}x^3+\frac{1}{3l^8}x^4+\frac{12}{l^7}x^3-\frac{4}{l^8}x^4+\frac{2}{l^9}x^5-\frac{12}{l^8}x^4+\frac{8}{l^9}x^5-\frac{2}{l^{10}}x^6\right]dx$$

$$= c.l^4EI_0\sin\omega t\left[\frac{2}{3l^6}l^3-\frac{4}{12l^7}l^4+\frac{1}{15l^8}l^5+\frac{12}{4l^7}l^4-\frac{8}{5l^8}l^5+\frac{2}{6l^9}l^6-\frac{12}{5l^8}l^5+\frac{8}{6l^9}l^6-\frac{2}{7l^{10}}l^7\right]$$

$$= c.l^4EI_0\frac{1}{l^3}\sin\omega t\left[\frac{2}{3}-\frac{1}{3}+\frac{1}{15}+3-\frac{8}{5}+\frac{1}{3}-\frac{12}{5}+\frac{4}{3}-\frac{2}{7}\right] = 0.797\,c.l^4EI_0\frac{1}{l^3}\sin\omega t \tag{E7.17}$$

Consider the second Eq. (E7.15) integration,

$$\int_0^l -p_0\left(1-\frac{x}{l}\right)\varnothing_j(x,t)dx = 0.0555\,p_0 l^3 \quad \text{From last integration} \tag{E7.18}$$

Consider the third integration in Eq. (E7.15),

$$\int_0^l m\,\ddot{w}\,\varnothing_j dx$$

$$= \int_0^l m\left[-\omega^2 l^2\right]\left[\left(\frac{x}{l}\right)^2 - \frac{2}{3}\left(\frac{x}{l}\right)^3 + \frac{1}{6}\left(\frac{x}{l}\right)^4\right]\sin\omega t\varnothing_j dx$$

$$= -m\omega^2 l^4 \int_0^l \left[\left(\frac{x}{l}\right)^2 - \frac{2}{3}\left(\frac{x}{l}\right)^3 + \frac{1}{6}\left(\frac{x}{l}\right)^4\right]^2 dx$$

$$= -m\omega^2 l^4 \int_0^l \left[\frac{x^4}{l^4} + \frac{4}{9}\frac{x^6}{l^6} + \frac{1}{36}\frac{x^8}{l^8} - \frac{4}{3}\frac{x^5}{l^5} + \frac{2}{6}\frac{x^6}{l^6} - \frac{4}{18}\frac{x^7}{l^7}\right]dx$$

$$= -m\omega^2 l^4\left[\frac{1}{5}\frac{l^5}{l^4} + \frac{4}{9*7}\frac{l^7}{l^6} + \frac{1}{9*36}\frac{l^9}{l^8} - \frac{4}{6*3}\frac{l^6}{l^5} + \frac{2}{6*7}\frac{l^7}{l^6} + \frac{2}{6*7}\frac{l^7}{l^6} - \frac{4}{3*6*8}\frac{l^8}{l^7}\right]$$

$$= -m\omega^2 l^5 * 0.159 * c \tag{E7.19}$$

Collecting the integration terms of Eqs. (E7.17)–(E7.19) gives,

$$c.l*0.797 + 0.0555\,p_0 l^3 - m\omega^2 l^5 * 0.159 * c = 0 \tag{E7.20}$$

From which,

$$c = \frac{0.0555\,_{p0}l^2}{(m\omega^2 l^4 * 0.159 - 0.797)}$$

Insert this value in Eq. (E7.14), the deflected is obtained as follows:

$$w(x,t) = \frac{0.0555\,p_0 l^2}{(m\omega^2 l^4 * 0.159 - 0.797)}.l^2\left[\left(\frac{x}{l}\right)^2 - \frac{2}{3}\left(\frac{x}{l}\right)^3 + \frac{1}{6}\left(\frac{x}{l}\right)^4\right]\sin\omega t \tag{E7.21}$$

2.4.5 Torsion of rectangular section using G.M

Example 2.8: Assuming,

$$\varnothing = \varnothing_0\left[1-\left(\frac{x}{a}\right)^2\right]\left[1-\left(\frac{y}{a}\right)^2\right]$$

Use Galerkin's method to obtain the maximum shearing stress and torsion constant for a square section bar of side $2a$. Take origin of coordinates at the center. How do you obtain the twisting moment?

Solution,
We have,

$$\varnothing(x,y) = \varnothing_0\left[1-\left(\frac{x}{a}\right)^2\right]\left[1-\left(\frac{y}{a}\right)^2\right]$$

For torsion problem:

$$\nabla^2\varnothing + 2G\theta = 0 \tag{E8.1}$$

On the basis of the given assumption Eq. (E8.1) becomes:

$$4\int_0^a\int_0^a\left\{\left(\frac{d^2}{dx^2}+\frac{d^2}{dy^2}\right)\left(1-\left(\frac{x}{a}\right)^2\right)\left(1-\left(\frac{y}{a}\right)^2\right)+2G\theta\right\}\varnothing_1 dx\,dy = 0 \tag{E8.2}$$

We have,

$$\frac{d^2}{dx^2}\left(1-\left(\frac{x}{a}\right)^2\right)\left(1-\left(\frac{y}{a}\right)^2\right) = \frac{d}{dx}\left(1-\left(\frac{y}{a}\right)^2\right)\left(-\frac{2x}{a^2}\right) = \left(1-\left(\frac{y}{a}\right)^2\right)\left(-\frac{2}{a^2}\right)$$

$$\frac{d^2}{dy^2}\left(1-\left(\frac{x}{a}\right)^2\right)\left(1-\left(\frac{y}{a}\right)^2\right) = \frac{d}{dy}\left(1-\left(\frac{x}{a}\right)^2\right)\left(-\frac{2y}{a^2}\right) = \left(1-\left(\frac{x}{a}\right)^2\right)\left(-\frac{2}{a^2}\right)$$

Substitute the above integrals in Eq. (E8.2) gives,

$$4\int_0^a\int_0^a\left\{\begin{array}{c}\left(-\frac{2}{a^2}\right)\left(1-\left(\frac{y}{a}\right)^2\right)^2\left(1-\left(\frac{x}{a}\right)^2\right)+\\ \left(1-\left(\frac{x}{a}\right)^2\right)^2\left(1-\left(\frac{y}{a}\right)^2\right)\left(-\frac{2}{a^2}\right)+2G\theta\varnothing_1\end{array}\right\}dx\,dy = 0 \tag{E8.3}$$

First part of integration of Eq. (E8.3)

$$\int_0^a\left[\left(-\frac{2}{a^2}\right)\left(1-\left(\frac{y}{a}\right)^2\right)^2\left(x-\frac{x^3}{3a^2}\right)\right]_0^a dy$$

$$= \int_0^a\left(-\frac{2}{a^2}\right)\left(a-\frac{a^3}{3a^2}\right)\left(1-\left(\frac{y}{a}\right)^2\right)^2 dy$$

$$= -\frac{2}{a^2}\frac{2}{3}a\int_0^a\left(1+\frac{y^4}{a^4}-2\frac{y^2}{a^2}\right)dy = -\frac{4}{3a}\left[y+\frac{y^5}{5a^4}-2\frac{y^3}{3a^2}\right]_0^a$$

$$= -\frac{4}{3a}\left[a+\frac{a^5}{5a^4}-2\frac{a^3}{3a^2}\right] = -\frac{4}{3a}\frac{8}{15}a = -\frac{32}{45} \tag{E8.4}$$

Similarly, the second term is integrated which gives $-\frac{32}{45}$: Then, collecting the integral terms of Eq. (E8.3) gives,

$$= 4\left(-\frac{32}{45}-\frac{32}{45}\right) = 4\left(-\frac{64}{45}\right)\varnothing_0^2$$

The third term integration is as follows,

$$4\int_0^a\int_0^a 2G\theta dx\, dy = 8G\theta\int_0^a\int_0^a\left(1-\left(\frac{x}{a}\right)^2\right)\left(1-\left(\frac{y}{a}\right)^2\right)dx\, dy$$

$$= 8G\theta\int_0^a\left(1-\left(\frac{y}{a}\right)^2\right)\left[x-\frac{x^3}{3a^3}\right]_0^a dy$$

$$= 8G\theta\varnothing_0\int_0^a\left(1-\left(\frac{y}{a}\right)^2\right)\left[a-\frac{a^3}{3a^3}\right]_0^a dy$$

$$= 8G\theta\left(\frac{2}{3}a\right)\left(\frac{2}{3}a\right) = \frac{32}{9}G\theta a^2\varnothing_0$$

Collecting the integrals of Eq. (E8.3) gives,
$-\frac{64}{45}4\varnothing_0^2 + \frac{32}{9}G\theta a^2\varnothing_0 = 0$, which results in,

$$\varnothing_0 = \frac{32}{9}\frac{45}{4*64}G\theta a^2 = 0.625G\theta a^2 \tag{E8.5}$$

$$\tau_{max} \approx \left(\frac{\partial\varnothing}{\partial x}\right)_{\substack{x=a\\ y=0}} \tag{E8.6}$$

Carrying the differentiation of Eq. (E8.1), and substitute the value of $\varnothing_0 = 0.625G\theta a^2$ yields,

$$\left(\frac{\partial\varnothing}{\partial x}\right)_{\substack{x=a\\ y=0}} = \varnothing_0\left[0-\frac{2x}{a^2}\right]\left[1-\left(\frac{y}{a}\right)^2\right] = \varnothing_0\left[0-\frac{2a}{a^2}\right][1-0] = -\varnothing_0\frac{2}{a} = -0.625G\theta a^2\frac{2}{a} = -1.25G\theta a$$

The twisting moment is given by:

$$M_t = 2\iint\varnothing dxdy = 2\iint 0.625G\theta a^2\left(1-\left(\frac{x}{a}\right)^2\right)\left(1-\left(\frac{y}{a}\right)^2\right)dxdy$$

Or,

$$k = \frac{20}{9}a^4$$

2.4.6 Application of energy methods to Lambda frame

Example 2.9: Each member of the lambda frame shown in Fig. 2.12, has length L, cross-sectional area A, modulus of elasticity E, and moment of inertia I. As a result of the external moment M applied at the vertex, the joint experiences displacement components (u,v) and a rotation θ, taking the extensions of the members into account, derive formulas for (u,v) by using the P.S.T.P.E.

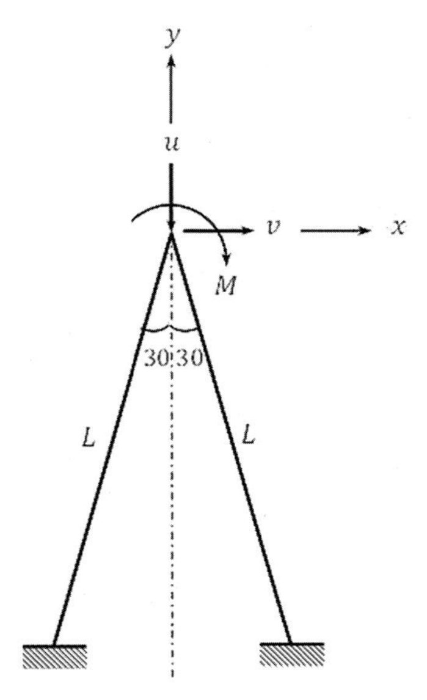

FIGURE 2.12 Lambda frame subjected to bending moment M.

Solution,

We have, the displacements in x and y are calculated as follows,

$$u(x,y) = -x\sin 30 + y\cos 30 \tag{E9.1}$$

$$v = x\cos 30 + y\sin 30 \tag{E9.2}$$

Assume,

$$y = a.x + b.x^2$$

And,

$$\frac{dy}{dx} = a + 2bx$$

Strain energy of the axial member is given by,

$$U = \int \frac{EA}{2}\left(\frac{dy}{dx}\right)^2 dx = \int \frac{EA}{2}\left(a^2 + 4abx + 4b^2x^2\right)dx$$

$$U = \frac{EA}{2}\left(a^2 l + 2abl^2 + \frac{4}{3}b^2 l^3\right) \tag{E9.3}$$

The potential energy due to bending moment M is given by,

$$\Omega = \int M\, d\theta$$

$$d\theta = \frac{dy}{dx}$$

Therefore,

$$\Omega = \int M(a + 2bx)\, dx = M\left(al + bl^2\right)$$

The total potential energy is given by,

$$V = U + \Omega$$

$$V = \frac{EA}{2}\left(a^2 l + 2abl^2 + \frac{4}{3}b^2 l^3\right) - M\left(al + bl^2\right)$$

And,

$$\frac{\partial V}{\partial a} = \frac{EA}{2}\left(2al + 2bl^2\right) - Ml = 0 \tag{E9.4}$$

$$\frac{\partial V}{\partial b} = \frac{EA}{2}\left(2al^2 + \frac{8}{3}bl^3\right) - Ml^2 = 0 \tag{E9.5}$$

Arrange Eqs. (E9.4) and (E9.5) in matrix form as follows,

$$\begin{bmatrix} EAl & EAl^2 \\ EAl^2 & \frac{8}{6}EAl^3 \end{bmatrix} \begin{bmatrix} a \\ b \end{bmatrix} = \begin{bmatrix} Ml \\ Ml^2 \end{bmatrix}$$

To find a, b,

$$\begin{bmatrix} a \\ b \end{bmatrix} = \begin{bmatrix} EAl & EAl^2 \\ EAl^2 & \frac{8}{6}EAl^3 \end{bmatrix}^{-1} \begin{bmatrix} Ml \\ Ml^2 \end{bmatrix}$$

Or,

$$\begin{bmatrix} a \\ b \end{bmatrix} = \frac{3}{(EA)^2 l^4} \begin{bmatrix} \frac{8}{6}EAl^3 & -EAl^2 \\ -EAl^2 & EAl \end{bmatrix} \begin{bmatrix} Ml \\ Ml^2 \end{bmatrix}$$

Which gives,

$$\begin{bmatrix} a \\ b \end{bmatrix} = \frac{3}{(EA)^2 l^4} \begin{bmatrix} \frac{1}{3}MEAl^4 \\ 0 \end{bmatrix} = \begin{bmatrix} \frac{M}{EA} \\ 0 \end{bmatrix}$$

To find x, y,

$$y = \frac{M}{EA}x$$

And,

$$x = \frac{EA}{M}y$$

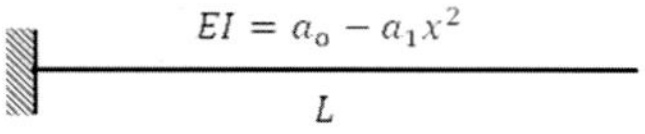

FIGURE 2.13 Modeling of airplane wing.

Therefore, using Eqs. (E9.1) and (E9.2),

$$u = -x \sin 30 + \frac{M}{EA} x \cos 30$$
$$v = \frac{EA}{M} y \cos 30 + y \sin 30 \quad \text{(E9.6)}$$

Example 2.10: An airplane wing shown in Fig. 2.13, is regarded as a cantilever beam of length L. The stiffness of the wing is assumed to be given by $EI = a_0 - a_1x^2$, where x is the distance of a cross section of the wing from the root and (a_0, a_1) are constants. The air load per unit length is approximated by $P = b_0 - b_1x$ where (b_0, b_1) are constants, the deflection is approximated by $y = Ax^2 + Bx^3$, where (A,B) are constants. Derive the expression for T.P.E. of the wing. Hence derive linear algebraic equations that determine the coefficients (A,B) by P.S.T.P.E.

Solution,

We have
$P = b_0 - b_1x$, and

$$y = Ax^2 + Bx^3 \quad \text{(E10.1a)}$$

Therefore,

$$\frac{dy}{dx} = 2Ax + 3Bx^2$$
$$\frac{d^2y}{dx^2} = 2A + 6Bx$$

The strain energy due to bending load is given by,

$$U = \int_0^L \frac{EI}{2}\left(\frac{d^2y}{dx^2}\right)^2 dx = \int_0^L \frac{(a_0 - a_1x^2)}{2}(4A^2 + 36B^2x^2 + 24ABx)dx \quad \text{(E10.1)}$$

Carrying out the integrations and simplifications of Eq. (E10.1) is achieved as follows,

$$U = \int_0^L (2A^2a_0 + 18B^2a_0x^2 + 12ABa_0x - 2A^2a_1x^2 - 18B^2a_1x^4 - 12ABa_1x^3)dx$$

$$U = 2A^2a_0L + 6B^2a_0L^3 + 6ABa_0L^2 - \frac{2}{3}A^2a_1L^3 - \frac{18}{5}B^2a_1L^5 - 3ABa_1L^4 \quad \text{(E10.2)}$$

The potential energy due to the applied load is given by,

$$\Omega = -\int_0^L P.y\, dx = -\int_0^L (b_0 - b_1x)(Ax^2 + Bx^3)dx \quad \text{(E10.3)}$$

Integrations and simplifications give,

$$\Omega = -\int_0^L \left(b_0Ax^2 - b_1Ax^3 + b_0Bx^3 - b_1Bx^4\right)dx$$

Or,

$$\Omega = -\frac{1}{3}b_0AL^3 + \frac{1}{4}b_1AL^4 - \frac{1}{4}b_0BL^4 + \frac{1}{5}b_1BL^5 \tag{E10.4}$$

And the total potential energy will be,

$$V = U + \Omega = V(A, B) \tag{E10.5}$$

Substitute the values of U and Ω from Eqs. (E10.2) and (E10.4) respectively results in,

$$V = \left(2a_0L - \frac{2}{3}a_1L^3\right)A^2 + \left(6a_0L^2 - 3a_1L^4\right)AB + \left(6a_0L^3 - \frac{18}{5}a_1L^5\right) \tag{E10.6}$$
$$B^2 - \left(\frac{1}{3}b_0L^3 - \frac{1}{4}b_1L^4\right)A - \left(\frac{1}{4}b_0L^4 - \frac{1}{5}b_1L^5\right)B$$

Minimization of T.P.E. with respect to A and B as follows,

$$\frac{\partial V}{\partial A} = \left(4a_0L - \frac{4}{3}a_1L^3\right)A + \left(6a_0L^2 - 3a_1L^4\right)B - \left(\frac{1}{3}b_0L^3 - \frac{1}{4}b_1L^4\right) = 0$$

$$\frac{\partial V}{\partial B} = \left(6a_0L^2 - 3a_1L^4\right)A + \left(12a_0L^3 - \frac{36}{5}a_1L^5\right)B - \left(\frac{1}{4}b_0L^4 - \frac{1}{5}b_1L^5\right) = 0 \tag{E10.7a}$$

Arranging the above two equations in matrix form gives,

$$\begin{bmatrix} \left(4a_0L - \frac{4}{3}a_1L^3\right) & \left(6a_0L^2 - 3a_1L^4\right) \\ \left(6a_0L^2 - 3a_1L^4\right) & \left(12a_0L^3 - \frac{36}{5}a_1L^5\right) \end{bmatrix} \begin{bmatrix} A \\ B \end{bmatrix} = \begin{bmatrix} \left(\frac{1}{3}b_0L^3 - \frac{1}{4}b_1L^4\right) \\ \left(\frac{1}{4}b_0L^4 - \frac{1}{5}b_1L^5\right) \end{bmatrix}$$

Or,

$$\begin{bmatrix} \left(48a_0 - 16a_1L^2\right) & \left(72a_0L - 36a_1L^3\right) \\ \left(120a_0 - 60a_1L^2\right) & \left(240a_0L - 144a_1L^3\right) \end{bmatrix} \begin{bmatrix} A \\ B \end{bmatrix} = \begin{bmatrix} \left(4b_0L^2 - 3b_1L^3\right) \\ \left(5b_0L^2 - 4b_1L^3\right) \end{bmatrix}$$

The solution results are,

$$A = \frac{\begin{vmatrix} \left(4b_0L^2 - 3b_1L^3\right) & \left(72a_0L - 36a_1L^3\right) \\ \left(5b_0L^2 - 4b_1L^3\right) & \left(240a_0L - 144a_1L^3\right) \end{vmatrix}}{\begin{vmatrix} \left(48a_0 - 16a_1L^2\right) & \left(120a_0 - 60a_1L^2\right) \\ \left(120a_0 - 60a_1L^2\right) & \left(240a_0L - 144a_1L^3\right) \end{vmatrix}} = \frac{[a_1]}{[a_0]}$$

Where,

$$[a_1] = 960a_0b_0L^3 + 432a_1b_1L^6 - 576b_0a_1L^5 - 480b_1a_0L^4 - 360a_0b_0L^3 - 144a_1b_1L^6 + 180b_0a_1L^5 + 288b_1a_0L^4$$

$$[a_1] = 600a_0b_0L^3 + 288a_1b_1L^6 - 396b_0a_1L^5 - 192b_1a_0L^4$$

$$[a_0] = 11520a_0^2L + 2304a_1^2L^5 - 6912a_0a_1L^3 - 3840a_0a_1L^3 - 8640a_0^2L - 2160a_1^2L^5 + 4320a_0a_1L^3 + 4320a_0a_1L^3$$

$$[a_0] = 2880a_0^2L + 144a_1^2L^5 - 2112a_0a_1L^3$$

Hence, the results of the coefficients A and B are as follows,

$$A = \frac{50a_0b_0L^2 + 24a_1b_1L^5 - 33b_0a_1L^4 - 16b_1a_0L^3}{240a_0^2 + 12a_1^2L^4 - 176a_0a_1L^2} \tag{E10.7}$$

And

$$B = \frac{\begin{vmatrix} (48a_0 - 16a_1L^2) & (4b_0L^2 - 3b_1L^3) \\ (120a_0 - 60a_1L^2) & (5b_0L^2 - 4b_1L^3) \end{vmatrix}}{[a_0]} = \frac{[a_2]}{[a_0]}$$

Where,

$$[a_2] = 240a_0b_0L^2 + 64a_1b_1L^5 - 192b_1a_0L^3 - 80b_0a_1L^4 - 480a_0b_0L^2 - 180a_1b_1L^5 + 360b_1a_0L^3 + 240a_1b_0L^4$$

$$[a_2] = -240a_0b_0L^2 - 116a_1b_1L^5 + 168b_1a_0L^3 + 160b_0a_1L^4$$

Therefore,

$$B = \frac{-60a_0b_0L - 29a_1b_1L^4 + 42a_0b_1L^2 + 40a_1b_0L^3}{720a_0^2 + 36a_1^2L^5 - 528a_0a_1L^3} \tag{E10.8}$$

Example 2.11: The tee-section cantilever shown in Fig. 2.14 has a span of $4m$ and is cut from a (914 mm × 305 mm × 253 mm) Universal Beam. The overall depth varies linearly from 306 mm at the free end to 612 mm at the built-in end. A concentrated load of 65 kN is carried at the free end and an anticlockwise couple of 120 kN.m is applied at mid-span.

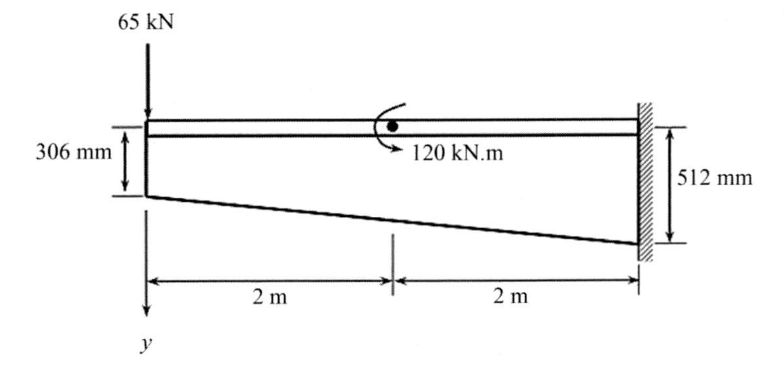

FIGURE 2.14 A tee-section cantilever beam.

A good approximation of the major axis second moment of area of the tee-section is given by:

$$I = I_0\left(1 + ax + bx^2\right)0 < x < 4 \text{ m}$$

Where,

$$I_0 = 9742 \text{ cm}^4 .a = 0.685 \text{ m}^{-1} \text{ and b} = 0.210 \text{ m}^{-2}$$

Estimate the maximum deflection Δ on the assumption that a suitable displacement function is: $y = \Delta\left(1 - \sin\frac{\pi x}{8}\right) \quad 0 < x < 4$
We have, $y' = -\frac{\pi\Delta}{8}\cos\frac{\pi x}{8}$

$$y'' = \frac{\pi^2\Delta}{64}\sin\frac{\pi x}{8}$$

The strain energy due to bending is given by,

$$U = \frac{E}{2}\int_0^4 I(y'')^2 dx \tag{E11.1}$$

Carrying out the substitutions and integrations of Eq. (E11.1) are achieved as follows,

$$U = \frac{\pi^4 E I_0 \Delta^2}{8192}\int_0^4 (1 + ax + bx^2)\sin^2\left(\frac{\pi x}{8}\right)dx$$

$$U = \frac{\pi^3 E I_0 \Delta^2}{1024}\int_0^{\pi/2}\left(\sin^2\theta + \frac{8a}{\pi}\theta\sin^2\theta + \frac{64b^2}{\pi^2}\theta^2\right)d\theta \tag{E11.2}$$

We have,

$$\int_0^{\pi/2}\sin^2\theta d\theta = \left[\frac{\theta}{2} - \frac{\sin^2\theta}{4}\right]_0^{\pi/2} = \frac{\pi}{4}$$

$$\int_0^{\pi/2}\theta\sin^2\theta d\theta = \left[\frac{\theta^2}{4} - \frac{\theta\sin 2\theta}{4} - \frac{\cos 2\theta}{8}\right]_0^{\pi/2} = \frac{\pi^2 + 4}{16}$$

$$\int_0^{\pi/2}\theta^2\sin^2\theta d\theta = \left[\frac{\theta^3}{6} - \frac{\theta^2}{4} - \frac{1}{8}\sin 2\theta - \theta\frac{\cos 2\theta}{4}\right]_0^{\pi/2} = \frac{\pi(\pi^2 + 6)}{48}$$

Substitute the above integrals in Eq. (E11.2) results,
$U = \frac{\pi^3 E I_0 \Delta^2}{1024}\left[\frac{\pi}{12}(3 + 6a + 16b) + \frac{2}{\pi}(a + b)\right]$
Hence, the strain energy U becomes,

$$U = 2190\Delta^2 \tag{E11.3}$$

The work done W_d is given by,

$$W_d = -65\Delta - 120\frac{\pi\Delta}{8\sqrt{2}}$$

Or,

$$W_d = -98.32\Delta \text{ KN.m} \tag{E11.4}$$

$$\text{Thetotalpotentialenergy } V = U + W_d \tag{E11.5}$$

Substitute the values of U and W_d from Eqs. (E11.3) and (E11.4) respectively in Eq. (E11.5) gives,

$$V = 2190\Delta^2 - 98.32\Delta \tag{E11.6}$$

>Minimization of total potential energy of Eq. (E11.6) with respect to Δ is carried out as follows,

$$\frac{\partial V}{\partial \Delta} = 4380\Delta - 98.32 = 0$$

Which yields,

$$\Delta = 22.4 \times 10^{-3} = 22.4 \text{ mm} \tag{E11.7}$$

Problems

P.2.1: Formulate Rayleigh-Ritz type analysis for the thermal stress analysis of plane disks, the thickness of which may vary. Consider that there is a central hole concentric with the outside diameter. Outline a CAD package for such components to be implemented on a microcomputer.

P.2.2: A loaded projection on a steel casting is modeled as a tapered cantilever as shown in Fig. P.2.2. Assuming the one parameter deflection function,

$$W(x) = C_1 x^2\left(3 - \frac{x}{l}\right)$$

Use the Ritz technique to obtain an approximation to the tip deflection. Comment on this choice of trial function. You may accept that,

$$U = \frac{1}{2}\int_0^l EI(w'')^2 dx$$

P.2.3: A circular shaft, diameter $2a$, has a pair of flats machined opposite each other, the distance across the flats is $2b$. Show that by suitable choice of coordinates,

$$\varnothing = \left(a^2 - x^2 - y^2\right)\left(b^2 - x^2\right)\sum a_{mn}x^m y^n$$

Is a reasonable stress function to use in the Rayleigh-Ritz technique.

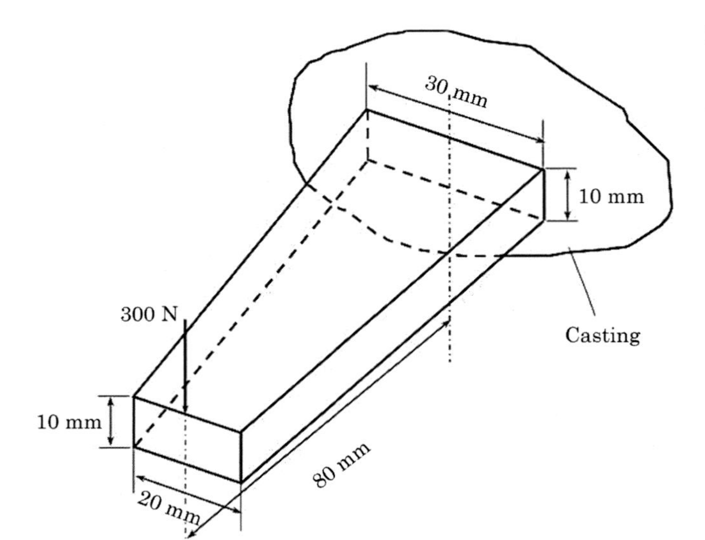

FIGURE P.2.2

P.2.4: The torsion of prismatic bars is described by

$\nabla^2\varnothing + 2G\theta = 0$ Over the section
$\varnothing = 0$ along the boundary
Where $\varnothing$ is the stress function related to the shearing stresses and torsion constant by:

$$\tau_{zx} = \frac{\partial\varnothing}{\partial y} \cdot \tau_{zy} = \frac{\partial\varnothing}{\partial x} \cdot K = \frac{2}{G\theta}\iint \varnothing dxdy$$

To obtain an approximate solution using a series representation:

$$\varnothing = \varnothing_n = \sum_{i=1}^{n} c_i f_i(x, y)$$

Show that Galerkin's method leads to a set of equations

$$\sum_{j=1}^{n} \alpha_{ij} c_j = \beta_i i = 1.2. \ldots n$$

For determining the constants c_j.
Show that the one term expansion

$$\varnothing = c_1 x\left(x^2 + y^2 - a^2\right)$$

Is admissible as a trial function for the section of Fig. 2.4.
Use Galerkin's method to determine c_1 and then obtain values for the maximum shearing stress and torsion constant.

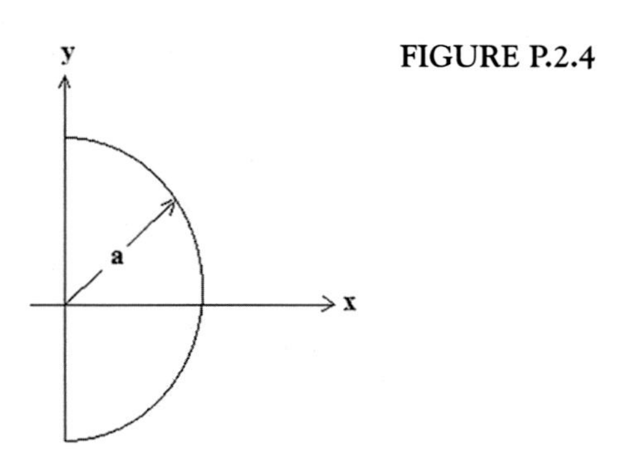

FIGURE P.2.4

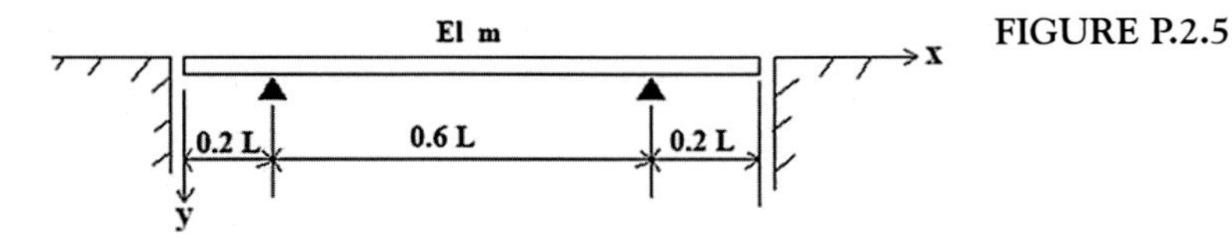

FIGURE P.2.5

The following information will prove useful (Fig. P.2.4),

1. $\int_0^a \left(a^2 - y^2\right)^{r/2} dy \int_0^{\pi/2} a^{(r+1)} \sin^{(r+1)} \theta d\theta$

2. $\int_0^{\pi/2} \sin^m \theta d\theta = \dfrac{(m-1)(m-3)\ldots..4.2}{m(m-2)\ldots\ldots\ldots\ldots..5.3}, \quad m = \text{odd}$

$\dfrac{(m-1)(m-3)\ldots..3.1}{m(m-2)\ldots\ldots\ldots\ldots..4.2} \cdot \dfrac{\pi}{2} m = \text{even}$

P.2.5: A bridge structure may be represented by a uniform beam, of length l, mass/unit length m and flexural rigidity EI supported at the one-fifth and four-fifth span positions as shown in Fig. P.2.5. The ends of the span are unsupported.

In order to compute the fundamental natural frequency the Rayleigh-Ritz method is to be used assuming a vibrating shape $y(x) = c_1\psi_1(x) + c_3\psi_3(x)$ where $\psi_1(x)$ and $\psi_3(x)$ are defined by

$$\psi_1(x) = \varnothing_1(x) + h_1, \psi_3(x) = \varnothing_3(x) + h_3$$

The functions $\varnothing_1(x)$ and $\varnothing_3(x)$ are the first and third Eigen-functions for a free-free beam and h_1 and h_3 are constants such as to make $\psi_1(x)$ and $\psi_3(x)$ zero at the support positions. With the help of the information tabulated below, calculate the fundamental natural frequency.

Comment on the suitability of the chosen functions $\psi_1(x)$ and $\psi_3(x)$ in this situation.

For a free-free beam:

$$\int_0^l \varnothing_n(x)dx = 0, \int_0^l \varnothing_n{}^2(x)dx = l, \int_0^l \varnothing_m(x)\varnothing_n(x)dx = 0$$

$$\int_0^l (\varnothing''_n(x))^2 dx = \beta_n^4 l, \int_0^l \varnothing''_m(x)\varnothing''_n(x)dx = 0$$

$$\beta_n^4 = \omega_n^2 \frac{m}{EI}, (\beta_1 l)^4 = 500.56, (\beta_3 l)^4 = 14617.63$$

x/l	$\varnothing_1(x)$	$\varnothing_3(x)$
0.0	2.0000	2.0000
0.2	0.1955	−1.2857
0.8	0.1955	−1.2857
1.0	2.0000	2.0000

P.2.6: To treat the problem of torsion of bars by the Rayleigh-Ritz process we may proceed in one of two ways. Visualize the analogous soap film and apply the potential energy principle to find an approximation to its deflexion. Work in terms of Prandtl's stress function and apply the complementary energy principle. Following the second approach, equation of compatibility result from $\delta v^* = 0$, where:

$$\delta v^* = \frac{l}{2G}\iint \left[\left(\frac{\partial \varnothing}{\partial x}\right)^2 + \left(\frac{\partial \varnothing}{\partial x}\right)^2 - 4G \propto \varnothing\right] dxdy$$

For a rectangular section $2a \times 2b$, obtain an approximate value for the torsion constant assuming a one term expansion for $\varnothing(x.y)$ in the Rayleigh-Ritz process. That is, assume:

$$\varnothing(x.y) = q_{11}\cos\frac{m\pi x}{2a}\cos\frac{n\pi y}{2b}$$

(Note: for a square section, side $2a$, Timoshenko and Goodier give the value $k = 2.2496a^4$ for the torsion constant). Compare with the value given by St. Venant's approximate formula in the Stress Analysis notes.

1. For the cantilever beam shown Fig. P.2.6A. The total potential energy is,

$$X = \frac{1}{2}\int_0^{\ell} EI\left(\frac{\partial^2 v}{\partial x^2}\right)dx - \int_0^{\ell} wvdx$$

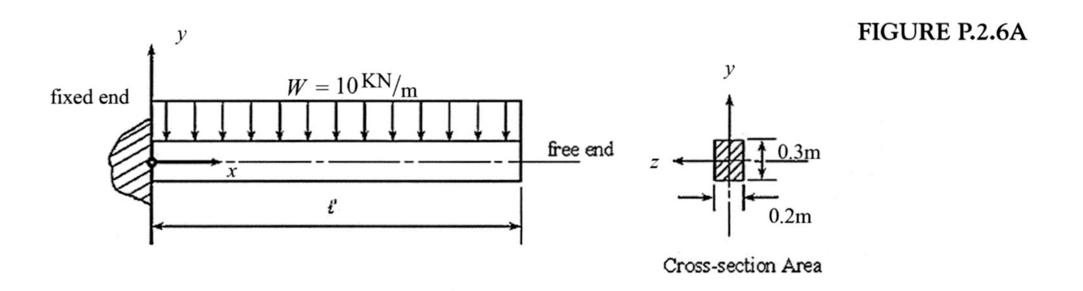

FIGURE P.2.6A

Where, E = Young's modulus. I = second moment of area about z-axis. v = vertical deflection. w = uniform distributed load.

Assuming a trial function solution as:

$$V(x) = \sum_{i=1}^{5} \alpha_i x^{i-1}$$

Find the unknown parameters ($\alpha_i . i = 1.5$) and the vertical displacement and rotation angle $\left(\frac{\partial v}{\partial x}\right)$ at $x = \ell$, for Rayleigh-Ritz method if $\ell = 1$ m and $E = 210\ \mathrm{GN/m^2}$.

2. A beam bonded onto an elastic base shown in Fig. P.2.6B. is described by:

$$EI\frac{d^4w}{dx^4} + kw = P$$

Or,

$$\frac{d^4w}{dx^4} + \mu^4 w = \frac{P}{EI}$$

k = foundation modulus, P = transverse load/unit length.

Such a beam is shown in Fig. P.2.6B.

The ends are rigidly clamped. Dividing the beam into two intervals and four intervals, taking advantage of symmetry and using Richardson's extrapolation, find the center deflection using finite differences.

P.2.7: A linearity elastic beam of non-uniform cross-section and span L is simply supported at each end. The beam carries a uniformly distributed vertical load of W per unit length. The beam has a horizontal centroid axis and the second moment of area (I) for bending in the vertical plane is given by:

$$I = I_0\left(1 + \pi \sin\frac{\pi x}{L}\right)$$

The origin of the coordinate axes is taken at the left-hand support. If the deflection shape of the beam is assumed to be given approximately by:

$$y(x) = \Delta \sin\frac{\pi x}{L}$$

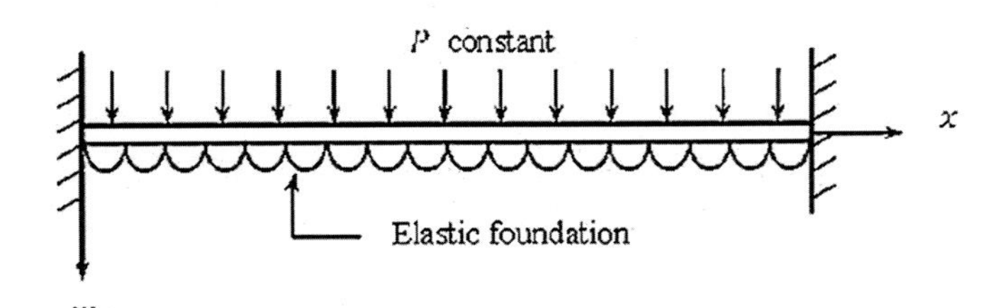

FIGURE P.2.6B

Show, by minimizing the potential energy, that:

$$\Delta = \frac{12}{11\pi^5}\frac{WL^4}{EI_0}$$

P.2.8: A simply supported variable thickness circular plate, radius a, carries a uniform pressure p. The bending stiffness is given by,

$$D = D_0\left(1 - \frac{r}{2a}\right)$$

Assuming that the deflection surface is approximated by,

$$w = w_0\left[1 - \left(\frac{r}{a}\right)^2\right]$$

Use the potential energy principle to show that,

$$w_0 = \frac{3}{32}\frac{pa^4}{(1+\nu)D_0}$$

CHAPTER

3

Application of energy methods to plate problems

In this chapter, the analysis of plates subjected to bending, stretching and buckling loading using the energy principles are presented. The variation principles and the direct methods are employed for different cases of plates. The presented examples cover the rectangular plates with different support conditions under bending loading and stretching plates in both x and y directions using the total potential energy principles. The displacement field, strains and stresses can be calculated. The chapter includes the application of direct methods in plate problems such as Galerkin's and Kantorovitch methods in plate bending problems. The resulted mathematical expressions are derived using the total potential energy principles. These expressions are derived for different loading cases and boundary conditions. The examples also cover the plate buckling cases and the derived expressions which are based upon the energy principles. At the end of the chapter, different plate problems are set to be solved using the derived expressions in this chapter.

3.1 Plate bending

Fig. 3.1 shows an element of a plate together with the resulted stresses due to bending.

In order to write the total potential energy of the plate element, we have to find the strain energy and the potential energy as follows,

1. To find **U** (strain energy stored)

Since we assume elements of the plate with plane stress condition, therefore we can write,

$$U^o = \frac{E}{2(1-\nu^2)}\left(\varepsilon_x^2 + \varepsilon_y^2 + 2\nu\varepsilon_x\varepsilon_y\right) + \frac{E}{4(1+\nu)}\gamma_{xy}^2 \tag{3.1}$$

where U^o is the strain energy per unit volume.

Since for the plate, the strain components are given by,

$$\varepsilon_x = -z\frac{\partial^2 w}{\partial x^2}, \quad \varepsilon_y = -z\frac{\partial^2 w}{\partial y^2}, \quad \gamma_{xy} = z\frac{\partial^2 w}{\partial x \partial y} \tag{3.2a}$$

Where $w(x,y)$ is the deflection. Note that all the terms carry usual meanings.

Energy Methods and Finite Element Techniques
DOI: https://doi.org/10.1016/B978-0-323-88666-6.00003-4

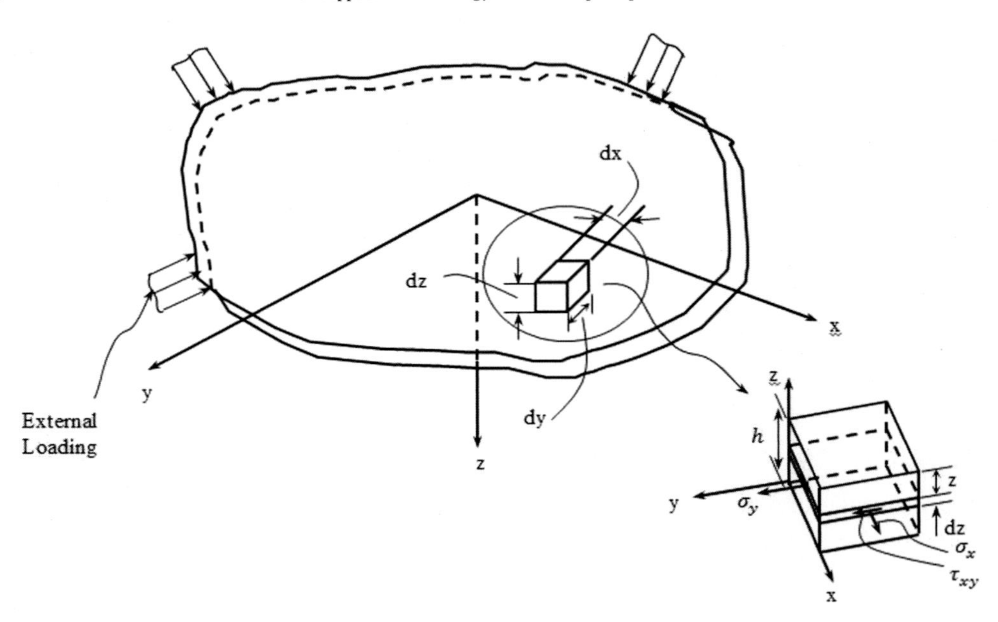

FIGURE 3.1 A plate element extracted from a loaded plane region.

And the total strain energy stored,

$$U = \int_{vol} U_o \, dvol = \iiint U_o \, dx \, dy \, dz$$

Or, using Eqs. (3.1) and (3.2a) gives,

$$U = \frac{D}{2} \iint \left\{ \left(\frac{\partial^2 w}{\partial x^2} + \frac{\partial^2 w}{\partial y^2} \right)^2 - 2(1-\nu) \left[\frac{\partial^2 w}{\partial x^2} \frac{\partial^2 w}{\partial y^2} - \frac{\partial^2 w}{\partial x \partial y} \right] \right\} dxdy \tag{3.2}$$

It may be shown that the term in square bracket can be discarded (because it makes zero contribution) in the following circumstances,

1. If the plate is of polygonal in plane form and the edge, is S.S or clamped.
2. If the plate is arbitrary plane form but the edge, is clamped.

If all edges of the plate are fixed, then the second term on the R.H.S. becomes zero, therefore,

$$U = \frac{D}{2} \iint \left\{ \left(\frac{\partial^2 w}{\partial x^2} + \frac{\partial^2 w}{\partial y^2} \right)^2 \right\} dx \, dy \tag{3.3}$$

The potential energy due to applied load Ω is,

$$\Omega = -\int_0^a \int_0^b P_z . w dxdy \tag{3.4}$$

The total P.E.V $= U + \Omega$ for a rectangular plate subjected to lateral load is obtained by the substitution of Eqs. (3.3) and (3.4) as follows,

$$V = \frac{1}{2}\int_0^a\int_0^b D\left\{\left(\frac{\partial^2 w}{\partial x^2}+\frac{\partial^2 w}{\partial y^2}\right)^2 - 2(1-\nu)\left[\frac{\partial^2 w}{\partial x^2}\frac{\partial^2 w}{\partial y^2} - \left(\frac{\partial^2 w}{\partial x \partial y}\right)^2\right]\right\}dx.dy - \int_0^a\int_0^b P_z.wdxdy \quad (3.5)$$

Example 3.1: For the simply supported rectangular plate under uniform pressure (Fig. 3.2), find the strain energy and the potential energy due to applied loading.

Solution:

Assume that the deflected shape of the plate is given by,

$$w(x,y) = \sum_{m=1}^{\infty}\sum_{n=1}^{\infty} a_{mn}\sin\frac{m\pi x}{a}\sin\frac{n\pi y}{b} \quad \text{(E1.1)}$$

The boundary conditions (B.Cs.) are S.S. gives as zero deflection at $x = 0$ and $x = a$, $w(0,y) = w(a,y) = 0$.

Now B.Cs.: $\frac{\partial^2 w}{\partial x^2}, \frac{\partial^2 w}{\partial y^2} = 0$ at edges are satisfied since the plate is S.S.

We observe that the assumption does indeed satisfy the geometric B.Cs.

The normal B.M implied $(\frac{\partial^2 w}{\partial x^2})$ and satisfy the mechanical (B.Cs) which give an efficient solution.

To find the strain energy of the plate U, we have,

$$\frac{\partial^2 w}{\partial x^2} = w_{,xx} = -\sum\sum a_{mn}\left(\frac{m\pi}{a}\right)^2\sin\frac{m\pi x}{a}\sin\frac{n\pi y}{b}$$

$$\frac{\partial^2 w}{\partial y^2} = w_{,yy} = -\sum\sum a_{mn}\left(\frac{n\pi}{b}\right)^2\sin\frac{m\pi x}{a}\sin\frac{n\pi y}{b}$$

Substitute above equations to $U_{\text{simplified}}$, Eq. (3.2), we find a typical term of the form,

$$\iint \sin\frac{m\pi x}{a}\sin\frac{n\pi y}{b}\sin\frac{p\pi x}{a}\sin\frac{q\pi y}{b}dxdy = \int_0^a \sin\frac{m\pi x}{a}\sin\frac{p\pi x}{a}dx\int_0^b \sin\frac{n\pi y}{b}\sin\frac{q\pi y}{b}dy$$
$$= 0 \text{ for } m \neq p,\ n \neq q$$

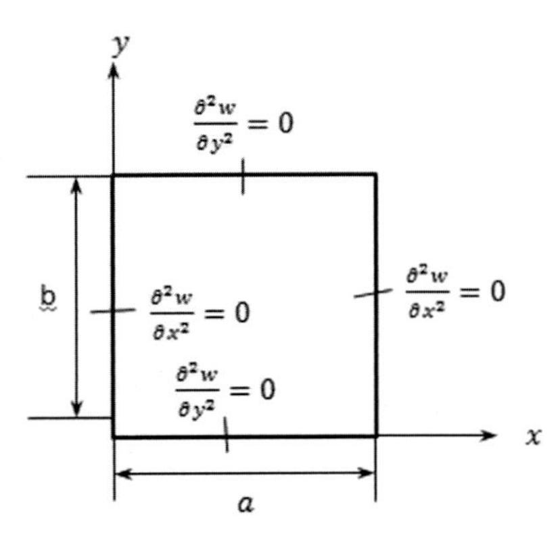

FIGURE 3.2 Simply supported plate (Note that the deflection $w(x, y) = 0$ all around).

Unless both $m = p$ and $n = q$ the terms disappear because of orthogonality properties, and the final form of the strain energy stored in the plate is given by,

$$U = \frac{\pi^4 Dab}{8} \sum \sum a_{mn}^2 \left[\frac{m^2}{a^2} + \frac{n^2}{b^2} \right]^2 \tag{E1.2}$$

To find the potential energy due to external applied load Ω, the pressure is like a point load so the P.E. due to this load is,

$$\Omega = - \iint p w(x, y) dx\, dy = - \iint \sum_{n=1}^{N} \sum_{m=1}^{N} a_{mn} \sin \frac{m\pi x}{a} \sin \frac{n\pi y}{b} dx\, dy \tag{E1.3}$$

To find the coefficient a_{mn},

The total P.E. of the plate $V = U + \Omega$ is obtained by using Eqs. (E1.2) and (E1.3) as follows,

$$\text{V} = \frac{\pi^4 Dab}{8} \sum \sum a_{mn}^2 \left[\frac{m^2}{a^2} + \frac{n^2}{b^2} \right]^2 - \iint \sum \sum a_{mn} \sin \frac{m\pi x}{a} \sin \frac{n\pi y}{b} dx\, dy$$

$$V = \frac{\pi^4 Dab}{8} \sum \sum a_{mn}^2 \left[\frac{m^2}{a^2} + \frac{n^2}{b^2} \right]^2 - 4 \sum \sum 4.a_{mn} \left(\frac{ab}{mn\pi^2} \right)$$

For equilibrium = 0,

That is,

$$\frac{\partial V}{\partial a_{mn}} = 0, m, n = 1, 2, \ldots$$

Therefore,

$$0 = \frac{\pi^4 Dab}{8} \sum \sum 2a_{mn} \left[\frac{m^2}{a^2} + \frac{n^2}{b^2} \right]^2 - 4 \sum \sum 4 \left(\frac{ab}{mn\pi^2} \right)$$

Hence,

$$a_{mn} = \sum \sum \frac{16}{\pi^2 D} \cdot \frac{a^3 b^3}{mn} \cdot \frac{1}{(b^2 m^2 + a^2 n^2)^2}$$

Having obtained a_{mn} the strain energy and the potential energy are calculated using Eqs. (E1.2) and (E1.3) respectively.

For the Clamped rectangular plate under uniform pressure, $p(x,y)$ as shown in (Fig. 3.3).

If we assume that the deflection is given by,

$$w(x, y) = \sum_{m}^{N} \sum_{n}^{N} C_{mn} \varnothing_m \varnothing_n$$

We must ensure that the geometric B.CS are satisfied.

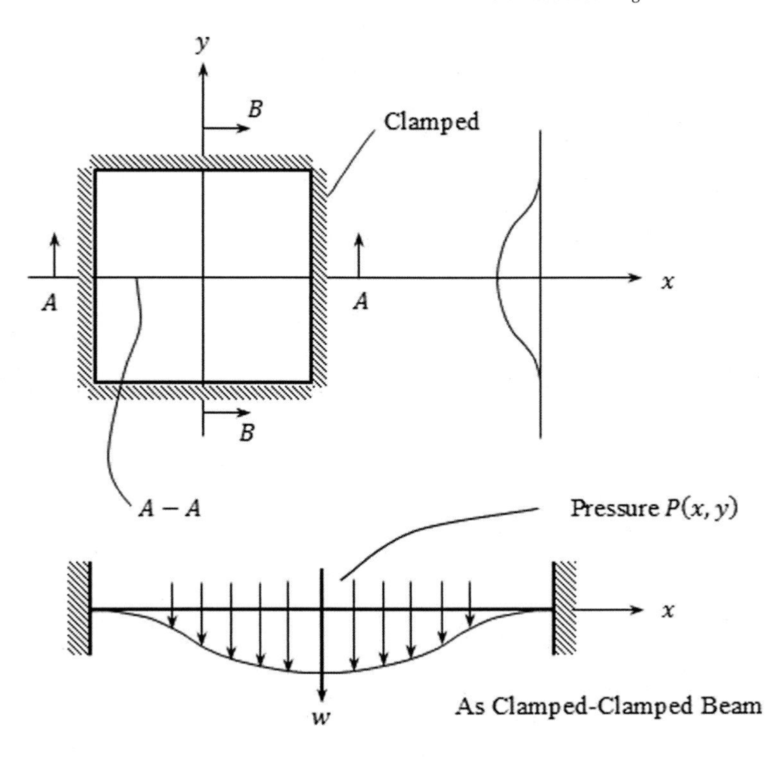

FIGURE 3.3 Clamped plate at all edges.

If we visualize clamped-clamped beam we can choose $\varnothing_m, \varnothing_n$ then if as a first approximation we use,

$$w(x,y) = c_{11}\left[1-\left(\frac{x}{a}\right)^2\right]^2\left[1-\left(\frac{y}{b}\right)^2\right]^2$$

For better accuracy,

$$w(x,y) = \sum\sum\left[1-\left(\frac{x}{a}\right)^2\right]^2\left[1-\left(\frac{y}{b}\right)^2\right]^2\left(c_1 + c_2x^2 + c_3y^2 + c_4x^2y^2 + \cdots\right)\text{etc}\ldots$$

3.2 Plate stretching

The plate is clamped on all edges and subjected to a tangential load X per surface in x-direction and Y per unit surface in y-direction as shown in Fig. 3.4.

Assume a displacement pattern,

$$u(x,y) = \sum_{m=1}^{\infty}\sum_{n=1}^{\infty} a_{mn}\sin\frac{m\pi x}{a}.\sin\frac{n\pi y}{b}$$

$$v(x,y) = \sum_{m=1}^{\infty}\sum_{n=1}^{\infty} b_{mn}\sin\frac{m\pi x}{a}.\sin\frac{n\pi y}{b} \tag{3.6}$$

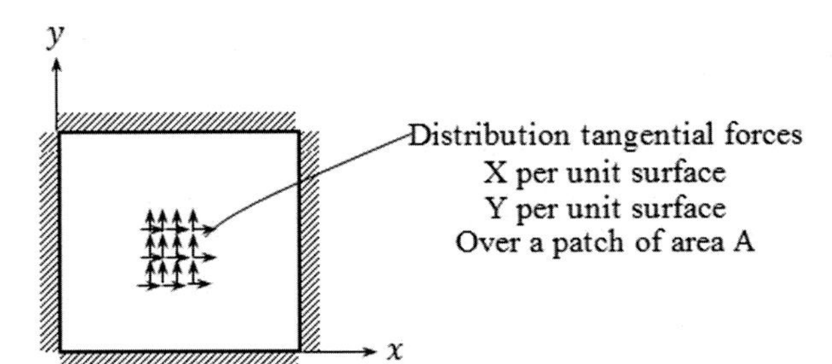

FIGURE 3.4 Clamped on all edges rectangular plate under stretching loads.

These satisfy the general B.Cs, the displacement at $x = 0$ and $x = a \equiv 0$ and same for y. The strain energy U for membrane forces is given by,

$$U = \int_0^b \int_0^a U_o t.dx.dy$$

And from Eq. (3.1),

$$U_o = \underbrace{\frac{E}{2(1-\nu^2)}\left[\varepsilon_x^2 + \varepsilon_y^2 + 2\nu\varepsilon_x\varepsilon_y\right] + \frac{E}{4(1+\nu)}\gamma_{xy}^2}_{\text{refer to energy method}} \tag{3.6a}$$

The potential energy due to given loading is,

$$\Omega = -\iint_{A_P} [X.u + Y.v]dx.dy \tag{3.6b}$$

Then, the total potential energy $V = U + \Omega$ is obtained by the substitution of the values of U and Ω given by Eqs. (3.6a) and (3.6b) respectively as follows,

$$V = \underbrace{\frac{E}{2(1-\nu^2)}\left[\varepsilon_x^2 + \varepsilon_y^2 + 2\nu\varepsilon_x\varepsilon_y\right] + \frac{E}{4(1+\nu)}\gamma_{xy}^2}_{\text{refer to energy method}} - \iint_{A_P} [X.u + Y.v]dx.dy$$

Using the minimization principles,

$$\frac{\partial V}{\partial a_{mn}} = 0, \quad \frac{\partial V}{\partial b_{mn}} = 0$$

From which the coefficients a_{mn} and b_{mn} are obtained. Having the displacement field, the strains, bending moments, and the stresses are obtained.

Example 3.2: Find the maximum deflection of a S.S. square plate under a lateral load in the form of a triangular prism by the energy method. Use R.R.M (Fig. 3.5).

Assume that the deflected shape is given by,

$$w(x,y) = \sum_{m=1}^{\infty}\sum_{n=1}^{\infty} w_{mn}\sin\frac{m\pi x}{a}.\sin\frac{n\pi y}{b}, (m = 1,3,5,\ldots \text{and } n = 1,3,5,\ldots) \tag{E2.1}$$

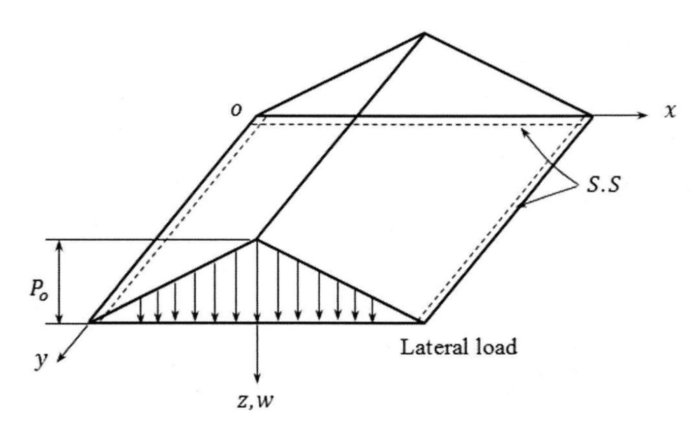

FIGURE 3.5 Square plate under lateral load of form triangular prism.

Mathematical expressions for lateral load are (Fig. 3.5),

$$P_z = \frac{2P_o x}{a} \text{ for } 0 < x < \frac{a}{2}$$

$$P_z = 2P_o - \frac{2P_o x}{a} \text{ for } \frac{a}{2} < x < a$$

The potential energy due to applied loading is given by,

$$\Omega = -\iint_A P_z(x, y).w(x, y)dA \tag{E2.2}$$

Substitute the value of $w(x,y)$ from Eq. (E2.1) into Eq. (E2.2) yields,

$$= -2\sum_m \sum_n \int_0^{b/2} \int_0^{a/2} \frac{2P_o x}{a} w_{mn} \sin\frac{m\pi x}{a}.\sin\frac{n\pi y}{b} dx.dy = -\sum_m \sum_n \frac{8P_o a^2}{m^2 n \pi^3} w_{mn} \sin\frac{m\pi}{2} \tag{E2.3}$$

The strain energy of the plate is given by,

$$U = \frac{D}{2}\int_0^b \int_0^a \sum_m \sum_n \left[w_{mn}\left(\frac{m^2\pi^2}{a^2} + \frac{n^2\pi^2}{a^2}\right) \sin\frac{m\pi x}{a}.\sin\frac{n\pi y}{b}\right]^2 dx.dy = \frac{D\pi^4 a^2}{8} \sum_m \sum_n w_{mn}^2 \left(\frac{m^2}{a^2} + \frac{n^2}{a^2}\right)^2 \tag{E2.4}$$

The total potential energy $V = U + \Omega$ is calculated from Eqs. (E2.3) and (E2.4) to give,

$$V = \frac{D\pi^4 a^2}{8} \sum_m \sum_n w_{mn}^2 \left(\frac{m^2}{a^2} + \frac{n^2}{a^2}\right)^2 - \sum_m \sum_n \frac{8P_o a^2}{m^2 n \pi^3} w_{mn} \sin\frac{m\pi}{2}$$

$\frac{\partial V}{\partial w_{mn}} = 0$, give for specific set of minimum values as follows,

$$\frac{D\pi^4 a^2}{4} w_{mn} \left(\frac{m^2}{a^2} + \frac{n^2}{a^2}\right)^2 - \frac{8P_o a^2}{m^2 n \pi^3} \sin\frac{m\pi}{2} = 0$$

From which we obtain,

$$w_{mn} = \frac{32P_o a^4}{m^2 n \pi^7 D (m^2 + n^2)^2} \tag{E2.5}$$

Then, the deflected shape is given by,

$$w(x,y) = \frac{32P_o a^4}{\pi^7 D} \sum_{m=1}^{\infty}\sum_{n=1}^{\infty} \frac{\sin\frac{m\pi}{2}}{m^2 n (m^2+n^2)^2} \sin\frac{m\pi x}{a} . \sin\frac{n\pi y}{b} \tag{E2.6}$$

$(m = 1, 3, 5, \ldots \text{and } n = 1, 3, 5, \ldots)$

The maximum deflection at $x = y = a/2$, using first three terms, $(m = n = 1), (m = 1, n = 3), (m = 3, n = 1)$,

$$w_{max} = 0.002625 \frac{P_o a^4}{D} \tag{E2.7}$$

$$\text{Exact, } w_{max} = 0.00253 \frac{P_o a^4}{D} \text{ (Theory of elasticity hand book)} \tag{E2.8}$$

3.3 Buckling of thin plates using energy method

The thin plate shown in Fig. 3.6 may buckle in a variety of modes depending upon its dimensions, the loading, and the method of support.

Consider the plate shown in Fig. 3.6 in which the plate is subjected to N_x/unitlength. Once the critical load is reached, the plate is incapable of supporting any farther load. We assume that the displacement is comprised of bending deflections only and that these are small in comparison with thickness of the plate.

The true deflected shape may be represented by the infinite double trigonometric series,

$$w(x,y) = \sum_{m=1}^{\infty}\sum_{n=1}^{\infty} a_{mn} \sin\frac{m\pi x}{a} . \sin\frac{n\pi y}{b} \tag{3.7}$$

Apply the method of the total potential energy, strain energy + potential energy due applied load as follows,

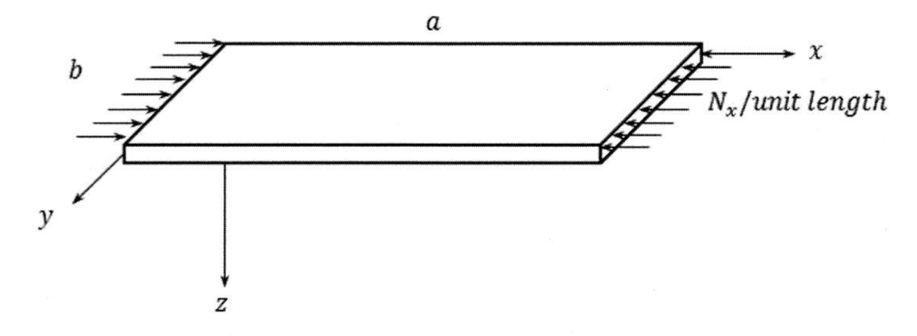

FIGURE 3.6 A simply supported plate subjected to an axial compression N_x/unitlength.

Applying Eq. (3.5) using the total potential energy due to N_x instead of P_z as follows,

$$U+V=\frac{1}{2}\int_0^a\int_0^b\left[D\left\{\left(\frac{\partial^2 w}{\partial x^2}+\frac{\partial^2 w}{\partial y^2}\right)^2-2(1-\nu)\left[\frac{\partial^2 w}{\partial x^2}\frac{\partial^2 w}{\partial y^2}-\left(\frac{\partial^2 w}{\partial x\partial y}\right)^2\right]\right\}-N_x\left(\frac{\partial w}{\partial x}\right)^2\right]dx.dy \tag{3.8}$$

On substituting for $w(x,y)$ from Eq. (3.7) in Eq. (3.8) and carrying out the integrations, we obtain,

$$\begin{aligned}U+V=\int_0^a\int_0^b\Bigg\{&\frac{D}{2}\sum_{m=1}^{\infty}\sum_{n=1}^{\infty}a_{mn}^2\left[\pi^4\left(\frac{m^2}{a^2}+\frac{n^2}{b^2}\right)^2\sin^2\frac{m\pi x}{a}.\sin^2\frac{n\pi y}{b}\right.\\&\left.-2(1-\nu)\frac{m^2n^2\pi^4}{a^2b^2}\left(\sin^2\frac{m\pi x}{a}.\sin^2\frac{n\pi y}{b}-\cos^2\frac{m\pi x}{a}.\cos^2\frac{n\pi y}{b}\right)\right]\\&-N_x\sum_{m=1}^{\infty}\sum_{n=1}^{\infty}a_{mn}^2\frac{m^2n^2}{a^2}\pi^2\cos^2\frac{m\pi x}{a}.\sin^2\frac{n\pi y}{b}\Bigg\}dxdy\end{aligned}$$

Or,

$$U+V=\frac{\pi^4abD}{8}\sum_{m=1}^{\infty}\sum_{n=1}^{\infty}a_{mn}^2\left(\frac{m^2}{a^2}+\frac{n^2}{b^2}\right)^2\sin^2\frac{m\pi x}{a}-N_x\sum_{m=1}^{\infty}\sum_{n=1}^{\infty}a_{mn}^2m^2 \tag{3.9a}$$

Note, the term multiplied by $2(1-\nu)$ integrates to zero and the mean value of $\sin^2$ or $\cos^2$ over a complete number of half waves is $\frac{1}{2}$.

The total potential energy of the plate has a stationary value in the neutral equilibrium of its buckled state (i.e., $N_x=N_{x,cr}$). Differentiating Eq. (3.9a) w.r.t each unknown coefficient a_{mn}, we have,

$$\frac{\partial(U+V)}{\partial a_{mn}}=\frac{\pi^4abD}{4}a_{mn}\left(\frac{m^2}{a^2}+\frac{n^2}{b^2}\right)^2-\frac{\pi^2b}{4a}N_{x,cr}.m^2.a_{mn}=0$$

Or,

$$N_{x,cr}=\pi^2a^2D\frac{1}{m^2}\left(\frac{m^2}{a^2}+\frac{n^2}{b^2}\right)^2 \tag{3.9}$$

The lowest value of the critical load evolves from some critical combination of the integers m and n, that is, the number of half waves in the x and y-directions.

$n=1$, Gives the minimum value, we may write,

$$N_{x,cr}=\pi^2a^2D\frac{1}{m^2}\left(\frac{m^2}{a^2}+\frac{1}{b^2}\right)^2 \tag{3.10}$$

or,

$$N_{x,cr}=\frac{k.\pi^2.D}{b^2} \tag{3.11}$$

k = Plate buckling coefficient, it is the minimum value, hence,

$$k = \left(\frac{m.b}{a} + \frac{a}{m.b}\right)^2 \tag{3.12}$$

Buckled load is minimum when $k = 4$ at values of $\frac{a}{b} = 1, 2, 3, \ldots$

The bucking coefficient k against the aspect ratio a/b plots are shown in Fig. 3.7 which shows the minimum buckling coefficient k for the simply supported plate under compressive loading.

The critical buckling stress $\sigma_{x,cr}$ is given by,

$$\sigma_{x,cr} = \frac{N_{x,cr}}{t} = \frac{k.\pi^2.E}{12(1-\nu^2)}\left(\frac{t}{b}\right)^2 \tag{3.13}$$

3.3.1 Inelastic buckling

For a plate having small values of b/t, the critical stress may exceed the elastic limit of the material of the plate. The effects of E and ν are usually included in a plasticity correction factor η. Therefore,

$$\sigma_{cr} = \frac{\eta.k.\pi^2.E}{12(1-\nu^2)}\left(\frac{t}{b}\right)^2 \tag{3.14}$$

$\eta = 1$ in the elastic region,

$$\eta = \frac{\left(1-\nu_e^2\right)}{\left(1-\nu_p^2\right)}\frac{E_s}{E}\left[\frac{1}{2} + \frac{1}{2}\left(\frac{1}{4} + \frac{3}{4}\frac{E_t}{E_s}\right)^{1/2}\right]$$

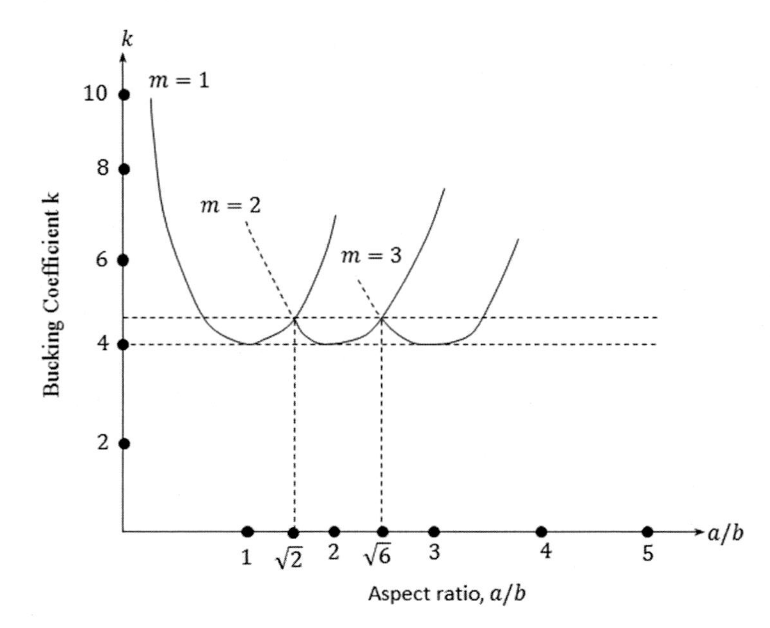

FIGURE 3.7 Simply supported plate subjected to axial loading aspect ratio (a/b) against buckling coefficient k.

E_t and E_s are the tangent modulus and secant modulus (stress/strain) of the plate in the inelastic region and ν_e and ν_p are Poisson's ratio in the elastic and inelastic ranges. These values are obtained from the tensile test in the plastic region.

3.3.2 Pure shear

Consider the plate shown in Fig. 3.8 which is subjected to N_{xy}

Required: the boundary conditions at the supported edges are satisfied. Suppose that the deflected shape is given by,

$$w(x,y) = \sum_{m=1}^{\infty}\sum_{n=1}^{\infty} a_{mn}\sin\frac{m\pi x}{a}.\sin\frac{n\pi y}{b} \tag{3.15a}$$

The stored strain energy is given by,

$$\Delta U = \frac{D}{2}\frac{\pi^2 ab}{4}\sum_{m=1}^{\infty}\sum_{n=1}^{\infty} a_{mn}^2\left(\frac{m^2}{a^2}+\frac{n^2}{b^2}\right)^2 \tag{3.15b}$$

The work done by the external shear forces is given by,

$$\Delta T = -N_{xy}\int_0^a\int_0^b \frac{\partial w}{\partial x}\frac{\partial w}{\partial y}dx.dy \tag{3.15c}$$

We have,

$\int_0^a \sin\frac{n\pi x}{a}\cos\frac{p\pi x}{a}dx = 0$ if $m \neq p$ is an even

$\int_0^a \sin\frac{n\pi x}{a}\cos\frac{p\pi x}{a}dx = \frac{2a}{\pi}\frac{m}{(m^2-p^2)}$ if $m \neq p$ is an odd

Now, we can write Eq. (3.15c) as follows,

$$\Delta T = -4N_{xy}\sum_m\sum_n\sum_p\sum_q a_{mn}a_{pq}\frac{m.n.p.q}{(m^2-p^2)(q^2-n^2)} \tag{3.15}$$

In which m, n, p, q are such integers that $m \neq p$ and $n \neq q$ are odd numbers.

Equating the work above produced by external forces to the strain energy, we obtain,

$$N_{xy} = -\frac{a.b.D\sum_{m=1}^{\infty}\sum_{n=1}^{\infty} a_{mn}^2\left(\frac{m^2\pi^2}{a^2}+\frac{n^2\pi^2}{b^2}\right)^2}{32\sum_m\sum_n\sum_p\sum_q a_{mn}a_{pq}\frac{m.n.p.q}{(m^2-p^2)(q^2-n^2)}} \tag{3.16}$$

It is necessary now to select such a system of constants a_{mn} and a_{pq} as to make N_{xy} a minimum by equating Eq. (3.16) to zero, the derivative of the expression (3.16) with

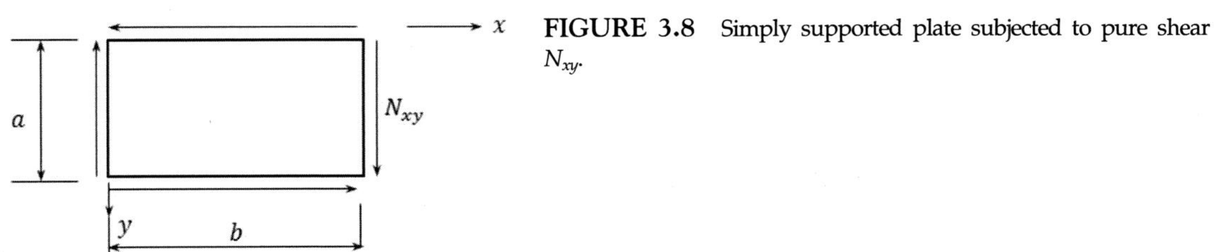

FIGURE 3.8 Simply supported plate subjected to pure shear N_{xy}.

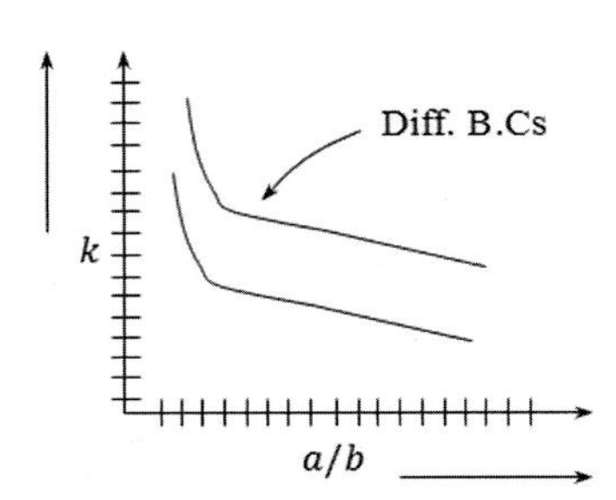

FIGURE 3.9 Aspect ratio against critical buckling coefficient for different boundary conditions.

respect to each of the coefficients a_{mn}, we obtain a system of homogenous linear equations in a_{mn}. This system can be divided into two groups, one containing constants a_{mn} for which $m+n$ odd and other for which $m+n$ even.

Using the notations,

$$\beta = \frac{a}{b}, \quad \lambda = -\frac{\pi^2}{32.\beta} \cdot \frac{\pi^2 D}{b^2 h.\tau_{cr}} \tag{3.17a}$$

And τ_{cr} is obtained for which τ_{cr} the smallest value, and occurs in the second group. $\left(\frac{a}{b} < 2 \text{ short plate}\right)$.

If the determinant of the system of equations is equated to zero, two terms are obtained as follows,

$$\lambda = \pm \frac{\beta^2}{9\left(1+\beta^2\right)^2} \tag{3.17b}$$

Using Eqs. (3.17a) and (3.17b), we can write,

$$\tau_{cr} = \pm \frac{9.\pi^2}{32} \frac{\left(1+\beta^2\right)^2}{\beta^3} \frac{\pi^2 D}{b^2 h} \tag{3.17}$$

The critical value of shearing stress does not depend upon the direction of stress.

Calculating λ and put it into τ_{cr} Eq. (3.17) gives, however, the buckling coefficient k against the aspect ratio of the plate is shown in Fig. 3.9 for different boundary conditions. The critical buckling shearing stress is given by,

$$\tau_{cr} = k \frac{\pi^2 D}{b^2 h} \tag{3.18}$$

Example 3.3: For the clamped square plate shown in Fig. 3.10 which is subjected to uniaxial $\left(N_c^a\right)$ or biaxial compression $\left(N_c^b\right)$, obtain the buckling loads using Ritz method and Galerkin's method:

Ritz method:

For the given boundary conditions assume,

$$w(x,y) = C_1\left(x^2-a^2\right)^2\left(y^2-b^2\right)^2 \tag{E3.1}$$

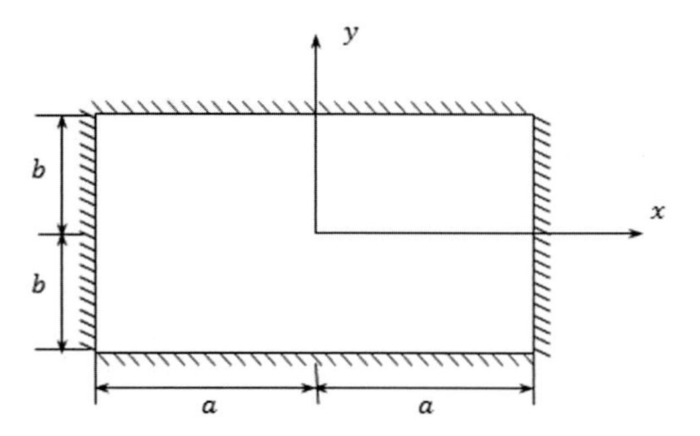

FIGURE 3.10 Clamped Plate Supported.

Which satisfy that the slope and deflection are zero all around edges and sides.
The strain energy stored for the clamped on all sides plate is given by,

$$U_b = \frac{D}{2}\int_{-b}^{b}\int_{-a}^{a}\left[\frac{\partial^2 w}{\partial x^2}+\frac{\partial^2 w}{\partial y^2}\right]^2 dxdy$$

$$U_b = \frac{D}{2}\int_{-b}^{b}\int_{-a}^{a}\left[\left(\frac{\partial^2 w}{\partial x^2}\right)^2 + 2\left(\frac{\partial^2 w}{\partial x^2}\right)\left(\frac{\partial^2 w}{\partial y^2}\right) + \left(\frac{\partial^2 w}{\partial y^2}\right)^2\right] dxdy \tag{E3.2}$$

We have,

$$\frac{\partial w}{\partial x} = 4C_1\left(y^2-b^2\right)^2\left(x^3-a^2x\right)$$

$$\frac{\partial^2 w}{\partial x^2} = 4C_1\left(y^2-b^2\right)^2\left(3x^2-a^2\right)$$

$$\frac{\partial w}{\partial y} = 4C_1\left(x^2-a^2\right)^2\left(y^3-b^2y\right)$$

$$\frac{\partial^2 w}{\partial y^2} = 4C_1\left(x^2-a^2\right)^2\left(3y^2-b^2\right)$$

Inserting the above integrations in Eq. (E3.2) gives,

$$U_b = \frac{DC_1^2}{2}\int_{-b}^{b}\int_{-a}^{a}\left[\begin{array}{c}16\left(y^2-b^2\right)^4\left(3x^2-a^2\right)^2+16\left(x^2-a^2\right)^4\left(3y^2-b^2\right)^2\\+32\left(y^2-b^2\right)^2\left(3x^2-a^2\right)\left(x^2-a^2\right)^2\left(3y^2-b^2\right)\end{array}\right]dxdy$$

Or,

$$U_b = \frac{DC_1^2}{2}\left[I_1+I_2+I_3\right] \tag{E3.3}$$

Where,

$$I_1 = 16\int_{-b}^{b}\int_{-a}^{a}\left(y^2-b^2\right)^4\left(3x^2-a^2\right)^2 dxdy$$

$$I_2 = 32\int_{-b}^{b}\int_{-a}^{a}(y^2-b^2)^2(3x^2-a^2)(x^2-a^2)^2(3y^2-b^2)dxdy$$

$$I_3 = 16\int_{-b}^{b}\int_{-a}^{a}(x^2-a^2)^4(3y^2-b^2)^2dxdy$$

Carrying out the above integrations, gives,

$$I_1 = 9161a^5b^9$$

$$I_2 = 14642a^7b^7$$

$$I_3 = 9161a^9b^5$$

Substitute the above integrations into Eq. (E3.3), gives

$$U_b = \frac{DC_1^2}{2}\left[9161a^5b^9 + 14642a^7b^7 + 9161a^9b^5\right] \tag{E3.4}$$

For a square plate, ($a = b$),
Therefore,

$$U_b = 16482Da^{14}C_1^2 \tag{E3.5}$$

The potential energy due to the applied load is given by,

$$\Omega = -\int_{-b}^{b}\int_{-a}^{a}N_{xx}\left(\frac{\partial w}{\partial x}\right)^2dxdy$$

Or,

$$= -16C_1^2N_{xx}\int_{-b}^{b}\int_{-a}^{a}(y^2-b^2)^4(x^3-a^2x)^2dxdy \tag{E3.6}$$

Carrying out the above integrations, gives,

$$\Omega = -1758a^7b^9C_1^2N_{xx} \tag{E3.7}$$

For a square plate, ($a = b$)

$$\Omega = -1758a^{16}C_1^2N_{xx} \tag{E3.8}$$

The total potential energy for a square plate $V = U + \Omega$ is obtained by substitution the expressions of U and Ω from Eqs. (E3.5) and (E3.8) are as follows,

$$V(C_1) = 16482Da^{14}C_1^2 - 1758a^{16}C_1^2N_{xx} \tag{E3.9}$$

The minimization of the total potential energy of Eq. (E3.9) is achieved as follows,

$$\frac{\partial V}{\partial C_1} = 0$$

Or,

$$32964Da^{14}C_1 - 3516a^{16}C_1N_{xx} = 0$$

Solving for N_{xx} for the critical load is,

$$N_{xxcr} = \frac{9.38D}{a^2} \text{ for a uniaxial compressive load.} \quad \text{(E3.10)}$$

Where,

For biaxial loading (N_{xx}) and (N_{yy}), the strain energy and the potential energy due to the applied load (N_{xx}) and (N_{yy}) are given as follows,

$$U_b = 16482Da^{14}C_1^2 \text{ (As it is given by Eq. (3.E3.5))} \quad \text{(E3.11)}$$

$$\Omega = -3516a^{16}C_1^2N_{cr} \text{ (Simply, expression of Eq. (3.E3.8) multiplied by 2)} \quad \text{(E3.12)}$$

The total potential energy $V = U + \Omega$ is obtained by the substitution of expressions of Eqs. (E3.11) and (E3.12) to yield,

$$V = 16482Da^{14}C_1^2 - 3516a^{16}C_1^2N_{cr} \quad \text{(E3.13)}$$

Minimization of the total potential energy given in Eq. (E3.13) is achieved as follows,

$$\frac{\partial V}{\partial C_1} = 0$$

Or,

$$32964Da^{14}C_1 - 7032a^{16}C_1N_{cr} = 0$$

From which,

$$N_{cr} = \frac{4.69D}{a^2} \text{ for a biaxial compressive load.} \quad \text{(E3.14)}$$

Galerkin's method:

We have, for plate,

$$\int_{-b}^{b}\int_{-a}^{a}\left[D\nabla^4 w + N_{xx}\frac{\partial^2 w}{\partial x^2}\right]\varnothing(x,y)\,dxdy = 0$$

Or,

$$\int_{-b}^{b}\int_{-a}^{a}\varnothing(x,y)D\nabla^4 w\,dxdy + \int_{-b}^{b}\int_{-a}^{a}N_{xx}\varnothing(x,y)\frac{\partial^2 w}{\partial x^2}dxdy = 0 \quad \text{(E3.15)}$$

Where,

$$\nabla^4 w = \frac{\partial^4 w}{\partial x^4} + 2\frac{\partial^4 w}{\partial x^2\partial y^2} + \frac{\partial^4 w}{\partial y^4}$$

In abbreviated form,

$$\text{or } DI_1 + N_{cr}I_2 = 0 \quad \text{(E3.16)}$$

Where,

$$I_1 = C_1 \int_{-b}^{b}\int_{-a}^{a} \left[24(y^2-b^2)^4(x^2-a^2)^2 + 32(3x^2-a^2)(x^2-a^2)^2(3y^2-b^2)(y^2-b^2)^2 + 24(x^2-a^2)^4(y^2-b^2)^2\right]dxdy$$

$$I_2 = 4C_1 \int_{-b}^{b}\int_{-a}^{a} (3x^2-a^2)(x^2-a^2)^2(y^2-b^2)^4 dxdy$$

After carrying out the above integrations, gives,

$$I_1 = 16626C_1a^9b^5$$

$$I_2 = 1153.3C_1a^7b^9$$

Substitute the values of I_1 and I_2 in Eq. (E3.16) gives,

$$16626C_1a^9b^5D = 1153.3C_1a^7b^9N_{cr}$$

$$N_{cr} = 14.416\frac{a^2D}{b^4}$$

From which,

$$N_{cr} = (16626C_1a^9b^5D)/(1153.3C_1a^7b^9)$$

For a square plate, $(a = b)$

$$N_{cr} = \frac{14.4D}{a^2} \text{ For uniaxial compressive load. } (N_{xx}) \tag{E3.17}$$

And,

$$N_{cr} = \frac{7.2D}{a^2} \text{ For biaxial compressive load. } (N_{xx}, N_{yy}) \tag{E3.18}$$

3.4 Application of Galerkin's method G.M. to plate bending

Suppose we wish to solve a linear (partial) differential equation,

$$L[w(x,y)] = f(x,y) \tag{3.19}$$

In a region over which $f(x,y)$ is prescribed and the boundary conditions are linear and homogenous. L stands for a linear differential operator, examples having relevance to stress analysis being Laplace or harmonic operator,

$$\nabla^4 = \frac{\partial^4}{\partial x^4} + 2\frac{\partial^4}{\partial x^2\partial y^2} + \frac{\partial^4}{\partial y^4}$$

The solution to linear differential equation, then,

$$L[w(x,y)] - f(x,y) = 0$$

If $w(x,y)$ is expressed in the form of a complete series of functions for the plate shown in Fig. 3.11,

$$w(x,y) = \sum_{n=1}^{\infty} C_k.\varnothing_k(x,y) \tag{3.20}$$

Satisfying the required boundary conditions, then the exactness of the solution could be expressed by the statement that in the region, the left hand side of solution to linear differential equation above is orthogonal to every term in series of functions.

That is,

$$\iint_D [L[w(x,y)] - f(x,y)]\varnothing_k(x,y)dxdy = 0, k = 1,2,\ldots$$

or,

$$\iint_D R(x,y)\varnothing_k(x,y)dxdy = 0$$

The approximated deflection equation,

$$w(x,y) = \left(c_1 + c_2x^2 + c_3y^2 + c_4x^2y^2 + \cdots\right)\left(x^2-a^2\right)^2\left(y^2-b^2\right)^2 \tag{3.21}$$

Which satisfies the boundary conditions.

For plate under bending in Galerkin's method,

$$\int_{-b}^{b}\int_{-a}^{a}\left[\nabla^4 w - \frac{P_o}{D}\right]\varnothing(x,y)dx.dy = 0 \tag{3.22}$$

Or,

$$\int_0^b\int_0^a \nabla^4 w.\varnothing(x,y)dx.dy = \int_0^b\int_0^a \frac{P_o}{D}\varnothing(x,y)dx.dy \tag{3.23a}$$

Or,

$$\int_0^b\int_0^a R(x,y)\varnothing_k(x,y).dx.dy = 0$$

And,

$$R(x,y) = \nabla^4 w - \frac{P_o}{D}$$

For the Plate Bending shown in Fig. 3.11

$$\nabla^4 w - \frac{P_o}{D} = 0$$

And, for clamped and uniformly distributed loading, $w = 0$, $\frac{\partial w}{\partial x} = 0$, all around.

Selecting the deflection function,

Assume one term,

$$w(x,y) = c_1\varnothing(x,y) \tag{3.23b}$$

Choose,

$$\varnothing(x,y) = (x^2-a^2)^2(y^2-b^2)^2 - \text{Deflection form} \tag{3.23c}$$

The deflection and the slopes, $\left|\frac{\partial w}{\partial x}\right|_{x=\pm a}$, $\left|\frac{\partial w}{\partial y}\right|_{y=\pm b}$ are specified for the plate, zero due to fixed edge boundary condition.

Therefore,

$$\frac{\partial^4 w}{\partial x^4} = 24(y^2-b^2)^2$$

$$\frac{\partial^4 w}{\partial y^4} = 24(x^2-a^2)^2$$

$$\frac{\partial^4 w}{\partial x^2 \partial y^2} = (12x^2-4a^2)(12y^2-4b^2)$$

Applying G.M, Eq. (3.25),

$$\int_{-b}^{b}\int_{-a}^{a} c_1\left\{\nabla^4(x^2-a^2)^2(y^2-b^2)^2 - \frac{P_o}{D}\right\}(x^2-a^2)^2(y^2-b^2)^2 dxdy = 0 \tag{3.23d}$$

In abbreviated form,

$$c_1 I_1 - \frac{P_o}{D} I_2 = 0$$

$$c_1 = \frac{P_o}{D}\frac{I_2}{I_1}$$

Carrying out all the integrations, we obtain,

$$c_1 = \frac{8}{225}\frac{P_o}{Db^4}\left[\frac{3072}{4725} + \frac{4096}{11025}\left(\frac{a}{b}\right)^2 + \frac{3072}{4725}\left(\frac{a}{b}\right)^4\right]^2$$

And,

$$w(x,y) = c_1(x^2-a^2)^2(y^2-b^2)^2$$

Or,

$$w(x,y) = \frac{8}{225}\frac{P_o}{Db^4}\left[\frac{3072}{4725} + \frac{4096}{11025}\left(\frac{a}{b}\right)^2 + \frac{3072}{4725}\left(\frac{a}{b}\right)^4\right]^2 (x^2-a^2)^2(y^2-b^2)^2 \tag{3.23e}$$

For a square plate, $a = b$

$$w_{\text{centre}} = 0.02b^2\frac{P_o a^4}{D} \tag{3.23f}$$

To improve the accuracy, take the deflected shape as follows,

$$w_{2(x,y)} = (c_1 + c_2 x^2)(x^2-a^2)^2(y^2-b^2)^2 \tag{3.23g}$$

We have,

$$\varnothing_1(x,y)=\frac{\partial w_2}{\partial c_1}=\left(x^2-a^2\right)^2\left(y^2-b^2\right)^2 \tag{3.23h}$$

$$\varnothing_2(x,y)=\frac{\partial w_2}{\partial c_2}=x^2\left(x^2-a^2\right)^2\left(y^2-b^2\right)^2 \tag{3.23i}$$

We may write the deflected shape in the form,

$$w_2(x,y)=\left(c_1+c_2x^2\right)\left(x^4-2x^2a^2+a^4\right)\left(y^4-2y^2b^2+b^4\right) \tag{3.23j}$$

We have,

$$\int_0^b\int_0^a \varnothing_1 R(x,y)dx.dy=0 \tag{3.23k}$$

$$\int_0^b\int_0^a \varnothing_2 R(x,y)dx.dy=0 \tag{3.23l}$$

We have,

$$R(x,y)=\left\{\begin{array}{l} c_1\left[24\left(y^2-b^2\right)^2+2\left(12x^2-4a^2\right)\left(12y^2-4b^2\right)+24\left(x^4-2x^2a^2+a^4\right)\right] \\ +c_2\left[\begin{array}{c}\left(360x^2-48a^2\right)\left(y^2-b^2\right)+24\left(x^6-2x^4a^2+a^4x^2\right)+ \\ 2\left(30x^4-24x^2a^2+2a^4\right)\left(12y^2-4b^2\right)\end{array}\right]-\dfrac{P_o}{D}\end{array}\right\} \tag{3.23m}$$

Substitute the expressions of $\varnothing_1$ and $\varnothing_2$ from Eqs. (3.23h) and (3.23i) respectively in the integration of Eq. (3.23m), simplification and arrange in matrix form gives,

$$\begin{bmatrix}0.81715175*10^{11} & 0.17038167*10^{12}\\ 0.18633288*10^{12} & 0.96143975*10^{13}\end{bmatrix}\begin{bmatrix}c_1\\ c_2\end{bmatrix}=\begin{bmatrix}0.27676627*10^{7}\\ 0.97880340*10^{7}\end{bmatrix}\frac{P_o}{D} \tag{3.23n}$$

Solving Eq. (3.23n) gives,

$$c_1=0.336838*10^{-4}\frac{P_o}{D},\ c_2=0.376875*10^{-6}\frac{P_o}{D}$$

Therefore Eq. (3.23m) becomes,

$$w_{2(x,y)}=\left(0.336838*10^{-4}+0.376875*10^{-6}x^2\right)\frac{P_o}{D}\left(x^2-a^2\right)^2\left(y^2-b^2\right)^2$$

At center, $x=y=0$

$$w_2=12.923359\frac{P_o}{D},\ \text{with an error}=+2.36\%$$

If we use the following assumption for the deflected shape,

$$w_3(x,y)=\left(c_1+c_2x^2+c_3y^2\right)\left(x^2-a^2\right)^2\left(y^2-b^2\right)^2$$

Or,

$$w_{3(x,y)}=c_1\left(x^2-a^2\right)^2\left(y^2-b^2\right)^2+c_2x^2\left(x^2-a^2\right)^2\left(y^2-b^2\right)^2+c_3y^2\left(x^2-a^2\right)^2\left(y^2-b^2\right)^2 \tag{3.23o}$$

Therefore,

$$\varnothing_1(x,y)=\frac{\partial w_3}{\partial c_1}=\left(x^2-a^2\right)^2\left(y^2-b^2\right)^2$$
$$\varnothing_2(x,y)=\frac{\partial w_3}{\partial c_2}=x^2\left(x^2-a^2\right)^2\left(y^2-b^2\right)^2 \tag{3.23p}$$
$$\varnothing_3(x,y)=\frac{\partial w_3}{\partial c_3}=y^2\left(x^2-a^2\right)^2\left(y^2-b^2\right)^2$$

We have,

$$\int_0^b\int_0^a \varnothing_1 R(x,y)dx.dy=0$$
$$\int_0^b\int_0^a \varnothing_2 R(x,y)dx.dy=0 \tag{3.23q}$$
$$\int_0^b\int_0^a \varnothing_3 R(x,y)dx.dy=0$$

Hence,

$$R(x,y)=\left\{\begin{array}{l} c_1\left[24\left(y^2-b^2\right)^2+2\left(12x^2-4a^2\right)\left(12y^2-4b^2\right)+24\left(x^2-a^2\right)^2\right]+ \\ c_2\left[\begin{array}{c}\left(360x^2-48a^2\right)\left(y^2-b^2\right)^2+24\left(x^6-2x^4a^2+a^4x^2\right)+ \\ 2\left(30x^4-24x^2a^2+2a^4\right)\left(12y^2-4b^2\right)\end{array}\right]+ \\ c_3\left[\begin{array}{c}\left(360y^2-48b^2\right)\left(x^2-a^2\right)+24\left(y^6-2y^4b^2+b^4y^2\right)+ \\ 2\left(30y^4-24y^2b^2+2b^4\right)\left(12x^2-4a^2\right)\end{array}\right]-\dfrac{P_o}{D}\end{array}\right\}$$

Substitution of expressions $\varnothing_1, \varnothing_2$ and $\varnothing_3$ from Eq. (3.23p) in integrations of Eq. (3.23q) and solve it numerically, three system of linear equations are obtained in matrix form as follows,

$$\begin{bmatrix} 0.181715*10^{11} & 0.17038*10^{12} & 0.17038*10^{12} \\ 0.186332*10^{12} & 0.96143*10^{13} & 0.4743*10^{12} \\ 0.186332*10^{12} & 0.4743*10^{12} & 0.96143*10^{13} \end{bmatrix}\begin{bmatrix} c_1 \\ c_2 \\ c_3 \end{bmatrix}=\begin{bmatrix} 0.2767*10^7 \\ 0.9788*10^7 \\ 0.9788*10^7 \end{bmatrix}\frac{P_o}{D} \tag{3.23r}$$

Solving of Eq. (3.23r) gives,

$$c_1=0.323125*10^{-4}\frac{P_o}{D}$$

$$c_2=0.373401*10^{-6}\frac{P_o}{D}$$

$$c_3=0.373401*10^{-6}\frac{P_o}{D}$$

Hence Eq. (3.23o) becomes,

$$w_3 = \left(0.323125 * 10^{-4} + 0.373401 * 10^{-6}(x^2 + y^2)\right)\left(x^2 - a^2\right)^2\left(y^2 - b^2\right)^2 \tag{3.23}$$

At center, $x = y = 0$

$$w_3(0,0) = 12.62207\frac{P_o}{D} \text{ and the error} = -0.02\%$$

3.5 Kantorovich method

In this approach, the problem of solving the partial differential equation is reduced to solving a set of ordinary different equation with a prospect of improved accuracy, R.R.M is used for functions in one or more variables, the Kantorovich method is appropriate for functions in more than one variable.

Suppose that $w(x,y)$ is an external to be determined in the region $-a \leq x \leq a, -b \leq y \leq b$ using R.R.M.,

$$w(x,y) = w_n(x,y) = c_1\varnothing_1(x,y) + c_2\varnothing_2(x,y) + \cdots + c_n\varnothing_n(x,y) \tag{3.24}$$

If we have a poor idea of how $w(x,y)$ varies with y and a good indication of how $w(x,y)$ vary with x. Under these conditions, assume that the deflection of the plate is given by,

$$w(x,y) = w_n(x,y) = f_1(y)\varnothing_1(x) + f_2(y)\varnothing_2(x) + \cdots + f_n(y)\varnothing_n(x) \tag{3.25}$$

$\varnothing$'s are functions of x alone and satisfy the forced B.Cs at $x = -a$ and $x = a$. The f's are functions of y and unknowns. Using variation principles V.P. of several functions of one variable, f' ssatisfy the boundary conditions along $y = \pm b$. In a more general form of Kantorovich, assume,

$$w_n(x,y) = \varnothing_o(x,y) + \sum_{k=1}^{n} f_k(y)\varnothing_k(x,y)$$

As an approximation to the deflected shape $w(x,y)$.

$\varnothing_o(x,y)$: is chosen to satisfy the B.Cs (Non-homogenous) while the remaining $\varnothing_k(x,y)$ are zero on the boundary.

In fact, for restricted shapes of boundary if we adjust the $f_k(y)$ to be zero along positions of the boundary, the $\varnothing_k(x,y)$ need not be zero there.

Example 3.4: The deflection $w(x,y)$ of a pre-tensioned membrane supporting a transverse pressure $P = T.\sin x.\sin y$ with T being the uniform pre-tension, is given by the stationary value of,

$$I(w) = \int_0^{\pi}\int_0^{1}\left[w_x^2 + w_y^2 + 2w \sin \pi x.\sin \pi y\right]dx.dy \tag{3.E4.1}$$

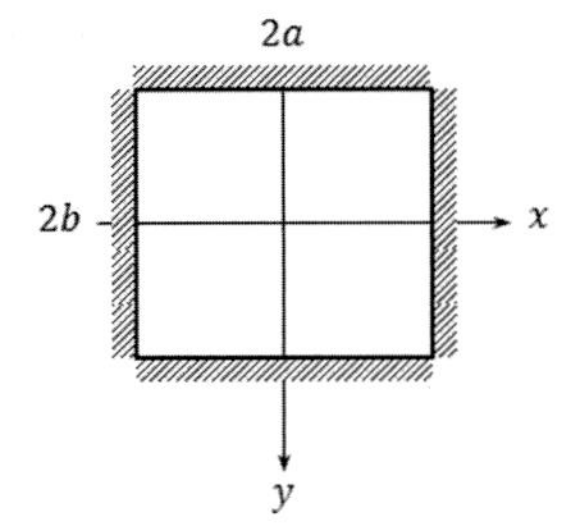

FIGURE 3.11 Clamped plate on all edges.

The B.Cs,

$$w=0 \quad at \quad x=0 \quad and \quad y=0,\pi$$

Assume that the deflected shape of the plate is given by,

$$w_1(x,y)=x(1-x)f(y) \tag{3.E4.2}$$

Determine an approximation to the deflected shape $w_1(x,y)$.

Solution:

Using the deflected shape given by Eq. (3.E4.2), carrying out the differentiations require by Eq. (3.E4.1) and insert them in this equation gives,

$$I(w_1)=\int_0^{\pi}\int_0^{1}\left[(1-2x)^2f^2+\left(x-x^2\right)^2f'^2+2\left(x-x^2\right)f\sin\pi x.\sin\pi y\right]dxdy \tag{3.E4.3}$$

We have,

$$I(w_1)=\int_0^{\pi}\left(\frac{f^2}{30}+\frac{f'^2}{3}+\left[\frac{8f}{\pi^3}\right]\sin y\right)dy \tag{3.E4.4}$$

Or,

$$I(w_1)=\int_0^{\pi}F\left(y,f,f'\right)dy \tag{3.E4.5}$$

For $I(w_1)$ to be stationary, f is given by the Euler-Lagrange equation,

$$\frac{\partial F}{\partial f}-\frac{d}{dy}\left(\frac{\partial F}{\partial f}\right)=0$$

Or,

$$f^n-10f=\left(\frac{120}{\pi^3}\right)\sin y \tag{3.E4.6}$$

Integrating Eq. (3.E4.6) gives,

$$w(y)=A\exp\left(\sqrt{10}y\right)+B\exp\left(-\sqrt{10}y\right)-\left(\frac{120}{\pi^3}\right)\sin y \tag{3.E4.7}$$

B.Cs, $y(0) = y(\pi) = 0$
Insert these conditions in Eq. (3.E4.7) gives, $A = B = 0$
First approximation,

$$w_1(x, y) = -0.3518x \quad (1 - x)\sin y$$

3.6 Application of Kantorovich's method to plate bending

Consider $\left(y^2 - b^2\right)^2$ to be the coordinate function $\varnothing_1(y)$ and use,

$$w_1(x, y) = c_1(x)\left(y^2 - b^2\right)^2 = c_1(x).\varnothing_1(y) \tag{3.26a}$$

Where,

$$\varnothing_1(y) = \left(y^2 - b^2\right)^2 \tag{3.26b}$$

$c_1(x)$: Unknown function
Thus for plate bending we have,

$$\int_{-a}^{a}\left[\int_{-b}^{b}\left(\nabla^4 w_1 - \frac{P_o}{D}\right)\varnothing_1(y)dy\right]dx = 0 \tag{3.26c}$$

Set the expression in brackets = 0, thus,

$$\int_{-b}^{b}\left(\nabla^4 w_1 - \frac{P_o}{D}\right)\varnothing_1(y)dy = 0 \tag{3.26d}$$

Substituting for w_1 from Eq. (3.26a) and $\varnothing_1$ from (3.26b), we have,

$$\int_{-b}^{b}\left\{\nabla^4\left[c_1(x)\left(y^2 - b^2\right)^2\right] - \frac{P_o}{D}\right\}\left(y^2 - b^2\right)^2 dy = 0$$

From which,

$$\int_{-b}^{b}\left[24c_1 + 2\left(12y^2 - 4b^2\right)\frac{d^2c_1}{dx^2} + \left(y^2 - b^2\right)^2\frac{d^4c_1}{dx^4} - \frac{P_o}{D}\right]\left(y^2 - b^2\right)^2 dy = 0 \tag{3.26e}$$

Integrating and multiplying through Eq. (3.26e) by $\frac{315}{256b^2}$, gives,

$$b^4\frac{d^4c_1}{dx^4} - 6b^2\frac{d^2c_1}{dx^2} + \frac{63}{2}c_1 = \frac{21}{6}\frac{P_o}{D} \tag{3.26f}$$

The characteristic equation of Eq. (3.26f) is,

$$b^4p^4 - 6b^2p^2 + \frac{63}{2} = 0 \tag{3.26g}$$

Solving Eq. (3.26g) for b^2p^2 with the quadratic formula,

$$b^2p^2 = 3 \pm i(4.75) \tag{3.26h}$$

The roots of Eq. (3.26h) are,

$$p_{1,2,3,4} = \frac{1}{b}(\pm\alpha \pm i\beta)$$

Where,

$$\alpha = 2.075$$

$$\beta = 1.143$$

The complementary solution of Eq. (3.26f) is,

$$(c_1)_c = B_1 e^{(\alpha+i\beta)x/b} + B_2 e^{(\alpha-i\beta)x/b} + B_3 e^{(-\alpha-i\beta)x/b} + B_4 e^{(-\alpha+i\beta)x/b} \tag{3.26i}$$

In terms of hyperbolic function,

$$(c_1)_c = A_1 \cosh\alpha\frac{x}{b}\cos\beta\frac{x}{b} + A_2 \cosh\alpha\frac{x}{b}\sin\beta\frac{x}{b} + A_3 \sinh\alpha\frac{x}{b}\sin\beta\frac{x}{b} + A_4 \sinh\alpha\frac{x}{b}\cos\beta\frac{x}{b} \tag{3.26j}$$

The particular solution of Eq. (3.26g) is,

$$(c_1)_p = \frac{2}{63}\frac{21}{16}\frac{P_o}{D} = \frac{1}{24}\frac{P_o}{D}$$

$c_1(x)$ will be an even function,
$A_2 = A_4 = 0$, therefore,

$$c_1(x) = A_1 \cosh\alpha\frac{x}{b}\cos\beta\frac{x}{b} + A_3 \sinh\alpha\frac{x}{b}\sin\beta\frac{x}{b} + \frac{1}{24}\frac{P_o}{D} \tag{3.26k}$$

B.Cs, Substitute the boundary conditions of deflection and the slope, $c(a) = c^{'}(a) = 0$, or edges $x = \pm a$, into Eq. (3.26k) gives,

$$A_1 \cosh\alpha\frac{a}{b}\cos\beta\frac{a}{b} + A_3 \sinh\alpha\frac{a}{b}\sin\beta\frac{a}{b} + \frac{P_o}{D} = 0 \tag{3.26l}$$

Or,

$$A_1\left(\frac{\alpha}{b}\sinh\alpha\frac{a}{b}\cos\beta\frac{a}{b} - \frac{\alpha}{b}\cosh\alpha\frac{a}{b}\sin\beta\frac{a}{b}\right) + A_3\left(-\frac{\beta}{b}\cosh\alpha\frac{a}{b}\sin\beta\frac{a}{b} + \frac{\beta}{b}\sinh\alpha\frac{a}{b}\cos\beta\frac{a}{b}\right) = 0 \tag{3.26m}$$

Solving Eqs. (3.26l) and (3.26m) gives,

$$A_1 = \frac{\gamma_1}{\gamma_o}\frac{1}{24}\frac{P_o}{D},$$

and

$$A_3 = \frac{\gamma_2}{\gamma_o}\frac{1}{24}\frac{P_o}{D}$$

Where,

$$\frac{a}{b} = \mu$$

$$\gamma_o = \beta.\text{sinh}\alpha\mu.\text{cosh}\alpha\mu + \alpha.\text{sin}\beta\mu.\text{cos}\beta\mu$$

$$\gamma_1 = -(\alpha.\text{cosh}\alpha\mu.\text{sin}\beta\mu + \beta.\text{sinh}\alpha\mu.\text{cos}\beta\mu)$$

$$\gamma_2 = .\text{sinh}\alpha\mu.\text{cos}\beta\mu - \beta.\text{cosh}\alpha\mu.\text{sin}\beta\mu$$

Therefore the solution becomes,

$$w(x,y) = \frac{1}{24}\frac{P_o}{D}\left[\left(\frac{\gamma_1}{\gamma_o}\text{cosh}\alpha\frac{x}{b}\text{cos}\beta\frac{x}{b} + \frac{\gamma_2}{\gamma_o}\text{sinh}\alpha\frac{x}{b}\text{sin}\beta\frac{x}{b}\right) + 1\right]\left(y^2 - b^2\right)^2 \quad (3.26\text{n})$$

At the center $x = y = 0$ and for square plate,

$$w_{\max} = \frac{1}{24}\frac{P_o}{D}\left(\frac{\gamma_1}{\gamma_o} + 1\right)a^4 = \frac{1}{24}\frac{P_o.a^4}{D}(0.479) = 0.01992\frac{P_o}{D}a^4 \quad (3.26\text{o})$$

Example 3.5: A simply supported rectangular plate of side ratio b/a carries a single central concentrated load P as shown in Fig. 3.12. If it is decided to use an energy method to give a solution, show that,

$$w = \sum_{m=1}^{\infty}\sum_{n=1}^{\infty} a_{mn}\text{sin}\frac{m\pi x}{a}\text{sin}\frac{n\pi y}{b}$$

is acceptable as an assumed deflection mode if the origin is taken at one corner.

Show that the center deflection is,

$$w_{\text{center}} = 0.173\frac{Pa^2}{Eh^3}$$

E = Young's modulus, h = plate thickness, Poisson's ratio $\nu = 0.3$, $\frac{b}{a} = 2$.

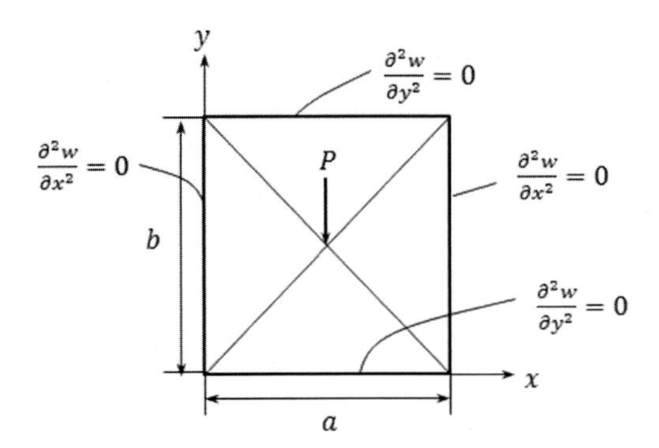

FIGURE 3.12 Simply supported plate on all edges and subjected to a central concentrated load.

Solution:
For a simply supported rectangular plate, assume that the deflected shape is given by,

$$w(x,y)=\sum_{m=1}^{\infty}\sum_{n=1}^{\infty}a_{mn}\sin\frac{m\pi x}{a}\sin\frac{n\pi y}{b} \tag{E5.1}$$

If the origin is taken at one corner as shown in Fig. 3.12, we have,

$$\left.\begin{matrix} w=0 \text{ at } x=0 \\ w=0 \text{ at } x=a \end{matrix}\right\}, \text{Also, } \left.\begin{matrix} w=0 \text{ at } y=0 \\ w=0 \text{ at } y=b \end{matrix}\right\} \text{S.S. plate} \tag{E5.2}$$

These geometric boundary conditions should be satisfied be Eq. (E5.1) of $w(x,y)$. Therefore,

$$w(x,y)=\sum_{m=1}^{\infty}\sum_{n=1}^{\infty}a_{mn}\sin\frac{m\pi x}{a}\sin\frac{n\pi y}{b}$$

Is an acceptable as an assumed deflection mode.
Strain energy stored U for the plate is,

$$U=\frac{D}{2}\iint\left\{\left(\frac{\partial^2 w}{\partial x^2}+\frac{\partial^2 w}{\partial y^2}\right)^2-2(1-\nu)\left[\frac{\partial^2 w}{\partial x^2}\frac{\partial^2 w}{\partial y^2}-\frac{\partial^2 w}{\partial x\partial y}\right]\right\}dx\ dy \tag{E5.3}$$

Since the edges of the plate are S.S, Eq. (E5.3) is simplified as follows,

$$U_{\text{Simplified}}=\frac{D}{2}\iint\left(\frac{\partial^2 w}{\partial x^2}+\frac{\partial^2 w}{\partial y^2}\right)^2 dx\ dy \tag{E5.4}$$

We have,

$$\frac{\partial^2 w}{\partial x^2}=w_{xx}=-\sum_{m=1}^{\infty}\sum_{n=1}^{\infty}a_{mn}\left(\frac{m\pi}{a}\right)^2\sin\frac{m\pi x}{a}\sin\frac{n\pi y}{b}$$

$$\frac{\partial^2 w}{\partial y^2}=w_{yy}=-\sum_{m=1}^{\infty}\sum_{n=1}^{\infty}a_{mn}\left(\frac{n\pi}{b}\right)^2\sin\frac{m\pi x}{a}\sin\frac{n\pi y}{b}$$

Hence,
$U=\frac{D}{2}\int_0^b\int_0^a\left[\sum_{m=1}^{\infty}\sum_{n=1}^{\infty}a_{mn}\left(\frac{m\pi}{a}\right)^2\sin\frac{m\pi x}{a}\sin\frac{n\pi y}{b}+\sum_{m=1}^{\infty}\sum_{n=1}^{\infty}a_{mn}\left(\frac{n\pi}{b}\right)^2\sin\frac{m\pi x}{a}\sin\frac{n\pi y}{b}\right]^2 dx\ dy$
Or,

$$U=\frac{D}{2}\int_0^b\int_0^a\left\{\sum_{m=1}^{\infty}\sum_{n=1}^{\infty}a_{mn}\sin\frac{m\pi x}{a}\sin\frac{n\pi y}{b}\left[\left(\frac{m\pi}{a}\right)^2+\left(\frac{n\pi}{b}\right)^2\right]\right\}^2 dx\ dy$$

$$=\frac{D}{2}\int_0^b\int_0^a\sum_{m=1}^{\infty}\sum_{n=1}^{\infty}a_{mn}^2\left(\left(\frac{m\pi}{a}\right)^2+\left(\frac{n\pi}{b}\right)^2\right)^2\sin^2\frac{m\pi x}{a}\sin^2\frac{n\pi y}{b}dx\ dy$$

$$=\frac{D}{2}\sum_{m=1}^{\infty}\sum_{n=1}^{\infty}a_{mn}^2\left(\left(\frac{m\pi}{a}\right)^2+\left(\frac{n\pi}{b}\right)^2\right)^2\left[\int_0^b\sin^2\frac{n\pi y}{b}dy.\int_0^a\sin^2\frac{m\pi x}{a}dx\right]$$

$$= \frac{D}{2}\sum_{m=1}^{\infty}\sum_{n=1}^{\infty}a_{mn}^2\left(\left(\frac{m\pi}{a}\right)^2+\left(\frac{n\pi}{b}\right)^2\right)^2\left(\frac{a}{2}\cdot\frac{b}{2}\right)$$

Simplification gives,

$$U = \frac{\pi^4 D.a.b}{8}\sum_{m=1}^{\infty}\sum_{n=1}^{\infty}a_{mn}^2\left(\frac{m^2}{a^2}+\frac{n^2}{b^2}\right)^2 \tag{E5.5}$$

The potential energy due to a central concentrated load P at $x=\frac{a}{2},\ y=\frac{b}{2}$ is,

$$\Omega = -\iint P.w.dx\ dy = -\iint P\left(\sum_{m=1}^{\infty}\sum_{n=1}^{\infty}a_{mn}\sin\frac{m\pi x}{a}\sin\frac{n\pi y}{b}\right)dxdy$$

Or,

$$= -P\sum_{m=1}^{\infty}\sum_{n=1}^{\infty}a_{mn}\iint\sin\frac{m\pi x}{a}\sin\frac{n\pi y}{b}dxdy \tag{E5.6}$$

The m, n parameters must be odd, because if they are even $\Omega = 0$.

Total potential energy $V = U + \Omega$ where U is given by Eq. (E5.5) and Ω is given by Eq. (E5.6), the result of substitution is,

$$= \frac{\pi^4 D.a.b}{8}\sum_{m=1}^{\infty}\sum_{n=1}^{\infty}a_{mn}^2\left(\frac{m^2}{a^2}+\frac{n^2}{b^2}\right)^2 - P\sum_{m=1}^{\infty}\sum_{n=1}^{\infty}a_{mn}\sin\frac{m\pi x}{a}\sin\frac{n\pi y}{b} \tag{E5.7}$$

Minimization of total potential energy is carried out as follows,

$$\frac{\partial V}{\partial a_{mn}} = 0$$

For Eq. (E5.7) gives,

$$\frac{\partial V}{\partial a_{mn}} = \frac{\pi^4 D.a.b}{8}2\sum_{m=1}^{\infty}\sum_{n=1}^{\infty}a_{mn}\left(\frac{m^2}{a^2}+\frac{n^2}{b^2}\right)^2 - P\sum_{m=1}^{\infty}\sum_{n=1}^{\infty}\sin\frac{m\pi x}{a}\sin\frac{n\pi y}{b} = 0$$

From which,

$$a_{mn} = \frac{P\sum\limits_{m=1}^{\infty}\sum\limits_{n=1}^{\infty}\sin\frac{m\pi x}{a}\sin\frac{n\pi y}{b}}{\frac{\pi^4 D.a.b}{4}\sum\limits_{m=1}^{\infty}\sum\limits_{n=1}^{\infty}\left(\frac{m^2}{a^2}+\frac{n^2}{b^2}\right)^2}$$

Hence, Eq. (E5.1) gives,

$$w(x,y) = \sum_{m=1}^{\infty}\sum_{n=1}^{\infty}\frac{P\sin\frac{m\pi x}{a}\sin\frac{n\pi y}{b}\sin\frac{m\pi x}{a}\sin\frac{n\pi y}{b}}{\frac{\pi^4 D.a.b}{4}\left(\frac{m^2}{a^2}+\frac{n^2}{b^2}\right)^2}$$

Or,

$$w(x,y)=\sum_{m=1}^{\infty}\sum_{n=1}^{\infty}\frac{4P}{\pi^4 D.a.b}\frac{\sin^2\frac{m\pi x}{a}\sin^2\frac{n\pi y}{b}}{\left(\frac{m^2}{a^2}+\frac{n^2}{b^2}\right)^2} \tag{E5.8}$$

We have,
$D=\frac{Eh^3}{12(1-\nu^2)}=\frac{Eh^3}{12*0.91}=\frac{Eh^3}{10.92}$, Bending stiffness of the plate using $\nu=0.3$.
Therefore the deflected shape is given by,

$$w(x,y)=\frac{4P}{\pi^4}\cdot\frac{10.92}{Eh^3}\cdot\frac{1}{2a^2}\sum_{m=1}^{\infty}\sum_{n=1}^{\infty}\frac{\sin^2\frac{m\pi x}{a}\sin^2\frac{n\pi y}{b}}{\left(\frac{m^2}{a^2}+\frac{n^2}{b^2}\right)^2} \tag{E5.9}$$

The deflection at the center, that is, $x=\frac{a}{2}, y=\frac{b}{2}$, is as follows,

$$w_{\text{centre}}=\frac{2P}{\pi^4}\cdot\frac{10.92}{Eh^3}\cdot\frac{1}{a^2}\sum_{m=1}^{\infty}\sum_{n=1}^{\infty}\frac{\sin^2\frac{m\pi}{2}\sin^2\frac{n\pi}{2}}{\left(\frac{m^2}{a^2}+\frac{n^2}{b^2}\right)^2} \tag{E5.10}$$

The m, n should be odd because of one of the even the deflection $w=0$ so we can collect the odd terms to get w_{centre} to a certain accuracy. If $m=1, n=1,3,5$ and $n=1, m=3,5$,
Hence, the central deflection is given by,

$$w_{\text{center}}=\frac{1}{4.46}\frac{P}{a^2Eh^3}\left\{\frac{1}{\left(\frac{1}{a^2}+\frac{1}{4a^2}\right)^2}+\frac{1}{\left(\frac{1}{a^2}+\frac{9}{4a^2}\right)^2}+\frac{1}{\left(\frac{1}{a^2}+\frac{25}{4a^2}\right)^2}+\frac{1}{\left(\frac{9}{a^2}+\frac{1}{4a^2}\right)^2}+\frac{1}{\left(\frac{25}{a^2}+\frac{1}{4a^2}\right)^2}+\cdots\right\}$$

$$w_{\text{center}}=\frac{1}{4.46}\frac{Pa^4}{a^2Eh^3}\left[\frac{1}{(1.25)^2}+\frac{1}{(3.25)^2}+\frac{1}{(7.25)^2}+\frac{1}{(9.25)^2}+\frac{1}{(25.25)^2}+\frac{1}{\left(1+\frac{49}{4}\right)^2}+\frac{1}{\left(1+\frac{81}{4}\right)^2}\right]$$

Or,

$$w_{\text{center}}=0.1837\frac{Pa^2}{Eh^3} \tag{E5.11}$$

Example 3.6: (a) A square plate of side a simply supported along its edges carries a uniformly distributed load of intensity P per unit surface area as shown in Fig. 3.13. Use the principle of stationary total potential energy and a double sine series deflection mode to prove that the center deflection is,

$$w_{\text{center}}=0.0442\frac{Pa^4}{Eh^3} \tag{E6.1a}$$

Where, $E=$ Young's modulus, $h=$ plate thickness, and Poisson's ratio $\nu=0.3$.

(b) A simply supported rectangular plate supports an arbitrary distributed load $p(x,y)$. Represent this load by a double Fourier half range sine series, then, assuming a deflection mode, $w=\sum_{m=1}^{\infty}\sum_{n=1}^{\infty}a_{mn}\sin\frac{m\pi x}{a}\sin\frac{m\pi y}{b}$, obtain an expression for the coefficient a_{mn}. Then show that the solution to (a) may obtained directly.

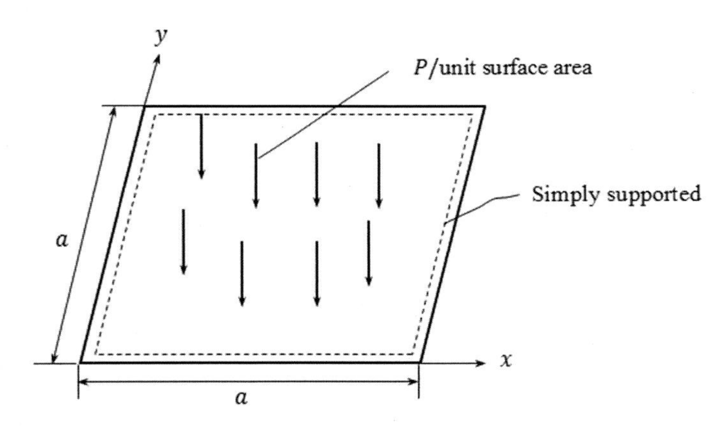

FIGURE 3.13 simply supported square plate subjected to a uniformly distributed load.

Solution:
(a) Let the deflected shape is given by,

$$w(x,y) = \sum_{m=1}^{\infty}\sum_{n=1}^{\infty} a_{mn}\sin\frac{m\pi x}{a}\sin\frac{n\pi y}{b} \tag{E6.1}$$

The strain energy stored, U Eq. (E5.5) of Example 3.5, is,

$$U = \frac{\pi^4 D.a.b}{8}\sum_{m=1}^{\infty}\sum_{n=1}^{\infty} a_{mn}^2\left(\frac{m^2}{a^2}+\frac{n^2}{b^2}\right)^2 \tag{E6.2}$$

The potential energy Ω due to external applied load (distributed load) P/unit surface area is given by,

$$\Omega = -\iint P.w(x,y).dx\,dy \tag{E6.3}$$

Substitution the value of $w(x,y)$ from Eq. (E6.1) into Eq. (E6.3) gives, for $b=a$,

$$= -\int_0^a\int_0^a P\sum_{m=1}^{\infty}\sum_{n=1}^{\infty} a_{mn}\sin\frac{m\pi x}{a}\sin\frac{n\pi y}{b}dxdy$$

Or,

$$= -P\sum_{m=1}^{\infty}\sum_{n=1}^{\infty} a_{mn}\int_0^a \sin\frac{m\pi x}{a}dx\int_0^a \sin\frac{n\pi y}{b}dy$$

$$= -P\sum_{m=1}^{\infty}\sum_{n=1}^{\infty} a_{mn}\left\{\frac{a}{m\pi}\left[-\cos\frac{m\pi x}{a}\right]_0^a\frac{a}{n\pi}\left[-\cos\frac{m\pi y}{a}\right]_0^a\right\}$$

$$= -P\sum_{m=1}^{\infty}\sum_{n=1}^{\infty} a_{mn}\left(\frac{a}{m\pi}\right)\left(\frac{a}{n\pi}\right)[\cos m\pi - 1][\cos n\pi - 1]$$

Then,

$$\Omega = -\mathrm{P}\sum_{m=1}^{\infty}\sum_{n=1}^{\infty} a_{mn}.4.\frac{a^2}{\pi^2 mn} = -\frac{4Pa^2}{\pi^2}\sum_{m=1}^{\infty}\sum_{n=1}^{\infty}\frac{a_{mn}}{mn}$$

For m, n even $\Omega = 0$, therefore m, n are odd. The total potential energy, $V = U + \Omega$ is obtained by substitution the expression of U using Eq. (E6.2) and the value of Ω from Eq. (E6.3) gives,

$$V = \frac{\pi^4 D.a.b}{8}\sum_{m=1}^{\infty}\sum_{n=1}^{\infty} a_{mn}^2\left(\frac{m^2}{a^2}+\frac{n^2}{b^2}\right)^2 - \frac{4Pa^2}{\pi^2}\sum_{m=1}^{\infty}\sum_{n=1}^{\infty}\frac{a_{mn}}{mn} \tag{E6.4}$$

Using the principles of total potential energy, $\delta V = 0$, the resulted equation is as follows,

$$\frac{\partial \mathrm{V}}{\partial a_{mn}} = \frac{\pi^4 D.a.b}{8}.2\sum_{m=1}^{\infty}\sum_{n=1}^{\infty} a_{mn}\left(\frac{m^2}{a^2}+\frac{n^2}{b^2}\right)^2 - \frac{4Pa^2}{\pi^2}\sum_{m=1}^{\infty}\sum_{n=1}^{\infty}\frac{1}{mn} = 0$$

Or,

$$a_{mn} = \frac{\frac{4Pa^2}{\pi^2}\cdot\frac{1}{mn}}{\frac{\pi^4 D.a.b}{4}\cdot\left(\frac{m^2}{a^2}+\frac{n^2}{b^2}\right)^2}(b=a), D = \frac{Eh^3}{10.92}$$

$$a_{mn} = \frac{4Pa^2 * 4 * 10.92}{\pi^2\pi^4 Eh^3 a^2}\left(\frac{1}{mn}\cdot\frac{1}{\left(\frac{m^2}{a^2}+\frac{n^2}{b^2}\right)^2}\right)$$

Simplification yields,

$$a_{mn} = \frac{16Pa^4 * 10.92}{\pi^6 Eh^3}\left(\frac{1}{mn}\cdot\frac{1}{(m^2+n^2)^2}\right) \tag{E6.5}$$

Hence the deflected shape given by Eq. (E6.1) becomes,

$$w(x,y) = \sum_{m=1}^{\infty}\sum_{n=1}^{\infty}\frac{8Pa^4 * 10.92}{\pi^6 Eh^3}\cdot\left(\frac{1}{mn}\cdot\frac{1}{(m^2+n^2)^2}\right)\sin\frac{m\pi x}{a}\sin\frac{n\pi y}{a}$$

Or,

$$w(x,y) = \frac{1}{5.5024}\cdot\frac{Pa^4}{Eh^3}\sum_{m=1}^{\infty}\sum_{n=1}^{\infty}\left(\frac{1}{mn}\cdot\frac{1}{(m^2+n^2)^2}\right)\sin\frac{m\pi x}{a}\sin\frac{n\pi y}{a} \tag{E6.6}$$

The central deflection can be calculated as follows,
w_{center}, $x = \frac{a}{2}$, $y = \frac{a}{2}$, and we have,
$\sin\frac{m\pi}{2}.\sin\frac{n\pi}{2} = -1 * -1 = 1, m, n$ should be odd otherwise $w_{\text{centre}} = 0$
Substitute the above values in Eq. (E6.6) and calculate $w(a/2, b/2)$ up to 5 terms give,

$$w_{\text{center}} = \frac{1}{5.5024}\cdot\frac{Pa^4}{Eh^3}\left(\frac{1}{1}\cdot\frac{1}{(1+1)^2}+\frac{1}{3}\cdot\frac{1}{(1+9)^2}+\frac{1}{5}\cdot\frac{1}{(1+25)^2}+\frac{1}{3}\cdot\frac{1}{(9+1)^2}+\frac{1}{5}\frac{1}{(1+25)^2}\right)$$

$$m = 3, 5,\ n = 1$$

The result is,

$$w_{\text{center}} = 0.0467\frac{Pa^4}{Eh^3} \tag{E6.7}$$

(b) For $P(x,y)$ to be represented by a double Fourier half range sine series and acted on the rectangular plate shown in Fig. 3.14,

$$P(x,y) = \sum\sum C_{mn}\sin\frac{m\pi x}{a}\sin\frac{n\pi y}{b} \tag{E6.8}$$

The assumption of $w(x,y)$ represented Eq. (E6.1) is repeated here,

$$w(x,y) = \sum\sum a_{mn}\sin\frac{m\pi x}{a}\sin\frac{n\pi y}{b} \tag{E6.9}$$

Form Eq. (E6.2),

$$U = \frac{\pi^4 D.a.b}{8}\sum_{m=1}^{\infty}\sum_{n=1}^{\infty}a_{mn}^2\left(\frac{m^2}{a^2}+\frac{n^2}{b^2}\right)^2 \tag{E6.10}$$

The potential energy due to the applied pressure is given by,

$$\Omega = -\iint P(x,y).w(x,y).dx\,dy \tag{E6.11}$$

Substitution of $w(x,y)$ form Eq. (E6.9) into Eq. (E6.11) gives,

$$= -\int_0^b\int_0^a \sum\sum C_{mn}\sin\frac{m\pi x}{a}\sin\frac{n\pi y}{b}.\sum\sum a_{mn}\sin\frac{m\pi x}{a}\sin\frac{n\pi y}{b}.dx\,dy$$

Or,

$$= -\sum\sum C_{mn}.\sum\sum a_{mn}\int_0^a \sin^2\frac{m\pi x}{a}dx.\int_0^b \sin^2\frac{n\pi x}{b}dx$$

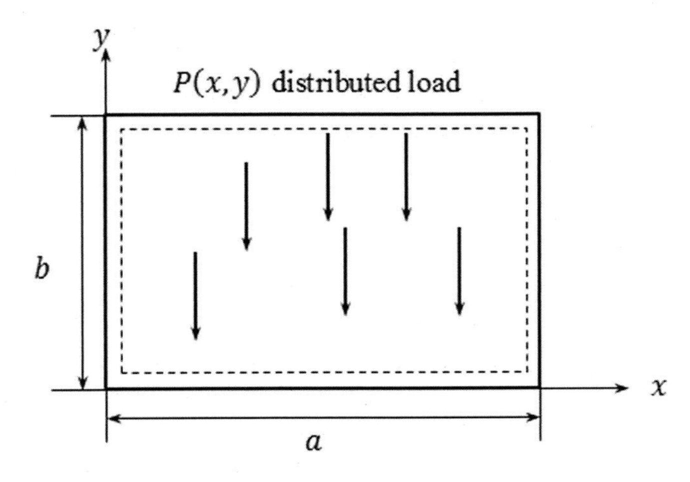

FIGURE 3.14 Rectangular plate simply supported on all edges subjected to a uniformly distributed load.

And simplification gives,

$$= -\sum\sum C_{mn}.\sum\sum a_{mn}.\frac{a}{2}.\frac{b}{2} = -\frac{a.b}{4}\sum\sum C_{mn}.\sum\sum a_{mn} \tag{E6.12}$$

The total potential energy $V = U + \Omega$ is calculated by the substitution of U from Eq. (E6.10) and Ω from Eq. (E6.12) gives,

$$= \frac{\pi^4 D.a.b}{8}\sum_{m=1}^{\infty}\sum_{n=1}^{\infty} a_{mn}^2\left(\frac{m^2}{a^2}+\frac{n^2}{b^2}\right)^2 - \frac{a.b}{4}\sum\sum C_{mn}.\sum\sum a_{mn} \tag{E6.13}$$

From T.P.E minimization principles $= 0$, this requires,

$$\frac{\partial V}{\partial a_{mn}} = \frac{\pi^4 D.a.b}{4}\sum_{m=1}^{\infty}\sum_{n=1}^{\infty} a_{mn}\left(\frac{m^2}{a^2}+\frac{n^2}{b^2}\right)^2 - \frac{a.b}{4}\sum\sum C_{mn}$$

From which,

$$a_{mn} = \frac{\frac{a.b}{4}C_{mn}}{\frac{\pi^4 D.a.b}{4}\left(\frac{m^2}{a^2}+\frac{n^2}{b^2}\right)^2} \tag{E6.14}$$

The plate bending stiffness is,

$$D = \frac{Eh^3}{12(1-\nu^2)}$$

Therefore Eq. (E6.14) becomes,

$$a_{mn} = C_{mn} * \frac{12(1-\nu^2)}{\pi^4 Eh^3}\frac{1}{\left(\frac{m^2}{a^2}+\frac{n^2}{b^2}\right)^2}$$

We have from Eq. (E6.8),

$$P(x,y) = \sum_{m=1}^{\infty}\sum_{n=1}^{\infty} C_{mn}\sin\frac{m\pi x}{a}\sin\frac{n\pi y}{b}$$

To find C_{mn} by using half rang sine series, we obtain,

$$C_{mn} = \frac{4}{ab}\int_0^b\int_0^a P(x,y)\sin\frac{m\pi x}{a}\sin\frac{n\pi y}{b}dx\,dy \tag{E6.15}$$

Since the plate in (E6.1) is square $a = b$ and $P =$ constant. Eq. (E6.15) becomes,

$$C_{mn} = \frac{4}{ab}\int_0^a\int_0^a P\sin\frac{m\pi x}{a}\sin\frac{n\pi y}{a}dx\,dy$$

Or,

$$= \frac{4P}{ab}\int_0^a \sin^2\frac{m\pi x}{a}dx.\int_0^a \sin^2\frac{n\pi x}{b}dx = \frac{4P}{ab}\left[-\frac{a}{m\pi}\cos\frac{m\pi x}{a}\right]\left[-\frac{a}{n\pi}\cos\frac{n\pi y}{a}\right]_0^a$$

Hence, simplification yields,

$$C_{mn} = \frac{4P}{a^2} \cdot \frac{a^2}{nm\pi^2} \overbrace{(\cos m\pi - 1)}^{-2} \overbrace{(\cos n\pi - 1)}^{-2}, \qquad \text{(E6.16)}$$

m and n should be odd to avoid zero value. This means that, $n = 1, 3, 5, 7, \ldots$, From Eq. (E6.1),

$$w(x,y) = \sum\sum a_{mn} \sin\frac{m\pi x}{a} \sin\frac{n\pi y}{b}$$

And, Eq. (E6.14) is,

$$a_{mn} = \frac{12(1-\nu^2)}{\pi^4 E h^3} \cdot \frac{1}{\left(\frac{m^2}{a^2} + \frac{n^2}{b^2}\right)^2} C_{mn}$$

Substitute the expression of C_{mn} given by Eq. (E6.16) into Eq. (E6.14) gives,

$$= \frac{12(1-\nu^2)}{\pi^4 E h^3} \bullet \left(\frac{m^2}{a^2} + \frac{n^2}{b^2}\right)^{-2} \bullet \frac{16P}{mn\pi^2} \qquad \text{(E6.17)}$$

Therefore the deflected shape is obtained by substitution of a_{mn} from Eq. (E6.17) into Eq. (E6.1) gives,

$$w(x,y) = \sum\sum \frac{12(1-\nu^2)}{\pi^4 E h^3} \left(\frac{m^2}{a^2} + \frac{n^2}{b^2}\right)^{-2} \frac{16P}{mn\pi^2} \sin\frac{m\pi x}{a} \sin\frac{n\pi y}{b}$$

The central deflection of the plate if $a = b$ is obtained as follows,

$$\text{Square plate} \left[\begin{array}{l} \text{at } x = \frac{a}{2} \\ \quad y = \frac{a}{2} \\ \quad b = a \end{array} \right.$$

Therefore,

$$w_{\text{center}} = \frac{12(1-\nu^2)}{\pi^6 E h^3} 16P \sum\sum \frac{1}{mn} \left(\frac{m^2}{a^2} + \frac{n^2}{a^2}\right)^{-2} \sin\frac{m\pi}{2} \sin\frac{n\pi}{2}$$

Simplification gives,

$$w_{\text{center}} = \frac{Pa^4}{Eh^3} \cdot \frac{174.72}{\pi^6} \sum\sum \frac{1}{mn} \left(\frac{m^2}{a^2} + \frac{n^2}{a^2}\right)^{-2} \sin\frac{m\pi}{2} \sin\frac{n\pi}{2}$$

Or,

$$w_{\text{center}} = \frac{Pa^4}{Eh^3} \cdot \frac{174.72}{\pi^6} \left(\frac{1}{4} - \frac{1}{300} + \frac{1}{5 * 26^2} - \frac{1}{300} + \frac{1}{5 * 26^2}\right), \quad m = 3, 5, \ n = 1$$

$$= 0.18173 \frac{Pa^4}{Eh^3} (0.243)$$

Hence,

$$w_{\text{center}} = 0.044328\frac{Pa^4}{Eh^3} \tag{E6.18}$$

This is the solution of Eq. (E6.1a) which is obtained directly as obtained in Eq. (E6.18)

Example 3.7: Determine the in-plane displacements at the center of the plate of Fig. 3.15. Using Rayleigh-Ritz method and assuming a displacement field,

$$u(x,y) = A\sin\frac{\pi x}{a}\sin\frac{\pi y}{b}$$
$$v(x,y) = B\sin\frac{\pi x}{a}\sin\frac{\pi y}{b}$$

Solution:
Displacement field,

$$u(x,y) = A\sin\frac{\pi x}{a}\sin\frac{\pi y}{b}$$
$$v(x,y) = B\sin\frac{\pi x}{a}\sin\frac{\pi y}{b} \tag{E7.1}$$

B.Cs.
$u = v = 0$ all around (plate of fixed sides)
Strain energy stored in stretched plate is given by the formula,

$$U = \int U_o.tdx\,dy$$
$$= \iint\left[\frac{Et}{2(1-\nu^2)}\left\{\varepsilon_x^2 + \varepsilon_y^2 + 2\nu\varepsilon_x\varepsilon_y\right\} + \frac{E}{4(1+\nu)}\gamma_{xy}^2\right]dx\,dy \tag{E7.2}$$

We have,
$\varepsilon_x = \frac{\partial u}{\partial x}, \varepsilon_y = \frac{\partial v}{\partial y},$ And since,

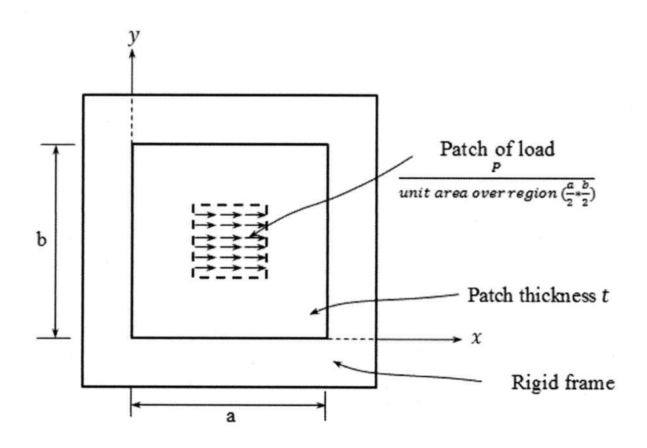

FIGURE 3.15 Rectangular plate clamped on all edges subjected to a patch load in x-direction.

$u(x,y) = A\sin\frac{\pi x}{a}\sin\frac{\pi y}{b}$, therefore,

$$\varepsilon_x = A\frac{\pi}{a}\cos\frac{\pi x}{a}\sin\frac{\pi y}{b}$$

$$\varepsilon_y = B\frac{\pi}{b}\sin\frac{\pi x}{a}\cos\frac{\pi y}{b}$$

$$\gamma_{xy} = A\frac{\pi}{b}\sin\frac{\pi x}{a}\cos\frac{\pi y}{b} + B\frac{\pi}{a}\cos\frac{\pi x}{a}\sin\frac{\pi y}{b}$$

$$\gamma_{xy} = A\frac{\pi}{b}\sin\frac{\pi x}{a}\cos\frac{\pi y}{b} + B\frac{\pi}{a}\cos\frac{\pi x}{a}\sin\frac{\pi y}{b},$$

and

$$\gamma_{xy}^2 = \left(A\frac{\pi}{b}\sin\frac{\pi x}{a}\cos\frac{\pi y}{b} + B\frac{\pi}{a}\cos\frac{\pi x}{a}\sin\frac{\pi y}{b}\right)^2$$

The above equations relate the strain components in terms of in plane displacement, are substituted in Eq. (E7.2) gives,

$$U = \iint\left[\frac{Et}{2(1-\nu^2)}\left\{\left(A\frac{\pi}{a}\cos\frac{\pi x}{a}\sin\frac{\pi y}{b}\right)^2 + \left(B\frac{\pi}{b}\sin\frac{\pi x}{a}\cos\frac{\pi y}{b}\right)^2 + 2\nu\left(A\frac{\pi}{a}\cos\frac{\pi x}{a}\sin\frac{\pi y}{b}B\frac{\pi}{b}\sin\frac{\pi x}{a}\cos\frac{\pi y}{b}\right)\right\}\right. \tag{E7.3}$$
$$\left. + \frac{E}{4(1+\nu)}\left(A\frac{\pi}{b}\sin\frac{\pi x}{a}\cos\frac{\pi y}{b} + B\frac{\pi}{a}\cos\frac{\pi x}{a}\sin\frac{\pi y}{b}\right)^2\right]dx\,dy$$

Potential energy of the applied load is,

$$\Omega = -\iint P.u(x,y)dx\,dy$$

Put the expression of $u(x,y)$ form Eq. (E7.1) gives,

$$= -\iint PA\sin\frac{\pi x}{a}\sin\frac{\pi y}{b}dx\,dy \tag{E7.4}$$

The total potential energy $V = U + \Omega$ is calculated by the substitution of U from Eq. (E7.3) and $\Omega$from Eq. (E7.4) gives,

$$V = \iint\left[\frac{Et}{2(1-\nu^2)}\left\{\left(A\frac{\pi}{a}\cos\frac{\pi x}{a}\sin\frac{\pi y}{b}\right)^2 + \left(B\frac{\pi}{b}\sin\frac{\pi x}{a}\cos\frac{\pi y}{b}\right)^2 + 2\nu\left(A\frac{\pi}{a}\cos\frac{\pi x}{a}\sin\frac{\pi y}{b}B\frac{\pi}{b}\sin\frac{\pi x}{a}\cos\frac{\pi y}{b}\right)\right\}\right. \tag{E7.5}$$
$$\left. + \frac{E}{4(1+\nu)}\left(A\frac{\pi}{b}\sin\frac{\pi x}{a}\cos\frac{\pi y}{b} + B\frac{\pi}{a}\cos\frac{\pi x}{a}\sin\frac{\pi y}{b}\right)^2\right]dx\,dy - \iint PA\sin\frac{\pi x}{a}\sin\frac{\pi y}{b}dx\,dy$$

Applying the minimization of T.P.E. which requires that, $\frac{\partial V}{\partial A} = 0$, therefore,

$$\frac{\partial V}{\partial A} = \iint\left[\frac{Et}{2(1-\nu^2)}\left\{2A\frac{\pi^2}{a^2}\cos^2\frac{\pi x}{a}\sin^2\frac{\pi y}{b} + 0 + 2\nu\left(B\frac{\pi^2}{ab}\cos\frac{\pi x}{a}\sin\frac{\pi x}{a}\sin\frac{\pi y}{b}\cos\frac{\pi y}{b}\right)\right\}\right.$$
$$\left. + \frac{E}{4(1+\nu)}\left(2A\frac{\pi^2}{b^2}\sin^2\frac{\pi x}{a}\cos^2\frac{\pi y}{b} + 2B\frac{\pi^2}{ab}\sin\frac{\pi x}{a}\cos\frac{\pi x}{a}\cos\frac{\pi y}{b}\sin\frac{\pi y}{b}\right)\right]dx\,dy - \iint P\sin\frac{\pi x}{a}\sin\frac{\pi y}{b}dx\,dy = 0$$
$$\tag{E7.6}$$

We have,

$$\iint 2A\frac{\pi^2}{a^2}\cos^2\frac{\pi x}{a}\sin^2\frac{\pi y}{b}\,dx\,dy =$$

$$= 2A\frac{\pi^2}{a^2}\int_0^b\int_0^a\frac{1}{2}\left(1+\cos\frac{2\pi x}{a}\right)\frac{1}{2}\left(1-\cos\frac{2\pi y}{b}\right)dx\,dy$$

$$= \frac{A}{2}\frac{\pi^2}{a^2}\int_0^b\left(1-\cos\frac{2\pi y}{b}\right)dy.\int_0^a\left(1+\cos\frac{2\pi x}{a}\right)dx$$

$$= \frac{A}{2}\frac{\pi^2}{a^2}\left[b-\frac{b}{2\pi}\left(\sin\frac{2\pi y}{b}\right)_0^b\right]\left[a-\frac{a}{2\pi}\left(\sin\frac{2\pi x}{a}\right)_0^a\right]$$

$$= \frac{A}{2}\frac{\pi^2}{a^2}b.a = \frac{A\pi^2 b}{2a} \tag{E7.7}$$

And,

$$B\frac{\pi^2}{ab}\iint\cos\frac{\pi x}{a}\sin\frac{\pi x}{a}\sin\frac{\pi y}{b}\cos\frac{\pi y}{b}\,dx\,dy =$$

$$= B\frac{\pi^2}{ab}\iint\frac{1}{2}\sin\frac{2\pi x}{a}.\frac{1}{2}\sin\frac{2\pi y}{b}\,dx\,dy$$

$$= B\frac{\pi^2}{4ab}\int_0^l\sin\frac{2\pi x}{a}dx.\int_0^l\sin\frac{2\pi y}{b}dy = 0 \tag{E7.8}$$

$$\iint A\frac{\pi^2}{b^2}\sin^2\frac{\pi x}{a}\cos^2\frac{\pi y}{b}\,dx\,dy =$$

$$= A\frac{\pi^2}{b^2}\int_0^a\frac{1}{2}\left(1-\cos\frac{2\pi x}{a}\right)dx.\int_0^b\frac{1}{2}\left(1+\cos\frac{2\pi y}{b}\right)dy$$

$$= \frac{A\pi^2}{4b^2}\left[a-\frac{a}{2\pi}\sin\frac{2\pi x}{a}\Big|_0^a\right]\left[b+\frac{b}{2\pi}\sin\frac{2\pi y}{b}\Big|_0^b\right]$$

$$= \frac{A\pi^2}{4b^2}4a.b = \frac{4.\pi^2 a}{4b}A \tag{E7.9}$$

Also,

$$\iint P.\sin\frac{\pi x}{a}\sin\frac{\pi y}{b}\,dx\,dy =$$

$$= P\int_{a/4}^{3a/4}\sin\frac{\pi x}{a}dx.\int_{b/4}^{3b/4}\sin\frac{\pi y}{b}dy = P\left[-\frac{a}{\pi}\cos\frac{\pi x}{a}\right]_{a/4}^{3a/4}.\left[-\frac{b}{\pi}\cos\frac{\pi y}{b}\right]_{b/4}^{3b/4}$$

$$= P\left[-\frac{a}{\pi}\left(-\frac{1}{\sqrt{2}}-\frac{1}{\sqrt{2}}\right)\right].\left[-\frac{b}{\pi}\left(-\frac{1}{\sqrt{2}}-\frac{1}{\sqrt{2}}\right)\right] = \frac{Pab}{\pi^2}\left(-\frac{2}{\sqrt{2}}\right)\left(-\frac{2}{\sqrt{2}}\right) = \frac{2ab}{\pi^2}P \tag{E7.10}$$

Insert these terms in $\frac{\partial V}{\partial A}$ given in Eq. (E7.6) gives,
$\frac{Et}{2(1-\nu^2)}\frac{A\pi^2 b}{2a} + \frac{E}{4(1+\nu)}\frac{4.\pi^2 a}{4b}A - \frac{2ab}{\pi^2}P = 0$, from which,

$$A\frac{Et}{2(1-\nu^2)}\left[\frac{\pi^2 b}{2a} + (1-\nu)\frac{\pi^2 a}{4b}\right] = \frac{2abP}{\pi^2}$$

Or,

$$A = \frac{4abP}{\pi^2}\frac{(1-\nu^2)}{Et}\frac{4ab}{(\pi^2 b.2b + (1-\nu)\pi^2 a^2)}$$

$$A = \frac{2*8a^2b^2P(1-\nu^2)}{\pi^2 Et}\frac{1}{\pi^2(2b^2 + (1-\nu)a^2)}$$

Hence,

$$A = \frac{2*8a^2b^2P(1-\nu^2)}{\pi^4 Et(2b^2 + (1-\nu)a^2)} \tag{E7.11}$$

Therefore the displacement in x-direction is, Eq. (E7.1),

$$u(x,y) = A\sin\frac{\pi x}{a}\sin\frac{\pi y}{b} = \frac{2*8a^2b^2P(1-\nu^2)}{\pi^4 Et(2b^2 + (1-\nu)a^2)}\sin\frac{\pi x}{a}\sin\frac{\pi y}{b}$$

At center of the plate $x = \frac{a}{2}, y = \frac{b}{2}$, we have,
$\sin\frac{\pi x}{a} = 1$ and $\sin\frac{\pi y}{b} = 1$, therefore the displacement u in the x-direction at the center of the plate is given by,

$$u_{\text{center}} = \frac{2*8a^2b^2P(1-\nu^2)}{\pi^4 Et(2b^2 + (1-\nu)a^2)} = \frac{16b^2P(1+\nu)}{\pi^4 Et\left(\frac{2b^2}{(1-\nu)a^2} + 1\right)} \tag{E7.12}$$

Similarly, the T.P.E principles require that $\frac{\partial V}{\partial B} = 0$, of Eq. (E7.5) gives,

$$\frac{\partial V}{\partial B} = 0 = \iint\left[\frac{Et}{2(1-\nu^2)}\left\{2B\left(\frac{\pi}{b}\sin\frac{\pi x}{a}\cos\frac{\pi y}{b}\right)^2 + 2\nu\left(A\frac{\pi}{a}\cos\frac{\pi x}{a}\sin\frac{\pi y}{b}\frac{\pi}{b}\sin\frac{\pi x}{a}\cos\frac{\pi y}{b}\right)\right\}\right. \\ \left. + \frac{2E}{4(1+\nu)}\left(A\frac{\pi}{b}\sin\frac{\pi x}{a}\cos\frac{\pi y}{b} + B\frac{\pi}{a}\cos\frac{\pi x}{a}\sin\frac{\pi y}{b}\right)\left(\frac{\pi}{a}\cos\frac{\pi x}{a}\sin\frac{\pi y}{b}\right)\right]dx\,dy \tag{E7.13}$$

We have,

$$\iint 2B\frac{\pi^2}{b^2}\sin^2\frac{\pi x}{a}\cos^2\frac{\pi y}{b}dx\,dy =$$

$$= 2B\frac{\pi^2}{b^2}\int_0^b\int_0^a\frac{1}{2}\left(1-\cos\frac{2\pi x}{a}\right)\frac{1}{2}\left(1+\cos\frac{2\pi y}{b}\right)dx\,dy$$

$$= 2B\frac{\pi^2}{b^2}\frac{1}{4}\int_0^b\left(1+\cos\frac{2\pi y}{b}\right)dy.\int_0^a\left(1-\cos\frac{2\pi x}{a}\right)dx$$

$$= \frac{1}{2}B\frac{\pi^2}{b^2}\left[b + \frac{b}{2\pi}\left(\sin\frac{2\pi y}{b}\right)_0^b\right]\left[a - \frac{a}{2\pi}\left(\sin\frac{2\pi x}{a}\right)_0^a\right]$$

$$= \frac{1}{2}B\frac{\pi^2}{b^2}b.a = \frac{1}{2}B\frac{\pi^2}{b}a \tag{E7.14}$$

And,

$$A\frac{\pi^2}{ab}\iint \cos\frac{\pi x}{a}\sin\frac{\pi x}{a}\sin\frac{\pi y}{b}\cos\frac{\pi y}{b}dx\,dy =$$

$$= A\frac{\pi^2}{ab}\iint \frac{1}{2}\sin\frac{2\pi x}{a}.\frac{1}{2}\sin\frac{2\pi y}{b}dx\,dy$$

$$= A\frac{\pi^2}{4ab}\int_0^l \sin\frac{2\pi x}{a}dx.\int_0^l \sin\frac{2\pi y}{b}dy = 0 \tag{E7.15}$$

Also,

$$\iint B\frac{\pi^2}{a^2}\cos^2\frac{\pi x}{a}\sin^2\frac{\pi y}{b}dx\,dy =$$

$$= B\frac{\pi^2}{a^2}\int_0^a \frac{1}{2}\left(1 + \cos\frac{2\pi x}{a}\right)dx.\int_0^b \frac{1}{2}\left(1 - \cos\frac{2\pi y}{b}\right)dy$$

$$= B\frac{\pi^2}{4a^2}\left[a + \frac{a}{2\pi}\sin\frac{2\pi x}{a}\Big|_0^a\right]\left[b - \frac{b}{2\pi}\sin\frac{2\pi y}{b}\Big|_0^b\right] = B\frac{\pi^2}{4a^2}a.b = B\frac{\pi^2}{4a}b \tag{E7.16}$$

Substitution of the expressions of Eq. (E7.14)–(E7.16) into Eq. (E7.13) gives, $B = 0$, where the other integrations are equal to zero.

Example 3.8: A simply supported rectangular plate of length (a) and width (b) carries a constant lateral pressure P. The deflection of the plate is approximated by the first term of a double sine series,

$$w(x, y) = w_o\sin\frac{\pi x}{a}\sin\frac{\pi y}{b}$$

In which (x, y) are rectangular coordinates with the axes coinciding with the two edges of the plate and w_o is a constant representing the deflection at the center of the plate. The strain energy of the plate due to bending is,

$$U = \frac{1}{2}D\iint\left(\frac{\partial^2 w}{\partial x^2} + \frac{\partial^2 w}{\partial y^2}\right)^2 dx\,dy$$

Where D is a constant called the "flexural rigidity." By the principle of stationary potential energy, derive a formula for w_o. Without using any approximation, derive the correct differential equation for w_o.

Solution:

For simply supported plate with constant lateral pressure, the deflection form and the strain energy stored are given as follows,

$$w(x,y) = w_o \sin\frac{\pi x}{a}\sin\frac{\pi y}{b} \quad \text{(E8.1)}$$

$$U = \frac{1}{2}D\iint\left(\frac{\partial^2 w}{\partial x^2} + \frac{\partial^2 w}{\partial y^2}\right)^2 dx\, dy \quad \text{(E8.2)}$$

To derive the correct differential equation for $w(x,y)$, we use Euler equation as follows,

$$V = \iint_R F(x, y, w, w_x, w_y, w_{xx}, w_{xy}, w_{yy})dx\, dy \quad \text{(E8.3)}$$

Where,

$w_x = \frac{\partial w}{\partial x}, w_{xy} = \frac{\partial^2 w}{\partial x \partial y} \ldots$ etc.

Then, the Euler equation becomes,

$$\frac{\partial F}{\partial w} - \frac{\partial}{\partial x}\left(\frac{\partial F}{\partial w_x}\right) - \frac{\partial}{\partial y}\left(\frac{\partial F}{\partial w_y}\right) + \frac{\partial^2}{\partial x^2}\left(\frac{\partial F}{\partial w_{xx}}\right) + \frac{\partial^2}{\partial y^2}\left(\frac{\partial F}{\partial w_{yy}}\right) + \frac{\partial^2}{\partial x \partial y}\left(\frac{\partial F}{\partial w_{xy}}\right) = 0$$

For the given case,

$$F(x,y) = \frac{D}{2}\left[w_{xx} + w_{yy}\right]^2 = \frac{D}{2}\left[w_{xx}^2 + w_{xx}w_{yy} + w_{yy}^2\right] \quad \text{(E8.4)}$$

Where,

$$\frac{\partial F}{\partial w} = 0, \frac{\partial F}{\partial w_x} = 0, \frac{\partial F}{\partial w_y} = 0, \frac{\partial F}{\partial w_{xy}} = 0$$

$$\frac{\partial F}{\partial w_{xx}} = \left[2w_{xx} + 2w_{yy}\right]\frac{D}{2}$$

$$\frac{\partial}{\partial x}\left(\frac{\partial F}{\partial w_{xx}}\right) = 2w_{xxx}\frac{D}{2}$$

$$\frac{\partial^2}{\partial x^2}\left(\frac{\partial F}{\partial w_{xx}}\right) = 2w_{xxxx}\frac{D}{2}$$

Similarly,

$$\frac{\partial^2}{\partial y^2}\left(\frac{\partial F}{\partial w_{yy}}\right) = 2w_{yyyy}\frac{D}{2}$$

Insert the above differentiations in Eq. (E8.4) gives,

$$D\left(w_{xxxx} + w_{yyyy}\right) = p$$

That is,

$$\frac{\partial^4 w}{\partial x^4} + \frac{\partial^4 w}{\partial y^4} = \frac{p}{D} \quad \text{(E8.5)}$$

To find an expression of w_o, using Eq. (E8.1),

$$w = w_o \sin\frac{\pi x}{a}\sin\frac{\pi y}{b}$$

The total potential energy $V = U + \Omega$ is obtained by calculating the strain energy U and the potential energy due to applied load Ω.

The strain energy stored in the plate is given by,

$$U = \frac{1}{2}D\int_0^a\int_0^b \left(w_{xx} + w_{yy}\right)^2 dx\,dy \tag{E8.6}$$

And the potential energy is,

$$\Omega = -\int_0^a\int_0^b pw(x,y)\,dx\,dy \tag{E8.7}$$

We have,

$$w_{xx} = -w_o\left(\frac{\pi}{a}\right)^2 \sin\frac{\pi x}{a}\sin\frac{\pi y}{b}$$

$$w_{yy} = -w_o\left(\frac{\pi}{b}\right)^2 \sin\frac{\pi x}{a}\sin\frac{\pi y}{b}$$

Substitute the above differentiations in Eq. (E8.6) gives,

$$U = \frac{D}{2}w_o^2\int_0^a\int_0^b \left[\left(\frac{\pi}{a}\right)^2 + \left(\frac{\pi}{b}\right)^2\right]^2 \sin^2\frac{\pi x}{a}\sin^2\frac{\pi y}{b}\,dx\,dy \tag{E8.8}$$

Using orthogonal property of sine function,

$$\int_0^a \sin^2\frac{\pi x}{a}\,dx = \frac{a}{2}$$

$$\int_0^b \sin^2\frac{\pi y}{b}\,dy = \frac{b}{2}$$

Substitute the above integrations into Eq. (E8.8) gives,

$$U = \frac{D}{2}w_o^2\left[\left(\frac{\pi}{a}\right)^2 + \left(\frac{\pi}{b}\right)^2\right]^2 \frac{ab}{4}$$

Or,

$$U = \frac{\pi^4 Dabw_o^2}{8}\left[\left(\frac{1}{a}\right)^2 + \left(\frac{1}{b}\right)^2\right]^2 \tag{E8.9}$$

To find Ω,

$$\Omega = -p\int_0^a\int_0^b w_o\sin\frac{\pi x}{a}\sin\frac{\pi y}{b}\,dx\,dy \tag{E8.10}$$

Where

$$\int_0^a \sin\frac{\pi x}{a}dx = \left(-\frac{a}{\pi}\right)\cos\frac{\pi x}{a}\Big|_0^a = -\frac{a}{\pi}[\cos\pi - \cos 0] = \frac{2a}{\pi}$$

$$\int_0^b \sin\frac{\pi y}{b}dy = \frac{2b}{\pi}$$

Substitute the above integrations into Eq. (E8.10) gives,

$$\Omega = -\frac{4pw_o ab}{\pi^2} \tag{E8.11}$$

$$V = \frac{\pi^4 Dabw_o^2}{8}\left[\left(\frac{1}{a}\right)^2 + \left(\frac{1}{b}\right)^2\right]^2 - \frac{4pw_o ab}{\pi^2}$$

P.S.T.P.E.

$\frac{\partial V}{\partial w_o} = 0$, which gives, $w_o = \frac{16p}{\pi^6 D\left[\left(\frac{1}{a}\right)^2 + \left(\frac{1}{b}\right)^2\right]^2}$

Example 3.9: For the square clamped edges and sides of the plate shown in Fig. 3.16, determine the displacements in x and y directions if it is subjected to the loading shown.

Solution:

$$u(x,y) = \sum\sum A_{mn}\sin\frac{m\pi x}{a}\sin\frac{n\pi y}{b}$$

$$v(x,y) = \sum\sum B_{mn}\sin\frac{m\pi x}{a}\sin\frac{n\pi y}{b} \tag{E9.1}$$

B.C, $= v = 0.0$, clamped plate all round

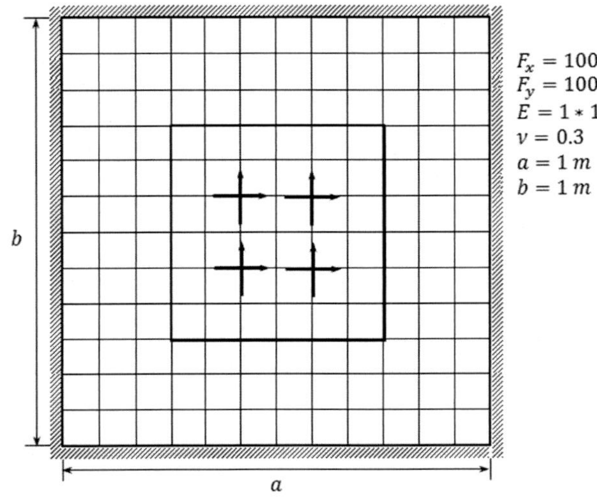

FIGURE 3.16 Clamped plate all round and subjected to forces in x and y directions.

By using potential energy principle, the displacement field can be obtained. Therefore $V = U + \Omega$.

For stretching plate, the strain energy stored is,

$$U = \iint_A \frac{Eh}{2(1-\nu^2)}\left\{(u_x + v_y)^2 - 2(1-\nu)\left[u_x v_y - \frac{1}{4}(u_y + v_x)^2\right]\right\}dx\,dy \tag{E9.2}$$

Where,

$$u_x(x,y) = \sum\sum A_{mn}\frac{m\pi}{a}\cos\frac{m\pi x}{a}\sin\frac{n\pi y}{b}$$

$$u_y(x,y) = \sum\sum A_{mn}\frac{n\pi}{b}\sin\frac{m\pi x}{a}\cos\frac{n\pi y}{b}$$

$$v_x(x,y) = \sum\sum B_{mn}\frac{m\pi}{a}\cos\frac{m\pi x}{a}\sin\frac{n\pi y}{b}$$

$$v_y(x,y) = \sum\sum B_{mn}\frac{n\pi}{b}\sin\frac{m\pi x}{a}\cos\frac{n\pi y}{b}$$

Where,

$$u_x = \frac{\partial u}{\partial x},\quad u_y = \frac{\partial u}{\partial y},\quad v_x = \frac{\partial v}{\partial x},\quad v_y = \frac{\partial v}{\partial y}$$

Therefore Eq. (E9.2) becomes,

$$U = \iint_A \frac{Eh}{2(1-\nu^2)}\left\{u_x^2 + v_y^2 + \frac{1}{2}(1-\nu)u_y^2 + \frac{1}{2}(1-\nu)v_x^2 + u_x v_y\right\}dA \tag{E9.3}$$

We have,

$$\iint_A u_x^2 dx\,dy = \int_0^b\int_0^a \sum\sum A_{mn}^2\left(\frac{m\pi}{a}\right)^2\cos^2\frac{m\pi x}{a}\sin^2\frac{n\pi y}{b}dx\,dy$$

$$= \frac{ab}{4}\sum\sum A_{mn}^2\left(\frac{m\pi}{a}\right)^2 = \frac{ab}{4}\frac{\pi^2}{a^2}\sum\sum A_{mn}^2 m^2$$

$$\iint_A v_y^2 dA = \int_0^b\int_0^a \sum\sum B_{mn}^2\left(\frac{n\pi}{b}\right)^2\sin^2\frac{m\pi x}{a}\cos^2\frac{n\pi y}{b}dx\,dy = \frac{ab}{4}\frac{\pi^2}{b^2}\sum\sum B_{mn}^2 n^2$$

$$\iint_A u_y^2 dA = \frac{ab}{4}\frac{\pi^2}{b^2}\sum\sum A_{mn}^2 n^2$$

$$\iint_A v_x^2 dA = \frac{ab}{4}\frac{\pi^2}{a^2}\sum\sum B_{mn}^2 m^2$$

$$\iint_A u_x v_y dA = \int_0^b\int_0^a \sum\sum A_{mn}B_{mn}\frac{n\pi}{b}\frac{m\pi}{a}\sin\frac{n\pi y}{b}\sin\frac{m\pi x}{a}\cos\frac{n\pi y}{b}\cos\frac{m\pi x}{a}dx\,dy = 0.0$$

Substitute the above derivatives in Eq. (E9.3) gives,

$$U = \frac{Eh}{2(1-\nu^2)}\left\{\begin{array}{c}\frac{ab}{4}\frac{\pi^2}{a^2}\sum\sum A_{mn}^2 m^2 + \frac{ab}{4}\frac{\pi^2}{b^2}\sum\sum B_{mn}^2 n^2 + \\ \frac{1}{2}(1-\nu)\frac{ab}{4}\frac{\pi^2}{b^2}\sum\sum A_{mn}^2 n^2 + \frac{1}{2}(1-\nu)\frac{ab}{4}\frac{\pi^2}{a^2}\sum\sum B_{mn}^2 m^2\end{array}\right\}$$

We have,

$$\frac{\partial U}{\partial A_{mn}} = \frac{Eh}{2(1-\nu^2)}\left\{\frac{m^2}{a^2} + \frac{(1-\nu)}{2}\frac{n^2}{b^2}\right\}\frac{\pi^2 ab}{2}A_{mn} \tag{E9.4}$$

$$\frac{\partial U}{\partial B_{mn}} = \frac{Eh}{2(1-\nu^2)}\left\{\frac{(1-\nu)}{2}\frac{m^2}{a^2} + \frac{n^2}{b^2}\right\}\frac{\pi^2 ab}{2}B_{mn} \tag{E9.5}$$

The potential energy of the applied load is given by,

$$\Omega = -\int_{-b/2}^{b/2}\int_{-a/2}^{a/2}\left(F_x.u + F_y.v\right)dx\,dy \tag{E9.6}$$

Substitute the expressions of $u(x,y)$ and $v(x,y)$ from Eq. (E9.1) into Eq. (E9.4) gives,

$$= -\int_{b/4}^{3b/4}\int_{a/4}^{3a/4}\left(F_x\sum\sum A_{mn}\sin\frac{m\pi x}{a}\sin\frac{n\pi y}{b} + F_y\sum\sum B_{mn}\sin\frac{m\pi x}{a}\sin\frac{n\pi y}{b}\right)dx\,dy$$

Carrying out the integrations in the above equation of the potential energy Ω yields,

$$\Omega = F_x\sum\sum A_{mn}\frac{4ab}{mn\pi^2}\cos\frac{m\pi}{4}\cos\frac{n\pi}{4}(-1)^{(m+n)/2} + F_y\sum\sum B_{mn}\frac{4ab}{mn\pi^2}\cos\frac{m\pi}{4}\cos\frac{n\pi}{4}(-1)^{(m+n)/2}$$

We have,

$$\frac{\partial\Omega}{dA_{mn}} = F_x\frac{4ab}{mn\pi^2}\sin\frac{m\pi}{4}\sin\frac{n\pi}{4}(-1)^{(m+n)/2} \tag{E9.7}$$

$$\frac{\partial\Omega}{dB_{mn}} = F_y\frac{4ab}{mn\pi^2}\sin\frac{m\pi}{4}\sin\frac{n\pi}{4}(-1)^{(m+n)/2} \tag{E9.8}$$

But for minimization of T.P.E., we have,

$$\frac{\partial \mathrm{V}}{\partial A_{mn}} = \frac{\partial \mathrm{U}}{d\partial} + \frac{\partial\Omega}{\partial A_{mn}} = 0 \tag{E9.9}$$

Substitution of the expressions of $\frac{\partial \mathrm{U}}{\partial A_{mn}}$ from Eq. (E9.4) and $\frac{\partial\Omega}{d\partial}$ from Eq. (E9.7) into Eq. (E9.9) gives,

$$\frac{\pi^2 Eab}{4(1-\nu^2)}\left[\frac{m^2}{a^2} + \frac{(1-\nu)}{2}\frac{n^2}{b^2}\right]A_{mn} + \frac{4(-1)^{(m+n)/2}abF_x}{\pi^2 mn}\sin\frac{m\pi}{4}\sin\frac{n\pi}{4} = 0$$

Hence,

$$A_{mn} = \frac{-4(-1)^{(m+n)/2} F_x . \sin\frac{m\pi}{4} \sin\frac{n\pi}{4}}{\frac{\pi^2 E}{4(1-\nu^2)}\left[\frac{m^2}{a^2} + \frac{(1-\nu)}{2}\frac{n^2}{b^2}\right] mn} \tag{E9.10}$$

Also, from T.P.E. principles,

$$\frac{\partial V}{\partial B_{mn}} = \frac{\partial U}{\partial B_{mn}} + \frac{\partial \Omega}{\partial B_{mn}} = 0 \tag{E9.11}$$

Substitution of the expressions of $\frac{\partial U}{\partial B_{mn}}$ from Eq. (E9.5) and $\frac{\partial \Omega}{\partial B}$ from Eq. (E9.8) into Eq. (E9.11) gives,

$$\frac{Eh}{2(1-\nu^2)}\left\{\frac{(1-\nu)}{2}\frac{m^2}{a^2} + \frac{n^2}{b^2}\right\}\frac{\pi^2 ab}{2} B_{mn} + F_y \frac{4ab}{mn\pi^2} \sin\frac{m\pi}{4} \sin\frac{n\pi}{4} (-1)^{(m+n)/2} = 0$$

Hence,

$$B_{mn} = \frac{-16(-1)^{(m+n)/2} . (1-\nu^2) F_y . \sin\frac{m\pi}{4} \sin\frac{n\pi}{4}}{\pi^2 E\left[\frac{(1-\nu)}{2}\frac{m^2}{a^2} + \frac{n^2}{b^2}\right] mn} \tag{E9.12}$$

Insert the expressions of Eqs. (E9.10) and (E9.12) in Eq. (E9.1) gives,

$$u(x,y) = \sum_{1,3,5,\ldots}^{\infty} \sum_{1,3,5,\ldots}^{\infty} A_{mn} \sin\frac{m\pi x}{a} \sin\frac{n\pi y}{b}$$

And,

$$v(x,y) = \sum_{1,3,5,\ldots}^{\infty} \sum_{1,3,5,\ldots}^{\infty} B_{mn} \sin\frac{m\pi x}{a} \sin\frac{n\pi y}{b}$$

If $b_1 = b$, $a_1 = a$

$$u(x,y) = \sum_{1,3,5,\ldots}^{\infty} \sum_{1,3,5,\ldots}^{\infty} \frac{-16.(1-\nu^2) F_x}{\pi^2 E\left[\frac{m^2}{a^2} + \frac{(1-\nu)}{2}\frac{n^2}{b^2}\right] mn} \sin\frac{m\pi x}{a} \sin\frac{n\pi y}{b}$$

$$v(x,y) = \sum_{1,3,5,\ldots}^{\infty} \sum_{1,3,5,\ldots}^{\infty} \frac{-16.(1-\nu^2) F_y}{\pi^2 E\left[\frac{m^2}{a^2}\frac{(1-\nu)}{2} + \frac{n^2}{b^2}\right] mn} \sin\frac{m\pi x}{a} \sin\frac{n\pi y}{b}$$

Example 3.10: A member in a ship structure consists of two edge beams and a thin sheet and is required, essentially, to transmit the longitudinal distributed forces $p(x)$ per unit length of edge beam as shown in Fig. 3.17. To describe approximately how the stresses are "diffused" into the sheet, the principle of stationary total potential energy may be used.

Assuming a displacement pattern in the sheet to be,

$$u = u_o \sin\frac{\pi x}{2a} \cosh\frac{k\pi y}{2b}, \quad v = 0 \tag{E10.1}$$

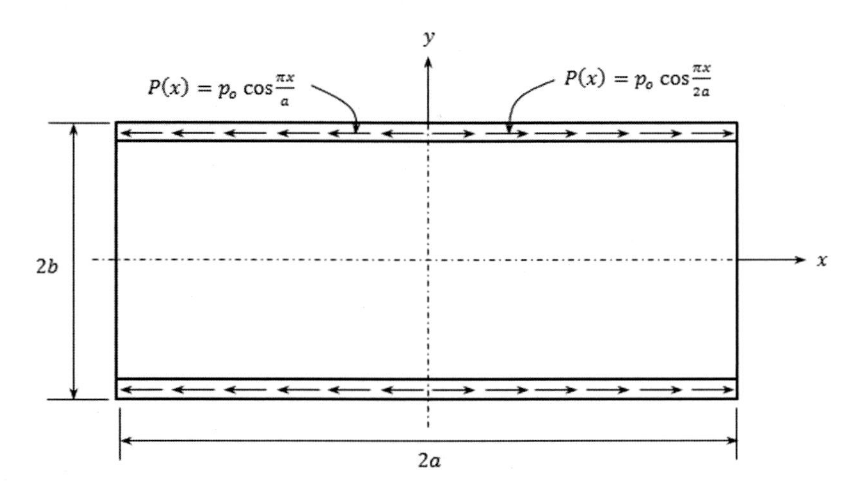

FIGURE 3.17 Sheet stiffened by two beams.

Where, $k = \sqrt{\frac{E}{G}}$ is a constant for the sheet. Determine a suitable value for u_o.

Take the material of the sheet and edge beams to have the same elastic constants.

You may accept that for a sheet in a state of plane stress, the strain energy density is given by,

$$U_o = \frac{E}{2(1-\nu^2)}\left[\varepsilon_x + \varepsilon_y + 2\nu\varepsilon_x\varepsilon_y\right] + \frac{E}{4(1+\nu)}\gamma_{xy}^2$$

Where, the symbols have the usual meanings.

Solution:

The displacement pattern of the sheet is given by,

$$u(x,y) = u_o \sin\frac{\pi x}{2a}\cosh\frac{k\pi y}{2b}, v = 0 \tag{E10.2}$$

Where,

$$k = \sqrt{\frac{E}{G}}$$

The problem is of plane stress, the strain energy density is given by,

$$U_o = \frac{E}{2(1-\nu^2)}\left[\varepsilon_x + \varepsilon_y + 2\nu\varepsilon_x\varepsilon_y\right] + \frac{E}{4(1+\nu)}\gamma_{xy}^2 \tag{E10.3}$$

Longitudinal distribution force per length of edge beam (Fig. 3.17) is given by,

$$P(x) = \begin{cases} p_o \cos\frac{\pi x}{2a} \text{for} 0 \le x \le a \\ -p_o \cos\frac{\pi x}{a} \text{for} -a \le x \le 0 \end{cases} \tag{E10.4}$$

Assume Hooke's law material with constant E and consider the edge beams, the strain energy and potential energy for beam$_{(1)}$ is given by,

$$U_{\text{beam}_{(1)}} = \int U_o dvol = \frac{1}{2}\int_{vol} E\ \varepsilon^2\ dvol = \frac{1}{2}EA\int \varepsilon_x^2 dx \quad \text{(E10.5)}$$

$$\Omega_{p_{\text{beam(1)}}} = -\int_{-a}^{a} p\ u\ dx \quad \text{(E10.6)}$$

The total potential energy for beam-1, is given by,

$$V_{\text{beam}_{(1)}} = U\text{beam}_1 + V\text{beam}_1 \quad \text{(E10.7)}$$

Substitute the values of U and Ω for beam-1, from Eqs. (E10.5) and (E10.6) respectively in Eq. (E10.7) gives,

$$V = \frac{1}{2}EA\int \varepsilon_x^2 dx - \int_{-a}^{a} p\ u\ dx\ 1^{\text{st}}\text{beam} \quad \text{(E10.8)}$$

Similarly for beam 2,

$$V_{\text{beam}_{(2)}} = \frac{1}{2}EA\int_{-a}^{a} \varepsilon_x^2 dx - \int_{-a}^{a} p\ u\ dx 2^{\text{nd}}\text{beam} \quad \text{(E10.9)}$$

The strain energy stored in the thin sheet is given by,

$$U = \int U_o dvol = \int_{-b}^{b}\int_{-a}^{a} U_o t\ dx\ dy \quad \text{(E10.10)}$$

Where, t is the sheet thickness.

The potential energy due tractions in the sheet is as follows,

$$\Omega = -\int T^n u\ ds$$

$$= -\left(\int_{-a}^{a} T_x^n utdx + \int_{a}^{-a} T_x^n u(-t)dx\right) = -2\int_{-a}^{a} T_x^n utdx \quad \text{(E10.11)}$$

We have,

$P(x)$ As a force $dF = pdx$

Or,

$dF = pdx$ Force per unit length $= T_x^n dxt$

Hence,

$T_x^n = \frac{P(x)}{t}$, and the potential energy is given by,

$$\Omega = -2\int_{-a}^{a} P\ u\ dx \quad \text{(E10.12)}$$

Therefore the total potential energy V is composed of strain energy $U +$ the potential energy due to tractions in the sheet Ω. Therefore,

$$V_{\text{sheet}} = \int_{-b}^{b}\int_{-a}^{a} U_o t\ dx\ dy - 2\int_{-a}^{a} P\ u\ dx \quad \text{(E10.13)}$$

Therefore,

$$\text{Total } V = V_{\text{beam}_{(1)}} + V_{\text{beam}_{(2)}} + V_{\text{sheet}}$$

This means that,

$$V = \underbrace{V_{\text{beam}_{(1)}}}_{U+\Omega} + \underbrace{V_{\text{beam}_{(2)}}}_{U+\Omega} + \underbrace{V_{\text{sheet}}}_{U+\Omega} \tag{E10.14}$$

Substitute the expressions of $V_{\text{beam}_{(1)}}$ from Eq. (E10.8), $V_{\text{beam}_{(2)}}$from Eq. (E10.9), and V_{sheet} from Eq. (E10.11) as follows,

$$V = \begin{pmatrix} \frac{1}{2}EA\int_{-a}^{a}\varepsilon_x^2\ dx - \int_{-a}^{a}p\ u\ dx|_{\text{beam}_{(1)}} + \\ \frac{1}{2}EA\int_{-a}^{a}\varepsilon_x^2\ dx - \int_{-a}^{a}p\ u\ dx|_{\text{beam}_{(2)}} + \\ \int_{-b}^{b}\int_{-a}^{a}U_o\ t\ dx\ dy - 2\int_{-a}^{a}P\ u\ dx|_{\text{sheet}} \end{pmatrix} \tag{E10.15}$$

The assumed displacement pattern and the strain components are as follows,

$$u = u_o \sin\frac{\pi x}{2a}\cosh\frac{k\pi y}{2b}, v = 0$$

$$\varepsilon_x = \frac{du}{dx} = u_o\frac{\pi}{2a}\cos\frac{\pi x}{2a}\cosh\frac{k\pi y}{2b}$$

$$\varepsilon_y = \frac{dv}{dy} = 0$$

$$\gamma_{xy} = \frac{du}{dy} + \frac{dv}{dx} = \frac{k\pi}{2b}u_o\sin\frac{\pi x}{2a}\sinh\frac{k\pi y}{2b}$$

The strain components and the displacement field are as follows,
For 1st beam

$$\varepsilon_x = \frac{\pi}{2a}u_o\cos\frac{\pi x}{2a}\cosh\frac{k\pi y}{2b}y = b$$

$$\varepsilon_x = \frac{\pi}{2a}u_o\cos\frac{\pi x}{2a}\cosh\frac{k\pi}{2}$$

$$u = u_o\sin\frac{\pi x}{2a}\cosh\frac{k\pi}{2}$$

For 2nd beam,

$$\varepsilon_x = \frac{\pi}{2a}u_o\cos\frac{\pi x}{2a}\cosh\frac{k\pi}{2}\frac{-b}{b} = \frac{\pi}{2a}u_o\cos\frac{\pi x}{2a}\cosh\frac{k\pi}{2}$$

$$u = u_o\sin\frac{\pi x}{2a}\cosh\frac{k\pi}{2}$$

Substitution the above corresponding strain components and the displacement u in the total V expression given in Eq. (E10.14) gives,

$$V=\left(\begin{array}{c} EA\int_{-a}^{a}\left(\frac{\pi}{2a}u_o\cos\frac{\pi x}{a}\cosh\frac{k\pi}{2}\right)^2dx-2\int_{-a}^{a}P(x)u_o\sin\frac{\pi x}{a}\cosh\frac{k\pi}{2}dx+\\ \int_{-b}^{b}\int_{-a}^{a}t\left[\frac{E}{2(1-\nu^2)}\frac{\pi}{2a}u_o\cos\frac{\pi x}{a}\cosh\frac{k\pi y}{2b}+0+0+\frac{E}{4(1-\nu^2)}\left(\frac{k\pi}{2b}u_o\sin\frac{\pi x}{2a}\sinh\frac{k\pi}{2b}\right)^2\right]dxdy-\\ 2\int_{-a}^{a}P(x)u_o\sin\frac{\pi x}{2a}\cosh\frac{k\pi}{2}dx \end{array}\right) \quad \text{(E10.16)}$$

We have,
$V=V(u_o)$, using the minimization principle of the T.P.E. requires that,

$$\delta V=0 \text{ so } \frac{\partial V}{\partial u_o}=0 \quad \text{(E10.17)}$$

Therefore Eq. (E10.16) becomes,

$$\frac{\partial V}{\partial u_o}=\left(\begin{array}{c} 2EA\left(\frac{\pi}{2a}\cosh\frac{k\pi}{2}\right)^2\int_{-a}^{a}u_o\cos^2\frac{\pi x}{a}dx+\\ \frac{tE}{2(1-\nu^2)}\frac{\pi}{2a}\int_{-b}^{b}\int_{-a}^{a}\cos\frac{\pi x}{2a}\cosh\frac{k\pi y}{2b}dx\,dy+\\ \frac{2tE}{4(1-\nu^2)}\left(\frac{k\pi}{2b}\right)^2\int_{-b}^{b}\int_{-a}^{a}\sin^2\frac{\pi x}{2a}\sinh^2\frac{k\pi y}{2b}dx\;dy+\\ 4P_o\cosh\frac{k\pi}{2}\int_{-a}^{a}\cos\frac{\pi x}{a}\sin\frac{\pi x}{2a}dx-\\ 4P_o\cosh\frac{k\pi}{2}\int_{0}^{a}\cos\frac{\pi x}{2a}\sin\frac{\pi x}{2a}dx \end{array}\right)=0 \quad \text{(E10.18)}$$

We have,

$$\int_{-a}^{a}\cos^2\frac{\pi x}{a}dx=a$$

$$\int_{0}^{a}\cos\frac{\pi x}{a}\sin\frac{\pi x}{2a}dx=\frac{2a}{\pi}\left[\frac{2a}{\pi}-\frac{2}{3}\right]$$

$$\int_{0}^{a}\cos\frac{\pi x}{2a}\sin\frac{\pi x}{2a}dx=\frac{a}{\pi}$$

$$\int_{-b}^{b}\int_{-a}^{a}\cos\frac{\pi x}{2a}\cosh\frac{k\pi y}{2b}dxdy=\frac{16ab}{b\pi^2}\sinh\frac{k\pi}{2}$$

$$\int_{-b}^{b}\int_{-a}^{a}\sin^2\frac{\pi x}{2a}\sinh^2\frac{k\pi y}{2b}dxdy=-ab+\frac{ab}{k\pi}\sin k\pi$$

Substitution the above integrations in Eq. (E10.18) results in,

$$\frac{\partial V}{\partial u_o} = \left(2EA\left(\frac{\pi}{2a}\cosh\frac{k\pi}{2}\right)^2 a.u_o + \frac{tE}{2(1-\nu^2)}\frac{\pi}{2a}\frac{16ab}{b\pi^2}\sinh\frac{k\pi}{2} + \frac{2tE}{4(1-\nu^2)}\left(\frac{k\pi}{2b}\right)^2\left(-ab+\frac{ab}{k\pi}\sin k\pi\right)\right.$$
$$\left. + 4P_o\cosh\frac{k\pi}{2}\frac{2a}{\pi}\left[\frac{2a}{\pi}-\frac{2}{3}\right] - 4P_o\frac{a}{\pi}\cosh\frac{k\pi}{2}\right)$$

So, $\frac{\partial V}{\partial u_o} = 0$ from this equation we can determine u_o,

$$u_o = \left(-\frac{1}{2EA\left(\frac{\pi}{2a}\cosh\frac{k\pi}{2}\right)^2 a}\frac{tE}{2(1-\nu^2)}\frac{\pi}{2a}\frac{16ab}{b\pi^2}\sinh\frac{k\pi}{2} + \frac{1}{2EA\left(\frac{\pi}{2a}\cosh\frac{k\pi}{2}\right)^2 a}\frac{2tE}{4(1-\nu^2)}\left(\frac{k\pi}{2b}\right)^2\left(ab-\frac{ab}{k\pi}\sin k\pi\right)\right.$$
$$\left. + \frac{4P_o}{2EA\left(\frac{\pi}{2a}\cosh\frac{k\pi}{2}\right)^2 a}\cosh\frac{k\pi}{2}\frac{2a}{\pi}\left(\frac{2}{3}-\frac{2a}{\pi}\right) + \frac{4P_o}{2EA\left(\frac{\pi}{2a}\cosh\frac{k\pi}{2}\right)^2 a}\frac{a}{\pi}\cosh\frac{k\pi}{2}\right)$$

Example 3.11: Fig. 3.18 shows an element of a machine structure is essentially a clamped square plate supporting a uniform transverse distributed load p per unit area. To determine the static deflection by means of the principle of stationary total potential energy, the deflection surface is approximated by,

$$w(x,y) = w_o\left(1-\cos\frac{2\pi x}{a}\right)\left(1-\cos\frac{2\pi y}{a}\right)$$

Where a is the length of the plate edge.

Explain why this is a reasonable shape to assume and why the stresses do not enjoy the same accuracy as the deflection. Determine w_o.

Solution:

The square bracket can be neglected since the edges are rigidly clamped. It is contributed by natural B.Cs only. For clamped on all edges,

$$w(x,y) = w_o\left(1-\cos\frac{2\pi x}{a}\right)\left(1-\cos\frac{2\pi y}{a}\right) \quad \text{(E11.1)}$$

$$\text{at } x=0,\ w=0,\ \text{at} x=a,\ w=0,\ \text{at } y=a,\ w=0$$

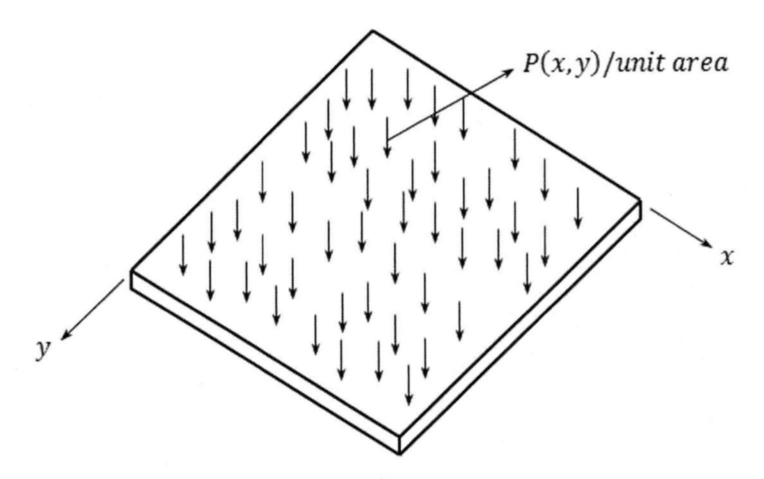

FIGURE 3.18 Clamped square plate.

Since it is satisfy the geometric B.Cs so it is a reasonable shape to assume at the clamped edges slopes are zero for example, when $x = a$,

$\frac{dw}{dx}$ = should be zero due to fixed edges.

The assumed deflected shape in Eq. (E11.1) can be written as follows,

$$w(x,y) = w_o\left[1 - \cos\frac{2\pi y}{a} - \cos\frac{2\pi x}{a} + \cos\frac{2\pi y}{a}\cos\frac{2\pi x}{a}\right] \quad \text{(E11.2)}$$

Applying the boundary conditions of the fixed edge, the slope at the fixed edge $x = a$ is zero, this means that,

$$\left.\frac{dw}{dx}\right|_{x=a} = w_o\left[\frac{2\pi}{a}\sin 2\pi - \frac{2\pi}{a}\cos\frac{2\pi y}{a}\sin 2\pi\right] = 0 \quad \text{(E11.3)}$$

The total potential energy,

$$V = U + \Omega \quad \text{(E11.4)}$$

Where,

$$U = \frac{D}{2}\iint\left(\frac{\partial^2 w}{\partial x^2} + \frac{\partial^2 w}{\partial y^2}\right)^2 \quad \text{(E11.5)}$$

We have,

$$\frac{\partial w}{\partial x} = w_o\left(1 - \cos\frac{2\pi y}{a}\right)\left(\frac{2\pi}{a}\right)\sin\frac{2\pi x}{a}$$

$$\frac{\partial^2 w}{\partial x^2} = w_o\left(1 - \cos\frac{2\pi y}{a}\right)\left(\frac{2\pi}{a}\right)^2\cos\frac{2\pi x}{a}$$

$$\frac{\partial^2 w}{\partial y^2} = w_o\left(1 - \cos\frac{2\pi x}{a}\right)\left(\frac{2\pi}{a}\right)^2\cos\frac{2\pi y}{a}$$

Therefore Eq. (E11.5) becomes,

$$U = \frac{D}{2}\iint w_o^2\left\{\left(1 - \cos\frac{2\pi y}{a}\right)\left(\frac{2\pi}{a}\right)^2\cos\frac{2\pi x}{a} + \left(1 - \cos\frac{2\pi x}{a}\right)\left(\frac{2\pi}{a}\right)^2\cos\frac{2\pi y}{a}\right\}^2 dx\,dy \quad \text{(E11.6)}$$

The potential energy due to applied load is given by,

$$\Omega = \iint - P(x,y)w(x,y)dxdy \quad \text{(E11.7)}$$

Substitute the expression of w from Eq. (E11.1) into Eq. (E11.2) gives,

$$= - P\iint w_o\left(1 - \cos\frac{2\pi x}{a}\right)\left(1 - \cos\frac{2\pi y}{a}\right)dx\,dy \quad \text{(E11.8)}$$

The total potential energy $V = U + \Omega$ is obtained by the substitution of the expression of U from Eq. (E11.6) and the expression of Ω from Eq. (E11.8) as follows,

$$= \frac{D}{2}\iint w_o^2\left\{\left(1-\cos\frac{2\pi y}{a}\right)\left(\frac{2\pi}{a}\right)^2\cos\frac{2\pi x}{a}+\left(1-\cos\frac{2\pi x}{a}\right)\left(\frac{2\pi}{a}\right)^2\cos\frac{2\pi y}{a}\right\}^2 dx\ dy$$
$$- P\iint w_o\left(1-\cos\frac{2\pi x}{a}\right)\left(1-\cos\frac{2\pi y}{a}\right)dx\ dy \quad \text{(E11.9)}$$

To find w_o, apply the T.P.E. minimization principles requires that $\frac{\partial V}{\partial w_o} = 0$ of Eq. (E11.9) gives,

$$0 = D\iint w_o\left\{\left(1-\cos\frac{2\pi y}{a}\right)\left(\frac{2\pi}{a}\right)^2\cos\frac{2\pi x}{a}+\left(1-\cos\frac{2\pi x}{a}\right)\left(\frac{2\pi}{a}\right)^2\cos\frac{2\pi y}{a}\right\}^2 dxdy$$
$$- P\iint\left(1-\cos\frac{2\pi x}{a}\right)\left(1-\cos\frac{2\pi y}{a}\right)dxdy \quad \text{(E11.10)}$$

Hence,

$$w_o = \frac{P\iint\left(1-\cos\frac{2\pi x}{a}\right)\left(1-\cos\frac{2\pi y}{a}\right)dx\ dy}{D\iint w_o\left\{\left(1-\cos\frac{2\pi y}{a}\right)\left(\frac{2\pi}{a}\right)^2\cos\frac{2\pi x}{a}+\left(1-\cos\frac{2\pi x}{a}\right)\left(\frac{2\pi}{a}\right)^2\cos\frac{2\pi y}{a}\right\}^2 dx\ dy} \quad \text{(E11.11)}$$

We have,

$$P\int_0^a\int_0^a\left(1-\cos\frac{2\pi x}{a}\right)\left(1-\cos\frac{2\pi y}{a}\right)dx\ dy$$
$$= P\int_0^a\left(1-\cos\frac{2\pi y}{a}\right)\left|x-\frac{a}{2\pi}\sin\frac{2\pi x}{a}\right|_0^a dy$$
$$= P\int_0^a\left(1-\cos\frac{2\pi y}{a}\right)\left(a-\frac{a}{2\pi}\sin 2\pi\right)dy$$
$$= Pa\left|y-\frac{a}{2\pi}\sin\frac{2\pi y}{a}\right|_0^a = Pa^2$$

Therefore substitute the above integrations into Eq. (E11.11) gives,

$$\int_0^a\int_0^a\left\{\begin{array}{c}\underbrace{\left(1-\cos\frac{2\pi y}{a}\right)^2\left(\frac{2\pi}{a}\right)^4\cos^2\frac{2\pi x}{a}}_{I_1}+\underbrace{\left(1-\cos\frac{2\pi x}{a}\right)^2\left(\frac{2\pi}{a}\right)^4\cos^2\frac{2\pi y}{a}}_{I_2}+\\ \underbrace{\left(\frac{2\pi}{a}\right)^4 2\left(1-\cos\frac{2\pi y}{a}\right)\left(1-\cos\frac{2\pi x}{a}\right)\cos\frac{2\pi y}{a}\cos\frac{2\pi x}{a}}_{I_3}\end{array}\right\}dx\ dy = I_1+I_2+I_3 \quad \text{(E11.12)}$$

Where,

$$I_1 = \int_0^a \int_0^a \left\{ \left(1 - \cos\frac{2\pi y}{a}\right)^2 \left(\frac{2\pi}{a}\right)^4 \cos^2\frac{2\pi x}{a} \right\} dxdy$$

$$= \int_0^a \left(1 - \cos\frac{2\pi y}{a}\right)^2 \left(\frac{2\pi}{a}\right)^4 \frac{1}{2}\left| x + \frac{a}{4\pi}\sin\frac{4\pi x}{a} \right|_0^a dy$$

$$= \left(\frac{2\pi}{a}\right)^4 \frac{a}{2}\int_0^a \left(1 - 2\cos\frac{2\pi y}{a} + \cos^2\frac{2\pi y}{a}\right)^2 dy$$

$$= \left(\frac{2\pi}{a}\right)^4 \frac{a}{2}\left| y - \frac{2a}{2\pi}\sin\frac{2\pi y}{a} + \frac{1}{2}\left(y + \sin\frac{4\pi y}{a}\right) \right|_0^a = \left(\frac{2\pi}{a}\right)^4 \frac{a}{2}\left| a + \frac{a}{2} \right| = \frac{12\pi^4}{a^2} \tag{E11.13}$$

Similarly,

$$I_2 = \int_0^a \int_0^b \left\{ \left(1 - \cos\frac{2\pi x}{a}\right)^2 \left(\frac{2\pi}{a}\right)^4 \cos^2\frac{2\pi y}{a} \right\} dx\ dy = \frac{12\pi^4}{a^2} \tag{E11.14}$$

$$I_3 = \int_0^a \int_0^a \left\{ \left(\frac{2\pi}{a}\right)^4 2\left(1 - \cos\frac{2\pi y}{a}\right)\left(1 - \cos\frac{2\pi x}{a}\right)\cos\frac{2\pi y}{a}\cos\frac{2\pi x}{a} \right\} dx\ dy$$

$$= \left(\frac{2\pi}{a}\right)^4 2\frac{a}{2}\int_0^a \left(1 - \cos\frac{2\pi y}{a}\right)\cos\frac{2\pi y}{a} dy = \left(\frac{2\pi}{a}\right)^4 \frac{a^2}{2} = \frac{8\pi^4}{a^2} \tag{E11.15}$$

Substitution the results of the above integrations I_1, I_2, and I_3 from Eqs. (E11.13)–(E11.15) respectively into Eq. (E11.11) and solving for w_o gives,

$$w_o = \frac{P.a^2}{D\left[2.\frac{12\pi^4}{a^2} + \frac{8\pi^4}{a^2}\right]} = \frac{P}{D}\frac{a^4}{32\pi^4}$$

Problems

P.3.1 Using the energy principles, find the center deflection for the elastic plate/spring systems of Fig. P.3.1A. Accept the results of Fig. P.3.1B and C. where, α_1, α_2 tabulated coeffts.

P.3.2 The rectangular plate of Fig. P.3.2 is simply supported all round its edges and supports a uniformly distributed transverse load p_o. Determine the deflection surface using Galerkin's method and assuming,

$$w_1(x,y) = c_1\varnothing_1(x,y) = c_1\left(x^2 - a^2\right)^2\left(y^2 - b^2\right)^2$$

b. Treat the plate of (a) by the Kantorovich method using the trial function,

$$w(x,y) = f(x)\left[1 - \left(\frac{2y}{b}\right)^2\right]$$

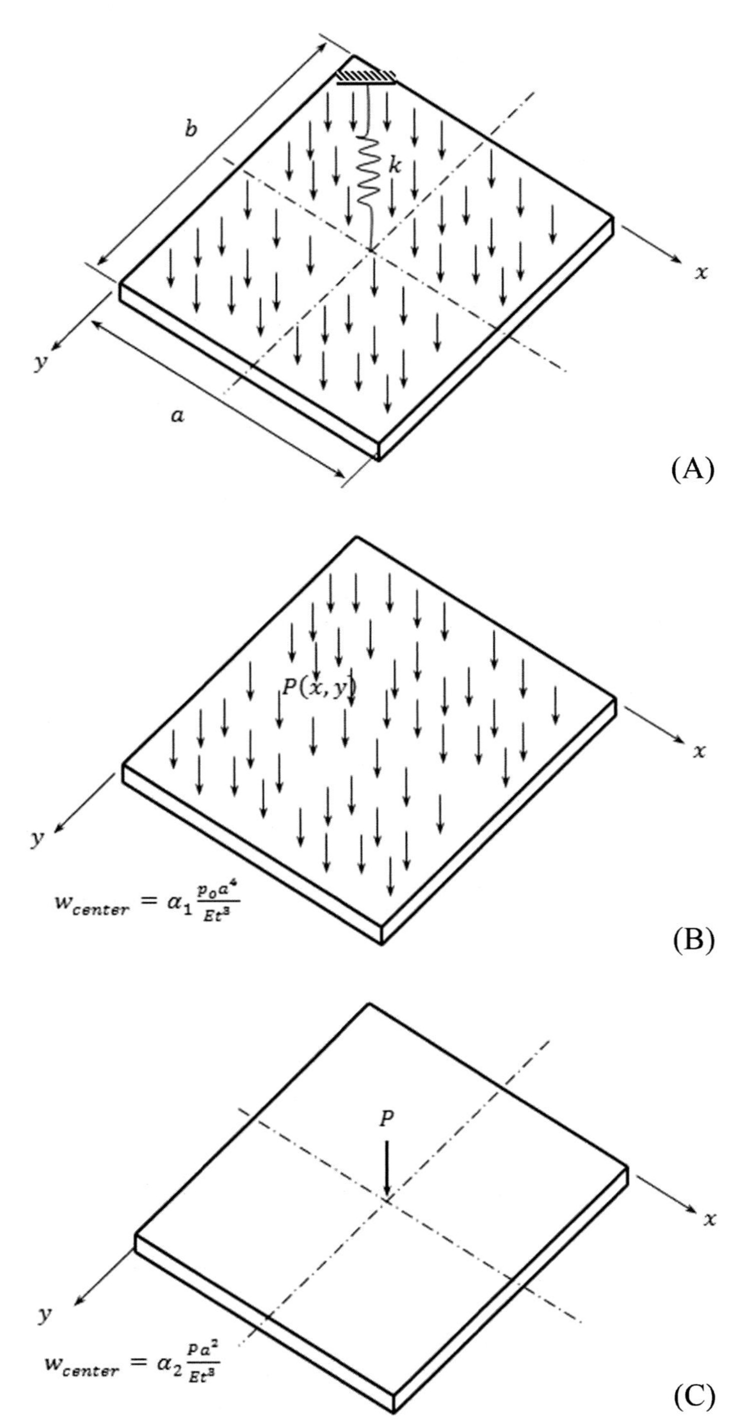

FIGURE P.3.1 Different plates configurations.

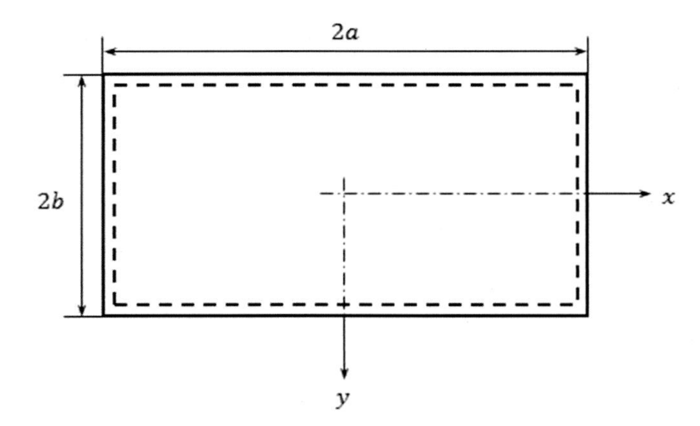

FIGURE P.3.2 Simply supported plate.

P.3.3 A rectangular steel plate of sides 300 mm × 200 mm is S.S. all round its edges and carries in a uniform pressure of $0.01N/\text{mm}^2$ over its entire surface. Show that, with the origin at one corner and the x-axis parallel to the long side,

$$w(x,y) = w_o \sin\frac{\pi x}{300}.\sin\frac{\pi y}{200}$$

Is suitable as a trial function to use in the Rayleigh-Ritz technique.
If the plate thickness is $t = 10$ mm, estimate the central deflection.
You may accept that for such a plate,

$$U = \frac{D}{2}\int_0^{200}\int_0^{300}\left[\frac{\partial^2 w}{\partial x^2} + \frac{\partial^2 w}{\partial y^2}\right]^2 dxdy$$

With,

$$D = \frac{Et^3}{12(1-\nu^2)}$$

ν = Poisson's ratio

P.3.4 The deflection at a point (x,y) on a plate due to a unit point load at a point (ξ, η) is called the influence function a (x, y, ξ, η). For a simply supported rectangular plate, sides $a \times b$, a deflection surface,

$$w(x,y) = \sum_{n=1}^{\infty}\sum_{n=1}^{\infty} a_{mn}\sin\frac{m\pi x}{a}\sin\frac{n\pi y}{b}$$

Use the potential energy principle to show that,

$$w(x,y,\xi,\eta) = \frac{4}{\pi^4 Dab}\sum_{n=1}^{\infty}\sum_{n=1}^{\infty}\left(\frac{1}{\left(\frac{m}{a}\right)^2 + \left(\frac{n}{b}\right)^2}\right)\sin\frac{m\pi x}{a}\sin\frac{n\pi y}{b}\sin\frac{m\pi\xi}{a}\sin\frac{n\pi\eta}{b}$$

Notice that $w(x,y,\xi,\eta) = w(\xi,\eta,x,y)$ as we would expect from the reciprocal theorem.

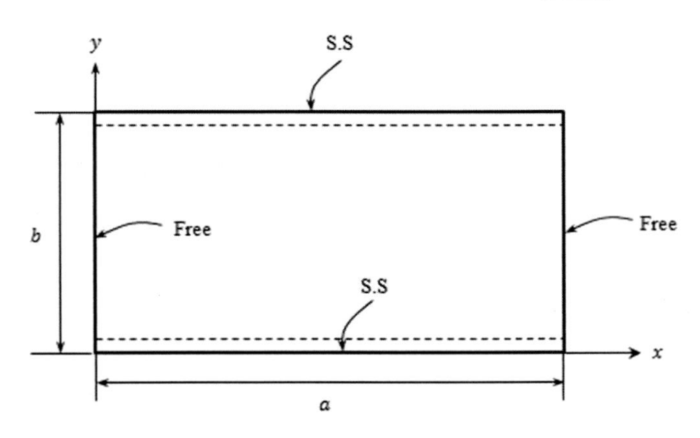

FIGURE P.3.5 Rectangular plate sides S.S. and ends free.

P.3.5 The Plate of Fig. P.3.5 supports a transverse pressure $P(x,y) = \tilde{P}(x).\tilde{P}(y)$ over its entire surface. $\tilde{P}(x)$ and $\tilde{P}(y)$ are respectively functions of x and y alone. If the total potential energy is given by,

$$V = \frac{D}{2}\iint\left\{\left(\nabla^2 w\right)^2 - 2(1-\nu)\left[\frac{\partial^2 w}{\partial x^2}\cdot\frac{\partial^2 w}{\partial y^2} - \left(\frac{\partial^2 w}{\partial x \partial y}\right)^2\right]\right\}dxdy - \iint P.w\,dx.dy$$

Find the deflection equation using on term by using Kantorovich method.
Calculate w_{max} for,

$$a = 400\text{ mm},\ \ b = 300\text{ mm},\ \ \tilde{P}(x) = 10\frac{N}{\text{mm}^2},\ \ \tilde{P}(y) = 5\frac{N}{\text{mm}^2},\ \ E = 70\times 10^3\frac{N}{\text{mm}^2}$$

$$\nu = 0.28,\ \ t = 5\text{ mm}$$

P.3.6 A simply supported variable thickness circular plate, radius a, carries a uniform pressure p. The bending stiffness is given by,

$$D = D_o\left(1 - \frac{r}{2a}\right)$$

Assuming that the deflection surface is approximated by,

$$w(r) = w_o\left[1 - \left(\frac{r}{a}\right)^2\right]$$

Use the potential energy principle to show that,

$$w_o = \frac{3}{32}\frac{pa^4}{(1+\nu)D_o}$$

P.3.7 To treat the problem of torsion of bars by Rayleigh-Ritz process we may proceed in one of ways,

1. Visualize the analogous soap film and apply the potential energy principle to find an approximation to its deflection.
2. Work in terms of Prandtl's stress function and apply the complementary energy principle.

Following the second approach, equations of compatibility result from,

$$\delta V^* = 0$$

Where,

$$V^* = \frac{l}{2G}\iint_{\text{section}}\left[\left(\frac{\partial \varnothing}{\partial x}\right)^2 + \left(\frac{\partial \varnothing}{\partial y}\right)^2 - 4G\alpha\phi\right]dx\,dy$$

For a rectangular section ($2a \times 2b$), obtain an approximate value for the torsion constant assuming a one term expansion for $\varnothing(x,y)$ in the Rayleigh-Ritz process. That is, assume,

$$\varnothing(x,y) = q_{11}\cos\frac{m\pi x}{2a}.\cos\frac{m\pi y}{2b}$$

Note, for a square section, side $2a$, Timoshenko and Goodier give the value $K = 2.2496a^4$ for the torsion constant. Compare with the value given by S_t Venant's approximate formula in the stress analysis notes.

P.3.8 A simply supported rectangular plate is side ratio b/a carries a single central concentrated load P. If it is decided to use an energy method, show that,

$$w(x,y) = \sum_{m=1}^{\infty}\sum_{n=1}^{\infty} a_{mn}\sin\frac{m\pi x}{a}\sin\frac{m\pi y}{b}$$

Is acceptable as an assumed deflection mode if the origin is taken at one corner.
Show that the center deflection is,

$$w_{\text{center}} = 0.173\frac{Pa^2}{Eh^3}$$

$E =$ Young's modulus, $h =$ plate thickness, Poisson's ratio $\nu = 0.3$

P.3.9 Asquare plate of side a simply supported along its edges carries a uniformly distributed load of intensity p per unit surface area. Use the principle of stationary total potential energy and a double sine series deflection mode to prove that the center deflection is,

$$w_{\text{center}} = 0.0442\frac{Pa^4}{Eh^3}$$

$E =$ Young's modulus, $h =$ plate thickness, Poisson's ratio $\nu = 0.3$

P.3.10 A simply supported rectangular plate supports an arbitrary distributed load $p(x,y)$. Represent this load by a double Fourier half rang sine series, then, assuming a deflection mode $w = \sum\sum a_{mn}\sin\frac{m\pi x}{a}\sin\frac{m\pi y}{b}$ obtain an expression for coefficient a_{mn} using the total potential energy principles.

P.3.11 The deflection at a point (*x*,*y*) on a plate due to a unit point load at a point (ξ,η) is called the influence function a (x,y,ξ,η). For a simply supported rectangular plate, sides $a \times b$, a deflection surface,

$$w = \sum_{n=1}^{\infty}\sum_{n=1}^{\infty} a_{mn}\sin\frac{m\pi x}{a}\sin\frac{n\pi y}{b}$$

Use the potential energy principle to show that,

$$w(x, y, \xi, \eta) = \frac{4}{\pi^4 Dab} \sum_{n=1}^{\infty} \sum_{n=1}^{\infty} \left(\frac{1}{\left(\frac{m}{a}\right)^2 + \left(\frac{n}{b}\right)^2} \right) \sin\frac{m\pi x}{a} \sin\frac{n\pi y}{b} \sin\frac{m\pi\xi}{a} \sin\frac{n\pi\eta}{b}$$

Notice that $w(x, y, \xi, \eta) = w(\xi, \eta, x, y)$ as we would expect from the reciprocal theorem.

P.3.12 A circular shaft, diameter $2a$, has a pair of flats machined opposite each other, the distance across the flats is $2b$. Show that by suitable choice of coordinates,

$$\varnothing(x, y) = \left(a^2 - x^2 - y^2\right)\left(b^2 - x^2\right) \sum a_{mn} x^m y^n$$

Is a reasonable stress function to use in the Rayleigh-Ritz technique?

Bibliography

[1] H.L. Langhaar, Energy Methods in Applied Mechanics, Wiley, New York, 1962.
[2] D.A. Saravanos, Integrated damping mechanics for thick composite laminates and plates, Journal of Applied Mechanics 61 (2) (1966) 375.
[3] J.S. Prezemieniecki, Theory of Matrix Structural Analysis, McGraw-Hill, New York, 1968.
[4] R.J. Roark, W.C. Young, Formulas for Stress and Strain, fifth ed., McGraw-Hill, New York, 1975.
[5] T.H. Richards, Energy Methods in Stress Analysis, Ellis Horwood Series in Engineering Science, Ellis Horwood, Chichester, 1977.
[6] M. Witts, K. Sobczyk, Dynamics response of laminated plates to random loading, Journal of Solids and Structures 16 (3) (1980) 231–238.
[7] J.N. Reddy, C.F. Liu, A higher–order shear deformation theory of laminated elastic shells, International Journal of Engineering Science 23 (3) (1985) 319–330.
[8] C. Lanczos, The Variational Principles of Mechanica, fourth ed., Dover, New York, 1986.
[9] G.W. Rowe, C.E.N. Sturgess, P. Hartley, I. Pillinger, Finite-element Plasticity and Metalforming Analysis, Cambridge University Press, Cambridge, 1991.
[10] M.J. Jweeg, S.Z. Said, Effect of rotational and geometric stiffness matrices on dynamic stresses and deformations of rotating blades, Journal of the Institution of Engineers, Mechanical Engineering Division 76 (1995) 29–38.
[11] E.J. Hearn, Mechanics of Materials, International Series on Materials Science and Technology I, Butterworth-Heinemann, Oxford, 1997.
[12] E.J. Hearn, Mechanics of Materials, International Series on Materials Science and Technology II, Butterworth-Heinemann, Oxford, 1997.
[13] J.N. Reddy, Mechanics of Laminated Composite Plates, Theory and Analysis, CRC Press, Boca Raton, 1997.
[14] A.C. Ugral, Stresses in Plates and Shells, seconded., McGraw-Hill, New York, 1999.
[15] J.N. Reddy, Theory and Analysis of Elastic Plates, Taylor & Francis, Philadelphia, 1999.
[16] M.J. Jweeg, S.K. Radi, Analysis of Composite Plates Subjected to Impact Loading, The National Engineering Conference, Kufa, Iraq, 2000.
[17] J.D. Hoffman, Numerical Methods for Engineers and Scientists, second ed.- Revised and Expanded, Marcel Dekker, New York, 2001.
[18] J.N. Reddy, Energy Principles and Variational Methods in Applied Mechanics, Wiley, New York, 2002.
[19] M.J. Jweeg, A.S. Hammood, M. Al-Waily, A suggestedanalytical solution of isotropic composite plate with crack effect, International Journal of Mechanical & Mechatronics Engineering 12 (5) (2012) 44–58.
[20] M.J. Jweeg, M. Al-Waily, A.A. Deli, Theoretical and numerical investigation of buckling of orthotropic hyper composite plates, International Journal of Mechanical & Mechatronics Engineering 15 (4) (2015) 1–12.
[21] M.J. Jweeg, Lecture Notes, Theory of Plates and Shells, Al-Nahrain University, University of Technology, and Military College of Engineering, Baghdad, Iraq, 2002.
[22] S.P. Timoshenko, J.M. Gere, Theory of Elastic Stability, second ed., McGraw-Hill, New York, 1961.
[23] V. Panc, Theory of Elastic Plates Plates, Noordhof, Leyden, 1975.
[24] E. Resissner, Reflections on the theory of elastic plates, Applied Mechanics Reviews 38 (11) (1985) 1453–1464.

CHAPTER

4

Energy methods in vibrations

In this chapter, the energy methods explained in the previous chapters are extended to deal with vibration problems and formulating the Eigen value problem. The methods will be presented and the formulation of total energy is achieved from which the frequency equations are deduced. Different examples are chosen to demonstrate the effectiveness of using energy methods in vibration analysis. The presented examples cover the application of Rsyleigh's principles to two degrees of freedom system, axial free vibration of beams, and cantilever beam bending vibrations. The frequency equations are formulated using Rayleigh Ritz from which the natural frequencies are obtained directly. The kinetic and strain energies were presented for plate bending applications presented in Chapter 3, Application of Energy Methods to Plate Problems. The applications cover rectangular plate shapes with different boundary conditions using Galerkin's, Galerkin's Vlasov and Ritz methods. The frequency equations will be derived for a vibrating taper beam and a non-uniform circular shaft carries a disk as practical applications. This chapter also includes problems covering different applications.

4.1 Rayleigh's method

Consider a continuous structure for example, a beam, let $y(y,t) = \varnothing(x).q(t)$, and assuming a harmonic function, then we can express,

$$\text{Strain energy}, V = \frac{1}{2}\bar{V}q^2 \tag{4.1}$$

$$\text{Kinetic energy}, T = \frac{1}{2}\bar{T}\dot{q}^2 \tag{4.2}$$

For a beam under bending, the strain energy is given by,

$$V = \frac{1}{2}EI\int_0^l y''^2 dx = \frac{1}{2}\left[EI\int_0^l \left[\varnothing''(x)\right]^2, dx\right]q^2 \tag{4.3}$$

DOI: https://doi.org/10.1016/B978-0-323-88666-6.00005-8

Where,

$$\bar{V} = EI\int_0^l \left[\varnothing''(x)\right]^2 dx$$

And the kinetic energy is given by,

$$T = \frac{1}{2}m\int_0^l \dot{y}^2 dx = \frac{1}{2}\left[m\int_0^l \varnothing^2(x)dx\right]\dot{q}^2.$$

Where,

$$\bar{T} = m\int_0^l \varnothing^2(x)dx$$

Where m, q, $\dot{q}$ are the mass per unit length, generalized displacement, and the generalized velocity, respectively.

If the system is conservative,

$$T_{\max} = V_{\max}$$

And since q is harmonic (Fig. 4.1), max. $q = \hat{q}$
Max. $\dot{q} = \omega\hat{q}$, this means that,

$$\frac{1}{2}\bar{T}\dot{q}^2_{\max} = \frac{1}{2}\bar{V}q^2_{\max}$$

$$\frac{1}{2}\bar{T}\omega^2\hat{q}^2 = \frac{1}{2}\bar{V}\hat{q}^2$$

Or,

$$\omega^2 = \frac{\bar{V}}{\bar{T}} \tag{4.4}$$

Where ω is the natural frequency.

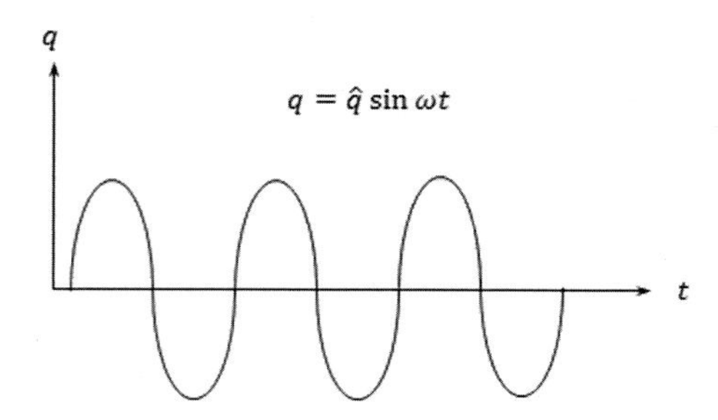

FIGURE 4.1 Displacement-time sinusoidal relation.

4.2 Rayleigh's energy theorem (R. Principle)

It is that the frequency of vibration of a conservative system vibrating in about an equilibrium position has a stationary (minimum) value with respect to vibrating shape in a neighborhood of fundamental mode.

Example 4.1: XXX

Consider the two degrees of freedom shown in Fig. 4.2, formulate the Eigen value problem.
Solution,
From Fig. 4.2, the potential and kinetic energy are given by,
Strain energy is given by,

$$2V = kq_1^2 + k(q_1 - q_2)^2 \tag{E1.1}$$

And, the potential energy is,

$$2T = m\dot{q}_1^2 + m\dot{q}_2^2 \tag{E1.2}$$

Let, the displacement field for the generalized displacements q_1 and q_2 are as follows,

$$q_1 = u_1 \sin \omega t, q_2 = u_2 \sin \omega t \tag{E1.3}$$

The velocity pattern for the generalized coordinate q_1 is as follows,

$$\dot{q}_1 = u_1\omega \cos \omega t, \text{and,} \tag{E1.4}$$

$$q_1(\text{max}) = u_1, \text{from which,} \tag{E1.5}$$

$$\dot{q}_1(\text{max}) = u_1\omega$$

Substitute the expressions of q_1, and q_2 form Eqs. (E1.3) and (E1.4) into Eq. (E1.1) gives,

$$2V = k\left[u_1^2 + (u_1 - u_2)^2\right] = u_2^2 k\left[\left(\frac{u_1}{u_2}\right)^2 + \left(\frac{u_1}{u_2} - 1\right)^2\right] \tag{E1.6}$$

Also, substitute the expressions of $\dot{q}_1$, and $\dot{q}_2$ form Eqs. (E1.3) and (E1.4) into Eq. (E1.2) gives,

$$2T_{\text{max}} = m\left[\omega^2 u_1^2 + \omega^2 u_2^2\right] = m\omega^2 u_2^2\left[\left(\frac{u_1}{u_2}\right)^2 - 1\right] \tag{E1.7}$$

We have,

$$2V_{\text{max}} = 2T_{\text{max}}$$

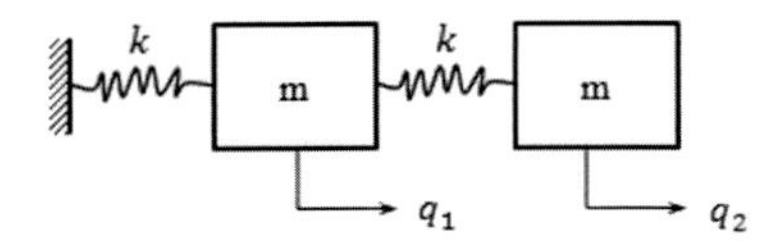

FIGURE 4.2 Two masses system.

Substitute their corresponding expressions from Eqs. (E1.6) and (E1.7) as follows,

$$u_2^2 k\left[\left(\frac{u_1}{u_2}\right)^2+\left(\frac{u_1}{u_2}-1\right)^2\right]=m\omega^2 u_2^2\left[\left(\frac{u_1}{u_2}\right)^2-1\right]$$

Hence,

$$\omega^2=\frac{k}{m}\frac{\left[\left(\frac{u_1}{u_2}\right)^2+\left(\frac{u_1}{u_2}-1\right)^2\right]}{\left[\left(\frac{u_1}{u_2}\right)^2-1\right]} \tag{E1.8}$$

The variation of natural frequency against the ratio $\left(\frac{u_1}{u_2}\right)$ is shown in Fig. 4.3 which indicate that the values of the natural frequencies are slightly higher than the exact value, $\omega_{\text{exact}} = 0.618$. The 'exact solution'.

Example 4.2: XXX

The simply supported beam shown in Fig. 4.4 is vibrating according to the shape given by.

$$y(x)=\left(\frac{x}{l}\right)\left(\frac{x}{l}-1\right) \tag{E2.1}$$

Use the Rayleigh's principle to find the natural frequency.
Solution,
Note, in a continuous problem, the geometric B.Cs,
$y = 0$ at $x = 0$ and $x = 1$ and the function prescribe the behavior is,

$$\psi(x)=\left(\frac{x}{l}\right)\left(\frac{x}{l}-1\right) \tag{E2.2}$$

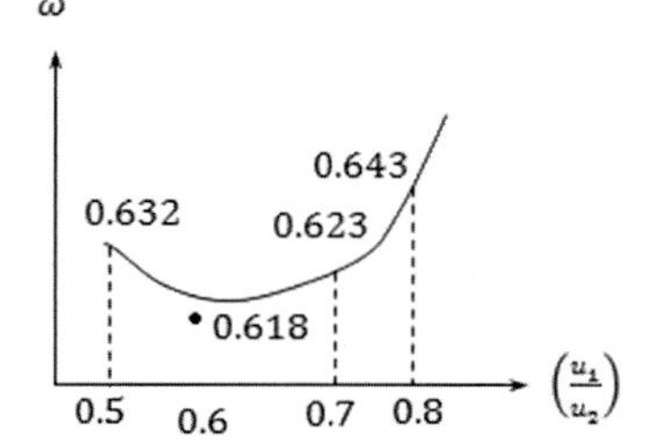

FIGURE 4.3 Frequency versus displacement ratio.

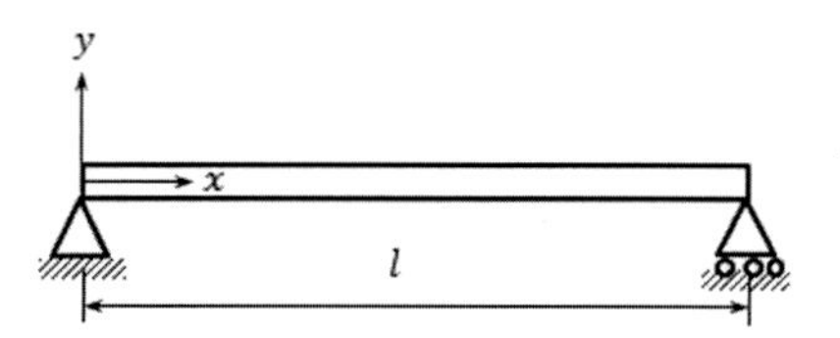

FIGURE 4.4 A simply supported beam.

From which,

$$\psi'' = 2/l^2 \tag{E2.3}$$

The potential and kinetic energy are given by,

$$\bar{V} = EI\int_0^l \psi''^2 dx = EI\int_0^l \frac{4}{l^4}dx = EI\frac{4}{l^3} \tag{E2.4}$$

And,

$$\bar{T} = m\int_0^l \psi^2 dx = m\int_0^l \left(\frac{x}{l}\right)^2\left(\frac{x}{l}-1\right)^2 dx = \frac{ml}{30} \tag{E2.5}$$

The natural frequency is given by,

$$\omega^2 = \frac{\bar{V}}{\bar{T}} = \frac{EI.4/l^3}{ml/30},$$

from which,

$$\omega = 10.95\sqrt{\frac{EI}{ml^4}}\text{rad/sec}$$

The exact solution from vibrations hand book, gives,

$$\omega = \text{exact for} \quad \text{S.S} = \pi^2\sqrt{\frac{EI}{ml^4}}$$

Which indicates a good agreement between Rayleigh's approximation and the exact solution with a percentage of discrepancy is %.

4.3 Rayleigh Ritz method (modified Ritz method)

Let,

$$\psi(x) = \sum_{i=1}^{N} a_i\psi_i(x) \tag{4.5}$$

a_i: are the generalized coordinates

We must to adjust vales of $a_1 \ldots a_N$ to make Rayleigh's Quotient a minimum,

$$Q = \frac{\bar{V}}{\bar{T}} = \frac{\bar{V}(a_1, \ldots, a_N)}{\bar{T}(a_1, \ldots, a_N)}$$

Therefore the minimum Q is given by $\frac{\partial Q}{\partial a_i} = 0, i = 1, 2, \ldots, N$
Hence,

$$Q = \frac{\overline{V}}{\overline{T}}, \frac{\partial Q}{\partial a_i} = \frac{\overline{T}\left(\partial \overline{V}/\partial a_i\right)}{\overline{T}^2} = 0 \tag{4.6}$$

Therefore,

$$\overline{T}\frac{\partial \overline{V}}{\partial a_i} = \overline{V}\frac{\partial \overline{T}}{\partial a_i}$$

Or,

$$\frac{\partial \overline{V}}{\partial a_i} = Q\frac{\partial \overline{T}}{\partial a_i}$$

Q here is our estimate of ω^2
The resulted equations will be in the form,

$$[A]\{a\} = \omega^2[B]\{a\} \tag{4.7}$$

The solution of the system of Eq. (4.7) gives the required natural frequencies.
For the cantilever beam shown in Fig. 4.5,
The fundamental frequency can be obtained by using energy methods as follows,

1. **Rayleigh's energy method,**
   We have from Eq. (4.6), the Rayleigh's Quotient is,

$$Q = \frac{\overline{V}[\psi]}{\overline{T}[\psi]} \tag{4.8}$$

For a uniform beam in flexure,

$$\overline{V}[\psi] = \frac{1}{2}EI\int_0^l \psi''^2 dx \tag{4.9}$$

$$\overline{T}[\psi] = \frac{1}{2}\rho A\int_0^l \psi^2 dx \tag{4.10}$$

a. Let us assume,

$$\psi = 1 - \cos\frac{\pi x}{2l} \tag{4.11}$$

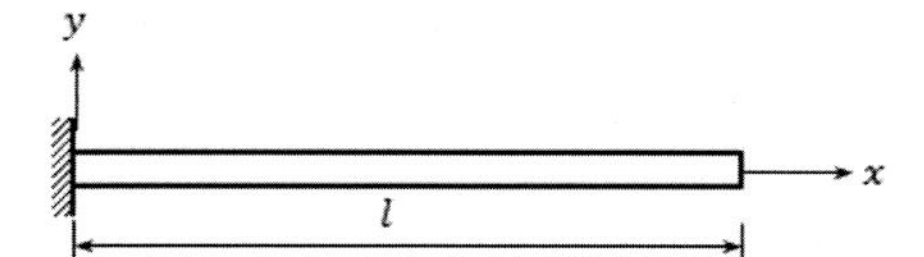

FIGURE 4.5 A cantilever beam.

This is an admissible function which satisfies the geometric boundary conditions but does not satisfy the requirement that the bending moment is zero at the free end of the beam, therefore,

$$\psi'' = \left(\frac{\pi}{2l}\right)^2 \cos\frac{\pi x}{2l}$$

And thus Eq. (4.9) becomes,

$$\bar{V}[\psi] = \frac{1}{2}EI\int_0^l \left(\frac{\pi}{2l}\right)^4 \cos^2\frac{\pi x}{2l}\,dx = \frac{EI\pi^4}{64l^3} \tag{4.12}$$

And Eq. (4.10) becomes,

$$\bar{T}[\psi] = \frac{1}{2}\rho A\int_0^l \left(1-\cos\frac{\pi x}{2l}\right)^2 dx = \frac{(3\pi-8)\rho A l}{4\pi} \tag{4.13}$$

Hence, Rayleigh's Quotient given in Eq. (4.8) is obtained by the substitution of $V[\psi]$ and $T[\psi]$ from Eqs. (4.12) and (4.13) respectively in Eq. (4.11) gives,

$$Q[\psi] = \frac{EI\pi^4}{64l^3}\cdot\frac{4\pi}{(3\pi-8)\rho A l} = 13.43\frac{EI}{\rho A l^4}$$

This value is 4.2% higher the exact value.

b. Let us assume,

$$\psi(x) = 3lx^2 - x^3. \tag{4.14}$$

This is the shape of the static deflection curve for a light beam with a concentrated mass at its tip. The function is an admissible function since it does not satisfy the condition that the shear force is zero at the free end of the beam. We have,

$$\psi'' = 6(l-x)$$

Hence Eq. (4.9) becomes,

$$\bar{V}[\psi] = \frac{1}{2}EI\int_0^l 36(l-x)^2 dx = 6EIl^3 \tag{4.15}$$

And Eq. (4.10) becomes,

$$\bar{T}[\psi] = \frac{1}{2}\rho A\int_0^l \left(3lx^2-x^3\right)^2 dx = \frac{33\rho A l^7}{70} \tag{4.16}$$

Hence, Rayleigh's Quotient given in Eq. (4.8) is obtained by the substitution of $V[\psi]$ and $T[\psi]$ from Eqs. (4.15) and (4.16) respectively in Eq. (4.11) gives,

$$Q[\psi] = 6EIl^3\frac{70}{33\rho A l^7} = 12.72\frac{EI}{\rho A l^4}$$

From which,

$$\omega_1^2 \leq 12.72\frac{EI}{\rho Al^4} \quad \text{or} \quad \omega_1 \leq 3.567\sqrt{\frac{EI}{\rho Al^4}}$$

This value is 1.46% higher than the exact value.

2. Rayleigh-Rita Method

Let the function $\varnothing$ be of the form,

$$\varnothing(x) = c_1\psi_1 + c_2\psi_2 \tag{4.17}$$

We have,

$$\bar{T}[\psi] = \frac{1}{2}\rho A\int_0^l (c_1\psi_1 + c_2\psi_2)^2 dx$$

Or,

$$= \frac{1}{2}\rho A\left\{c_1^2\int_0^l \psi_1^2 + 2c_1c_2\int_0^l \psi_1\psi_2 dx + c_2^2\int_0^l \psi_2^2\right\}$$

And,

$$\frac{\partial\bar{T}}{\partial c_1} = \frac{1}{2}\rho A\left\{2c_1\int_0^l \psi_1^2 + 2c_2\int_0^l \psi_1\psi_2 dx\right\} \tag{4.18}$$

$$\frac{\partial\bar{T}}{\partial c_2} = \frac{1}{2}\rho A\left\{2c_1\int_0^l \psi_1\psi_2 dx + 2c_2\int_0^l \psi_2^2\right\} \tag{4.19}$$

Similarly for $\frac{\partial\bar{V}}{\partial c_1}, \frac{\partial\bar{V}}{\partial c_2}$ leads to an equation of the form,

$$EI[a]\{c\} = \lambda^2\rho A[b]\{c\}$$

Where,

$$a_{11} = \int_0^l \psi''^2 dx, a_{12} = \int_0^l \psi_1''\psi_2'' dx \ldots \text{etc.}$$

$$b_{11} = \int_0^l \psi_1^2 dx, b_{12} = \int_0^l \psi_1\psi_2 dx \ldots \text{etc.}$$

Let us assume a series of the form,

$$\varnothing(x) = c_1\left(6l^2x^2 - 4lx^3 + x^4\right) + c_2\left(20l^3x^2 - 10l^3x^2 + x^5\right) \tag{4.20}$$

It this case ψ_1 is the deflected shape due to U.D.L and ψ_2 is the deflected shape due to the triangular load distribution. Thus ψ_1 and ψ_2 satisfy both the geometric and natural boundary conditions,

$$b_{11} = \int_0^l \psi_1^2 dx = l^9 \frac{104}{45}$$

$$b_{12} = b_{21} = \int_0^l \psi_1 \psi_2 dx = l^{10} \frac{2644}{315} \ldots \text{etc.}$$

Similarly, with all other coefficients. This leads to an Eigen value problem of the form,

$$EIl^5 \begin{bmatrix} \frac{144}{5} & 104l \\ 104l & \frac{2640}{7} l^2 \end{bmatrix} \begin{Bmatrix} c_1 \\ c_2 \end{Bmatrix} = \rho A l^9 \lambda^2 \begin{bmatrix} \frac{104}{45} & \frac{2644}{315} l \\ \frac{2644}{315} l & \frac{21128}{603} l^2 \end{bmatrix} \begin{Bmatrix} c_1 \\ c_2 \end{Bmatrix}$$

Solution of this Eigen value problem yields,

$$\lambda_1^2 = 12.37 \frac{EI}{\rho A l^4}, \lambda_2^2 = 518.8 \frac{EI}{\rho A l^4}$$

Therefore the natural frequencies are,

$$\omega_1 \leq 3.517 \sqrt{\frac{EI}{\rho A l^4}}$$

$$\omega_2 \leq 22.7 \sqrt{\frac{EI}{\rho A l^4}}$$

The first natural frequency is very accurate and the second natural frequency is 3% higher than the exact value.

Example 4.3: Show that if the deflection shape of a beam is assumed to be,

$$w(x) = \sum_{i=1}^{n} c_i . \varnothing_i(x) \tag{E3.1}$$

Then, Galerkin's method leads to,

$$\left([K] - \omega^2 [M]\right) \{\hat{q}\} = \{0\} \tag{E3.2}$$

Where $[K]$ is the stiffness matrix, $[M]$ is the mass matrix and $\{\hat{q}\}$ is the generalized coordinates.

For determining natural frequencies, the stiffness and mass matrices are calculated as follows,

$$k_{ij} = k_{ji} = \int_0^l EI \varnothing_i'' \varnothing_j'' dx \tag{E3.3}$$

$$m_{ij} = m_{ji} = \int_0^l m \varnothing_i \varnothing_j dx \tag{E3.4}$$

The mass per unit length is m and ω^2, EI are the natural frequency and the beam bending stiffness.

Solution,

We have from Eq. (E3.1),

$$w(x) = \sum_{i=1}^{n} c_i.\varnothing_i(x)$$

To be depend on time let,

$$w(x,t) = \sum_{i=1}^{n} c_i.\varnothing_i(x) \sin \omega t \tag{E3.5}$$

The disturbing force f is the inertia force $\text{I.F} = m\ddot{w}$ (E3.6)

G. method,

G.M. states that,

$$\int_0^l \left\{ L\left[\sum_{i=1}^{n} c_i.\varnothing_i(x)\right] - f \right\} \varnothing_j dx = 0 \tag{E3.7}$$

Where,

$$L = \frac{d^2}{dx^2}\left(EI \frac{d^2}{dx^2}\right)$$

Using the assumption of $w(x,t)$ in Eq. (E3.6) and substitute into Eq. (E3.7) gives,

$$\int_0^l \left\{ \frac{d^2}{dx^2} EI \frac{d^2}{dx^2} \left[\sum_{i=1}^{n} c_i.\varnothing_i(x) \sin \omega t\right] - m\omega^2 \sum_{i=1}^{n} c_i.\varnothing_i(x) \sin \omega t \right\} \varnothing_j dx = 0$$

Or,

$$\int_0^l \left\{ \left[\frac{d^2}{dx^2}\left(EI \sum_{i=1}^{n} c_i.\varnothing_i''(x)\right)\right] \varnothing_j - \left[m\omega^2 \sum_{i=1}^{n} c_i.\varnothing_i(x)\right] \varnothing_j \right\} dx = 0 \tag{E3.8}$$

Let,

$$\int_0^l \frac{d^2}{dx^2}\left(EI c_i.\varnothing_i''\right) \varnothing_j dx = k_{ij} = k_{ji}, \text{ stiffness coefficients} \tag{E3.9}$$

And let,

$$\int_0^l mc_i.\varnothing_i.\varnothing_j dx = m_{ij} = m_{ji}, \text{ mass coefficients} \tag{E3.10}$$

Form Eq. (E3.9),

$$k_{ij} = \int_0^l \underbrace{\frac{d^2}{dx^2}(EIc_i.\varnothing''_i)}_{dv} \underbrace{\varnothing_j}_{u} dx$$

Or,

$$k_{ij} = \underbrace{\left[\varnothing_j \frac{d}{dx}(EI.\varnothing''_i)\right]_0^l}_{\substack{=0 \\ \text{since } \varnothing''_i = 0 \text{ at } x=0|x=l}} - \int_{dx}^{d} \frac{d}{dx}\left(EI.\varnothing''_i \varnothing'_j\right) dx$$

Hence,

$$k_{ij} = -\int_0^l \frac{d}{dx}\left(EI.\varnothing''_i \varnothing'_j\right)$$

Carrying out the integrations of this equation gives,

$$= -\left|\varnothing'_j EI.\varnothing''_i\right|_0^l - \int_0^l EI.\varnothing''_i \varnothing''_j dx$$

In matrix form, the EVP will be in a form,

$$\left[\left[k_{ij}\right] - \omega^2\left[m_{ij}\right]\right]\{c\} = 0 \tag{E3.11}$$

The solution of the EVP (E3.11) results in the natural frequencies, and from this the mode shapes are obtained.

4.4 Plate applications

Consider a vibrating plate, the kinetic energy is given by,

$$T = \frac{1}{2}\iint_A \bar{m}(x,y)\left[\frac{\partial w(x,y,t)}{\partial t}\right]^2 dxdy \tag{4.21}$$

Or in terms of polar coordinates,

$$T = \frac{1}{2}\iint_A \bar{m}(r,\theta)\left[\frac{\partial w(r,\theta,t)}{\partial t}\right] r.dr.d\theta \tag{4.22}$$

$\bar{m}$: mass of the plate per unit area,

The total potential energy is given by,

$$V = \underset{\text{/strain energy}}{U} + \underset{\text{P.E}}{\Omega} + \underset{\text{K.E}}{T} \tag{4.23}$$

The strain energy and the potential energy are obtained as derived in Chapter 3, Application of Energy Methods to Plate Problems.

4.4.1 Rayleigh's method

To determine the lowest natural frequency, a similar procedure achieved in the determination of natural frequencies in beam problems is adopted here. This means that,

$$U_{b_{\max}} = T_{\max} \tag{4.24}$$

Assuming that the plate is undergoing harmonic oscillations,

$$w(x, y, t) = W(x, y)\sin \omega t \tag{4.25}$$

Where, $W(x, y)$: Shape function and ω is the un-known natural circular frequency.

Using Eq. (4.28), the kinetic energy is written as follows,

$$T = \frac{\omega^2}{2}\cos^2\omega t \iint_A \bar{m}\,(x, y)\, W^2(x, y)dxdy \tag{4.26}$$

We have,

$$\iint_A \bar{m}\,(x, y)\, W^2(x, y)dxdy \tag{4.27}$$

The K.E is at a maximum when the velocity of the plate is maximum which occurs when $w(x, y, t)$ is zero. This will be true if $\sin \omega t = 0$, this means that, $\omega t = n\pi (n = 0, 1, \ldots)$.

Therefore the maximum kinetic energy is given by,

$$T_{\max} = \frac{\omega^2}{2} \iint_A \bar{m}\,(x, y)\, W^2(x, y)dxdy = \frac{\omega^2}{2} g \tag{4.28}$$

Where $g = \iint_A \bar{m}(x, y)\, W^2(x, y)dxdy$

The bending part of the strain energy is maximum when the deflection of the plate is maximum. This occurs when $\omega t = 1$, this means that,

$$\omega t = \left(n + \frac{1}{2}\right)\pi, (n = 0, 1, 2, 3, \ldots) \tag{4.29}$$

The strain energy stored in the plate is given by,

$$U_{b_{\max}} = \frac{1}{2} \iint_A D\left\{ \left(\frac{\partial^2 w}{\partial x^2} + \frac{\partial^2 w}{\partial y^2}\right)^2 - 2(1 - \nu)\left(\frac{\partial^2 w}{\partial x^2}\frac{\partial^2 w}{\partial y^2} - \left(\frac{\partial^2 w}{\partial x \partial y}\right)^2\right) \right\} dxdy \tag{4.30}$$

In applying Rayleigh's method, we select an appropriate deflection shape, $w(x,y,t)$, then we equate $U_{b_{\max}} = T_{\max}$, results in,

$$\omega^2 = \frac{2U_{b_{\max}}}{\iint_A \bar{m}(x,y) W^2(x,y) dxdy} \tag{4.31}$$

The approximate natural frequencies calculated from Rayleigh's method are always higher than the exact values, since we have arbitrarily stiffened the plate by assuming a modal shape, thus increasing its frequencies.

4.4.2 Ritz method

In this method, the extended Rayleigh's method is achieved by including more than one parameter in the expression of the shape function. In this way, not only a more accurate value for the lowest natural frequency can be obtained, but also additional information concerning the higher frequencies and mode shapes.

Assume the shape function,

$$W(x,y) = c_1\psi_1(x,y) + c_2\psi_2(x,y) + \ldots + c_n\psi_n(x,y) \tag{4.32}$$

Where, $\psi_1(x,y), \psi_2(x,y), \ldots, \psi_n(x,y)$ are the appropriate displacement functions, which individually satisfy at least the geometric boundary conditions. Therefore in this respect,

$$\frac{\partial V}{\partial c_i} = 0, (i = 1, 2, 3 \ldots, n) \tag{4.33}$$

Where,

$$V = U_{b_{\max}} - T_{\max}$$

Or,

$$\frac{\partial U_b}{\partial c_1} - \frac{1}{2}\omega^2 \frac{\partial g}{\partial c_1} = 0$$

$$\frac{\partial U_b}{\partial c_2} - \frac{1}{2}\omega^2 \frac{\partial g}{\partial c_2} = 0$$

$$\cdots\cdots\cdots$$

$$\frac{\partial U_b}{\partial c_n} - \frac{1}{2}\omega^2 \frac{\partial g}{\partial c_n} = 0 \tag{4.34}$$

This will reduce the problem to an eigenvalue and eigenvector.

4.4.3 Galerkin-Vlasov method

Assume a shape function in the form of,

$$W(x,y) = c_1\psi_1(x,y) + c_2\psi_2(x,y) + \cdots + c_n\psi_n(x,y) \tag{4.35}$$

Which, satisfies term by term, all boundary conditions. For the plate problem, we have,

$$\iint_A \left[D\nabla^2\nabla^2 W(x,y) - \omega^2 \bar{m}\,(x,y)\right]\psi_1(x,y)dxdy = 0$$

$$\iint_A \left[D\nabla^2\nabla^2 W(x,y) - \omega^2 \bar{m}\,(x,y)\right]\psi_2(x,y)dxdy = 0$$

$$\cdots\cdots\cdots$$

$$\iint_A \left[D\nabla^2\nabla^2 W(x,y) - \omega^2 \bar{m}\,(x,y)\right]\psi_n(x,y)dxdy = 0 \tag{4.36}$$

Where D is the bending stiffness of the plate and the other terms carry usual meaning.

Let us represent the shape function $W(x,y)$ in an infinite series,

$$W(x,y) = \sum_n \sum_m W_{mn}\psi_{mn}(x,y) \tag{4.37}$$

$W_{mn} = 1$ normalized amplitudes of the free vibration modes and $\psi_{mn}(x,y)$ is the product of the eigenfunctions of lateral beam vibrations.

$$\psi_{mn}(x,y) = X_m(x).Y_n(y)\text{satisfy the B.Cs} \tag{4.38}$$

$X_m(x)$: the mth mode of a freely vibrating uniform beam, with a span a.
$Y_n(y)$: the nth node of a beam of length b.

Insert the above definitions into Eq. (4.36) gives,

$$\iint_A \left[D\nabla^2\nabla^2\psi_{mn}(x,y) - \omega^2 \bar{m}\,\psi_{mn}(x,y)\right]\psi_{mn}(x,y)dxdy \tag{4.39}$$

m, n can be take the values $1, 2, 3, \ldots$. By introducing the notations,

$$I_1 = \iint_A \left[\psi_{mn}(x,y)\nabla^2\nabla^2\psi_{mn}(x,y)\right]dxdy$$

$$I_2 = \iint_A \psi_{mn}^2 dxdy \tag{4.40}$$

An approximate analytical expression for the frequencies of the free flexural vibration of plates of uniform thickness is obtained,

$$\omega_{mn}^2 = \frac{I_1}{I_2}\frac{D}{\bar{m}} \tag{4.41}$$

For the rectangular plate,

$$I_1 = \int_0^b \int_0^a \left\{X_m(x).Y_n(y)\nabla^2\nabla^2\left[X_m(x).Y_n(y)\right]\right\}dxdy$$

$$= \int_0^b \int_0^a \left[X_m''''(x)Y_n(y) + 2X_m''(x)Y_n''(y) + X_m(x)Y_n''''(y)\right]X_m(x)Y_n(y)dxdy \tag{4.42}$$

$$I_2 = \iint_A \psi_{mn}^2(x,y)dxdy$$

$$I_2 = \int_0^b \int_0^a X_m^2(x)Y_n^2(y)dxdy \tag{4.43}$$

The integrations in Eqs. (4.41) and (4.43) can be broken into the following integrations,

$$I_3 = \int_0^a X_m^2(x)dx$$

$$I_4 = \int_0^b Y_n^2(y)dy$$

$$I_5 = \int_0^a X_m''''(x)X_m(x)dx$$

$$I_6 = \int_0^a X_m''(x)X_m(x)dx$$

$$I_7 = \int_0^b Y_n''(y)Y_n(y)dy$$

$$I_8 = \int_0^b Y_n''''(y)Y_n(y)dy \tag{4.44}$$

Using the expressions in Eq. (4.44), the circular frequencies of free vibration of plates become,

$$\omega_{mn} = \sqrt{\frac{I_4 I_5 + 2I_6 I_7 + I_3 I_8}{I_3 I_4}}\sqrt{\frac{D}{\overline{m}}} \tag{4.45}$$

Example 4.4: Find the lowest natural frequency of a rectangular plate clamped on all edges shown in Fig. 4.6 using the Ritz method and check the results by the Galerkin's method.

1. Ritz Method

Assume a suitable infinite series expression of the deflection which satisfies the geometrical boundary conditions and closely approximates the shape of the first mode of vibration, we use,

$$W(x,y) = C_1\psi_1(x,y) = C_1\left(x^2 - a^2\right)^2\left(y^2 - b^2\right)^2 \tag{E4.1}$$

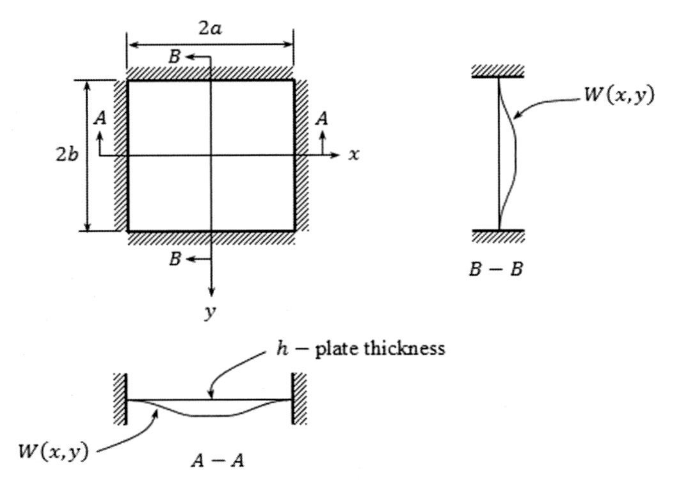

FIGURE 4.6 A rectangular plate fixed in all edges.

(first term only) static deflection formulas of uniformly distributed load for a beam clamped at both ends.

For this case, the strain energy of the plate is given by,

$$U_b = \frac{D}{2} \int_{-b}^{b} \int_{-a}^{a} \left(\frac{\partial^2 w}{\partial x^2} + \frac{\partial^2 w}{\partial y^2} \right)^2 dxdy$$

$$= C_1 D \frac{2^8}{5^2} a^9 b^9 \left(\frac{1}{a^4} + \frac{2^8}{3^2 * 7^2} \frac{1}{a^2 b^2} + \frac{1}{b^4} \right) \tag{E4.2}$$

And the kinetic energy is given by,

$$T_{\max} = \frac{1}{2} \omega_1^2 g = \frac{1}{2} \omega_1^2 \bar{m} \int_{-b}^{b} \int_{-a}^{a} W^2(x,y) dxdy = \frac{1}{2} \omega_1^2 \bar{m} \frac{2^4}{5^2} a^9 b^9$$

Where,

$$g = \frac{1}{2} \omega_1^2 \bar{m} \int_{-b}^{b} \int_{-a}^{a} W^2(x,y) dxdy \tag{E4.3}$$

Substitution the expression of W(x,y) of Eq. (E4.1) into Eq. (E4.3) gives,

$$g = \frac{1}{2} \omega_1^2 \bar{m} \int_{-b}^{b} \int_{-a}^{a} C_1 \left(x^2 - a^2 \right)^4 \left(y^2 - b^2 \right)^4 dxdy \tag{E4.4}$$

From minimization of the total energy,

$$\frac{\partial U_b}{\partial C_1} - \frac{1}{2} \omega_1^2 \frac{\partial g}{\partial C_1} = 0$$

Thus,

$$\frac{\partial U_b}{\partial C_1} = D\frac{2^8}{5^2}a^9b^9\left(\frac{1}{a^4} + \frac{2^8}{3^2 * 7^2}\frac{1}{a^2b^2} + \frac{1}{b^4}\right)$$

And,

$$\frac{\partial g}{\partial C_1} = \frac{2}{2}\omega_1^2\bar{m}\int_{-b}^{b}\int_{-a}^{a} C_1\left(x^2-a^2\right)^4\left(y^2-b^2\right)^4 \quad dxdy \tag{E4.5}$$

Carrying out the integrations of Eq. (E4.5), gives

$$\omega_1 = \sqrt[4]{\left(\frac{1}{a^4} + \frac{2^8}{3^2 * 7^2}\frac{1}{a^2b^2} + \frac{1}{b^4}\right)}\sqrt{\frac{D}{\bar{m}}}$$

Hence,

$$\omega_1 = \frac{9.09}{a^2}\sqrt{\frac{D}{\bar{m}}}a$$

This value agrees with the exact value with a percentage of discrepancy (1.04%).

2. Galerkin's Method

The equation of motion for plate bending is given by,

$$\int_{-b}^{b}\int_{-a}^{a}\left[D\nabla^2\nabla^2 W(x,y) - \omega_1^2\bar{m}\,W(x,y)\right]W(x,y)dxdy = 0 \tag{E4.6}$$

Using the shape function given by (E4.2), gives,

$$\int_{-b}^{b}\int_{-a}^{a} C_1\left\{D\left[24\left(y^2-b^2\right)^2 + 32\left(3x^2-b^2\right)^2 + 24\left(x^2-a^2\right)^2\right]\right.$$

$$\left. - \omega_1^2\bar{m}\left(x^2-a^2\right)^2\left(y^2-b^2\right)^2\right\}\left(x^2-a^2\right)^2\left(y^2-b^2\right)^2 dxdy = 0 \tag{E4.7}$$

Which after the evaluation of the definite integrals, becomes,

$$C_1\left\{\frac{D * 2^{15}}{3^2 * 5^2 * 7}\left[a^9b^5 + \frac{4}{7}a^7b^7 + a^5b^9\right] - \omega^2\frac{2^{16} * \bar{m}}{3^4 * 5^2 * 7^2}a^9b^9\right\}$$

Hence, the natural frequency is given by,

$$\omega = \sqrt[3]{\frac{7}{2}\left(\frac{1}{a^4} + \frac{4}{7}\frac{1}{a^2b^2} + \frac{1}{b^4}\right)}\sqrt{\frac{D}{\bar{m}}} \tag{E4.8}$$

The lowest circular frequency of square plates ($a = b$) is,

$$\omega = \frac{9}{a^2}\sqrt{\frac{D}{\bar{m}}}$$

Exact value of the natural frequency is,

$$\omega_{\text{exact}} = \frac{8.9965}{a^2}\sqrt{\frac{D}{\bar{m}}} \tag{E4.9}$$

A good agreement is obtained between the Galerkin's method and the exact solution.
Alternative deflection assumption using G.M.

$$W(x,y) = \left(C_1 + C_2x^2\right)\left(x^2-a^2\right)^2\left(y^2-b^2\right)^2 \tag{E4.10}$$

Using this shape function for the estimation of the natural frequencies in a procedure similar to that used for the estimation of the fundamental natural frequency by using this shape function (E4.10) and carrying out the differentiations as required by Eq. (E4.6) and rearranging to obtain the 2 unknowns C_1 and C_2. We have,

$$R(x,y) = D\nabla^4 W(x,y) - \omega^2\bar{m}W(x,y) \tag{E4.11}$$

And for G.M. it is required that,

$$R(x,y)\varnothing_1 = 0 \tag{E4.12}$$

$$R(x,y)\varnothing_2 = 0 \tag{E4.13}$$

Where,

$$\varnothing_1(x,y) = \frac{\partial W(x,y)}{\partial C_1} = \left(x^2-a^2\right)^2\left(y^2-b^2\right)^2 \tag{E4.14}$$

$$\varnothing_2(x,y) = \frac{\partial W(x,y)}{\partial C_2} = x^2\left(x^2-a^2\right)^2\left(y^2-b^2\right)^2 \tag{E4.15}$$

Substitution of $R(x,y)$ from Eq. (E4.11), $\varnothing_1(x,y)$ from Eq. (E4.14), $\varnothing_2(x,y)$ from Eq. (E4.15) into Eqs. (E4.12) and (E4.13) respectively as follows,

$$D\nabla^4 W(x,y) - \omega^2\,\bar{m}\,W(x,y).\left(x^2-a^2\right)^2\left(y^2-b^2\right)^2 = 0 \tag{E4.16}$$

$$D\nabla^4 W(x,y) - \omega^2\bar{m}W(x,y).\quad x^2\left(x^2-a^2\right)^2\left(y^2-b^2\right)^2 = 0 \tag{E4.17}$$

Carrying out the substitution of Eq. (E4.10) and differentiations required in Eqs. (E4.16) and (E4.17) and arranging them in matrix form gives,

$$D\begin{bmatrix} 0.81715*10^{11} & 0.1703816*10^{12} \\ 0.18633288*10^{12} & 0.96143*10^{13} \end{bmatrix}\begin{bmatrix} C_1 \\ C_2 \end{bmatrix}$$

$$= \omega^2 \bar{m} \begin{bmatrix} 0.6298*10^{12} & 0.14311*10^{13} \\ 0.143117*10^{13} & 0.82338*10^{13} \end{bmatrix} \begin{bmatrix} C_1 \\ C_2 \end{bmatrix} \tag{E4.18}$$

Comparison Eq. (E4.18) with EVP, gives,

$$D[K]\{C\} = \omega^2 \bar{m}[M]\{C\}$$

Or,

$$\frac{D}{\omega^2 \bar{m}}[K]\{C\} = [M]\{C\} \tag{E4.19}$$

Pre- multiplying Eq. (E4.19) by $[K]^{-1}$ gives,

$$\frac{D}{\omega^2 \bar{m}}[K]^{-1}[K]\{C\} = [K]^{-1}[M]\{C\}$$

Hence,

$$[K]^{-1}[M]\{C\} = \frac{1}{\omega^2}[I]$$

Or,

$$[A][M]\{C\} - \frac{1}{\omega^2}\frac{D}{\bar{m}} = 0$$

Where,

$$[A] = [K]^{-1} = \begin{bmatrix} 1.2752*10^{11} & -2.26002*10^{-13} \\ -2.471604*10^{-13} & 1.083407*10^{-13} \end{bmatrix}$$

Solving the EVP gives,

$$\omega_1 = 0.3601918\sqrt{\frac{D}{m}}, \omega_2 = 1.3608307\sqrt{\frac{D}{m}}$$

Where ω_1, and ω_2 are the first and the second natural frequencies.

Example 4.5: A cantilever beam shown in Fig. 4.7 is tapered such that,

$$EI = EI_o\left[1 - \frac{1}{2}\left(\frac{x}{l}\right)^2\right] \tag{E5.1}$$

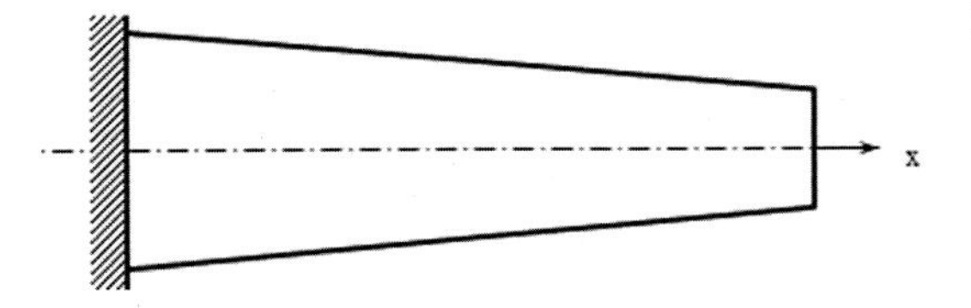

FIGURE 4.7 A cantilevered tapered beam.

Where x is measured from the clamped end. the beam caries a load having intensity,

$$p = p_o\left(1 - \frac{x}{l}\right) \tag{E5.2}$$

Assuming a deflection beam,

$$w(x, y) = c.l^2\left[\left(\frac{x}{l}\right)^2 - \frac{2}{3}\left(\frac{x}{l}\right)^3 + \frac{1}{6}\left(\frac{x}{l}\right)^4\right] \tag{E5.3}$$

1. Use Galerkin's method to obtain an expression for the tip deflection.
2. Determine an approximate value for the first natural frequency of beam over vibration of the cantilever using the dynamic equilibrium.

Solution,

1. for the deflection beam,

$$\frac{d^2}{dx^2}\left(EI\frac{d^2y}{dx^2}\right) = P \tag{E5.4}$$

Where,
$L[w] = f$ Disturbing function, and,

$$EI = EI_o\left[1 - \frac{1}{2}\left(\frac{x}{l}\right)^2\right], p = p_o\left(1 - \frac{x}{l}\right) \tag{E5.5}$$

The deflected shape in Eq. (E5.1) is,

$$w(x) = c.l^2\left[\left(\frac{x}{l}\right)^2 - \frac{2}{3}\left(\frac{x}{l}\right)^3 + \frac{1}{6}\left(\frac{x}{l}\right)^4\right]$$

Where,

$$\varnothing_j(x) = l^2\left[\left(\frac{x}{l}\right)^2 - \frac{2}{3}\left(\frac{x}{l}\right)^3 + \frac{1}{6}\left(\frac{x}{l}\right)^4\right] \tag{E5.6}$$

Using G.M. which states that,

$$\int_0^l \left\{L\left[\sum_{k=1}^{n} c_k \varnothing_k\right] - P\right\}\varnothing_j dx = 0 \tag{E5.7}$$

Using Eqs. (E5.1), (E5.4), and (E5.5) into Eq. (E5.7) gives,

$$c.l^2 \int_0^l \frac{d^2}{dx^2}\left(\underbrace{EI_0\left[1 - \frac{1}{2}\left(\frac{x}{l}\right)^2\right]\frac{d^2}{dx^2}\left[\left(\frac{x}{l}\right)^2 - \frac{2}{3}\left(\frac{x}{l}\right)^3 + \frac{1}{6}\left(\frac{x}{l}\right)^4\right]}_{\text{first part}} - \underbrace{p_0\left(1 - \frac{x}{l}\right)}_{\text{second part}}\right)\varnothing_j dx = 0 \tag{E5.8}$$

We have,

$$\frac{d}{dx}\left(\frac{2x}{l^2}-2\frac{x^2}{l^3}+\frac{4}{6}\frac{x^3}{l^4}\right)=\frac{2}{l^2}-4\frac{x}{l^3}+2\frac{x^2}{l^4} \tag{E5.9}$$

Therefore the first part becomes,

$$\frac{d^2}{dx^2}EI_0\left[1-\frac{1}{2}\left(\frac{x}{l}\right)^2\right]\left(\frac{2}{l^2}-4\frac{x}{l^3}+2\frac{x^2}{l^4}\right)$$

Or,

$$=EI_0\frac{d^2}{dx^2}\left(\frac{2}{l^2}-4\frac{x}{l^3}+2\frac{x^2}{l^4}-\frac{x^2}{l^4}+2\frac{x^3}{l^5}-\frac{x^4}{l^6}\right)=EI_0\left(\frac{4}{l^4}-\frac{2}{l^4}+12\frac{x}{l^5}-12\frac{x^2}{l^6}\right)$$

$$=EI_0\left(\frac{2}{l^4}+12\frac{x}{l^5}-12\frac{x^2}{l^6}\right)=EI_0\varnothing_j\left(\frac{2}{l^4}+12\frac{x}{l^5}-12\frac{x^2}{l^6}\right) \tag{E5.10}$$

Substitution of the expression of $\varnothing_j$ from Eq. (E5.6) into Eq. (E5.10) gives,

$$=EI_0\left(c.l^2\right)\left[\left(\frac{x}{l}\right)^2-\frac{2}{3}\left(\frac{x}{l}\right)^3+\frac{1}{6}\left(\frac{x}{l}\right)^4\right]\left[\frac{2}{l^4}+12\frac{x}{l^5}-12\frac{x^2}{l^6}\right] \tag{E5.11}$$

First part of the Eq. (E5.8)

$$\int_0^l \underbrace{EI_0\frac{d^2}{dx^2}\left(\frac{2}{l^2}-4\frac{x}{l^3}+2\frac{x^2}{l^4}-\frac{x^2}{l^4}+2\frac{x^3}{l^5}-\frac{x^4}{l^6}\right)}_{EI_0\left(\frac{4}{l^4}-\frac{2}{l^4}+12\frac{x}{l^5}-12\frac{x^2}{l^6}\right)}\varnothing_j dx \tag{E5.12}$$

Substitute the expression of Eq. (E5.6) into Eq. (E5.12) gives,

$$\int_0^l EI_0c.l^2\left\{\frac{2}{l^4}\left(\frac{x}{l}\right)^2+12\frac{x^3}{l^7}-12\frac{x^4}{l^8}-\frac{4}{3}\left(\frac{x}{l}\right)^3\frac{1}{l^4}-8\frac{x^4}{l^8}+8\frac{x^5}{l^9}+\frac{1}{3l^4}\left(\frac{x}{l}\right)^4+2\frac{x^5}{l^9}-\frac{x^6}{l^{10}}\right\}dx$$

Integration gives,

$$=EI_0c^2l^4\left|\left[\frac{2}{l^4}\frac{x^3}{3.l^3}+\frac{12}{4}\frac{x^4}{l^7}-\frac{12}{5}\frac{x^5}{l^8}-\frac{4}{4x^3}\frac{x^4}{l^7}-\frac{8}{5}\frac{x^5}{l^8}+\frac{8}{6}\frac{x^6}{l^9}+\frac{1}{12l^4}\frac{x^5}{l^4}+\frac{2}{6}\frac{x^6}{l^9}-\frac{2}{7}\frac{x^7}{l^{10}}\right]\right|_0^l$$

Hence,

$$=EI_0c^2l^4\left\{\frac{2}{3}\frac{1}{l^3}+\frac{3}{l^3}-\frac{12}{5}\frac{1}{l^3}-\frac{1}{3}\frac{1}{l^3}-\frac{8}{5}\frac{1}{l^3}+\frac{8}{6}\frac{1}{l^3}+\frac{1}{12}\frac{1}{l^3}+\frac{1}{3}\frac{1}{l^3}-\frac{2}{7}\frac{1}{l^3}\right\}$$

$$=EI_0c^2l^4\frac{1}{l^3}\left\{\frac{2}{3}+3-\frac{12}{5}-\frac{8}{5}+\frac{4}{3}+\frac{1}{12}-\frac{2}{7}\right\}$$

$$=EI_0c^2l^4\frac{1}{l^3}\left[\frac{6}{3}-\frac{20}{5}+3+\frac{1}{12}-\frac{2}{7}\right]=EI_0c^2l\left[1+\frac{1}{12}-\frac{2}{7}\right]=EI_0c^2l\left[\frac{12*7+7-24}{12*7}\right]$$

The result of first part of Eq. (E5.6) is,

$$= EI_0c^2l*0.797 \tag{E5.13}$$

To find the integration of second term in Eq. (E5.8),

$$\int_0^l p_0\left(1-\frac{x}{l}\right)\varnothing_j dx \tag{E5.14}$$

The substitution of the expression of Eq. (E5.6) into Eq. (E5.14) gives,

$$c.l^2p_0\int_0^l\left(1-\frac{x}{l}\right)\left(\left(\frac{x}{l}\right)^2-\frac{2}{3}\left(\frac{x}{l}\right)^3+\frac{1}{6}\left(\frac{x}{l}\right)^4\right)dx$$

Multiplication the terms in brackets gives,

$$c.l^2p_0\int_0^l\left(\left(\frac{x}{l}\right)^2-\frac{2}{3}\left(\frac{x}{l}\right)^3+\frac{1}{6}\left(\frac{x}{l}\right)^4-\left(\frac{x}{l}\right)^3+\frac{2}{3}\left(\frac{x}{l}\right)^4-\frac{1}{6}\left(\frac{x}{l}\right)^5\right)dx$$

Carrying out the above integration gives,

$$= c.l^2p_0\left[\frac{x^3}{3.l^2}-\frac{2}{12}\frac{x^4}{l^3}+\frac{1}{30}\frac{x^5}{l^4}-\frac{1}{4}\frac{x^4}{l^3}+\frac{2}{15}\frac{x^5}{l^4}-\frac{1}{36}\frac{x^6}{l^5}\right]_0^l$$

Or,

$$= c.l^2p_0\left[\frac{1}{3}l-\frac{2}{12}l+\frac{1}{30}l-\frac{1}{4}l+\frac{2}{15}l-\frac{1}{36}l\right] = c.l^3p_0\left[\frac{1}{3}-\frac{2}{12}+\frac{1}{30}-\frac{1}{4}+\frac{2}{15}-\frac{1}{36}\right]$$

The final result of the integration of the second part is,

$$= c.l^3p_0\frac{10}{180} = 0.0555\,c.l^3p_0 \tag{E5.15}$$

Therefore the expressions of Eqs. (E5.13) and (E5.15) are substituted into Eq. (E5.8) to give,

$$EI_0c^2l*0.797 - 0.0555\,c.l^3p_0 = 0$$

From which, $c = \frac{0.0555}{0.797}\frac{p_0l^2}{EI_0}$
Therefore Eq. (E5.1) becomes,

$$w(x) = 0.073\frac{p_0l^4}{EI_0}\left[\left(\frac{x}{l}\right)^2-\frac{2}{3}\left(\frac{x}{l}\right)^3+\frac{1}{6}\left(\frac{x}{l}\right)^4\right]$$

The tip deflection is,

$$w(l) = w_{\text{tip}} = 0.073\frac{p_0l^4}{EI_0}\left(1-\frac{2}{3}+\frac{1}{6}\right)$$

$$= 0.0365\frac{p_0l^4}{EI_0}$$

2. Deflection of beam. Using the dynamic equilibrium,

$$\frac{\partial^2}{\partial x^2}\left(EI\frac{\partial^2 w}{\partial x^2}\right) = m\frac{\partial^2 w}{\partial t^2} \tag{E5.16}$$

We have,

$$w(x,t) = c.l^2\left[\left(\frac{x}{l}\right)^2 - \frac{2}{3}\left(\frac{x}{l}\right)^3 + \frac{1}{6}\left(\frac{x}{l}\right)^4\right]\sin\omega t \tag{E5.17a}$$

And,

$$\frac{\partial^2 w}{\partial t^2} = -\omega^2 c.l^2\left[\left(\frac{x}{l}\right)^2 - \frac{2}{3}\left(\frac{x}{l}\right)^3 + \frac{1}{6}\left(\frac{x}{l}\right)^4\right]\sin\omega t$$

Therefore Eq. (E5.16) becomes,

$$\left(\begin{array}{c} c.l^2\dfrac{d^2}{dx^2}\displaystyle\int_0^l\left\{EI_0\left(1-\frac{1}{2}\left(\tfrac{x}{l}\right)^2\right)\frac{d^2}{dx^2}\left[\left(\tfrac{x}{l}\right)^2-\frac{2}{3}\left(\tfrac{x}{l}\right)^3+\frac{1}{6}\left(\tfrac{x}{l}\right)^4\right]\right\}+ \\ \displaystyle\int_0^l m\omega^2 c.l^2\left[\left(\tfrac{x}{l}\right)^2-\frac{2}{3}\left(\tfrac{x}{l}\right)^3+\frac{1}{6}\left(\tfrac{x}{l}\right)^4\right]\varnothing_j dx \end{array}\right) = 0 \tag{E5.17b}$$

Using Galerkin's method to find fundamental natural frequency, we use,

$$\int_0^l\left[\frac{\partial^2}{\partial x^2}\left(EI\frac{\partial^2 w}{\partial x^2}\right) + m\frac{\partial^2 w}{\partial t^2}\right]\varnothing_j dx = 0 \tag{E5.18}$$

For the vibration of beam, the equation of motion is,

$$\frac{\partial^2}{\partial x^2}\left(EI\frac{\partial^2 w}{\partial x^2}\right) = p(x,t) - m\frac{\partial^2 w}{\partial t^2} \tag{E5.19}$$

By using Galerkin's method,

$$L[w] = f$$

Where,

$$EI = EI_0\left[1 - \frac{1}{2}\left(\frac{x}{l}\right)^2\right], p = p_0\left(1 - \frac{x}{l}\right).\sin\omega t \tag{E5.20}$$

Let,

$$w(x,t) = c.l^2\left[\left(\frac{x}{l}\right)^2 - \frac{2}{3}\left(\frac{x}{l}\right)^3 + \frac{1}{6}\left(\frac{x}{l}\right)^4\right]\sin\omega t$$

And,

$$\ddot{w(x,t)} = -\omega^2 c.l^2\left[\left(\frac{x}{l}\right)^2 - \frac{2}{3}\left(\frac{x}{l}\right)^3 + \frac{1}{6}\left(\frac{x}{l}\right)^4\right]\sin\omega t \tag{E5.21}$$

Therefore,

$$\int_0^l \left\{L\left[\sum_{k=1}^{n} c_k \varnothing_k - p + m\frac{\partial^2 w}{\partial t^2}\right]\right\}\varnothing_j dx = 0 \tag{E5.22}$$

Where,

$$L = \frac{\partial^2}{\partial x^2}\left(EI\frac{\partial^2}{\partial x^2}\right)$$

Therefore Eq. (E5.17b) becomes,

$$c.l^2\int_0^l \left(\frac{d^2}{dx^2}\left\{EI_0\left(1-\frac{1}{2}\left(\frac{x}{l}\right)^2\right)\frac{d^2}{dx^2}\left[\left(\frac{x}{l}\right)^2 - \frac{2}{3}\left(\frac{x}{l}\right)^3 + \frac{1}{6}\left(\frac{x}{l}\right)^4\right]\right\}\sin\omega t - p_0\left(1-\frac{x}{l}\right).\sin\omega t + m\ddot{w}\right)\varnothing_j dx = 0 \tag{E5.23}$$

We have,

$$\frac{d^2}{dx^2}\left[\left(\frac{x}{l}\right)^2 - \frac{2}{3}\left(\frac{x}{l}\right)^3 + \frac{1}{6}\left(\frac{x}{l}\right)^4\right]$$

$$= \frac{d}{dx}\left(\frac{2x}{l^2} - 2\frac{x^2}{l^3} + \frac{4}{6}\frac{x^3}{l^4}\right) = \frac{2}{l^2} - 4\frac{x}{l^3} + 2\frac{x^2}{l^4}$$

Consider first part of Eq. (E5.23),

$$c.l^2\int_0^l \left[\frac{d^2}{dx^2}EI_0\left(1-\frac{1}{2}\left(\frac{x}{l}\right)^2\right)\left(\frac{2}{l^2} - 4\frac{x}{l^3} + 2\frac{x^2}{l^4}\right)\right]\varnothing_j dx$$

Or,

$$= c.l^2\int_0^l \left[\frac{d^2}{dx^2}EI_0\left(\frac{2}{l^2} - 4\frac{x}{l^3} + 2\frac{x^2}{l^4} - \frac{x^2}{l^4} + 2\frac{x^3}{l^5} - \frac{x^4}{l^6}\right)\right]\varnothing_j dx$$

Simplification gives,

$$= c.l^2\int_0^l EI_0\left[\frac{4}{l^4} - \frac{2}{l^4} + 12\frac{x}{l^5} - 12\frac{x^2}{l^6}\right]\varnothing_j dx \tag{E5.24}$$

And

$$\varnothing_j(x,t) = l^2\left[\left(\frac{x}{l}\right)^2 - \frac{2}{3}\left(\frac{x}{l}\right)^3 + \frac{1}{6}\left(\frac{x}{l}\right)^4\right]\sin\omega t \tag{E5.25}$$

Substitute the expression in Eq. (E5.25) into Eq. (E5.24) gives,

$$= c.l^4 \sin \omega t \int_0^l EI_0 \left[\frac{2}{l^4} + 12\frac{x}{l^5} - 12\frac{x^2}{l^6}\right]\left[\left(\frac{x}{l}\right)^2 - \frac{2}{3}\left(\frac{x}{l}\right)^3 + \frac{1}{6}\left(\frac{x}{l}\right)^4\right] dx$$

Multiplication and simplification give,

$$= c.l^4 EI_0 \sin \omega t \int_0^l \left[\frac{2}{l^6}x^2 - \frac{4}{3l^3}x^3 + \frac{1}{3l^8}x^4 + \frac{12}{l^7}x^3 - \frac{4}{l^8}x^4 + \frac{2}{l^9}x^5 - \frac{12}{l^8}x^4 + \frac{8}{l^9}x^5 - \frac{2}{l^{10}}x^6\right] dx$$

Integrations give,

$$= c.l^4 EI_0 \sin \omega t \left[\frac{2}{3l^6}l^3 - \frac{4}{12l^7}l^4 + \frac{1}{15l^8}l^5 + \frac{12}{4l^7}l^4 - \frac{8}{5l^8}l^5 + \frac{2}{6l^9}l^6 - \frac{12}{5l^8}l^5 + \frac{8}{6l^9}l^6 - \frac{2}{7l^{10}}l^7\right]$$

Or,

$$= c.l^4 EI_0 \frac{1}{l^3} \sin \omega t \left[\frac{2}{3} - \frac{1}{3} + \frac{1}{15} + 3 - \frac{8}{5} + \frac{1}{3} - \frac{12}{5} + \frac{4}{3} - \frac{2}{7}\right]$$

$$= 0.797\ c.l^4 EI_0 \frac{1}{l^3} \sin \omega t \tag{E5.26}$$

Consider the second integration part of Eq. (E5.23),

$$\int_0^l -p_0\left(1 - \frac{x}{l}\right)\varnothing_j dx = 0.0555 p_0 l^3, \text{From last integration} \tag{E5.27}$$

Consider the third integration part of Eq. (E5.23),

$$\int_0^l m\ddot{w} \quad \varnothing_j dx$$

Substitution the expression of $\ddot{w}(x,t)$ from Eq. (E5.21) into the above equation gives,

$$= \int_0^l m\left[-\omega^2 l^2\right]\left[\left(\frac{x}{l}\right)^2 - \frac{2}{3}\left(\frac{x}{l}\right)^3 + \frac{1}{6}\left(\frac{x}{l}\right)^4\right] \sin \omega t \varnothing_j\ dx$$

And the substitution of $\varnothing_j$ from Eq. (E5.25) into the above equation gives,

$$= -m\omega^2 l^4 \int_0^l \left[\left(\frac{x}{l}\right)^2 - \frac{2}{3}\left(\frac{x}{l}\right)^3 + \frac{1}{6}\left(\frac{x}{l}\right)^4\right]^2 dx$$

Or,

$$= -m\omega^2 l^4 \int_0^l \left[\frac{x^4}{l^4} + \frac{4}{9}\frac{x^6}{l^6} + \frac{1}{36}\frac{x^8}{l^8} - \frac{4}{3}\frac{x^5}{l^5} + \frac{2}{6}\frac{x^6}{l^6} - \frac{4}{18}\frac{x^7}{l^7}\right] dx$$

Carrying out the integrations give,

$$= -m\omega^2 l^4\left[\frac{1}{5}\frac{l^5}{l^4}+\frac{4}{9*7}\frac{l^7}{l^6}+\frac{1}{9*36}\frac{l^9}{l^8}-\frac{4}{6*3}\frac{l^6}{l^5}+\frac{2}{6*7}\frac{l^7}{l^6}+\frac{2}{6*7}\frac{l^7}{l^6}-\frac{4}{3*6*8}\frac{l^8}{l^7}\right]$$

Or,

$$= -m\omega^2 l^5 * 0.159 * c \tag{E5.28}$$

Substitution the expressions of Eqs. (E5.26), (E5.27), and (E5.28) into Eq. (E5.23) gives,

$$c.l*0.797 + 0.0555\, p_0 l^3 - m\omega^2 l^5 * 0.159 * c = 0$$

Or,

$$c = \frac{0.0555\, p_0 l^2}{(m\omega^2 l^4 * 0.159 - 0.797)} \tag{E5.29}$$

Therefore the deflected shape is obtained by substitution the value of c from Eq. (E5.29) into Eq. (5.17a) as follows,

$$w(x,t) = \frac{0.0555\, p_0 l^2}{(m\omega^2 l^4 * 0.159 - 0.797)}.l^2\left[\left(\frac{x}{l}\right)^2 - \frac{2}{3}\left(\frac{x}{l}\right)^3 + \frac{1}{6}\left(\frac{x}{l}\right)^4\right]\sin\omega t \tag{E5.30}$$

Example 4.6: Show that if the deflection shape of a beam is assumed to be,

$$w(x) = \sum_{i=1}^{n} c_i.\varnothing_i(x) \tag{E6.1}$$

Then, Galerkin's method leads to,

$$\left([K] - \omega^2[M]\right)\{\hat{q}\} = \{0\} \tag{E6.2}$$

The Eigen Value Problem for determining the natural frequencies, and,

$$k_{ij} = k_{ji} = \int_0^l EI\varnothing_i''\varnothing_j''\, dx \tag{E6.3}$$

$$m_{ij} = m_{ji} = \int_0^l m\varnothing_i\varnothing_j dx \tag{E6.4}$$

The mass per unit length is m and ω^2, EI have the usual meanings.
Solution,
We have from Eq. (E6.1),

$$w(x) = \sum_{i=1}^{n} \text{To be depend on time let,}$$

To be depend on time let,

$$w(x,t)=\sum_{i=1}^{n}c_i.\varnothing_i(x)\sin\omega t \tag{E6.5}$$

$f=$ disturbing function $=I.F=m\ddot{w}$
Galerkin's method requires that,

$$\int_0^l\left\{L\left[\sum_{i=1}^{n}c_i.\varnothing_i(x)\right]-f\right\}\varnothing_j dx=0 \tag{E6.6}$$

Where,

$$L=\frac{d^2}{dx^2}\left(EI\frac{d^2}{dx^2}\right)$$

Therefore Eq. (E6.6) becomes,

$$\int_0^l\left\{\frac{d^2}{dx^2}EI\frac{d^2}{dx^2}\left[\sum_{i=1}^{n}c_i.\varnothing_i(x)\sin\omega t\right]-m\omega^2\sum_{i=1}^{n}c_i.\varnothing_i(x)\sin\omega t\right\}\varnothing_j dx=0$$

Or,

$$\int_0^l\left\{\left[\frac{d^2}{dx^2}\left(EI\sum_{i=1}^{n}c_i.\varnothing_i''(x)\right)\right]\varnothing_j-\left[m\omega^2\sum_{i=1}^{n}c_i.\varnothing_i(x)\right]\varnothing_j\right\}dx=0 \tag{E6.7}$$

Let,

$$\int_0^l\frac{d^2}{dx^2}\left(EIc_i.\varnothing_i''\right)\varnothing_j dx=k_{ij}=k_{ji},\ \text{stiffness coefficient.} \tag{E6.8}$$

And let,

$$\int_0^l mc_i.\varnothing_i.\varnothing_j dx=m_{ij}=m_{ji},\ \text{mass coefficient.} \tag{E6.9}$$

From Eqs. (E6.8) and (E6.9), we have,

$$k_{ij}=\int_0^l\underbrace{\frac{d^2}{dx^2}(EIc_i.\varnothing''_i)}_{dv}\ \underbrace{\varnothing_j}_{u}\ dx$$

$$k_{ij} = \underbrace{\left[\varnothing_j \frac{d}{dx}(EI.\varnothing''_i)\right]_0^l}_{\substack{=0 \\ \text{since } \varnothing''_i = 0 \text{ at } x=0|x=l}} - \int_0^l \frac{d}{dx}\left(EI.\varnothing''_i \varnothing'_j\right) dx$$

Or,

$$k_{ij} = -\int_0^l \frac{d}{dx}\left(EI.\varnothing''_i \varnothing'_j\right)$$

$$= -\left|\varnothing'_j EI.\varnothing''_i\right|_0^l - \int_0^l EI.\varnothing''_i \varnothing''_j dx$$

$$\left[\left[k_{ij}\right] - \omega^2\left[m_{ij}\right]\right]\{c\} = 0 \tag{E6.10}$$

Eq. (E6.10) represents the Eigen value problem and the solution gives the natural frequencies of the problem.

4.5 Application to the governing differential equation of plates

The differential equation governing the pure bending of plates subjected to lateral static loading is to be found in numerous books pertaining to the analysis of plates. It is attributed to Lagrange and its development is presented in detail by Timoshenko.

$$\frac{\partial^4 w(x,y)}{\partial x^4} + 2\frac{\partial^4 w(x,y)}{\partial x^2 \partial y^2} + \frac{\partial^4 w(x,y)}{\partial y^4} = \frac{p}{D}$$

In free vibration of plate will be no surface loading (p) there will be, however, exist an inertial body force which must be taken into consideration. Thus we obtain the governing differential equation,

$$D\left[\frac{\partial^4 w}{\partial x^4} + 2\frac{\partial^4 w}{\partial x^2 \partial y^2} + \frac{\partial^4 w}{\partial y^4}\right] + \rho h \frac{\partial^2 w}{\partial t^2} = 0 \tag{4.46}$$

Strain Energy

$$V = \int_0^a \int_0^a \frac{Eh}{2(1-\nu^2)}\left[\varepsilon_x^2 + \varepsilon_y^2 + 2\nu\varepsilon_x\varepsilon_y + \frac{1}{2}(1-\nu)\varepsilon_{xy}^2\right] dydx$$

$$= \frac{D}{2}\int_0^a\int_0^a \left[\left(\frac{\partial^2 w}{\partial x^2}\right)^2 + \left(\frac{\partial^2 w}{\partial y^2}\right)^2 + 2\nu\left(\frac{\partial^2 w}{\partial x^2}\right)\left(\frac{\partial^2 w}{\partial y^2}\right) + 2(1-\nu)\left(\frac{\partial^2 w}{\partial x \partial y}\right)^2\right] dydx \tag{4.47}$$

Where,
$D = \frac{Eh^3}{12(1-\nu^2)}$, Bending stiffness of the plate.
Kinetic Energy

$$T = \frac{1}{2}\rho h \int_0^a \int_0^b \left(\frac{\partial w}{\partial t}\right)^2 dydx \tag{4.48}$$

Functions and Boundary Conditions
It is assumed that,

$$w(x,y) = \sum_{i=1}^{I}\sum_{j=1}^{I} c.\phi_i(x).\psi_j(y) \tag{4.49}$$

$\phi_i(x)$ and $\psi_j(y)$ must satisfy any geometric boundary condition on the edges ($x=0$, $x=a$) and ($y=0$, $y=a$), respectively.

Consider $\phi_i(x)$ and $\psi_j(y)$ for plate are the same appropriate beam shape function which satisfy all these conditions.

For simply supported edges in x and y,

$$\phi(x) = \sin\frac{\pi x}{a} \tag{4.50}$$

$$\psi(y) = \sin\frac{\pi y}{a} \tag{4.51}$$

These functions are normalized and orthogonal,

$$\int_0^a \phi_i.\phi_n = \begin{cases} 0 & \text{if } i \neq n \\ \frac{a}{2} & \text{if } i = n \end{cases} \tag{4.52}$$

Also,

$$\int_0^a \psi_j.\psi_n = \begin{cases} 0 & \text{if } j \neq n \\ \frac{a}{2} & \text{if } j = n \end{cases} \tag{4.53}$$

For fixed edges in x and y which satisfy the geometric boundary conditions,

$$\phi(x) = \left(1 - \cos\frac{\pi x}{a}\right) \tag{4.54}$$

$$\psi(x) = \left(1 - \cos\frac{\pi y}{a}\right) \tag{4.55}$$

$$\int_0^a \phi_i.\phi_n = \begin{cases} 0 & \text{if } i \neq n \\ \frac{3a}{2} & \text{if } i = n \end{cases} \tag{4.56}$$

$$\int_0^a \psi_j.\psi_n = \begin{cases} 0 & \text{if } j \neq n \\ \frac{3a}{2} & \text{if } j = n \end{cases} \tag{4.57}$$

Case 1: First we take simply supported plate,
In this case we assume,

$$\phi_i(x) = \sin\frac{i\pi x}{a}, \psi_j(x) = \sin\frac{i\pi y}{a} \tag{4.58}$$

Then,

$$w(x,y) = \sum_{i=1}^{3}\sum_{j=1}^{3} C_i.\phi_i.\psi_j \tag{4.59}$$

We can arrange this assumption for 3 d.o.f as follows,

$$[w(x,y)] = \begin{bmatrix} \phi_1\psi_1 & \phi_1\psi_2 & \phi_1\psi_3 \\ \phi_2\psi_1 & \phi_2\psi_2 & \phi_2\psi_3 \\ \phi_3\psi_1 & \phi_3\psi_2 & \phi_3\psi_3 \end{bmatrix} \begin{bmatrix} c_1 \\ c_2 \\ c_3 \end{bmatrix} = [w]c \tag{4.60}$$

Where,

$$\phi_1(x) = \sin\frac{\pi x}{a}, \psi_1(y) = \sin\frac{\pi y}{a}$$

$$\phi_2(x) = \sin\frac{2\pi x}{a}, \psi_2(y) = \sin\frac{2\pi y}{a}$$

$$\phi_3(x) = \sin\frac{3\pi x}{a}, \psi_3(y) = \sin\frac{3\pi y}{a}$$

Note that,

$$\frac{\partial^2\varnothing_1}{\partial x^2} = -\left(\frac{\pi}{a}\right)^2 \sin\frac{\pi x}{a} = -\left(\frac{\pi}{a}\right)^2\phi_1, \frac{\partial^2\psi_1}{\partial y^2} = -\left(\frac{\pi}{a}\right)^2 \sin\frac{\pi y}{a} = -\left(\frac{\pi}{a}\right)^2\psi_1$$

$$\frac{\partial^2\varnothing_2}{\partial x^2} = -4\left(\frac{\pi}{a}\right)^2 \sin\frac{2\pi x}{a} = -4\left(\frac{\pi}{a}\right)^2\phi_2, \frac{\partial^2\psi_2}{\partial y^2} = -4\left(\frac{\pi}{a}\right)^2 \sin\frac{2\pi y}{a} = -4\left(\frac{\pi}{a}\right)^2\psi_2 \tag{4.61}$$

$$\frac{\partial^2\varnothing_3}{\partial x^2} = -9\left(\frac{\pi}{a}\right)^2 \sin\frac{3\pi x}{a} = -9\left(\frac{\pi}{a}\right)^2\phi_3, \frac{\partial^2\psi_3}{\partial y^2} = -9\left(\frac{\pi}{a}\right)^2 \sin\frac{3\pi y}{a} = -9\left(\frac{\pi}{a}\right)^2\psi_3$$

By the same way, we can write,

$$\frac{\partial^4\varnothing_1}{\partial x^4} = \left(\frac{\pi}{a}\right)^4\phi_1, \frac{\partial^4\psi_1}{\partial y^4} = \left(\frac{\pi}{a}\right)^4\psi_1$$

$$\frac{\partial^4\varnothing_2}{\partial x^4} = 16\left(\frac{\pi}{a}\right)^4\phi_2, \frac{\partial^4\psi_2}{\partial y^4} = 16\left(\frac{\pi}{a}\right)^4\psi_2 \tag{4.62}$$

$$\frac{\partial^4 \varnothing_3}{\partial x^4} = 81\left(\frac{\pi}{a}\right)^4 \phi_3, \frac{\partial^4 \psi_3}{\partial y^4} = 81\left(\frac{\pi}{a}\right)^4 \psi_3$$

For fixed edges, we use the following conditions,

$$\frac{\partial^2 \varnothing_1}{\partial x^2} = \left(\frac{\pi}{a}\right)^2 \cos\frac{\pi x}{a} = \left(\frac{\pi}{a}\right)^2 \phi_1', \frac{\partial^2 \psi_1}{\partial y^2} = \left(\frac{\pi}{a}\right)^2 \cos\frac{\pi y}{a} = \left(\frac{\pi}{a}\right)^2 \psi_1' \tag{4.63}$$

$$\frac{\partial^4 \varnothing_1}{\partial x^4} = -\left(\frac{\pi}{a}\right)^4 \phi_1', \frac{\partial^4 \psi_1}{\partial y^4} = -\left(\frac{\pi}{a}\right)^4 \psi_1'$$

4.5.1 Rayleigh-Ritz

Strain energy,
The strain energy of the plate under bending is given by,

$$V = \frac{D}{2}\int_0^a\int_0^a \left[\left(\frac{\partial^2 w}{\partial x^2}\right)^2 + \left(\frac{\partial^2 w}{\partial y^2}\right)^2 + 2\nu\left(\frac{\partial^2 w}{\partial x^2}\right)\left(\frac{\partial^2 w}{\partial y^2}\right) + 2(1-\nu)\left(\frac{\partial^2 w}{\partial x \partial y}\right)^2\right] dydx \tag{4.64}$$

Carrying out the differentiations of Eq. (4.59) are achieved as follows,

$$\frac{\partial^2 w}{\partial x^2} = -\left(\frac{\pi}{a}\right)^2 \begin{bmatrix} \phi_1\psi_1 & \phi_1\psi_2 & \phi_1\psi_3 \\ 4\phi_2\psi_1 & 4\phi_2\psi_2 & 4\phi_2\psi_3 \\ 9\phi_3\psi_1 & 9\phi_3\psi_2 & 9\phi_3\psi_3 \end{bmatrix} \begin{bmatrix} c_1 \\ c_2 \\ c_3 \end{bmatrix} = -\left(\frac{\pi}{a}\right)^2 [A]C \tag{4.65}$$

$$\frac{\partial^2 w}{\partial y^2} = -\left(\frac{\pi}{a}\right)^2 \begin{bmatrix} \phi_1\psi_1 & 4\phi_1\psi_2 & 9\phi_1\psi_3 \\ \phi_2\psi_1 & 4\phi_2\psi_2 & 9\phi_2\psi_3 \\ \phi_3\psi_1 & 4\phi_3\psi_2 & 9\phi_3\psi_3 \end{bmatrix} \begin{bmatrix} c_1 \\ c_2 \\ c_3 \end{bmatrix} = -\left(\frac{\pi}{a}\right)^2 [B]C \tag{4.66}$$

$$\frac{\partial^2 w}{\partial x \partial y} = \left(\frac{\pi}{a}\right)^2 \begin{bmatrix} \phi_1'\psi_1' & 2\phi_1'\psi_2' & 3\phi_1'\psi_3' \\ 2\phi_2'\psi_1' & 4\phi_2'\psi_2' & 6\phi_2'\psi_3' \\ 3\phi_3'\psi_1' & 6\phi_3'\psi_2' & 6\phi_3'\psi_3' \end{bmatrix} \begin{bmatrix} c_1 \\ c_2 \\ c_3 \end{bmatrix} = \left(\frac{\pi}{a}\right)^2 [D]C \tag{4.67}$$

Where,

$$\phi_1' = \frac{d\phi_1}{dx} = \cos\frac{\pi x}{a}$$

$$\psi_1' = \frac{d\psi_1}{dx} = \cos\frac{\pi y}{a}$$

$$\left(\frac{\partial^2 w}{\partial x^2}\right)\left(\frac{\partial^2 w}{\partial y^2}\right) = C^t[A]^t[B]C$$

So we can write strain energy Eq. (4.64) in other form as follows,

$$V = \left(\frac{D}{2}\left(\frac{\pi}{a}\right)^4 \int_0^a \int_0^a \begin{bmatrix} \{C\}^t[A]^t[A]\{C\} + \{C\}^t[B]^t[B]\{C\} + \\ 2\nu C^t[A]^t[B]\{C\} + 2(1-\nu)C^t[D]^t[D]\{C\} \end{bmatrix} dxdy\right)$$

Then,

$$V = \frac{D}{2}\left(\frac{\pi}{a}\right)^4 \{C\}^t \int_0^a \int_0^a \left[[A]^t[A] + [B]^t[B] + 2\nu[A]^t[B] + 2(1-\nu)[D]^t[D]\right] dxdy\{C\}$$

Minimizing V gives,

$$\frac{\partial V}{\partial C} = D\left(\frac{\pi}{a}\right)^4 \int_0^a \int_0^a \left[[A]^t[A] + [B]^t[B] + 2\nu[A]^t[B] + 2(1-\nu)[D]^t[D] \quad \right] dxdy\{C\} \tag{4.68}$$

Now, we perform the integration and remembering that ϕ and ψ are normalized and orthogonal. The integrations in Eq. (4.68) as follows,

First integration part of Eq. (4.68), we have,

$$D\left(\frac{\pi}{a}\right)^4 \int_0^a \int_0^a [A]^t[A] \quad \{C\}dxdy = \begin{bmatrix} \phi_1\psi_1 & 4\phi_2\psi_1 & 9\phi_3\psi_1 \\ \phi_1\psi_2 & 4\phi_2\psi_2 & 9\phi_3\psi_2 \\ \phi_1\psi_3 & 4\phi_2\psi_3 & 9\phi_3\psi_3 \end{bmatrix} \begin{bmatrix} \phi_1\psi_1 & \phi_1\psi_2 & \phi_1\psi_3 \\ 4\phi_2\psi_1 & 4\phi_2\psi_2 & 4\phi_2\psi_3 \\ 9\phi_3\psi_1 & 9\phi_3\psi_2 & 9\phi_3\psi_3 \end{bmatrix} \begin{bmatrix} c_1 \\ c_2 \\ c_3 \end{bmatrix}$$

Carrying out the matrix multiplications of $[A]^t[A]$ (Eq. 4.65) and substitutions of their expressions and the integrations of the above multiplications give,

$$= D\frac{\pi^4}{a^2} \begin{bmatrix} 24.5 & 0 & 0 \\ 0 & 24.5 & 0 \\ 0 & 0 & 24.5 \end{bmatrix} \begin{bmatrix} c_1 \\ c_2 \\ c_3 \end{bmatrix} \tag{4.69}$$

Second integration part of Eq. (4.68), we have,

$$D\left(\frac{\pi}{a}\right)^4 \int_0^a \int_0^a [B]^t[B][C]dxdy = \begin{bmatrix} \phi_1\psi_1 & \phi_2\psi_1 & \phi_3\psi_1 \\ 4\phi_1\psi_2 & 4\phi_2\psi_2 & 4\phi_3\psi_2 \\ 9\phi_1\psi_3 & 9\phi_2\psi_3 & 9\phi_3\psi_3 \end{bmatrix} \begin{bmatrix} \phi_1\psi_1 & 4\phi_1\psi_2 & 9\phi_1\psi_3 \\ \phi_2\psi_1 & 4\phi_2\psi_2 & 9\phi_2\psi_3 \\ \phi_3\psi_1 & 4\phi_3\psi_2 & 9\phi_3\psi_3 \end{bmatrix} \begin{bmatrix} c_1 \\ c_2 \\ c_3 \end{bmatrix}$$

Carrying out the matrix multiplications $[B]^t[B]$ (Eq. 4.66) and substitutions of their expressions and the integrations of the above multiplications give,

$$= D\frac{\pi^4}{a^2} \begin{bmatrix} 0.75 & 0 & 0 \\ 0 & 12.0 & 0 \\ 0 & 0 & 60.75 \end{bmatrix} \begin{bmatrix} c_1 \\ c_2 \\ c_3 \end{bmatrix} \tag{4.70}$$

Third integration part of Eq. (4.68), we have,

$$D\left(\frac{\pi}{a}\right)^4 \int_0^a \int_0^a [A]^t[B][C]dxdy = \begin{bmatrix} \phi_1\psi_1 & 4\phi_2\psi_1 & 9\phi_3\psi_1 \\ \phi_1\psi_2 & 4\phi_2\psi_2 & 9\phi_3\psi_2 \\ \phi_1\psi_3 & 4\phi_2\psi_3 & 9\phi_3\psi_3 \end{bmatrix} \begin{bmatrix} \phi_1\psi_1 & 4\phi_1\psi_2 & 9\phi_1\psi_3 \\ \phi_2\psi_1 & 4\phi_2\psi_2 & 9\phi_2\psi_3 \\ \phi_3\psi_1 & 4\phi_3\psi_2 & 9\phi_3\psi_3 \end{bmatrix} \begin{bmatrix} c_1 \\ c_2 \\ c_3 \end{bmatrix}$$

Carrying out the matrix multiplications $[A]^t[B]$ get use of Eqs. (4.65) and (4.66) and substitutions of their expressions and the integrations of the above multiplications give,

$$=D\frac{\pi^4}{a^2}\begin{bmatrix}3.5 & 0 & 0\\ 0 & 14 & 0\\ 0 & 0 & 31.5\end{bmatrix}\begin{bmatrix}c_1\\ c_2\\ c_3\end{bmatrix} \tag{4.71}$$

Fourth integration part of Eq. (4.68), we have,

$$D\left(\frac{\pi}{a}\right)^4\int_0^a\int_0^a [D]^t[D][C]dxdy = \begin{bmatrix}\phi_1'\psi_1' & 2\phi_2'\psi_1' & 3\phi_3'\psi_1'\\ 2\phi_1'\psi_2' & 4\phi_2'\psi_2' & 6\phi_3'\psi_2'\\ 3\phi_1'\psi_3' & 6\phi_2'\psi_3' & 6\phi_3'\psi_3'\end{bmatrix}\begin{bmatrix}\phi_1'\psi_1' & 2\phi_1'\psi_2' & 3\phi_1'\psi_3'\\ 2\phi_2'\psi_1' & 4\phi_2'\psi_2' & 6\phi_2'\psi_3'\\ 3\phi_3'\psi_1' & 6\phi_3'\psi_2' & 6\phi_3'\psi_3'\end{bmatrix}\begin{bmatrix}c_1\\ c_2\\ c_3\end{bmatrix}$$

Carrying out the matrix multiplications $[D]^t[D]$ (Eq. 4.67) and substitutions of their expressions and the integrations of the above multiplications give,

$$=D\frac{\pi^4}{a^2}\begin{bmatrix}3.5 & 0 & 0\\ 0 & 14 & 0\\ 0 & 0 & 31.5\end{bmatrix}\begin{bmatrix}c_1\\ c_2\\ c_3\end{bmatrix} \tag{4.72}$$

Substitution of Eqs. (4.69), (4.70), (4.71), and (4.72) into Eq. (4.68) give,

$$\frac{\partial V}{\partial C}=D\frac{\pi^4}{a^2}\begin{bmatrix}32.25 & 0 & 0\\ 0 & 64.5 & 0\\ 0 & 0 & 148.25\end{bmatrix}\begin{bmatrix}c_1\\ c_2\\ c_3\end{bmatrix} \tag{4.73}$$

Kinetic Energy

$$T=\frac{1}{2}\rho h\int_0^a\int_0^b\left(\frac{\partial w}{\partial t}\right)^2 dxdy=\frac{1}{2}\rho h\int_0^a\int_0^b \{C\}^t[w]^t[w]\{C\}dxdy \tag{4.74}$$

Minimizing T gives,

$$\frac{\partial T}{\partial C}=\rho h\int_0^a\int_0^b [w]^t[w]\{C\}dxdy \tag{4.75}$$

The derivative of the displacement vector given in Eq. (4.60) is substituted in Eq. (4.75) to give,

$$=\rho h\int_0^a\int_0^b\begin{bmatrix}\phi_1\psi_1 & \phi_2\psi_1 & \phi_3\psi_1\\ \phi_1\psi_2 & \phi_2\psi_2 & \phi_3\psi_2\\ \phi_1\psi_3 & \phi_2\psi_3 & \phi_3\psi_3\end{bmatrix}\begin{bmatrix}\phi_1\psi_1 & \phi_1\psi_2 & \phi_1\psi_3\\ \phi_2\psi_1 & \phi_2\psi_2 & \phi_2\psi_3\\ \phi_3\psi_1 & \phi_3\psi_2 & \phi_3\psi_3\end{bmatrix}dxdy$$

Multiplications of the above matrices and substitution of the $\phi_{'s}$ and $\psi_{'s}$ from Eq. (4.58) give,

$$=\frac{3\rho h a^2}{4}\begin{bmatrix}1 & 0 & 0\\ 0 & 1 & 0\\ 0 & 0 & 1\end{bmatrix}\begin{bmatrix}c_1\\ c_2\\ c_3\end{bmatrix} \tag{4.76}$$

Now

$$\frac{\partial V}{\partial C} - \omega^2 \frac{\partial T}{\partial C} = 0 \tag{4.77}$$

Substitute the expression $\frac{\partial V}{\partial C}$ from Eq. (4.73) and $\frac{\partial T}{\partial C}$ from Eq. (4.76) into Eq. (4.77) gives,

$$\left[D\frac{\pi^4}{a^2}\begin{bmatrix} 32.25 & 0 & 0 \\ 0 & 64.5 & 0 \\ 0 & 0 & 148.25 \end{bmatrix} - \frac{3\rho h a^2}{4}\omega^2 \begin{bmatrix} 1 & 0 & 0 \\ 0 & 1 & 0 \\ 0 & 0 & 1 \end{bmatrix} \right] \begin{bmatrix} c_1 \\ c_2 \\ c_3 \end{bmatrix}$$

Carrying out the above integrations and arrange the results in a standard EVP form and setting $\lambda = \frac{3\rho h a^2}{4}\omega^2$ gives,

$$\begin{bmatrix} (32.25 - \lambda) & 0 & 0 \\ 0 & (64.5 - \lambda) & 0 \\ 0 & 0 & (148.25 - \lambda) \end{bmatrix} = 0$$

The solution of the above matrix is that the determinate = 0, Then,

$$\lambda_1 = 32.25, \lambda_2 = 64.5, \lambda_3 = 148.25$$

Where,

$$\lambda = \omega^2 \frac{3\rho h a^4}{4D\pi^4}$$

Then, the natural frequencies are given by,

$$\omega_1 = 6.55744\sqrt{\frac{D\pi^4}{\rho h a^4}}$$

$$\omega_2 = 9.27362\sqrt{\frac{D\pi^4}{\rho h a^4}}$$

$$\omega 3 = 14.07\sqrt{\frac{D\pi^4}{\rho h a^4}} \tag{4.78}$$

4.5.2 Galerkin's method

The governing differential equation,

$$D\left[\frac{\partial^4 w}{\partial x^4} + 2\frac{\partial^4 w}{\partial x^2 \partial y^2} + \frac{\partial^4 w}{\partial y^4}\right] + \rho h \frac{\partial^2 w}{\partial t^2} = R \tag{4.79}$$

Where, R is residual.

Now, by **Galerkin's method,**

$$\int_0^a \int_0^a R.w(x,y)dx\ dy = 0$$

Or,

$$\int_0^a \int_0^a \left\{ D\left[\frac{\partial^4 w}{\partial x^4} + 2\frac{\partial^4 w}{\partial x^2 \partial y^2} + \frac{\partial^4 w}{\partial y^4}\right] + \rho h \frac{\partial^2 w}{\partial t^2} \right\} w(x,y)dxdy = 0 \tag{4.80}$$

We have from Eqs. (4.65), (4.66), and (4.67) as follows,

$$\frac{\partial^4 w}{\partial x^4} = \left(\frac{\pi}{a}\right)^4 \begin{bmatrix} \varphi_1\psi_1 & \varphi_1\psi_2 & \varphi_1\psi_3 \\ 16\varphi_2\psi_1 & 16\varphi_2\psi_2 & 16\varphi_2\psi_3 \\ 81\varphi_3\psi_1 & 81\varphi_3\psi_2 & 81\varphi_3\psi_3 \end{bmatrix} \begin{bmatrix} c_1 \\ c_2 \\ c_3 \end{bmatrix} = \left(\frac{\pi}{a}\right)^4 [A]\{C\} \tag{4.81}$$

$$\frac{\partial^4 w}{\partial y^4} = \left(\frac{\pi}{a}\right)^4 \begin{bmatrix} \varphi_1\psi_1 & 16\varphi_1\psi_2 & 81\varphi_1\psi_3 \\ \varphi_2\psi_1 & 16\varphi_2\psi_2 & 81\varphi_2\psi_3 \\ \varphi_3\psi_1 & 16\varphi_3\psi_2 & 81\varphi_3\psi_3 \end{bmatrix} \begin{bmatrix} c_1 \\ c_2 \\ c_3 \end{bmatrix} = \left(\frac{\pi}{a}\right)^4 [B]\{C\} \tag{4.82}$$

$$\frac{\partial^4 w}{\partial x^2 \partial y^2} = \left(\frac{\pi}{a}\right)^4 \begin{bmatrix} \varphi_1\psi_1 & 4\varphi_1\psi_2 & 9\varphi_1\psi_3 \\ 4\varphi_2\psi_1 & 16\varphi_2\psi_2 & 36\varphi_2\psi_3 \\ 9\varphi_3\psi_1 & 36\varphi_3\psi_2 & 81\varphi_3\psi_3 \end{bmatrix} \begin{bmatrix} c_1 \\ c_2 \\ c_3 \end{bmatrix} = \left(\frac{\pi}{a}\right)^4 [D]\{C\} \tag{4.83}$$

So, put the above derivatives into Eq. (4.80) gives,

$$\int_0^a \int_0^a \left\{ D\left(\frac{\pi}{a}\right)^4 \left[[w]^t[A]\{C\} + 2[w]^t[D]\{C\} + [w]^t[B]\{C\}\right] - \omega^2 \rho h [w]^t[w]\{C\} \right\} dxdy = 0$$

Or,

$$= \int_0^a \int_0^a [w]^t \left\{ \left(\frac{\pi}{a}\right)^4 [[A] + 2[D] + [B]] - \omega^2 \rho h [w] \right\} [C] dxdy = 0 \tag{4.84}$$

Then, carrying out the addition of matrices by using Eq. (4.60) for [w(x,y)] matrix, Eq. (4.81) for $[A]$ matrix, Eq. (4.82) for $[B]$ matrix and Eq. (4.83) for $[D]$ matrix into Eq. (4.84) gives,

$$\iint \begin{bmatrix} \phi_1\psi_1 & \phi_2\psi_1 & \phi_3\psi_1 \\ \phi_1\psi_2 & \phi_2\psi_2 & \phi_3\psi_2 \\ \phi_1\psi_3 & \phi_2\psi_3 & \phi_3\psi_3 \end{bmatrix} \left\{ \left(\frac{\pi}{a}\right)^4 \left\{ \begin{bmatrix} \varphi_1\psi_1 & \varphi_1\psi_2 & \varphi_1\psi_3 \\ 16\varphi_2\psi_1 & 16\varphi_2\psi_2 & 16\varphi_2\psi_3 \\ 81\varphi_3\psi_1 & 81\varphi_3\psi_2 & 81\varphi_3\psi_3 \end{bmatrix} \right.\right.$$

$$\left. + 2\begin{bmatrix} \varphi_1\psi_1 & 4\varphi_1\psi_2 & 9\varphi_1\psi_3 \\ 4\varphi_2\psi_1 & 16\varphi_2\psi_2 & 36\varphi_2\psi_3 \\ 9\varphi_3\psi_1 & 36\varphi_3\psi_2 & 81\varphi_3\psi_3 \end{bmatrix} + \begin{bmatrix} \varphi_1\psi_1 & 16\varphi_1\psi_2 & 81\varphi_1\psi_3 \\ \varphi_2\psi_1 & 16\varphi_2\psi_2 & 81\varphi_2\psi_3 \\ \varphi_3\psi_1 & 16\varphi_3\psi_2 & 81\varphi_3\psi_3 \end{bmatrix} \right\}$$

$$\left. - \omega^2 \rho h \begin{bmatrix} \phi_1\psi_1 & \phi_1\psi_2 & \phi_1\psi_3 \\ \phi_2\psi_1 & \phi_2\psi_2 & \phi_2\psi_3 \\ \phi_3\psi_1 & \phi_3\psi_2 & \phi_3\psi_3 \end{bmatrix} \right\} \begin{bmatrix} c_1 \\ c_2 \\ c_3 \end{bmatrix} dx\ dy = 0$$

Simplification gives,

$$\iint \begin{bmatrix} \phi_1\psi_1 & \phi_2\psi_1 & \phi_3\psi_1 \\ \phi_1\psi_2 & \phi_2\psi_2 & \phi_3\psi_2 \\ \phi_1\psi_3 & \phi_2\psi_3 & \phi_3\psi_3 \end{bmatrix} \left\{ \left(\frac{\pi}{a}\right)^4 \begin{bmatrix} 4\varphi_1\psi_1 & 25\varphi_1\psi_2 & 100\varphi_1\psi_3 \\ 25\varphi_2\psi_1 & 64\varphi_2\psi_2 & 169\varphi_2\psi_3 \\ 100\varphi_3\psi_1 & 189\varphi_3\psi_2 & 324\varphi_3\psi_3 \end{bmatrix} \right.$$

$$\left. - \omega^2 \rho h \begin{bmatrix} \phi_1\psi_1 & \phi_1\psi_2 & \phi_1\psi_3 \\ \phi_2\psi_1 & \phi_2\psi_2 & \phi_2\psi_3 \\ \phi_3\psi_1 & \phi_3\psi_2 & \phi_3\psi_3 \end{bmatrix} \right\} \begin{bmatrix} c_1 \\ c_2 \\ c_3 \end{bmatrix} dx\, dy = 0$$

Carrying out the above integrations and arrange the results in a standard EVP form gives,

$$= \left\{ \frac{D\pi^4}{a^2} \begin{bmatrix} 32.25 & 0 & 0 \\ 0 & 64.5 & 0 \\ 0 & 0 & 148.5 \end{bmatrix} - \frac{\omega^2 3\rho h a^2}{4} \begin{bmatrix} 1 & 0 & 0 \\ 0 & 1 & 0 \\ 0 & 0 & 1 \end{bmatrix} \right\} \begin{bmatrix} c_1 \\ c_2 \\ c_3 \end{bmatrix} = 0$$

Arrange the above EVP in the following form by substitution of $\lambda = \frac{\omega^2 3\rho h a^2}{4}$,

$$= \begin{bmatrix} (32.25 - \lambda) & 0 & 0 \\ 0 & (64.5 - \lambda) & 0 \\ 0 & 0 & (148.5 - \lambda) \end{bmatrix} \begin{bmatrix} c_1 \\ c_2 \\ c_3 \end{bmatrix} = 0$$

Solving the above system of equations using the standard procedure by setting the determinant of the above (3×3) matrix gives,

$$\lambda_1 = 32.25, \lambda_2 = 64.5, \lambda_3 = 148.5$$

From which the natural frequencies will be as follows,

$$\omega_1 = 6.55744\sqrt{\frac{D\pi^4}{\rho h a^2}}$$

$$\omega_2 = 9.27362\sqrt{\frac{D\pi^4}{\rho h a^2}} \tag{4.85}$$

$$\omega_3 = 14.07\sqrt{\frac{D\pi^4}{\rho h a^2}}$$

Case 2:

We take plate fixed in $y = 0$ and $y = a$ and simply supported in $x = 0$ and $x = a$ in this case $\psi(y) = 1 - \cos\frac{\pi x}{a}$ and $\phi(x)$ will be as before, $\phi(x) = \sin\frac{\pi x}{a}$.

Rayleigh-Ritz

1. Strain energy stored in the plate is given by,

$$V = \frac{D}{2}\int_0^a\int_0^a \left[\left(\frac{\partial^2 w}{\partial x^2}\right)^2 + \left(\frac{\partial^2 w}{\partial y^2}\right)^2 + 2\nu\left(\frac{\partial^2 w}{\partial x^2}\right)\left(\frac{\partial^2 w}{\partial y^2}\right) + 2(1-\nu)\left(\frac{\partial^2 w}{\partial x \partial y}\right)^2 \right] dydx$$

Where,

$$\frac{\partial^2 w}{\partial x^2} = -\left(\frac{\pi}{a}\right)^2 \begin{bmatrix} \phi_1\psi_1 & \phi_1\psi_2 & \phi_1\psi_3 \\ 4\phi_2\psi_1 & 4\phi_2\psi_2 & 4\phi_2\psi_3 \\ 9\phi_3\psi_1 & 9\phi_3\psi_2 & 9\phi_3\psi_3 \end{bmatrix} \begin{bmatrix} c_1 \\ c_2 \\ c_3 \end{bmatrix} = -\left(\frac{\pi}{a}\right)^2 [A]\{C\} \tag{4.86}$$

$$\frac{\partial^2 w}{\partial y^2} = -\left(\frac{\pi}{a}\right)^2 \begin{bmatrix} \phi_1\psi_1 & 4\phi_1\psi_2 & 9\phi_1\psi_3 \\ \phi_2\psi_1 & 4\phi_2\psi_2 & 9\phi_2\psi_3 \\ \phi_3\psi_1 & 4\phi_3\psi_2 & 9\phi_3\psi_3 \end{bmatrix} \begin{bmatrix} c_1 \\ c_2 \\ c_3 \end{bmatrix} = -\left(\frac{\pi}{a}\right)^2 [B]\{C\} \tag{4.87}$$

$$\frac{\partial^2 w}{\partial x \partial y} = \left(\frac{\pi}{a}\right)^2 \begin{bmatrix} \phi_1'\psi_1' & 2\phi_1'\psi_2' & 3\phi_1'\psi_3' \\ 2\phi_2'\psi_1' & 4\phi_2'\psi_2' & 6\phi_2'\psi_3' \\ 3\phi_3'\psi_1' & 6\phi_3'\psi_2' & 6\phi_3'\psi_3' \end{bmatrix} \begin{bmatrix} c_1 \\ c_2 \\ c_3 \end{bmatrix} = \left(\frac{\pi}{a}\right)^2 [D]\{C\} \tag{4.88}$$

Equation of strain energy using Eqs. (4.86), (4.87), and (4.88) can be written as follows,

$$V = \frac{D}{2}\left(\frac{\pi}{a}\right)^4 \int_0^a \int_0^a \begin{bmatrix} \{C\}^t[A]^t[A]\{C\} + \{C\}^t[B]^t[B]\{C\} - \\ 2\nu\{C\}^t[A]^t[B]\{C\} + 2(1-\nu)\{C\}^t[D]^t[D]\{C\} \end{bmatrix} dxdy$$

Minimizing V in the above equation gives,

$$\frac{\partial V}{\partial C} = \begin{pmatrix} D\iint \left(\frac{\pi}{a}\right)^4 [A]^t[A]\{C\}dx\,dy + D\iint \left(\frac{\pi}{a}\right)^4 [B]^t[B]\{C\}dx\,dy - \\ D\iint 2\nu\left(\frac{\pi}{a}\right)^4 [A]^t[B]Cdx\,dy + D\iint 2(1-\nu)\left(\frac{\pi}{a}\right)^4 [D]^t[D]Cdx\,dy \end{pmatrix} \tag{4.89}$$

First integral part of Eq. (4.89) using matrix $[A]$ given in Eq. (4.86) is,

$$\left(\frac{\pi}{a}\right)^4 D \iint [A]^t[A]\{C\}dx\,dy =$$

$$\left(\frac{\pi}{a}\right)^4 D \iint \begin{bmatrix} \phi_1\psi_1 & 4\phi_2\psi_1 & 9\phi_3\psi_1 \\ \phi_1\psi_2 & 4\phi_2\psi_2 & 9\phi_3\psi_2 \\ \phi_1\psi_3 & 4\phi_2\psi_3 & 9\phi_3\psi_3 \end{bmatrix} \begin{bmatrix} \phi_1\psi_1 & \phi_1\psi_2 & \phi_1\psi_3 \\ 4\phi_2\psi_1 & 4\phi_2\psi_2 & 4\phi_2\psi_3 \\ 9\phi_3\psi_1 & 9\phi_3\psi_2 & 9\phi_3\psi_3 \end{bmatrix}$$

$$dxdy = \frac{D\pi^4}{a^2} \begin{bmatrix} 73.5 & 0 & 0 \\ 0 & 73.5 & 0 \\ 0 & 0 & 73.5 \end{bmatrix} \begin{bmatrix} c_1 \\ c_2 \\ c_3 \end{bmatrix} \tag{4.90}$$

Second integral part using Eq. (4.86) for Matrix $[A]$ and Eq. (4.87) for matrix $[B]$ is,

$$-2\nu\left(\frac{\pi}{a}\right)^4 D \iint [A]^t[B][C]dxdy =$$

$$-2\nu\left(\frac{\pi}{a}\right)^4 D \iint \begin{bmatrix} \phi_1\psi_1 & 4\phi_2\psi_1 & 9\phi_3\psi_1 \\ \phi_1\psi_2 & 4\phi_2\psi_2 & 9\phi_3\psi_2 \\ \phi_1\psi_3 & 4\phi_2\psi_3 & 9\phi_3\psi_3 \end{bmatrix} \begin{bmatrix} \phi_1'\psi_1' & 4\phi_1'\psi_2' & 9\phi_1'\psi_3' \\ \phi_2'\psi_1' & 4\phi_2'\psi_2' & 9\phi_2'\psi_3' \\ \phi_3'\psi_1' & 4\phi_3'\psi_2' & 9\phi_3'\psi_3' \end{bmatrix}$$

$$dxdy = \frac{2\nu D\pi^4}{a^2} \begin{bmatrix} 3.5 & 0 & 0 \\ 0 & 14 & 0 \\ 0 & 0 & 31.5 \end{bmatrix} \begin{bmatrix} c_1 \\ c_2 \\ c_3 \end{bmatrix} \tag{4.91}$$

Third integral part of Eq. (4.89) using Eq. (4.87) is,

$$\left(\frac{\pi}{a}\right)^4 D \iint [B]^t[B]\{C\}dxdy =$$

$$\left(\frac{\pi}{a}\right)^4 D \iint \begin{bmatrix} \phi_1'\psi_1' & \phi_2'\psi_1' & \phi_3'\psi_1' \\ 4\phi_1'\psi_2' & 4\phi_2'\psi_2' & 4\phi_3'\psi_2' \\ 9\phi_1'\psi_3' & 9\phi_2'\psi_3' & 9\phi_3'\psi_3' \end{bmatrix} \begin{bmatrix} \phi_1'\psi_1' & 4\phi_1'\psi_2' & 9\phi_1'\psi_3' \\ \phi_2'\psi_1' & 4\phi_2'\psi_2' & 9\phi_2'\psi_3' \\ \phi_3'\psi_1' & 4\phi_3'\psi_2' & 9\phi_3'\psi_3' \end{bmatrix}$$

$$dxdy = \frac{D\pi^4}{a^2} \begin{bmatrix} 0.75 & 0 & 0 \\ 0 & 12 & 0 \\ 0 & 0 & 60.75 \end{bmatrix} \begin{bmatrix} c_1 \\ c_2 \\ c_3 \end{bmatrix} \tag{4.92}$$

Fourth integral part of Eq. (4.89) using Eq. (4.88) is,

$$2(1-\nu)\left(\frac{\pi}{a}\right)^4 D \iint [D]^t[D]\{C\}dx\, dy =$$

$$2(1-\nu)\left(\frac{\pi}{a}\right)^4 D \iint \begin{bmatrix} \phi_1'\psi_1' & 2\phi_2'\psi_1' & 3\phi_3'\psi_1' \\ 2\phi_1'\psi_2' & 9\phi_2'\psi_2' & 6\phi_3'\psi_2' \\ 3\phi_1'\psi_3' & 6\phi_2'\psi_3' & 9\phi_3'\psi_3' \end{bmatrix} \begin{bmatrix} \phi_1'\psi_1' & 2\phi_1'\psi_2' & 3\phi_1'\psi_3' \\ 2\phi_2'\psi_1' & 4\phi_2'\psi_2' & 6\phi_2'\psi_3' \\ 3\phi_3'\psi_1' & 6\phi_3'\psi_2' & 9\phi_3'\psi_3' \end{bmatrix}$$

$$dxdy = \frac{2(1-\nu)\pi^4}{a^2} \begin{bmatrix} 3.5 & 0 & 0 \\ 0 & 14 & 0 \\ 0 & 0 & 31.5 \end{bmatrix} \tag{4.93}$$

Therefore substitution of the above four part integrals in Eq. (4.89) gives,

$$\frac{\partial V}{\partial C} = \frac{D\pi^4}{a^2} \left(\begin{matrix} \begin{bmatrix} 73.5 & 0 & 0 \\ 0 & 73.5 & 0 \\ 0 & 0 & 73.5 \end{bmatrix} + \begin{bmatrix} 0.75 & 0 & 0 \\ 0 & 12 & 0 \\ 0 & 0 & 60.75 \end{bmatrix} + \\ 2\nu \begin{bmatrix} 3.5 & 0 & 0 \\ 0 & 14 & 0 \\ 0 & 0 & 31.5 \end{bmatrix} + 2(1-\nu) \begin{bmatrix} 3.5 & 0 & 0 \\ 0 & 14 & 0 \\ 0 & 0 & 31.5 \end{bmatrix} \end{matrix} \right) \begin{bmatrix} c_1 \\ c_2 \\ c_3 \end{bmatrix}$$

And, by simplification of the above resulted system of equations give,

$$\frac{\partial V}{\partial C} = \frac{D\pi^4}{a^2} \begin{bmatrix} 81.25 & 0 & 0 \\ 0 & 113.5 & 0 \\ 0 & 0 & 197.25 \end{bmatrix} \begin{bmatrix} c_1 \\ c_2 \\ c_3 \end{bmatrix} \tag{4.94}$$

2. Kinetic energy T is given by,

$$T = \frac{1}{2}\rho h \iint \left(\frac{\partial w}{\partial t}\right)^2 dxdy$$

Or.

$$T = \frac{1}{2}\rho h \iint [C]^t [w]^t [w] C dxdy$$

Insert the matrix $[w]$ given in Eq. (4.60) in the above matrix and use the minimization principles give,

$$\frac{\partial T}{\partial C} = \rho h \iint [w]^t [w] \quad [C]$$

$$dxdy = \rho h \int_0^a \int_0^a \begin{bmatrix} \phi_1\psi_1 & \phi_2\psi_1 & \phi_3\psi_1 \\ \phi_1\psi_2 & \phi_2\psi_2 & \phi_3\psi_2 \\ \phi_1\psi_3 & \phi_2\psi_3 & \phi_3\psi_3 \end{bmatrix} \begin{bmatrix} \phi_1\psi_1 & \phi_1\psi_2 & \phi_1\psi_3 \\ \phi_2\psi_1 & \phi_2\psi_2 & \phi_2\psi_3 \\ \phi_3\psi_1 & \phi_3\psi_2 & \phi_3\psi_3 \end{bmatrix}$$

$$dxdy = \frac{3\rho h a^2}{4} \begin{bmatrix} 1 & 0 & 0 \\ 0 & 1 & 0 \\ 0 & 0 & 1 \end{bmatrix} \begin{bmatrix} c_1 \\ c_2 \\ c_3 \end{bmatrix} \tag{4.95}$$

Now,

$$\frac{\partial V}{\partial C} - \omega^2 \frac{\partial T}{\partial C} = 0$$

Using Eqs. (4.94) and (4.95) and arrange in a standard EVP as follows,

$$\left\{ \frac{D\pi^4}{a^2} \begin{bmatrix} 81.25 & 0 & 0 \\ 0 & 113.5 & 0 \\ 0 & 0 & 197.25 \end{bmatrix} - \frac{3\omega^2 \rho h a^2}{4} \begin{bmatrix} 1 & 0 & 0 \\ 0 & 1 & 0 \\ 0 & 0 & 1 \end{bmatrix} \right\} \begin{bmatrix} c_1 \\ c_2 \\ c_3 \end{bmatrix} = 0$$

Using the substitution $\lambda = \frac{3\omega^2 \rho h a^2}{4}$ in the above equation gives,

$$= \begin{bmatrix} (81.25 - \lambda) & 0 & 0 \\ 0 & (113.5 - \lambda) & 0 \\ 0 & 0 & (197.25 - \lambda) \end{bmatrix} \begin{bmatrix} c_1 \\ c_2 \\ c_3 \end{bmatrix} = 0$$

To solve the above resulted system of questions, the determinant of the matrix $(3 \times 3) = 0$, the roots of the resulted equation are,

$$\lambda_1 = 81.25, \lambda_2 = 113.5, \lambda_3 = 197.25$$

From which, the natural frequencies are as follows,

$$\omega_1 = 10.408\sqrt{\frac{D\pi^4}{\rho h a^4}}$$

$$\omega_2 = 12.3\sqrt{\frac{D\pi^4}{\rho h a^4}} \tag{4.96}$$

$$\omega_3 = 16.217\sqrt{\frac{D\pi^4}{\rho h a^4}}$$

Galerkin's method,
The strain energy is given by,

$$\int_0^a\int_0^a\left\{D\left[\frac{\partial^4 w}{\partial x^4}+2\frac{\partial^4 w}{\partial x^2\partial y^2}+\frac{\partial^4 w}{\partial y^4}\right]+\rho h\frac{\partial^2 w}{\partial t^2}\right\}wdxdy=0 \tag{4.97}$$

The derivatives in the strain energy formula are achieved as follows,

$$\frac{\partial^4 w}{\partial x^4}=\left(\frac{\pi}{a}\right)^4\begin{bmatrix}\varphi_1\psi_1 & \varphi_1\psi_2 & \varphi_1\psi_3\\ 16\varphi_2\psi_1 & 16\varphi_2\psi_2 & 16\varphi_2\psi_3\\ 81\varphi_3\psi_1 & 81\varphi_3\psi_2 & 81\varphi_3\psi_3\end{bmatrix}\begin{bmatrix}c_1\\c_2\\c_3\end{bmatrix}$$

$$\frac{\partial^4 w}{\partial y^4}=-\left(\frac{\pi}{a}\right)^4\begin{bmatrix}\phi_1'\psi_1' & 16\phi_1'\psi_2' & 81\phi_1'\psi_3'\\ \phi_2'\psi_1' & 16\phi_2'\psi_2' & 81\phi_2'\psi_3'\\ \phi_3'\psi_1' & 16\phi_3'\psi_2' & 81\phi_3'\psi_3'\end{bmatrix}\begin{bmatrix}c_1\\c_2\\c_3\end{bmatrix}$$

$$\frac{\partial^4 w}{\partial x^2\partial y^2}=-\left(\frac{\pi}{a}\right)^4\begin{bmatrix}\phi_1'\psi_1' & 4\phi_1'\psi_2' & 9\phi_1'\psi_3'\\ 4\phi_2'\psi_1' & 16\phi_2'\psi_2' & 36\phi_2'\psi_3'\\ 9\phi_3'\psi_1' & 36\phi_3'\psi_2' & 81\phi_3'\psi_3'\end{bmatrix}\begin{bmatrix}c_1\\c_2\\c_3\end{bmatrix}$$

The φ's, ψ's, ϕ's, and ψ's are as defined in the previous cases of plate examples. Hence,

$$\iint D\left(\frac{\pi}{a}\right)^4\left\{\begin{bmatrix}\varphi_1\psi_1 & \varphi_1\psi_2 & \varphi_1\psi_3\\ 16\varphi_2\psi_1 & 16\varphi_2\psi_2 & 16\varphi_2\psi_3\\ 81\varphi_3\psi_1 & 81\varphi_3\psi_2 & 81\varphi_3\psi_3\end{bmatrix}-\begin{bmatrix}\phi_1'\psi_1' & 16\phi_1'\psi_2' & 81\phi_1'\psi_3'\\ \phi_2'\psi_1' & 16\phi_2'\psi_2' & 81\phi_2'\psi_3'\\ \phi_3'\psi_1' & 16\phi_3'\psi_2' & 81\phi_3'\psi_3'\end{bmatrix}-\right.$$
$$\left.2\begin{bmatrix}\phi_1'\psi_1' & 4\phi_1'\psi_2' & 9\phi_1'\psi_3'\\ 4\phi_2'\psi_1' & 16\phi_2'\psi_2' & 36\phi_2'\psi_3'\\ 9\phi_3'\psi_1' & 36\phi_3'\psi_2' & 81\phi_3'\psi_3'\end{bmatrix}-\omega^2\rho h\begin{bmatrix}\varphi_1\psi_1 & \varphi_1\psi_2 & \varphi_1\psi_3\\ \varphi_2\psi_1 & \varphi_2\psi_2 & \varphi_2\psi_3\\ \varphi_3\psi_1 & \varphi_3\psi_2 & \varphi_3\psi_3\end{bmatrix}\right\}\begin{bmatrix}c_1\\c_2\\c_3\end{bmatrix}\begin{bmatrix}\varphi_1\psi_1 & \varphi_1\psi_2 & \varphi_1\psi_3\\ \varphi_2\psi_1 & \varphi_2\psi_2 & \varphi_2\psi_3\\ \varphi_3\psi_1 & \varphi_3\psi_2 & \varphi_3\psi_3\end{bmatrix}dxdy \tag{4.98}$$

Or,

$$D\left(\frac{\pi}{a}\right)^4\iint\begin{bmatrix}\phi_1\psi_1 & \phi_2\psi_1 & \phi_3\psi_1\\ \phi_1\psi_2 & \phi_2\psi_2 & \phi_3\psi_2\\ \phi_1\psi_3 & \phi_2\psi_3 & \phi_3\psi_3\end{bmatrix}\left\{\begin{bmatrix}\varphi_1\psi_1 & \varphi_1\psi_2 & \varphi_1\psi_3\\ 16\varphi_2\psi_1 & 16\varphi_2\psi_2 & 16\varphi_2\psi_3\\ 81\varphi_3\psi_1 & 81\varphi_3\psi_2 & 81\varphi_3\psi_3\end{bmatrix}\right.$$
$$\left.-\begin{bmatrix}3\phi_1'\psi_1' & 24\phi_1'\psi_2' & 99\phi_1'\psi_3'\\ 9\phi_2'\psi_1' & 48\phi_2'\psi_2' & 153\phi_2'\psi_3'\\ 19\phi_3'\psi_1' & 88\phi_3'\psi_2' & 243\phi_3'\psi_3'\end{bmatrix}\right\}dxdy$$
$$-\omega^2\rho h\iint\begin{bmatrix}\phi_1\psi_1 & \phi_2\psi_1 & \phi_3\psi_1\\ \phi_1\psi_2 & \phi_2\psi_2 & \phi_3\psi_2\\ \phi_1\psi_3 & \phi_2\psi_3 & \phi_3\psi_3\end{bmatrix}\begin{bmatrix}\varphi_1\psi_1 & \varphi_1\psi_2 & \varphi_1\psi_3\\ \varphi_2\psi_1 & \varphi_2\psi_2 & \varphi_2\psi_3\\ \varphi_3\psi_1 & \varphi_3\psi_2 & \varphi_3\psi_3\end{bmatrix}dxdy=0$$

Carrying out the multiplications and the required integrations and simplifications give,

$$= \frac{D\pi^4}{a^2}\begin{bmatrix} 73.5 & 0 & 0 \\ 0 & 73.5 & 0 \\ 0 & 0 & 73.5 \end{bmatrix} + \frac{D\pi^4}{a^2}\begin{bmatrix} 7.75 & 0 & 0 \\ 0 & 40 & 0 \\ 0 & 0 & 123.25 \end{bmatrix} - \frac{3\omega^2 \rho h a^2}{4}\begin{bmatrix} 1 & 0 & 0 \\ 0 & 1 & 0 \\ 0 & 0 & 1 \end{bmatrix} = 0 \quad (4.99)$$

Arrange Eq. (4.99) in a standard EVP and simplifications give,

$$= \frac{D\pi^4}{a^2}\begin{bmatrix} 21.25 & 0 & 0 \\ 0 & 113.5 & 0 \\ 0 & 0 & 197.25 \end{bmatrix}\begin{bmatrix} c_1 \\ c_2 \\ c_3 \end{bmatrix} - \frac{3\omega^2 \rho h a^2}{4}\begin{bmatrix} 1 & 0 & 0 \\ 0 & 1 & 0 \\ 0 & 0 & 1 \end{bmatrix}\begin{bmatrix} c_1 \\ c_2 \\ c_3 \end{bmatrix} = 0 \quad (4.100)$$

Or,

$$([k] - \omega^2[m])c = 0$$

By substitution of $\omega^2 = \lambda = \frac{3\omega^2 \rho h a^4}{4D\pi^4}$, therefore,

$$\begin{bmatrix} (21.25 - \lambda) & 0 & 0 \\ 0 & (113.5 - \lambda) & 0 \\ 0 & 0 & (197.25 - \lambda) \end{bmatrix} = 0$$

$$\lambda_1 = 81.25, \lambda_2 = 113.5, \lambda_3 = 197.25$$

Hence, the natural frequencies are as follows,

$$\omega_1 = 10.408\sqrt{\frac{D\pi^4}{\rho h a^4}}$$

$$\omega_2 = 12.3\sqrt{\frac{D\pi^4}{\rho h a^4}} \quad (4.101)$$

$$\omega_3 = 16.217\sqrt{\frac{D\pi^4}{\rho h a^4}}$$

Example 4.7: (a) The simply supported plate, Fig. 4.8, all around its edges and supports a uniformly distributed load P_0. Determine the deflection surface using the total potential energy principles,

$$w = C_1\left(x^2 - a^2\right)^2\left(y^2 - b^2\right)^2$$

(b) Treat the prob. In (a) by Rayleigh-Ritz method to calculate the fundamental natural frequency by assuming a reasonable vibrating shape.

Solution,

$$w(x, y) = C_1\left(x^2 - a^2\right)^2\left(y^2 - b^2\right)^2 \quad (E7.1)$$

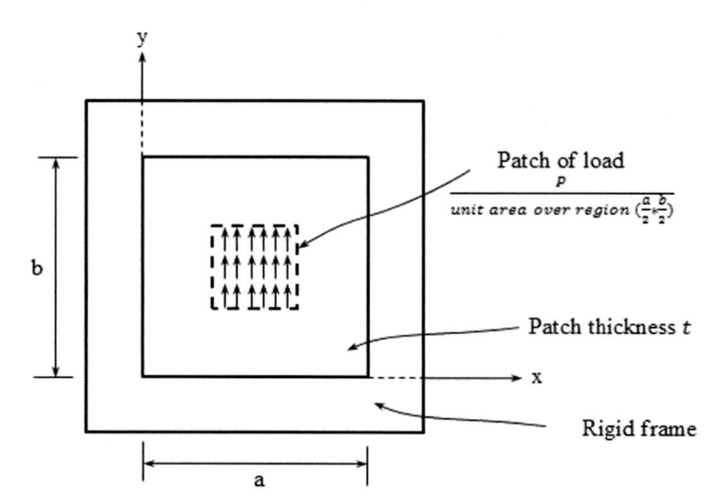

FIGURE 4.8 A simply supported plate subjected to patch load.

To find deflection constant, C_1,
We use P.S.T.P.E.

$$U = \frac{D}{2}\int_{-a}^{a}\int_{-b}^{b}\left(\frac{\partial^2 w}{\partial x^2} + \frac{\partial^2 w}{\partial y^2}\right)^2 dxdy$$

From Eq. (E7.1), we have,

$$\frac{\partial w}{\partial x} = 4C_1 x\left(x^2 - a^2\right)\left(y^2 - b^2\right)^2 = 4C_1\left(x^3 - a^2 x\right)\left(y^2 - b^2\right)^2$$

$$\frac{\partial^2 w}{\partial x^2} = 4C_1\left(3x^2 - a^2\right)\left(y^2 - b^2\right)^2$$

$$\frac{\partial^2 w}{\partial y^2} = 4C_1\left(3y^2 - b^2\right)\left(x^2 - a^2\right)^2$$

Then,

$$U = \frac{16DC_1^2}{2}\int_{-a}^{a}\int_{-b}^{b}\left(\left(3x^2 - a^2\right)\left(y^2 - b^2\right)^2 + \left(3y^2 - b^2\right)\left(x^2 - a^2\right)^2\right)^2 dxdy$$

$$= -\int_{-a}^{a}\int_{-b}^{b} q_0 w(x,y)\,dxdy$$

$$= -\int_{-a}^{a}\int_{-b}^{b} q_0 C_1\left(x^2 - a^2\right)^2\left(y^2 - b^2\right)^2 dxdy \quad \text{(E7.2)}$$

$$V = U + \Omega$$

$$V=\frac{16DC_1^2}{2}\int_{-a}^{a}\int_{-b}^{b}\left(\left(3x^2-a^2\right)\left(y^2-b^2\right)^2+\left(3y^2-b^2\right)\left(x^2-a^2\right)^2\right)^2dx\,dy-q_oC_1\int_{-a}^{a}\int_{-b}^{b}\left(x^2-a^2\right)^2\left(y^2-b^2\right)^2dxdy \tag{E7.3}$$

To find C_1,

$$\frac{\partial V}{\partial C_1}=0$$

$$16DC_1\int_{-a}^{a}\int_{-b}^{v}\left(\left(3x^2-a^2\right)\underbrace{\left(y^2-b^2\right)^2}_{\left(y^4-2y^2b^2+b^4\right)}+\left(3y^2-b^2\right)\underbrace{\left(x^2-a^2\right)^2}_{\left(x^4-2x^2a^2+a^4\right)}\right)^2dxdy$$

$$-q_o\int_{-a}^{a}\int_{-b}^{b}\left(x^2-a^2\right)^2\left(y^2-b^2\right)^2dx\,dy=0 \tag{E7.4}$$

Carrying out the integration of Eq. (E7.4) we have,

$$2\int_0^a\int_0^b\left(\left(3x^2-a^2\right)\left(y^2-b^2\right)^2+\left(3y^2-b^2\right)\left(x^2-a^2\right)^2\right)^2dxdy$$

$$=2\int_0^a\int_0^b\left[\underbrace{\left(3x^2-a^2\right)^2\left(y^2-b^2\right)^4}_{I_1}+\underbrace{2\left(3x^2-a^2\right)\left(3y^2-b^2\right)\left(y^2-b^2\right)^2\left(x^2-a^2\right)^2}_{0}+\underbrace{\left(3y^2-b^2\right)^2\left(x^2-a^2\right)^4}_{I_2}\right]dxdy$$

The potentials energy is given by,

$$\Omega=-2\int_{-a}^{a}\int_{-b}^{b}C_1q_o\left(x^2-a^2\right)^2\left(y^2-b^2\right)^2dxdy \tag{E7.5}$$

The integrations values are as follows,

$$\int_0^a\left(x^2-a^2\right)^2dx=\left[\frac{x^5}{5}-2\frac{a^2x^3}{3}+a^4x\right]_0^a=\frac{8}{15}a^5$$

$$\int_0^b\left(y^2-b^2\right)^2dy=\frac{8}{15}b^5$$

Therefore,

$$\Omega=-\frac{128}{15}C_1q_0a^5b^5$$

And,

$$\frac{\partial \Omega}{\partial C_1} = -\frac{128}{15} q_0 a^5 b^5$$

We have,

$$I_1 = \int_0^a \left(3x^2 - a^2\right)^2 dx = \int_0^a \left(9x^4 - 6a^2x^2 + a^4\right) dx = \frac{4}{5} a^5$$

Similarly,

$$\int_0^b \left(3y^2 - b^2\right)^2 dy = \frac{4}{5} b^5$$

$\int_0^b \left(y^2 - b^2\right)^4 dy = \frac{128}{315} b^4$ see following work on Ritz,

$$\int_0^a \left(x^2 - a^2\right)^4 dx = \frac{128}{315} a^4$$

$$\int_0^b \left(y^2 - b^2\right)^2 dy = \int_0^b \left[y^4 - 2b^2y^2 + b^4\right] dy = \left[\frac{y^5}{5} - 2\frac{b^2y^3}{3} + b^4 y\right]_0^b = \frac{8}{15} b^5$$

$$\int_0^a \left(x^2 - a^2\right)^2 dx = \frac{8}{15} a^5$$

$$\int_0^a \left(3x^2 - a^2\right) dx = \left[x^3 - a^2 x\right]_0^a = a^3 - a^3 = 0 \tag{E7.6}$$

$$\frac{\partial U_b}{\partial C_1} = 16DC_1 \left[\frac{4}{5} a^5 \frac{128}{315} b^9 + \frac{4}{5} b^5 \frac{128}{315} a^9\right] - 5.2DC_1 \left[a^5 b^9 + a^9 b^5\right] \tag{E7.7a}$$

Therefore Eq. (E7.4) becomes,

$$5.2DC_1 \left[a^5 b^9 + a^9 b^5\right] = 8.5 q_0 a^5 b^5$$

From which,

$$C_1 = \frac{1.641 q_0 a^5 b^5}{D\left[a^5 b^9 + a^9 b^5\right]}$$

Or,

$$C_1 = \frac{1.641 q_0}{D\left[b^4 + a^4\right]}$$

The max. deflection at $x = 0, y = 0$ by using the value of C_1 in Eq. (E7.1) gives,

$$w_{\max} = \frac{1.641 q_0 a^4 b^4}{D\left[b^4 + a^4\right]} \tag{E7.7b}$$

1. Ritz Method,

$$w(x,y) = C_1\left(x^2-a^2\right)^2\left(y^2-b^2\right)^2 \tag{E7.7}$$

$$g = \overline{m} \iint w^2(x,y)\,dxdy$$

mass per unit area

$$U = \frac{D}{2}\int_0^a\int_0^b\left(\frac{\partial^2 w}{\partial x^2}+\frac{\partial^2 w}{\partial y^2}\right)^2 dxdy$$

Ritz Method,

$$\frac{\partial U_{b_{\max}}}{\partial c_1} - \frac{1}{2}w^2\frac{\partial g}{\partial C_1} = 0$$

w = lowest natural frequency.

Galerkin's Method, Assume that,

$$w(x,y) = C_1\psi_{mn}(x,y) = C_1 X_m(x)Y_n(y) \tag{E7.8}$$

Where,

$$X_m(x) = \left(x^2-a^2\right)^2$$

And,

$$Y_n(y) = \left(y^2-b^2\right)^2$$

We have,

$$\iint_A\left[D\nabla^2\nabla^2 w(x,y) - \bar{m}\omega^2 w(x,y)\right]w(x,y)\,dxdy$$

$$\iint_A\left[D(\nabla^4(X_mY_n)) - \bar{m}\omega^2 X_mY_n\right]X_mY_n\,dxdy = 0$$

Or,

$$\nabla^4(X_mY_n) = X_m''''Y_n + 2X_m''Y_n'' + Y_n''''X_m$$

$$\iint_A\left[D\left(X_m''''Y_n^2X_m + 2X_m''Y_n''X_mY_n + Y_n''''X_m^2Y_n\right) - \bar{m}\omega^2X_m^2Y_n^2\right]dxdy = 0$$

Define,

$$I_1 = \iint_A\left(X_m''''Y_n^2X_m + 2X_m''Y_n''X_mY_n + Y_n''''X_m^2Y_n\right)dxdy$$

$$I_2 = \iint_A X_m^2Y_n^2\,dxdy$$

$$\omega^2 = \frac{I_1}{I_2}\frac{D}{\overline{m}} \tag{E7.9}$$

We have,

$$X_m' = 2(2x)\left(x^2 - a^2\right) \; 4x\left(x^2 - a^2\right) \; 4\left(x^3 - a^2x\right)$$

$$X_m'' = 4\left(3x^2 - a^2\right)$$

$$X_m''' = 24x$$

$$X_m'''' = 24$$

$$Y_n'' = 4\left(3y^2 - b^2\right)$$

$$Y_n'''' = 24$$

Define,

$$I_3 = \int_{-a}^{a} X_m^2 X_m dx$$

$$I_4 = \int_{-b}^{b} Y_n^2 dy$$

$$I_5 = \int_{-a}^{a} X_m'''' X_m dx$$

$$I_6 = \int_{-a}^{a} X_m'' X_m dx$$

$$I_7 = \int_{-b}^{b} Y_n'' Y_n dy$$

$$I_8 = \int_{-b}^{b} Y_n'''' Y_n dy$$

$$\frac{I_1}{I_2} = \frac{I_4 I_5 + 2 I_6 I_7 + I_3 I_8}{I_3 I_4}$$

$$I_3 = \int_{-a}^{a} \left(x^2 - a^2\right)^4 dx = \int_{-a}^{a} \left[x^4 - 2a^2x^2 + a^4\right]^2 dx = \frac{256}{315}a^9$$

Similarly,

$$I_4 = \frac{256}{315}b^9$$

$$I_5 = 24\int_{-a}^{a} \left(x^2 - a^2\right)^2 dx = 24\int_{-a}^{a} \left(x^4 - 2a^2x^2 + a^4\right) dx = \frac{384}{15}a^5$$

Similarly,

$$I_8 = \frac{384}{15}b^5$$

$$I_6 = \int_{-a}^{a} X_m'' X_m dx = 8\int_{-a}^{a} (3x^2 - a^2)(x^2 - a^2)^2 dx$$
$$= 8\int_{-a}^{a} (3x^2 - a^2)(x^4 - 2a^2x^2 + a^4)dx = -\frac{256}{105}a^7$$

Similarly,

$$I_7 = -\frac{256}{105}b^7$$

$$\frac{I_1}{I_2} = \frac{\frac{256}{315}b^9.\frac{384}{15}a^5 + 2.\left(\frac{256}{105}\right)^2 a^7.b^7 + \frac{256}{315}a^9.\frac{384}{15}b^5}{\left(\frac{256}{315}\right)^2 a^9 b^9} = \frac{31.5}{a^4} + \frac{18}{a^2b^2} + \frac{31.5}{b^4} \tag{E7.10}$$

From Eq. (E7.9),

$$\omega^2 = \frac{I_1}{I_2}\frac{D}{\bar{m}}$$

$$\omega = \sqrt{\frac{31.5}{a^4} + \frac{18}{a^2b^2} + \frac{31.5}{b^4}}\sqrt{\frac{D}{\bar{m}}}$$

Ritz Method,
We have,

$$g = \bar{m}C_1^2 \int_{-a}^{a}\int_{-b}^{b} (x^2 - a^2)^4 (y^2 - b^2)^4 dxdy \tag{E7.11}$$

And,

$$\int_{-a}^{a} (x^2 - a^2)^4 dx = \frac{256}{315}a^9$$

$$\int_{-b}^{b} (y^2 - b^2)^4 dy = \frac{256}{315}b^9$$

Therefore,

$$g = \bar{m}C_1^2 \left(\frac{256}{315}\right)^2 a^9 b^9$$

And,

$$\frac{\partial g}{\partial C_1} = 2\bar{m}C_1 \left(\frac{256}{315}\right)^2 a^9 b^9 \tag{E7.12}$$

After finding U_b, Find $\frac{\partial U_b}{\partial C_1}$,

$$\text{Ritz} \quad \frac{\partial U_{b_{\max}}}{\partial C_1} - \frac{1}{2}\omega^2 \frac{\partial g}{\partial C_1} = 0$$

From which the natural frequency is obtained, by using Eqs. (E7.7a) and (E7.12) as follows,

$$\omega^2 = \left(16DC_1\left[\frac{4}{5}a^5\frac{128}{315}b^9 + \frac{4}{5}b^5\frac{128}{315}a^9\right] - 5.2DC_1\left[a^5b^9 + a^9b^5\right]\right) \Big/ \left(\overline{m}C_1\left(\frac{256}{315}\right)^2 a^9b^9\right) \quad \text{(E7.13)}$$

Example 4.8: Assuming that the plate deflects according to,

$$w = A\cos\omega t\left(1 + \cos\frac{\pi x}{a}\right)\left(1 + \cos\frac{\pi y}{a}\right) \quad \text{(E8.1)}$$

Establish the equations of free vibration for the system.
The plate is of thickness h and the material had density ρ.
It will be recalled that Lagrange's equations of motion are,

$$\frac{d}{dt}\left(\frac{dT}{d\dot{q}_1}\right) - \frac{\partial U}{\partial q_1} = 0 \quad \text{(E8.2)}$$

It will be convenient to denote the vertical displacement of the center of the plate and of M as $q_1$and q_2 respectively.

Solution,

It's permissible to discard the terms in square brackets. If the edges are rigidly clamped, this is because that the square bracket only contribute to the line integral along the boundary. This can either be the clamped to the edge in the natural boundary conditions. In this clamped edge case the boundary conditions are prescribed, therefore this square bracket term matches no contribution to the natural boundary conditions (boundary displacement w and dw/dn are prescribed all around the clamped edge of the plate).

From Eq. (E8.1),

$$w(x,y) = A\cos\omega t\left(1 + \cos\frac{\pi x}{a}\right)\left(1 + \cos\frac{\pi y}{a}\right)$$

Therefore,

$$\frac{\partial w}{\partial x} = A\cos\omega t\left(1 + \cos\frac{\pi y}{a}\right)\left(-\frac{\pi}{a}\sin\frac{\pi x}{a}\right)$$

$$\frac{\partial^2 w}{\partial x^2} = A\cos\omega t\left(1 + \cos\frac{\pi y}{a}\right)\left(-\left(\frac{\pi}{a}\right)^2\cos\frac{\pi x}{a}\right)$$

Similarly,

$$\frac{\partial^2 w}{\partial y^2} = A \cos \omega t\left(1 + \cos \frac{\pi x}{a}\right)\left(-\left(\frac{\pi}{a}\right)^2 \cos \frac{\pi y}{a}\right)$$

And,

$$\left[\frac{\partial^2 w}{\partial x^2} + \frac{\partial^2 w}{\partial y^2}\right]^2 = (A \cos \omega t)^2\left[\left(\frac{\pi}{a}\right)^2 \cos \frac{\pi x}{a}\left(1+ \cos \frac{\pi y}{a}\right) + \left(\frac{\pi}{a}\right)^2(-1)\left(1+ \cos \frac{\pi x}{a}\right) \cos \frac{\pi y}{a}\right]^2$$

Or,

$$= (A \cos \omega t)^2\left(\frac{\pi}{a}\right)^4\left\{ \cos \frac{\pi x}{a}\left(1+ \cos \frac{\pi y}{a}\right) + \cos \frac{\pi y}{a}\left(1+ \cos \frac{\pi x}{a}\right)\right\}^2$$

$$= A^2 \cos^2\omega t\left(\frac{\pi}{a}\right)^4\left\{\cos^2 \frac{\pi x}{a} + \cos^2 \frac{\pi x}{a} \cos^2 \frac{\pi y}{a} + 2 \cos \frac{\pi y}{a} \cos^2 \frac{\pi x}{a} + \cos^2 \frac{\pi y}{a} + \cos^2 \frac{\pi y}{a} \cos^2 \frac{\pi x}{a}\right.$$

$$\left. + 2 \cos^2 \frac{\pi y}{a} \cos \frac{\pi x}{a} + 2 \cos \frac{\pi x}{a}\left(1 + \cos \frac{\pi y}{a}\right) \cos \frac{\pi y}{a}\left(1 + \cos \frac{\pi x}{a}\right)\right\} \quad \text{(E8.3)}$$

Simplification Eq. (E8.3) given,

$$= A^2 \cos^2\omega t\left(\frac{\pi}{a}\right)^4\left\{\cos^2 \frac{\pi x}{a} + \cos^2 \frac{\pi y}{a} + 2 \cos \frac{\pi x}{a} \cos \frac{\pi y}{a}\right.$$

$$\left. + 4 \cos^2 \frac{\pi x}{a} \cos \frac{\pi y}{a} + 4 \cos \frac{\pi x}{a} \cos^2 \frac{\pi y}{a} + 4 \cos^2 \frac{\pi x}{a} \cos^2 \frac{\pi y}{a}\right\} \quad \text{(E8.4)}$$

Now, the strain energy of the plate is given by,

$$U = \frac{D}{2}\int_{-a}^{a}\int_{-a}^{a}\left(\frac{\partial^2 w}{\partial x^2} + \frac{\partial^2 w}{\partial y^2}\right)^2 dx\, dy \quad \text{(E8.5)}$$

We have,

$$\int_{-a}^{a} \cos \frac{\pi x}{a} dx = 0$$

$$\int_{-a}^{a} \cos^2 \frac{\pi x}{a} dx = a$$

Therefore,

$$U = \frac{D}{2}\left(\frac{\pi}{a}\right)^4 A^2 \cos^2\omega t \int_{-a}^{a}\left[a + a \cos^2 \frac{\pi y}{a} + 0 + 4a \cos^2 \frac{\pi y}{a} + 4a \cos^2 \frac{\pi y}{a}\right] dy \quad \text{(E8.6)}$$

Or,

$$= \frac{D}{2}\left(\frac{\pi}{a}\right)^4 A^2 \cos^2\omega t\{2a^2 + a^2 + 4a^2\} = \frac{D}{2}\left(\frac{\pi}{a}\right)^4 A^2 \cos^2\omega t.7a^2$$

Therefore,

$$U = \frac{7}{2}a^2D\left(\frac{\pi}{a}\right)^4 A^2\cos^2\omega t = \frac{7}{32}a^2D\left(\frac{\pi}{a}\right)^4 q_1^2 \tag{E8.7}$$

The central deflection,

$$w_{\text{center}} = A\cos\omega t\left(1+\cos\frac{\pi 0}{a}\right)\left(1+\cos\frac{\pi 0}{a}\right) = 4A\cos\omega t = q_1$$

$$\dot{q}_1 = -4A\omega\sin\omega t$$

$$q_2 = \text{Vertical displacement of the mass}\quad M.$$

$$\text{K.E of the mass} = \frac{1}{2}M\dot{q}_2^2$$

$$\text{P.E of springs} = \frac{1}{2}k(q_1-q_2)^2$$

$$\text{P.E of the plate} = \frac{1}{2}\int_{-a}^{a}\int_{-a}^{a}\rho h dxdy\left(\frac{dw}{dt}\right)^2 \tag{E8.8}$$

Or,

$$= \frac{1}{2}\int_{-a}^{a}\int_{-a}^{a}\rho h\left[-\omega A\sin\omega t\left(1+\cos\frac{\pi x}{a}\right)\left(1+\cos\frac{\pi y}{a}\right)\right]^2 dxdy$$

$$= \frac{1}{2}\rho h\omega^2 A^2\sin\omega t\iint\left[\left(1+2\cos\frac{\pi x}{a}+\cos^2\frac{\pi x}{a}\right)\left(1+2\cos\frac{\pi y}{a}+\cos^2\frac{\pi y}{a}\right)\right]dx\,dy$$

$$= \frac{1}{2}\rho h\omega^2 A^2\sin\omega t\iint\left(1+2\cos\frac{\pi y}{a}+\cos^2\frac{\pi y}{a}+2\cos\frac{\pi x}{a}+4\cos\frac{\pi x}{a}\cos\frac{\pi y}{a}+2\cos\frac{\pi x}{a}\cos^2\frac{\pi y}{a}\right.$$

$$\left.+\cos^2\frac{\pi x}{a}2\cos^2\frac{\pi x}{a}\cos\frac{\pi y}{a}+\cos^2\frac{\pi x}{a}\cos^2\frac{\pi y}{a}\right)dxdy$$

Carrying out the integration gives,

$$= \frac{1}{2}\rho h\omega^2 A^2\left[4a^2+2a^2+2a^2+a^2\right] = \frac{9}{2}\rho h a^2\omega^2 A^2\sin\omega t \tag{E8.9}$$

We have,

$$\dot{q}_1 = -4A\omega\sin\omega t$$

$$\dot{q}_1^2 = 16\omega^2 A^2\sin^2\omega t$$

Therefore,

$$\text{K.E of plate} = \frac{9}{32}\rho h a^2\dot{q}_1^2 \tag{E8.10}$$

We have,

$$V = U + \text{PE} = \frac{7a^2D}{32}\left(\frac{\pi}{4}\right)^2 q_1^2 + \overbrace{\frac{1}{2}k(q_1 - q_2)^2}^{\text{for spring}}$$

$$T = \underbrace{\frac{1}{2}M\dot{q}_2^2}_{\text{for } M} + \frac{9}{32}\rho h a^2 \dot{q}_1^2$$

We have,

$$\frac{\partial T}{\partial \dot{q}_1} = \frac{9}{16}\rho h a^2 \dot{q}_1, \frac{\partial T}{\partial \dot{q}_2} = M\dot{q}_2$$

$$\frac{\partial V}{\partial q_1} = \frac{7a^2D}{16}\left(\frac{\pi}{4}\right)^2 q_1 + k(q_1 - q_2)$$

$$\frac{\partial V}{\partial q_2} = -k(q_1 - q_2)$$

The equation of motion (free vibration), is,

$$\frac{9}{16}\rho h^2 \ddot{q}_2 - \frac{7a^2D}{16}\left(\frac{\pi}{4}\right)^2 q_1 + k(q_1 - q_2) = 0$$

In matrix form,

$$[M]\{\ddot{q}\} - [K]\{q\} = 0 \tag{E8.11}$$

Example 4.9: A simply supported non uniform circular shaft of length $2L$ carries a disk of mass $M = m_0L/a$ at mid span. The non-uniformity is symmetrical about mid span and varies such that the shaft diameter is proportional to $\sqrt{\left(1 - \frac{x}{2L}\right)}$ for $0 \le x \le L$ where the origin of the co-ordinates is taken at the center as shown in Fig. 4.9.

To obtain an estimate for the fundamental natural frequency the Rayleigh-Ritz method is to be used. Assuming a deflected shape,

$$\varnothing(x) = c_1 \cos\frac{\pi x}{2L} + c_3 \cos\frac{3\pi x}{2L} \tag{E9.1}$$

Establish the eigenvalue problem in terms of $L.E, I_0$, the second moment of area of the shaft at the mid span, and m_0, the mass/unit length of the shaft at the mid span. You may accept the following integrals,

$$\int_0^L \cos\frac{n\pi x}{L}dx = 0$$

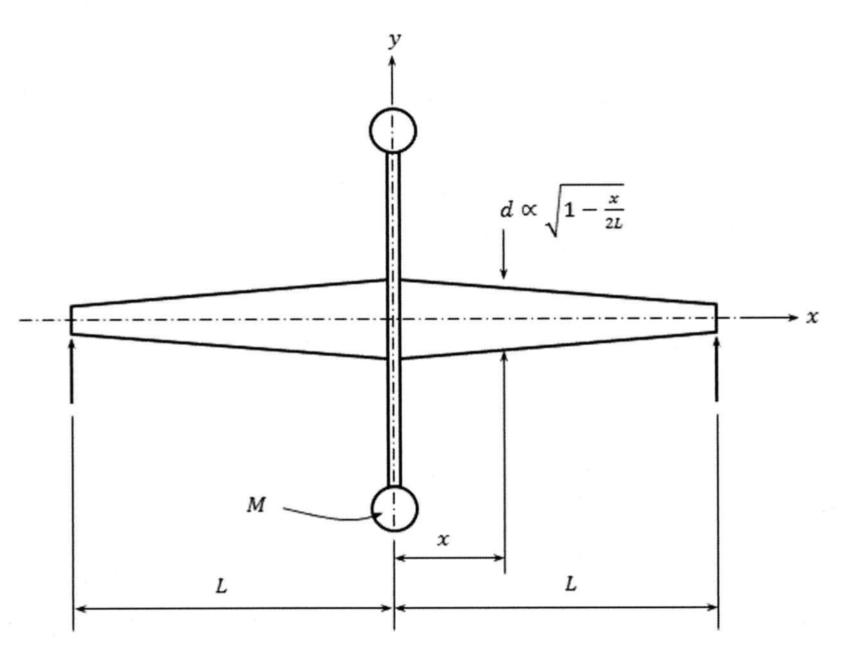

FIGURE 4.9 Non-uniform circular shaft carries a disk.

$$\int_0^L \frac{x}{L}\cos\frac{n\pi x}{L}dx = \begin{cases} 0 & \text{if } n \text{ is even} \\ -\frac{2L}{n^2\pi^2} & \text{if } n \text{ is odd} \end{cases}$$

$$\int_0^L \left(\frac{x}{L}\right)^2 \cos\frac{n\pi x}{L}dx = \begin{cases} \frac{2L}{n^2\pi^2} & \text{if } n \text{ is even} \\ -\frac{2L}{n^2\pi^2} & \text{if } n \text{ is odd} \end{cases} \tag{E9.2}$$

Solution,
From Eq. (E9.1),

$$\varnothing(x) = c_1 \cos\frac{\pi x}{2L} + c_3 \cos\frac{3\pi x}{2L}$$

And,

$$d \propto \sqrt{1-\frac{x}{2L}}\, 0 \le x \le L, dx = k\sqrt{1-\frac{x}{2L}} \tag{E9.3}$$

I_o = Second moment of area of the shaft at the mid span.
We have,

$$\varnothing(x) = c_1 \cos\frac{\pi x}{2L} + c_3 \cos\frac{3\pi x}{2L}$$

We have,

$$I_{xx} = \int_0^R 2\pi r.drr^2 = \frac{\pi}{32}d^4$$

$$I_0 = \frac{\pi}{32}d_0^4$$

$$d = k\sqrt{1 - \frac{x}{2L}}$$

At $=0, d = d_0 = k\sqrt{1-0}, k = d_0$

$$d = d_0\sqrt{1 - \frac{x}{2L}},$$

$$I = \frac{\pi}{32}\left[d_0\sqrt{1-\frac{x}{2L}}\right]^4$$

$$I = I_0\left(1-\frac{x}{2L}\right)^2 \quad \text{(E9.4)}$$

And,

$$\varnothing'(x) = -\frac{\pi}{2L}c_1 \sin\frac{\pi x}{2L} - \frac{3\pi}{2L}c_3 \sin\frac{3\pi x}{2L}$$

$$\varnothing''(x) = -\left(\frac{\pi}{2L}\right)^2 c_1 \cos\frac{\pi x}{2L} - \left(\frac{3\pi}{2L}\right)^2 c_3 \cos\frac{3\pi x}{2L} \quad \text{(E9.5)}$$

Let,

$$w(x) = \varnothing(x)\sin\Omega t$$

$$\dot{w}(x) = \varnothing(x)\Omega\cos\Omega t$$

$$\ddot{w}(x) = -\varnothing(x)\Omega^2\sin\Omega t$$

$$m_0dx = \frac{\pi}{4}d_0^2, mdx = \frac{\pi}{4}d^2dx \quad \text{(E9.6)}$$

$$m = m_0\left(1 - \frac{x}{2L}\right)$$

Free vibration,

$$V_{\text{shaft}} = \frac{1}{2}\int_{-L}^{L} EI(x)\left[\phi''(x)\right]^2 dx \quad \text{(E9.7)}$$

$$\bar{V}_{\text{shaft}} = \frac{1}{2}\int_{-L}^{L} EI(x)\left[\phi''(x)\right]^2 dx$$

$$T_{\text{shaft}} = \frac{1}{2}\int_{-L}^{L} m(x)dx(\dot{w})^2 \tag{E9.8}$$

$$\bar{T}_{\text{shaft}} = \frac{1}{2}\int_{-L}^{L} M_0\left(1-\frac{x}{2L}\right)dx\varnothing^2$$

$$T_{\text{disk}} = \frac{1}{2}M\left(\dot{w}^2\right)_{x=0} \tag{E9.9}$$

$$\dot{T}_{\text{disk}} = \frac{1}{2}M\varnothing_{x=0}^2 = \frac{1}{2}M(c_1+c_3)^2$$

Therefore,

$$\bar{T} = \int_0^l m_0\left(c_1\cos\frac{\pi x}{2L}+c_3\cos\frac{3\pi x}{2L}\right)^2\left(1-\frac{x}{2L}\right)dx + \frac{1}{2}m_0.\frac{1}{4}(c_1+c_3)^2 \tag{E9.10}$$

And,

$$\bar{V} = \int_0^L EI(x)\left[\phi''(x)\right]^2 dx = \int_0^L EI(x)\left[\left(\frac{\pi}{2L}\right)^2 c_1\cos\frac{\pi x}{2L}+\left(\frac{3\pi}{2L}\right)^2 c_3\cos\frac{3\pi x}{2L}\right]^2 dx \tag{E9.11}$$

We have,

$$\frac{\partial\bar{V}}{\partial c_1} = \int_0^L 2EI_0\left(1-\frac{x}{2L}\right)^2\left(\left(\frac{\pi}{2L}\right)^2 c_1\cos\frac{\pi x}{2L}+\left(\frac{3\pi}{2L}\right)^2 c_3\cos\frac{3\pi x}{2L}\right)\left(\left(\frac{\pi}{2L}\right)^2\cos\frac{\pi x}{2L}\right)dx$$

$$= \left(\begin{matrix} 2EI_0\int_0^L\left(1-\frac{x}{2L}\right)^2\left[\left(\frac{\pi}{2L}\right)^4\cos^2\frac{\pi x}{2L}\right]dxc_1 + \\ 2EI_0\int_0^L\left(1-\frac{x}{2L}\right)^2\left[\left(\frac{\pi}{2L}\right)^2\left(\frac{3\pi}{2L}\right)^2\cos\frac{\pi x}{2L}\cos\frac{3\pi x}{2L}\right]dxc_3 \end{matrix}\right) = A_{11}c_1 + A_{12}c_3 \tag{E9.12}$$

Where,

$$A_{11} = 2EI_0\int_0^L\left(1-\frac{x}{2L}\right)^2\left[\left(\frac{\pi}{2L}\right)^4\cos^2\frac{\pi x}{2L}\right]dx = EI_0\left(\frac{\pi}{2L}\right)^4 L(0.046)$$

$$A_{12} = 2EI_0 \int_0^L \left(1 - \frac{x}{2L}\right)^2 \left[\left(\frac{\pi}{2L}\right)^2 \left(\frac{3\pi}{2L}\right)^2 \cos\frac{\pi x}{2L} \cos\frac{3\pi x}{2L}\right] dx = 0.093\ EI_0 \frac{\pi^4}{L^3}$$

And,

$$\frac{\partial \bar{T}}{\partial c_1} = m_0 \left\{ 2\int_0^l \left(1 - \frac{x}{2L}\right)\left(c_1 \cos\frac{\pi x}{2L} + c_3 \cos\frac{3\pi x}{2L}\right) \cos\frac{\pi x}{2L} dx + \frac{m_0 L}{4}(c_1 + c_3)^2 \right\} = B_{11}c_1 + B_{12}c_3 \tag{E9.13}$$

Where,

$$B_{11} = 2m_0 \int_0^L \left(1 - \frac{x}{2L}\right) \cos^2\frac{\pi x}{2L} dx + \frac{m_o L}{4} = 1.101\ m_0 L$$

And,

$$B_{12} = 2m_0 \int_0^L \left(1 - \frac{x}{2L}\right) \cos\frac{3\pi x}{2L} \cos\frac{\pi x}{2L} dx + \frac{m_o L}{4} = 0.453\ m_0 L$$

$$\frac{\partial \bar{V}}{\partial c_2} = A_{21}c_1 + A_{22}c_3$$

$$= \int_0^L 2EI_0\left(1 - \frac{x}{2L}\right)^2 \left(\left(\frac{\pi}{2L}\right)^2 c_1 \cos\frac{\pi x}{2L} + \left(\frac{3\pi}{2L}\right)^2 c_3 \cos\frac{3\pi x}{2L}\right)\left(\left(\frac{3\pi}{2L}\right)^2 \cos\frac{3\pi x}{2L}\right) dx \tag{E9.14}$$

Where,

$$A_{21} = 2EI_0 \int_0^L \left(1 - \frac{x}{L} + \frac{1}{4}\left(\frac{x}{L}\right)^2\right)\left(\frac{\pi}{2L}\frac{3\pi}{2L}\right)^2 \cos\frac{\pi x}{2L} \cos\frac{3\pi x}{2L} dx$$

And,

$$A_{22} = 2EI_0 \int_0^L \left(1 - \frac{x}{L} + \frac{1}{4}\left(\frac{x}{L}\right)^2\right)\left(\frac{3\pi}{2L}\right)^4 \cos^2\frac{3\pi x}{2L} dx$$

$$A_{21} = A_{12} = 0.093\ EI_0 \frac{\pi^4}{L^3}$$

$$A_{22} = 3.039 \frac{\pi^4}{L^3} EI_0$$

Eqs. (E9.12) and (E9.14) arrange in matrix form,

$$\frac{\pi^4}{L^3}EI_0\begin{bmatrix}0.046 & 0.093\\ 0.093 & 3.039\end{bmatrix}\begin{bmatrix}c_1\\ c_2\end{bmatrix}=\begin{bmatrix}\dfrac{\partial\bar{V}}{\partial c_1}\\ \dfrac{\partial\bar{V}}{\partial c_2}\end{bmatrix}$$

And,

$$\frac{\partial\bar{T}}{\partial c_3}=m_0\left\{2\int_0^l\left(1-\frac{x}{2L}\right)\left(c_1\cos\frac{\pi x}{2L}+c_3\cos\frac{3\pi x}{2L}\right)\cos\frac{3\pi x}{2L}dx+\frac{m_oL}{4}(c_1+c_3)\right\}$$
$$=B_{21}c_1+B_{22}c_3 \tag{E9.15}$$

Where,

$$B_{21}=2m_0\int_0^l\left(1-\frac{x}{2L}\right)\cos\frac{\pi x}{2L}\cos\frac{3\pi x}{2L}dx+\frac{m_0L}{4}=0.453\ m_0L$$

And,

$$B_{22}=2m_0\int_0^l\left(1-\frac{x}{2L}\right)\cos^2\frac{3\pi x}{2L}dx+\frac{m_0L}{4}=1.011\ m_0L$$

Arrange Eqs. (E9.13) and (E9.15) in matrix form, gives,

$$\begin{Bmatrix}\dfrac{\partial\bar{T}}{\partial c_1}\\ \dfrac{\partial\bar{T}}{\partial c_3}\end{Bmatrix}=m_0L\begin{bmatrix}1.101 & 0.453\\ 0.453 & 1.011\end{bmatrix}\begin{bmatrix}c_1\\ c_3\end{bmatrix}$$

Or,

$$\frac{\pi^4}{L^3}EI_0\begin{bmatrix}0.046 & 0.093\\ 0.093 & 3.039\end{bmatrix}\begin{bmatrix}c_1\\ c_2\end{bmatrix}=\Lambda^2m_0L\begin{bmatrix}1.101 & 0.453\\ 0.453 & 1.011\end{bmatrix}\begin{bmatrix}c_1\\ c_3\end{bmatrix}$$
$$\begin{bmatrix}0.046 & 0.093\\ 0.093 & 3.039\end{bmatrix}\begin{bmatrix}c_1\\ c_2\end{bmatrix}=\Lambda^2\frac{m_0L}{\frac{\pi^4}{L^3}EI_0}\begin{bmatrix}1.101 & 0.453\\ 0.453 & 1.011\end{bmatrix}\begin{bmatrix}c_1\\ c_3\end{bmatrix} \tag{E9.16}$$

This is the frequency equation.

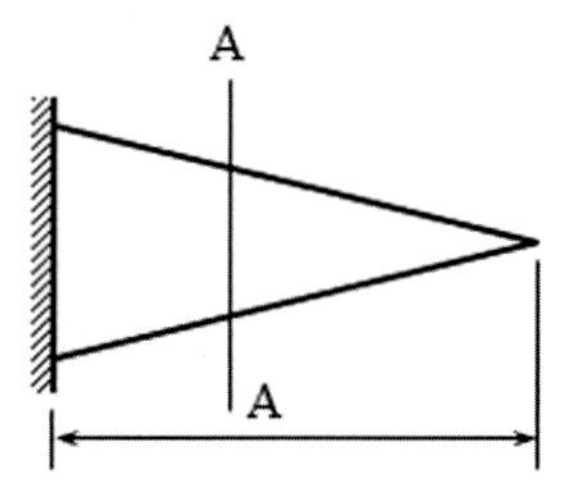

FIGURE 4.10 Simplified model for propeller blade.

Example 4.10: An aircraft propeller blade shown in Fig. 4.10, may be represented, for the purpose of approximate vibration analysis, by a tapered beam. The cross sectional area of the beam at a distance x from the root is given by,

$$A_x = A_0\left(1 - \frac{x}{L}\right) \tag{E10.1}$$

And the second moment of area is given by,

$$I_x = I_0\left(1 - \frac{x}{L}\right)^2 \tag{E10.2}$$

Where, A_0 is the cross sectional area at the root, I_0 is the second moment of area at the root and L is the length of the beam. The density of the material of the propeller is ρ and the propeller is assumed to be rigidly built into the hub.

In order to compute the fundamental natural frequency of the propeller blade the Rayleigh-Ritz method is to be used assuming a vibrating shape,

$$y(x) = c_1\psi_1(x) + c_2\psi_2(x) \tag{E10.3}$$

Where,

$$\psi_1(x) = \left(\frac{x}{L}\right)^2 \text{and} \quad \psi_2(x) = \left(\frac{x}{L}\right)^3 \tag{E10.4}$$

Calculate the fundamental natural frequency.

Explain very briefly the characteristics which distinguish the following functions when applied to the Rayleigh-Ritz method of vibration analysis,

1. An admissible function.
2. A comparison function.
3. An Eigen function.

Solution,
We have,

$$A_x = A_0\left(1 - \frac{x}{L}\right) \text{that is at } x = l, A_x = 0 \tag{E10.5}$$

$$I_x = I_0\left(1 - \frac{x}{L}\right)^2$$

$$\rho = density$$

We have from Eq. (E10.2),

$$y(x) = c_1\psi_1(x) + c_2\psi_2(x)$$

And,

$$\psi_1(x) = \left(\frac{x}{L}\right)^2, \psi_2(x) = \left(\frac{x}{L}\right)^3$$

And,

$$y(x) = c_1\left(\frac{x}{L}\right)^2 + c_2\left(\frac{x}{L}\right)^3$$

$$y' = 2c_1\frac{x}{L^2} + 3c_2\frac{x^2}{L^2}$$

$$y'' = \frac{2c_1}{L^2} + 6c_2\frac{x}{L^2}$$

By Rayleigh-Ritz method, the potential and kinetic energy are given by,

$$V = \frac{1}{2}\bar{V}q^2, T = \frac{1}{2}\bar{T}\dot{q}^2 \tag{E10.6}$$

We have,

$$\bar{V} = \int_0^l EIy''^2 dx = E\int_0^l I_0\left(1-\frac{x}{L}\right)^2\left(\frac{2c_1}{L^2} + 6c_2\frac{x}{L^3}\right)^2 dx$$

$$\bar{V} = EI_0\int_0^l \left(1-\frac{x}{L}\right)^2\left(\frac{2c_1}{L^2} + 6c_2\frac{x}{L^3}\right)^2 dx \tag{E10.7}$$

And,

$$\bar{T} = \int_0^l \rho A y^2 dx$$

$$\bar{T} = \rho\int_0^l A_0\left(1-\frac{x}{L}\right)\left(c_1\left(\frac{x}{L}\right)^2 + c_2\left(\frac{x}{L}\right)^3\right)^2 dx$$

Or,

$$\bar{T} = \rho A_0\int_0^l \left(1-\frac{x}{L}\right)^2\left(\frac{c_1x^2}{L^2} + \frac{c_2x^3}{L^3}\right)^2 dx$$

And,

$$\frac{\partial\bar{V}}{\partial c_1} = EI_0\int_0^l \left(1-\frac{x}{L}\right).2.\left(\frac{2c_1}{L^2} + 6c_2\frac{x}{L^3}\right)\frac{2}{L^2}dx$$

$$\frac{\partial\bar{V}}{\partial c_1} = \frac{4EI_o}{L^2}\int_0^l \left(1-\frac{x}{L}\right)\left(\frac{2c_1}{L^2} + 6c_2\frac{x}{L^3}\right)dx \tag{E10.8}$$

And,

$$\frac{\partial \bar{V}}{\partial c_2} = EI_0 \int_0^l \left(1 - \frac{x}{L}\right).2.\left(\frac{2c_1}{L^2} + 6c_2\frac{x}{L^3}\right).6.\frac{x}{L^3}dx$$

$$\frac{\partial \bar{V}}{\partial c_2} = \frac{12EI_0}{L^3}\int_0^l \left(1 - \frac{x}{L}\right)\left(\frac{2c_1}{L^2} + 6c_2\frac{x}{L^3}\right)dx \tag{E10.9}$$

Then,

$$\frac{\partial \bar{T}}{\partial c_1} = \rho A_0 \int_0^l \left(1 - \frac{x}{L}\right)^2 2\left(\frac{c_1x^2}{L^2} + \frac{c_2x^3}{L^3}\right)\frac{x^2}{L^2}dx$$

Or,

$$\frac{\partial \bar{T}}{\partial c_1} = \frac{2.\rho A_0}{L^2}\int_0^l \left(1 - \frac{x}{L}\right)^2\left(\frac{c_1x^4}{L^2} + \frac{c_2x^5}{L^3}\right)dx \tag{E10.10}$$

And,

$$\frac{\partial \bar{T}}{\partial c_2} = \rho A_0 \int_0^l \left(1 - \frac{x}{L}\right)^2 2\left(\frac{c_1x^2}{L^2} + \frac{c_2x^3}{L^3}\right)\frac{x^3}{L^3}dx$$

$$\frac{\partial \bar{T}}{\partial c_2} = \frac{2.\rho A_0}{L^3}\int_0^l \left(1 - \frac{x}{L}\right)^2\left(\frac{c_1x^5}{L^2} + \frac{c_2x^6}{L^3}\right)dx \tag{E10.11}$$

The Eigen Value problem is as follows,

$$[a]\{c\} = \Lambda^2[b]\{c\}$$

Simplifications Eq. (E10.8) gives,

$$\frac{\partial \bar{V}}{\partial c_1} = \frac{4EI_0}{L^2}\int_0^l \left(1 - \frac{x}{L}\right)\frac{2}{L^2}c_1dx + \frac{4EI_0}{L^2}\int_0^l \left(1 - \frac{x}{L}\right)\frac{6x}{L^3}c_2dx \tag{E10.12}$$

And from Eq. (E10.9) gives,

$$\frac{\partial \bar{V}}{\partial c_2} = \frac{12EI_0}{L^3}\int_0^l \left(1 - \frac{x}{L}\right)\frac{2x}{L^2}c_1dx + \frac{12EI_0}{L^3}\int_0^l \left(1 - \frac{x}{L}\right)\frac{6x^2}{L^3}c_2dx \tag{E10.13}$$

And from Eqs. (E10.10) and (E10.11) gives,

$$\frac{\partial \bar{T}}{\partial c_1} = \frac{2.\rho A_0}{L^2}\int_0^l \left(1-\frac{x}{L}\right)^2 \frac{x^4}{L^2} c_1 dx + \frac{2.\rho A_o}{L^2}\int_0^l \left(1-\frac{x}{L}\right)^2 \frac{x^5}{L^3} c_2 dx \tag{E10.14}$$

And,

$$\frac{\partial \bar{T}}{\partial c_2} = \frac{2.\rho A_0}{L^3}\int_0^l \left(1-\frac{x}{L}\right)^2 \frac{x^5}{L^2} c_1 dx + \frac{2.\rho A_0}{L^3}\int_0^l \left(1-\frac{x}{L}\right)^2 \frac{x^6}{L^3} c_2 dx \tag{E10.15}$$

Array Eqs. (E10.12)–(E10.15) gives,

$$\begin{bmatrix} \left(\frac{4EI_0}{L^2}\int_0^l \left(1-\frac{x}{L}\right)\frac{2}{L^2}dx\right) & \left(\frac{4EI_0}{L^2}\int_0^l \left(1-\frac{x}{L}\right)\frac{6x}{L^3}dx\right) \\ \frac{12EI_0}{L^3}\int_0^l \left(1-\frac{x}{L}\right)\frac{2x}{L^2}dx & \frac{12EI_0}{L^3}\int_0^l \left(1-\frac{x}{L}\right)\frac{6x^2}{L^3}dx \end{bmatrix} \begin{Bmatrix} c_1 \\ c_2 \end{Bmatrix} =$$

$$\frac{2.\rho A_0}{L^4}\begin{bmatrix} \left(\int_0^l \left(1-\frac{x}{L}\right)^2 x^4 dx\right) & \left(\frac{1}{l}\int_0^l \left(1-\frac{x}{L}\right)^2 x^5 dx\right) \\ \left(\frac{1}{l}\int_0^l \left(1-\frac{x}{L}\right)^2 x^5 dx\right) & \left(\frac{1}{l^2}\int_0^l \left(1-\frac{x}{L}\right)^2 x^6 c_2 dx\right) \end{bmatrix} \begin{Bmatrix} c_1 \\ c_2 \end{Bmatrix} \tag{E10.16}$$

Or,

$$[b]^{-1}[a]\{c\} = \Lambda\{c\}$$

Iteration for $[b]^{-1}[a]$ by an trial vector gives $\Lambda = \omega$,

$$\frac{\partial \bar{T}}{\partial c_1} = \frac{1}{\omega^2}\frac{\partial \bar{V}}{\partial c_1}$$

$$\frac{\partial \bar{T}}{\partial c_2} = \frac{1}{\omega^2}\frac{\partial \bar{V}}{\partial c_2}$$

$$\begin{bmatrix} \frac{\partial \bar{T}}{\partial c_1} \\ \frac{\partial \bar{T}}{\partial c_2} \end{bmatrix} = \frac{1}{\omega^2}\left[\frac{\partial \bar{V}}{\partial c}\right] \tag{E10.17}$$

The solution of Eqs. (E10.16) or (E10.17) gives the natural frequency.

Problems

P.4.1: (a) A square plate, side a, is clamped all round its edges. To estimate the fundamental natural frequency, assume a displacement model,

$$w(x,y) = q_1(t)\left(1 + \cos\frac{2\pi x}{a}\right)\left(1 + \cos\frac{2\pi y}{a}\right)$$

With the origin at the center. Confirm that this is indeed admissible for use in the Rayleigh-Ritz process and find the lowest natural frequency.

(b) Check the result of (a) by the Galerkin's method.

P.4.2: (a) A cantilever beam of uniform width b and length l tapers uniformly from a depth h at the clamped end to $\frac{h}{5}$ at the free end. Use the Rayleigh-Ritz process to estimate the first and second natural frequencies. Note that the exact solution is,

$$\omega_1 = 1.239\sqrt{(Eh^2/\rho l^4)},$$

$$\omega_2 = 4.545\sqrt{(Eh^2/\rho l^4)}$$

(b) If the beam has a concentrated mass equal to half the beam mass added to the 'free' end, determine the new value of the first two frequencies.

P.4.3: Find the lowest natural frequency of a rectangular plate ($2a$*$2b$) clamped on all edges using Rayleigh-Ritz method and check the results by the Galerkin's method.

P.4.4: in Problem P.4.1 If the plate is stiffened by compact ribs attached along the symmetry line $x = 0$ and $y = 0$. Each rib has stiffness EI and distributed mass $\frac{m}{\text{unit length}}$. Assuming

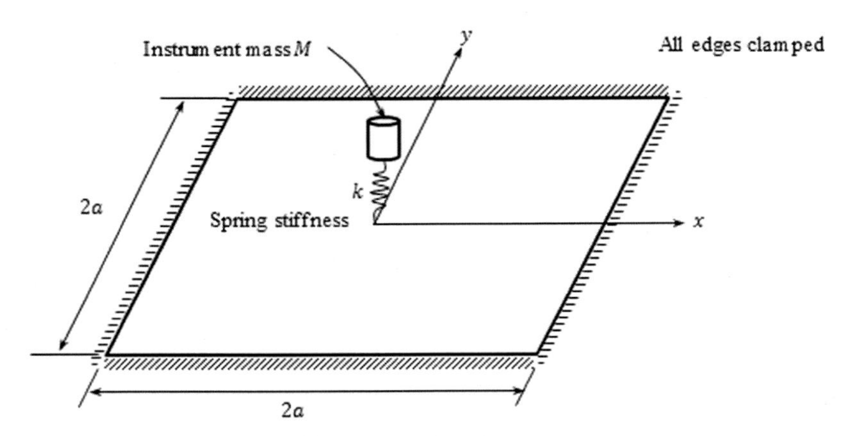

FIGURE P.4.4 A square plate supported by a spring of stiffness k.

that the first mode shape is not significantly affected by this design change, deduce a correction factor to apply the result of Problem P.4.1(a) to this new situation.

(c) Interpret the plate of (a) as a model for the base of a dynamical test rig. Equipment having mass M is supported by a spring of stiffness k at the center of the plate as shown in Fig. P.4.4 Determine the eigenvalue problem giving the two natural frequencies of the system.

P.4.5: The natural frequencies of transverse vibration of a cantilever beam are increased if a distributed axial load is applied. (Turbine blades are similarly affected). Proceed as follows to estimate the effect using,

$$w(x) = \sum_{i=1}^{n} q_1 . N_i(x)$$

In Fig. P.4.5, the centrifugal load dF_c has a potential energy,

$$d\Omega = \left(\bar{m}(x)\omega^2 x\, dx\right)\frac{1}{2}\int_0^x \left(\frac{\partial w}{\partial \xi}\right)^2 d\xi$$

So that,

$$\Omega = \frac{\omega^2}{2}\int_0^l \left(\bar{m}(x)x\left(\frac{\partial w}{\partial \xi}\right)^2 d\xi\right)dx$$

Use Hamilton's principle to generate the eigenvalue problem corresponding to this system. Express it in terms suitable for handling in a CAD package.

Confirm that, compared to the "static" situation, each stiffness coefficient is augmented to give,

$$k_{ij} = \int_0^l EI(x)N_i''(x)N_j''(x)dx + \omega^2\int_0^l \left(\bar{m}(x)x\int_0^x N_i'(\xi)N_j'(\xi)d\xi\right)dx$$

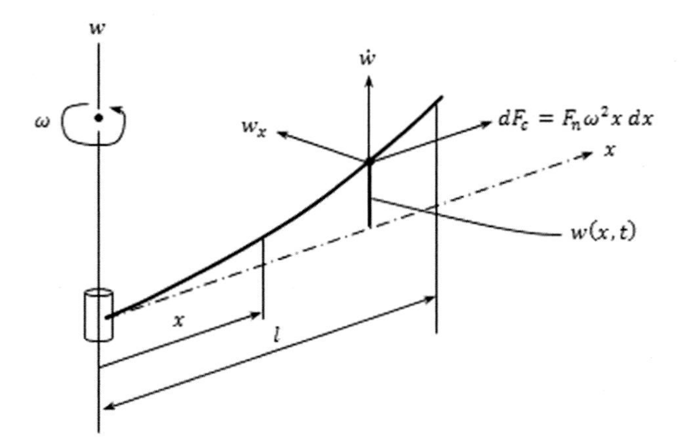

FIGURE P.4.5 A cantilever beam.

And each mass coefficient,

$$m_{ij} = \int_0^l \bar{m}\ N_i(x)N_j(x)dx + \omega^2 \int_0^l \bar{m}\ x^2 dx$$

P.4.6: A circular plate of radius a, mass/unit area ρh and flexural rigidity D is clamped along its edge. At the center is clamped a vertical blade length l, mass/unit length m and such a cross section that bending occurs most readily in the $r-y$ plane defined by $\theta = 0$ as shown in Fig. P.4.6, the appropriate bending stiffness being EI.

It is desired to compute the frequency at which the plate and blade assembly will vibrate when the plate vibrates in just one diametric mode alone. Assuming the plate vibrates a shape,

$$w(r,\theta,t) = \left[\left(r^3 - 2ar^2 + a^2 r\right)\cos\theta\right] q_1(t)$$

And the blade vibrates with a shape,

$$y(x,t) = \left[\frac{3}{x^2} - x^3\right] q_2(t)$$

Establish the equation of free vibrations for the system.
You may accept that for a circular plate with clamped edges,

$$U = \frac{D}{2}\int_0^{2\pi}\int_0^a \left(\frac{\partial^2 w}{\partial r^2} + \frac{1}{r}\frac{\partial w}{\partial r} + \frac{1}{r}\frac{\partial^2 w}{\partial \theta^2}\right)^2 r d\theta dr$$

$$V = \frac{\rho h}{2}\int_0^{2\pi}\int_0^a \dot{w}^2 r d\theta dr$$

Establish the equation of free vibrations for the system.

P.4.7: A heavy uniform shaft of moment of inertia/unit length I and torsional stiffness GJ is fixed at one and carries inertia $I_0 = Il/2$.

In order to estimate the fundamental natural frequency of torsional oscillation of this system, the Rayleigh-Ritz method is to used assuming a vibrating shape

$$f(x) = a_1 \frac{x}{l} + a_2 \frac{x^2}{l^2}$$

Where l is the length of the shaft and x is measured from the clamped end.
Calculate this first natural frequency and explain why shape function used is suitable.
An alternative approach to this problem is to use Rayleigh's Method, together with the shape function

$$g(x) = \frac{x}{l} - \frac{\beta x^3}{6l}$$

FIGURE P.4.6

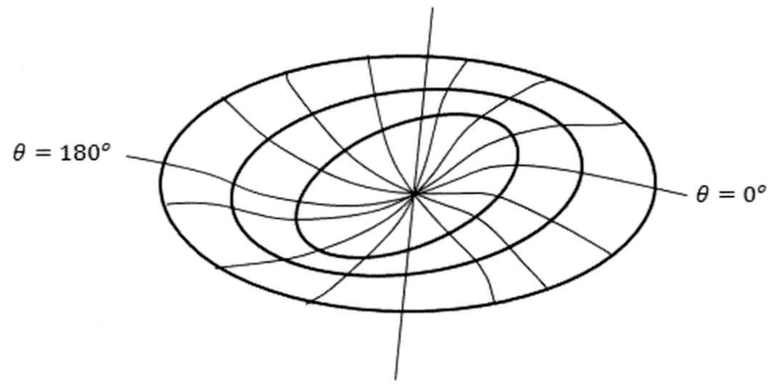

Vibration shape of plate

(A)

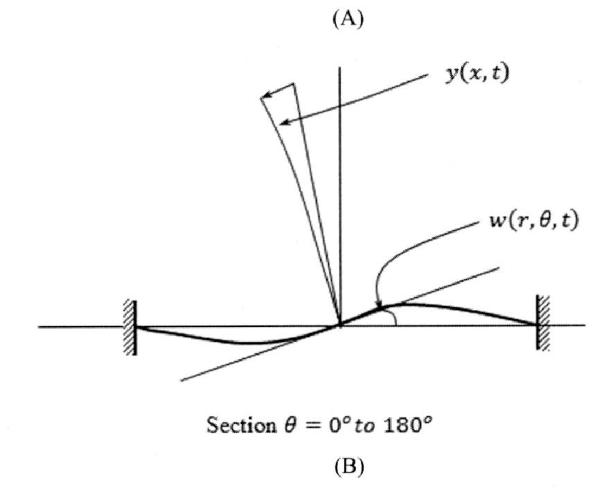

Section $\theta = 0°$ to $180°$

(B)

where $\beta = I\omega^2/GJ$

From an equation of the form

$$\omega^2 A = B$$

Where A and B are both function of ω^2 and hence develop a cubic equation in λ (where $\lambda = \omega^2 Il^2/GJ$). Do not solve this equation.

If according to this latter approach, $\lambda = 1.1601$, comment on the relative merits of the frequencies predicted by two methods.

P.4.8: A turbine blade is modeled as a cantilever of uniform depth and a breadth which varies as $b = b_0 e^{x/l}$ where l is the length of the cantilever and x is measured from the free end. Using the method of lumped parameters divide the beam into 3 sections and concentrate the mass of each section to the center of the section.

The flexibility influence function for this cantilever is

$$a(x.\zeta) = \frac{l^3}{EI_0}\left[e^{-x/l}\left(2 + \frac{x}{l} - \frac{\zeta}{l}\right) - e^{-1}\left\{5 - 2\left(\frac{x}{l} + \frac{\zeta}{l}\right) + \frac{x\zeta}{l^2}\right\}\right] x > \zeta$$

Where x and ζ are measured from the free end and I_0 is the second moment of area of beam at the free end. Develop the eigenvalue problem in terms of EI_0, l, and m_0, the mass/unit length at the free end, and starting with the vector

$\{u\}^t = [19\ 8\ 1]$ iterate twice only to obtain an estimate from ω_1.

From this algebraic eigenvalue problem formulation obtain an Dunkerley's equation and then proceed to Morris's modification. Using the Dunkerley-Morris equation obtain an estimate for ω_1.

You may accept the following integrals

$$\int xe^{ax}dx = e^{ax}\left(\frac{x}{a} - \frac{1}{a^2}\right)$$

$$\int x^2e^{ax}dx = e^{ax}\left(\frac{x^2}{a} - \frac{2}{a^2} + \frac{2x}{a^3}\right)$$

P.4.9: A pump is supported on a large rectangular plate simply supported at its edges. The pump is small relative to the plate and may thus be considered as a concentrated mass fixed to the plate at (p,q) as shown in Fig. P.4.9. The plate is of thickness h, side length a and b (where $b = 0.6a$) and density ρ. The pump mass is one quarter the plate mass.

An out of balance force $F \sin \Omega t$, where $\Omega^2 = 1800\frac{D}{\rho h a^4}$ acts transverse to the plane of the plate at the position of the pump, setting the plate into transverse vibration.

Assuming a vibrating shape,

$$w(x,y,t) = \phi_1(x)\psi(y)q_1(t) + \phi_2(x)\psi(y)q_2(t)$$

Where,

$$\phi_1(x) = \sin\frac{\pi x}{a}$$

$$\phi_2(x) = \sin\frac{2\pi x}{a}$$

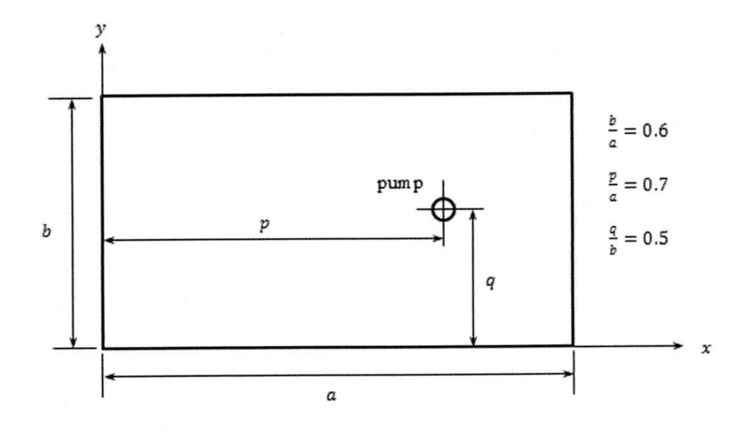

FIGURE P.4.9

$$\psi(y) = \sin \frac{\pi y}{b}$$

Determine the response of the plate at the position of the pump.

You may accept that the strain energy stored in a plate is,

$$U = \frac{D}{2}\int_0^a \int_0^b \left\{\frac{\partial^2 w}{\partial y^2} + \frac{\partial^2 w}{\partial x^2}\right\}^2 dxdy$$

P.4.10: In order to estimate the fundamental natural frequency of a simply supported rectangular plate, it is proposed to use the lumped parameter method. To this end, it is decided to determine the influence function for the plate by an application of principle of stationary total potential energy. Explain why

$$w(x,y) = \sum_{i=1}^{2}\sum_{j=1}^{2} a_{ij}\varnothing_i(x)\varnothing_j(y)$$

Where

$$\varnothing_1(x) = \frac{x}{a} - \left(\frac{x}{a}\right)^2 . \varnothing_1(y) = \frac{y}{b} - \left(\frac{y}{b}\right)^2$$

$$\varnothing_2(x) = \frac{x}{a} - 3\left(\frac{x}{a}\right)^3 + 2\left(\frac{x}{a}\right)^3 . \varnothing_2(y) = \frac{y}{b} - 3\left(\frac{y}{b}\right)^3 + 2\left(\frac{y}{b}\right)^3$$

It is a reasonable shape to assume for the deflection surface if the origin of coordinates is taken at one corner of a plate having sides of length a and b.

If $a = 2$ and $b = 3$ show that the influence function is

$$a(x, y, \zeta, \eta) = \frac{1}{2}\sum_{m=1}^{2}\sum_{n=1}^{2}\frac{1}{\varnothing_{mn}}\varnothing_m(x)\varnothing_\mu(y)\varnothing_m(\zeta)\varnothing_\mu(\eta)$$

Where

$$\varnothing_{11} = 0 \cdot 5815/12$$
$$\varnothing_{12} = 1 \cdot 0362/12$$
$$\varnothing_{21} = 0 \cdot 3254/12$$
$$\varnothing_{22} = 0 \cdot 1313/12$$

to proceed to determine the fundamental natural frequencies for this plate the subdivision this plate is used with the mass of each portion concentrated at its center, the mass per unit surface area is ρ. Using the coordinates $q_1 \ldots q_6$ of the plate, write down the mass matrix and complete the flexibility matrix $[a]$ given that,

$$[a] = \frac{1}{10^3 D}\begin{bmatrix} 16 \cdot 97 & 14 \cdot 36 & -1 \cdot 02 & 8 \cdot 13 & 10 \cdot 82 & * \\ & 25 \cdot 85 & 14 \cdot 36 & 10 \cdot 82 & 19 \cdot 49 & 10 \cdot 82 \\ & & 16 \cdot 97 & * & 10 \cdot 82 & 8 \cdot 13 \\ \text{symm} & & & 16 \cdot 97 & 14 \cdot 36 & -1 \cdot 02 \\ & & & & 25 \cdot 85 & 14 \cdot 36 \\ & & & & & 16 \cdot 97 \end{bmatrix}$$

Where the a_{ij}'s have been obtained from the $a(x, y, \zeta, \eta)$'s appropriately.

Obtain an approximation to the fundamental mode shape and natural frequency. If you use iteration, write down what you consider to be a reasonable trial vector and iterate twice only.

P.4.11: For the Example 3.11, How would you extend the above analysis to describe the dynamic behavior of the plate? Determine an approximate value for the lowest frequency of free vibration if the load is removed and the mass/unit area of the plate is m.

P.4.12: For the rectangular plate shown in Fig. P.4.12,

1. Obtain the maximum deflection if it is subjected to a uniformly distributed load (q_o) N/mm^2.
2. If $q = 0$, find the lowest natural frequency using Ritz method and check the result by the Galerkin method. Assume,

$$W(x, y) = C_1 (x^2 - a^2)^2 (y^2 - b^2)^2$$

P.4.13: A simply supported rectangular plate is stiffened by a single rib which divided the plate as shown in Fig. P.4.13. To calculate the fundamental natural frequency of the stiffened plate using Rayleigh-Ritz method a suitable deflection shape is assumed to be

$$w(x, y, t) = \varnothing(x, y) \sin \omega t$$

Where $\varnothing(x, y) = \sin \frac{\pi x}{a} \left[A_1 \sin \frac{\pi y}{b} + A_2 \sin \frac{3\pi y}{b} \right]$

Using the coordinates shown, justify this assumed deflection shape.

If the flexural rigidities of the beam and plate EI and D respectively, ρ is the mass/unit area of the plate and m is the mass/unit length of the beam apply the Rayleigh-Ritz method and evaluate all the appropriate integrals and set up, but no solve, the eigenvalue problem. From this eigenvalue problem deuce the fundamental natural frequency of an unstiffened simply supported beam.

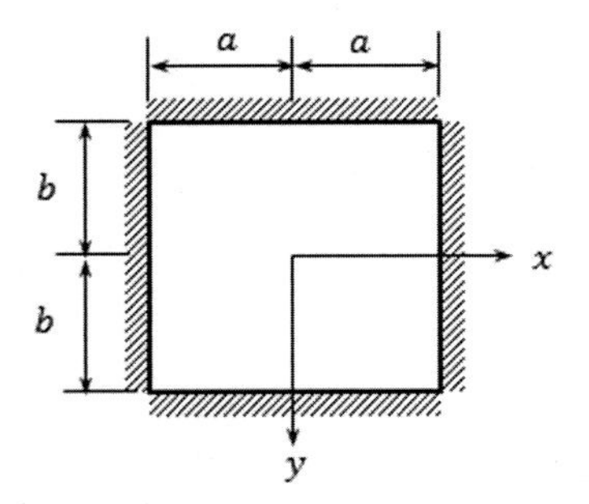

FIGURE P.4.12 A rectangular plate fixed around all edges.

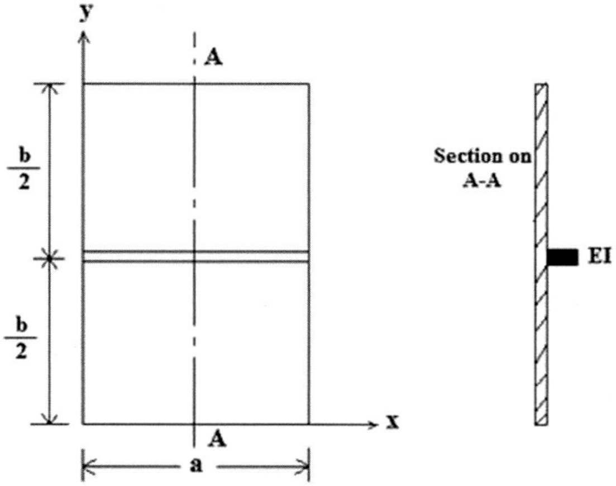

FIGURE P.4.13 A stiffened rectangular plate.

You may accept that for the assumed deflection shape given above the maximum energy stored in the plate alone is

$$-U = \frac{abD}{8}\left[A_1^2\left(\frac{1}{a^2}+\frac{1}{b^2}\right)^2 + A_2^2\left(\frac{1}{a^2}+\frac{9}{b^2}\right)^2\right]$$

P.4.14: A cantilever beam of uniform width (b) and length l tapers uniformly from a depth h at the clamped end to $h/5$ at the free end. Use the Rayleigh-Ritz process to estimate the first and second natural frequencies. Note that the exact solution is:

$$\omega_1 = 1.239\sqrt{(Eh^2/\rho l^4)}, \omega_2 = 4.545\sqrt{(Eh^2/\rho l^4)}$$

P.4.15: If the beam of Problem P.4.14 has a concentrated mass equal to half the beam mass added to the "free" end, determine the new value the first two frequencies.

P.4.16: (a) A square plate, side a, is clamped all round its edges. To estimate the fundamental natural frequency, assume a displacement model.

$$w(x.y) = q_1(t)\left(1 + \cos\frac{2\pi x}{a}\right)\left(1 + \cos\frac{2\pi y}{a}\right)$$

With origin at the center. Confirm that this is, indeed admissible for use in the Rayleigh-Ritz process.

(b) If the plate in (a) is stiffened by compact ribs attached along the symmetry lines $x = 0$ and $y = 0$. Each rib has a stiffness EI and distributed mass m/unit length. Assuming that the first mode shape is not significantly affected by this design change, deduce a correction factor to apply the result of (a) to this new situation.

P.4.17: The bending strain energy in a flat plate is given by,

$$U = \frac{D}{2}\iint\left\{\left(\frac{\partial^2 w}{\partial x^2}+\frac{\partial^2 w}{\partial y^2}\right)^2 - 2(1-\nu)\left[\frac{\partial^2 w}{\partial x^2}\frac{\partial^2 w}{\partial y^2} - \left(\frac{\partial^2 w}{\partial x \partial y}\right)^2\right]\right\}dxdy$$

Where the symbols have the usual meanings.

Why is it permissible to discard the term in square brackets if the edges are rigidly clamped?

An element of machine structure is essentially a clamed square plate supporting a uniform transverse distributed load p per unit area.

P.4.18: The specification for a certain vibrating table for environmental testing requires that it shall have a maximum difference of displacement across its surface within prescribed limits. As part of design work, it is necessary to consider a constant thickness square plate, clamped along its edges of length $2a$. In operation, the plate edges experience a force transverse displacement $d = d_0 \sin \Omega t$.

To obtain an approximate solution, assume that the plate has a deformed shape given by:

$$w = w_0(t)\left[1 - \frac{x^2}{a^2}\right]\left[1 - \frac{y^2}{a^2}\right] \quad \text{(P19.1)}$$

Where $w_0(t)$ is the Centre deflection. Show that w_0 is given by the ordinary differential equation,

$$\frac{d^2 w_0}{dt^2} + \lambda^2 w_0 = 1.7226\, \Omega^2 d_0 \cos \Omega t \quad \text{(P19.2)}$$

Where, $\lambda^2 = 81D/\text{ma}^2$. D is the flexural rigidity of the plate and m is its mass per unit surface.

Deduce an expression for a suitable plate thickness h if $\{[w(a) - w(0)]/w(0)\}$ is limited to 0.05 during steady vibrations.

You may accept that the strain energy in the plate corresponding to deformation (P19.1) is given by:

$$U = 53.5\frac{Dw_0^2}{2a^2} \quad \text{(P19.3)}$$

and that the Lagrange's equations are

$$\frac{d}{dt}\left(\frac{\partial T}{\partial q_i}\right) + \frac{\partial U}{\partial q_i} = Q_i i = 1.2.\cdots \quad \text{(P19.4)}$$

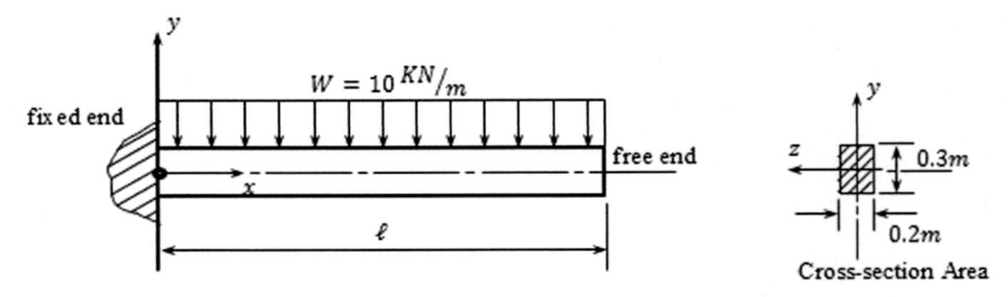

FIGURE P.4.19

P.4.19: (a) For the cantilever beam shown in Fig. P.4.19 if the total potential energy is:

$$X = \frac{1}{2}\int_0^{\ell} EI\left(\frac{\partial^2 v}{\partial x^2}\right)dx - \int_0^{\ell} wv dx$$

Where, E = Young's modulus. I = second moment of area about z-axis. v = vertical deflection. w = uniform distributed load.

Assuming a trial function solution as:

$$V(x) = \sum_{i=1}^{5} \alpha_i x^{i-1}$$

Find the unknown parameters ($\alpha_i . i = 1.5$) and the vertical displacement and rotation angle $\left(\frac{\partial v}{\partial x}\right)$ at $x = \ell$, for Rayleigh-Ritz method if $\ell = 1\ m$ and $E = 210\ \text{GN/m}^2$.

P.4.20: The specification for a certain vibrating table for environmental testing requires that it shall have a maximum difference of displacement across its surface within prescribed limits. As part of design work, it is necessary to consider a constant thickness circular plate, clamped along its edges at radius a. In operation, the plate edges experience a force transverse displacement $d = d_0 \sin \Omega t$.

To obtain an approximate solution, assume that the plate has a deformed shape given by:

$$w = w_0(t)\left[1 - \frac{r^2}{a^2}\right]^2 \qquad \text{(P20.1)}$$

Where, $w_0(t)$ is the center deflection. Show that, w_0 is given by the following ordinary differential equation,

$$\frac{d^2 w_0}{dt^2} + \mu^2 w_0 = \frac{5}{3}\ \Omega^2 d_0 \cos \Omega t \qquad \text{(P20.2)}$$

Where, $\mu^2 = 32\ Dw_0/3\ ma^4$. D is the flexural rigidity of the plate and m is its mass per unit surface.

Deduce an expression for a suitable plate thickness h if $\{[w(a) - w(0)]/w(0)\}$ is limited to 0.05 during steady vibrations. You may accept that the strain energy in the plate corresponding to deformation (P20.1) is given by:

$$U = \frac{32\pi}{3}\frac{Dw_0^2}{a^2} \tag{P20.3}$$

And that the Lagrange's equations are

$$\frac{d}{dt}\left(\frac{\partial T}{\partial q_i}\right) + \frac{\partial U}{\partial q_i} = Q_i i = 1.2.\cdots \tag{P20.4}$$

P.4.21: Using Rayleigh's principle, calculate the fundamental frequency of longitudinal vibrations for the system shown in Fig. P.4.21. The mass/unit length is given by

$$m(x) = \frac{6m}{5}\left(1 - \frac{x}{3L}\right),$$

the stiffness by $EA(x) = \frac{6}{5}EI\left(1 - \frac{x}{3L}\right)$, and the stiffness of the end spring by $k = EA/6.25L$. Use the Eigen-function for a uniform rod fixed at $x = 0$ and free at $x = L$ that is, $\psi = \sin\frac{\pi x}{L}$.

What type of function is this in these circumstances? If you use the Eigen-function for a uniform rod fixed at $x = 0$ with a spring of $k = EA/6.25\,L$ at $x = L$ that is, $\psi = \sin 1.6887x/L$ would you expect a better result? If $k \to 0$ compare the nature frequency with that previously computed for the system (Fig. P.4.21).

P.4.22: Calculate the first and second natural frequencies of a non-uniform cantilever if mass $\rho A = \rho A_0\left(1 - \frac{x}{L}\right)$, flexural rigidity $EI = EI_0\left(1 - \frac{x}{L}\right)$ where x is measured from the clamped end. Use the trial function, $\Phi = c_1\psi_1 + c_2\psi_2$

P.4.23: Calculate the fundamental natural frequency of a square plate clamped along one pair of opposite edges and simply supported along the other pair of opposite edges.

P.4.24: A heavy uniform cantilever of length L, flexural rigidity EI, has a concentrated mass m fixed to its tip. If m is twice as heavy as the beam calculate the natural frequency of the system using the trial functions,

$$\psi = 1 - \cos\frac{x}{2L}, \psi = \frac{1}{6EI}\left(3Lx^2 + x^3\right)$$

What shape does the second function represent.

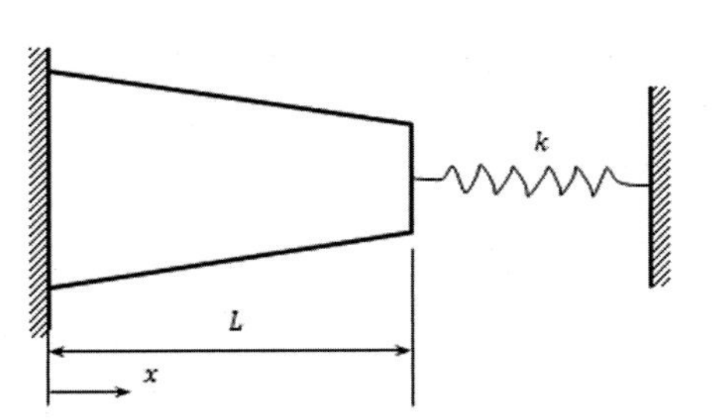

FIGURE P.4.21

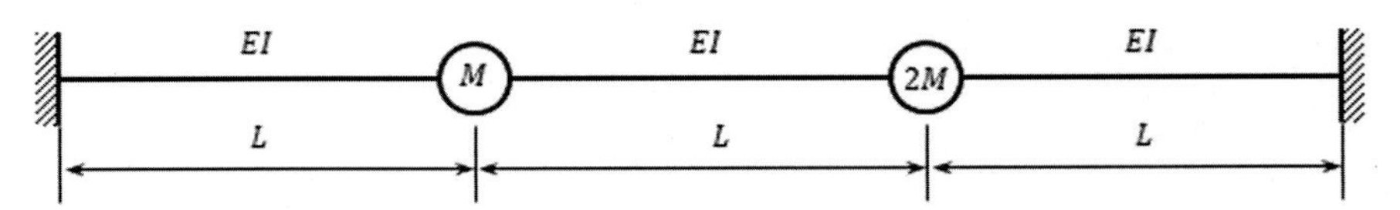

FIGURE P.4.25

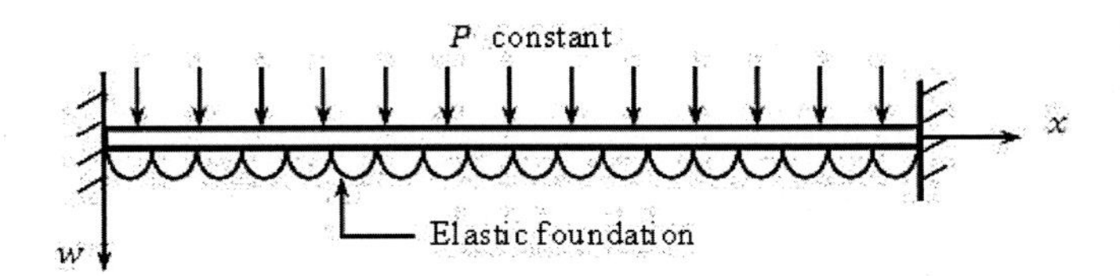

FIGURE P.4.26 XXXXXX.

P.4.25: The system shown in Fig. P.4.25 demonstrates that $\omega^2 = 3.4\ EI/ML^3$ is a good estimate for the fundamental frequency. What is the approximate modal shape when vibrating in the first mode? Put the transfer matrix equation into the Delta matrix form and shown that

$\omega^2 = 3.4\ EI/ML^3$ is a natural frequency using the Delta matrices?

P.4.26: A beam bonded onto an elastic base is described by:

$$EI\frac{d^4w}{dx^4} + kw = P$$

Or,

$$\frac{d^4w}{dx^4} + \mu^4 w = \frac{P}{EI}$$

k = foundation modulus, P = transverse load/unit length. Such a beam is shown in Fig. P.4.26.

The ends are rigidly clamped. Dividing the beam into two intervals and four intervals, taking advantage of symmetry and using the total potential energy principles, find the central deflection.

Bibliography

[1] Z. Kodal, Numerical Analysis, second ed., Chapman and Hall, 1961.
[2] S.P. Timoshenko, J.M. Gere, Theory of Elastic Stability, second ed., McGraw-Hill, New York, 1961.
[3] H.L. Langhaar, Energy Methods in Applied Mechanics, John Wiley and Sons Inc, 1962.
[4] R.J. Melosh, Basis for derivation of matrices for the direct stiffness method, Journal of the American Institute of Aeronautics and Astronautics I (7) (1963) 1631–1637.
[5] R.W. Clough, J. Penzien, Dynamics of Structures, McGraw-Hill, New York, 1975.
[6] R.J. Roark, W.C. Young, Formulas for Stress and Strain, fifth ed., McGraw-Hill, New York, 1975.
[7] T.H. Richards, Energy Methods in Stress Analysis, Ellis Horwood Series in Engineering Science, 1977.

[8] M. Witts, K. Sobczyk, Dynamics response of laminated plates to random loading, Journal of Solids and Structures 16 (3) (1980).
[9] M.J. Jweeg, Static bending deflections and flexural frequencies of prismatic cantilever tapered beams by Galerkin's method, A Scientific Journal, College of Engineering, University of Baghdad 4 (3) (1997) 13–24.
[10] K.J. Bathe, E.N. Dvorkin, A four-node, plate bending element based on Mindlin/Reissner plate theory and mixed interpolation, International Journal for Numerical Methods in Engineering 21 (1985) 367–383.
[11] C. Lanczos, The Variational Principles of Mechanica, fourth ed., Dover, New York, 1986.
[12] W. Weaver Jr., S.P. Timoshenko, D.H. Young, Vibration Problems in Engineering, fifth ed., John Wiley and Sons Inc, New York, 1990.
[13] M.J. Jweeg, S.K. Radi, Analysis of Composite Plates Subjected to Impact Loading, The National Eng. Conference, Kufa, Iraq, 2000.
[14] J.N. Reddy, Energy Principles and Variational Methods in applied Mechanics, John Wiley and Sons Inc, 2002.
[15] M.J. Jweeg, A.S. Hammood, M. Al-Waily, A. Suggested, Analytical solution of isotropic composite plate with crack effect, International Journal of Mechanical & Mechatronics Engineering 12 (5) (2012).
[16] M.A. Al-Shammari, M. Al-Waily, Theoretical and numerical vibration investigation study of orthotropic hyper composite plate structure, International Journal of Mechanical & Mechatronics Engineering 14 (6) (2014).
[17] M.J. Jweeg, M. Al-Waily, A.A. Deli, Theoretical and numerical investigation of buckling of orthotropic hyper composite plates, International Journal of Mechanical & Mechatronics Engineering 15 (4) (2015).
[18] M.J. Jweeg, A. Suggested, Analytical solution for vibration of honeycombs sandwich combined plate structure, International Journal of Mechanical & Mechatronics Engineering 16 (2) (2016).
[19] M. Al-Waily, K.K. Resan, A.H. Al-Wazir, Z.A.A. Abud Ali, Influences of glass and carbon powder reinforcement on the vibration response and characterization of an isotropic hyper composite materials plate structure, International Journal of Mechanical & Mechatronics Engineering 17 (6) (2017).
[20] M.A. Al-Shammari, M. Al-Waily, Analytical investigation of buckling behavior of honeycombs sandwich combined plate structure, International Journal of Mechanical and Production Engineering Research and Development 8 (4) (2018) 771–786.
[21] E.N. Abbas, M.J. Jweeg, M. Al-Waily, Analytical and numerical investigations for dynamic response of composite plates under various dynamic loading with the influence of carbon multi-wall tube nano materials, International Journal of Mechanical & Mechatronics Engineering 18 (6) (2018) 1–10.
[22] M. Al-Waily, M.A. Al-Shammari, M.J. Jweeg, An analytical investigation of thermal buckling behavior of composite plates reinforced by carbon nano particles, Engineering Journal 24 (3) (2020).
[23] I.A. Hussain, M.J. Jweeg, Optimum thickness of symmetric angle-ply laminates to eliminate extensional and bending stiffnesses, Journal of the Institution of Engineering 74 (1993) 99–104.

PART II

Finite Element Method

CHAPTER

5

Introduction to finite element method: bar and beam applications

This chapter presents an introduction to the finite element method. The formulations of bar and beam problems are derived depending upon the energy principles presented in the previous chapters. The finite element steps are explained in detail. The first step is the structure discretization which includes the number, type, size, and arrangements of elements used in the finite element simulation. The displacement model selection is the next step to obtain an approximate solution which is taken in this respect as a polynomial. Depending upon the displacement model, the element stiffness matrix and the load vector are derived for both types of structural elements, bar and beam applications. The assembly of the system stiffness matrix and the system load vector are constructed for the formulation of the equilibrium equations. The boundary conditions are applied to obtain the reduced system stiffness and load vector. The solution of the equilibrium equations for the displacement, then calculation of strains and stresses is presented in detail for the selected examples using different support conditions sand loading scheme. Finally, selected problems and the relevant references are given.

5.1 Bar extension

Consider a tapered bar under a simple uniaxial loading system as shown in Fig. 5.1A. The material obeys Hooke's law. The first step is to define the finite element mesh. In this case, an element is merely a finite portion of length of the bar so that the discretization is a linear sequence of elements interconnected at their end faces only.

The mesh defines element and nodes and these have been numbered on a global or system scale as element 1 to n and nodes 1 to $(n + 1)$ as shown in Fig. 5.1B, the line modeling is shown in Fig. 5.1C. We know from the usual engineering (or mechanics of materials) theory of bars that, provided the taper is not too severe, the deformation is characterized by the axial displacement $u(x)$ along the bar length with a constant cross-section. Then on a global scale we may label the nodal displacements as u_1, u_2, ... u_{n+1}, being quantities which characterize the deformation of the bar as a whole with the global or system reference frame XOY. When concentrating our attention on a typical element, it is convenient

Energy Methods and Finite Element Techniques
DOI: https://doi.org/10.1016/B978-0-323-88666-6.00009-5

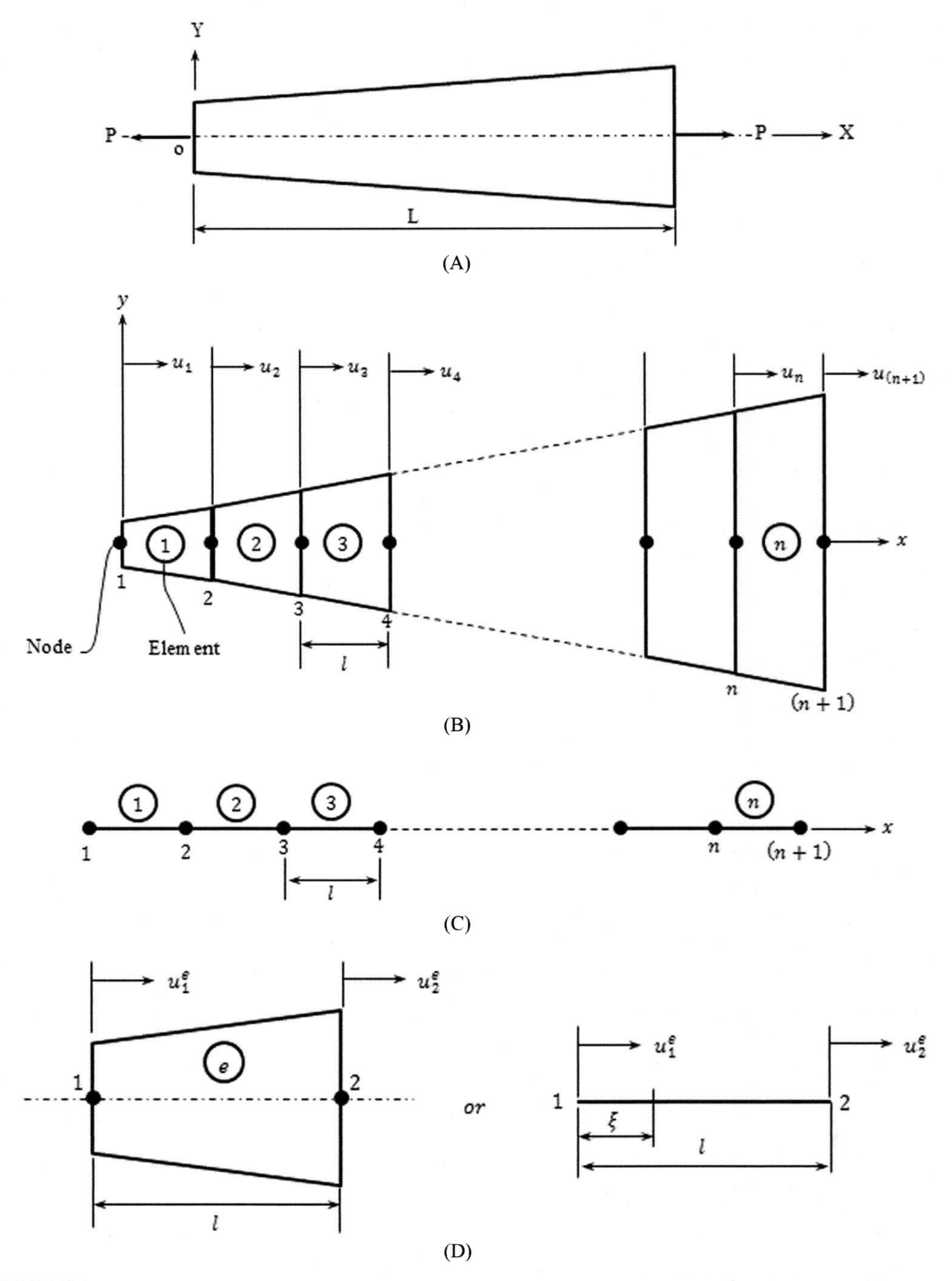

FIGURE 5.1 (A) A uni-axially tapered stressed bar, (B) The finite element modeling, (C) line modeling of the bar, No elements n and $n + 1$ nodes, (D) Typical element 2 nodes.

to introduce a coordinate ξ local to it and to label its external nodes as 1 and 2, then on a local or element scale, deformation is characterized by the element nodal displacements u_1^e and u_2^e, as shown in Fig. 5.1D.

1. **Idealization**, let the bar be considered as assemblage of n elements we have only axial displacement at any point in the element.
2. **Selecting a displacement model**,

$$u^e(\xi) = a_0 + a_1\xi = \begin{bmatrix} 1 & \xi \end{bmatrix} \begin{Bmatrix} a_0 \\ a_1 \end{Bmatrix}, \text{or} \tag{5.1}$$

$$u^e(\xi) = [f(\xi)]\{a\} \tag{5.2}$$

Using the boundary conditions that the displacement $= u_1^e$ at $\xi = 0$ and u_2^e at $\xi = l$ and inserting in Eq. (5.1) gives:

$$\begin{Bmatrix} u_1^e \\ u_2^e \end{Bmatrix} = \begin{bmatrix} 1 & 0 \\ 1 & l \end{bmatrix} \begin{Bmatrix} a_0 \\ a_1 \end{Bmatrix}$$

$$u^e(\xi) = \left(1 - \frac{\xi}{l}\right) u_1^e + \frac{\xi}{l} u_2^e$$

Or,

$$u^e(\xi) = \begin{bmatrix} \left(1 - \frac{\xi}{l}\right) & \frac{\xi}{l} \end{bmatrix} \begin{Bmatrix} u_1^e \\ u_2^e \end{Bmatrix}$$

Or,

$$u(\xi) = \underbrace{[N(\xi)]}_{\text{Shape Function}} \{u\}^e \tag{5.3}$$

Where,

$[N(\xi)] = \begin{bmatrix} \left(1 - \frac{\xi}{l}\right) & \frac{\xi}{l} \end{bmatrix}$ is the shape function for a uniaxial extension element.

3. **Element stiffness matrix**, the element stiffness matrix can be derived from the principle of minimum potential energy.

$$\text{Potential Energy} = \text{Strain Energy} - \text{Work Done by Ext. Forces } V = U - \Omega \tag{5.4}$$

On the basis of displacement equation above the strain ε^e(strain within element) $= \frac{du}{d\xi}$

In the finite element literature $N(\xi)$ is often referred to as a shape function, note that u_1^e and u_2^e are element nodal displacements as distinct from $u_1, u_2, \ldots\ldots u_{n+1}$ which are global or system nodal displacements.

$$\varepsilon = \frac{dU^e}{d\xi} = \left[\frac{dN(\xi)}{d\xi}\right]\{u\}^e = [B]\{u\}^e \tag{5.5}$$

Where,

$$[B] = \left[\frac{dN(\xi)}{d\xi}\right] \tag{5.6}$$

And,

$$[N(\xi)] = \left[\left(1 - \frac{\xi}{l}\right) \ \frac{\xi}{l}\right] \tag{5.7}$$

Then,

$$\left[\frac{dN(\xi)}{d\xi}\right] = \left[-\frac{1}{l} \ \frac{1}{l}\right]$$

Therefore,

$$[B] = \left[-\frac{1}{l} \ \frac{1}{l}\right] \tag{5.8}$$

Or,

$$\{\varepsilon\} = \left[-\frac{1}{l} \ \frac{1}{l}\right]\begin{Bmatrix} u_1^e \\ u_2^e \end{Bmatrix}$$

Since Hooke's law for uniaxial stresses is $\sigma = E\varepsilon$ or in matrix form,

$$\{\sigma\} = [D]\{\varepsilon\} \tag{5.9}$$

Note, in general $\{\sigma\}$ and $\{\varepsilon\}$ are column listings of stress and strain components; correspondingly, $[D]$ is a symmetric matrix of elastic constants (Young's modulus), they are all of order (1×1).

The strain energy of a uniaxial bar is represented by the area of the triangle shown in Fig. 5.2.

$$\text{Area} = \text{Strain Energy} = \frac{1}{2}\sigma\varepsilon = \frac{1}{2}E\varepsilon^2$$

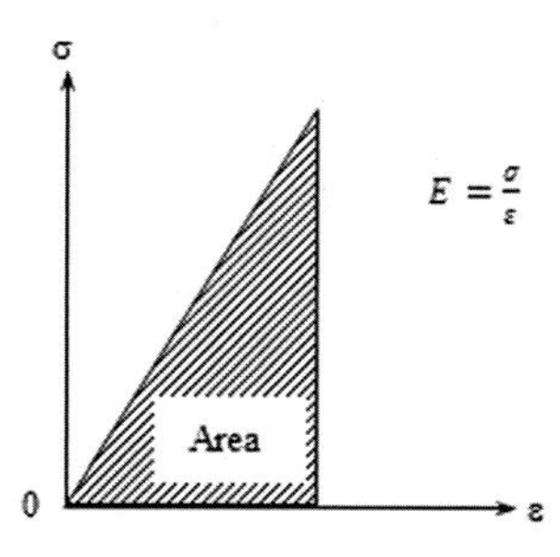

FIGURE 5.2 Stress—Strain diagram (Linear part).

$$U^e = \int_e \frac{1}{2} E\varepsilon^2 dvol. = \int_e \frac{1}{2} \sigma\varepsilon dvol.$$

$$U^e = \int_e \frac{1}{2} \{\varepsilon\}^t [D] \{\varepsilon\} dvol$$

$$U^e = \frac{1}{2} \{u\}^{et} \left(\int_e [B]^t [D][B] dvol. \right) \{u\}^e$$

Evaluating the triple matrix product and carrying out the integration, yields a matrix of constant coefficients so that,

$$U^e = \frac{1}{2} \{u\}^{et} [k]^e \{u\}^e$$

The matrix,

$$[k]^e = \int_e [B]^t [D][B] dvol.$$

is the element stiffness matrix. It is symmetric and is of order 2×2.

In this particular case after carrying out the triple integration, the element stiffness becomes:

$$[k]^e = \frac{EA}{l} \begin{bmatrix} 1 & -1 \\ -1 & 1 \end{bmatrix} \tag{5.10}$$

To find the total strain energy stored in the bar U,

$$U = \sum U^e$$

It is convenient to express the equations in matrix formulation,

$$\{\tilde{u}\} = \begin{Bmatrix} \{u\}^1 \\ \{u\}^2 \\ \vdots \\ \{u\}^n \end{Bmatrix}$$

$$\{\tilde{k}\} = \begin{bmatrix} [k]^1 & 0 & 0 & \cdots & 0 \\ 0 & [k]^2 & 0 & \cdots & 0 \\ \vdots & \vdots & \vdots & & \vdots \\ 0 & 0 & 0 & & [k]^n \end{bmatrix}$$

And,

$$U = \frac{1}{2} \{\tilde{u}\}^t \{\tilde{k}\} \{\tilde{u}\}$$

Now the elements of the vector $\{u\}^e$ are not completely independent because $u_2^1 = u_1^2 = u_2$, $u_2^2 = u_1^3 = u_3$ etc.

We can collect all these compatibility statements together in matrix as,

$$\begin{Bmatrix} u_1^1 \\ u_2^1 \\ \cdots \\ u_1^2 \\ u_2^2 \\ \cdots \\ \vdots \\ u_1^n \\ u_2^n \end{Bmatrix} = \begin{Bmatrix} \{u\}^1 \\ \cdots \\ \{u\}^2 \\ \cdots \\ \vdots \\ \{u\}^n \end{Bmatrix} = \begin{bmatrix} 1 & 0 & 0 & 0 & \cdots \\ 0 & 1 & 0 & 0 & \cdots \\ \cdots & \cdots & \cdots & \cdots & \cdots \\ 0 & 1 & 0 & 0 & \cdots \\ 0 & 0 & 1 & 0 & \cdots \\ \cdots & \cdots & \cdots & \cdots & \cdots \\ \vdots & \vdots & \vdots & \vdots & \vdots \end{bmatrix} \begin{Bmatrix} u_1 \\ u_2 \\ \vdots \\ u_{n+1} \end{Bmatrix}$$

Or,

$$\begin{Bmatrix} \{u\}^1 \\ \{u\}^2 \\ \vdots \\ \{u\}^n \end{Bmatrix} = \begin{bmatrix} [c]^1 \\ [c]^2 \\ \vdots \\ [c]^n \end{bmatrix} \begin{Bmatrix} u_1 \\ u_2 \\ \vdots \\ u_{n+1} \end{Bmatrix}$$

Or,

$$\{\tilde{u}\} = [c]\{u\}$$

Where,

$[c] \equiv$ (Connection Matrix) or (Compatibility Matrix)

Therefore,

$$U = \frac{1}{2}\{u\}^t\left([c]^t\{\tilde{k}\}[c]\right)\{u\}$$

Or,

$$U = \frac{1}{2}\{u\}^t[k]\{u\} \quad (5.11a)$$

Where, $[K] \equiv$ System stiffness matrix

$$[K] = [c]^t[\tilde{k}][c]$$

Potential Energy of the applied loads (Suppose Concentrated Load at the R.H.S.),

$$\Omega = -u_1p_1 - u_2p_2 - u_3p_3 - \cdots\cdots - u_{n+1}p_{n+1} = -\{u\}^t\{P\} \quad (5.11b)$$

Total Potential Energy,

$$V = U + \Omega$$

Using Eqs. (5.11a) and (5.11b) gives,

$$V = \frac{1}{2}\{u\}^t[K]\{u\} - \{u\}^t\{P\}$$

From equilibrium $\delta V = 0$ so that,

$$\delta V = \frac{1}{2}\left(\{\delta u\}^t[K]\{u\} + \{u\}^t[K]\{\delta u\}\right) - \{\delta u\}^t\{P\}$$

Or,

$$\delta V = \{\delta u\}^t([K]\{u\} - \{P\})$$

Since, $\{\delta u\}^t$ arbitrary $\neq 0$
Therefore,

$$[K]\{u\} - \{P\} = 0$$

Or,

$$\{P\} = [K]\{u\} \tag{5.11}$$

Example 5.1: The bar shown in Fig. 5.3 is of constant thickness and linearly varying width. It is subjected to a variety of boundary conditions. Using only two uniform elements and a linear displacement modal, determine the displacements and stresses at the nodes.

Idealization: Discretize the given into two elements (three nodes) as shown in Fig. 5.4. Note that the section areas are taken as average between A_0 and 1.5 A_0 for the first element and between 1.5 A_0 and 2 A_0 for the second element.

Therefore,

$$[k]^1 = \frac{5}{4}\frac{A_0E}{l}\begin{bmatrix} 1 & -1 \\ -1 & 1 \end{bmatrix} \text{And,} \tag{E1.1}$$

$$[k]^2 = \frac{7}{4}\frac{A_0E}{l}\begin{bmatrix} 1 & -1 \\ -1 & 1 \end{bmatrix} \tag{E1.2}$$

Or, we can resolve the example as follows,

$$p_1 = k\delta = k(u_1 - u_2) = ku_1 - ku_2 \tag{E1.3}$$

$$p_2 = k\delta = k(u_2 - u_1) = ku_2 - ku_1 \tag{E1.4}$$

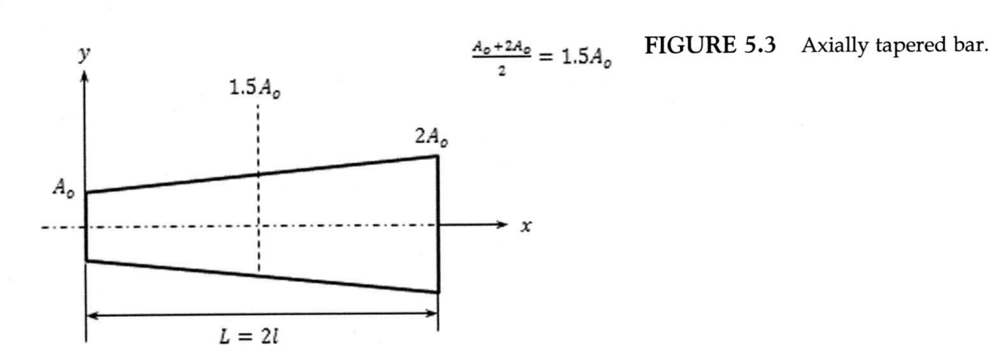

FIGURE 5.3 Axially tapered bar.

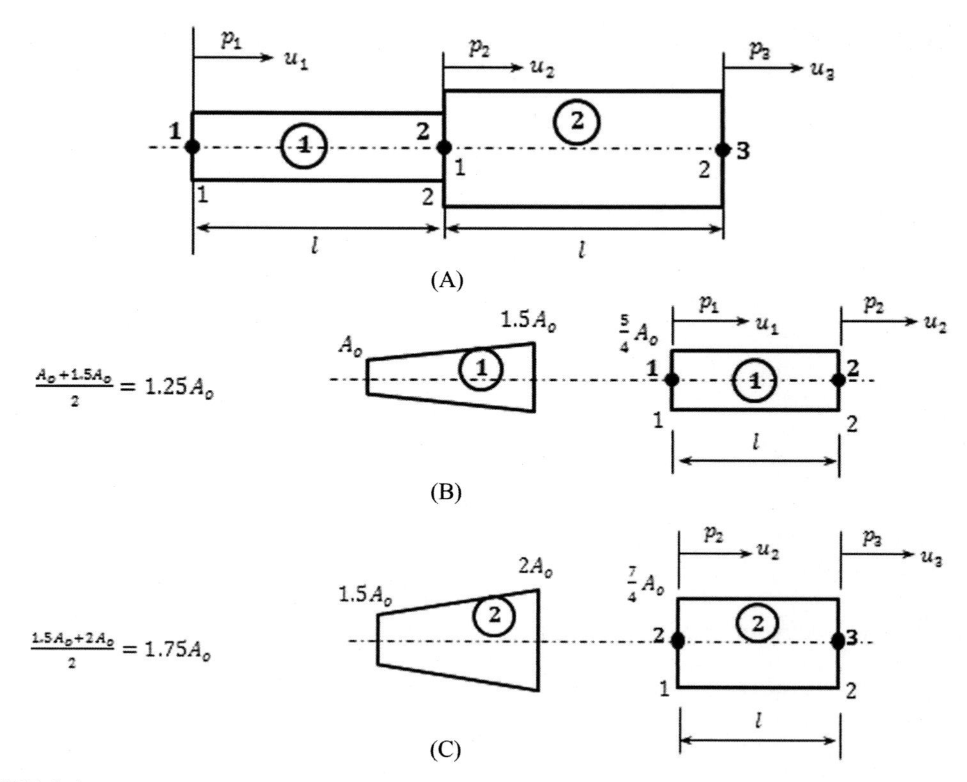

FIGURE 5.4 (A) Finite element modeling, (B) Element no. 1, (C) Element no 2.

Arranging Eqs. (E1.3) and (E1.4) in matrix form gives,

$$\begin{Bmatrix} p_1 \\ p_2 \end{Bmatrix} = \begin{bmatrix} k & -k \\ -k & k \end{bmatrix} \begin{Bmatrix} u_1 \\ u_2 \end{Bmatrix}$$

But,

$$k = \frac{AE}{l} = \frac{5}{4}\frac{A_0 E}{l}, \text{therefore,}$$

$$\begin{Bmatrix} p_1 \\ p_2 \end{Bmatrix} = \frac{AE}{l} \begin{bmatrix} 1 & -1 \\ -1 & 1 \end{bmatrix} \begin{Bmatrix} u_1 \\ u_2 \end{Bmatrix},$$

Where,

$$[k]^1 = \frac{5}{4}\frac{A_0 E}{l} \begin{bmatrix} 1 & -1 \\ -1 & 1 \end{bmatrix} \tag{E1.5}$$

Similarly,

$$[k]^2 = \frac{7}{4}\frac{A_0 E}{l} \begin{bmatrix} 1 & -1 \\ -1 & 1 \end{bmatrix} \tag{E1.6}$$

$[k]^1$ and $[k]^2$ in Eqs. (E1.5) and (E1.6) are similar to that obtained in (E1.1) and (E1.2). For compatibility, we must have,

$$\begin{Bmatrix} u_1^1 \\ u_2^1 \\ u_1^2 \\ u_2^2 \end{Bmatrix} = \begin{bmatrix} 1 & 0 & 0 \\ 0 & 1 & 0 \\ 0 & 1 & 0 \\ 0 & 0 & 1 \end{bmatrix} \begin{Bmatrix} u_1 \\ u_2 \\ u_3 \end{Bmatrix}$$

Or,

$$\{\tilde{u}\} = [c]\{u\}$$

Now, to assemble the system stiffness matrix $[K]$,

$$[K] = [c]^t[\tilde{k}][c] \tag{E1.7}$$

Where,
$[\tilde{k}]$, element or local stiffness matrix
$[K]$, assembled or system stiffness matrix.
The nodal connection matric $[c]$ is given by,

$$[c] = \begin{bmatrix} 1 & 0 & 0 \\ 0 & 1 & 0 \\ 0 & 1 & 0 \\ 0 & 0 & 1 \end{bmatrix},$$

And,

$$[c]^t = \begin{bmatrix} 1 & 0 & 0 & 0 \\ 0 & 1 & 1 & 0 \\ 0 & 0 & 0 & 1 \end{bmatrix}, [\tilde{k}] = \begin{bmatrix} [k]^1 & 0 \\ 0 & [k]^2 \end{bmatrix} \tag{E1.8}$$

Inserting the values of the element matrices as derived in Eqs. (E1.1) and (E1.2) in Eq. (E1.8) gives,

$$[\tilde{k}] = \begin{bmatrix} \left(\frac{5A_0E}{4l}\right) & \left(\frac{-5A_0E}{4l}\right) & 0 & 0 \\ \left(\frac{-5A_0E}{4l}\right) & \left(\frac{5A_0E}{4l}\right) & 0 & 0 \\ 0 & 0 & \left(\frac{7A_0E}{4l}\right) & \left(\frac{-7A_0E}{4l}\right) \\ 0 & 0 & \left(\frac{-7A_0E}{4l}\right) & \left(\frac{7A_0E}{4l}\right) \end{bmatrix}$$

Or, using Eq. (E1.7) gives,

$$[K] = \begin{bmatrix} 1 & 0 & 0 & 0 \\ 0 & 1 & 1 & 0 \\ 0 & 0 & 0 & 1 \end{bmatrix} \begin{bmatrix} \left(\frac{5A_0E}{4l}\right) & \left(\frac{-5A_0E}{4l}\right) & 0 & 0 \\ \left(\frac{-5A_0E}{4l}\right) & \left(\frac{5A_0E}{4l}\right) & 0 & 0 \\ 0 & 0 & \left(\frac{7A_0E}{4l}\right) & \left(\frac{-7A_0E}{4l}\right) \\ 0 & 0 & \left(\frac{-7A_0E}{4l}\right) & \left(\frac{7A_0E}{4l}\right) \end{bmatrix} \begin{bmatrix} 1 & 0 & 0 \\ 0 & 1 & 0 \\ 0 & 1 & 0 \\ 0 & 0 & 1 \end{bmatrix}$$

Carrying out the multiplications gives,

$$[K] = \begin{bmatrix} 1 & 0 & 0 & 0 \\ 0 & 1 & 1 & 0 \\ 0 & 0 & 0 & 1 \end{bmatrix} \begin{bmatrix} \left(\frac{5A_0E}{4l}\right) & \left(\frac{-5A_0E}{4l}\right) & 0 \\ \left(\frac{-5A_0E}{4l}\right) & \left(\frac{5A_0E}{4l}\right) & 0 \\ 0 & \left(\frac{7A_0E}{4l}\right) & \left(\frac{-7A_0E}{4l}\right) \\ 0 & \left(\frac{-7A_0E}{4l}\right) & \left(\frac{7A_0E}{4l}\right) \end{bmatrix}$$

$$[K] = \begin{bmatrix} \left(\frac{5A_0E}{4l}\right) & \left(\frac{-5A_0E}{4l}\right) & 0 \\ \left(\frac{-5A_0E}{4l}\right) & \left[\left(\frac{5A_0E}{4l}\right) + \left(\frac{7A_0E}{4l}\right)\right] & \left(\frac{-7A_0E}{4l}\right) \\ 0 & \left(\frac{-7A_0E}{4l}\right) & \left(\frac{7A_0E}{4l}\right) \end{bmatrix} = \frac{A_0E}{l} \begin{bmatrix} \left(\frac{5}{4}\right) & \left(\frac{-5}{4}\right) & 0 \\ \left(\frac{-5}{4}\right) & \left(\frac{12}{4}\right) & \left(\frac{-7}{4}\right) \\ 0 & \left(\frac{-7}{4}\right) & \left(\frac{7}{4}\right) \end{bmatrix} \tag{E1.9}$$

Or,

$$[K] = \begin{bmatrix} k_{11}^1 & k_{12}^1 & 0 \\ k_{21}^1 & (k_{22}^1 + k_{11}^2) & k_{12}^2 \\ 0 & k_{21}^2 & k_{22}^2 \end{bmatrix}$$

Where k_{11}^1 is the element $k(1,1)$ for the first element, etc.

We have,

$$\{p\} = [K]\{u\} \tag{E1.10}$$

Using Eq. (E1.9) into Eq. (E1.10) gives,

$$\frac{A_0E}{4l}\begin{bmatrix} 5 & -5 & 0 \\ -5 & 12 & -7 \\ 0 & -7 & 7 \end{bmatrix}\begin{Bmatrix} u_1 \\ u_2 \\ u_3 \end{Bmatrix} = \begin{Bmatrix} p_1 \\ p_2 \\ p_3 \end{Bmatrix} \tag{E1.11}$$

$[K]$, is singular and cannot be inverted, that is, the structure is not adequate supported in space and as before we find $[K]^{-1}$, we must impose a displacement boundary conditions as shown in Fig. 5.5.

Suppose L.H.S Clamped, $p_3 = P, u_1 = 0$.

Therefore,

$$\frac{A_0E}{4l}\begin{bmatrix} 5 & -5 & 0 \\ -5 & 12 & -7 \\ 0 & -7 & 7 \end{bmatrix}\begin{Bmatrix} 0 \\ u_2 \\ u_3 \end{Bmatrix} = \begin{Bmatrix} p_1 \\ 0 \\ P \end{Bmatrix} \tag{E1.12}$$

Then the reduced Eq. (E1.12) becomes,

$$\frac{A_0E}{4l}\begin{bmatrix} 12 & -7 \\ -7 & 7 \end{bmatrix}\begin{Bmatrix} u_2 \\ u_3 \end{Bmatrix} = \begin{Bmatrix} 0 \\ P \end{Bmatrix},$$

Or,

$$\frac{12A_0E}{4l}u_2 - \frac{7A_0E}{4l}u_3 = 0$$

$$\frac{12A_0E}{4l}u_2 = \frac{7A_0E}{4l}u_3,$$

Which gives,

$$u_2 = 0.583u_3$$

And,

$$\frac{-7A_0E}{4l}u_2 + \frac{7A_0E}{4l}u_3 = P$$

$$\frac{-7A_0E}{4l}0.583u_3 + \frac{7A_0E}{4l}u_3 = P$$

Then,

$$u_3 = \frac{1.37Pl}{A_0E}, \text{and}, u_2 = \frac{0.8Pl}{A_0E}$$

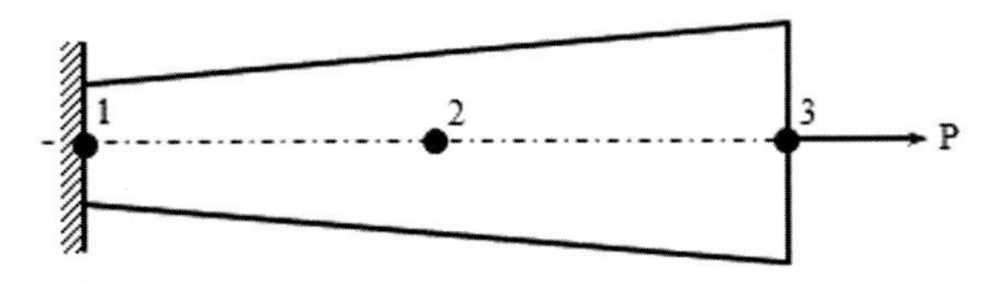

FIGURE 5.5 A Cantilever bar subjected to an end axial load P, two elements and three nodes.

The reactions are obtained using the unreduced equilibrium equations (E1.12). To determine the reaction at node 1, the first equation is as follows:

$$\frac{A_0E}{4l}(5*"0"-5*u_2+"0"u_3)=P_1$$

Insert the value of u_2 and simplify gives,

$$p_1=-P$$

To evaluate the stresses in elements, the strains are calculated using Eq. (1.8) as follows,

$$\{\varepsilon\}=[B]\{u\}^e=\begin{bmatrix}-\frac{1}{l} & \frac{1}{l}\end{bmatrix}\{u\}^e$$

Substitute the values of the displace vector $\{u\}^e$ gives,

$$\varepsilon_1=\begin{bmatrix}-\frac{1}{l} & \frac{1}{l}\end{bmatrix}\begin{Bmatrix}0\\ \left(\frac{0.8Pl}{A_oE}\right)\end{Bmatrix}=\frac{0.8P}{A_0E}$$

$$\varepsilon_2=\begin{bmatrix}-\frac{1}{l} & \frac{1}{l}\end{bmatrix}\begin{Bmatrix}\left(\frac{0.8Pl}{A_0E}\right)\\ \left(\frac{1.37Pl}{A_0E}\right)\end{Bmatrix}=-\frac{0.8P}{A_0E}+\frac{1.37P}{A_0E}=\frac{0.57P}{A_0E}$$

Then, the stresses are calculated using Hooke's law:

$$\sigma=E\varepsilon=E[B]\{u\}^e$$

For the element No. 1,

$$\sigma_1=E\varepsilon_1=E[B]\{u\}^1=E\begin{bmatrix}-\frac{1}{l} & \frac{1}{l}\end{bmatrix}\begin{Bmatrix}0\\ \left(\frac{0.8Pl}{A_0E}\right)\end{Bmatrix}$$

$$\sigma_1=\frac{0.8P}{A_0}$$

Similarly, for the element No. 2,

$$\sigma_2=E\begin{bmatrix}-\frac{1}{l} & \frac{1}{l}\end{bmatrix}\begin{Bmatrix}\left(\frac{0.8Pl}{A_0E}\right)\\ \left(\frac{1.37Pl}{A_0E}\right)\end{Bmatrix} \tag{E1.13}$$

$$\sigma_2=\frac{0.571P}{A_0} \tag{E1.14}$$

Using the mechanics of materials concept, the stresses evaluated by using $\sigma = \frac{P}{A}$ (axial loading) are (Fig. 5.6):

1. On the R.H.E ($A = 2A_0$),

$$\sigma = \frac{P}{A} = \frac{P}{2A_0} = \frac{0.5P}{A_0}$$

2. On the left hand end ($A = A_0$)

$$\sigma = \frac{P}{A_0}$$

3. At the center of the element No. 1 ($A = 1.5A_0$)

$$\sigma = \frac{P}{1.5A_0} = \frac{0.666P}{A_0}$$

$$\sigma \ \sigma_1$$

4. At the center of the element No. 2 ($A = 2A_0$)

$$\sigma = \frac{P}{2A_0} = \frac{0.5P}{A_0} \tag{E1.15}$$

$$\sigma \ \sigma_2$$

The results of Eqs. (E1.14) and (E1.15) are in good agreement, which means that the finite element results are in good agreement with those obtained analytically.

5.2 Equivalent nodal forces of the axially distributed loading

The axially distributed loading should be converted to an axially concentrated loading at the nodes. Consider a portion of length l and it is required to convert the uniformly axially loaded member into an equivalent nodal axial forces acted at the nodes 1 and 2ss shown in Fig. 5.7.

Potential energy of the applied loads (suppose distributed load),

$$\Omega_d^e = -\int_0^l P(\xi)u d\xi \tag{5.12}$$

But,

$$u(\xi) = [N(\xi)]\{u\}^e$$

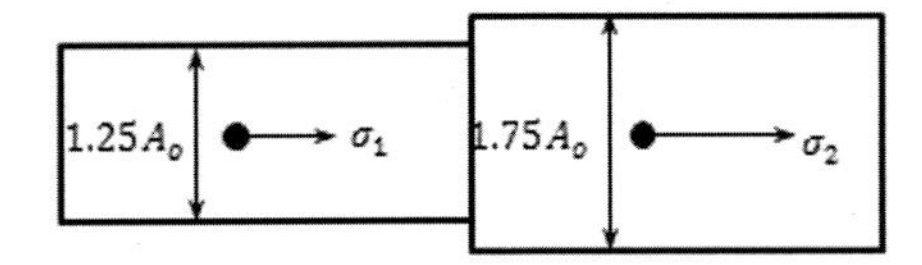

FIGURE 5.6 Element stresses positions.

We have,

$$\Omega_d^e = -\int_0^l P(\xi)[N(\xi)]\{u\}^e d\xi$$

Or,

$$= -\int_0^l P(\xi)\{u\}^{et}[N(\xi)]^t d\xi$$

$$= -\{u\}^t \int_0^l P(\xi)[N(\xi)]^t d\xi \tag{5.13a}$$

Eq. (1.13a) can be arranged as follows,

$$= -\{u\}^t \int_0^l P(\xi) \left\{ \begin{array}{c} \left(1-\frac{\xi}{l}\right) \\ \left(\frac{\xi}{l}\right) \end{array} \right\} d\xi$$

Or,

$$= -\{u\}^t \left\{ \begin{array}{c} \tilde{P}_1 \\ \tilde{P}_2 \end{array} \right\}$$

And,

$$\Omega_d^e = -\{u\}^t \{\tilde{P}_d\}^e$$

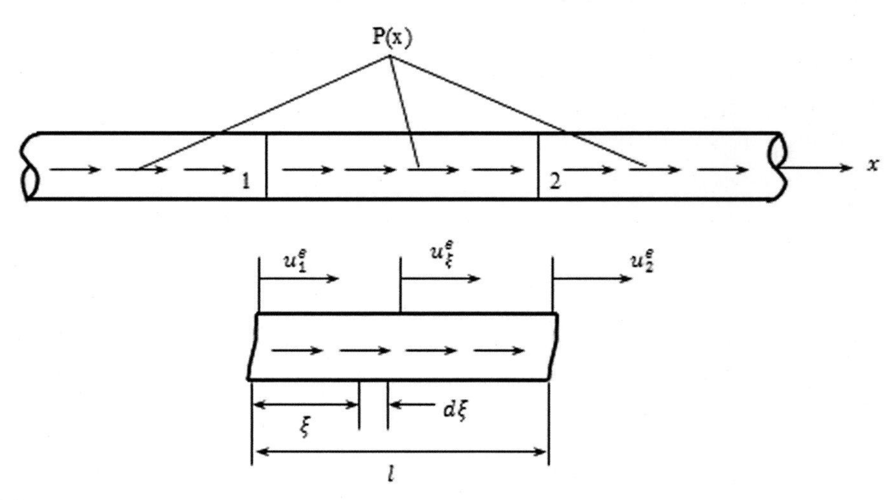

FIGURE 5.7 Uniformly axially loaded bar.

Where,

$$\{P\}^e = \int_0^l P(\xi)[N(\xi)]^t d\xi \tag{5.13}$$

Note that,

$$[N(\xi)] = \left[\left(1 - \frac{\xi}{l}\right) \quad \left(\frac{\xi}{l}\right) \right]$$

$$[N(\xi)]^t = \left\{ \begin{matrix} \left(1 - \frac{\xi}{l}\right) \\ \left(\frac{\xi}{l}\right) \end{matrix} \right\}$$

$$\Omega_d = \sum \Omega_d^e \tag{5.14a}$$

Denote,

$$\{\tilde{\tilde{P}}\} = \left\{ \begin{matrix} \{\tilde{P}\}^1 \\ \{\tilde{P}\}^2 \end{matrix} \right\}$$

$$\Omega_d = -\{\tilde{u}\}^t\{\tilde{\tilde{P}}\} = -\{\tilde{u}\}^t\left([c]^t\{\tilde{\tilde{P}}\}\right)$$

Or,

$$\Omega_d = -\{u\}^t\{P\}$$

The total potential energy is thus,

$$V = \frac{1}{2}\{u\}^t[k]\{u\} - \{u\}^t(\{P_c\} + \{P_d\}) \tag{5.14}$$

Where P_c and P_d are the concentrated and distributed loading respectively.

Example 5.2: The line element of Fig. 5.8 has an internal node 3 in addition, to the external nodes 1 and 2. Show that an axial displacement model, ($u = a_0 + a_1\xi + a_2\xi^2$) can be written in the form ($u = [N]\{u\}^e$), where,

$$[N] = \begin{bmatrix} N_1 & N_2 & N_3 \end{bmatrix} = \left[\left(1 - 3\frac{\xi}{l} + 2\frac{\xi^2}{l^2}\right) \quad \left(-\frac{\xi}{l} + 2\frac{\xi^2}{l^2}\right) \quad \left(4\frac{\xi}{l} - 4\frac{\xi^2}{l^2}\right) \right]$$

And,

$$\{u\}^t = \begin{bmatrix} u_1 & u_2 & u_3 \end{bmatrix}$$

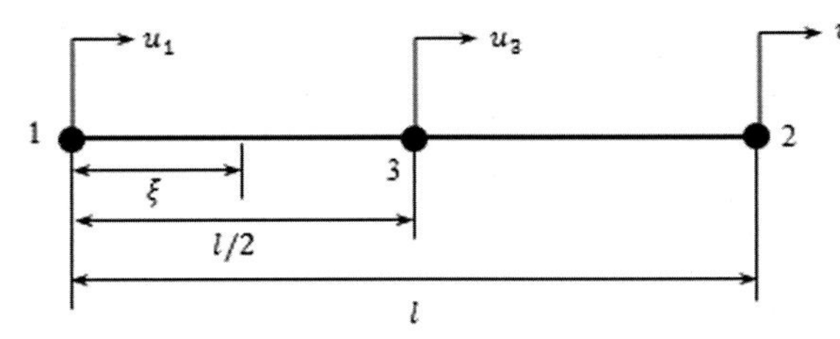

FIGURE 5.8 Line element with three nodes.

By assuming that,

$$u(\xi) = a_0 + a_1\xi + a_2\xi^2 \tag{E2.1}$$

At,

$$\xi = 0 \longrightarrow u(\xi) = a_0 = u_1$$

At,

$$\xi = \frac{l}{2} \longrightarrow u(\xi) = a_0 + a_1\frac{l}{2} + a_2\frac{l^2}{4} = u_3$$

At,

$$\xi = l \longrightarrow u(\xi) = a_0 + a_1 l + a_2 l^2 = u_2$$

We can write the above equations as,

$$\begin{aligned} a_0 + 0 + 0 &= u_1 \\ a_0 + a_1 l + a_2 l^2 &= u_2 \\ a_0 + a_1\frac{l}{2} + a_2\frac{l^2}{4} &= u_3 \end{aligned} \tag{E2.2}$$

Arranging Eq. (E2.2) in matrix form gives,

$$\begin{Bmatrix} u_1 \\ u_2 \\ u_3 \end{Bmatrix} = \begin{bmatrix} 1 & 0 & 0 \\ 1 & l & l^2 \\ 1 & \left(\frac{l}{2}\right) & \left(\frac{l^2}{4}\right) \end{bmatrix} \begin{Bmatrix} a_0 \\ a_1 \\ a_2 \end{Bmatrix} \tag{E2.3}$$

Or,

$$\{u\} = [A]\{a\}$$

Where,

$$[A] = \begin{bmatrix} 1 & 0 & 0 \\ 1 & l & l^2 \\ 1 & \left(\frac{l}{2}\right) & \left(\frac{l^2}{4}\right) \end{bmatrix}$$

And,

$$\{a\} = [A]^{-1}\{u\} \tag{E2.4}$$

Or,

$$[A]^{-1} = \frac{\mathrm{adj}[A]}{|A|} \tag{E2.5a}$$

$$|A| = \frac{l^3}{4} - \frac{l^3}{2} = -\frac{l^3}{4}$$

$$\mathrm{Cof}[A] = (-1)^{i+j}|\mathrm{Co} - \mathrm{Factor}|$$

And,

$$\mathrm{adj}[A] = [\mathrm{Cof}[A]]^t$$

Now,

$$\mathrm{Cof}[A] = \begin{bmatrix} -\frac{l^3}{4} & 3\frac{l^2}{4} & -\frac{l}{2} \\ 0 & \frac{l^2}{4} & -\frac{l}{2} \\ 0 & -l^2 & l \end{bmatrix}$$

And,

$$\mathrm{adj}[A] = \begin{bmatrix} -\frac{l^3}{4} & 0 & 0 \\ 3\frac{l^2}{4} & \frac{l^2}{4} & -l^2 \\ -\frac{l}{2} & -\frac{l}{2} & l \end{bmatrix}$$

Therefore using Eq. (E5.2a) gives,

$$[A]^{-1} = \begin{bmatrix} 1 & 0 & 0 \\ \left(-\frac{3}{l}\right) & \left(-\frac{1}{l}\right) & \left(\frac{4}{l}\right) \\ \left(\frac{2}{l^2}\right) & \left(\frac{2}{l^2}\right) & \left(\frac{-4}{l^2}\right) \end{bmatrix}$$

And, using Eq. (2.4) gives,

$$\begin{Bmatrix} a_0 \\ a_1 \\ a_2 \end{Bmatrix} = \begin{bmatrix} 1 & 0 & 0 \\ \left(-\frac{3}{l}\right) & \left(-\frac{1}{l}\right) & \left(\frac{4}{l}\right) \\ \left(\frac{2}{l^2}\right) & \left(\frac{2}{l^2}\right) & \left(\frac{-4}{l^2}\right) \end{bmatrix} \begin{Bmatrix} u_1 \\ u_2 \\ u_3 \end{Bmatrix} \tag{E2.5}$$

The assumption of the displacement field Eq. (E2.1) is given by,

$$u(\xi) = a_0 + a_1\xi + a_2\xi^2$$

Or,

$$\{u\} = \begin{bmatrix} 1 & \xi & \xi^2 \end{bmatrix} \begin{Bmatrix} a_0 \\ a_1 \\ a_2 \end{Bmatrix} \tag{E2.6}$$

Using Eq. (E2.5) in Eq. (E2.6) gives,

$$\{u\} = \begin{bmatrix} 1 & \xi & \xi^2 \end{bmatrix} \begin{bmatrix} 1 & 0 & 0 \\ \left(-\frac{3}{l}\right) & \left(-\frac{1}{l}\right) & \left(\frac{4}{l}\right) \\ \left(\frac{2}{l^2}\right) & \left(\frac{2}{l^2}\right) & \left(\frac{-4}{l^2}\right) \end{bmatrix} \begin{Bmatrix} u_1 \\ u_2 \\ u_3 \end{Bmatrix} \tag{E2.7}$$

Or,

$$= \begin{bmatrix} \left(1 - 3\frac{\xi}{l} + 2\frac{\xi^2}{l^2}\right) & \left(-\frac{\xi}{l} + 2\frac{\xi^2}{l^2}\right) & \left(4\frac{\xi}{l} - 4\frac{\xi^2}{l^2}\right) \end{bmatrix} \begin{Bmatrix} u_1 \\ u_2 \\ u_3 \end{Bmatrix} \tag{E2.8}$$

We have,

$$\{u\} = \begin{bmatrix} N_1 & N_2 & N_3 \end{bmatrix} \{u\}^e$$

$$\{u\} = [N]\{u\}^e$$

Therefore,

$$N_1 = 1 - \frac{3}{l}\xi + \frac{2}{l^2}\xi^2$$

$$N_2 = -\frac{1}{l}\xi + \frac{2}{l^2}\xi^2$$

$$N_3 = \frac{4}{l}\xi - \frac{4}{l^2}\xi^2 \tag{E2.9}$$

Note: that the derivation of higher order line elements may be derived using the above procedure using higher order displacement model.

5.3 Temperature effects—application to axially loaded problems

If the distribution of the change in temperature $\Delta T(x)$ is known, as in Fig. 5.9, then the strain due to this temperature changes can be treated as initial strain ε_0 so,

$$\varepsilon_0 = \alpha.\Delta T \tag{5.15}$$

Where, α, coefficient of thermal expansion, ΔT is the temperature difference.

$$\sigma = E(\varepsilon - \varepsilon_0) \tag{5.16a}$$

The strain energy per unit volume is, (area under the curve),

$$U = \frac{1}{2}\sigma(\varepsilon - \varepsilon_o) \tag{5.16}$$

Or,

$$U = \frac{1}{2}(\varepsilon - \varepsilon_0)^t E(\varepsilon - \varepsilon_0)$$

The total strain energy U_{total} in the structure is,

$$U_t = \frac{1}{2}\int_e (\varepsilon - \varepsilon_0)^t E(\varepsilon - \varepsilon_0)Adx$$

Or in matrix form,

$$U_t = \frac{1}{2}\int_e (\{\varepsilon\}^t - \{\varepsilon_0\}^t)[D](\{\varepsilon\} - \{\varepsilon_0\})Adx \tag{5.17}$$

Carrying out the multiplications in Eq. (1.17) gives,

$$U_t = \frac{1}{2}\int_e \{\varepsilon\}^t[D]\{\varepsilon\}Adx - \frac{1}{2}\int_e \{\varepsilon_0\}^t[D]\{\varepsilon\}Adx - \frac{1}{2}\int_e \{\varepsilon\}^t[D]\{\varepsilon_0\}Adx + \frac{1}{2}\int_e \{\varepsilon_0\}^t[D]\{\varepsilon_0\}Adx$$

Or,

$$U_t = \frac{1}{2}\int_e \{\varepsilon\}^t[D]\{\varepsilon\}Adx - \int_e \{\varepsilon\}^t[D]\{\varepsilon_0\}Adx + \frac{1}{2}\int_e \{\varepsilon_0\}^t[D]\{\varepsilon_0\}Adx$$

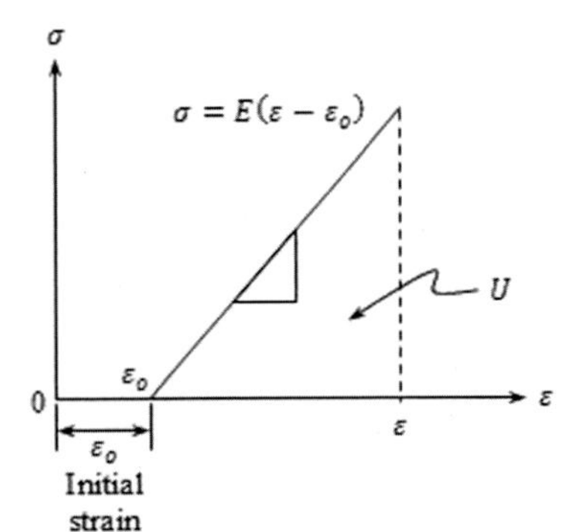

FIGURE 5.9 Temperature effects with an initial strain.

But,

$$\{\varepsilon\} = [B]\{u\}\{F\}^e \quad (5.18a)$$

We have,

$$U^e = \frac{1}{2}\{u\}^{et}[k]\{u\}^e - \{u\}^{et}\{F\}^e + \frac{1}{2}\int \{\varepsilon_0\}^t[D]\{\varepsilon_0\}Adx$$

Or,

$$U = \sum_r \frac{1}{2}\{u\}^t(E_rA_r\int_{-1}^{+1}[B]^t[B]d\xi)\{u\} - \underbrace{\sum r\,\{u\}^t E_rA_r\frac{l}{2}{}^r\varepsilon_0\int_{-1}^{+1}[B]^t d\xi}_{LoadVector} + \underbrace{\sum_r \frac{1}{2}E_rA_r\frac{l}{2}{}^r\varepsilon\frac{2}{0}}_{Constant}$$

Where,

$$\{F\}^e = E_rA_r\frac{l_r}{2}\varepsilon_0\int_{-1}^{+1}[B]^t d\xi \quad (5.18)$$

But,

$$\varepsilon_0 = \alpha.T$$

And,

$$[B] = \begin{bmatrix} -\frac{1}{l} & \frac{1}{l} \end{bmatrix}$$

Therefore,

$$\{F\}^e_{\text{th}} = E_rA_r\alpha\Delta T\begin{Bmatrix} -1 \\ +1 \end{Bmatrix}$$

ΔT, average change in temperature within the element.

$\{F\}^e_{\text{thermal}}$, can be assembled with body force, traction force and point load vector.

$$\underset{\text{global}}{\{F\}} = \sum\left(\underset{\text{Traction Force}}{T} + \underset{\text{Body Force}}{F_b} + \underset{\text{Thermal}}{F_{\text{th}}} + \underset{\text{Concentrated}}{P_c}\right) \quad (5.19)$$

The stress in each element is,

$$\sigma = E([B]\{u\} - \alpha.\Delta T) = E[B]\{u\} - E\alpha.\Delta T$$

$$= E\frac{1}{l}\begin{bmatrix} -1 & 1 \end{bmatrix}\{u\} - E\alpha.\Delta T \quad (5.20)$$

Example 5.3: Axial load member is subjected to loading as shown in Fig. 5.10 and an increase in temperature $\Delta T = 80°\text{C}$. Determine the displacements, stresses, and support reactions. The specifications are shown in Table 5.1.

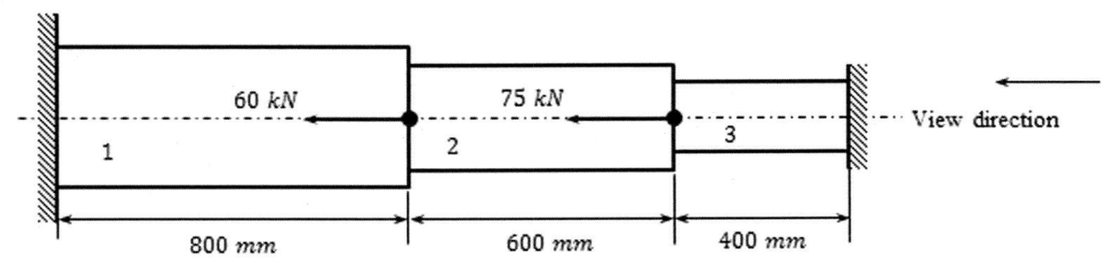

FIGURE 5.10 A uniaxial loading problem.

TABLE 5.1 Geometry and material properties.

Member	1	2	3
Materials	Bronze	Aluminum	Steel
Area (mm^2)	2400	1200	600
Modulus of elasticity E (GPa)	83	70	200
Coefficient of linear expansion α (°C)	$18.9 * 10^{-6}$	$23 * 10^{-6}$	$11.7 * 10^{-6}$

Solution,

We have for axial loading conditions

$$[k]^e = \frac{EA}{l}\begin{bmatrix} 1 & -1 \\ -1 & 1 \end{bmatrix}, \text{so}$$

$$[k]^1 = \frac{83 * 10^3 * 2400}{800}\begin{bmatrix} 1 & -1 \\ -1 & 1 \end{bmatrix}\text{N/mm} = 10^3\begin{bmatrix} 249 & -249 \\ -249 & 249 \end{bmatrix}$$

$$[k]^2 = \frac{70 * 10^3 * 1200}{600}\begin{bmatrix} 1 & -1 \\ -1 & 1 \end{bmatrix}\text{N/mm} = 10^3\begin{bmatrix} 140 & -140 \\ -140 & 140 \end{bmatrix}$$

$$[k]^3 = \frac{200 * 10^3 * 600}{400}\begin{bmatrix} 1 & -1 \\ -1 & 1 \end{bmatrix}\text{N/mm} = 10^3\begin{bmatrix} 300 & -300 \\ -300 & 300 \end{bmatrix} \tag{E3.1}$$

The system stiffness, matrix $[K]$ is obtained either by inspection or overlapping method. In both ways, we obtain the system stiffness matrix $[K]$ as follows:

$$[K] = 10^3\begin{bmatrix} 249 & -249 & 0 & 0 \\ -249 & 249+140 & -140 & 0 \\ 0 & -140 & 140+300 & -300 \\ 0 & 0 & -300 & 300 \end{bmatrix}$$

$$[K] = 10^3\begin{bmatrix} 249 & -249 & 0 & 0 \\ -249 & 389 & -140 & 0 \\ 0 & -140 & 440 & -300 \\ 0 & 0 & -300 & 300 \end{bmatrix} \tag{E3.2}$$

The thermal load is given by,

$$\{F_{st}\}_{th} = E.\alpha.A.\Delta T\begin{Bmatrix} -1 \\ 1 \end{Bmatrix}, \text{global}$$

$$\{F_{br}\}^1_{th} = 83 * 10^3 * 18.9 * 10^{-6} * 2400 * 80\begin{Bmatrix} -1 \\ 1 \end{Bmatrix}\begin{matrix} 1 \\ 2 \end{matrix} = \begin{Bmatrix} -301,190.4 \\ +301,190.4 \end{Bmatrix}\begin{matrix} 1 \\ 2 \end{matrix} \tag{E3.3}$$

$$\{F_{al}\}^2_{th} = 70 * 10^3 * 23 * 10^{-6} * 1200 * 80\begin{Bmatrix} -1 \\ 1 \end{Bmatrix}\begin{matrix} 2 \\ 3 \end{matrix} = \begin{Bmatrix} -154,560 \\ +154,560 \end{Bmatrix}\begin{matrix} 2 \\ 3 \end{matrix} \tag{E3.4}$$

$$\{F_{st}\}^3_{th} = 200 * 10^3 * 11.7 * 10^{-6} * 600 * 80\begin{Bmatrix} -1 \\ 1 \end{Bmatrix}\begin{matrix} 3 \\ 4 \end{matrix} = \begin{Bmatrix} -112,320 \\ +112,320 \end{Bmatrix}\begin{matrix} 3 \\ 4 \end{matrix} \tag{E3.5}$$

$$\{F_{th}\} = \begin{Bmatrix} -301,190.4 \\ 301,190.4 - 154,560 \\ 154,560 - 112,320 \\ 112,320 \end{Bmatrix}\begin{matrix} 1 \\ 2 \\ 3 \\ 4 \end{matrix} \tag{E3.6}$$

$$\{F_{th}\} = \begin{Bmatrix} -301,190.4 \\ 301,190.4 - 154,560 + 60,000 \\ 154,560 - 112,320 + 75,000 \\ 112,320 \end{Bmatrix} = \begin{Bmatrix} -301,190.4 \\ 152,630.4 \\ 117,240 \\ 112,320 \end{Bmatrix} \tag{E3.7}$$

The equilibrium equations are,

$$10^3\begin{bmatrix} 249 & -249 & 0 & 0 \\ -249 & 389 & -140 & 0 \\ 0 & -140 & 440 & -300 \\ 0 & 0 & -300 & 300 \end{bmatrix}\begin{Bmatrix} u_1 = 0 \\ u_2 \\ u_3 \\ u_4 = 0 \end{Bmatrix} = \begin{Bmatrix} F_1 \\ F_2 \\ F_3 \\ F_4 \end{Bmatrix} = \begin{Bmatrix} -301,190.4 \\ 152,630.4 \\ 117,240 \\ 112,320 \end{Bmatrix} \tag{E3.8}$$

The reduced equations are obtained by deleting first row and first column which gives,

$$10^3\begin{bmatrix} 389 & -140 \\ -140 & 440 \end{bmatrix}\begin{Bmatrix} u_2 \\ u_3 \end{Bmatrix} = \begin{Bmatrix} 152,630.4 \\ 117,240 \end{Bmatrix}$$

Solving gives,

$$\begin{Bmatrix} u_2 \\ u_3 \end{Bmatrix} = \frac{10^{-3}}{(389 * 440 - 140^2)}\begin{bmatrix} 440 & 140 \\ 140 & 389 \end{bmatrix}\begin{Bmatrix} 152,630.4 \\ 117,240 \end{Bmatrix} \tag{E3.9}$$

$$u_2 = 0.558$$

$$u_3 = 0.461 \tag{E3.10}$$

To find the element stresses,

$$\sigma = \frac{E}{l}\begin{bmatrix} -1 & 1 \end{bmatrix}\{u\} - E.\alpha.\Delta T$$

$$\sigma_1 = \frac{83 * 10^3}{800}\begin{bmatrix} -1 & 1 \end{bmatrix}\begin{Bmatrix} 0 \\ u_2 \end{Bmatrix} - 83 * 10^3 * 18.9 * 10^{-6} * 80 = 103.75u_2 - 125.49 = -148.44 \text{ MPa}$$

$$\sigma_2 = \frac{70 * 10^3}{600}\begin{bmatrix} -1 & 1 \end{bmatrix}\begin{Bmatrix} u_2 \\ u_3 \end{Bmatrix} - 70 * 10^3 * 23 * 10^{-6} * 80 = 116.6u_3 - 116.6u_2 - 128.8 = -102.54 \text{ MPa}$$

$$\sigma_3 = \frac{200 * 10^3}{400}\begin{bmatrix} -1 & 1 \end{bmatrix}\begin{Bmatrix} u_3 \\ 0 \end{Bmatrix} - 200 * 10^3 * 11.7 * 10^{-6} * 80 = -500u_3 - 187.2 = -189.2 \text{ MPa}$$

(E3.11)

5.4 Application to the beam bending

The beam bending problem will be presented using the standard procedure explained in the previous articles in this chapter using the energy principles. The equilibrium equations are constructed using the polynomial assumption of the displacement model.

1. **Selecting a displacement model**, we have to take into account (continuity) compatibility, both deflection and slope at an element between nodes. To achieve this, we have to assume a polynomial at least three degrees for modeling the element shown in Fig. 5.11 which has two nodes with two degrees of freedom at each node.

$$w(\xi) = a_0 + a_1\xi + a_2\xi^2 + a_3\xi^3 \tag{5.21}$$

It is convenient to express ($a_0 \longrightarrow a_3$) in terms of the elements of the element nodal displacements.

Where,

$$\text{At}, \xi = 0 \longrightarrow w(0) = u_1^e, \frac{dw}{d\xi} = u_2^e \ldots\ldots \text{etc.}$$

Inserting these boundary conditions in Eq. (5.21), ($a_0 \longrightarrow a_3$) can be found, from which we can write:

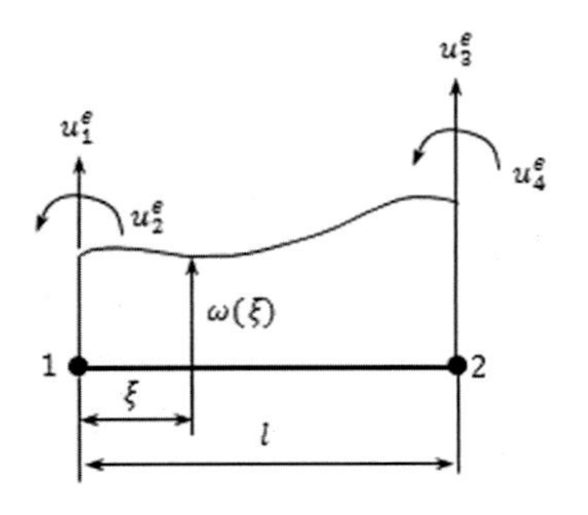

FIGURE 5.11 Beam element have 2 nodes and 4 d.o.f.

$$w(\xi) = \left[\left(1 - 3\frac{\xi^2}{l^2} + 2\frac{\xi^3}{l^3}\right) \quad \left(\xi - 2\frac{\xi^2}{l} + \frac{\xi^3}{l^2}\right) \quad \left(3\frac{\xi^2}{l^2} - 2\frac{\xi^3}{l^3}\right) \quad \left(-\frac{\xi^2}{l} + \frac{\xi^3}{l^2}\right) \right] \begin{Bmatrix} u_1 \\ u_2 \\ u_3 \\ u_4 \end{Bmatrix}^e$$

Or,

$$w(\xi) = \left[N_1(\xi) \quad N_2(\xi) \quad N_3(\xi) \quad N_4(\xi) \right] \begin{Bmatrix} u_1 \\ u_2 \\ u_3 \\ u_4 \end{Bmatrix}^e \tag{5.22}$$

Where,

$$N_1(\xi) = \left(1 - 3\frac{\xi^2}{l^2} + 2\frac{\xi^3}{l^3}\right)$$

$$N_2(\xi) = \left(\xi - 2\frac{\xi^2}{l} + \frac{\xi^3}{l^2}\right)$$

$$N_3(\xi) = \left(3\frac{\xi^2}{l^2} - 2\frac{\xi^3}{l^3}\right)$$

$$N_4(\xi) = \left(-\frac{\xi^2}{l} + \frac{\xi^3}{l^2}\right) \tag{5.23}$$

Or,

$$w(\xi) = \underbrace{[N(\xi)]}_{\text{shape function}} \quad \{u\}^e \tag{5.24a}$$

2. Formulating the stiffness equilibrium equations

To find the strain energy for a beam element U^e is given by,

$$U^e = \int_0^l \frac{EI}{2}\left(\frac{d^2\omega}{d\xi^2}\right)^2 d\xi \tag{5.24}$$

But,

$$w'' = \left[N''(\xi)\right]\{u\}^e = \underset{1*4}{[B]}\,\underset{4*1}{\{u\}}$$

Where,

$$[B] = \left[N''(\xi)\right] = \left[\frac{d^2N(\xi)}{d\xi^2}\right], \text{the strain} - \text{displacement relationship} \tag{5.25a}$$

Then denoting,

$$EI = \underset{1*1}{[D]} \text{elasticity matrix}$$

Where, EI is the bending stiffness of the beam. Using Eq. (5.24) gives,

$$U^e = \frac{1}{2}\int_0^l \{u\}^{et}[B]^t[D][B]\{u\}^e d\xi = \frac{1}{2}\{u\}^{et}\left(\int_0^l \underbrace{\underset{4*1^t}{[B]}\ \underset{1*1}{[D]}\ \underset{1*4}{[B]}}_{(4*4)} d\xi\right)\{u\}^e \tag{5.25b}$$

Or,

$$U^e = \frac{1}{2}\{u\}^{e\ t}_{1*4}[k]^e_{4*4}\{u\}^e_{4*1}, \text{scalar quantity}$$

Where,

$$[k]^e = \int_0^l [B]^t[D][B]d\xi \tag{5.25}$$

Carrying out the triple multiplication of Eq. (5.25) gives,

$$[k]^e = \frac{EI}{l^3}\begin{bmatrix} 12 & 6l & -12 & 6l \\ 6l & 4l^2 & -6l & 2l^2 \\ -12 & -6l & 12 & -6l \\ 6l & 2l^2 & -6l & 4l^2 \end{bmatrix} \tag{5.26}$$

3. To find Ω (potential energy of the external loads),
   a. **Concentrated load or moment**, for the elements shown in Fig. 5.12,

$$\Omega_c = -p_1u_1 - p_2u_2 - \ldots - p_nu_n$$

$$\Omega_c = -\{u\}^t[p]p, \text{concentrated load or moment.}$$

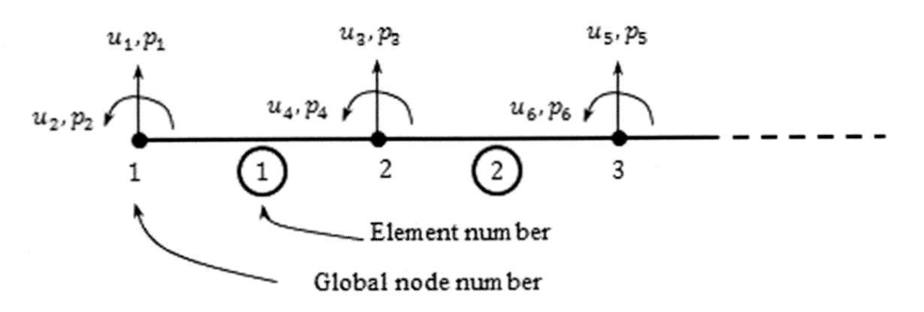

FIGURE 5.12 Finite Element Modeling of the beam, 2 elements and 3 nodes.

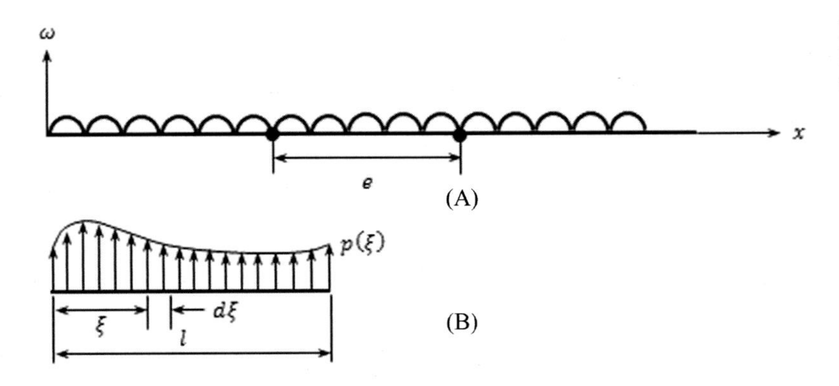

FIGURE 5.13 (A) Distributed loading. (B) Typical element.

b. Distributed loads,

The distributed loads shown in Fig. 5.13 should be converted to concentrated load or moment. Using the potential energy principles, we have,

$$\Omega_d^e = -\int_0^l p w(\xi) d\xi \tag{5.27}$$

Using Eq. (5.24a) gives,

$$\Omega_d = -\int_0^l p(\xi)\{u\}^{et}[N]^t d\xi = -\{u\}^{et}\int_0^l p(\xi)[N(\xi)]^t d\xi$$

Where, $N(\xi)$ is gives by Eq. (5.21).

Or,

$$\Omega_d = -\{u\}^t\{p_d\}$$

Where,

$$\{p_d\} = -\int_0^l p(\xi)[N(\xi)]^t d\xi = \int_0^l \left\{ \begin{matrix} \left(\xi - \dfrac{\xi^3}{l^2} + \dfrac{1}{2}\dfrac{\xi^4}{l^3}\right) \\ \left(\dfrac{1}{2}\xi^2 - \dfrac{2}{3}\dfrac{\xi^3}{l} + \dfrac{1}{4}\dfrac{\xi^4}{l^2}\right) \\ \left(\dfrac{\xi^2}{l^2} - \dfrac{2}{4}\dfrac{\xi^4}{l^3}\right) \\ \left(-\dfrac{1}{3}\dfrac{\xi^3}{l} + \dfrac{1}{4}\dfrac{\xi^4}{l^2}\right) \end{matrix} \right\} p(\xi) d\xi$$

$$\{p_e\}^e = \begin{Bmatrix} \frac{pl}{2} \\ \frac{pl^2}{2} \\ \frac{pl}{2} \\ -\frac{pl^2}{12} \end{Bmatrix} = \begin{Bmatrix} P_1 \\ P_2 \\ P_3 \\ P_4 \end{Bmatrix} \text{Equivalent nodal forces and moments.} \tag{5.28}$$

These are shown in Fig. 5.14.

Note: If we had merely shared the statically load, we would have lost the couple terms $\pm\frac{pl^2}{2}$ which the work calculation showed to be necessary.

In the system equilibrium equations,

$$[K]\{u\} = \{P\}$$

The system load vector includes the effects of concentrated and distributed in addition, to thermal effects. In case of distributed and concentrated loading, the force vector is composed of the concentrated forces or moments and the equivalent nodal forces and moments. The addition is achieved in the respective degrees of freedom.

$$\{P\} = \{P_d\} + \{P_c\} \tag{E5.28a}$$

Example 5.4: A propped cantilever beam is loaded as shown in Fig. 5.15, By using the two element representation deduce the stiffness equilibrium equations.

Solution:

By inspection, we can write the equilibrium equations:

$$\begin{bmatrix} k_{11}^1 & k_{12}^1 & k_{13}^1 & k_{14}^1 & 0 & 0 \\ k_{21}^1 & k_{22}^1 & k_{23}^1 & k_{24}^1 & 0 & 0 \\ k_{31}^1 & k_{32}^1 & (k_{33}^1 + k_{11}^2) & (k_{34}^1 + k_{12}^2) & k_{13}^2 & k_{14}^2 \\ k_{41}^1 & k_{42}^1 & (k_{43}^1 + k_{21}^2) & (k_{44}^1 + k_{22}^2) & k_{23}^2 & k_{24}^2 \\ 0 & 0 & k_{31}^2 & k_{32}^2 & k_{33}^2 & k_{34}^2 \\ 0 & 0 & k_{41}^2 & k_{42}^2 & k_{43}^2 & k_{44}^2 \end{bmatrix} \begin{Bmatrix} u_1 = 0 \\ u_2 = 0 \\ u_3 \\ u_4 \\ u_5 = 0 \\ u_6 \end{Bmatrix} = \begin{Bmatrix} p_1 = ? \\ p_2 = ? \\ p_3 \\ p_4 \\ p_5 = ? \\ p_6 \end{Bmatrix} \tag{E4.1}$$

Where k_{11}^1 is the stiffness at the location $k(1,1)$ for the element No. 1, ... etc.

Eq. (E4.1) can be solved as it stands since the system has not been located in space, Fig. 5.16 shows that in fact $(u_1 = u_2 = u_5 = 0)$ and (p_1, p_2, p_5) are thus unknown reactions.

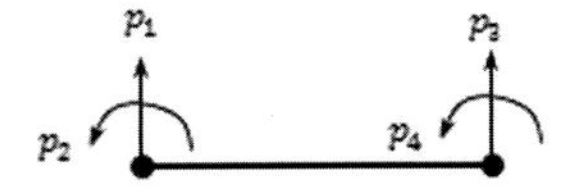

FIGURE 5.14 Equivalent nodal forces and moments for the typical beam bending element.

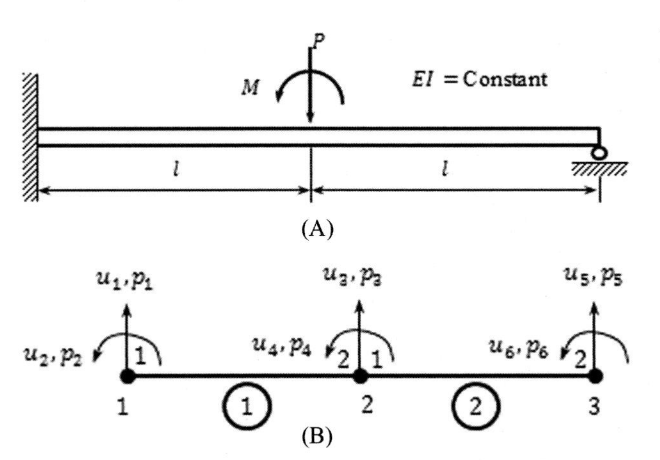

FIGURE 5.15 (A) Propped cantilever beam, (B) F.E. modeling—2 elements.

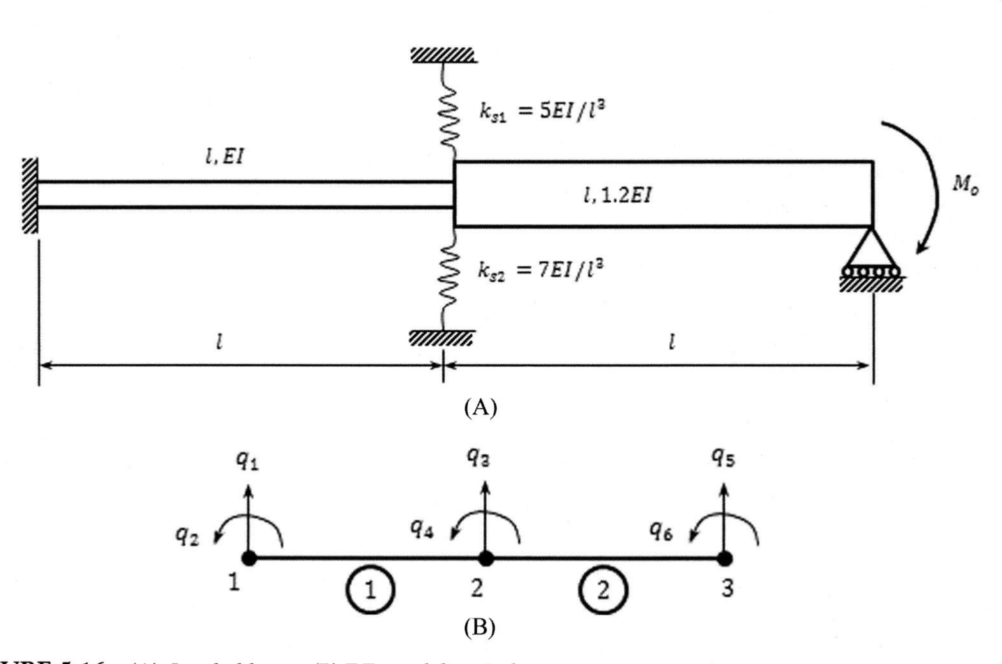

FIGURE 5.16 (A): Loaded beam, (B) F.E. modeling 2 elements.

Deleting the rows and columns corresponding to $u_1, u_2,$ and u_5 in $[k^e]$, we obtain the reduced stiffness matrix as follows:

$$\begin{bmatrix} (k_{33}^1 + k_{11}^2) & (k_{34}^1 + k_{12}^2) & k_{14}^2 \\ (k_{43}^1 + k_{21}^2) & (k_{44}^1 + k_{22}^2) & k_{24}^2 \\ k_{41}^2 & k_{42}^2 & k_{44}^2 \end{bmatrix} \begin{Bmatrix} u_3 \\ u_4 \\ u_6 \end{Bmatrix} = \begin{Bmatrix} -P \\ M \\ 0 \end{Bmatrix} \tag{E4.2a}$$

Then, by substituting the values of element stiffness matrices, using Eq. (5.26), we can write:

$$\frac{EI}{l^3}\begin{bmatrix} 24 & 0 & 6l \\ 0 & 8l^2 & 2l^2 \\ 6l & 2l^2 & 4l^2 \end{bmatrix}\begin{Bmatrix} u_3 \\ u_4 \\ u_6 \end{Bmatrix} = \begin{Bmatrix} -P \\ M \\ 0 \end{Bmatrix} \tag{E4.2}$$

Solving Eq. (E4.2a) given,

$$\begin{aligned} u_3 &= \frac{l^3}{EI}\left(\frac{M}{32L} - \frac{7P}{96}\right) \\ u_4 &= \frac{l^3}{EI}\left(\frac{5M}{32L^2} - \frac{P}{32L}\right) \\ u_6 &= \frac{l^3}{EI}\left(\frac{P}{8L} - \frac{M}{8L^2}\right) \end{aligned} \tag{E4.3}$$

Example 5.5: Formulate the global matrix equation $[K]\{q\} = \{Q\}$ for the models shown in Fig. 5.16A and B with the fewest possible degrees of freedom.

Solution:

Using the fewest possible degrees of freedom, the beam is modeled as shown in Fig. 5.16B, 2 elements and 6 unconstrained degrees of freedom. The element stiffness matrix for the beam element is given by:

$$[k]^e = \frac{EI}{l^3}\begin{bmatrix} 12 & 6l & -12 & 6l \\ 6l & 4l^2 & -6l & 2l^2 \\ -12 & -6l & 12 & -6l \\ 6l & 2l^2 & -6l & 4l^2 \end{bmatrix} = \frac{EI}{l^3}[A]\text{say} \tag{E5.1}$$

We have,

$$[k]^1 = \frac{EI}{l^3}[A]$$

$$[k]^2 = \frac{1.2EI}{l^3}[A]$$

The effective stiffness of the springs attached at node 2 is given by,

$$k_{\text{spring}} = k_{s_1} + k_{s_2} = \frac{5EI}{l^3} + \frac{7EI}{l^3}$$

Or,

$$k_{\text{spring}} = \frac{12EI}{l^3} \tag{E5.2}$$

By inspection, we can write the system stiffness matrix as follows:

$$[K] = \begin{bmatrix} k_{11}^1 & k_{12}^1 & k_{13}^1 & k_{14}^1 & 0 & 0 \\ k_{21}^1 & k_{22}^1 & k_{23}^1 & k_{24}^1 & 0 & 0 \\ k_{31}^1 & k_{32}^1 & (k_{33}^1 + k_{11}^2 + k_s) & (k_{34}^1 + k_{12}^2) & k_{13}^2 & k_{14}^2 \\ k_{41}^1 & k_{42}^1 & (k_{43}^1 + k_{21}^2) & (k_{44}^1 + k_{22}^2) & k_{23}^2 & k_{24}^2 \\ 0 & 0 & k_{31}^2 & k_{32}^2 & k_{33}^2 & k_{34}^2 \\ 0 & 0 & k_{41}^2 & k_{42}^2 & k_{43}^2 & k_{44}^2 \end{bmatrix}$$

Note,

$$\{q\} = \begin{Bmatrix} q_1 = 0 \\ q_2 = 0 \\ q_3 \\ q_4 \\ q_5 = 0 \\ q_6 \end{Bmatrix}, \text{displacement vector}$$

Inserting the values of the element stiffness matrix for each element, we obtain,

$$[K] = \frac{EI}{l^3} \begin{bmatrix} 12 & 6l & -12 & 6l & 0 & 0 \\ 6l & 4l^2 & -6l & 2l^2 & 0 & 0 \\ -12 & -6l & (12 + (1.2 * 12) + 12) & (-6l + (1.2 * 6l)) & (-12 * 1.2) & (6l * 1.2) \\ 6l & 2l^2 & (-6l + (1.2 * 6l)) & (4l^2 + 4l^2 * 1.2) & (-6l * 1.2) & (2l^2 * 1.2) \\ 0 & 0 & (-12 * 1.2) & (-6l * 1.2) & (12 * 1.2) & (-6l * 1.2) \\ 0 & 0 & (6l * 1.2) & (2l^2 * 1.2) & (-6l * 1.2) & (4l^2 * 1.2) \end{bmatrix} \tag{E5.3}$$

There reduced stiffness matrix is obtained by deleting the rows and the corresponding columns for each d.o.f., which constrained the nodal loading are,

The nodal loading is,

$$p_1 = p_2 = p_5 = ?,$$

$$p_6 = -M_0$$

$$p_3 = p_4 = 0$$

The results in the reduced stiffness matrix as follows,

$$[k] = [3 \times 3] \tag{E5.4}$$

Then, the reduced stiffness matrix, which is obtained by deleting the corresponding rows and columns for the constrained degrees of freedom that is, first, second and fifth rows and columns from the system stiffness matrix in Eq. (E6.3), is as follows

$$\frac{EI}{l^3} \begin{bmatrix} 38.4 & 1.2l & 7.2l \\ 1.2l & 8.8l^2 & 2.4l^2 \\ 7.2l & 2.4l^2 & 4.8l^2 \end{bmatrix} \begin{Bmatrix} q_3 \\ q_4 \\ q_6 \end{Bmatrix} = \begin{Bmatrix} 0 \\ 0 \\ -M_0 \end{Bmatrix} \tag{E5.5}$$

Or,

$$[K]\{q\} = \{Q\}$$

Solving Eq. (E6.4) gives,

$$q_3 = \frac{L^3}{EI}\left(\frac{21M_0}{340l}\right)$$

$$q_4 = \frac{L^3}{EI}\left(\frac{29M_0}{340l^2}\right)$$

$$q_6 = \frac{L^3}{EI}\left(-\frac{701M_0}{2040l^2}\right)$$

You may obtain the reaction by using the unconstrained equilibriums equations.

Example 5.6: For the loaded beam shown in Fig. 5.17A with the using the generalized coordinates shown on the Fig. 5.17B, formulate the equilibrium equations.

Solution:

Using the generalized coordinates, that is, the unconstrained degrees of freedom as shown in Fig. 5.17B

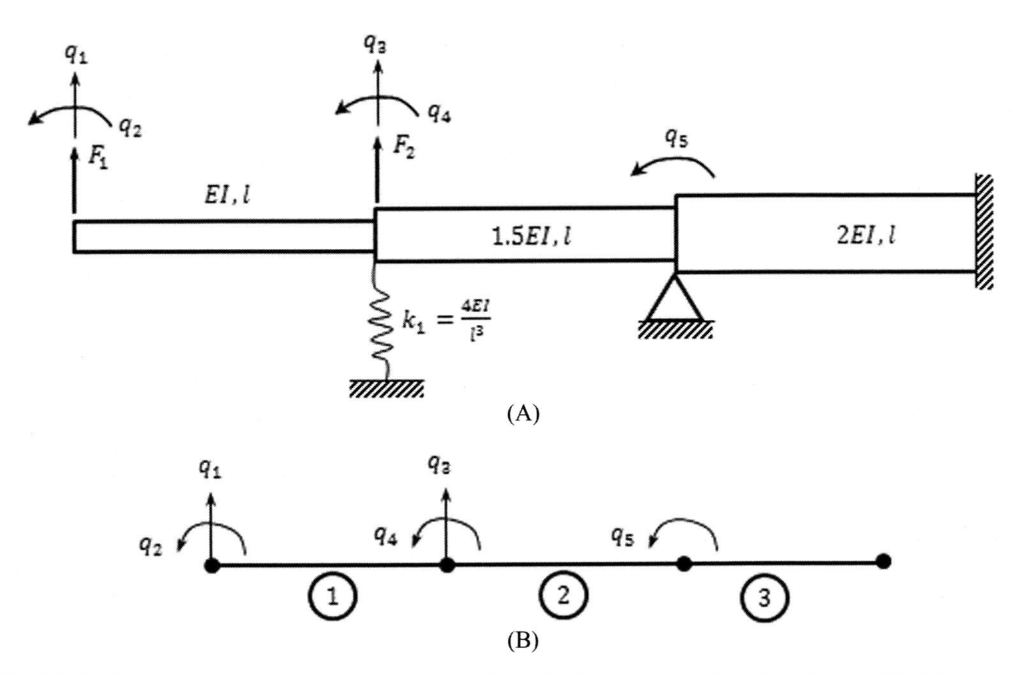

FIGURE 5.17 (A) Loaded beam with the generalized displacements and applied forces, (B) F.E. modeling with generalized coordinates.

For the beam element, the element stiffness matrix is given by,

$$[k]^e = \frac{EI}{l^3}\begin{bmatrix} 12 & 6l & -12 & 6l \\ 6l & 4l^2 & -6l & 2l^2 \\ -12 & -6l & 12 & -6l \\ 6l & 2l^2 & -6l & 4l^2 \end{bmatrix} \tag{E6.1}$$

The stiffness of the spring attached to node 2 is:

$$k_{\text{spring}} = \frac{4EI}{l^3} \tag{E6.2}$$

The element stiffness matrix for elements No. 1, No. 2, and No. 3 as follows,

$$[k]^1 = \frac{EI}{l^3}\begin{bmatrix} 12 & 6l & -12 & 6l \\ 6l & 4l^2 & -6l & 2l^2 \\ -12 & -6l & 12 & -6l \\ 6l & 2l^2 & -6l & 4l^2 \end{bmatrix} \tag{E6.3}$$

$$[k]^2 = \frac{EI}{l^3}\begin{bmatrix} 18 & 9l & -18 & 9l \\ 9l & 6l^2 & -9l & 3l^2 \\ -18 & -9l & 18 & -9l \\ 9l & 3l^2 & -9l & 6l^2 \end{bmatrix} \tag{E6.4}$$

$$[k]^3 = \frac{EI}{l^3}\begin{bmatrix} 24 & 12l & -24 & 12l \\ 12l & 8l^2 & -12l & 4l^2 \\ -24 & -12l & 24 & -12l \\ 12l & 4l^2 & -12l & 8l^2 \end{bmatrix} \tag{E6.5}$$

By using the standard assembly procedure and adding the contribution of the stiffness of each element. Note that the spring stiffness is added to the $K(3,3)$ which corresponds to the degree of freedom in which the spring is attached. The system stiffness matrix will be as follows:

$$[K] = \begin{bmatrix} k_{11}^1 & k_{12}^1 & k_{13}^1 & k_{14}^1 & 0 \\ k_{21}^1 & k_{22}^1 & k_{23}^1 & k_{24}^1 & 0 \\ k_{31}^1 & k_{32}^1 & (k_{33}^1 + k_{11}^2 + k_s) & (k_{34}^1 + k_{12}^2) & k_{14}^2 \\ k_{41}^1 & k_{42}^1 & (k_{43}^1 + k_{21}^2) & (k_{44}^1 + k_{22}^2) & k_{24}^2 \\ 0 & 0 & k_{41}^2 & k_{42}^2 & (k_{44}^2 + k_{22}^3) \end{bmatrix} \tag{E6.6}$$

With substitution of element stiffness matrix, we can write the stiffness equations as follows:

$$\frac{EI}{l^3}\begin{bmatrix} 12 & 6l & -12 & 6l & 0 \\ 6l & 4l^2 & -6l & 2l^2 & 0 \\ -12 & -6l & (12+18+4) & (-6l+9l) & 9l \\ 6l & 2l^2 & (-6l+9l) & (4l^2+6l^2) & 3l^2 \\ 0 & 0 & 9l & 3l^2 & (6l^2+8l^2) \end{bmatrix}\begin{Bmatrix} q_1 \\ q_2 \\ q_3 \\ q_4 \\ q_5 \end{Bmatrix} = \begin{Bmatrix} F_1 \\ 0 \\ F_2 \\ 0 \\ 0 \end{Bmatrix}$$

Example 5.7: For the loaded beam shown in Fig. 5.18 with the indicated unconstrained coordinates, formulate the system stiffness equations.

Solution:

The finite element modeling is shown in Fig. 5.18B with 2 elements, 3 nodes, and 6 d.o.f. As before, the element stiffness matrix for a typical element may written as follows:

$$[k]^e = \frac{EI}{l^3}\begin{bmatrix} 12 & 6l & -12 & 6l \\ 6l & 4l^2 & -6l & 2l^2 \\ -12 & -6l & 12 & -6l \\ 6l & 2l^2 & -6l & 4l^2 \end{bmatrix} \tag{E7.1}$$

The element stiffness matrix for the elements 1 and 2 are,

$$[k]^1 = \frac{EI}{l^3}\begin{bmatrix} 12 & 6l & -12 & 6l \\ 6l & 4l^2 & -6l & 2l^2 \\ -12 & -6l & 12 & -6l \\ 6l & 2l^2 & -6l & 4l^2 \end{bmatrix} \tag{E7.2}$$

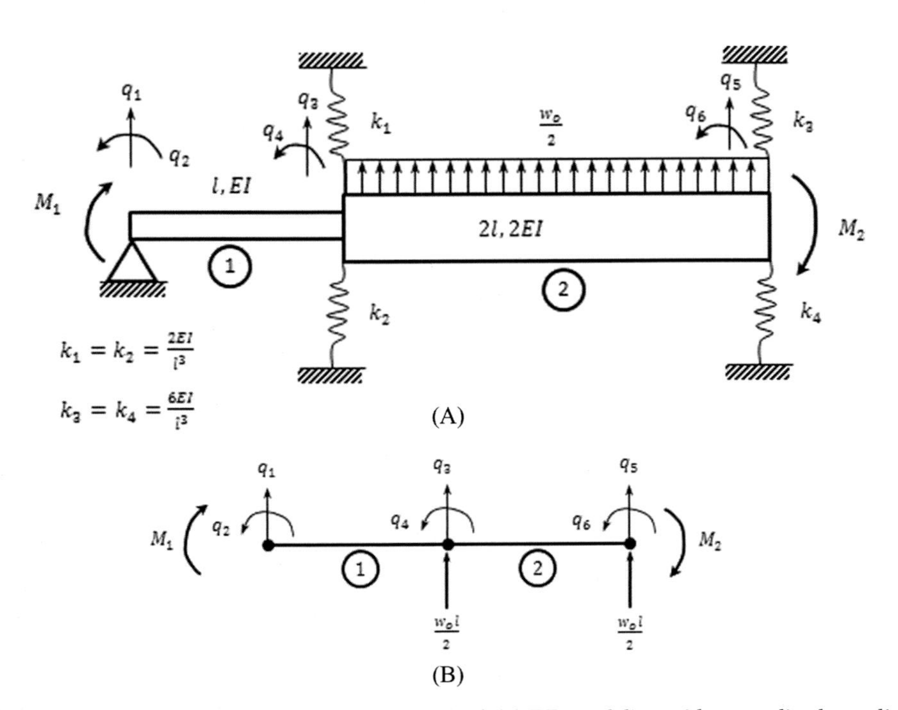

FIGURE 5.18 (A) Loaded beam with unconstrained d.o.f. (B) F.E. modeling with generalized coordinates.

$$[k]^2 = \frac{EI}{l^3}\begin{bmatrix} 3 & 1.5l & -3 & 1.5l \\ 1.5l & l^2 & -1.5l & 0.5l^2 \\ -3 & -1.5l & 3 & -1.5l \\ 1.5l & 0.5l^2 & -1.5l & l^2 \end{bmatrix} \tag{E7.3}$$

The assembly of the system stiffness matrix is obtained using the overlapping method and adding the effects of the springs at the corresponding degrees of freedom at which the springs are attached.

$$[K]_{sys} = \begin{bmatrix} k_{11}^1 & k_{12}^1 & k_{13}^1 & k_{14}^1 & 0 & 0 \\ k_{21}^1 & k_{22}^1 & k_{23}^1 & k_{24}^1 & 0 & 0 \\ k_{31}^1 & k_{32}^1 & \left(k_{33}^1 + k_{11}^2 + k_{S1,2}\right) & \left(k_{34}^1 + k_{12}^2\right) & k_{13}^2 & k_{14}^2 \\ k_{41}^1 & k_{42}^1 & \left(k_{43}^1 + k_{21}^2\right) & \left(k_{44}^1 + k_{22}^2\right) & k_{23}^2 & k_{24}^2 \\ 0 & 0 & k_{31}^2 & k_{32}^2 & \left(k_{33}^2 + k_{S3,4}\right) & k_{34}^2 \\ 0 & 0 & k_{41}^2 & k_{42}^2 & k_{43}^2 & k_{44}^2 \end{bmatrix} \tag{E7.4}$$

Substituting the values of the element stiffness matrix for each element and the springs stiffness in Eq. (E8.4), the system stiffness matrix becomes,

$$[K]_{sys} = \frac{EI}{l^3}\begin{bmatrix} 12 & 6l & -12 & 6l & 0 & 0 \\ 6l & 4l^2 & -6l & 2l^2 & 0 & 0 \\ -12 & -6l & 19 & -4.5l & -3 & 1.5l \\ 6l & 2l^2 & -4.5l & 5l^2 & -1.5l & 0.5l^2 \\ 0 & 0 & -3 & -1.5l & 15 & -1.5l \\ 0 & 0 & -1.5l & 0.5l^2 & -1.5l & l^2 \end{bmatrix} \tag{E7.5}$$

But, $[K]\{q\} = \{Q\}$ unconstrained, therefore,

$$\frac{EI}{l^3}\begin{bmatrix} 12 & 6l & -12 & 6l & 0 & 0 \\ 6l & 4l^2 & -6l & 2l^2 & 0 & 0 \\ -12 & -6l & 19 & -4.5l & -3 & 1.5l \\ 6l & 2l^2 & -4.5l & 5l^2 & -1.5l & 0.5l^2 \\ 0 & 0 & -3 & -1.5l & 15 & -1.5l \\ 0 & 0 & 1.5l & 0.5l^2 & -1.5l & l^2 \end{bmatrix}\begin{Bmatrix} q_1 \\ q_2 \\ q_3 \\ q_4 \\ q_5 \\ q_6 \end{Bmatrix} = \begin{Bmatrix} Q_1 \\ Q_2 \\ Q_3 \\ Q_4 \\ Q_5 \\ Q_6 \end{Bmatrix} \tag{E7.6}$$

Since,

$$q_1 = 0, Q_1 = p_1 = ?, Q_2 = -M_1 = p_2 = ?, Q_3 = p_3 = \frac{w_0 l}{2}$$

$$Q_4 = 0, Q_5 = p_5 = \frac{w_0 l}{2}, Q_6 = -M_2 \tag{E7.7}$$

Then, by using the above values, the system stiffness equations may be written as follows:

$$\frac{EI}{l^3}\begin{bmatrix} 4l^2 & -6l & 2l^2 & 0 & 0 \\ -6l & 19 & -4.5l & -3 & 1.5l \\ 2l^2 & -4.5l & 5l^2 & -1.5l & 0.5l^2 \\ 0 & -3 & -1.5l & 15 & -1.5l \\ 0 & 1.5l & 0.5l^2 & -1.5l & l^2 \end{bmatrix}\begin{Bmatrix} q_2 \\ q_3 \\ q_4 \\ q_5 \\ q_6 \end{Bmatrix} = \begin{Bmatrix} -M_1 \\ \frac{w_0 l}{2} \\ 0 \\ \frac{w_0 l}{2} \\ -M_2 \end{Bmatrix} \quad \text{(E7.8)}$$

Example 5.8: Deduce the equilibrium equations $[K]\{x\} = \{F\}$ which are satisfied by the displacement, components at the nodes of the continuous beam shown in Fig. 5.19.

Solution:

The finite element modeling of the continuous beam shown in Fig. 5.19A is shown in Fig. 5.19B with 4 elements, 5 nodes and 10 d.o.f. which represents the fewest degrees of freedom to model the beam.

The element stiffness matrix is given by:

$$[k]^e = \frac{EI}{l^3}\begin{bmatrix} 12 & 6l & -12 & 6l \\ 6l & 4l^2 & -6l & 2l^2 \\ -12 & -6l & 12 & -6l \\ 6l & 2l^2 & -6l & 4l^2 \end{bmatrix} \quad \text{(E8.1)}$$

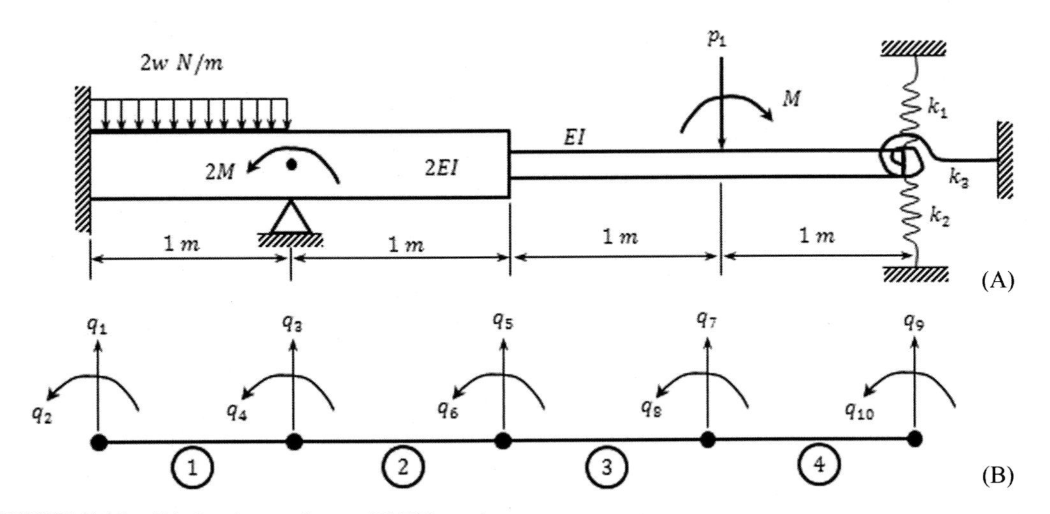

FIGURE 5.19 (A) Continuous beam, (B) F.E. mesh.

The element stiffness matrix for elements 1, 2, 3, and 4 are as follows:

$$[k]^1 = EI\begin{bmatrix} 24 & 12l & -24 & 12l \\ 12l & 8l^2 & -12l & 4l^2 \\ -24 & -12l & 24 & -12l \\ 12l & 4l^2 & -12l & 8l^2 \end{bmatrix} = [k]^2 \tag{E8.2}$$

$$[k]^3 = EI\begin{bmatrix} 12 & 6l & -12 & 6l \\ 6l & 4l^2 & -6l & 2l^2 \\ -12 & -6l & 12 & -6l \\ 6l & 2l^2 & -6l & 4l^2 \end{bmatrix} = [k]^4 \tag{E8.3}$$

Or, using the geometry of Fig. 1.20, gives,

$$[k]^1 = [k]^2 = EI\begin{bmatrix} 24 & 12 & -24 & 12 \\ 12 & 8 & -12 & 4 \\ -24 & -12 & 24 & -12 \\ 12 & 4 & -12 & 8 \end{bmatrix} \tag{E8.4}$$

And,

$$[k]^3 = [k]^4 = EI\begin{bmatrix} 12 & 6 & -12 & 6 \\ 6 & 4 & -6 & 2 \\ -12 & -6 & 12 & -6 \\ 6 & 2 & -6 & 4 \end{bmatrix} \tag{E8.5}$$

For this beam, we have,

$$q_1 = q_2 = q_3 = 0, Q_4 = 2M, Q_5 = 0, Q_6 = 0, Q_7 = -p_1, Q_8 = -M, Q_9 = 0, Q_{10} = 0 \tag{E8.6}$$

By inspection, the unconstrained global stiffness matrix is given by:

$$[K]_{\text{sys}} = \begin{bmatrix} k_{11}^1 & k_{12}^1 & k_{13}^1 & k_{14}^1 & 0 & 0 & 0 & 0 & 0 & 0 \\ k_{21}^1 & k_{22}^1 & k_{23}^1 & k_{24}^1 & 0 & 0 & 0 & 0 & 0 & 0 \\ k_{31}^1 & k_{32}^1 & \begin{pmatrix} k_{33}^1 + \\ k_{11}^2 \end{pmatrix} & \begin{pmatrix} k_{34}^1 + \\ k_{12}^2 \end{pmatrix} & k_{13}^2 & k_{14}^2 & 0 & 0 & 0 & 0 \\ k_{41}^1 & k_{42}^1 & \begin{pmatrix} k_{43}^1 + \\ k_{21}^2 \end{pmatrix} & \begin{pmatrix} k_{44}^1 + \\ k_{22}^2 \end{pmatrix} & k_{23}^2 & k_{24}^2 & 0 & 0 & 0 & 0 \\ 0 & 0 & k_{31}^2 & k_{32}^2 & \begin{pmatrix} k_{33}^2 + \\ k_{11}^3 \end{pmatrix} & \begin{pmatrix} k_{34}^2 + \\ k_{12}^3 \end{pmatrix} & k_{13}^3 & k_{14}^3 & 0 & 0 \\ 0 & 0 & k_{41}^2 & k_{42}^2 & \begin{pmatrix} k_{43}^2 + \\ k_{21}^3 \end{pmatrix} & \begin{pmatrix} k_{44}^2 + \\ k_{22}^3 \end{pmatrix} & k_{23}^3 & k_{24}^3 & 0 & 0 \\ 0 & 0 & 0 & 0 & k_{31}^3 & k_{32}^3 & \begin{pmatrix} k_{33}^3 + \\ k_{11}^4 \end{pmatrix} & \begin{pmatrix} k_{34}^3 + \\ k_{12}^4 \end{pmatrix} & k_{13}^4 & k_{14}^4 \\ 0 & 0 & 0 & 0 & k_{41}^3 & k_{42}^3 & \begin{pmatrix} k_{43}^3 + \\ k_{21}^4 \end{pmatrix} & \begin{pmatrix} k_{44}^3 + \\ k_{22}^4 \end{pmatrix} & k_{23}^4 & k_{24}^4 \\ 0 & 0 & 0 & 0 & 0 & 0 & k_{31}^4 & k_{32}^4 & \begin{pmatrix} k_{33}^4 + \\ k_1 + \\ k_2 \end{pmatrix} & k_{34}^4 \\ 0 & 0 & 0 & 0 & 0 & 0 & k_{41}^4 & k_{42}^4 & k_{43}^4 & \begin{pmatrix} k_{44}^4 + \\ k_3 \end{pmatrix} \end{bmatrix} \tag{E8.7}$$

By inserting the values of the element stiffness matrices in the system stiffness matrix, the equilibrium equations are given by:

$$EI\begin{bmatrix} 16 & -12 & 4 & 0 & 0 & 0 & 0 \\ -12 & 36 & -6 & -12 & 6 & 0 & 0 \\ 4 & -6 & 12 & -6 & 2 & 0 & 0 \\ 0 & -12 & -6 & 24 & 0 & -12 & 6 \\ 0 & 6 & 2 & 0 & 8 & -6 & 2 \\ 0 & 0 & 0 & -12 & -6 & (12+k_1+k_2) & -6 \\ 0 & 0 & 0 & 6 & 2 & -6 & (4+k_3) \end{bmatrix}\begin{Bmatrix} q_4 \\ q_5 \\ q_6 \\ q_7 \\ q_8 \\ q_9 \\ q_{10} \end{Bmatrix} = \begin{Bmatrix} (2M) \\ 0 \\ 0 \\ -p_1 \\ -M \\ 0 \\ 0 \end{Bmatrix} \quad \text{(E8.8)}$$

5.5 Inclined bar element

Consider the inclined bar element shown in Fig. 5.20, before and after deformation positions.

Fig. 5.20 shows the inclined bar in local and global positions. We have,

$$\sin\theta = \frac{x_1}{q_2}, \text{and}$$

$$x_1 = q_2 \sin\theta, \text{therefore}$$

$$\cos\theta = \frac{x_2}{q_1}. \text{ Also}$$

$$x_2 = q_1 \cos\theta, \text{and}$$

$$q_1' = x_1 + x_2$$

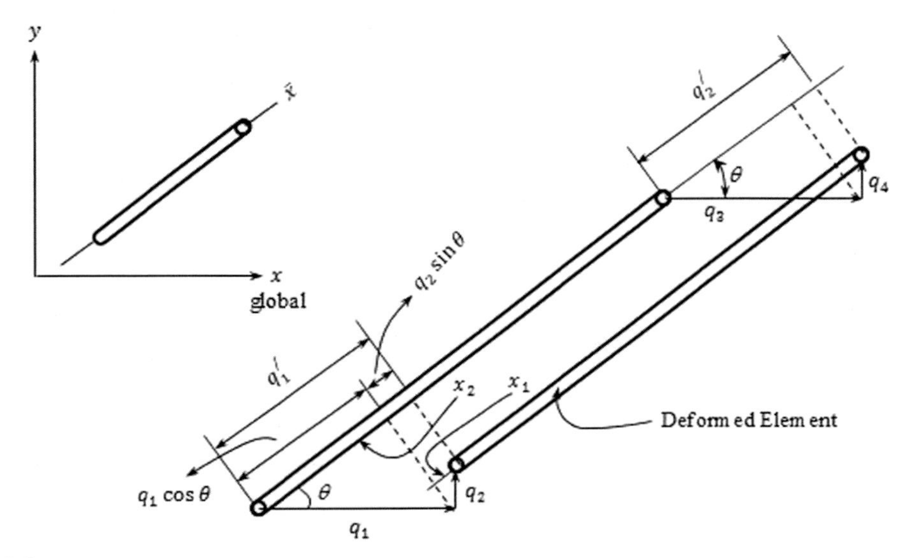

FIGURE 5.20 An inclined bar element.

Therefore,

$$q_1^{'} = q_1\cos\theta + q_2\sin\theta$$

$$q_2^{'} = q_3\cos\theta + q_4\sin\theta \tag{5.29a}$$

The element displacement vector in the global coordinates system is,

$$\{q\} = \begin{Bmatrix} q_1 \\ q_2 \\ q_3 \\ q_4 \end{Bmatrix} \tag{5.29b}$$

Let, $l = \cos\theta$and$m = \sin\theta$, therefore,

$$\{q^{'}\} = [L]\{q\} \tag{5.29c}$$

Where, $[L]$ = Transformation matrix.
$[L]$ is given by,

$$[L] = \begin{bmatrix} l & m & 0 & 0 \\ 0 & 0 & l & m \end{bmatrix} \tag{5.29}$$

$\{q^{'}\}$ is obtained by using Eqs. (5.29c) and (5.29), as follows,

$$\{q^{'}\} = \begin{bmatrix} l & m & 0 & 0 \\ 0 & 0 & l & m \end{bmatrix} \begin{Bmatrix} q_1 \\ q_2 \\ q_3 \\ q_4 \end{Bmatrix} = \begin{Bmatrix} lq_1 + mq_2 \\ lq_3 + mq_4 \end{Bmatrix}$$

Therefore,

$$\begin{Bmatrix} q_1^{'} \\ q_2^{'} \end{Bmatrix} = \begin{Bmatrix} q_1\cos\theta + q_2\sin\theta \\ q_3\cos\theta + q_4\sin\theta \end{Bmatrix}$$

To calculate $l$and$m$,
From Fig. 5.21

$$\cos\theta \text{ or } l = \frac{x_2 - x_1}{l_e}$$

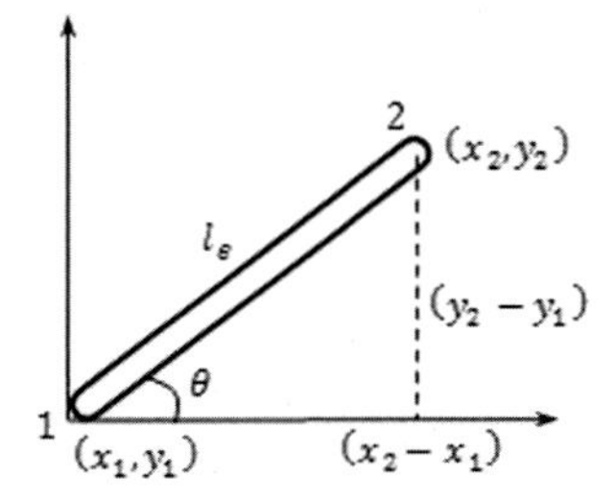

FIGURE 5.21 An inclined bar element.

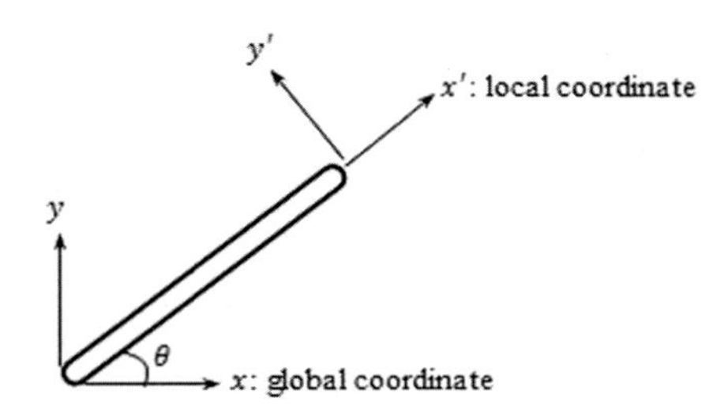

FIGURE 5.22 An Inclined bar with local and global coordinate system.

$$\sin\theta \text{ or } m = \frac{y_2 - y_1}{l_e}$$

$$l_e = \sqrt{(x_2 - x_1)^2 + (y_2 - y_1)^2}$$

Element stiffness matrix in its local position (Fig. 5.22),

$$[k']^e = \frac{E_r A_r}{l_r}\begin{bmatrix} 1 & -1 \\ -1 & 1 \end{bmatrix} \tag{5.30}$$

It is a truss element one-dimensional when viewed in the coordinate system.
$[k]^e$, in global coordinate system.

$$\text{Potential energy} \frac{1}{2}(kx^2) \tag{5.31a}$$

The strain energy is given by,

$$U^e = \frac{1}{2}\{q'\}^t[k']^e\{q'\} \text{in local coordinate system.} \tag{5.31b}$$

And,

$$\{q'\} = [L]\{q\} \tag{5.31c}$$

Therefore,

$$U^e = \frac{1}{2}\{q\}^t[L]^t[k']^e[L]\{q\} \tag{5.31d}$$

And the strain energy in global coordinate system,

$$U^e = \frac{1}{2}\{q\}^t[k]^e\{q\}$$

And the element stiffness matrix in global directions is given by,

$$[k]^e = [L]^t[k'][L] \tag{5.31e}$$

Therefore,

$$[k]^e = \frac{A_r E_r}{l_r}\begin{bmatrix} l^2 & lm & -l^2 & -lm \\ lm & m^2 & -lm & -m^2 \\ -l^2 & -lm & l^2 & lm \\ -lm & -m^2 & lm & m^2 \end{bmatrix} \tag{5.31}$$

Stress calculation, the stress σ in a truss element is given by,

$$\sigma = E_r.\varepsilon = E_r.\frac{q_2' - q_1'}{l_r} \tag{5.32}$$

Or,

$$\sigma = \frac{E_r}{l_r}\begin{bmatrix} -1 & 1 \end{bmatrix}\begin{Bmatrix} q_1' \\ q_2' \end{Bmatrix} \tag{5.33}$$

The above equation can be written in terms of the global displacement $\{q\}$ using the transformation, given in Eq. (5.29c),

$$\{q'\} = [L]\{q\}$$

Therefore insert Eq. (5.29c) into Eq. (5.33) gives,

$$\sigma = \frac{E_r}{l_r}\begin{bmatrix} -1 & 1 \end{bmatrix}[L]\{q\} = \frac{E_r}{l_r}\begin{bmatrix} -1 & 1 \end{bmatrix}\begin{bmatrix} l & m & 0 & 0 \\ 0 & 0 & l & m \end{bmatrix}\{q\}$$

Or,

$$\sigma = \frac{E_r}{l_r}\begin{bmatrix} -l & -m & l & m \end{bmatrix}\{q\} \tag{5.34}$$

Thermal Effects

For the thermal effects of an inclined element as shown in Fig. 5.22,
The element thermal load given by Eq. (5.35) as follows,

$$\{F\}^e_{\text{th}} = E.A.\alpha.\Delta T\begin{Bmatrix} -1 \\ 1 \end{Bmatrix} \tag{5.35}$$

Expressing in $x' - y'$ directions will be as follows,

$$\underbrace{\begin{Bmatrix} q_1 \\ q_2 \end{Bmatrix}^{t'}\{F'\}_{\text{th}}}_{\text{local}} = \underbrace{\{q\}^t\{F\}_{\text{th}}}_{\text{global}} \tag{5.36}$$

We have,

$$\{q'\}^t = [L]^t\{q\}^t \tag{5.37}$$

Where $[L]$ is the transformation matrix, therefore the thermal load in the local direction is given by,

$$\{F\}'_{\text{th}} = E.A.\alpha.\Delta T \begin{Bmatrix} -l \\ -m \\ l \\ m \end{Bmatrix} \tag{5.38}$$

Where,

$$l = \cos\theta \text{ and } m = \sin\theta$$

Example 5.9: Use the derived $[k]^e$ Eqs. (5.30) and (5.31) to establish the matrix equilibrium equations for the frame shown in Fig. 5.23,

Solution,

The modeling may be achieved by using 3 elements, 2 vertical beam-bar, and 1 horizontal beam-bar with six degrees' freedom (unconstrained)

By using global coordinates directly,

$$[k]_{\text{bar}} = \frac{EA}{l}\begin{bmatrix} 1 & -1 \\ -1 & 1 \end{bmatrix} \tag{E9.1}$$

The element stiffness matrix for the bar element lying horizontally.

And the element stiffness matrix for the inclined bar is given by Eq. (5.31),

$$[k]_{\text{beam}} = \frac{EI}{l^3}\begin{bmatrix} 12 & 6l & -12 & 6l \\ 6l & 4l^2 & -6l & 2l^2 \\ -12 & -6l & 12 & -6l \\ 6l & 2l^2 & -6l & 4l^2 \end{bmatrix} \tag{E9.2}$$

For element (2), it is a combination of bar and beam element (Fig. 5.24),

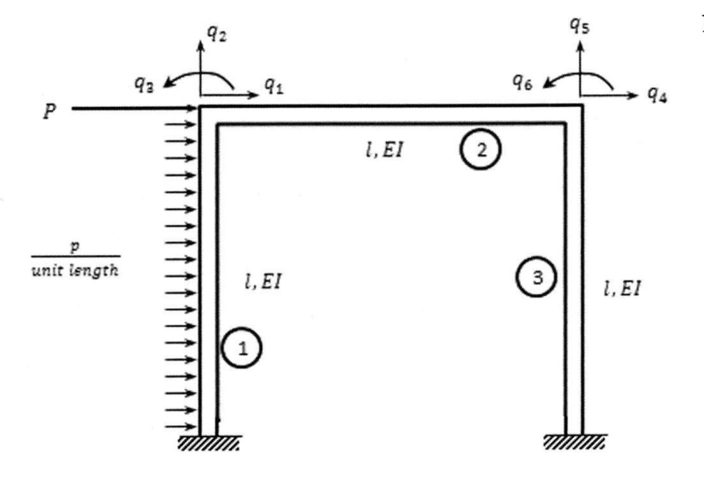

FIGURE 5.23 Frame problem.

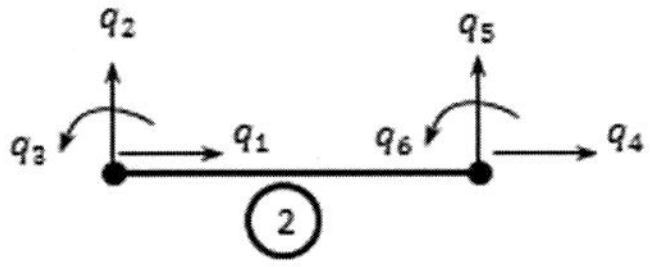

FIGURE 5.24 Element no. 2.

$$[k]^{(2)} = \frac{EI}{l^3}\overset{\begin{matrix} q_2 & \quad q_3 & \quad q_5 & \quad q_6 \end{matrix}}{\begin{bmatrix} 12 & 6l & -12 & 6l \\ 6l & 4l^2 & -6l & 2l^2 \\ -12 & -6l & 12 & -6l \\ 6l & 2l^2 & -6l & 4l^2 \end{bmatrix}} + \frac{AE}{l}\overset{\begin{matrix} q_1 & \quad q_4 \end{matrix}}{\begin{bmatrix} 1 & -1 \\ -1 & 1 \end{bmatrix}} \tag{E9.3}$$

Or,

$$= \frac{EI}{l^3}\overset{\begin{matrix} q_1 & q_2 & q_3 & q_4 & q_5 & q_6 \end{matrix}}{\begin{bmatrix} 0 & 0 & 0 & 0 & 0 & 0 \\ 0 & 12 & 6l & 0 & -12 & 6l \\ 0 & 6l & 4l^2 & 0 & -6l & 2l^2 \\ 0 & 0 & 0 & 0 & 0 & 0 \\ 0 & -12 & -6l & 0 & 12 & -6l \\ 0 & 6l & 2l^2 & 0 & -6l & 4l^2 \end{bmatrix}} + \frac{AE}{l}\overset{\begin{matrix} q_1 & q_2 & q_3 & q_4 & q_5 & q_6 \end{matrix}}{\begin{bmatrix} 1 & 0 & 0 & -1 & 0 & 0 \\ 0 & 0 & 0 & 0 & 0 & 0 \\ 0 & 0 & 0 & 0 & 0 & 0 \\ -1 & 0 & 0 & 1 & 0 & 0 \\ 0 & 0 & 0 & 0 & 0 & 0 \\ 0 & 0 & 0 & 0 & 0 & 0 \end{bmatrix}}$$

Which gives,

$$[k]^{(2)} = \begin{bmatrix} \frac{AE}{l} & 0 & 0 & -\frac{AE}{l} & 0 & 0 \\ 0 & 12\frac{EI}{l^3} & 6\frac{EI}{l^2} & 0 & -12\frac{EI}{l^3} & 6\frac{EI}{l^2} \\ 0 & 6\frac{EI}{l^2} & 4\frac{EI}{l} & 0 & -6\frac{EI}{l^2} & 2\frac{EI}{l} \\ -\frac{AE}{l} & 0 & 0 & \frac{AE}{l} & 0 & 0 \\ 0 & -12\frac{EI}{l^3} & -6\frac{EI}{l^2} & 0 & 12\frac{EI}{l^3} & -6\frac{EI}{l^2} \\ 0 & 6\frac{EI}{l^2} & 2\frac{EI}{l} & 0 & -6\frac{EI}{l^2} & 4\frac{EI}{l} \end{bmatrix} \tag{E9.4}$$

Since element 1 and 3 are identical (exactly) then (Fig. 5.25),

$$[k]^{(1)} = [k]^{(3)}$$

$$[k]^{(1)} = [k]^{(3)} = \frac{EI}{l^3}\overset{\begin{matrix} q_1 & \quad q_3 & \quad q_4 & \quad q_6 \end{matrix}}{\begin{bmatrix} 12 & 6l & -12 & 6l \\ 6l & 4l^2 & -6l & 2l^2 \\ -12 & -6l & 12 & -6l \\ 6l & 2l^2 & -6l & 4l^2 \end{bmatrix}} + \frac{AE}{l}\overset{\begin{matrix} q_2 & \quad q_5 \end{matrix}}{\begin{bmatrix} 1 & -1 \\ -1 & 1 \end{bmatrix}} \tag{E9.5}$$

Or,

$$[k]^{(1)} = [k]^{(3)} = \frac{EI}{l^3}\begin{bmatrix} 12 & 0 & 6l & -12 & 0 & 6l \\ 0 & 0 & 0 & 0 & 0 & 0 \\ 6l & 0 & 4l^2 & -6l & 0 & 2l^2 \\ -12 & 0 & -6l & 12 & 0 & -6l \\ 0 & 0 & 0 & 0 & 0 & 0 \\ 6l & 0 & 2l^2 & -6l & 0 & 4l^2 \end{bmatrix} + \frac{AE}{l}\begin{bmatrix} 0 & 0 & 0 & 0 & 0 & 0 \\ 0 & 1 & 0 & 0 & -1 & 0 \\ 0 & 0 & 0 & 0 & 0 & 0 \\ 0 & 0 & 0 & 0 & 0 & 0 \\ 0 & -1 & 0 & 0 & 1 & 0 \\ 0 & 0 & 0 & 0 & 0 & 0 \end{bmatrix}$$

Which gives,

$$[k]^{(1)} = [k]^{(3)} = \begin{bmatrix} 12\frac{EI}{l^3} & 0 & 6\frac{EI}{l^2} & -12\frac{EI}{l^3} & 0 & 6\frac{EI}{l^2} \\ 0 & \frac{AE}{l} & 0 & 0 & -\frac{AE}{l} & 0 \\ 6\frac{EI}{l^2} & 0 & 4\frac{EI}{l} & -6\frac{EI}{l^2} & 0 & 2\frac{EI}{l} \\ -12\frac{EI}{l^3} & 0 & -6\frac{EI}{l^2} & 12\frac{EI}{l^3} & 0 & -6\frac{EI}{l^2} \\ 0 & -\frac{AE}{l} & 0 & 0 & \frac{AE}{l} & 0 \\ 6\frac{EI}{l^2} & 0 & 2\frac{EI}{l} & -6\frac{EI}{l^2} & 0 & 4\frac{EI}{l} \end{bmatrix} \tag{E9.6}$$

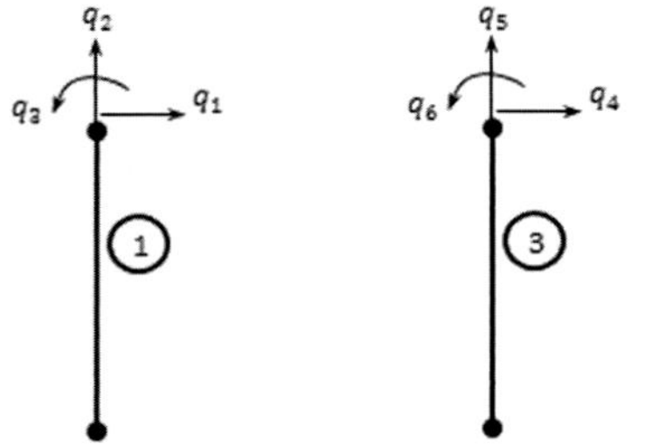

FIGURE 5.25 Elements 1 and 3.

The system stiffness matrix is formed by inspection according to the corresponding degrees of freedom which results in the following form,

$$[K]_{sys} = \begin{bmatrix} \left(12\frac{EI}{l^3}+\frac{AE}{l}\right) & 0 & 6\frac{EI}{l^2} & \left(-12\frac{EI}{l^3}-\frac{AE}{l}\right) & 0 & 6\frac{EI}{l^2} \\ 0 & \left(12\frac{EI}{l^3}+\frac{AE}{l}\right) & 6\frac{EI}{l^2} & 0 & \left(-12\frac{EI}{l^3}-\frac{AE}{l}\right) & 6\frac{EI}{l^2} \\ 6\frac{EI}{l^2} & 6\frac{EI}{l^2} & 8\frac{EI}{l} & -6\frac{EI}{l^2} & -6\frac{EI}{l^2} & 4\frac{EI}{l} \\ \left(-12\frac{EI}{l^3}-\frac{AE}{l}\right) & 0 & -6\frac{EI}{l^2} & \left(\frac{AE}{l}+\frac{12EI}{l^3}\right) & \frac{6EI}{l^2} & \frac{6EI}{l^2} \\ 0 & \left(-12\frac{EI}{l^3}-\frac{AE}{l}\right) & -6\frac{EI}{l^2} & \frac{6EI}{l^2} & \left(\frac{12EI}{l^3}+\frac{AE}{l}\right) & -\frac{6EI}{l^2} \\ 6\frac{EI}{l^2} & 6\frac{EI}{l^2} & 4\frac{EI}{l} & \frac{6EI}{l^2} & -\frac{6EI}{l^2} & 8\frac{EI}{l} \end{bmatrix} \quad \text{(E9.7)}$$

The load vector is,

$$Q = \left\{ \left(\frac{pl}{2}+p\right) \quad 0 \quad 0 \quad 0 \quad 0 \quad 0 \right\}^t, q = \{ q_1 \quad q_2 \quad q_3 \quad q_4 \quad q_5 \quad q_6 \}^t \quad \text{(E9.8)}$$

Example 5.10: Generate the overall stiffness matrix for the pin-jointed plane frame of Fig. 5.26.

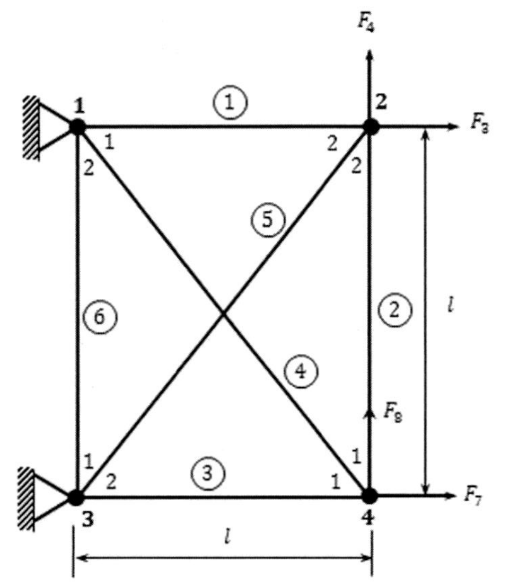

FIGURE 5.26 Pin jointed frame.

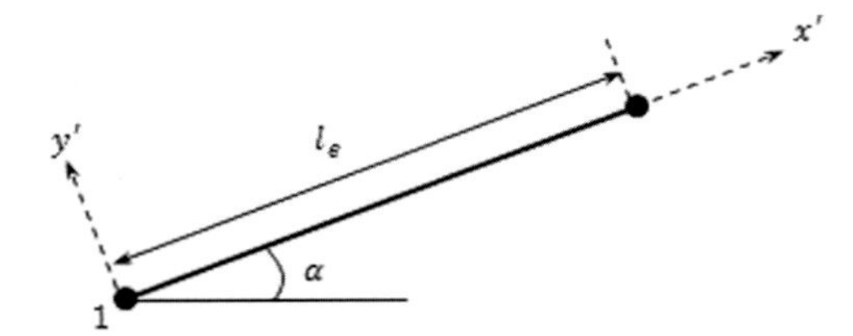

FIGURE 5.27 Typical inclined element.

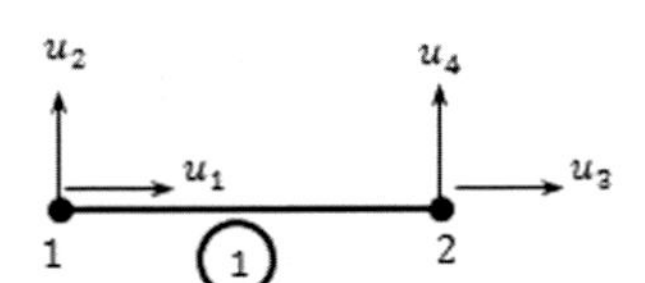

FIGURE 5.28 Element no. 1.

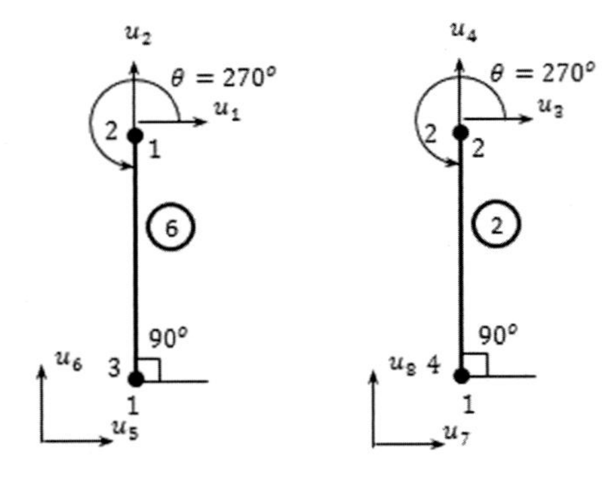

FIGURE 5.29 Elements 2 and 6.

As in the previous example, the element stiffness matrix for the bar in global direction and for the inclined bar shown in Fig. 5.27 in global direction respectively are given by,

$$[k]^e = \frac{AE}{l_e}\begin{bmatrix} 1 & -1 \\ -1 & 1 \end{bmatrix}$$

$$[k]^e = \frac{A_rE_r}{l_e}\begin{bmatrix} l^2 & lm & -l^2 & -lm \\ lm & m^2 & -lm & -m^2 \\ -l^2 & -lm & l^2 & lm \\ -lm & -m^2 & lm & m^2 \end{bmatrix} \tag{E10.1}$$

Element (1) shown in Fig. 5.28, $l = \cos\alpha, m = \sin\alpha, (\alpha = 0), l = 1, m = 0,$

$$\begin{matrix} & & 1234 & \\ [k]^1 = \dfrac{AE}{l_e} & \begin{bmatrix} 1 & 0 & -1 & 0 \\ 0 & 0 & 0 & 0 \\ -1 & 0 & 1 & 0 \\ 0 & 0 & 0 & 0 \end{bmatrix} & \begin{matrix} 1 \\ 2 \\ 3 \\ 4 \end{matrix} \end{matrix} \tag{E10.2}$$

Element (2) and (6) shown in Fig. 5.29, $(\alpha = 90^o), l = 0, m = 1,$

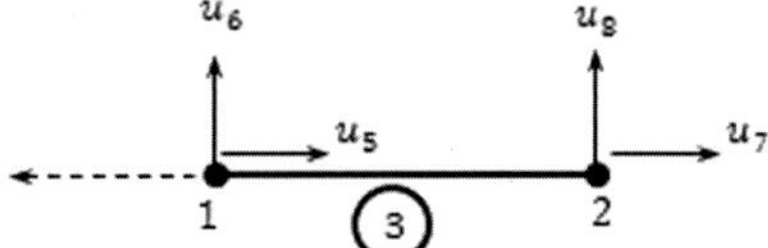

FIGURE 5.30 Element no. 3.

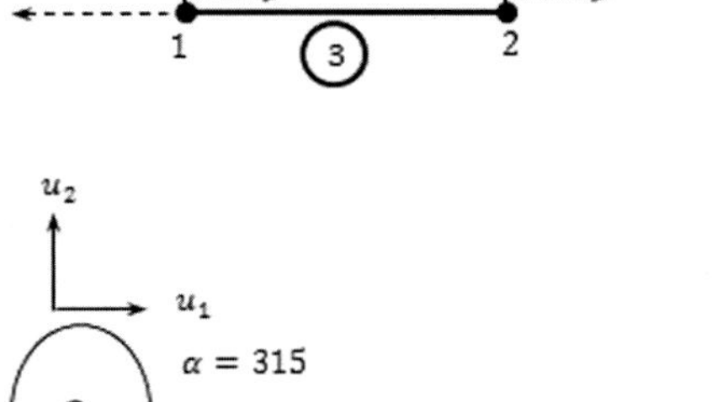

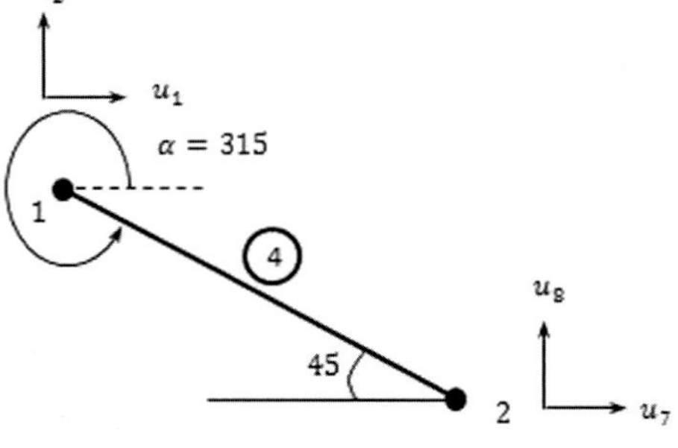

FIGURE 5.31 Element no. 4.

$$[k]^2 = [k]^6 = \frac{AE}{l_e}\begin{array}{c} \scriptstyle 6\to1 \quad 2 \quad 5 \quad 6 \\ \scriptstyle 2\to3478 \\ \begin{bmatrix} 0 & 0 & 0 & 0 \\ 0 & 1 & 0 & -1 \\ 0 & 0 & 0 & 0 \\ 0 & -1 & 0 & 1 \end{bmatrix}\end{array} \begin{array}{cc} \overset{2}{\downarrow} & \overset{6}{\downarrow} \\ 3 & 1 \\ 4 & 2 \\ 7 & 5 \\ 8 & 6 \end{array} \tag{E10.3}$$

Element (3) shown in Fig. 5.30, ($\alpha = 0$)

$$[k]^3 = \frac{AE}{l_e}\begin{array}{c} \scriptstyle 5678 \\ \begin{bmatrix} 1 & 0 & -1 & 0 \\ 0 & 0 & 0 & 0 \\ -1 & 0 & 1 & 0 \\ 0 & 0 & 0 & 0 \end{bmatrix}\end{array} \begin{array}{c} 5 \\ 6 \\ 7 \\ 8 \end{array} \tag{E10.4}$$

For element (4) shown in Fig. 5.31, ($\alpha = 315^o$), Length element $\left(\sqrt{2}l\right), l = \frac{1}{\sqrt{2}}, m = -\frac{1}{\sqrt{2}},$

$$[k]^4 = \frac{\sqrt{2}AE}{4l_e}\begin{array}{c} \scriptstyle 1278 \\ \begin{bmatrix} 1 & -1 & -1 & 1 \\ -1 & 1 & 1 & -1 \\ -1 & 1 & 1 & -1 \\ 1 & -1 & -1 & 1 \end{bmatrix}\end{array} \begin{array}{c} 1 \\ 2 \\ 7 \\ 8 \end{array} \tag{E10.5}$$

For element (5) shown in Fig. 5.32, ($\alpha = 45^o$), $l = \frac{1}{\sqrt{2}}, m = \frac{1}{\sqrt{2}},$

$$[k]^5 = \frac{\sqrt{2}AE}{4l_e}\begin{array}{c} \scriptstyle 5634 \\ \begin{bmatrix} 1 & 1 & -1 & -1 \\ 1 & 1 & -1 & -1 \\ -1 & -1 & 1 & 1 \\ -1 & -1 & 1 & 1 \end{bmatrix}\end{array} \begin{array}{c} 5 \\ 6 \\ 3 \\ 4 \end{array} \tag{E10.6}$$

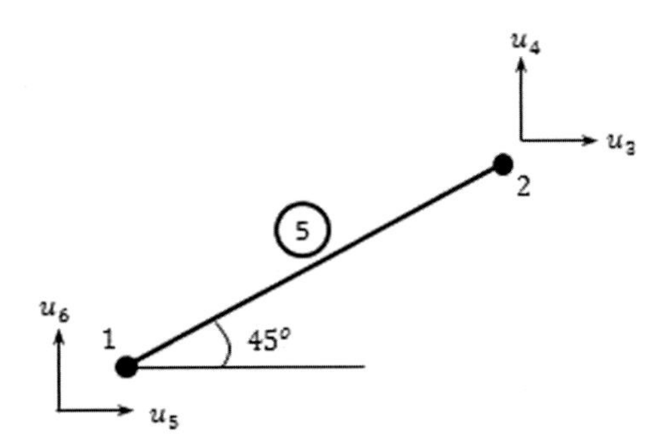

FIGURE 5.32 Element no. 5.

Having obtained the stiffness matrix for each element, the unconstrained system stiffness matrix is obtained by inspection and will be of size (8 × 8) as follows,

12345678

$$[K]_{sys}=\frac{AE}{L}\begin{bmatrix} \left(1+\frac{\sqrt{2}}{4}\right) & -\frac{\sqrt{2}}{4} & -1 & 0 & 0 & 0 & -\frac{\sqrt{2}}{4} & \frac{\sqrt{2}}{4} \\ -\frac{\sqrt{2}}{4} & \left(1+\frac{\sqrt{2}}{4}\right) & 0 & 0 & 0 & -1 & \frac{\sqrt{2}}{4} & -\frac{\sqrt{2}}{4} \\ -1 & 0 & \left(1+\frac{\sqrt{2}}{4}\right) & \frac{\sqrt{2}}{4} & -\frac{\sqrt{2}}{4} & -\frac{\sqrt{2}}{4} & 0 & 0 \\ 0 & 0 & \frac{\sqrt{2}}{4} & \left(1+\frac{\sqrt{2}}{4}\right) & -\frac{\sqrt{2}}{4} & -\frac{\sqrt{2}}{4} & 0 & -1 \\ 0 & 0 & -\frac{\sqrt{2}}{4} & \frac{\sqrt{2}}{4} & \left(1+\frac{\sqrt{2}}{4}\right) & \frac{\sqrt{2}}{4} & -1 & 0 \\ 0 & -1 & -\frac{\sqrt{2}}{4} & -\frac{\sqrt{2}}{4} & \frac{\sqrt{2}}{4} & \left(1+\frac{\sqrt{2}}{4}\right) & 0 & 0 \\ -\frac{\sqrt{2}}{4} & \frac{\sqrt{2}}{4} & 0 & 0 & -1 & 0 & \left(1+\frac{\sqrt{2}}{4}\right) & -\frac{\sqrt{2}}{4} \\ \frac{\sqrt{2}}{4} & -\frac{\sqrt{2}}{4} & 0 & -1 & 0 & 0 & -\frac{\sqrt{2}}{4} & \left(1+\frac{\sqrt{2}}{4}\right) \end{bmatrix} \tag{E10.7}$$

The displacement vector is,

$$\{u\}=\begin{Bmatrix} u_1=0 \\ u_2=0 \\ u_3 \\ u_4 \\ u_5=0 \\ u_6=0 \\ u_7 \\ u_8 \end{Bmatrix} \tag{E110.8}$$

$u_1 = u_2 = u_5 = u_6 = 0$, Specified degrees of freedom. Using theses constraints, the reduced stiffness matrix a gives in Eq. (E11.7) is

$$[K]_{\text{sys}} = \frac{AE}{L}\begin{bmatrix} 1.353 & 0.353 & 0 & 0 \\ 0.353 & 1.353 & 0 & -1 \\ 0 & 0 & 1.353 & -0.353 \\ 0 & -1 & -0.353 & 1.353 \end{bmatrix} \tag{E10.9}$$

$$[K]\{u\} = \{F\}$$

And the final constrained system equilibrium equations are,

$$\frac{AE}{L}\begin{bmatrix} 1.353 & 0.353 & 0 & 0 \\ 0.353 & 1.353 & 0 & -1 \\ 0 & 0 & 1.353 & -0.353 \\ 0 & -1 & -0.353 & 1.353 \end{bmatrix}\begin{Bmatrix} u_3 \\ u_4 \\ u_7 \\ u_8 \end{Bmatrix} = \begin{Bmatrix} F_3 \\ F_4 \\ F_7 \\ F_8 \end{Bmatrix} \tag{E10.10}$$

Example 5.11: For the truss shown in Fig. 5.33 a horizontal load of $P = 20,000$ N is applied in the x-direction at node (2),

1. Write down the element stiffness matrix $[k]^e$ for each element.
2. Assemble the system stiffness matrix $[k]$.
3. Using the elimination approach, solve for $\{u\}$.
4. Evaluate the stress in elements (2) and (3).
5. Determine the reaction force at node (2) in the y-direction.
6. $E = 300 * 10^3 \ \text{N/mm}^2, A = 200 \text{ mm}^2$ for each member, $p = 20,000$ N.

Solution,

1. As in the previous example,

$$[k']^e = \frac{EA}{l}\begin{bmatrix} 1 & -1 \\ -1 & 1 \end{bmatrix} \tag{E11.1}$$

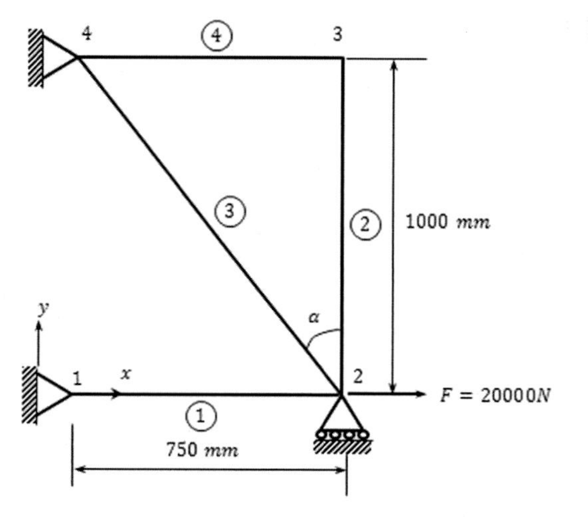

FIGURE 5.33 Pin jointed plane frame.

$$[L] = \begin{bmatrix} l & m & 0 & 0 \\ 0 & 0 & l & m \end{bmatrix} \text{Transformation matrix}$$

And we have the element stiffness matrix for the bar in global direction is given by,

$$[k]^e = [L]^t \left[k'\right]^e [L]$$

$$[k]^e = \frac{EA}{l} \begin{bmatrix} l & 0 \\ m & 0 \\ 0 & l \\ 0 & m \end{bmatrix} \begin{bmatrix} 1 & -1 \\ -1 & 1 \end{bmatrix} \begin{bmatrix} l & m & 0 & 0 \\ 0 & 0 & l & m \end{bmatrix}$$

$$[k]^e = \frac{EA}{l} \begin{bmatrix} l^2 & lm & -l^2 & -lm \\ lm & m^2 & -lm & -m^2 \\ -l^2 & -lm & l^2 & lm \\ -lm & -m^2 & lm & m^2 \end{bmatrix} \quad \text{(E11.2)}$$

In general, for the typical element shown in Fig. 5.34,
For elements (1) and (4) shown in Figs. 5.35 and 5.36 respectively, ($\theta = 0$),

$$l = \cos 0 = 1, m = \sin 0 = 0$$

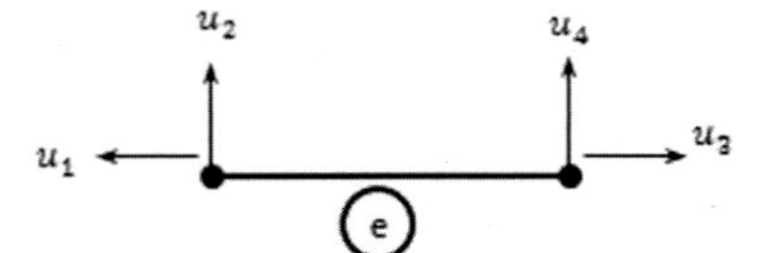

FIGURE 5.34 Typical beam element.

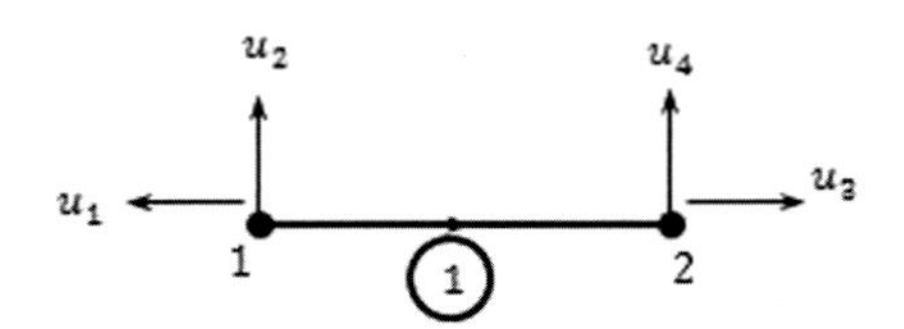

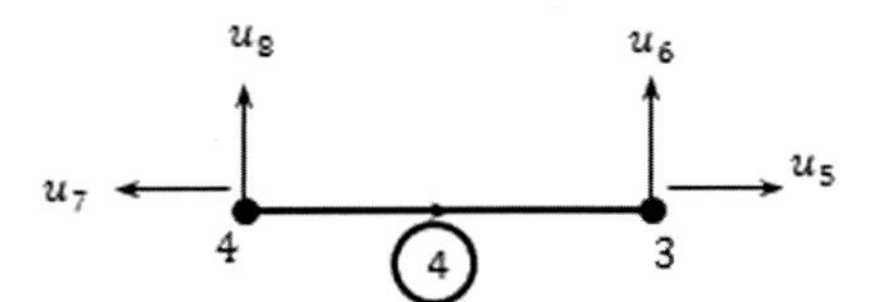

FIGURE 5.35 Element No. 1. and element No. 4

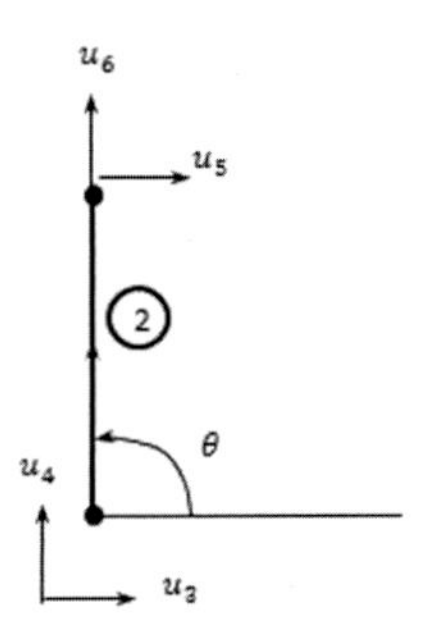

FIGURE 5.36 Element No. 2.

$$[k]^1 = \frac{300*10^3*200}{750}\overset{1234}{\begin{bmatrix} 1 & 0 & -1 & 0 \\ 0 & 0 & 0 & 0 \\ -1 & 0 & 1 & 0 \\ 0 & 0 & 0 & 0 \end{bmatrix}} \tag{E11.3}$$

$$[k]^1 = 10^3\overset{1234}{\begin{bmatrix} 80 & 0 & -80 & 0 \\ 0 & 0 & 0 & 0 \\ -80 & 0 & 80 & 0 \\ 0 & 0 & 0 & 0 \end{bmatrix}} \tag{E11.4}$$

$$[k]^4 = \frac{300*10^3*200}{750}\overset{7856}{\begin{bmatrix} 1 & 0 & -1 & 0 \\ 0 & 0 & 0 & 0 \\ -1 & 0 & 1 & 0 \\ 0 & 0 & 0 & 0 \end{bmatrix}} \tag{E11.5}$$

$$[k]^4 = 10^3\overset{7856}{\begin{bmatrix} 80 & 0 & -80 & 0 \\ 0 & 0 & 0 & 0 \\ -80 & 0 & 80 & 0 \\ 0 & 0 & 0 & 0 \end{bmatrix}} \tag{E11.6}$$

For element (2) shown in Fig. 5.37, ($\theta = 90^o$),

$$l = \cos 90 = 0, m = \sin 90 = 1$$

$$[k]^2 = \frac{300*10^3*200}{1000}\overset{3456}{\begin{bmatrix} 0 & 0 & 0 & 0 \\ 0 & 1 & 0 & -1 \\ 0 & 0 & 0 & 0 \\ 0 & -1 & 0 & 1 \end{bmatrix}} \tag{E11.7}$$

$$[k]^2 = 10^3\overset{3456}{\begin{bmatrix} 0 & 0 & 0 & 0 \\ 0 & 60 & 0 & -60 \\ 0 & 0 & 0 & 0 \\ 0 & -60 & 0 & 60 \end{bmatrix}} \tag{E11.8}$$

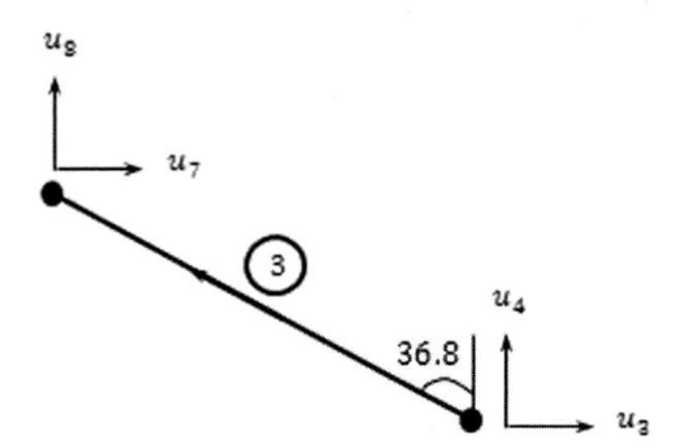

FIGURE 5.37 Element No. 3.

For element (3) shown in Fig. 5.38,

$$\alpha = tan^{-1}\left(\frac{750}{1000}\right) = 36.8^o$$

Therefore,

$$\theta = 90(270) + 36.8 = 126.8^o$$

$$l = \cos 126.8 = -0.6, m = \sin 126.8 = 0.8$$

$$\text{length} = \sqrt{(750)^2 + (1000)^2} = 1250 \text{ mm}$$

$$[k]^3 = \frac{300 * 10^3 * 200}{1250} \overset{3478}{\begin{bmatrix} 0.36 & 0.48 & -0.36 & -0.48 \\ 0.48 & 0.64 & -0.48 & -0.64 \\ -0.36 & -0.48 & 0.36 & 0.48 \\ -0.48 & -0.64 & 0.48 & 0.64 \end{bmatrix}} \tag{E11.9}$$

$$[k]^3 = 10^3 \overset{3478}{\begin{bmatrix} 17.28 & 23.04 & -17.28 & -23.04 \\ 23.04 & 30.72 & -23.04 & -30.72 \\ -17.28 & -23.04 & 17.28 & 23.04 \\ -23.04 & -30.72 & 23.04 & 30.72 \end{bmatrix}} \tag{E11.10}$$

2. The system stiffness matrix is obtained by using the inspection using the similar procedure employed in the previous examples.

$$[K]_{sym.} = 10^3 \begin{bmatrix} 80 & 0 & -80 & 0 & 0 & 0 & 0 & 0 \\ 0 & 0 & 0 & 0 & 0 & 0 & 0 & 0 \\ -80 & 0 & (80+17.28) & 23.04 & 0 & 0 & -17.28 & -23.04 \\ 0 & 0 & 23.04 & (60+30.72) & 0 & -60 & -23.04 & -30.04 \\ 0 & 0 & 0 & 0 & 80 & 0 & -80 & 0 \\ 0 & 0 & 0 & -60 & 0 & 60 & 0 & 0 \\ 0 & 0 & -17.28 & -23.04 & -80 & 0 & (17.28+80) & 23.04 \\ 0 & 0 & -23.04 & -30.72 & 0 & 0 & 23.04 & 30.72 \end{bmatrix} \tag{E11.11}$$

3. $\{P\} = [K]\{u\}$

$$\{u\} = \begin{Bmatrix} u_1 = 0 \\ u_2 = 0 \\ u_3 \\ u_4 = 0 \\ u_5 \\ u_6 \\ u_7 = 0 \\ u_8 = 0 \end{Bmatrix}, \text{The unconstrained displacement vector.} \tag{E11.12}$$

And,

$$\{P\} = \left\{ \begin{array}{c} p_1 = ? \\ p_2 = ? \\ p_3 = 20,000 \\ p_4 = ? \\ p_5 = 0 \\ p_6 = 0 \\ p_7 = ? \\ p_8 = ? \end{array} \right\}, \text{The unconstrained load vector.} \tag{E11.13}$$

Employing the boundary conditions by deleting the rows and the corresponding columns corresponds to the constrained degree of freedom, then,

$$[K]_{\text{sys}} = 10^3 \begin{bmatrix} (80 + 17.28) & 0 & 0 \\ 0 & 80 & 0 \\ 0 & 0 & 60 \end{bmatrix} \tag{E11.14}$$

Or, the constrained equations will be as follows,

$$10^3 \begin{bmatrix} 97.28 & 0 & 0 \\ 0 & 80 & 0 \\ 0 & 0 & 60 \end{bmatrix} \left\{ \begin{array}{c} u_3 \\ u_5 \\ u_6 \end{array} \right\} = \left\{ \begin{array}{c} 20,000 \\ 0 \\ 0 \end{array} \right\} \tag{E11.15}$$

And solving gives,

$$97,280 u_3 = 20,000$$

$$u_3 = \frac{20,000}{97,280} = 0.205 \text{ mm}$$

Or,

$$u_5 = u_6 = 0$$

Or,

$$[K]\{u\} = \{P\}$$

$$\{u\} = [k]^{-1}\{P\}$$

Or,

$$\left\{ \begin{array}{c} u_3 \\ u_5 \\ u_6 \end{array} \right\} = \begin{bmatrix} 0.00001028 & 0 & 0 \\ 0 & 0.0000125 & 0 \\ 0 & 0 & 0.0000166 \end{bmatrix} \left\{ \begin{array}{c} 20,000 \\ 0 \\ 0 \end{array} \right\}$$

$$u_3 = 0.00001028 * 20,000 = 0.205 \text{ mm}$$

$$u_5 = u_6 = 0$$

4. Stress calculation (in members (2) and (3)),

$$\sigma = \frac{E}{l} \begin{bmatrix} -l & -m & l & m \end{bmatrix} \{u\} \tag{E11.16}$$

For element (2) we have already,

$$(\theta = 90^{o}), l = 0, m = 1, L = 1000 \text{ mm}$$

We can write,

$$\{u\} = \begin{Bmatrix} u_3 \\ u_4 \\ u_5 \\ u_6 \end{Bmatrix} = \begin{Bmatrix} 0.205 \\ 0 \\ 0 \\ 0 \end{Bmatrix} \quad \text{(E11.17)}$$

The stress can be calculated using Eq. (E12.16) as follows,

$$\sigma_2 = \frac{300 * 10^3}{1000} \underset{(1*4)}{\begin{bmatrix} 0 & -1 & 0 & 1 \end{bmatrix}} \underset{(4*1)}{\begin{Bmatrix} 0.205 \\ 0 \\ 0 \\ 0 \end{Bmatrix}}$$

Or,

$$\sigma_2 = 0$$

For element (3),

$$(\theta = 126.8^{o}), l = -0.6, m = 0.8, L = 1250 \text{ mm}$$

And the stress in element (3) is,

$$\sigma_3 = \frac{300 * 10^3}{1250} \begin{bmatrix} 0.6 & -0.8 & -0.6 & 0.8 \end{bmatrix} \begin{Bmatrix} u_3 = 0.205 \\ u_4 = 0 \\ u_7 = 0 \\ u_8 = 0 \end{Bmatrix} = 240 * (0.6 * 0.205)$$

Or,

$$\sigma_3 = 29.52 \text{ N/mm}^2$$

5. To find the reaction force (P_4) in node (2) we must return to general equation of the form,

$$[K]\{u\} = \{P\}$$

we have,

$$p_3 = p_5 = p_6 = 0$$

The reaction for p_4 is obtained as follows,

$$10^3\begin{bmatrix}0 & 0 & 23.04 & 90.72 & 0 & -60 & -23.04 & -30.72\end{bmatrix}\begin{Bmatrix}0\\0\\0.205\\0\\0\\0\\0\\0\end{Bmatrix}=p_4 \tag{E11.18}$$

Therefore,

$$p_4 = 10^3 * 23.04 * 0.205 = 4723.2\text{ N}$$

The reactions at the supports can be obtained in a similar manner to that employed for obtaining p_4 by using the equilibrium equations$[k]\{u\} = \{P\}$, and selecting the corresponding row of the required reaction. Also, the reactions are,

$$10^3\begin{bmatrix}80 & 0 & -80 & 0 & 0 & 0 & 0 & 0\\0 & 0 & 0 & 0 & 0 & 0 & 0 & 0\\-80 & 0 & (80+17.28) & 23.04 & 0 & 0 & -17.28 & -23.04\\0 & 0 & 23.04 & (60+30.72) & 0 & -60 & -23.04 & -30.04\\0 & 0 & 0 & 0 & 80 & 0 & -80 & 0\\0 & 0 & 0 & -60 & 0 & 60 & 0 & 0\\0 & 0 & -17.28 & -23.04 & -80 & 0 & (17.28+80) & 23.04\\0 & 0 & -23.04 & -30.72 & 0 & 0 & 23.04 & 30.72\end{bmatrix}\begin{Bmatrix}0\\0\\0.205\\0\\0\\0\\0\\0\end{Bmatrix}=\begin{Bmatrix}p_1\\p_2\\20,000\\4723.2\\0\\0\\p_7\\p_8\end{Bmatrix} \tag{E11.19}$$

Therefore,

$$p_1 = -16,400\text{ N}$$

$$p_2 = 0$$

$$p_4 = 4723.2\text{ N}$$

$$p_7 = -3542.4\text{ N},$$

and,

$$p_8 = -4723.2\text{ N}$$

Example 5.12: Find all parameters in the previous example for the truss configuration shown in Fig. 5.38,

$$[k]^e = \frac{EA}{l}\begin{bmatrix}l^2 & lm & -l^2 & -lm\\lm & m^2 & -lm & -m^2\\-l^2 & -lm & l^2 & lm\\-lm & -m^2 & lm & m^2\end{bmatrix} \tag{E12.1}$$

$$l = \cos\theta, m = \sin\theta$$

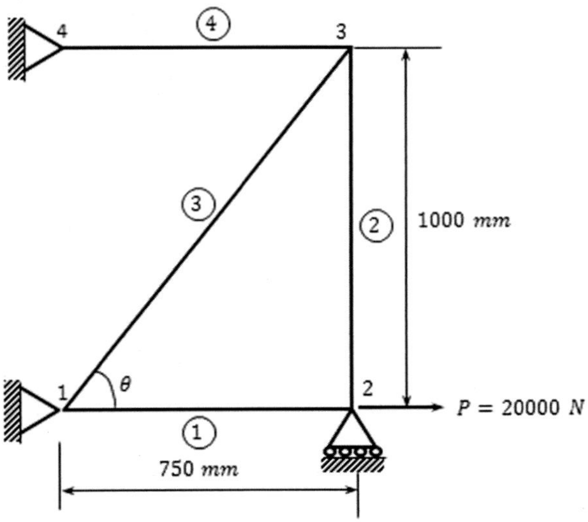

FIGURE 5.38 Pin jointed plane frame.

$$A = 200\ \text{mm}^2, E = 200 * 10^3 \quad \text{N/mm}^2$$

$$tan^{-1}\left(\frac{1000}{750}\right) = 53^{o}$$

The finite elements of the pin jointed frame are shown in Fig. 5.39.
Element (1),

$$[k]^1 = \frac{200 * 200 * 10^3}{750}\overset{1234}{\begin{bmatrix} 1 & 0 & -1 & 0 \\ 0 & 0 & 0 & 0 \\ -1 & 0 & 1 & 0 \\ 0 & 0 & 0 & 0 \end{bmatrix}}$$

Or,

$$[k]^1 = 10^3\overset{1234}{\begin{bmatrix} 53.3 & 0 & -53.3 & 0 \\ 0 & 0 & 0 & 0 \\ -53.3 & 0 & 53.3 & 0 \\ 0 & 0 & 0 & 0 \end{bmatrix}} \tag{E12.2}$$

Element (2),

$$[k]^2 = \frac{200 * 200 * 10^3}{1000}\overset{3456}{\begin{bmatrix} 0 & 0 & 0 & 0 \\ 0 & 1 & 0 & -1 \\ 0 & 0 & 0 & 0 \\ 0 & -1 & 0 & 1 \end{bmatrix}}$$

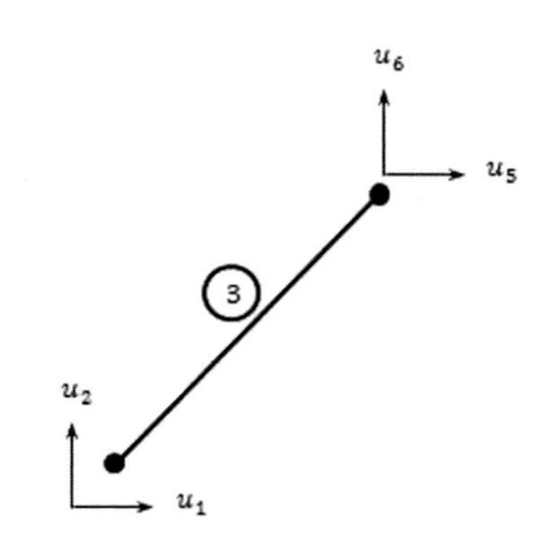

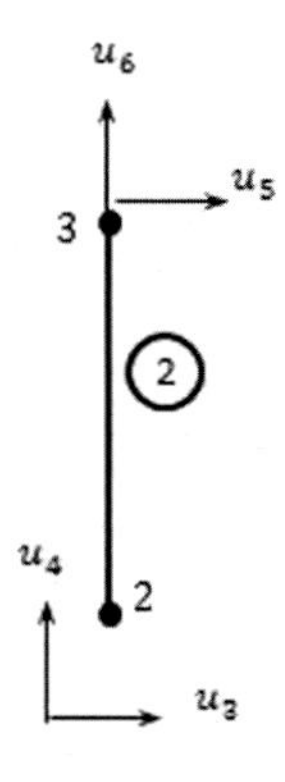

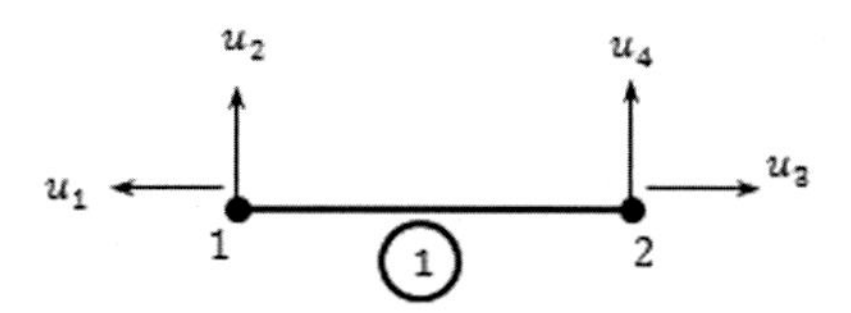

FIGURE 5.39 Elements 1, 2, 3 and 4

Element no.	θ^o	l	m	L (mm)
1	0	1	0	750
2	90	0	1	1000
3	53	0.6	0.8	1250
4	180	−1	0	750

Or,

$$[k]^2 = 10^3 \overset{3456}{\begin{bmatrix} 0 & 0 & 0 & 0 \\ 0 & 40 & 0 & -40 \\ 0 & 0 & 0 & 0 \\ 0 & -40 & 0 & 40 \end{bmatrix}} \tag{E12.3}$$

Element (3),

$$[k]^3 = \frac{200*200*10^3}{1250}\overset{1256}{\begin{bmatrix} 0.36 & 0.48 & -0.36 & -0.48 \\ 0.48 & 0.64 & -0.48 & -0.64 \\ -0.36 & -0.48 & 0.36 & 0.48 \\ -0.48 & -0.64 & 0.48 & 0.64 \end{bmatrix}}$$

Or,

$$[k]^3 = 10^3\overset{1256}{\begin{bmatrix} 11.52 & 15.36 & -11.52 & -15.36 \\ 15.36 & 20.48 & -15.36 & -20.48 \\ -11.52 & -15.36 & 11.52 & 15.36 \\ -15.36 & -20.48 & 15.36 & 20.48 \end{bmatrix}} \tag{E12.4}$$

Element (4),

$$[k]^4 = \frac{200*200*10^3}{750}\overset{5678}{\begin{bmatrix} 1 & 0 & -1 & 0 \\ 0 & 0 & 0 & 0 \\ -1 & 0 & 1 & 0 \\ 0 & 0 & 0 & 0 \end{bmatrix}}$$

Or,

$$[k]^4 = 10^3\overset{5678}{\begin{bmatrix} 53.3 & 0 & -53.3 & 0 \\ 0 & 0 & 0 & 0 \\ -53.3 & 0 & 53.3 & 0 \\ 0 & 0 & 0 & 0 \end{bmatrix}} \tag{E12.5}$$

The system stiffness matrix is obtained by using the inspection method and will be as follows,

$$[K]_{sys} = 10^3\begin{bmatrix} \begin{pmatrix} 53.3+ \\ 11.52 \end{pmatrix} & 15.36 & -53.3 & 0 & -11.52 & -15.36 & 0 & 0 \\ 15.36 & 20.48 & 0 & 0 & -15.36 & -20.48 & 0 & 0 \\ -53.3 & 0 & 53.3 & 0 & 0 & 0 & 0 & 0 \\ 0 & 0 & 0 & 40 & 0 & -40 & 0 & 0 \\ -11.52 & -15.36 & 0 & 0 & \begin{pmatrix} 11.52+ \\ 53.3 \end{pmatrix} & 15.36 & -53.3 & 0 \\ -15.36 & -20.48 & 0 & -40 & 15.36 & \begin{pmatrix} 40+ \\ 20.48 \end{pmatrix} & 0 & 0 \\ 0 & 0 & 0 & 0 & -53.3 & 0 & 53.3 & 0 \\ 0 & 0 & 0 & 0 & 0 & 0 & 0 & 0 \end{bmatrix} \tag{E12.6}$$

Or,

$$[K]_{sys} = 10^3 \begin{bmatrix} 64.82 & 15.36 & -53.3 & 0 & -11.52 & -15.36 & 0 & 0 \\ 15.36 & 20.48 & 0 & 0 & -15.36 & -20.48 & 0 & 0 \\ -53.3 & 0 & 53.3 & 0 & 0 & 0 & 0 & 0 \\ 0 & 0 & 0 & 40 & 0 & -40 & 0 & 0 \\ -11.52 & -15.36 & 0 & 0 & 64.82 & 15.36 & -53.3 & 0 \\ -15.36 & -20.48 & 0 & -40 & 15.36 & 60.48 & 0 & 0 \\ 0 & 0 & 0 & 0 & -53.3 & 0 & 53.3 & 0 \\ 0 & 0 & 0 & 0 & 0 & 0 & 0 & 0 \end{bmatrix} \quad \text{(E12.7)}$$

But,

$$u_1 = u_2 = u_4 = u_7 = u_8 = 0 \text{ Constrained degrees of freedom} \quad \text{(E12.8)}$$

$$F_1 = F_2 = F_4 = F_5 = F_6 = F_7 = F_8 = 0, \text{F3} = 20,000 \text{ N} \quad \text{(E12.9)}$$

$$[K]\{u\} = \{F\}$$

$$\begin{bmatrix} 53,300 & 0 & 0 \\ 0 & 64,820 & 15,360 \\ 0 & 15,360 & 60,480 \end{bmatrix} \begin{Bmatrix} u_3 \\ u_5 \\ u_6 \end{Bmatrix} = \begin{Bmatrix} 20,000 \\ 0 \\ 0 \end{Bmatrix}$$

Or,

$$u_3 = \frac{20,000}{53,300} = 0.3745 \text{ mm}$$

$$u_5 = u_6 = 0 \quad \text{(E12.10)}$$

We have from Eq. (E12.10),

$$\sigma = \frac{E}{L}\begin{bmatrix} -l & -m & l & m \end{bmatrix}\{u\} \quad \text{(E12.11)}$$

Or,

$$\{u\}^2 = \begin{Bmatrix} u_3 = 0.3745 \\ u_4 = 0 \\ u_5 = 0 \\ u_6 = 0 \end{Bmatrix}$$

And,

$$\sigma_2 = \frac{200 * 10^3}{1000}\begin{bmatrix} 0 & -1 & 0 & 1 \end{bmatrix}\begin{Bmatrix} 0.3745 \\ 0 \\ 0 \\ 0 \end{Bmatrix} \quad \text{(E12.12)}$$

Or,

$$\sigma_2 = 0$$

And,

$$\sigma_3 = \frac{200 * 10^3}{1250}\begin{bmatrix}-0.6 & -0.8 & 0.6 & 0.8\end{bmatrix}\begin{Bmatrix}0\\0\\0\\0\end{Bmatrix} \quad \text{(E12.13)}$$

Or,

$$\sigma_3 = 0.0\ \mathrm{N/mm^2}$$

To find the vertical reaction at node (2) return to general matrix and choose the 4th row.

$$R_4 = \begin{bmatrix}0 & 0 & 0 & 40,000 & 0 & -40,000 & 0 & 0\end{bmatrix}\begin{Bmatrix}u_1 = 0\\u_2 = 0\\u_3 = 0.3745\\u_4 = 0\\u_5 = 0\\u_6 = 0\\u_7 = 0\\u_8 = 0\end{Bmatrix} \quad \text{(E12.14)}$$

Problems

P.5.1: For the beam shown in Fig. P.5.1

1. Use the finite element method to prove that the maximum displacement V_{max} is given by:

$$V_{max} = \frac{FL^3}{192EI}$$

2. What are the reaction forces and moments at the fixed end a and c?
3. If the supports are not rigid, and have:
   a. Translation stiffness $= 2 \times 10^{10}\ \mathrm{N/m^2}$
   b. Rotational stiffness $= 1.5 \times 10^{10}\ \mathrm{N.m/rad}$

Find the value of the maximum displacement and the displacement and rotation at the supports.

$$E = 70 * 10^3\ \mathrm{N/mm^2}, L = 2.5\ \mathrm{m}, I = 160\ \mathrm{cm^4}, \text{and} F = 4 \times 10^4\ \mathrm{N}$$

Note: use $[k]^e$, as,

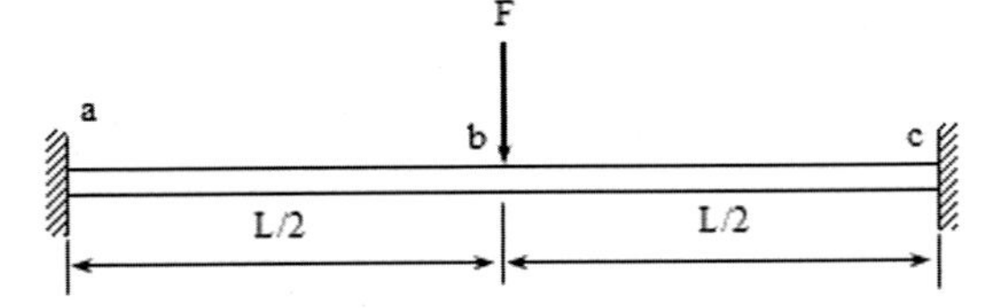

FIGURE P.5.1

$$[k]^e = \frac{EI}{l^3}\begin{bmatrix} 12 & 6l & -12 & 6l \\ 6l & 4l^2 & -6l & 2l^2 \\ -12 & -6l & 12 & -6l \\ 6l & 2l^2 & -6l & 4l^2 \end{bmatrix}$$

P.5.2: A plain shaft has one end supported in a 'long' bearing whilst the other end is carried in a 'short' ball bearing. Due to some misalignment problems it is necessary to examine the influence of a concentrated couple and transverse forces applied at mid span as shown in Fig. P.5.2A. The unit is modeled as a propped cantilever beam as shown. By using a two element discretization, with three degrees of freedom as in Fig. P.5.2B, establish, by hand, the matrix stiffness equilibrium equation for the system. If both bearing were 'long', how would the equation be modified?

P.5.3: A bar element supports a uniformly distributed load p/unit length. Using the displacement model given in Eq. (5.1), deduce the element load vector using work basis principles.

P.5.4: The system shown in Fig. P.5.4, is used as a vehicle for demonstrating how elements having a suitable mixture of characteristics can be accommodated systematically in a finite element formulation. Use three elements and observe that,

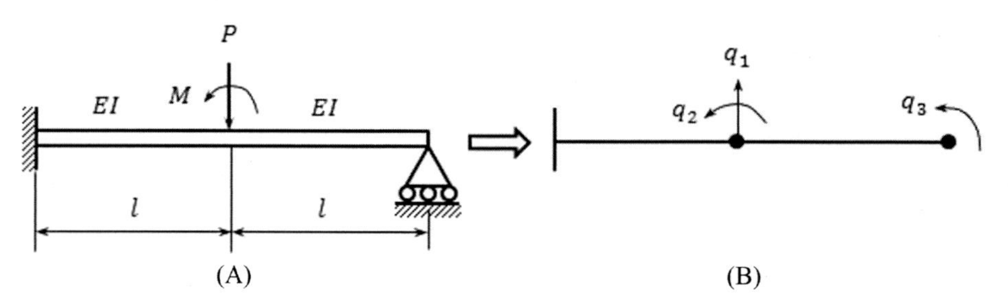

FIGURE P.5.2.A AND FIGURE P.5.2.B

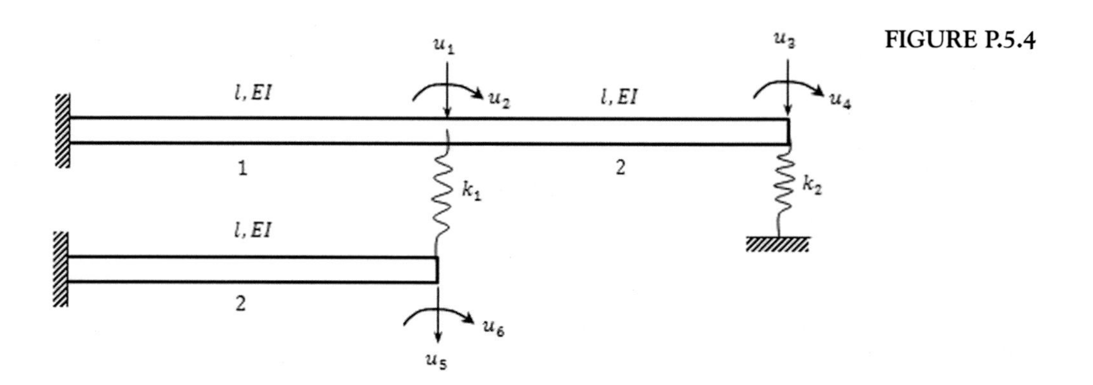

FIGURE P.5.4

FIGURE P.5.5

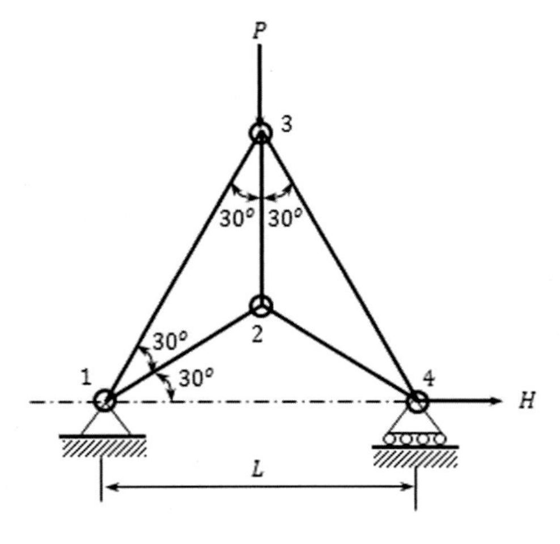

$$U = \sum U_{\text{beam}} \sum U_{\text{springs}}$$

Deduce the system stiffness matrix.

P.5.5: Find the ratio H/P for which the displacement of node 4 of the plane pin-jointed frame shown in Fig. P.5.5. is zero and for that case give the displacements of nodes 2 and 3.

(All members equal axial rigidity EA).

P.5.6: Set up the structure stiffness matrix for the structure shown in Fig. P.5.6 in terms of the individual member stiffness. For all bars and beams $EA = 0.8 * 10^{10}$ N, and for all beams $EI = 12 * 10^{10}$ N.mm^2. Length of each of the vertical members = 0.25 m and the length of each of the horizontal members = 1.0 m. Use the finite element modeling with a fewer number of degrees of freedom.

P.5.7: In the plane truss shown in Fig. P.5.7, bar 1 a temperature rise of 40°C in addition, to the concentrated load $Q_1 = 50$ kN, formulate the reduced system stiffness equations.

A small project may be achieved by employing the finite element steps with the aid of detailed flow charts should be constructed for this purpose using 6 elements, 4 nodes and 8 d.o.f as follows:

1. Write down the element stiffness matrix for each element using Eq. (5.31).
2. Formulate the system stiffness matrix using the inspection method.
3. Formulate the load vector due to thermal and mechanical loads (Section 5.4).
4. Apply the boundary conditions at the supports 1 and 4 (hinged supports).
5. Solve the equilibrium to find the displacements at the joints.
6. Find the strains and stresses at the frame members.
7. Calculate the internal forces at the members (axial bars).

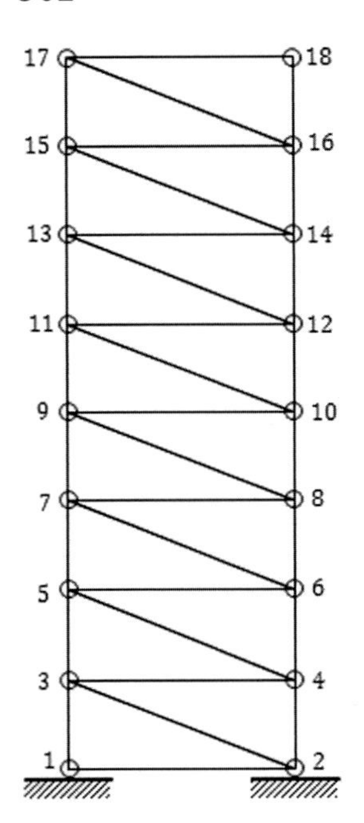

FIGURE P.5.6

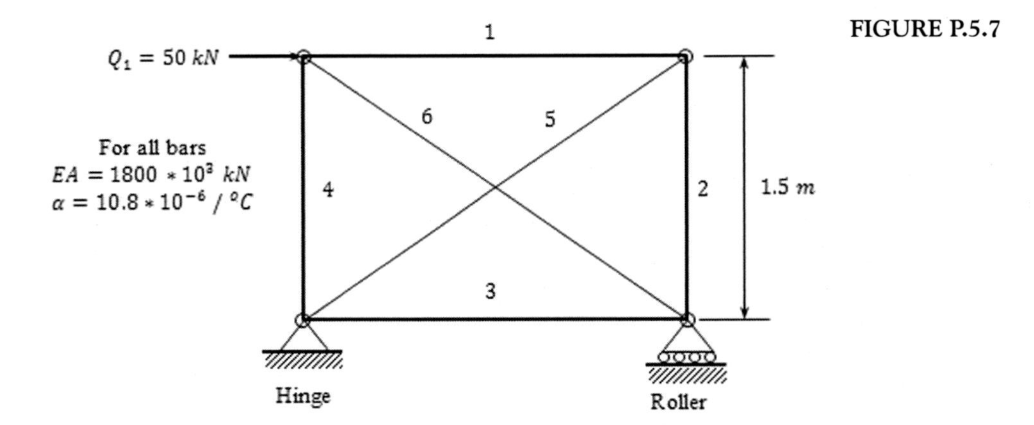

FIGURE P.5.7

Apply the energy principles explained in Part I of this book to find the internal forces in the members. Compare the results obtained in using both methods.

P.5.8: For the structure shown in Fig. P.5.8, nodes 1 and 4 are fixed to the foundation. Bars 1–3 and 2–4 are hinged to the nodes. Obtain the reduced stiffness matrix of the structure.

P.5.9: The grid framework shown in the Fig. P.5.9 is rigidly fixed to a stiff wall at nodes 1 and 2. Obtain the reduced stiffness matrix of this structure.

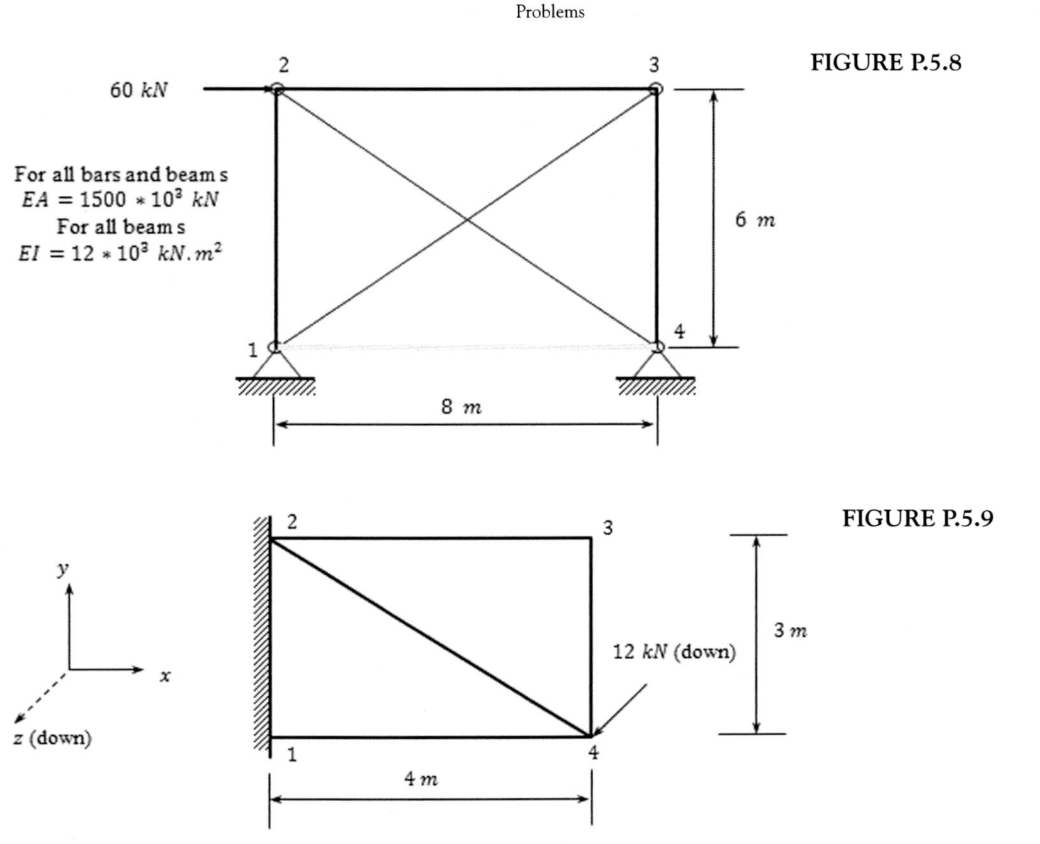

FIGURE P.5.8

FIGURE P.5.9

P.5.10: The tapered rod shown in Fig. P.5.10A is subjected to body force f/unit volume, traction T/unit length and the concentrated load P. Determine the global force vector assuming that the bar is divided into three elements. $P = 20$ kN. Use 3 elements and 4 nodes as shown in the finite element model shown in Fig. P.5.10B

P.5.11: (a) Deduce the matrix equations which are satisfied by all the displacement components at the nodes of the continuous beam of Fig. P.5.11. $W = 12$ kN, $M = 5$ kN.m, $L = 1$ m and $EI = 3.2 * 10^5$ N.m^2, and $L = 1$ m.

Accepting that, for a certain choice of $[N(\xi)]$, $[K]^e$ for a uniform element is given by,

$$[k]^l = \frac{EI}{l^3}\begin{bmatrix} 12 & 6l & -12 & 6l \\ & 4l^2 & -6l & 2l^2 \\ & & 12 & -6l \\ \text{Sym.} & & & 4l^2 \end{bmatrix}$$

(b) Write a computer program flow chart to generate the system stiffness matrix and the load vector for the problem in (a). the generality of the program is needed in order to be used for different types of finite element meshes with different geometrical properties.

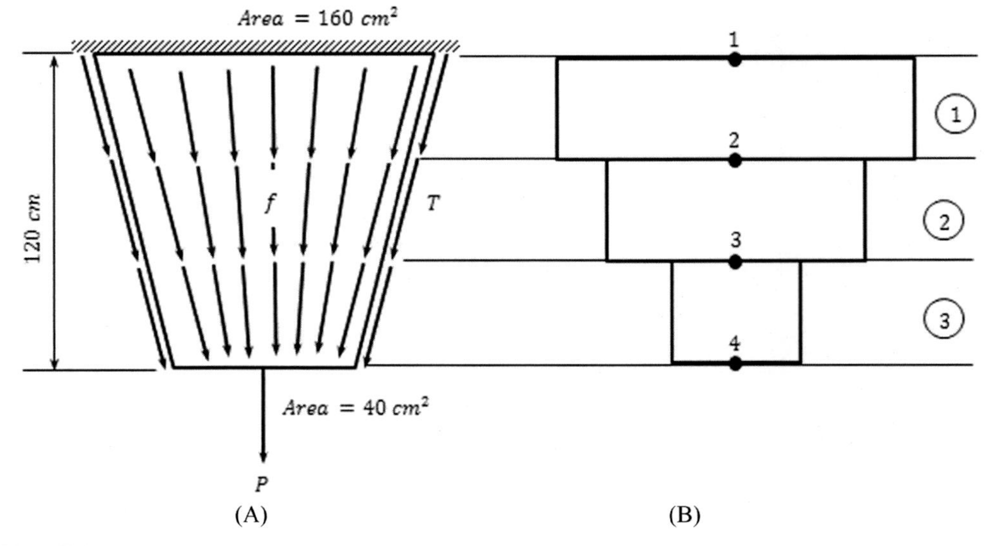

FIGURE P.5.10

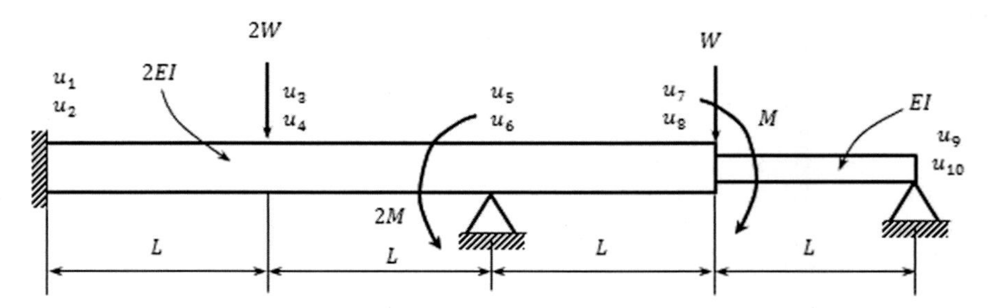

FIGURE P.5.11

P.5.12: For the tapered plate shown in Fig. P.5.12, calculate the extension due to F, using the 'Finite element method' $F = 5000$ N, $E = 210{,}000$ N/mm^2, $\nu = 0.3$.

P.5.13: The 2-node bar element, shown in Fig. P.5.13, has been used in one-dimensional stress analysis,

1. Derive element equations, if the value of the elastic modulus E vary linearly as,

$$E(x) = E_1 N_1 + E_2 N_2$$

Where, E_1 and E_2 are the values of the modulus of elasticity at node 1 and 2, respectively N_1 and N_2 are the shape functions

2. Consider the column in Fig. P.5.14B, with linearly varying of 'E' and 'A'. length 'L' of each element is equal to 12cm.

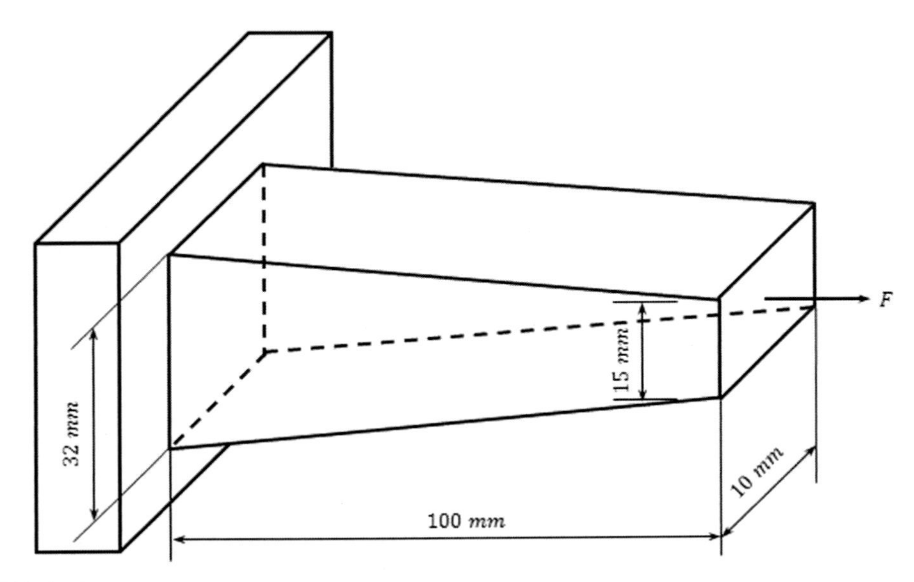

FIGURE P.5.12

a. Assembly the equations for the whole structure if the element stiffness matrix is,

$$[k]^e = \lambda \begin{bmatrix} 1 & -1 \\ -1 & 1 \end{bmatrix}$$

Where,

$$\lambda = \frac{1}{6L} [A_1 \quad E_1 \quad A_2 \quad E_3] \begin{bmatrix} 0 & 1 & 0 & \frac{1}{2} \\ 1 & 0 & \frac{1}{2} & 0 \\ 0 & \frac{1}{2} & 0 & 1 \\ \frac{1}{2} & 0 & 1 & 0 \end{bmatrix} \begin{bmatrix} A_1 \\ E_1 \\ A_2 \\ E_2 \end{bmatrix} = \text{Scalar}$$

b. Explain how the boundary conditions can be introduced, and find the unknown displacements.

Evaluate the strain in each element.
Hint,

$$\Omega = U - W = \frac{1}{L}\int_0^L AE\left(\frac{du}{dy}\right)^2 dy - \int_0^l P(y)\left(\frac{du}{dy}\right)dy = \text{min, with usual notations. 2}$$

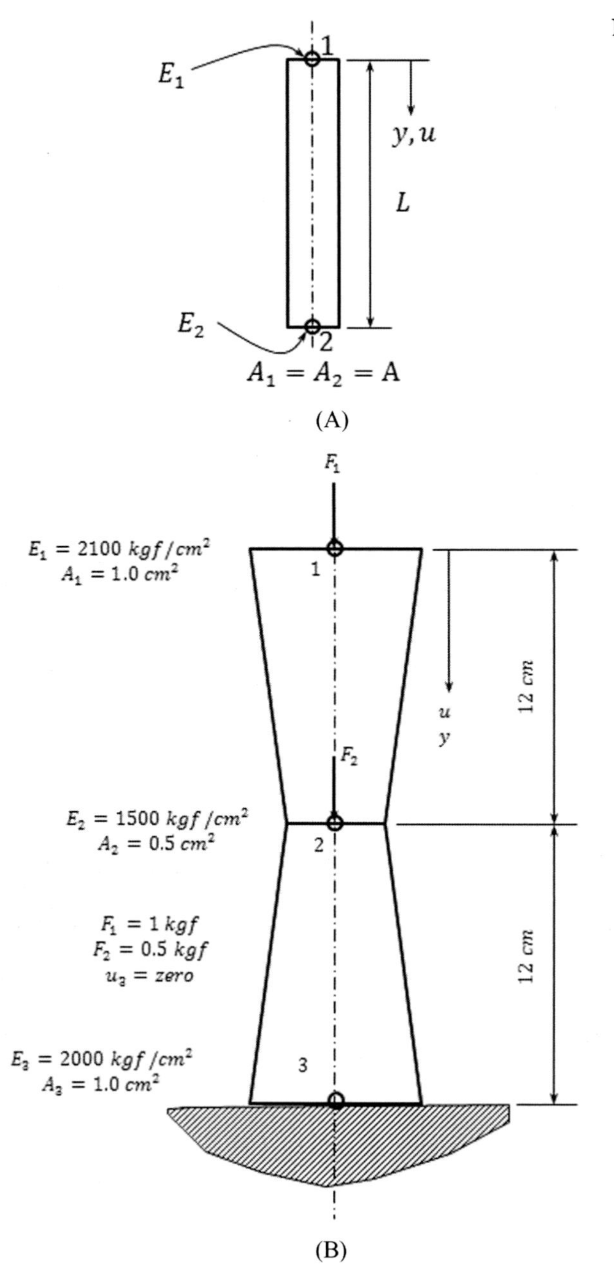
1
E_1
y, u
L
E_2
2
$A_1 = A_2 = A$
(A)
F_1
$E_1 = 2100\ kgf/cm^2$
$A_1 = 1.0\ cm^2$
1
u
y
12 cm
F_2
$E_2 = 1500\ kgf/cm^2$
$A_2 = 0.5\ cm^2$
2
$F_1 = 1\ kgf$
$F_2 = 0.5\ kgf$
$u_3 = zero$
12 cm
$E_3 = 2000\ kgf/cm^2$
$A_3 = 1.0\ cm^2$
3
(B)

FIGURE P.5.13

P.5.14: Consider a system with 4 DOFs. The system equation is written as follows,

$$\begin{bmatrix} K_{11} & K_{12} & K_{13} & K_{14} \\ K_{21} & K_{22} & K_{23} & K_{24} \\ K_{31} & K_{32} & K_{33} & K_{34} \\ K_{41} & K_{42} & K_{43} & K_{44} \end{bmatrix} \begin{Bmatrix} u_1 \\ u_2 \\ u_3 \\ u_4 \end{Bmatrix} = \begin{Bmatrix} F_1 \\ F_2 \\ F_3 \\ F \end{Bmatrix} or\ [K]\{u\} = \{F\}$$

Where, K is 4×4 stiffness matrix, now assume that the displacement boundary condition is,

$$\begin{cases} u_1 = 0 \\ u_4 = \Delta \end{cases}$$

Apply this displacement boundary condition to the system equations to reduce the size of the stiffness matrix and write down your result. In this situation, identify which components of the traction boundary vector F must be given and which components do not.

P.5.15: Modify the global stiffness matrix for the problem shown in Fig. P.5.15, by considering the constraint equation corresponding to the skew boundary condition at support b which is shown in Fig. P.5.15. Solve for the remaining degrees of freedom.

$$K = \begin{bmatrix} 4.75 & -4.763 & -3.0 \\ -4.763 & 21.25 & -0.804 \\ +3.0 & -0.804 & 8.0 \end{bmatrix}$$

$$x = \begin{bmatrix} u_B \\ v_B \\ \varphi_B \end{bmatrix}, F = \begin{bmatrix} 0 \\ -10 \\ 0 \end{bmatrix}$$

P.5.16: Formulate the equilibrium equations $[k]\{x\} = \{F\}$, which are satisfied by the displacement components of the nodes of the continuous beam shown in Fig. P.5.17, refer to example 5.9. $P_1 = 12$ kN, $w = 4$ kN/m, $k_1 = \frac{5EI}{l^3} = 3.5$ kN/m, $k_2 = \frac{3EI}{l^3} = 3.5$ kN/m, $k_3 = \frac{5EI}{l^3} = 4$ kN.m/rad. Use 4 beam elements which represents the minimum number of degrees of freedom to be used in this problem (Fig. P.5.16)

$$[k]^e = \frac{EI}{l^3} \begin{bmatrix} 12 & 6l & -12 & 6l \\ & 4l^2 & -6l & 2l^2 \\ & & 12 & 6l \\ \text{Sym.} & & & 4l \end{bmatrix}$$

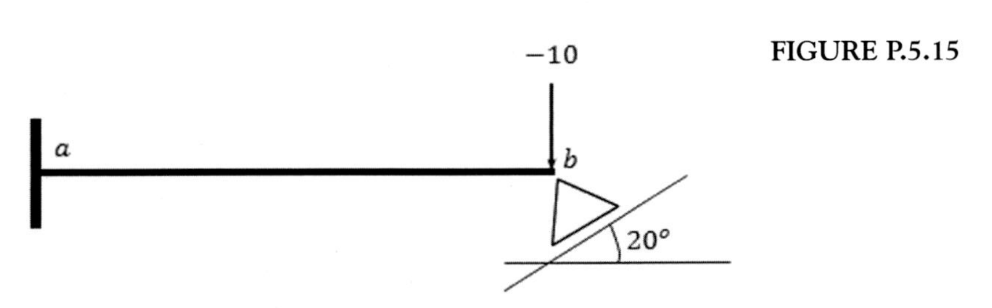

FIGURE P.5.15

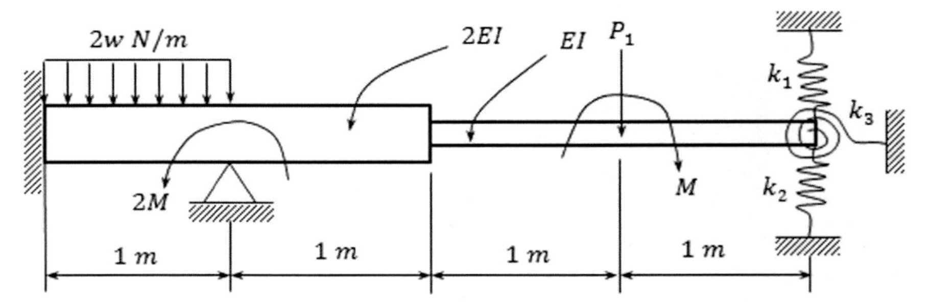

FIGURE P.5.16

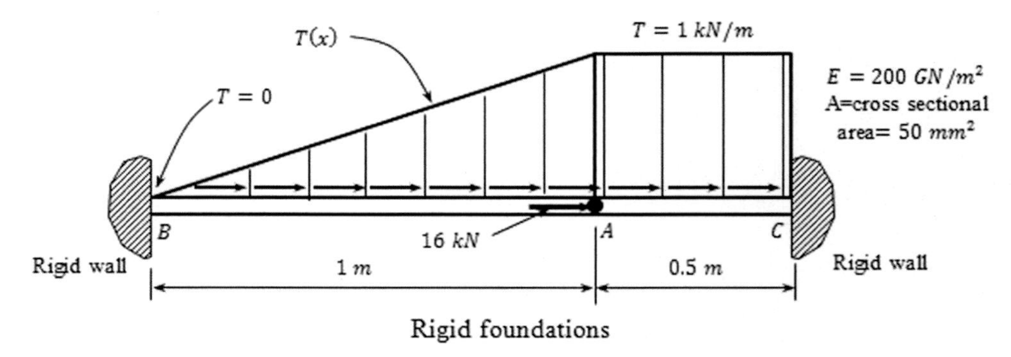

FIGURE P.5.17A

P.5.17: Steel rod is attached to rigid walls at each end. the rod is subjected to distributed loading $T(x)$ and concentrated force of 16 kN, as shown in Fig. P.5.17A and B. Use a suitable number of 2-noded elements to calculate,

1. The displacement at point (A).
2. The stress distribution in zone 'AB'.
3. Prove that the FEM results satisfies the equilibrium conditions.
4. If the rod is attached to elastic foundation at ends B and C with elastic stiffness $k_B = k_C = 3.5$ N/m, determine the reactions and displacements at the ends.

Hint, Finite element equation for 2-node bar element is,

$$\frac{AE}{l}\begin{bmatrix} 1 & -1 \\ -1 & 1 \end{bmatrix}\begin{bmatrix} u_1 \\ u_2 \end{bmatrix} = \begin{bmatrix} F_1 \\ F_2 \end{bmatrix} + \frac{l}{6}\begin{bmatrix} 2 & 1 \\ 1 & 2 \end{bmatrix}\begin{bmatrix} T_1 \\ T_2 \end{bmatrix}, \text{with usual notations.}$$

P.5.18: Develop a microcomputer design package for a component which is essentially a beam, simply supported at its ends, and with the possibility of an additional linear spring

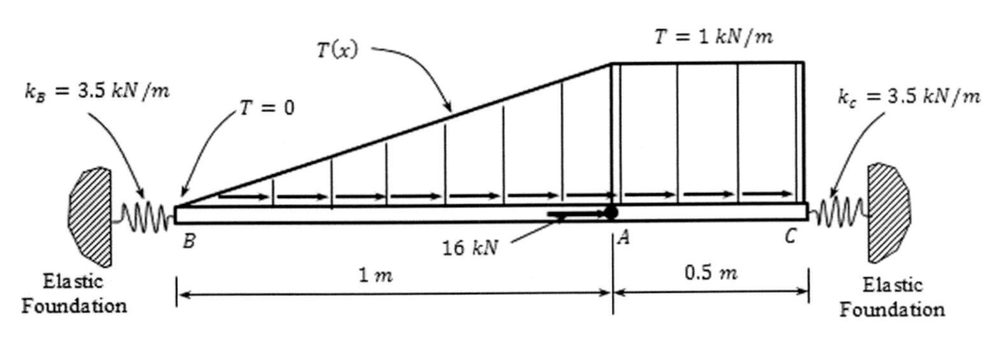

FIGURE P.5.17B

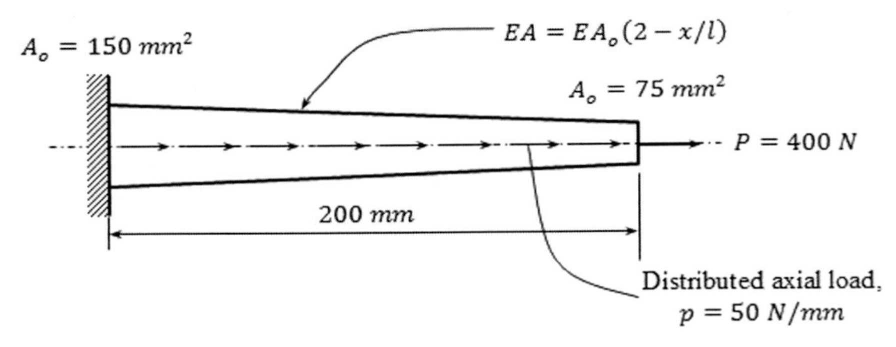

FIGURE P.5.19

at mid span. A variety of loads, including end thrust is possible. Formulate the model on the basis of sine series and the potential energy principle.

P.5.19: (a) An element in a special purpose machine (Fig. P.5.19) is essentially a tapered bar. It is held at one end carries a large number of axial forces applied at short intervals along its entire length. For finite element analysis, it is modeled into the system of figure below.

Using only 3 uniform elements, generate $[k]\{q\} = \{Q\}$ by hand and solve using the computer program provided. Compare with the 'exact' solution. Modify the equations to suit the case when the right hand end of the bar is clamped instead of supporting the 400 N end load.

P.5.20: For the structure shown in Fig. P.5.20 find the maximum displacement caused by the heavy weight $W = 5$ metric tons. b and c are hinges using 2 finite elements. Compare the results with the exact solution employing strength of materials principles.

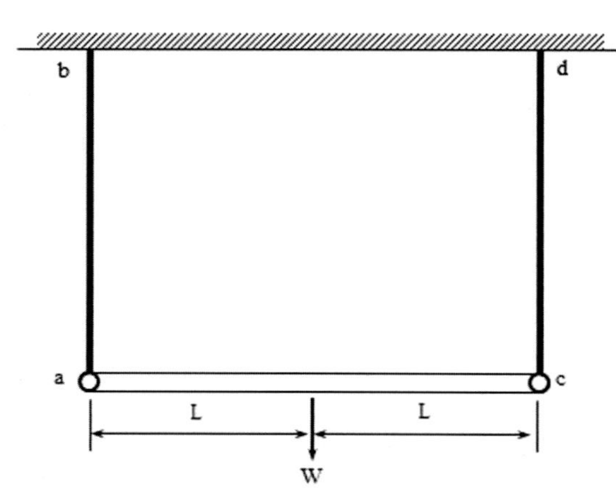

FIGURE P.5.20

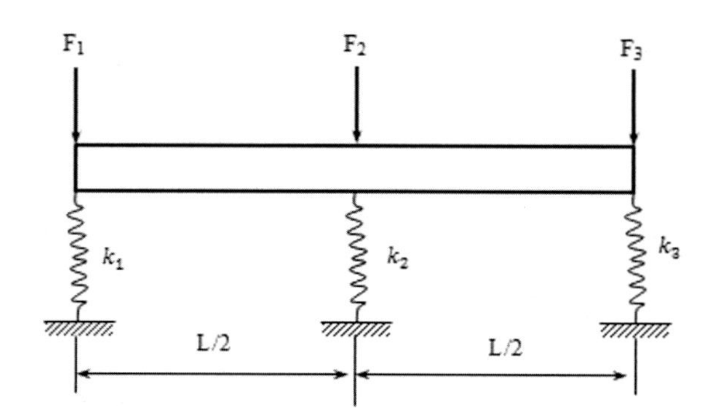

FIGURE P.5.21

Beam	Length (m)	A (cm^2)	I (cm^4)	E (N/m^2)
ab	1	25	52.08	7×10^{10}
bc	2	36	108.00	2×10^{11}
cd	1	25	52.08	7×10^{10}

P.5.21: A beam on an elastic foundation (which resists vertical displacement only) has been modeled as shown in Fig. P.5.21 where:

Total beam length = 4.0 m
Second moment of area for the beam
cross section = 1.6×10^{-6} m^4
Young's modulus for the beam material = 2×10^{11} N/m^2

$$F_1 = F_3 = 1 \times 10^4 \text{ N}$$

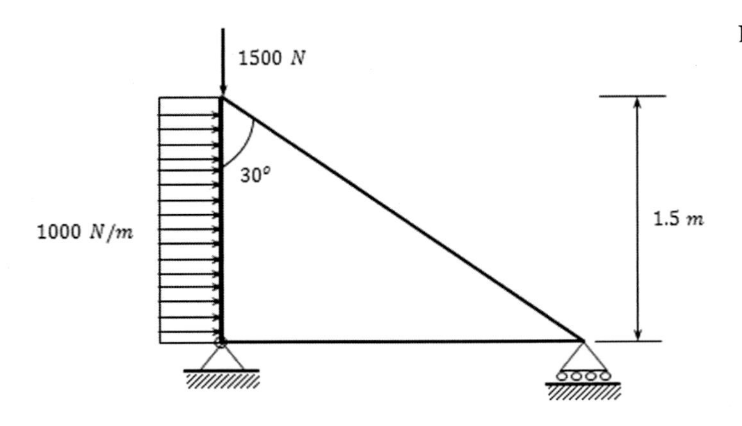

FIGURE P.5.22

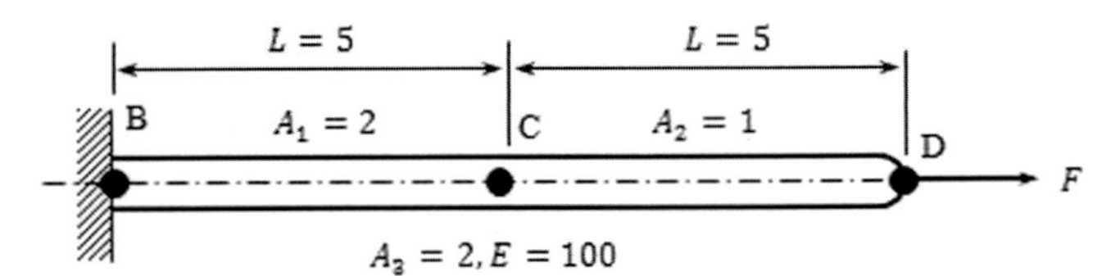

FIGURE P.5.23

$$F_2 = 4 \times 10^4 \text{ N}$$

$$k_1 = k_3 = 1.7 \times 10^5 \text{ N/m}, k_2 = 3.4 \times 10^5 \text{ N/m}$$

Use the finite element method to:

1. find the maximum displacement of the beam
2. the reaction at the foundation.

P.5.22: For the frame shown in Fig. P.5.22.

1. Construct the system load vector $\{P\}$ corresponds to the unconstrained d.o.f.
2. construct the constrained system stiffness matrix $[k]$.
3. Calculate the nodal displacement.

P.5.23: (a) For the following uniaxial multiple bar structure shown in Fig. P.5.23, if a force $F = 8$ is applied on the node D (at the right side of the structure), find the nodal displacement, element strains and stresses and nodal reactions.

(b) For the same structure, if the node D is displaced by -0.3 (which means node D moves toward the left for 0.3 unit), find the nodal displacement, element strain and stresses and nodal reactions.

(c) Denote the force $F = 8$ at node D in (a) as $F(a)$ and the corresponding displacement at node D as $u(a)$. Denote the displacement $u = -0.3$ at node D as $u(b)$ and the corresponding reaction force at node D as $F(b)$. Verify whether $\frac{F^{(a)}}{u^{(a)}} = \frac{F^{(b)}}{u^{(b)}}$ or not.

P.5.24: If in example 5.12 a downward force 20, 000 N is acting as node 2, the support 4 is settled by 3 mm downward, write down the equilibrium equations stating clearly the procedure of solving such types of problems. The truss is shown in Fig. P.5.24.

Note: Use the penalty function presented in Appendix B to solve the resulted equations.

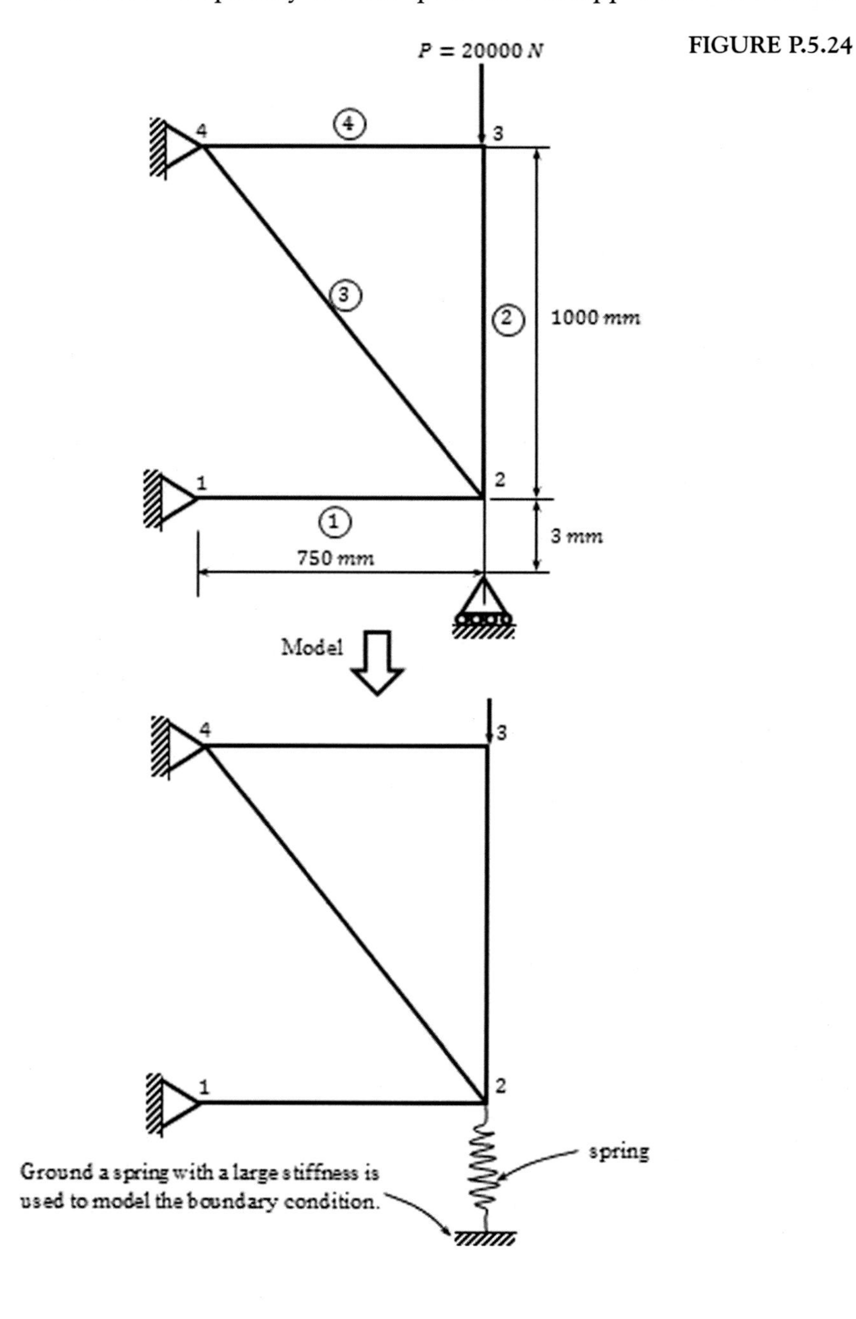

FIGURE P.5.24

Bibliography

[1] T.R. Chandrupatla, A.D. Belegundu, Introduction to Finite Elements in Engineering, Prentice-Hall, 2009.
[2] H.L. Langhaar, Energy Methods in applied Mechanics, Wiley, 1962.
[3] R.J. Melosh, Basis for berivation of matrices for the direct stiffness method, Journal of the American Institute of Aeronautics and Astronautics I (7) (1963) 1631–1637.
[4] K.J. Bathe, Finite Element Procedures, first ed., Prentice-Hall, Inc, 2006.
[5] K.J. Bathe, Finite Element Procedures, second ed., Prentice-Hall, Inc, 2014.
[6] J.S. Prezemieniecki, Theory of Matrix Structural Analysis, McGraw–Hill, 1968.
[7] L.J. Segerlind, Applied Finite Element Analysis, Wiley, New York, 1976.
[8] T.H. Richards, Energy Methods in Stress Analysis, Ellis Horwood Series in Engineering Science, 1977.
[9] E. Hinton, D.R. Owen, An Introduction to Finite Element Computations, Pineridge Press, 1979.
[10] Y.Y. Hsieh, Elementary Theory of Structures, second ed., Prentice-Hall, Englewood, 1982.
[11] J.M. Gere, W. Weaver, Analysis of Framed Structures, second ed., Van Nostrand, 1982.
[12] M.J. Jweeg, Application of Finite Element Analysis to Rotating Fan Impellers (Doctoral Thesis), Aston University, 1983.
[13] W. Weaver Jr., P.R. Johnston, Finite Element for Structural Analysis, Prentice-Hall, Inc, 1984.
[14] L.J. Segerling, Applied Finite Element Analysis, John Wilry and Sons, Inc, 1984.
[15] Internal Report, Finite Element Metohd, Granfield Institute of Technology, School of Mechanical Engineering, 1985.
[16] D.S. Burnett, Finite Element Analysis, Addison_Wesley Publishing Company, 1987.
[17] W. Weaver Jr., R.J. Paul, Structural Dynamics by Finite Elements, Prentice-Hall, Inc, New Jersey, 1987.
[18] M.J. Jweeg, S.Z. Said, Effect of rotational and geometric stiffness matrices on dynamic stresses and deformations of rotating blades, Journal of the Institution of Engineers, Mechanical Engineering Division 76 (1995) 29–38.
[19] R.D. Cook, Finite Element Modeling for Stress Analysis, John Wiley and Sons, Inc, 1995.
[20] R.T. Fenner, Finite Element Methods for Beginners, Imperial College Press, 1996.
[21] K.J. Bathe, Finite Element Procedures, Prentice-Hill, Englewood, 1996.
[22] E.J. Hearn, Mechanics of Materials, International Series on Materials Science and Technology I, 1997.
[23] E.J. Hearn, Mechanics of Materials, International Series on Materials Science and Technology II, 1997.
[24] S. Moaveni, Finite Element Analysis, Theory and Application With Ansys, Prentice-Hall, Inc, 1999.
[25] S.S. Rao, The Finite Element Method in Engineering, fourth ed., Elsevier Science and Technology, 2004.
[26] M.J. Jweeg, M. Al-Waily, K.K. Al-Kinani, Lecture Notes, Al-Nahrain University, Al-Mustansirya University, Unversity of Kufa.

CHAPTER

6

Two-dimensional problems: application of plane strain and stress

In this chapter, the two-dimensional problems will be modeled employing the finite element method and choosing the simplest type of elements to present the procedure of discretization and apply the finite element formulations in-plane strain and stress. The formulation covers the derivation of the element stiffness matrix for the constant stain triangulate element and the system stiffness matrix presentation for some selected examples. The isoparametric quadrilateral element of four nodes is also presented in detail. The displacement model is assumed, and the shape function in terms of the intrinsic coordinates is used to derive the element stiffness matrix. The determination of the equivalent nodal forces is explained, and some cases are presented, which include different loading schemes. At the end of the chapter, problems are included, containing different types of two-dimensional problems under different support conditions and loading types.

6.1 Two-dimensional modeling: triangular elements

The standard procedure steps will be employed as follows:

1. The definition of the finite element mesh used to model the region.

The region will be modeled into triangular elements with three nodes and six degrees of freedom for each element as shown in Fig. 6.1.

2. Selection of a displacement model: The simplest assumption of the displacement mode is to assume:

$$
\begin{aligned}
u^e(x,y) &= a_0 + a_1 x + a_2 y \\
v^e(x,y) &= a_3 + a_4 x + a_5 y
\end{aligned}
\tag{6.1}
$$

Energy Methods and Finite Element Techniques
DOI: https://doi.org/10.1016/B978-0-323-88666-6.00004-6

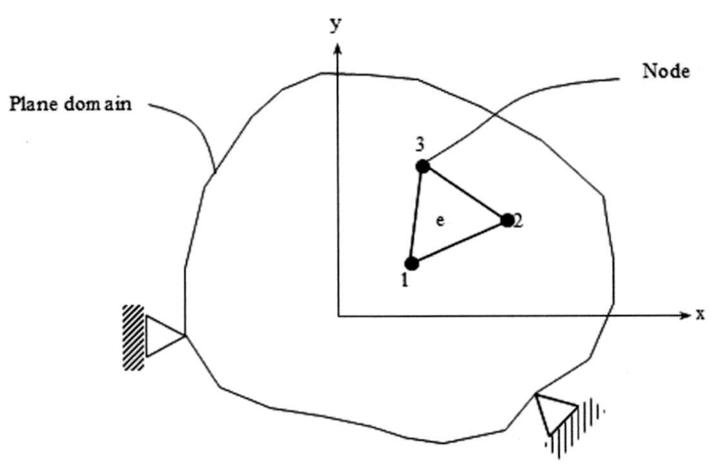

FIGURE 6.1 Two-dimensional region.

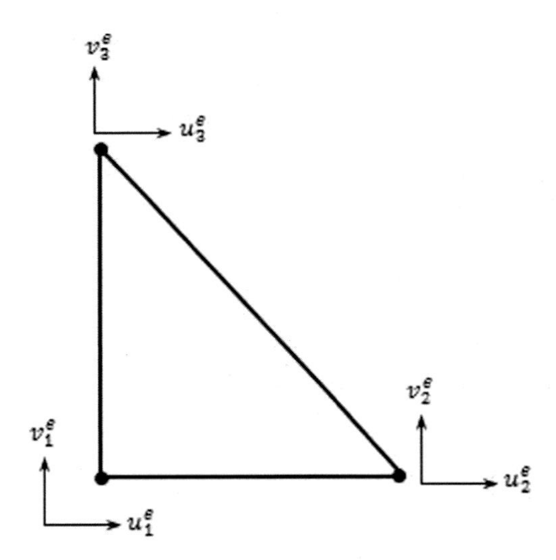

FIGURE 6.2 Typical finite element three nodes and six degrees of freedom.

a_0, a_1, a_2, a_3, a_4, and a_5 represent the generalized coordinates (G.C.) for the element. u and v are the displacements in x and y directions, respectively, at each node as shown in Fig. 6.2. We can write:

$$\{u\}^e = \begin{Bmatrix} u(x,y) \\ v(x,y) \end{Bmatrix} = [N(x,y)]\{u\}_0^e \tag{6.2}$$

Or,

$$\underset{2\times1}{\begin{Bmatrix} u \\ v \end{Bmatrix}^e} = \begin{bmatrix} \underset{(2\times2)}{[N_1]} & \underset{(2\times2)}{[N_2]} & \underset{(2\times2)}{[N_3]} \end{bmatrix} \underset{(6\times1)}{\begin{Bmatrix} \begin{Bmatrix} u_1 \\ v_1 \end{Bmatrix} \\ \begin{Bmatrix} u_2 \\ v_2 \end{Bmatrix} \\ \begin{Bmatrix} u_3 \\ v_3 \end{Bmatrix} \end{Bmatrix}}$$

$$\text{In fact } [N_1] = \underset{\uparrow - \text{scalar function}}{N_1} \begin{bmatrix} 1 & 0 \\ 0 & 1 \end{bmatrix}.$$

Where,

$$N_1 = \frac{1}{2\Delta}(a_1 + b_1 x + c_1 y) \tag{6.4a}$$

Δ is the area of the element.

$$a_1 = x_2 y_3 - x_3 y_2, b_1 = y_2 - y_3, c_1 = x_3 - x_2$$

$N_2(x,y)$ and $N_3(x,y)$ are found by cyclic permutations and will be derived in Section 6.1.1 (constant triangle element), hence,

$$N_2 = \frac{1}{2\Delta}(a_2 + b_2 x + c_2 y)$$

And,

$$N_3 = \frac{1}{2\Delta}(a_3 + b_3 x + c_3 y)$$

Where,

$$a_2 = x_3 y_1 - x_1 y_3, b_2 = y_3 - y_1, c_2 = x_1 - x_3$$

And,

$$a_3 = x_1 y_2 - x_2 y_1, b_3 = y_1 - y_2, c_3 = x_2 - x_1$$

3. Determining the stiffness equilibrium equations,

a. Expressing for the strain components, consider plane stress, relevant strain component is,

$$\varepsilon_x = \frac{\partial u}{\partial x}, \varepsilon_y = \frac{\partial v}{\partial y}, \gamma_{xy} = \frac{\partial u}{\partial y} + \frac{\partial v}{\partial x} \tag{6.4}$$

Hooke's law,

$$\begin{Bmatrix} \sigma_x \\ \sigma_y \\ \tau_{xy} \end{Bmatrix} = \frac{E}{1-\nu^2} \begin{bmatrix} 1 & \nu & 0 \\ \nu & 1 & 0 \\ 0 & 0 & \frac{1-\nu}{2} \end{bmatrix} \begin{Bmatrix} \varepsilon_x \\ \varepsilon_y \\ \gamma_{xy} \end{Bmatrix} \tag{6.5}$$

Or,

$$\{\sigma\}^e = [D]\{\varepsilon\}^e \tag{6.6}$$

If we wish to consider the plane strain, the only change will be in the form of matrix $[D]$, now,

$$\{\varepsilon\}^e = \begin{bmatrix} \frac{\partial}{\partial x} & 0 \\ 0 & \frac{\partial}{\partial y} \\ \frac{\partial}{\partial y} & \frac{\partial}{\partial x} \end{bmatrix} \begin{Bmatrix} u \\ v \end{Bmatrix} = [\partial]\{u\}^e \tag{6.7}$$

But,

$$\{u\}^e = [N]\{u\}_0^e \tag{6.8}$$

Then,

$$\underset{(3\times1)}{\{\varepsilon\}^e} = \underset{(3\times2)}{[\partial]}\ \underset{(2\times6)}{[N]}\ \underset{(6\times1)}{\{u\}_0^e}$$

$$\{\varepsilon\}^e = [B]\{u\}_0^e \tag{6.9}$$

Where, $[B] = \underset{(3\times2)(2\times6)}{[\partial]\ [N]}$

b. To find U^e (strain energy stored in the element), and using Eq. (6.9) gives,

$$U^e = \frac{1}{2}\iint_e \{\varepsilon\}^t[D]\{\varepsilon\}\, t\, dx\, dy = \frac{1}{2}\{u\}_0^{et}\left(\iint_e \underbrace{\underset{(6\times3)}{[B]^t}\ \underset{(3\times3)}{[D]}\ \underset{(3\times6)}{[B]}}_{(6\times6)}\ t\, dx\, dy\right)\{u\}_0^e$$

c. To find U, total strain energy

$$U = \sum_e U^e \tag{6.11}$$

Using the notation $\{\tilde{u}\} = [c]\{u\}$, for compatibility ... etc., we find,

$$U = \frac{1}{2}\{u\}^t[k]\{u\} \tag{6.12}$$

Where,

$$[K] = [c]^t\left[\tilde{k}\right][c]$$

d. To find Ω, total potential energy due to applied loads.

The applied forces could be comprising concentrated forces applied at the nodes, body forces having intensity per unit area, Fig. 6.3,

i. Finding Ω^e,

$$\Lambda(\Omega^e) = -\left(\overline{u}.\overline{T}n\right)(t.ds),\ or,\ = -\left(\{u\}_0^{et}[N]^t\left\{\overset{n}{T}\right\}\right)t.ds$$

Where, $\overset{n}{T}$ is the surface traction per unit area.

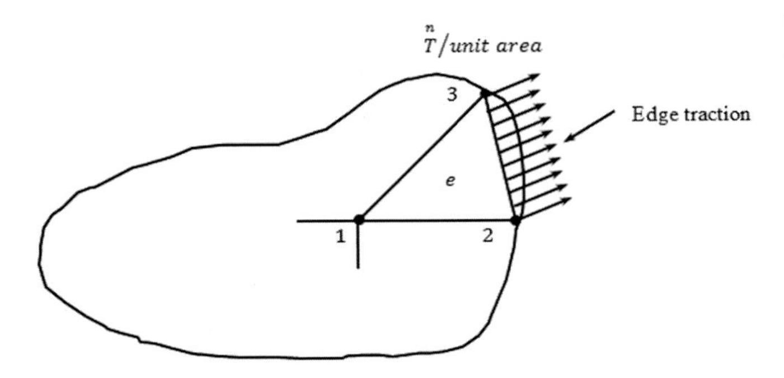

FIGURE 6.3 Two-dimensional region with surface traction.

Or,

$$= -\int_{sT} \{u\}^{et}_{sT}[N]^t\left\{\overset{n}{T}\right\}t\,ds = -\underset{(1\times 6)}{\{u\}^{et}_0}\left(\int_{sT}[N]^t_{(6\times 2)}\underset{(6\times 2)}{[\overset{n}{T}]}\;t\cdot ds\right) = -\{u\}^{et}_0\{P\}^e_0 \tag{6.13}$$

ii. Finding the potential energy due to applied load Ω,

$$\Omega = \sum_e \Omega^e$$

Or,

$$\Omega = -\{u\}^t\left([c]\{\tilde{P}\}\right)$$

$$\text{That is}, \Omega = -\{u\}^t\{P\} \tag{6.14}$$

iii. To find the total potential energy V, using Eqs. (6.12) and (6.14) we have,

$$V = \frac{1}{2}\{u\}^t K\{u\} - \{u\}\{P\}$$

For equilibrium, $\delta V = 0$, so that,

$$[K]\{u\} = \{P\} \tag{6.15}$$

As previously derived in Chapter 5, Introduction to Finite Element Method (F.E.M.): Bar and Beam Applications.

6.1.1 Constant strain triangle element

Consider the region shown in Fig. 6.4 with a triangle mesh and a typical element, Assume that the displacement model, Eq. (6.1) is repeated here as follows,

$$u^e(x,y) = a_0 + a_1x + a_2y$$

$$v^e(x,y) = a_3 + a_4x + a_5y \tag{6.16}$$

Or,

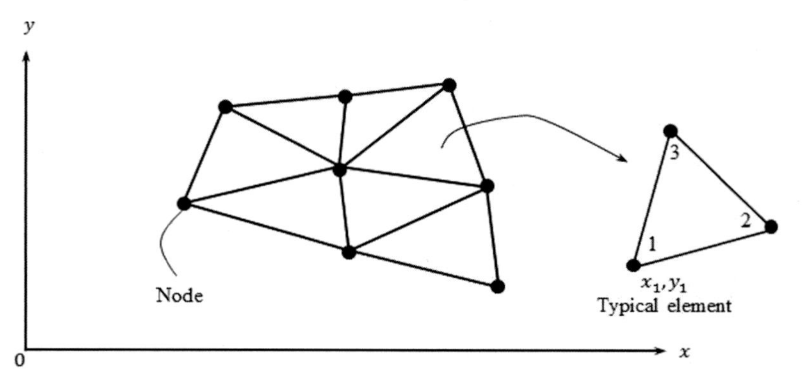

FIGURE 6.4 Constant strain element.

$$\begin{Bmatrix} u \\ v \end{Bmatrix}^e = \begin{bmatrix} 1 & x & y & 0 & 0 & 0 \\ 0 & 0 & 0 & 1 & x & y \end{bmatrix} \begin{bmatrix} a_0 \\ a_1 \\ a_2 \\ a_3 \\ a_4 \\ a_5 \end{bmatrix} \tag{6.17}$$

Eq. (6.17) can be written as,

$$\{u\}^e = [f(x,y)]\{a\} \tag{6.18}$$

Each element has six degrees of freedom, two at each node.
Eq. (6.18) can be written in extended form as,

$$\begin{bmatrix} u_1 \\ v_1 \\ u_2 \\ v_2 \\ u_3 \\ v_3 \end{bmatrix} = \begin{bmatrix} 1 & x_1 & y_1 & 0 & 0 & 0 \\ 0 & 0 & 0 & 1 & x_1 & y_1 \\ 1 & x_2 & y_2 & 0 & 0 & 0 \\ 0 & 0 & 0 & 1 & x_2 & y_2 \\ 1 & x_3 & y_3 & 0 & 0 & 0 \\ 0 & 0 & 0 & 1 & x_3 & y_3 \end{bmatrix} \begin{bmatrix} a_0 \\ a_1 \\ a_2 \\ a_3 \\ a_4 \\ a_5 \end{bmatrix} \tag{6.19}$$

Or,

$$\{u\}_0^e = [A]\{a\}$$

Therefore

$$\{a\} = [A]^{-1}\{u\}_0^e \tag{6.20}$$

Inserting Eq. (6.20) into Eq. (6.18) gives,

$$\{u\}^e = [f(x,y)][A]^{-1}\{u\}_0^e = [N(x,y)]\{u\}_0^e \tag{6.21}$$

Where,

$$[N(x,y)] = [\,[N(x,y)]_1 \quad [N(x,y)]_2 \quad [N(x,y)]_3\,]$$

And,

$$[N(x,y)]_1 = \frac{1}{2\Delta}(a_1 + b_1x + c_1y)[I] = \frac{1}{2\Delta}[N_1(x,y)][I] \tag{6.22}$$

Where,

$$a_1 = x_2y_3 - x_3y_2$$
$$b_1 = y_2 - y_3$$
$$c_1 = x_3 - x_2$$

Where Δ is the triangle area.

The other components are obtained by cyclic change of the subscripts, and, $[I]$, unit matrix.

Δ: element area

$$\Delta = \frac{1}{2}\begin{vmatrix} 1 & 1 & 1 \\ x_1 & x_2 & x_3 \\ y_1 & y_2 & y_3 \end{vmatrix} \tag{6.23}$$

The strain components are,

$$\begin{Bmatrix} \varepsilon_x \\ \varepsilon_y \\ \gamma_{xy} \end{Bmatrix} = \begin{Bmatrix} \frac{\partial u}{\partial x} \\ \frac{\partial v}{\partial y} \\ \frac{\partial u}{\partial y} + \frac{\partial v}{\partial x} \end{Bmatrix} = \begin{bmatrix} \frac{\partial}{\partial x} & 0 \\ 0 & \frac{\partial}{\partial y} \\ \frac{\partial}{\partial y} & \frac{\partial}{\partial x} \end{bmatrix} \begin{Bmatrix} u \\ v \end{Bmatrix} \tag{6.24}$$

Or,

$$\{\varepsilon\} = [\partial]\{u\}_0^e \tag{6.25}$$

Using Eq. (6.24) gives,

$$\{\varepsilon\} = [\partial][N(x,y)]\{u\}_0^e = [B]\{u\}_0^e$$

The strain–displacement relationship $[B]$ includes constant terms as follows,

$$\{\varepsilon\} = \frac{1}{2\Delta}\begin{bmatrix} b_1 & 0 & b_2 & 0 & b_3 & 0 \\ 0 & c_1 & 0 & c_2 & 0 & c_3 \\ c_1 & b_1 & c_2 & b_2 & c_3 & b_3 \end{bmatrix}\{u\}_0^e \tag{6.26}$$

Where,

$$[B] = \frac{1}{2\Delta}\begin{bmatrix} b_1 & 0 & b_2 & 0 & b_3 & 0 \\ 0 & c_1 & 0 & c_2 & 0 & c_3 \\ c_1 & b_1 & c_2 & b_2 & c_3 & b_3 \end{bmatrix}$$

Using Hooke's law,

$$\{\sigma\} = [D]\{\varepsilon\} \tag{6.27}$$

Also,

$$\{\sigma\} = \left\{ \sigma_x \quad \sigma_y \quad \tau_{xy} \right\}^t,$$

and,

$$[D] = \frac{E}{(1-\nu^2)}\begin{bmatrix} 1 & \nu & 0 \\ \nu & 1 & 0 \\ 0 & 0 & \dfrac{1-\nu}{2} \end{bmatrix} \text{for plane stress} \tag{6.28a}$$

Or,

$$[D] = \frac{E}{(1+\nu)(1-\nu)}\begin{bmatrix} (1-\nu) & \nu & 0 \\ \nu & (1-\nu) & 0 \\ 0 & 0 & \left(\dfrac{1-\nu}{2}\right) \end{bmatrix} \text{for plane strain} \tag{6.28b}$$

6.1.2 Loading conditions

The loading should be converted into an equivalent concentrated force at the nodes in x- and y-directions. The forces may be,

1. Body force $\{F_b\}$ per unit volume.

Where, the potential energy stored in the element is given by,

$$\Omega_b^e = -\iint_e \{u\}_0^{et}\{F_b\}tdxdy = -\{u\}_0^{et}\iint_e [N]^t\{F_b\}tdxdy = -\{u\}_0^{et}\{F_b\}^e$$

And,

$$\{F_b\}_0^e = \left\{ F_x \quad F_y \right\}^t$$

2. Surface force, Ω_s per unit area, where, the potential energy stored in the element is given by,

$$\Omega_s^e = -\iint_{st} \{u\}_{st}^{et}\{T_s\}tds = -\{u\}_0^{et}\iint_{st} [N]^t\{T_s\}tds = -\{u\}_0^{et}\{F_s\}^e$$

Where,

$$\{F_s\}^e = \{\, F_x \quad F_y \,\}^t$$

3. Concentrated forces may be applied directly into x- and y-directions,

$$\{F_s\}^e = \{\, F_x \quad F_y \,\}^t$$

Therefore

$$\{F\}^e = \{\, F_{x_1} \quad F_{y_1} \quad F_{x_2} \quad F_{y_2} \quad F_{x_3} \quad F_{y_3} \,\}^t \tag{6.28}$$

And, $F_{x_1}, F_{y_1}, \ldots, F_{y_3}$ Each contains all the acted components of loading on the R.H.S. of the equilibrium equations.

Example 6.1: Formulate the constrained stiffness matrix and load vector. You may use the triangular elements as a discretization element and consider only the G.C. for the problem shown in Fig. 6.5, where t is the plate thickness.

Solution,

The cantilever beam shown in Fig. 6.5A can be modeled by taking advantage of symmetry (geometry and loading) using 4-elements as shown in Fig. 6.5B. The typical elements are shown in Fig. 6.5C.

The stiffness equations may be assembled by inspection,

$$\begin{array}{c} \\ q_1 \\ q_2 \\ q_3 \\ q_4 \\ q_5 \end{array}\begin{array}{c} \begin{array}{ccccc} q_1 & q_2 & q_3 & q_4 & q_5 \end{array} \\ \begin{bmatrix} k^1_{33} & k^1_{34} & 0 & k^1_{35} & k^1_{36} \\ k^1_{43} & k^1_{44} & 0 & k^1_{45} & k^1_{46} \\ 0 & 0 & (k^2_{11} + k^3_{11} + k^4_{33}) & (k^2_{15} + k^3_{13}) & (k^2_{16} + k^3_{14}) \\ k^1_{53} & k^1_{54} & (k^2_{15} + k^3_{13}) & (k^3_{33} + k^2_{55} + k^1_{55}) & (k^1_{56} + k^3_{34} + k^2_{56}) \\ k^1_{63} & k^1_{64} & (k^2_{61} + k^3_{41}) & (k^1_{56} + k^3_{34} + k^2_{56}) & (k^1_{66} + k^2_{66} + k^3_{44}) \end{bmatrix} \end{array} \begin{Bmatrix} q_1 \\ q_2 \\ q_3 \\ q_4 \\ q_5 \end{Bmatrix} = \begin{Bmatrix} \frac{p\,b.t}{2} \\ 0 \\ 0 \\ 0 \\ 0 \end{Bmatrix}$$

As previously stated, the element for example $(k^2_{61} + k^3_{41})$ means $k(6,1)$ location from the element stiffness matrix of the second element $+\,k(4,1)$ location from the element stiffness matrix for the third element ... etc.

$k(5,5)$ means the contribution of the stiffness matrix from the three elements 1, 2, and 3 as $k(6,6), k(6,6)$, and $k(4,4)$, respectively.

Example 6.2: a. For the triangular element shown in Fig. 6.6A, derives the element stiffness matrix using the constant strain element concept.

b. Nodes (1) and (3) of the element in Fig. 6.6B lie on the edge of a plate where tractions are prescribed. Suppose the tractions consist of pressure and upward shear varying

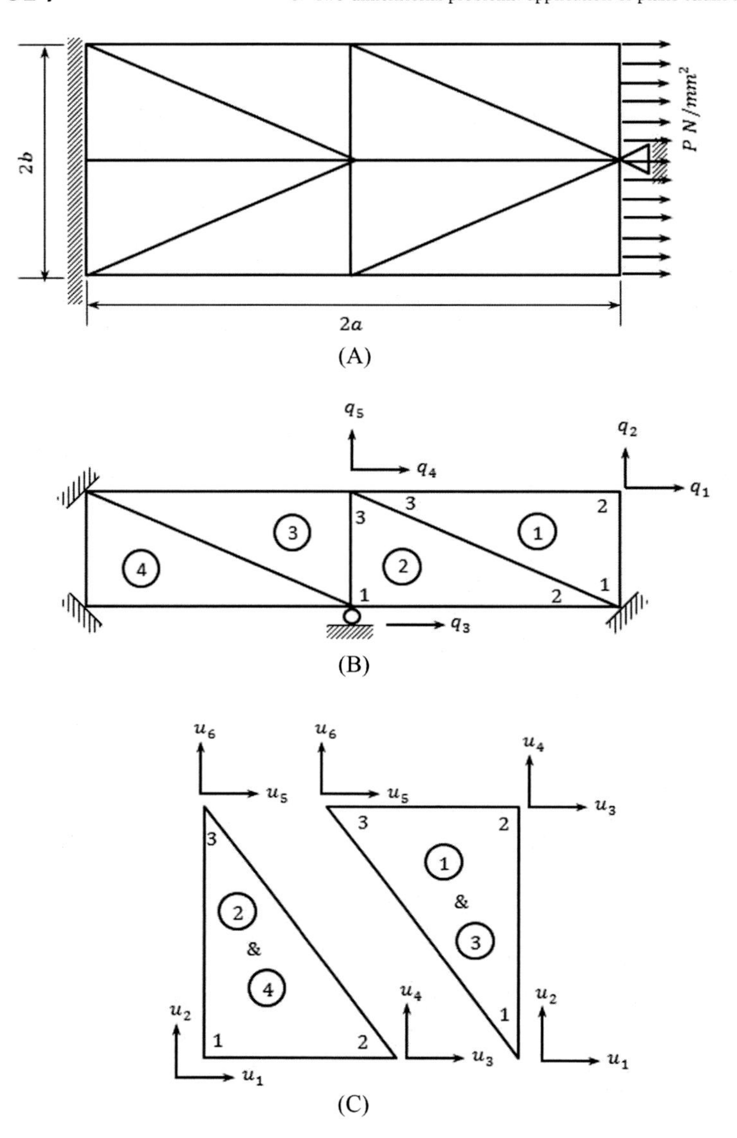

FIGURE 6.5 (A) A cantilever plate under axial loading. (B) F.E. modeling. (C)Typical elements.

linearly from P_1 to P_3 and S_1 to S_3, respectively, per unit surface between nodes (1) and (3). Determine the element load vector.

Solution,

For the triangular element shown in Fig. 6.6A using Eq. (6.4a):

$$a_1 = x_2y_3 - x_3y_2 = 1 - 0, a_1 = 1, a_2 = x_3y_1 - x_1y_3 = 0 - 0, a_2 = 0$$

$$a_3 = x_1y_2 - x_2y_1 = 0 - 0, a_3 = 0$$

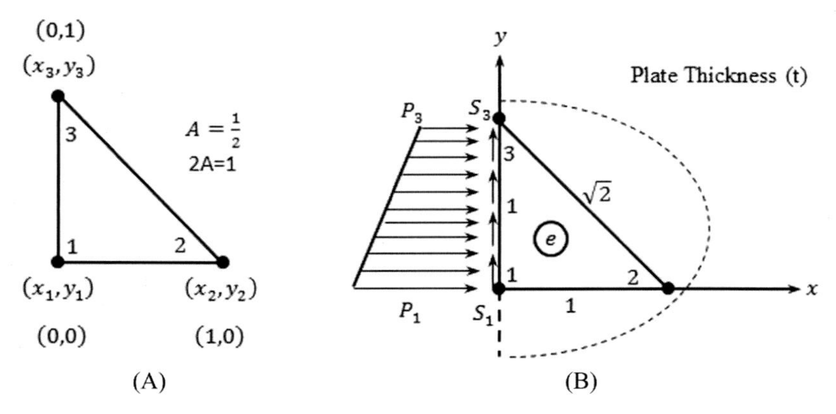

FIGURE 6.6 Triangular element with loaded edge: (A) typical element and (B) traction on the edge of the plate element.

$$c_1 = x_3 - x_2 = 0 - 1, c_1 = -1$$

$$c_2 = x_1 - x_3 = 0 - 0, c_2 = 0$$

$$c_3 = x_2 - x_1 = 1 - 0, c_3 = 1$$

$$b_1 = y_2 - y_3 = 0 - 1 = -1$$

$$b_2 = y_3 - y_1 = 1 - 0 = 1$$

$$b_3 = y_1 - y_2 = 0 - 0 = 0 \tag{E2.1a}$$

Again using Eq. (6.4a) results in the following shape functions:

$$N_1 = (a_1 + b_1 x + c_1 y)/2\Delta, N_1 = 1 - y - x$$

$$N_2 = (a_2 + b_2 x + c_2 y)/2\Delta, N_2 = x$$

$$N_3 = (a_3 + b_3 x + c_3 y)/2\Delta, N_3 = y$$

Using Eq. (2.1a) gives,
Arrange in matrix form gives,

$$[N] = \begin{bmatrix} N_1 & 0 & N_2 & 0 & N_3 & 0 \\ 0 & N_1 & 0 & N_2 & 0 & N_3 \end{bmatrix}$$

Or,

$$[N] = \begin{bmatrix} (1-y-x) & 0 & x & 0 & y & 0 \\ 0 & (1-y-x) & 0 & x & 0 & y \end{bmatrix}$$

Therefore using Eq. (2.1a), the strain–displacement relationship becomes:

$$[B] = \begin{bmatrix} -1 & 0 & 1 & 0 & 0 & 0 \\ 0 & -1 & 0 & 0 & 0 & 1 \\ -1 & -1 & 0 & 1 & 1 & 0 \end{bmatrix}$$

And,

$$[k]^e = 2\Delta t[B]^t[D][B] \tag{E2.1}$$

Carrying out the multiplications of matrices in (E2.1) gives,

$$[k]^e = \frac{Et}{2(1-\nu^2)} \begin{bmatrix} \frac{(1-\nu)}{2} & \frac{(1+\nu)}{2} & -1 & \frac{(1-\nu)}{2} & -\frac{(1-\nu)}{2} & -\nu \\ & \frac{(3-\nu)}{2} & -\nu & \frac{(1-\nu)}{2} & \frac{(1-\nu)}{2} & -1 \\ & & 1 & 0 & 0 & \nu \\ & & & \frac{(1-\nu)}{2} & \frac{(1-\nu)}{2} & 0 \\ & \text{Sym.} & & & \frac{(1-\nu)}{2} & 0 \\ & & & & & 1 \end{bmatrix} \tag{E2.2}$$

From the similarity of triangles in Fig. 6.7C,

$$\frac{(P-P_3)}{(P_1-P_3)} = \frac{(1-y)}{1}$$

$$P - P_3 = (P_1 - P_3)(1-y)$$

Or,

$$P = (P_1 - P_3)(1-y) + P_3 \tag{E2.3}$$

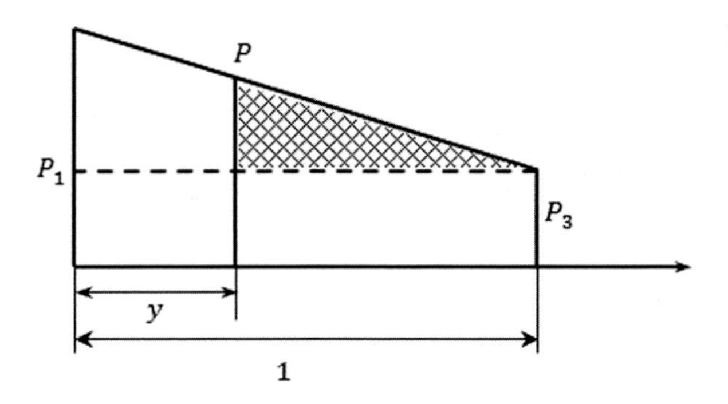

FIGURE 6.7 Pressure loading representation.

Simplifying gives,

$$P = P_1 - P_3 - P_1 y + P_3 y + P_3$$
$$P = P_1 - P_1 y + P_3 y$$

For the shear loading,

$$S = S_1 - S_1 y + S_3 y \tag{E2.4}$$

We have,

$$[N]^t = \begin{bmatrix} (1-y) & 0 \\ 0 & (1-y) \\ 0 & 0 \\ 0 & 0 \\ y & 0 \\ 0 & y \end{bmatrix}$$

The loading is composed of P and S,

$$\{F\} = \begin{Bmatrix} P \\ S \end{Bmatrix}$$

We have, for the traction loading,

$$\{T\} = \int_0^1 [N]^t \{F\} t dy$$

Or,

$$\{T\} = \int_0^1 \begin{Bmatrix} (1-y) & 0 \\ 0 & (1-y) \\ 0 & 0 \\ 0 & 0 \\ y & 0 \\ 0 & y \end{Bmatrix} \begin{Bmatrix} P \\ S \end{Bmatrix} dy = \int_0^1 \begin{Bmatrix} P(1-y) \\ S(1-y) \\ 0 \\ 0 \\ Py \\ Sy \end{Bmatrix} dy \tag{E2.5}$$

Using Eqs. (E2.3) and (E2.4) in Eq. (E2.5) gives,

$$\{T\} = \int_0^1 \begin{Bmatrix} (P_1 - P_1 y + P_3 y)(1-y) \\ S_1 - S_1 y + S_3 y(1-y) \\ 0 \\ 0 \\ (P_1 - P_1 y + P_3 y) y \\ (S_1 - S_1 y + S_3 y) y \end{Bmatrix} dy$$

Or,

$$\{T\} = \int_0^1 \begin{Bmatrix} P_1 - P_1 y + P_3 y - P_1 y + P_1 y^2 - P_3 y^2 \\ S_1 - S_1 y + S_3 y - S_1 y + S_1 y^2 - S_3 y^2 \\ 0 \\ 0 \\ P_1 y - P_1 y^2 + P_3 y^2 \\ S_1 y - S_1 y^2 + S_3 y^2 \end{Bmatrix} dy$$

$$\{T\} = \int_0^1 \begin{Bmatrix} P_1 + (P_3 - 2P_1)y + (P_1 - P_3)y^2 \\ S_1 + (S_3 - 2S_1)y + (S_1 - S_3)y^2 \\ 0 \\ 0 \\ P_1 y + (P_3 - P_1)y^2 \\ S_1 y + (S_3 - S_1)y^2 \end{Bmatrix} dy \tag{E2.6}$$

Carrying out the integration in Eq. (E2.6) gives,

$$\{T\} = \begin{Bmatrix} P_1 y\Big|_0^1 + (P_3 - 2P_1)\dfrac{y^2}{2}\Big|_0^1 + (P_1 - P_3)\dfrac{y^3}{3}\Big|_0^1 \\ S_1 y\Big|_0^1 + (S_3 - 2S_1)\dfrac{y^2}{2}\Big|_0^1 + (S_1 - S_3)\dfrac{y^3}{3}\Big|_0^1 \\ 0 \\ 0 \\ P_1 \dfrac{y^2}{2}\Big|_0^1 + (P_3 - P_1)\dfrac{y^3}{3}\Big|_0^1 \\ S_1 \dfrac{y^2}{2}\Big|_0^1 + (S_3 - S_1)\dfrac{y^3}{3}\Big|_0^1 \end{Bmatrix}$$

And carrying out the integrations inside the parentheses gives,

$$\{T\} = \begin{Bmatrix} P_1 + \dfrac{1}{2}(P_3 - 2P_1) + \dfrac{1}{3}(P_1 - P_3) \\ S_1 + \dfrac{1}{2}(S_3 - 2S_1) + \dfrac{1}{3}(S_1 - S_3) \\ 0 \\ 0 \\ \dfrac{1}{2}P_1 + \dfrac{1}{3}(P_3 - P_1) \\ \dfrac{1}{2}S_1 + \dfrac{1}{3}(S_3 - S_1) \end{Bmatrix} \tag{E2.7}$$

Simplifications of Eq. (E2.7) gives, the equivalent nodal forces vector due to P and S is,

$$\{T\} = \begin{Bmatrix} \left(\frac{P_1}{3} + \frac{P_3}{6}\right) \\ \left(\frac{S_1}{3} + \frac{S_3}{6}\right) \\ 0 \\ 0 \\ \left(\frac{P_1}{6} + \frac{P_3}{3}\right) \\ \left(\frac{S_1}{6} + \frac{S_3}{3}\right) \end{Bmatrix}$$

Example 6.3: Determine the system stiffness matrix and load vector for the finite formulation of the plane stress problem in Fig. 6.8. Choose the necessary G.C. $E = 207 \quad \text{GPa}, \nu = 0.3$.

Solution,

In this problem, and due to symmetry we can choose only a quarter of the region. Also, to illustrate the procedure of finding the coefficients a_i, b_i, and c_i from which the shape functions can be obtained, consider the element No. 4.

$$\Delta = \frac{100}{2} = 50, 2\Delta = 100$$

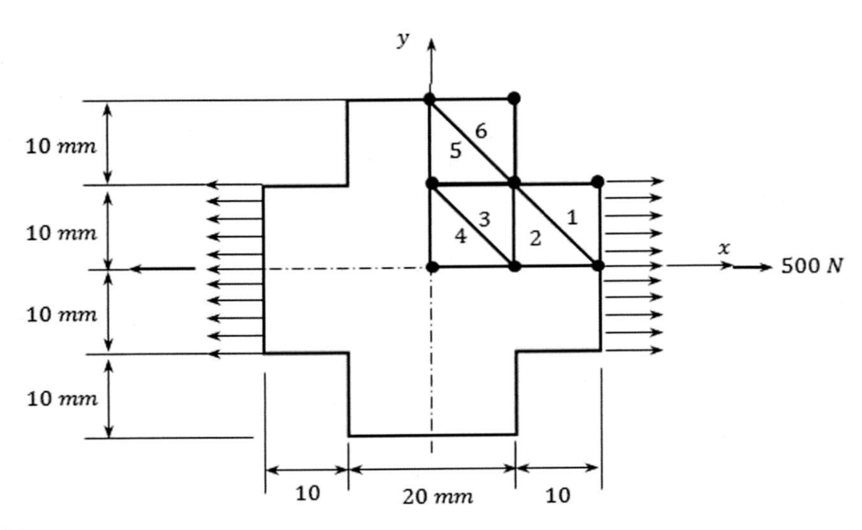

FIGURE 6.8 Plane stress region.

From the similarity of triangles,

$$\frac{1}{2\Delta} = \frac{1}{100}, x_1 = 0, x_2 = 10 \quad \text{mm}, x_3 = 0$$

$$y_1 = 0, y_2 = 0, y_3 = 10 \quad \text{mm}$$

$$a_1 = x_2y_3 - x_3y_2$$

$$a_1 = 100, y_2 = 0, b_1 = y_2 - y_3, b_1 = -10$$

$$c_1 = x_3 - x_2,$$

$$a_2 = x_3y_1 - x_1y_3, c_1 = -10 \quad a_2 = 0, b_2 = y_3 - y_1, b_2 = 10$$

$$c_2 = x_1 - x_3, c_2 = 0, a_3 = x_1y_2 - x_2y_1, a_3 = 0$$

$$b_3 = y_1 - y_2, b_3 = 0, c_3 = x_2 - x_1, c_3 = 10 \tag{E3.1a}$$

The typical elements of the quarter of the plane stress region shown in Fig. 6.9 are shown in Fig. 6.10.

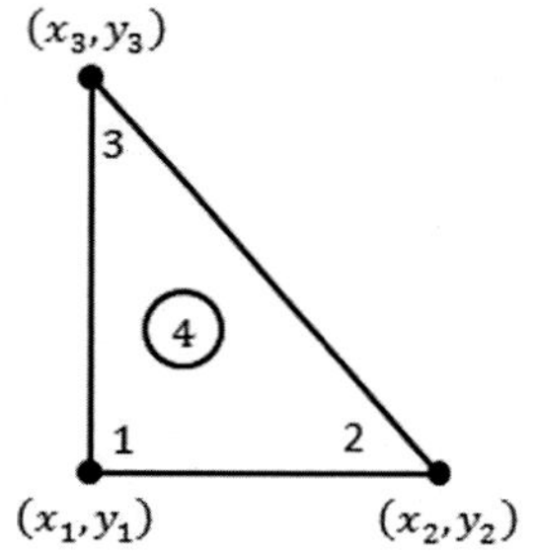

FIGURE 6.9 Typical discretization element.

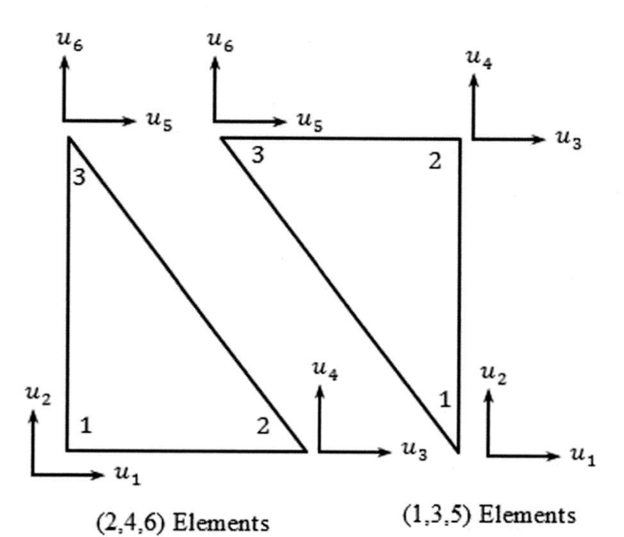

FIGURE 6.10 Finite elements, the local numbering sequence.

Also, the quarter of the plane stress region is shown in Fig. 6.11. The corresponding generalized degrees of freedom $q_1 - q_9$ are shown in this figure.

We have, from Eq. (E3.1a),

$$N_1(x,y) = \frac{(a_1 + b_1x + c_1y)}{2\Delta}$$
$$N_1(x,y) = \frac{(100 - 10x - 10y)}{100}$$
$$N_1(x,y) = 1 - \frac{x}{10} - \frac{y}{10}$$

Similarly,

$$N_2(x,y) = \frac{(a_2 + b_2x + c_2y)}{2\Delta}$$
$$N_2(x,y) = \frac{(0 + 10x + 0)}{100}$$
$$N_2(x,y) = \frac{x}{10}$$

And,

$$N_3(x,y) = \frac{(a_3 + b_3x + c_3y)}{2\Delta}$$
$$N_3(x,y) = \frac{(0 + 0 + 10y)}{100}$$
$$N_3(x,y) = \frac{y}{10} \quad \text{(E3.1b)}$$

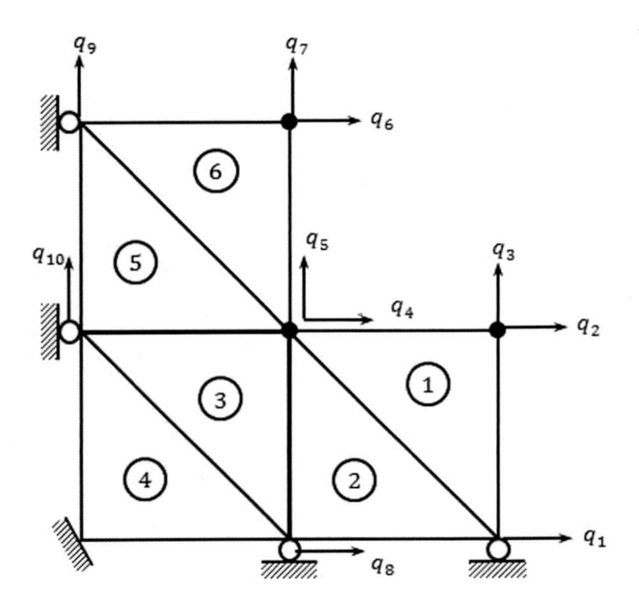

FIGURE 6.11 F.E. modeling.

Therefore

$$[N(x,y)] = [\,[N_1(x,y)] \quad [N_2(x,y)] \quad [N_3(x,y)]\,]$$

Or,

$$[N(x,y)] = \begin{bmatrix} N_1 & 0 & N_2 & 0 & N_3 & 0 \\ 0 & N_1 & 0 & N_2 & 0 & N_3 \end{bmatrix}$$

Using Eq. (E3.1b) gives,

$$[N(x,y)] = \frac{1}{10}\begin{bmatrix} (10-x-y) & 0 & x & 0 & y & 0 \\ 0 & (10-x-y) & 0 & x & 0 & y \end{bmatrix}$$

Strain−displacement relationship matrix $[B]$ is given by,

$$[B] = [\partial][N(x,y)] = \begin{bmatrix} \frac{\partial}{\partial x} & 0 \\ 0 & \frac{\partial}{\partial y} \\ \frac{\partial}{\partial y} & \frac{\partial}{\partial x} \end{bmatrix} \begin{bmatrix} N_1 & 0 & N_2 & 0 & N_3 & 0 \\ 0 & N_1 & 0 & N_2 & 0 & N_3 \end{bmatrix}$$

Using the above shape functions N_1, N_2, and N_3 given in Eq. (E3.1b) as follows,

$$[B] = \frac{1}{10}\begin{bmatrix} \frac{\partial}{\partial x} & 0 \\ 0 & \frac{\partial}{\partial y} \\ \frac{\partial}{\partial y} & \frac{\partial}{\partial x} \end{bmatrix} \begin{bmatrix} (10-x-y) & 0 & x & 0 & y & 0 \\ 0 & (10-x-y) & 0 & x & 0 & y \end{bmatrix}$$

Carrying out the multiplications gives,

$$[B] = \frac{1}{10}\begin{bmatrix} -1 & 0 & 1 & 0 & 0 & 0 \\ 0 & -1 & 0 & 0 & 0 & 1 \\ -1 & -1 & 0 & 1 & 1 & 0 \end{bmatrix} \qquad \text{(E3.1c)}$$

And the transpose of $[B]$ matrix is,

$$[B]^t = \frac{1}{10}\begin{bmatrix} -1 & 0 & -1 \\ 0 & -1 & -1 \\ 1 & 0 & 0 \\ 0 & 0 & 1 \\ 0 & 0 & 1 \\ 0 & 1 & 0 \end{bmatrix} \qquad \text{(E3.1d)}$$

The elasticity matrix for plane problems is given by Eq. (28a),

$$[D] = \frac{E}{(1-\nu^2)}\begin{bmatrix} 1 & \nu & 0 \\ \nu & 1 & 0 \\ 0 & 0 & \frac{(1-\nu)}{2} \end{bmatrix}$$

The element stiffness matrix is given by,

$$[k]^e = \int_0^y \int_0^x [B]^t[D][B]t dxdy$$

And by substitution of the matrices involved in the calculation of element stiffness matrix and achieving the multiplications inside the integrals becomes,

$$[k]^e = \frac{1}{100}\int_0^y \int_0^x \begin{bmatrix} -1 & 0 & -1 \\ 0 & -1 & -1 \\ 1 & 0 & 0 \\ 0 & 0 & 1 \\ 0 & 0 & 1 \\ 0 & 1 & 0 \end{bmatrix} \frac{E}{(1-\nu^2)}\begin{bmatrix} 1 & \nu & 0 \\ \nu & 1 & 0 \\ 0 & 0 & \frac{(1-\nu)}{2} \end{bmatrix} \begin{bmatrix} -1 & 0 & 1 & 0 & 0 & 0 \\ 0 & -1 & 0 & 0 & 0 & 1 \\ -1 & -1 & 0 & 1 & 1 & 0 \end{bmatrix} dxdy$$

Or,

$$[k]^e = \frac{E}{91}\int\limits_0^y\int\limits_0^x \begin{bmatrix} -1 & 0 & -1 \\ 0 & -1 & -1 \\ 1 & 0 & 0 \\ 0 & 0 & 1 \\ 0 & 0 & 1 \\ 0 & 1 & 0 \end{bmatrix} \begin{bmatrix} -1 & -\nu & 1 & 0 & 0 & \nu \\ -\nu & -1 & \nu & 0 & 0 & 1 \\ \left(\frac{\nu-1}{2}\right) & \left(\frac{\nu-1}{2}\right) & 0 & \left(\frac{1-\nu}{2}\right) & \left(\frac{1-\nu}{2}\right) & 0 \end{bmatrix} dxdy \tag{E3.1e}$$

Simplification of Eq. (E3.1e) gives,

$$[k]^e = \frac{E}{91}\int_0^y \int_0^x \begin{bmatrix} \left(1-\left(\frac{\nu-1}{2}\right)\right) & \left(\nu-\left(\frac{\nu-1}{2}\right)\right) & -1 & \left(\frac{\nu-1}{2}\right) & \left(\frac{\nu-1}{2}\right) & -\nu \\ \left(\nu-\left(\frac{\nu-1}{2}\right)\right) & \left(1-\left(\frac{\nu-1}{2}\right)\right) & -\nu & \left(\frac{\nu-1}{2}\right) & \left(\frac{\nu-1}{2}\right) & -1 \\ -1 & -\nu & 1 & 0 & 0 & \nu \\ \left(\frac{\nu-1}{2}\right) & \left(\frac{\nu-1}{2}\right) & 0 & \left(\frac{1-\nu}{2}\right) & \left(\frac{1-\nu}{2}\right) & 0 \\ \left(\frac{\nu-1}{2}\right) & \left(\frac{\nu-1}{2}\right) & 0 & \left(\frac{1-\nu}{2}\right) & \left(\frac{1-\nu}{2}\right) & 0 \\ -\nu & -1 & \nu & 0 & 0 & 1 \end{bmatrix} dxdy$$

Inserting the values of ν gives,

$$[k]^e = \frac{E}{91}\int_0^y\int_0^x \begin{bmatrix} 1.35 & 0.65 & -1 & -0.35 & -0.35 & -0.3 \\ 0.65 & 1.35 & -0.3 & -0.35 & -0.35 & -1 \\ -1 & -0.3 & 1 & 0 & 0 & 0.3 \\ -0.35 & -0.35 & 0 & 0.35 & 0.35 & 0 \\ -0.35 & -0.35 & 0 & 0.35 & 0.35 & 0 \\ -0.3 & -1 & 0.3 & 0 & 0 & 1 \end{bmatrix} dxdy \quad \text{(E3.1f)}$$

We have,

$$\int\int dxdy = \int x|_0^{10} dy = 10y|_0^{10} = 100$$

Carrying out the multiplications and integrations of Eq. (E3.1f) gives,

$$[k] = \frac{E}{91}\begin{bmatrix} 135 & 65 & -100 & -35 & -35 & -30 \\ 65 & 135 & -30 & -35 & -35 & -100 \\ -100 & -30 & 100 & 0 & 0 & 30 \\ -35 & -35 & 0 & 35 & 35 & 0 \\ -35 & -35 & 0 & 35 & 35 & 0 \\ -30 & -100 & 30 & 0 & 0 & 100 \end{bmatrix}$$

Note that the derivate was achieved according to the assigned local node numbering sequence for the elements (2,4,6) and the elements (1,3,5) local numbering as shown in Fig. 6.9.

$$[k]_{sys} = \begin{bmatrix} \begin{pmatrix} k_{11}^1 + \\ k_{33}^2 \end{pmatrix} & k_{13}^1 & k_{14}^1 & \begin{pmatrix} k_{15}^1 + \\ k_{35}^2 \end{pmatrix} & \begin{pmatrix} k_{16}^1 + \\ k_{36}^2 \end{pmatrix} & 0 & 0 & k_{31}^2 & 0 & 0 \\ & k_{33}^1 & k_{34}^1 & k_{35}^1 & k_{36}^1 & 0 & 0 & 0 & 0 & 0 \\ & & k_{44}^1 & k_{45}^1 & k_{46}^1 & 0 & 0 & 0 & 0 & 0 \\ & & & \begin{pmatrix} k_{55}^1 + k_{55}^2 + \\ k_{33}^5 + k_{33}^3 + \\ k_{11}^6 \end{pmatrix} & \begin{pmatrix} k_{56}^1 + k_{56}^2 + \\ k_{34}^5 + k_{34}^3 + \\ k_{12}^6 \end{pmatrix} & k_{13}^6 & k_{14}^6 & \begin{pmatrix} k_{51}^2 + \\ k_{31}^3 \end{pmatrix} & \begin{pmatrix} k_{36}^5 + \\ k_{16}^6 \end{pmatrix} & \begin{pmatrix} k_{36}^5 + \\ k_{32}^3 \end{pmatrix} \\ & & & & \begin{pmatrix} k_{66}^1 + k_{66}^2 + \\ k_{44}^3 + k_{44}^5 + \\ k_{22}^6 \end{pmatrix} & k_{23}^6 & k_{24}^6 & \begin{pmatrix} k_{61}^2 + \\ k_{41}^3 \end{pmatrix} & \begin{pmatrix} k_{46}^5 + \\ k_{26}^6 \end{pmatrix} & \begin{pmatrix} k_{46}^5 + \\ k_{42}^3 \end{pmatrix} \\ & & & & & k_{33}^6 & k_{34}^6 & 0 & k_{36}^6 & 0 \\ & & & & & & k_{44}^6 & 0 & k_{46}^6 & 0 \\ & & Sym & & & & & \begin{pmatrix} k_{11}^2 + \\ k_{11}^3 + \\ k_{33}^4 \end{pmatrix} & 0 & \begin{pmatrix} k_{16}^3 + \\ k_{36}^4 \end{pmatrix} \\ & & & & & & & & \begin{pmatrix} k_{66}^6 + \\ k_{66}^5 \end{pmatrix} & k_{62}^3 \\ & & & & & & & & & \begin{pmatrix} k_{66}^4 + \\ k_{22}^5 + \\ k_{66}^3 \end{pmatrix} \end{bmatrix} \quad \text{(E3.1)}$$

Then,

$$[k]_{sys} = \begin{bmatrix} 235 & -100 & -35 & -35 & 0 & 0 & 0 & -100 & 0 & 0 \\ & 100 & 0 & 0 & 30 & 0 & 0 & 0 & 0 & 0 \\ & & 35 & 35 & 0 & 0 & 0 & 0 & 0 & 0 \\ & & & 405 & 65 & -100 & -35 & -135 & 0 & 0 \\ & & & & 405 & -30 & -35 & -65 & -100 & -35 \\ & & & & & 100 & 0 & 30 & 0 & 0 \\ & & & & & & 35 & 0 & 0 & 0 \\ & & & & & & & 370 & 0 & 0 \\ & & \text{Sym.} & & & & & & 200 & -100 \\ & & & & & & & & & 335 \end{bmatrix} \quad \text{(E3.2)}$$

The load vector corresponds to the G.C. is,

$$\{Q\} = \begin{Bmatrix} \frac{wl}{2} \\ \frac{wl}{2} \\ 0 \\ 0 \\ 0 \\ 0 \\ 0 \\ 0 \\ 0 \\ 0 \end{Bmatrix}, \text{and, } \{q\} = \begin{Bmatrix} q_1 \\ q_2 \\ q_3 \\ q_4 \\ q_5 \\ q_6 \\ q_7 \\ q_8 \\ q_9 \\ q_{10} \end{Bmatrix} \quad \text{(E3.3)}$$

6.2 Derivation of the 4-node quadrilateral element, formulation of the element equations

Step-1, define the nodal parameters.
The nodal displacement vector can be defined as follows, Fig. 6.12,

$$\{\delta\} = \{ u_1 \quad v_1 \quad u_2 \quad v_2 \quad u_3 \quad v_3 \quad u_4 \quad v_4 \}^t \quad (6.29)$$

Note that, the displacement component along the z-axis is neglected.
And, the force vector is,

$$\{F\} = \{ F_{x_1} \quad F_{y_1} \quad F_{x_2} \quad F_{y_2} \quad F_{x_a} \quad F_{y_b} \quad F_{x_4} \quad F_{y_4} \}^t \quad (6.30)$$

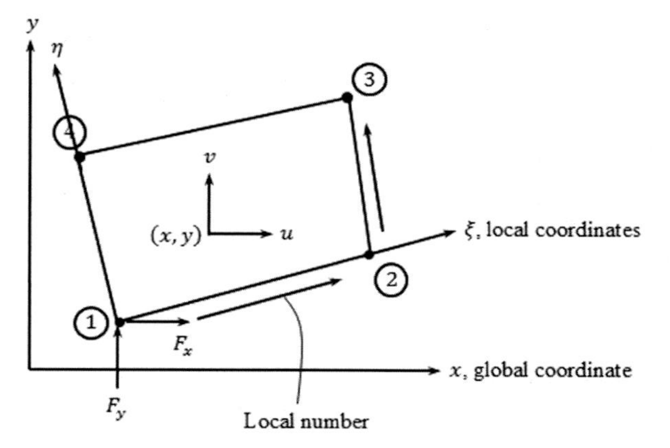

FIGURE 6.12 4-Node quadrilateral element.

Step-2,

Express the displacement components at any point in terms of nodal displacement and shape functions.

$$u(x,y) = u_1N_1(x,y) + u_2N_2(x,y) + u_3N_3(x,y) + u_4N_4(x,y)$$
$$v(x,y) = v_1N_1(x,y) + v_2N_2(x,y) + v_3N_3(x,y) + v_4N_4(x,y) \tag{6.31}$$

Furthermore, it is required to formulate the shape function.

For the case of the 4-node quadrilateral element, assume,

$$u_1 = \alpha_1 + \alpha_2x_1 + \alpha_3y_1 + \alpha_4x_1y_1$$
$$u_2 = \alpha_1 + \alpha_2x_2 + \alpha_3y_2 + \alpha_4x_2y_2$$
$$u_3 = \alpha_1 + \alpha_2x_3 + \alpha_3y_3 + \alpha_4x_3y_3$$
$$u_4 = \alpha_1 + \alpha_2x_4 + \alpha_3y_4 + \alpha_4x_4y_4 \tag{6.32}$$

Or,

$$\begin{bmatrix} 1 & x_1 & y_1 & x_1y_1 \\ 1 & x_2 & y_2 & x_2y_2 \\ 1 & x_3 & y_3 & x_3y_3 \\ 1 & x_4 & y_4 & x_4y_4 \end{bmatrix} \begin{bmatrix} \alpha_1 \\ \alpha_2 \\ \alpha_3 \\ \alpha_4 \end{bmatrix} = \begin{bmatrix} u_1 \\ u_2 \\ u_3 \\ u_4 \end{bmatrix}, \text{four unknowns} \quad \alpha_1, \alpha_2, \alpha_3, \alpha_4 \tag{6.33}$$

By substitution $(\alpha_1, \alpha_2, \alpha_3, \alpha_4)$ back into $u(x,y)$ expression, the shape functions can be deduced.

The significant difficulty is that the shape functions depend upon the nodal coordinates. For every subdomain, the previous procedure of shape function derivation should be repeated in addition to the complexity of the integration involved to calculate the element stiffness matrix.

It is useful to employ a local system which is simple, unique and independent of the global system such a system is known as an intrinsic system and its coordinates are the intrinsic coordinates ξ and η.

One excellent idea is to transform the quadrilateral element into a square of unit side length, as shown in Fig. 6.13.

The quadrilateral element shape functions are (This will be derived in Chapter 12: Shape Function Determinations and Numerical Integration),

$$\begin{aligned} N_1 &= (1-\xi)(1-\eta) \\ N_2 &= \xi(1-\eta) \\ N_3 &= \xi\eta \\ N_4 &= (1-\xi)\eta \end{aligned} \tag{6.34}$$

And hence,

$$\begin{aligned} u(\xi,\eta) &= \sum_{i=1}^{4} u_i N_i(\xi,\eta) \\ v(\xi,\eta) &= \sum_{i=1}^{4} v_i N_i(\xi,\eta) \end{aligned} \tag{6.35}$$

The problem now reduces to one of obtaining the equations of the transformation,

$$\begin{aligned} x &= x(\xi,\eta) \\ y &= y(\xi,\eta) \end{aligned} \tag{6.36}$$

Another idea is to assume that x and y have one field functions and defined as follows,

$$\begin{aligned} x(\xi,\eta) &= x_1N_1 + x_2N_2 + x_3N_3 + x_4N_4 \\ y(\xi,\eta) &= y_1N_1 + y_2N_2 + y_3N_3 + y_4N_4 \end{aligned} \tag{6.37}$$

Such transformation is known as "Isoparametric Transformation."

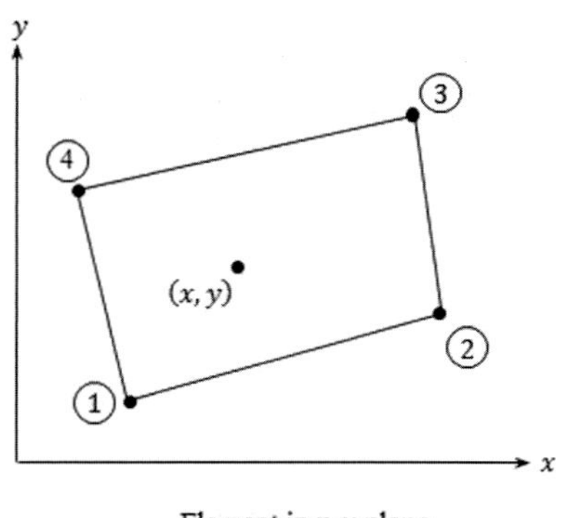

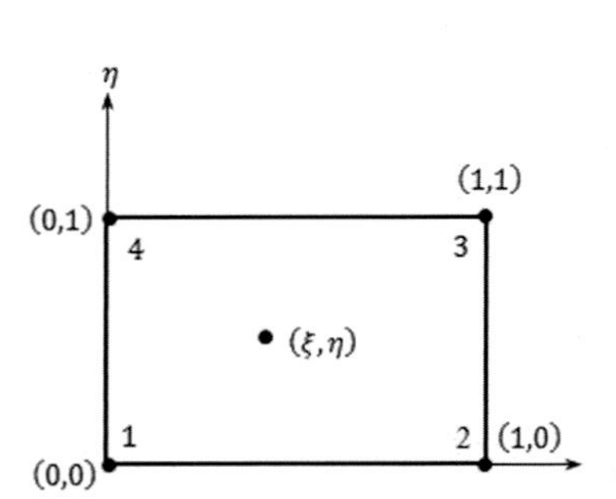

FIGURE 6.13 4-Node quadrilateral element.

Finally, it can be defined as the displacement vectors $\{q\}$ at any point (x,y) as follows,

$$q^t = \begin{bmatrix} N_1 & 0 & N_2 & 0 & N_3 & 0 & N_4 & 0 \\ 0 & N_1 & 0 & N_2 & 0 & N_3 & 0 & N_4 \end{bmatrix} \delta \tag{6.38}$$

Step-3

Express the strain components at any point in terms of nodal displacement and the shape function. Defining,

$$\delta^t = \{ u_1 \quad v_1 \quad u_2 \quad v_2 \quad u_3 \quad v_3 \quad u_4 \quad v_4 \} \tag{6.39}$$

It can be shown that,

$$\begin{aligned} u(x,y) &= [N_1 \quad 0 \quad N_2 \quad 0 \quad N_3 \quad 0 \quad N_4 \quad 0]\delta \\ v(x,y) &= [0 \quad N_1 \quad 0 \quad N_2 \quad 0 \quad N_3 \quad 0 \quad N_4]\delta \end{aligned} \tag{6.40}$$

We have,

$$\varepsilon_x = \frac{\partial u}{\partial x} = \begin{bmatrix} \frac{\partial N_1}{\partial x} & 0 & \frac{\partial N_2}{\partial x} & 0 & \frac{\partial N_3}{\partial x} & 0 & \frac{\partial N_4}{\partial x} & 0 \end{bmatrix} \delta$$

$$\varepsilon_y = \frac{\partial u}{\partial y} = \begin{bmatrix} 0 & \frac{\partial N_1}{\partial y} & 0 & \frac{\partial N_2}{\partial y} & 0 & \frac{\partial N_3}{\partial y} & 0 & \frac{\partial N_4}{\partial y} \end{bmatrix} \delta$$

And,

$$\gamma_{xy} = \frac{\partial u}{\partial y} + \frac{\partial v}{\partial x} = \begin{bmatrix} \frac{\partial N_1}{\partial y} & \frac{\partial N_1}{\partial x} & \frac{\partial N_2}{\partial y} & \frac{\partial N_2}{\partial x} & \frac{\partial N_3}{\partial y} & \frac{\partial N_3}{\partial x} & \frac{\partial N_4}{\partial y} & \frac{\partial N_4}{\partial x} \end{bmatrix} \tag{6.41}$$

Hence, it can be deduced that,

$$\underset{3\times1}{\{\varepsilon\}} = \underset{3\times8}{[B]}\,\underset{8\times1}{\{\delta\}}$$

We have,

$$\{\varepsilon\} = \{ \varepsilon_x \quad \varepsilon_y \quad \gamma_{xy} \}$$

And,

$$[B] = \begin{bmatrix} \frac{\partial N_1}{\partial x} & 0 & \frac{\partial N_2}{\partial x} & 0 & \frac{\partial N_3}{\partial x} & 0 & \frac{\partial N_4}{\partial x} & 0 \\ 0 & \frac{\partial N_1}{\partial y} & 0 & \frac{\partial N_2}{\partial y} & 0 & \frac{\partial N_3}{\partial y} & 0 & \frac{\partial N_4}{\partial y} \\ \frac{\partial N_1}{\partial y} & \frac{\partial N_1}{\partial x} & \frac{\partial N_2}{\partial y} & \frac{\partial N_2}{\partial x} & \frac{\partial N_3}{\partial y} & \frac{\partial N_3}{\partial x} & \frac{\partial N_4}{\partial y} & \frac{\partial N_4}{\partial x} \end{bmatrix}$$

Since the shape functions are expressed in terms of the intrinsic coordinates ξ, η it is useful to deduce the Cartesian derivatives in terms of intrinsic derivatives. We have,

$$\frac{\partial N_i}{\partial \xi} = \frac{\partial N_i}{\partial x}\frac{\partial x}{\partial \xi} + \frac{\partial N_i}{\partial y}\frac{\partial y}{\partial \xi}$$

$$\frac{\partial N_i}{\partial \eta} = \frac{\partial N_i}{\partial x}\frac{\partial x}{\partial \eta} + \frac{\partial N_i}{\partial y}\frac{\partial y}{\partial \eta} \tag{6.42}$$

In matrix form,

$$\begin{bmatrix} \frac{\partial N_i}{\partial \xi} \\ \frac{\partial N_i}{\partial \eta} \end{bmatrix} = \left[J\left(\frac{x,y}{\xi,\eta}\right) \right] \begin{bmatrix} \frac{\partial N_i}{\partial x} \\ \frac{\partial N_i}{\partial y} \end{bmatrix} \tag{6.43}$$

Where,

$\left[J\left(\frac{x,y}{\xi,\eta}\right)\right] = \begin{bmatrix} \frac{\partial x}{\partial \xi} & \frac{\partial y}{\partial \xi} \\ \frac{\partial x}{\partial \eta} & \frac{\partial y}{\partial \eta} \end{bmatrix}$ is the Jacobian matrix

Finally,

$$\begin{bmatrix} \frac{\partial N_i}{\partial x} \\ \frac{\partial N_i}{\partial y} \end{bmatrix} = [J]^{-1}\left(\frac{x,y}{\xi,\eta}\right) \begin{bmatrix} \frac{\partial N_i}{\partial \xi} \\ \frac{\partial N_i}{\partial \eta} \end{bmatrix} \tag{6.44}$$

And it can be shown that,

$$[J]^{-1} = \frac{1}{|J|} \begin{bmatrix} \frac{\partial y}{\partial \eta} & -\frac{\partial y}{\partial \xi} \\ -\frac{\partial x}{\partial \eta} & \frac{\partial x}{\partial \xi} \end{bmatrix}$$

$$|J| = \begin{vmatrix} \frac{\partial x}{\partial \xi} & \frac{\partial y}{\partial \xi} \\ \frac{\partial x}{\partial \eta} & \frac{\partial y}{\partial \eta} \end{vmatrix} = \frac{\partial x}{\partial \xi}\frac{\partial y}{\partial \eta} - \frac{\partial y}{\partial \xi}\frac{\partial x}{\partial \eta} \tag{6.45}$$

6.3 Parallelogramic element

In some two-dimensional representations, we may use a quadrilateral elements 4-nodes with eight degrees of freedom. The element shown in Fig. 6.14 (parallelogram element) may be used.

The side 1–4 is parallel to the side 2–3, and the side 1–2 is parallel to the side 4–3 and the intrinsic coordinates (local coordinate system) are attached on the element with directions cosines respective to the global coordinate system xy.

From parallelism,

$$x_3 - x_2 = x_4 - x_1$$

$$y_3 - y_2 = y_4 - y_1$$

or,

$$x_3 = -x_1 + x_2 + x_4$$

$$y_3 = -y_1 + y_2 + y_4 \tag{6.46}$$

Appling the isoparametric transformation and considering the 4-nodes quadrilateral element shown in Fig. 6.15, we have,

$$x = \sum_{i=1}^{4} x_i N_i \tag{6.47a}$$

Using Eqs. (6.34) and (6.46) gives,

$$\begin{aligned} x(\xi,\eta) &= (1-\xi-\eta+\xi\eta)x_1 + \xi(1-\eta)x_2 + \xi\eta(-x_1+x_2+x_4) + (\eta-\xi\eta)x_4 \\ &= (1-\xi-\eta)x_1 + \xi x_2 + \eta x_4 \end{aligned}$$

Or,

$$x(\xi,\eta) = x_1 + \xi(x_2 - x_1) + \eta(x_4 - x_1) \tag{6.47}$$

Similarly, it can be shown that,

$$y(\xi,\eta) = y_1 + \xi(y_2 - y_1) + \eta(y_4 - y_1) \tag{6.48}$$

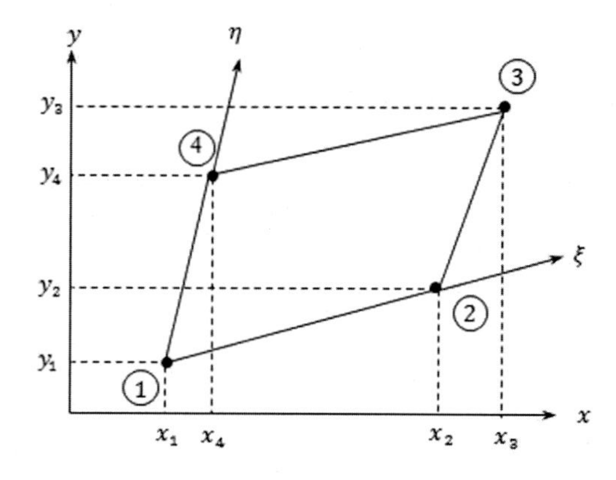

FIGURE 6.14 Parallelogramic element.

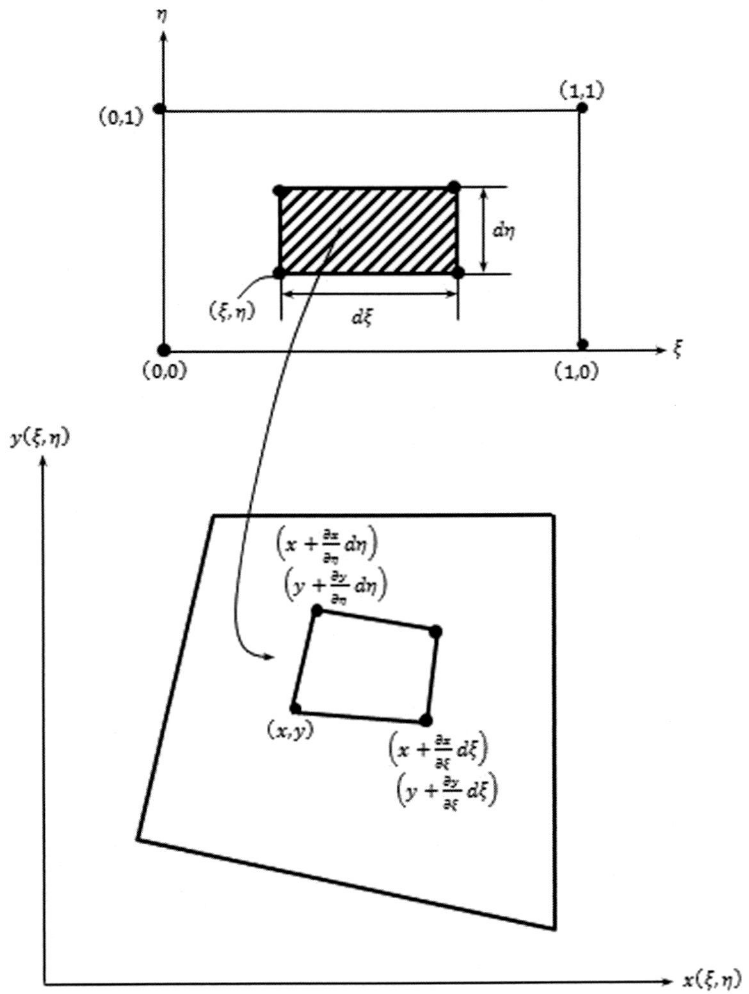

FIGURE 6.15 4-Node quadrilateral element.

The Jacobian matrix is,

$$[J] = \begin{bmatrix} \dfrac{\partial x}{\partial \xi} & \dfrac{\partial y}{\partial \xi} \\ \dfrac{\partial x}{\partial \eta} & \dfrac{\partial y}{\partial \eta} \end{bmatrix} = \begin{bmatrix} (x_2 - x_1) & (y_2 - y_1) \\ (x_4 - x_1) & (y_4 - y_1) \end{bmatrix} \tag{6.49}$$

Hence,

$$|J| = \begin{vmatrix} (x_2 - x_1) & (y_2 - y_1) \\ (x_4 - x_1) & (y_4 - y_1) \end{vmatrix} = \begin{vmatrix} 1 & x_1 & y_1 \\ 1 & x_2 & y_2 \\ 1 & x_4 & y_4 \end{vmatrix} = A(\text{the area of the element})$$

$$[J]^{-1} = \frac{1}{A}\begin{bmatrix} (y_4 - y_1) & -(y_2 - y_1) \\ -(x_4 - x_1) & (x_2 - x_1) \end{bmatrix}$$

$$\begin{bmatrix} \dfrac{\partial N_i}{\partial x} \\ \dfrac{\partial N_i}{\partial y} \end{bmatrix} = \frac{1}{A}\begin{bmatrix} (y_4 - y_1) & -(y_2 - y_1) \\ -(x_4 - x_1) & (x_2 - x_1) \end{bmatrix}\begin{bmatrix} \dfrac{\partial N_i}{\partial \xi} \\ \dfrac{\partial N_i}{\partial \eta} \end{bmatrix} \tag{6.50}$$

Or explicitly,

$$\frac{\partial N_i}{\partial x} = \frac{1}{A}\left[(y_4 - y_1)\frac{\partial N_i}{\partial \xi} - (y_2 - y_1)\frac{\partial N_i}{\partial \eta}\right]$$

$$\frac{\partial N_i}{\partial y} = \frac{1}{A}\left[-(x_4 - x_1)\frac{\partial N_i}{\partial \xi} + (x_2 - x_1)\frac{\partial N_i}{\partial \eta}\right] \tag{6.51}$$

Step-4,

Express the stress components at any point in terms of nodal displacements and shape functions,

$$\{\sigma\} = \{\sigma_x \quad \sigma_y \quad \tau_{xy}\}$$

$$\{\sigma\} = [D]\{\varepsilon\} = [D][B]\{\delta\} \tag{6.52}$$

Where, $[D]$ is the stress–strain matrix

From the theory of elasticity,

$$\varepsilon_x = \frac{1}{E}\left[\sigma_x - \nu(\sigma_y + \sigma_z)\right]$$

$$\varepsilon_y = \frac{1}{E}\left[\sigma_y - \nu(\sigma_z + \sigma_x)\right]$$

$$\varepsilon_z = \frac{1}{E}\left[\sigma_z - \nu(\sigma_x + \sigma_y)\right]$$

$$\gamma_{xy} = \frac{2(1+\nu)}{E}\tau_{xy}$$

$$\gamma_{yz} = \frac{2(1+\nu)}{E}\tau_{yz}$$

$$\gamma_{zx} = \frac{2(1+\nu)}{E}\tau_{zx} \tag{6.53}$$

1. Plane stress,

$$\sigma_z = \tau_{yz} = \tau_{zx} = 0$$

$$\varepsilon_x = \frac{1}{E}\left[\sigma_x - \nu\sigma_y\right]$$

$$\varepsilon_y = \frac{1}{E}\left[\sigma_y - \nu\sigma_x\right]$$

$$\begin{bmatrix} \varepsilon_x \\ \varepsilon_y \end{bmatrix} = \frac{1}{E}\begin{bmatrix} 1 & -\nu \\ -\nu & 1 \end{bmatrix}\begin{bmatrix} \sigma_x \\ \sigma_y \end{bmatrix} \tag{6.54}$$

Hence,

$$\begin{bmatrix} \sigma_x \\ \sigma_y \end{bmatrix} = E\begin{bmatrix} 1 & -\nu \\ -\nu & 1 \end{bmatrix}^{-1}\begin{bmatrix} \varepsilon_x \\ \varepsilon_y \end{bmatrix}$$

It can be shown that,

$$\begin{bmatrix} 1 & -\nu \\ -\nu & 1 \end{bmatrix}^{-1} = \frac{1}{(1-\nu^2)}\begin{bmatrix} 1 & \nu \\ \nu & 1 \end{bmatrix}$$

Also,

$$\tau_{xy} = \frac{E}{2(1+\nu)}\gamma_{xy} \tag{6.55}$$

Hence, it can be shown that,

$$\{\sigma\} = [D]\{\varepsilon\}$$

Where,

$$[D] = \frac{E}{(1-\nu^2)}\begin{bmatrix} 1 & \nu & 0 \\ \nu & 1 & 0 \\ 0 & 0 & \left(\dfrac{1-\nu}{2}\right) \end{bmatrix}$$

2. Plane strain

$$\varepsilon_z = \gamma_{yz} = \gamma_{zx} = 0$$

$$\varepsilon_z = 0 = \frac{1}{E}\left[\sigma_z - \nu\left(\sigma_x + \sigma_y\right)\right]$$

That is,

$$\sigma_z = \nu(\sigma_x + \sigma_y)$$

$$\varepsilon_x = \frac{1}{E}\left[\sigma_x - \nu(\sigma_y + \sigma_z)\right] = \frac{1}{E}\left[\sigma_x - \nu\{\sigma_y + \nu(\sigma_x + \sigma_y)\}\right] = \frac{1}{E}\left[(1-\nu^2)\sigma_x - \nu(1+\nu)\sigma_y\right]$$

$$= \frac{1+\nu}{E}\left[(1-\nu)\sigma_x - \nu\sigma_y\right]$$

Similarly,

$$\varepsilon_y = \frac{1+\nu}{E}\left[(1-\nu)\sigma_y - \nu\sigma_x\right]$$

Or,

$$\begin{bmatrix} \varepsilon_x \\ \varepsilon_y \end{bmatrix} = \frac{1+\nu}{E}\begin{bmatrix} (1-\nu) & -\nu \\ -\nu & (1-\nu) \end{bmatrix}\begin{bmatrix} \sigma_x \\ \sigma_y \end{bmatrix} \tag{6.56}$$

And it can be shown that,

$$\{\sigma\} = [D]\{\varepsilon\}$$

Where,

$$[D] = \frac{E}{(1+\nu)(1-2\nu)}\begin{bmatrix} (1-\nu) & \nu & 0 \\ \nu & (1-\nu) & 0 \\ 0 & 0 & \dfrac{(1-2\nu)}{2} \end{bmatrix} \tag{6.57}$$

Step-5,
Express the total potential energy of the element in terms of nodal displacements,

$$V = U - W$$

$$U = \frac{1}{2}\iiint_{\text{element}} \sigma^t \varepsilon dxdydz$$

$$W = \{\delta\}^t\{F\} \tag{6.58}$$

Where,

$$\{F\} = F_1\delta_1 + F_2\delta_2 + F_3\delta_3$$

From the previous steps,

$$\{\varepsilon\} = [B]\{\delta\},$$

and,

$$\{\sigma\} = [D][B]\{\delta\}$$

$$\{\sigma\}^t = \{\delta\}^t[B]^t[D]^t \tag{6.59}$$

Where,

$$[[D]\{\varepsilon\}]^t = \{\varepsilon\}^t[D]^t = ([B]\{\delta\})^t[D]^t = \{\delta\}^t[B]^t[D]^t$$

But, $[D]$ is a symmetric matrix,

$$[D] = [D]^t$$

Therefore

$$\{\sigma\}^t = \{\delta\}^t[B]^t[D] \tag{6.60}$$

Hence, the strain energy is given by,

$$U = \frac{1}{2}\iiint_{\text{element}} \{\delta\}^t[B]^t[D][B]\{\delta\}dxdydz = \frac{1}{2}\{\delta\}^t\left[\iiint [B]^t[D][B]dxdydz\right]\{\delta\} \tag{6.61}$$

And,

$$V = \frac{1}{2}\{\delta\}^t\left[\iiint \underbrace{[B]^t}_{8\times 3}\ \underbrace{[D]}_{3\times 3}\ \underbrace{[B]}_{3\times 8}\ dx\,dy\,dz\right]\{\delta\} - \{\delta\}^t\{F\} \tag{6.62}$$

Step-6,
Apply the minimum total potential energy theorem,

$$\frac{\partial \chi}{\partial \delta} = 0 = \left(\iiint [B]^t[D][B]dxdydz\right)\{\delta\} - \{F\} \tag{6.63}$$

Or,

$$[k]^e\{\delta\} = \{F\}$$

Where,

$$[k]^e = \iiint_{\text{element}} [B]^t[D][B]dxdydz$$

Since $[B]$ is the function of x and y only, it can be deduced that,

$$[k]^e = \iint_{\text{element}} t[B]^t[D][B]dxdy \tag{6.64}$$

Where, t = element thickness along the z-direction.
If an intrinsic element of area is plotted in the $x-y$ plane as shown in Fig. 6.15, then,

$$dA = \begin{vmatrix} 1 & -x & -y \\ 1 & x + \dfrac{dx}{d\xi}\xi & y + \dfrac{dy}{d\xi}\xi \\ 1 & x + \dfrac{dx}{d\eta}\eta & y + \dfrac{dy}{d\eta}\eta \end{vmatrix} = \begin{vmatrix} 1 & x & y \\ 0 & \dfrac{dx}{d\xi}\xi & \dfrac{dy}{d\xi}\xi \\ 0 & \dfrac{dx}{d\eta}\eta & \dfrac{dy}{d\eta}\eta \end{vmatrix}$$

$$dA = \begin{vmatrix} \frac{dx}{d\xi} & \frac{dy}{d\xi} \\ \frac{dx}{d\eta} & \frac{dy}{d\eta} \end{vmatrix} d\xi d\eta = \left| J\left(\frac{x,y}{\xi,\eta}\right) \right| d\xi d\eta \tag{6.65}$$

Hence,

$$[k]^e = \int_0^1 \int_0^1 t[B]^t[D][B]|J| d\xi d\eta \tag{6.66}$$

Eq. (6.66) is integrated numerically, which will be presented in detail in Chapter 12, Shape Function Determinations and Numerical Integration.

Problems

P.6.1 Determine the constrained system stiffness matrix and load vector for the finite formulation of the plane stress problem of Fig. P.6.1. Choose the necessary G.C. for the shown mesh (4 constant strain elements). Use the derived element stiffness matrix for the constant stain triangular element presented in Section 6.1.1. Use, $a = 20\ \text{mm}, E = 200 * 10^3\ \text{N/mm}^2, \nu = 0.3, A = 25\ \text{mm}^2, \text{and}\ t = 1\ \text{mm}$

$$k_b = \frac{EA}{l}\begin{bmatrix} 1 & -1 \\ -1 & 1 \end{bmatrix}$$

P.6.2a. Develop the expression for the traction force vector of the element loaded as shown in Fig. P.6.2a. The general form of the shape function is,

$$N_i = \frac{1}{4}(1 + \xi\xi_i)(1 + \eta\eta_i)$$

b. Derive the strain–displacement relationship $[B]$ for the element shown in Fig. P.6.2b at $\xi = 0$ and $\eta = 0$. Use the same function for both displacement (u, v) and coordinates (x,y). Consider the element to be used in a plane stress problem.

P.6.3 Fig. P.6.3 shows a local detail of a simplex triangle finite element mesh. The element is at that part of a 1-mm-thick plate edge where its normal unit vector is at 45 degrees to the OX axis. The surface traction has components $T_x^n = 20\ \text{N/mm}^2$ and $T_y^n = 30\ \text{N/mm}^2$. Using a standard work calculation, show that,

$$\{P\}_0^e t = \begin{bmatrix} 0 & 0 & 100\sqrt{2} & 150\sqrt{2} & 100\sqrt{2} & 150\sqrt{2} \end{bmatrix}$$

P.6.4 Determine the constrained stiffness matrix and load vector by choosing the minimum number of the required G.C. for the finite element formulation of the plane stress problem of Fig. P.6.4 using the indicated node numbering. The element stiffness matrix for the triangular elements calculated in example 6.3 may formulate the system stiffness matrix.

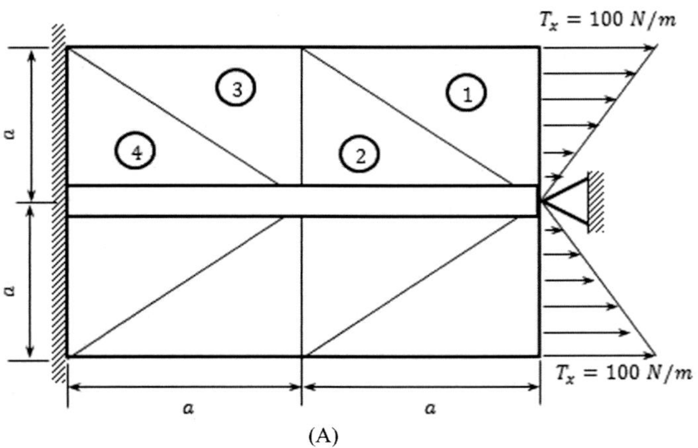

FIGURE P.6.1

(A)

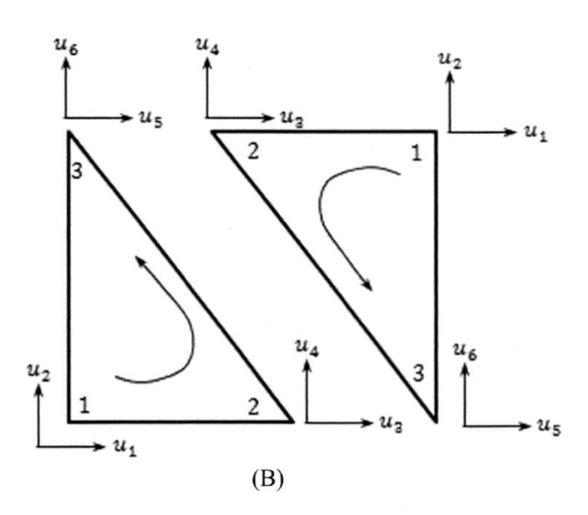

(B)

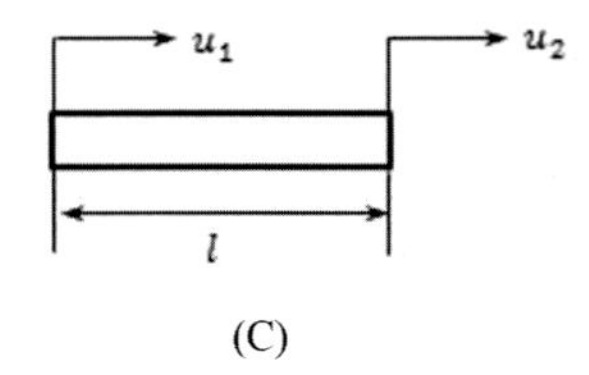

(C)

Bar cross-section = 4 mm^2

$$\text{Plate thickness} = 1 \text{ mm}$$

$$\nu = 0.3, E = 207 \text{ kN/mm}^2$$

What can be said about the results obtained from this coarse mesh? What are special problems attached to the reentrant corners?

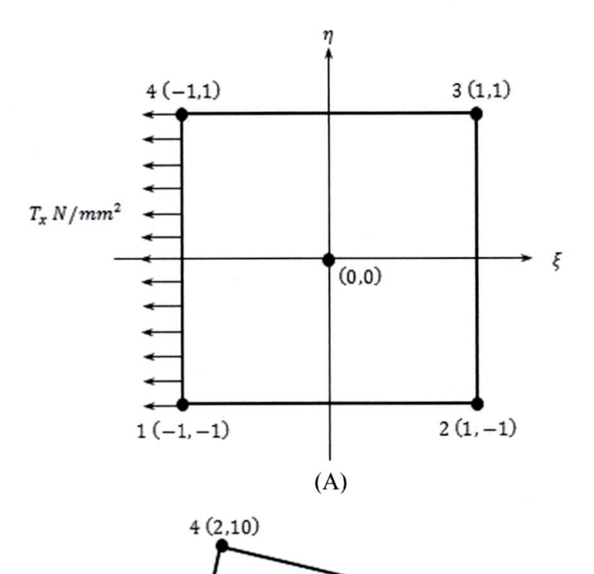

FIGURE P.6.2 (a) 4-Nodes quadrilateral element subjected to traction loading and (b) 4-nodes quadrilateral element.

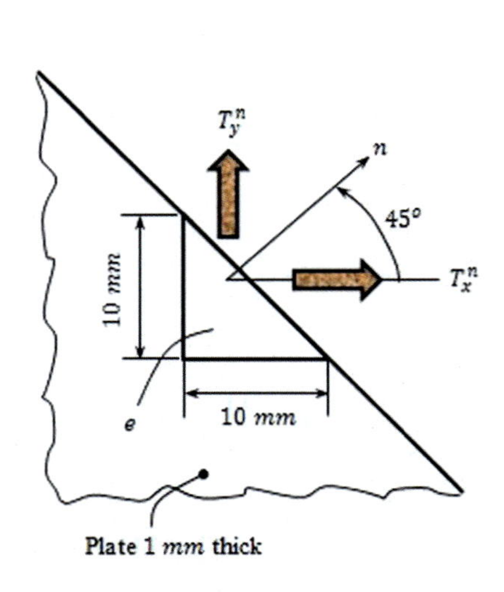

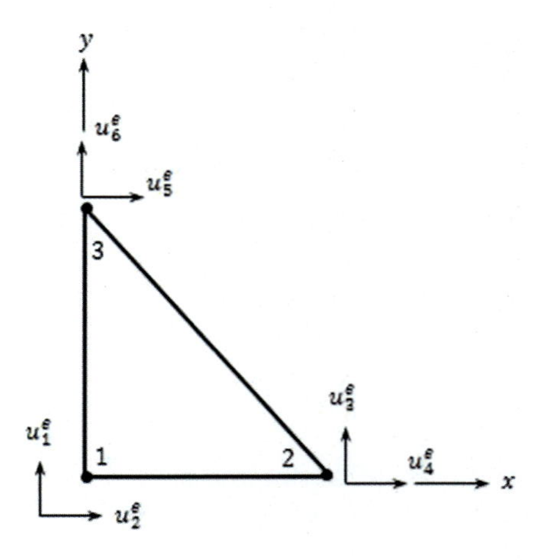

FIGURE P.6.3

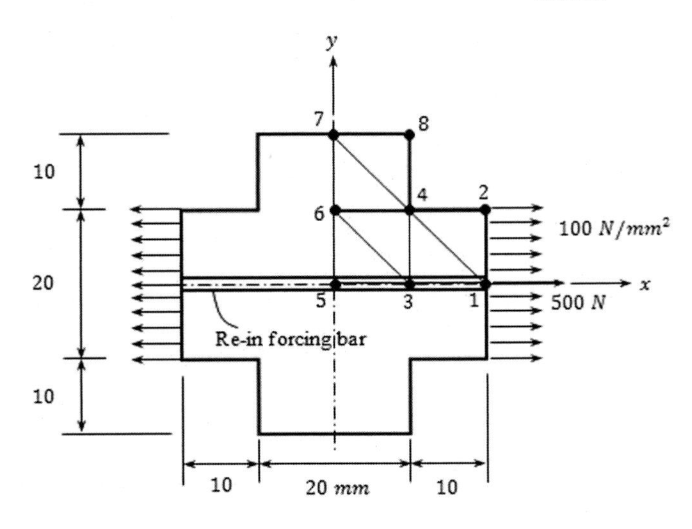

FIGURE P.6.4

P.6.5 In the finite element analysis of plane stress problem, an element displacement model is assumed in the form,

$$\{u(x,y)\} = [N(x,y)]\{u\}_0^e$$

Where, $\{u\}_0^e$ is a vector of element nodal displacement and $[N(x,y)]$ is a shape function matrix.

Show, without all the algebraic details, that the strain energy stored in a typical element is,

$$U^e = \frac{1}{2}\{u\}_0^{e^t}[k]^e\{u\}_0^e$$

If a particular situation is discretized to have N degrees of freedom, with G.C. $q_1, \ldots q_N$, state the essential consideration which allows the total strain energy stored to be expressed as,

$$U = \frac{1}{2}\{q\}^t[K]\{q\}$$

Show, again without all the algebraic details, how $[K]$ is generated from all the $[k]^e$.

In the system of Fig. P.6.5a, the applied load P is diffused into the plate support by way of the reinforcing bar A.B., cross-sectional area $A_b = 4.t^4$, which is continuously bonded to the plate, thickness $t = 0.1$, all along A.B. The plate is supported rigidly along CD while travel in the horizontal direction is completely prevented at corners E and F. If the coarse mesh of constant strain triangles of Fig. P.6.5b is used, explain why only a five degrees of freedom discretization need be analyzed and briefly discuss how representative of reality is the predicted stress field.

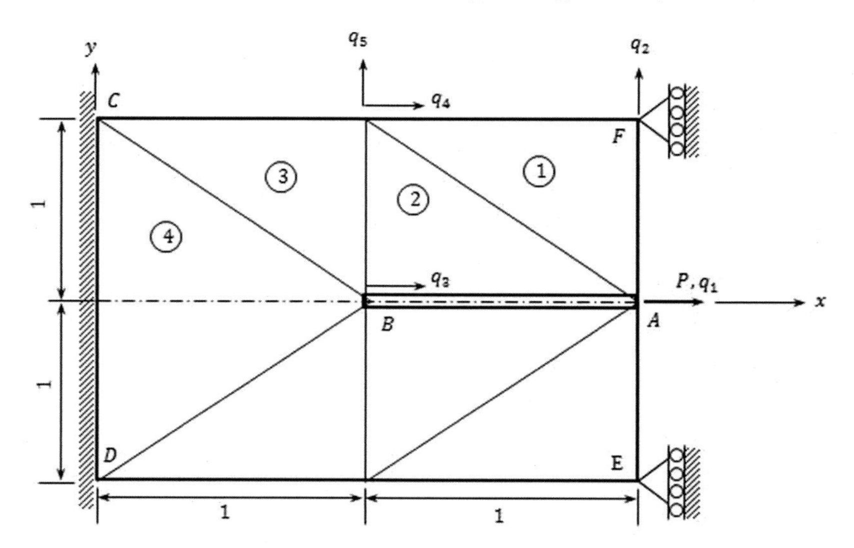

FIGURE P.6.5 (A and B).

By inspection or otherwise, generate the matrix equilibrium equation for the assemblage, given that for the elements of Fig. P.6.5b.

$$[k]^1 = [k]^2 = 0.55Et\begin{bmatrix} 1.35 & 0.65 & -1.00 & 0.35 & -0.35 & -0.30 \\ & 1.35 & -0.30 & -0.35 & -0.35 & -1.00 \\ & & 1.00 & 0 & 0 & 0.30 \\ & & & 0.35 & 0.35 & 0 \\ & \text{Sym.} & & & 0.35 & 0 \\ & & & & & 1.00 \end{bmatrix}$$

$$[k]^b = \frac{EA}{l}\begin{bmatrix} 1 & -1 \\ -1 & 1 \end{bmatrix}$$

P.6.6 A plate in-plane stress is constrained such that displacements describe the plate's deformation behavior in one direction.

1. Derive the stiffness matrix for a rectangular element of this plate. The deformation is described by displacements in x-direction $u = u(x,y)$. The displacements in the y-direction are zero $v = 0$.

   The plate is shown in Fig. P.6.6a and has a constant thickness and isotropic material.

2. A plate of constant thickness is compressed between two rigid boundaries, as shown in Fig. P.6.6b. No slipping takes place between the plate and the constraining boundaries. Using the finite element in (*i*) gives a solution for deformation and stresses in this plate.

   Plate thickness $t = 0.025$ m, elastic modulus $E = 200$ GN/m^2, Poisson's ratio $\nu = 0.3$.

P.6.7 In a plane strain problem, the structure is divided into some triangular elements. A typical element is shown in Fig. P.6.7,

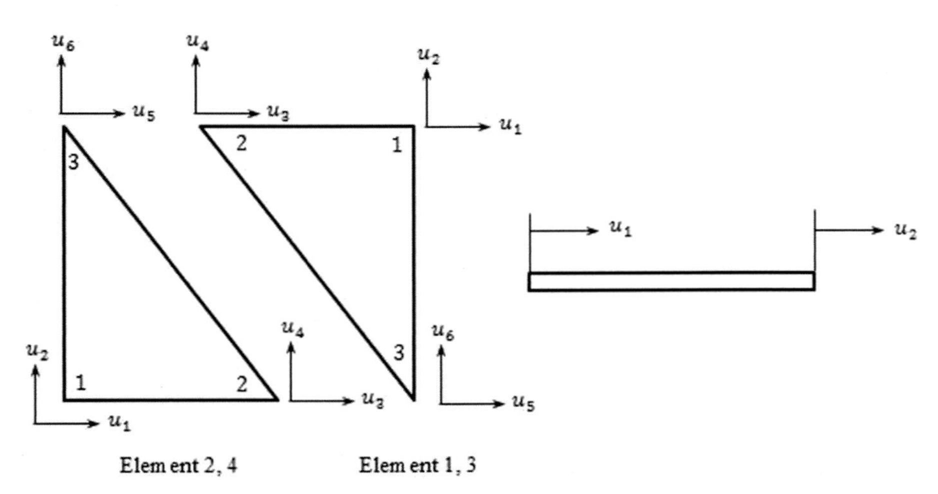

FIGURE P.6.6 A.

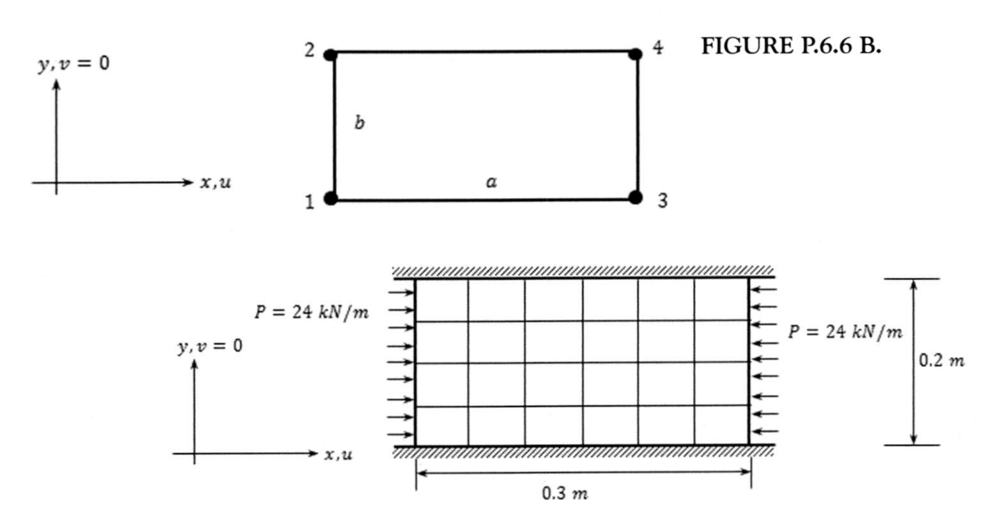

FIGURE P.6.6 B.

FIGURE P.6.7

1. Find out the stiffness matrix of the element. Take thickness of element as 2 mm, modulus of elasticity = 70 GPa, and Poisson's ratio $\nu = 0.3$.
2. Find out the load vector for the loads $p_x = 10$ MPa and $p_y = 15$ MPa.

P.6.8 For the element shown in Fig. P.6.8, how do you obtain the equivalent load vector if used in the plane stress case? Draw a flow chart showing the procedure of writing a computer subroutine for this purpose.

P.6.9 Consider the 3-node triangle element shown in Fig. P.6.9 with a linear distributed pressure from 1 MPa at node k to 4 MPa at node j.

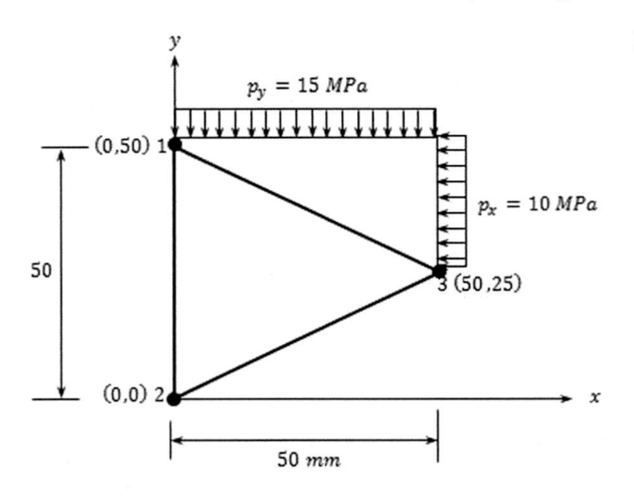

FIGURE P.6.8

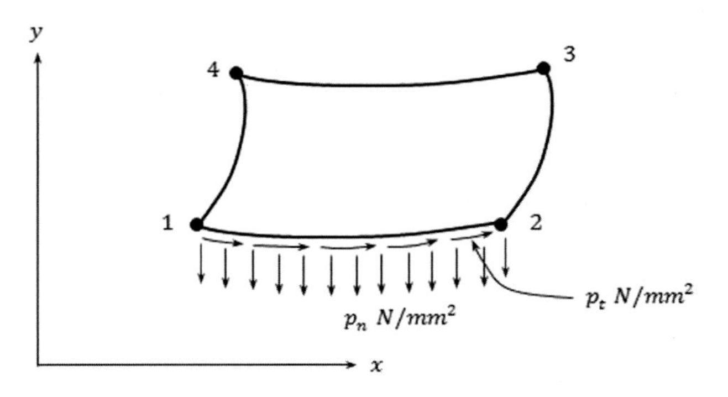

FIGURE P.6.9

TABLE P.6.9A

Node	x-Coordinate	y-Coordinate
i	0	1
j	6	0
k	3	4

Here E is Young's modulus, ν is the Poisson's ratio, and t is the thickness. The nodal coordinates are given as follows (unit: m), Table P.6.9a:

1. Given the following nodal displacements (unit: cm), Table P.6.9b:

   Compute the strain components ε_{xx}.

2. A linearly distributed pressure (i.e., the loading direction is perpendicular to the surface) is applied on the edge jk (Fig. P.6.9), with end pressure values of $P_k = 1$ MPa, $P_j = 4$ MPa, what are the equivalent nodal forces (in x and y directions)?

TABLE P.6.9B

Node	x-Coordinate	y-Coordinate
i	0.15	0.1
j	0.15	−0.1
k	0.15	0.2

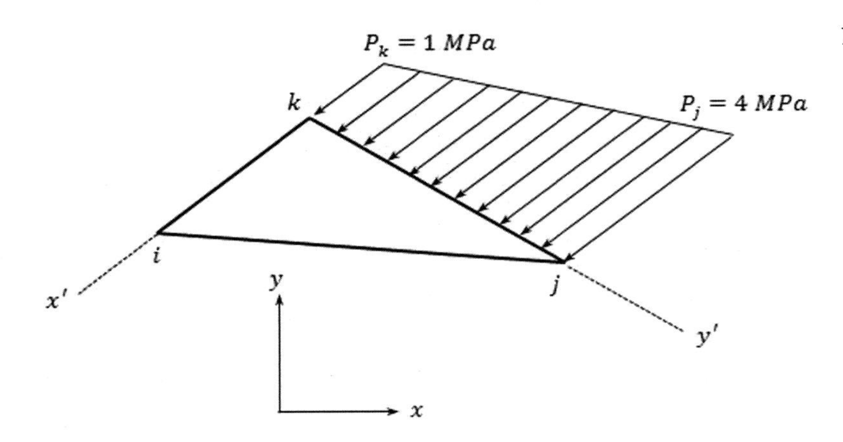

FIGURE P.6.10

Note that,

$$N_i = x'$$

$$N_j = y'$$

$$N_k = 1 - x' - y'$$

P.6.10 To solve the problem of Fig. P.6.10, by finite elements, explain why it is only necessary to consider the discretized quadrant. Explain why the four simplex triangle discretization only has seven degrees of freedom. Show, by inspection, that the system stiffness matrix equilibrium equation is,

$$\begin{bmatrix} (k_{11}^1 + k_{33}^2) & k_{13}^1 & k_{14}^1 & k_{31}^2 & (k_{15}^1 + k_{35}^2) & (k_{16}^1 + k_{36}^2) & 0 \\ k_{31}^1 & k_{33}^1 & k_{34}^1 & 0 & k_{35}^1 & k_{36}^1 & 0 \\ k_{41}^1 & k_{43}^1 & k_{44}^1 & 0 & k_{45}^1 & k_{46}^1 & 0 \\ k_{13}^2 & 0 & 0 & (k_{11}^2 + k_{11}^3 + k_{33}^4) & (k_{15}^2 + k_{13}^3) & (k_{16}^2 + k_{14}^3) & (k_{16}^3 + k_{36}^4) \\ (k_{51}^1 + k_{53}^2) & k_{53}^1 & k_{54}^1 & (k_{51}^2 + k_{31}^3) & (k_{55}^1 + k_{55}^2 + k_{33}^3) & (k_{56}^1 + k_{56}^2 + k_{34}^3) & k_{36}^3 \\ (k_{61}^1 + k_{63}^2) & k_{63}^1 & k_{64}^1 & (k_{61}^2 + k_{41}^3) & (k_{65}^1 + k_{65}^2 + k_{43}^3) & (k_{66}^1 + k_{66}^2 + k_{44}^3) & k_{46}^3 \\ 0 & 0 & 0 & (k_{61}^3 + k_{63}^4) & k_{63}^3 & k_{64}^3 & (k_{66}^3 + k_{66}^4) \end{bmatrix} \begin{Bmatrix} q_1 \\ q_2 \\ q_3 \\ q_4 \\ q_5 \\ q_6 \\ q_7 \end{Bmatrix} = \begin{Bmatrix} p/2 \\ 0 \\ 0 \\ 0 \\ 0 \\ 0 \\ 0 \end{Bmatrix}$$

Note, use the typical F.E. models of Fig. P.6.5b.

P.6.11 A plate in two-dimensional stresses is constrained such that the plate will have displacements in one direction only [$u = u(x,y)$ in $x-$direction and $v = 0$ in the $y-$direction]. The plate is made of isotropic elastic material and has a constant thickness.

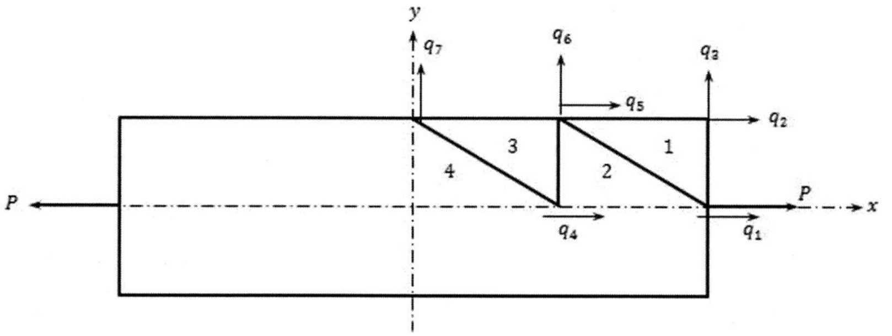

FIGURE P.6.11 A.

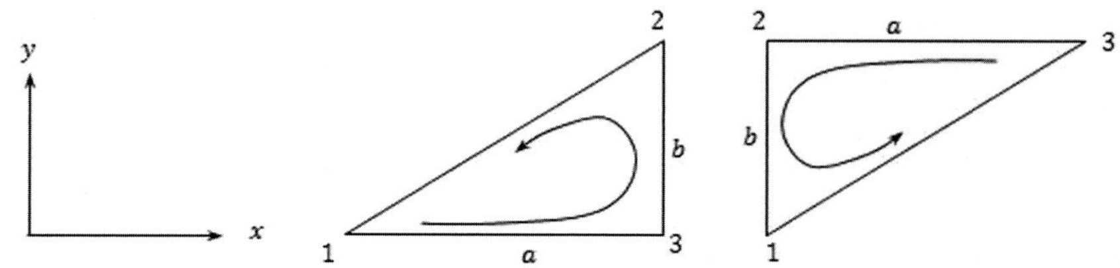

FIGURE P.6.11 B.

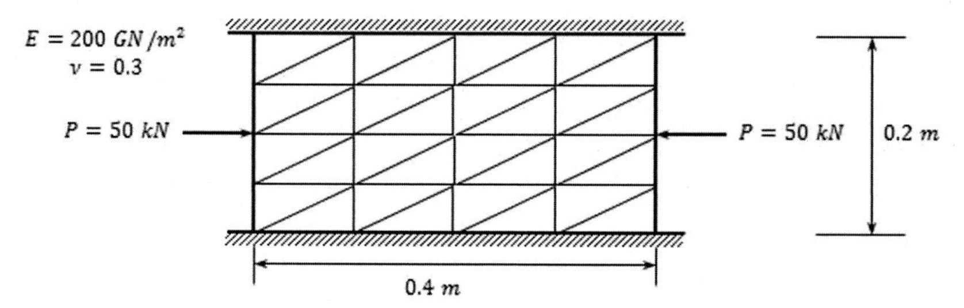

FIGURE P.6.12

1. Derive the stiffness matrices for the right-angled triangular elements (1) and (2) as shown in Fig. P.6.11a.
2. A rectangular plate $0.4 * 0.2$ m and thickness 0.02 m is constrained between two rigid boundaries (Fig. P.6.11b) such that no slip will take place. The plate is subjected to two equal and opposite forces $P = 50$ kN at the ends of the center line (as shown). For the constraining conditions described above, describe the method of solution.

P.6.12 Determine the reduced system stiffness matrix and the load vector for the finite element formulation for the problem shown in Fig. P.6.12 use G.C. and the node numbers indicated in the drawing.

$$E = 200 * 10^3 \ \ \text{N/mm}^2$$

$$t = 2 \ \ \text{mm}$$

Bar sectional area $A = 200$ mm^2

For bar element,

$$[k]^b = \frac{EA}{l}\begin{bmatrix} 1 & -1 \\ -1 & 1 \end{bmatrix}$$

For the 4-node quadrilateral element,

$$[k]^e = \begin{bmatrix} 487 & -156.1 & -546.8 & 270.5 & -95.93 & 12.71 & 155.72 & -127.1 \\ & 566.52 & 294.04 & -873 & 2.15 & 82.4 & -140 & 224.1 \\ & & 834.2 & -350.4 & 3.85 & -90.16 & -291.2 & 105.9 \\ & & & 1771 & -66.71 & -183.7 & 146.5 & -704.6 \\ & & & & 222.7 & -37.21 & -130.6 & 101.7 \\ & & & & & 191.2 & 114.7 & -69 \\ & & & & & & 266.62 & -101.1 \\ & & & & & & & 580 \end{bmatrix} \text{N/m}$$

P.6.13 a. For the element shown in Fig. P.6.13, determine the surface traction vectors if $t = 20$ mm.

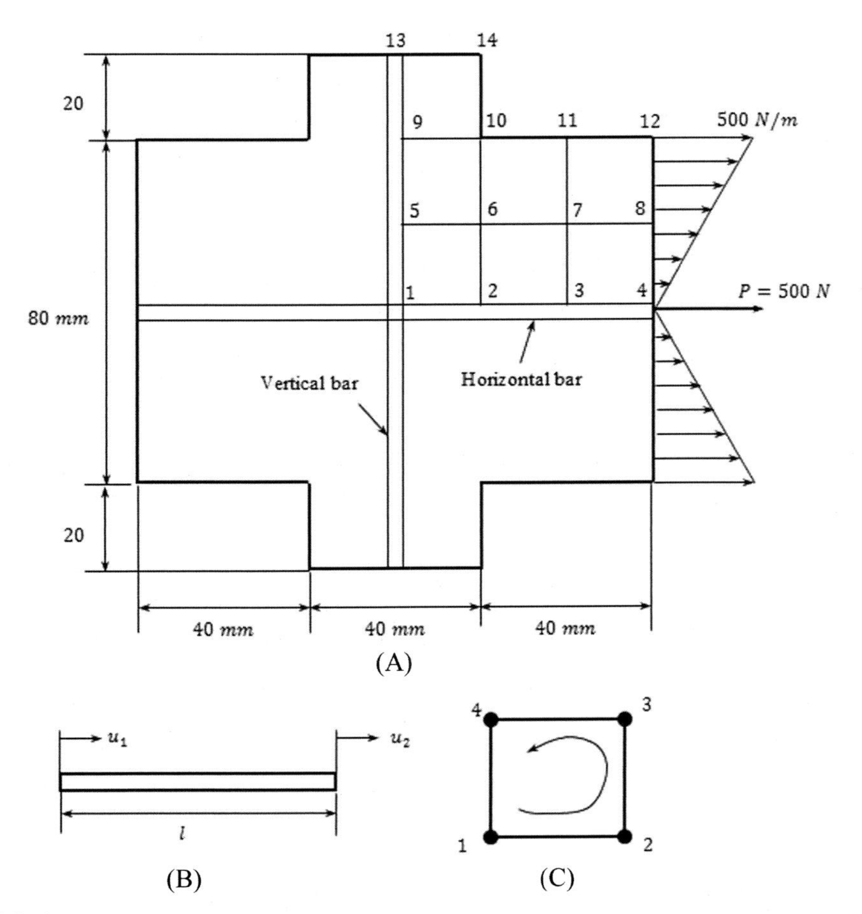

FIGURE P.6.13

b. Find the stresses if the element happens to represent the system if the nodes 1 and 2 are restrained, $E = 200$ GPa.

$$[k]^e = \frac{2E}{65.52}\begin{bmatrix} 50.4 & 9.1 & -25.2 & 5.6 & -25.2 & -14.7 \\ & 21.15 & 7.4 & -1.8 & -16.2 & -19.35 \\ & & 21.6 & -10.4 & 3.6 & 3.0 \\ & & & 21.6 & 4.8 & -19.8 \\ & \text{Sym.} & & & 21.6 & 11.7 \\ & & & & & 39.15 \end{bmatrix}$$

P.6.14 Explain why, if $[N(x,y)]$ is assumed linear, a 3-node triangle element is suitable and show that $[k]^e$ is of order (6×6).

Fig. P.6.14 shows a critical portion of a plane structure. The tie bar B.D. with a cross-sectional area $A_b = 4t^2$, Young modulus E and Poisson's ratio $\nu = 0.3$. The whole subassembly is firmly bonded to complete rigid support. For a first approximation solution, the F.E. mesh shown is established. Exploiting symmetry and explaining the process involved, apply the total potential energy principle to determine the system's stiffness equilibrium equations.

You may accept that for the right-angle triangle and bar elements of Fig. P.6.14, the stiffness matrices appropriate to the given element nodal orderings are:

$$[k]^1 = [k]^2 = \frac{Et}{2(1-\nu)}\begin{bmatrix} \frac{3-\nu}{2} & \frac{1+\nu}{2} & -1 & \frac{1-\nu}{2} & -\frac{1-\nu}{2} & -\nu \\ & \frac{3-\nu}{2} & -\nu & -\frac{1-\nu}{2} & -\frac{1-\nu}{2} & -1 \\ & & 1 & 0 & 0 & 0 \\ & & & \frac{1-\nu}{2} & \frac{1-\nu}{2} & 0 \\ & \text{sym.} & & & \frac{1-\nu}{2} & 0 \\ & & & & & 1 \end{bmatrix}$$

$$[k]^b = \frac{EA}{l}\begin{bmatrix} 1 & -1 \\ -1 & 1 \end{bmatrix}$$

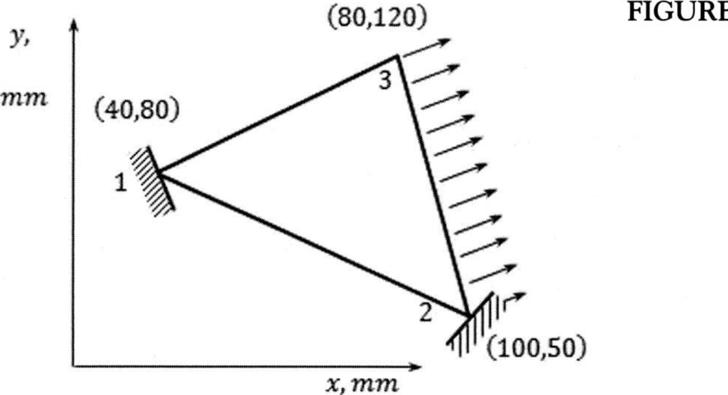

FIGURE P.6.14

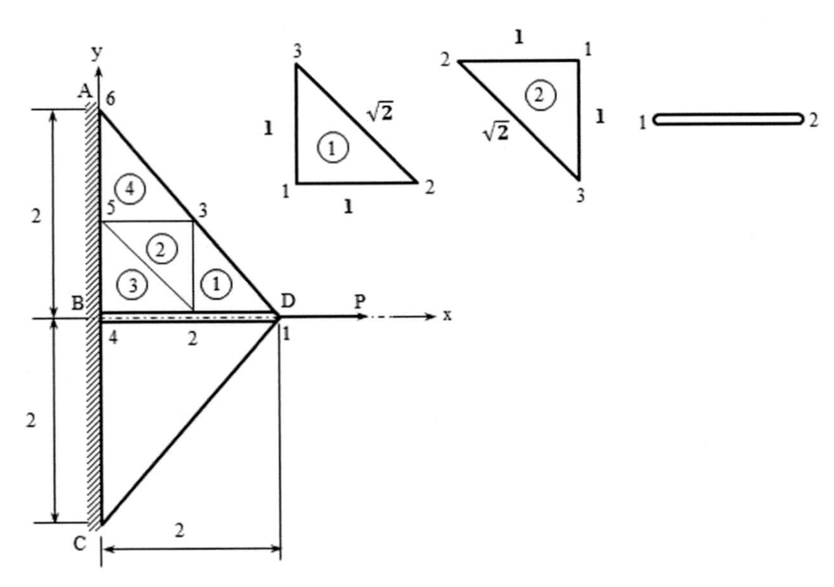

FIGURE P.6.15

Note: The element stiffness matrix for the elements 1,3,4 are identical. The local node numbering indicated in the typical elements should be considered in the assembly of the system stiffness matrix.

P.6.15 For the 8-node quadrilateral isoperimetric plane element shown in Fig. P.6.15, establish the following:

1. The displacement field in terms of the element's natural coordinates ($\xi - \eta$).
2. All the shape functions $N_i (i = 1{-}8)$.
3. The location and weighting factors of the 3×3 Gaussian integration points. (Indicate these points on the element.)
4. If the nodal displacements for this element are tabulated below, calculate the displacements u and v at Gauss point $(+a, +a)$, where $a > 0$.

P.6.16 a. For the isoperimetric formulation of the triangular element, the Cartesian coordinates and the shape functions are:

$$x = (x_1 - x_3)\zeta + (x_2 - x_3)\eta + x_3$$

$$y = (y_1 - y_3)\zeta + (y_2 - y_3)\eta + y_3$$

$$N_1 = \zeta,\ N_2 = \eta,\ \text{and}\ N_3 = 1 - \zeta - \eta$$

Develop the Jacobian if the field variables (u,v) and (x,y) are all functions of ξ and η.

b. Derive the strain–displacement matrix $[B]$ if required to be used as a plane stress element (Fig. P.6.16).

P.6.17 The problem shown in Fig. P.6.17a has been modeled by FEM, where:

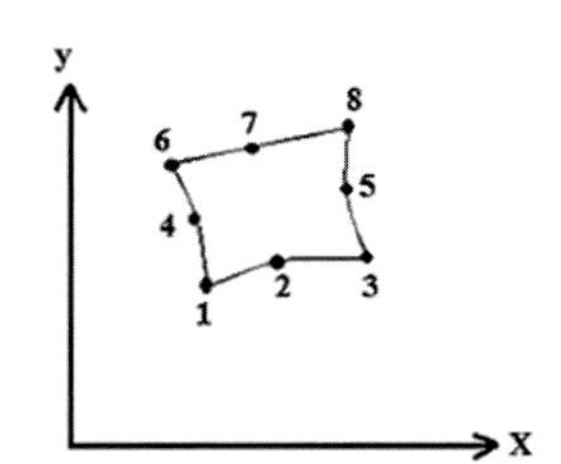

Node	u x 10^{-3}	v
1	1	0
2	3	4
3	-2	5
4	3	1
5	1	-3
6	5	-2
7	4	3
8	0	1

FIGURE P.6.16

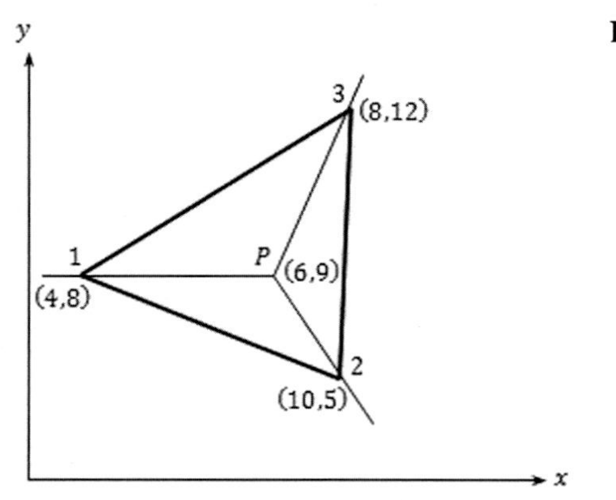

FIGURE P.6.17 (a).

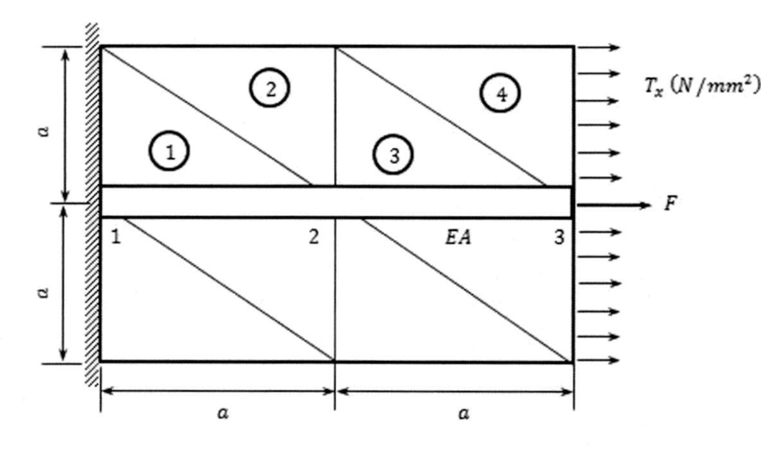

FIGURE P.6.17 (b) Elements 1 and 3.

$$a = 1\ \text{m},\ E = 2 \times 10^{11} N/m^2,\ v = 0.3$$

$$T_x^n = 100\ N/\text{mm}^2,\ F = 200\ \text{KN},\ \text{structure thickness} = 0.1\ m,\ \text{and}\ A = 25\ \text{mm}^2.$$

The elements of the stiffness matrix are given in Figs. P.6.17b and P.6.17c. Using the previous date, assemble and write the first four equations that relate the whole domain's nodal displacement with F_{x1}, F_{y1}, F_{x2}, and F_{y2} corresponding to G.C.

Note: use the G.C. stating from node 1 (Table P.6.17a).

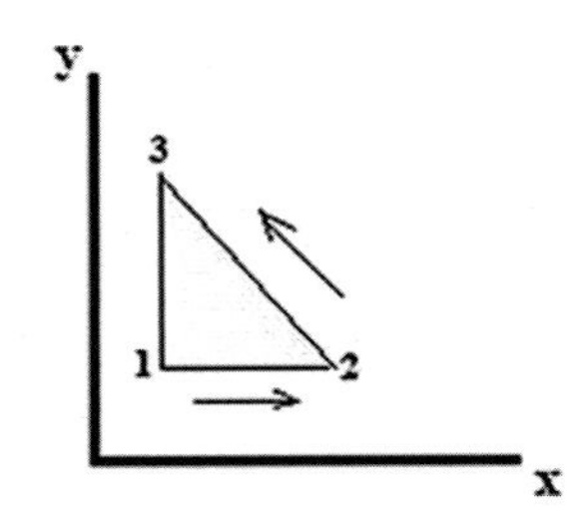

FIGURE P.6.17 (c) Elements 2 and 4.

TABLE P.6.17A

1.48	0.714	−1.09	−0.384	−0.384	−0.329	× 10^10
	1.48	−0.329	−0.384	−0.384	−1.09	
		1.09	0.000	0.000	0.329	
			0.384	0.384	0.000	
				0.384	0.000	
					1.098	

TABLE P.6.17B

0.384	0.000	−0.384	−0.384	0.000	0.384	× 10^10
	1.098	−0.329	−1.090	0.329	0.000	
		1.48	0.714	−1.09	−0.384	
			1.48	−0.329	−0.384	
				1.09	0.000	
					0.384	

This is valid for elements 1 and 3 with local numbers as shown.
Table P.6.17b
This is valid for elements 2 and 4 with local nodal numbers as shown.

$$[k]^e = \frac{EA}{l}\begin{bmatrix} 1 & -1 \\ -1 & 1 \end{bmatrix}$$

P.6.18 The problem shown in Fig. P.6.18 is solved employing FEM using three 4-noded isoperimetric elements. If the element stiffness matrix for the element I is given in Table P.6.18a, and the F.E. solution is as shown in Table P.6.18b, find

1. The reactions at nodes 1 and 4 and prove that the structure is in static equilibrium.
2. The strain and stress components at node one if the shape functions for the element is

$$N_1 = (1-\xi)(1-\eta), N_2 = \xi(1-\eta), N_3 = \xi\eta, N_4 = \eta(1-\xi)$$

FIGURE P.6.18

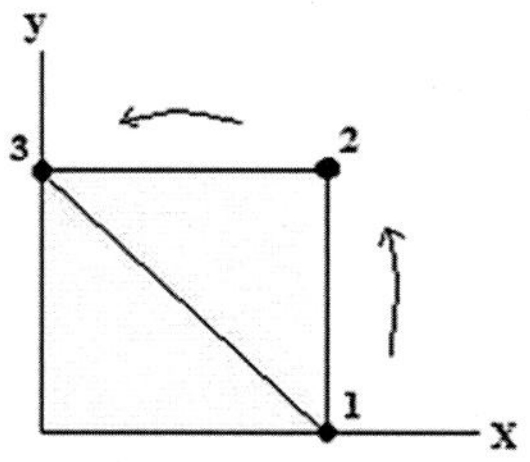

TABLE P.6.18A Element stiffness matrix.

9.8901	3.5714	−6.044	−0.2747	−4.9451	−3.5714	1.0989	0.2747	$k_e = 10^9X$
3.5714	9.8901	0.2747	1.0989	−3.5714	−4.9451	−0.2747	−6.044	
−6.044	0.2747	9.8901	−3.5714	1.0989	−0.2749	−4.9451	3.5714	
−0.2747	1.0989	−3.5714	8.8901	0.2747	−6.044	3.5714	−4.9451	
−4.9451	−3.5714	1.0989	0.2747	9.8901	3.5714	−6.044	−0.2747	
−3.5714	−4.9451	−0.2747	−6.044	3.5714	9.8901	0.2747	1.0989	
1.0989	−0.2747	−4.9451	3.5714	−6.044	0.2747	9.8901	−3.5714	
0.2747	−6.044	3.5714	−4.9451	−0.2747	1.0989	−3.5714	9.8901	

TABLE P.6.18B Finite element solution.

Node	1	2	3	4	5	6	7	8
X (m)	0.0	3.0	3.0	0.0	1.0	2.0	2.0	1.0
Y (m)	0.0	0.0	1.0	1.0	0.0	0.0	1.0	1.0
u (m)	0.000	−0.01835	0.01805	0.00	−0.01012	−0.01615	0.01620	0.01010
v (m)	0.000	−0.07929	−0.07787	0.00	−0.01273	−0.04148	−0.04172	−0.01269

With usual notations. The element stiffness matrix for the elements II and III are the same as element stiffness matrix for the element No. 1 considering the sequence local numbering of the elements. The typical element is shown in Fig. P.6.18a.

P.6.19 For the plane shown in Fig. P.6.19, node two settles 6 mm downward, determine the forces and stresses in each element due to the settlement. You may use the sequence of numbering and the element stiffness matrix of P.6.18 with the sequence of numbering as shown in Fig. P.6.18b and c.

$$E = 70 * 10^3 \quad \text{N/mm}^2, \nu = 0.28, A = 1500 \quad \text{mm}^2$$

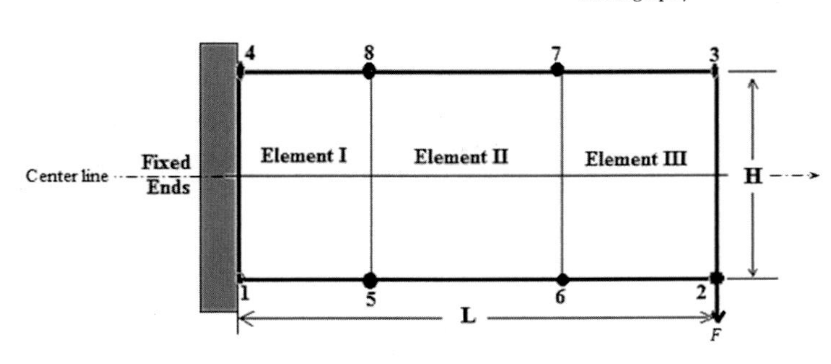

FIGURE P.6.18 A

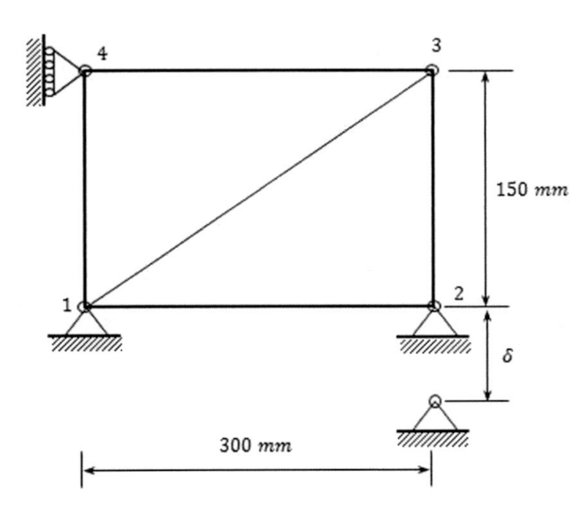

FIGURE P.6.19

Bibliography

[1] C.A. Brebbia, J.J. Conner, Fundamentals of Finite Element Techniques for Structural Engineers, Butterworths, London, 1975.

[2] R.W. Clough, The Finite Element Method in Plane Stress Analysis, Proceedings of the Second ASCE Conference on Electronic Computation, Pittsburgh (1960) 345–378.

[3] J.L. Batoz, K.J. Bathe, L.W. Ho, A study of three-node triangular plate bending elements, International Journal for Numerical Methods in Engineering 15 (1980) 1771–1812.

[4] L.A. Ying, Some special interpolation formulae for triangular and quadrilateral elements, International Journal for Numerical Methods in Engineering 18 (1982) 959–966.

[5] H.L. Langhaar, Energy Methods in Applied Mechanics, John Wiley and Sons, Inc, 1962.

[6] R.J. Melosh, Basis for derivation of matrices for the direct stiffness method, Journal of the American Institute of Aeronautics and Astronautics I (7) (1963) 1631–1637.

[7] B. Irons, S. Ahmad, Techniques of Finite Elements, Wiley, New York, 1980.

[8] D.R.J. Owen, E. Hinton, A simple Guide for Finite Elements, Pineridge Press, Swansea, 1980.

[9] J. Robinson, Understanding Finite Element Stress Analysis, Robinson and Associates, Wimborne, 1973.

[10] B.M. Irons, Engineering applications of numerical integrations in stiffness methods, Journal of the American Institute of Aeronautics and Astronautics 4 (11) (1966) 2035–2037.

[11] J.L. Tocher, B.J. Hertz, Higher order finite elements for plane stress, Journal of the Engineering Mechanics Division, Proceedings of the American Society of Civil Engineers 93 (EM4) (1967) 149–174.

[12] I. Ergatoudis, B.M. Irons, O.C. Zienkiewicz, Curved isoparametric quadrilateral elements for finite element analysis, International Journal of Solids and Structures 4 (1968) 31–42.
[13] J.S. Prezemieniecki, Theory of Matrix Structural Analysis, McGraw – Hill, 1968.
[14] Y. Yamada, Dynamic Analysis of Civil Engineering Structures, Recent Advances in Matrix Methods of Structural Analysis and Design, University of Alabama Press, Alabama, 1970, pp. 487–512.
[15] C.S. Desai, J.F. Abel, Introduction to the Finite Element Method, Van Nostrand, Reinhold, New York, 1972.
[16] N.A. Mahmood, M.J. Jweeg, A.A. Hussain, A study of localized buckling in thin tensioned plates with holes, Engineering and Technology 13 (13) (1994) 133–153. University of Technology.

CHAPTER

7

Torsion problem

The torsion problem of circular and noncircular sections is presented in this chapter. The derived equations are based upon the formulation of total potential energy due to twisting loading. The twisted section region is used such as triangular element with a single degree of freedom at each node. The line element of two nodes with one degree of freedom at each node was used for modeling the one dimensional applications. The standard procedure of the finite element formulation was explained in detail. Different examples are chosen for this purpose, line and plane finite elements are used in the chosen examples. Having obtained the torsion stress function from, the shear stress values are calculated at each discretized element's center. Different problems reflected that the torsion concept is included at the end of the chapter.

7.1 Total potential energy

The total potential energy of the torsion problem is given by:

$$V^* = \frac{1}{2G}\iint\left[\left(\frac{\partial\varnothing}{\partial x}\right)^2 + \left(\frac{\partial\varnothing}{\partial y}\right)^2 - 4G\alpha\varnothing\right]dx\,dy \tag{7.1}$$

Where $\varnothing(x,y)$ is the torsion stress function, α is the angle of twist, and G is the modulus of rigidity.

Note that $\varnothing = 0$ on the boundary of the domain, as shown in Fig. 7.1.

Assume,

Assuming $\phi(x,y)$ varies linearly with each element.

$$\varnothing^e(x,y) = \begin{bmatrix} 1 & x & y \end{bmatrix}\begin{Bmatrix} a_0 \\ a_1 \\ a_2 \end{Bmatrix} = [f(x,y)]\{a\} \tag{7.2}$$

Considering each node in the element shown in Fig. 7.1 in turn,

$$\begin{Bmatrix} \varnothing_1 \\ \varnothing_2 \\ \varnothing_3 \end{Bmatrix}^e = \begin{bmatrix} 1 & x_1 & y_1 \\ 1 & x_2 & y_2 \\ 1 & x_3 & y_3 \end{bmatrix}\begin{Bmatrix} a_0 \\ a_1 \\ a_2 \end{Bmatrix} \tag{7.3}$$

Energy Methods and Finite Element Techniques
DOI: https://doi.org/10.1016/B978-0-323-88666-6.00014-9

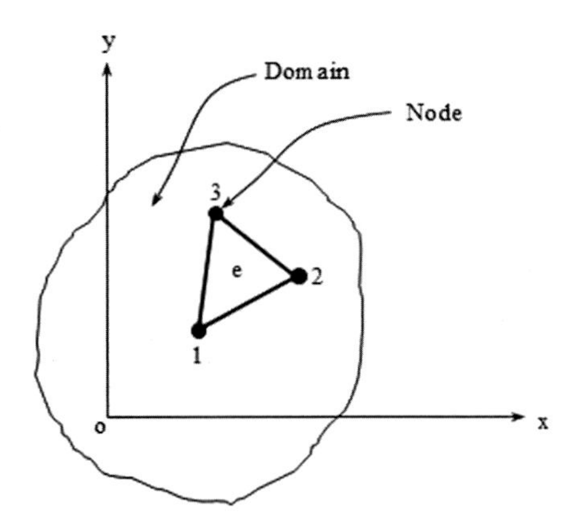

FIGURE 7.1 Irregular section under torsion and the typical discretized element.

Or in matrix form,

$$\{\varnothing\}_0^e = [A]\{a\} \tag{7.4}$$

So,

$$\begin{aligned}\varnothing^e = [f(x,y)][A]^{-1}\{\varnothing\}_0^e &= \begin{bmatrix} N_1(x,y) & N_2(x,y) & N_3(x,y) \end{bmatrix}\{\varnothing\}_0^e \\ &= [N(x,y)]\{\varnothing\}_0^e \end{aligned} \tag{7.5}$$

Where $N(x,y)$ is the shape function. The same shape function for the triangular element of the three nodes explained in detail in Chapter 5, Introduction to Finite Element Method: Bar and Beam Applications, can be used here as follows,

$$[N(x,y)]_1 = \frac{1}{2\Delta}(a_1 + b_1 x + c_1 y)[I] = \frac{1}{2\Delta}N_1(x,y)[I] \tag{7.6}$$

Where,

$$\begin{aligned} a_1 &= x_2 y_1 - x_1 y_2 \\ b_1 &= y_2 - y_3 \\ c_1 &= x_3 - x_2 \end{aligned} \tag{7.7}$$

The other constants $a_2, b_2, \ldots\ldots\ldots, c_3$ are given by the cyclic change of the subscripts, and,

$$\Delta = \frac{1}{2}\begin{vmatrix} 1 & 1 & 1 \\ x_1 & x_2 & x_3 \\ y_1 & y_2 & y_3 \end{vmatrix} = \text{area of the element.}$$

Now,

$$\frac{\partial \varnothing^e}{\partial x} = \left[\frac{\partial N}{\partial x}\right]\{\varnothing\}_0^e = [B_x]\{\varnothing\}_0^e$$

$$\frac{\partial \varnothing^e}{\partial y} = \left[\frac{\partial N}{\partial y}\right]\{\varnothing\}_0^e = \left[B_y\right]\{\varnothing\}_0^e \tag{7.8}$$

Where $[B_x]$, $\left[B_y\right]$ Are the derivatives of the shape function concerning z, and y respectively.

Then for a typical element, the total potential energy,

$$V^{*e} = \frac{1}{2G}\iint\left(\{\varnothing\}_0^{et}[B_x]^t[B_x]\{\varnothing\}_0^e + \{\varnothing\}_0^{et}\left[B_y\right]^t\left[B_y\right]\{\varnothing\}_0^e - 4G\alpha\{\varnothing\}_0^{et}[N]^t\right)dxdy \tag{7.9}$$

Or,

$$\frac{V^{*e}}{l} = \frac{1}{2}\{\varnothing\}_0^{et}\left(\frac{1}{G}\iint_e\left([B_x]^t[B_x] + \left[B_y\right]^t\left[B_y\right]\right)dxdy\right)\{\varnothing\}_0^e - 2\alpha\{\varnothing\}_0^{et}\left(\iint_e [N]^t dx\, dy\right)$$

$$\frac{V^{*e}}{l} = \frac{1}{2}\{\varnothing\}_0^{et}[h]^e\{\varnothing\}_0^e - \{\varnothing\}_0^{et}\{u\} \tag{7.10}$$

Where, $\varnothing^e$ varying linearly as supposed above, and,

$$h_{ij}^e = \frac{1}{4G\Delta}\left(b_i b_j + c_i c_j\right),\ \ u_i^e = -\frac{2\alpha\Delta}{3} \tag{7.11a}$$

Where h_{ij}^e are the elements of the torsional stiffness matrix.

For continuity,

$$\{\tilde{\varnothing}\} = \left\{\{\varnothing\}_0^e\right\} = [c]\{\varnothing\}$$

$$\{\tilde{u}\} = \left\{\{u\}_0^e\right\} = [c]\{\varnothing\}$$

Where $[c]$ is the connection matrix between global and local coordinates.

$$[\tilde{h}] = \begin{bmatrix} [h]^1 & 0 & 0 & \cdots \\ 0 & [h]^2 & 0 & \cdots \\ \vdots & \vdots & \vdots & \end{bmatrix} \tag{7.11}$$

The total complementary energy, Eq. (7.1),

$$\frac{V^*}{l} = \sum_e \frac{V^{*e}}{l} = \frac{1}{2}\{\tilde{\varnothing}\}^t[\tilde{h}]\{\tilde{\varnothing}\} - \{\tilde{\varnothing}\}^t[\tilde{u}] = \frac{1}{2}\{\varnothing\}^t\{c\}^t[\tilde{h}][c]\{\varnothing\} - \{\tilde{\varnothing}\}^t[c][\tilde{u}]$$

Or,

$$\frac{V^*}{l} = \frac{1}{2}\{\varnothing\}^t[H]\{\varnothing\} - \{\varnothing\}^t\{u\} \tag{7.12}$$

Where $[H] = \{c\}^t[\tilde{h}][c]$, and $\{u\} = [c][\tilde{u}]$ Where $[H]$ is the system torsion stiffness matrix.

For equilibrium, $\delta V = 0$, which gives,

$$[H]\{\phi\} = \{u\} \tag{7.13}$$

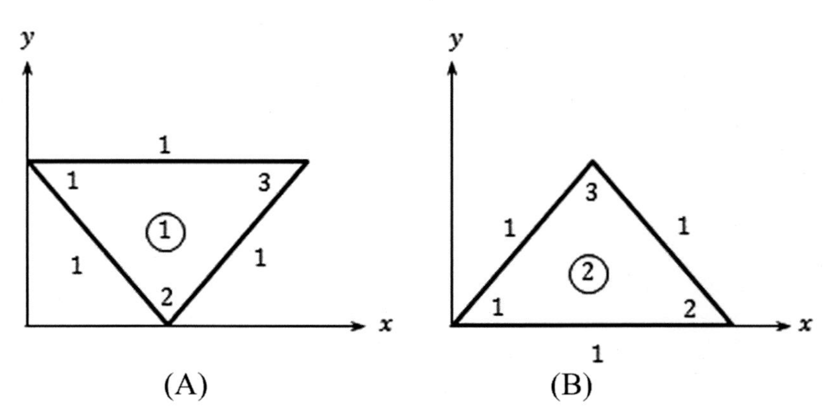

FIGURE 7.2 Triangular elements, (A) Element No. 1, (B) Element No. 2.

Example 7.1: Determine the coefficient matrices for the elements, Fig. 7.2A and B based on linear interpolation for the torsion stress function.

Solution,

1. For element No. 1, we have,

$$N_1 = \frac{1}{2\Delta}(a_1 + b_1x + c_1y),$$

where,

$$x_1 = 0,\ y_1 = \frac{\sqrt{3}}{2}$$

$$x_2 = \frac{1}{2},\ y_2 = 0$$

$$x_3 = 1,\ y_3 = \frac{\sqrt{3}}{2},$$

and,

$$\Delta = \frac{1}{2} * 1 * \frac{\sqrt{3}}{2} = \frac{\sqrt{3}}{4}$$

$$\frac{1}{4G\Delta} = \frac{1}{4G\frac{\sqrt{3}}{4}} = \frac{1}{\sqrt{3}G}$$

Using Eq. (7.7) gives,

$$a_1 = x_2y_1 - x_1y_2 = \frac{\sqrt{3}}{4}$$

$$b_1 = y_2 - y_3 = -\frac{\sqrt{3}}{2}$$

$$c_1 = x_3 - x_2 = -\frac{1}{2}$$

$$a_2 = x_3y_2 - x_2y_3 = -\frac{\sqrt{3}}{4}$$

$$b_2 = y_3 - y_1 = 0$$

$$c_2 = x_1 - x_3 = -1$$

$$a_3 = x_1y_3 - x_3y_1 = -\frac{\sqrt{3}}{2}$$

$$b_3 = y_1 - y_2 = \frac{\sqrt{3}}{2}$$

$$c_3 = x_2 - x_1 = \frac{1}{2}$$

From Eq. (7.11a), the torsional stiffness elements are given by,

$$h_{ij} = \frac{1}{4G\Delta}\left(b_ib_j + c_ic_j\right),$$

therefore,

$$h_{11} = \frac{1}{6\sqrt{3}},\ h_{12} = -\frac{1}{12\sqrt{3}},\ h_{13} = -\frac{1}{12\sqrt{3}},\ h_{22} = \frac{1}{6\sqrt{3}},\ h_{23} = -\frac{1}{12\sqrt{3}},\ h_{33} = \frac{1}{6\sqrt{3}}$$

$$[h]^e = \frac{1}{G\sqrt{3}}\begin{bmatrix} 1 & -\frac{1}{2} & -\frac{1}{2} \\ -\frac{1}{2} & 1 & -\frac{1}{2} \\ -\frac{1}{2} & -\frac{1}{2} & 1 \end{bmatrix} = \frac{1}{2\sqrt{3}G}\begin{bmatrix} 2 & -1 & -1 \\ -1 & 2 & -1 \\ -1 & -1 & 2 \end{bmatrix} * \frac{\sqrt{3}}{\sqrt{3}} = \frac{\sqrt{3}}{6G}\begin{bmatrix} 2 & -1 & -1 \\ -1 & 2 & -1 \\ -1 & -1 & 2 \end{bmatrix} \tag{7E1.1}$$

2. For element No. 2, we have,

$$x_1 = 0,\ y_1 = 0$$

$$x_2 = 1,\ y_2 = 0$$

$$x_3 = \frac{1}{2},\ y_3 = \frac{\sqrt{3}}{2}$$

$$\Delta = \frac{\sqrt{3}}{4}$$

Again using Eq. (7.7),

$$a_1 = x_2y_1 - x_1y_2 = 0$$

$$b_1 = y_2 - y_3 = -\frac{\sqrt{3}}{2}$$

$$c_1 = x_3 - x_2 = -\frac{1}{2}$$

$$a_2 = x_3y_2 - x_2y_3 = -\frac{\sqrt{3}}{2}$$

$$b_2 = y_3 - y_1 = \frac{\sqrt{3}}{2}$$

$$c_2 = x_1 - x_3 = -\frac{1}{2}$$

$$a_3 = x_1y_3 - x_3y_1 = 0$$

$$b_3 = y_1 - y_2 = 0$$

$$c_3 = x_2 - x_1 = 1$$

From Eq. (7.11a), the elements of torsional stiffness are, $h_{ij} = \frac{1}{4GD}\left(b_ib_j + c_ic_j\right)$, therefore,

$$h_{11} = \frac{1}{\sqrt{3}G},\ h_{12} = \frac{1}{\sqrt{3}G}\left(-\frac{1}{2}\right),\ h_{13} = \frac{1}{\sqrt{3}G}\left(-\frac{1}{2}\right),\ h_{22} = \frac{1}{\sqrt{3}G},\ h_{23} = \frac{1}{\sqrt{3}G}\left(-\frac{1}{2}\right),\ h_{33} = \frac{1}{\sqrt{3}G},$$

Therefore,

$$[h]^e = \frac{1}{G\sqrt{3}}\begin{bmatrix} 1 & -\frac{1}{2} & -\frac{1}{2} \\ -\frac{1}{2} & 1 & -\frac{1}{2} \\ -\frac{1}{2} & -\frac{1}{2} & 1 \end{bmatrix} * \frac{2\sqrt{3}}{2\sqrt{3}} = \frac{\sqrt{3}}{6G}\begin{bmatrix} 2 & -1 & -1 \\ -1 & 2 & -1 \\ -1 & -1 & 2 \end{bmatrix} \tag{7E1.2}$$

This mean, $[h]^{e1} = [h]^{e2}$

Example 7.2: Using the results of previous torsion stiffness matrix equations for the matrix compatibility equation for the torsion of the section shown in Fig. 7.3 is given by,

$$\begin{vmatrix} -6 & 1 & 1 & 0 \\ 1 & -6 & 1 & 1 \\ 1 & 1 & -6 & 1 \\ 0 & 1 & 1 & -6 \end{vmatrix}\begin{Bmatrix} \varnothing_4 \\ \varnothing_7 \\ \varnothing_8 \\ \varnothing_{11} \end{Bmatrix} = -3G\alpha\begin{Bmatrix} 1 \\ 1 \\ 1 \\ 1 \end{Bmatrix}$$

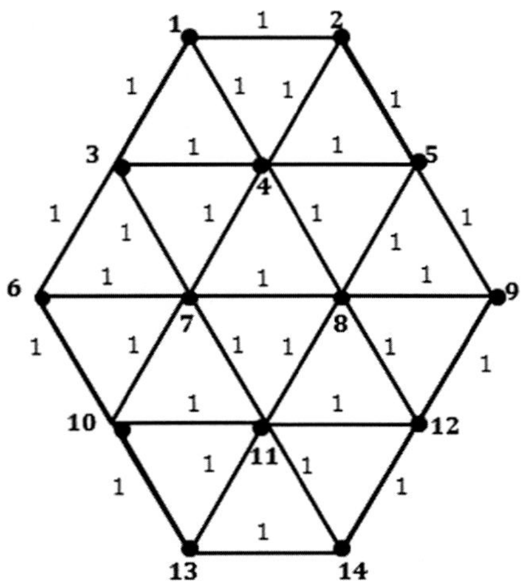

FIGURE 7.3 A shaft section.

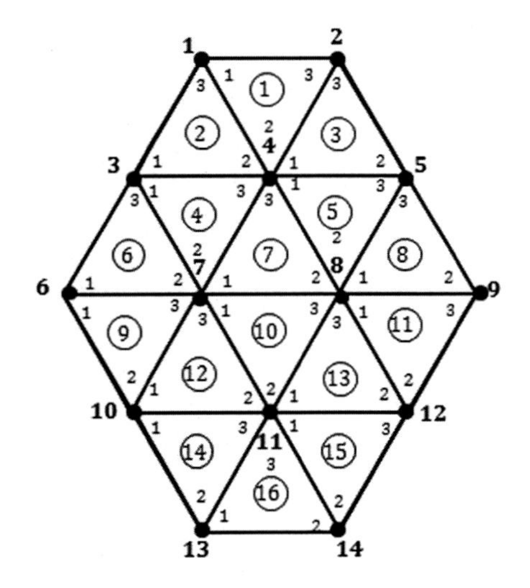

FIGURE 7.4 Finite element mesh.

The results of the element torsion stiffness matrix presented in Example 7.1 are given by,

$$[h]^e = \frac{\sqrt{3}}{6G}\begin{bmatrix} 2 & -1 & -1 \\ -1 & 2 & -1 \\ -1 & -1 & 2 \end{bmatrix}$$

And, from the obtained results of $[h]^e$ for the two types of elements 1 and 2, the above $[h]^e$ can be used for each element in the finite element mesh shown in Fig. 7.4 with the assigned local numbering for each element, from which we can construct the global stiffness matrix equations with a number of unconstrained degree of freedom equals to 14. The global equilibrium equations are in the following form:

$$[H]\{\varnothing\} = \{u\}$$

The global torsional stiffness elements are calculated according to the procedure explained in Chapter 2, Direct Methods. The stiffness at each node is obtained from the contribution of the stiffness for each element at that node, for example,

$H(2,2)$: the stiffness of the 1 and 3 have a contribution at this element with the stiffness $h(3,3)$ for element No. 1 and $h(3,3)$ for element No. 3 according to the assigned local numbering. Therefore by substitution the stiffness value for each element given in Eq. (7E1.2) gives,

$$H(3,3) = \frac{\sqrt{3}}{6G}(2+2) = \frac{\sqrt{3}}{6G}(4)$$

Another example, to obtain the value of $H(11,11)$, the stiffness of the elements 10, 12, 13, 14, 15, and 16 have a contribution in the stiffness of the node 11 according to the assigned local numbering on which the derivative of the torsion stiffness matrix is obtained. Therefore,

$H(11,11) = h(2,2)$ for element $10 + h(2,2)$ for element $12 + h(1,1)$ for element $13 + h(3,3)$ for element $14 + h(3,3)$ for element $15 + h(3,3)$ for element 16.

Substitution the corresponding values of the stiffness elements given in Eq. (E1.2) gives,

$$H(11,11) = \frac{\sqrt{3}}{6G}(2+2+2+2+2+2) = \frac{\sqrt{3}}{6G}(12)$$

Finally, to find $H(10,11)$, two elements contribute to stiffness, elements 12, and 14. From Fig. 7.4 and according to the local numbering, $H(10,1) = h(1,2)$ for element No. $12 + h(1,3)$ for element No. 14. Their corresponding values were obtained from Eq. (E1.2).

$$\mathrm{H}(10,1) = \frac{\sqrt{3}}{6G}(-1-1) = \frac{\sqrt{3}}{6G}(-2)$$

The other elements of the unconstrained $[H]$ matrix of size (14,14) are obtained and written as follows,

$$[H]=\frac{\sqrt{3}}{6G}\begin{bmatrix} 4 & -1 & -1 & -2 & 0 & 0 & 0 & 0 & 0 & 0 & 0 & 0 & 0 & 0 \\ -1 & 4 & 0 & -2 & -1 & 0 & 0 & 0 & 0 & 0 & 0 & 0 & 0 & 0 \\ -1 & 0 & 6 & -2 & 0 & -1 & -2 & 0 & 0 & 0 & 0 & 0 & 0 & 0 \\ -2 & -2 & -2 & 12 & -2 & 0 & -2 & -2 & 0 & 0 & 0 & 0 & 0 & 0 \\ 0 & -1 & 0 & -2 & 6 & 0 & 0 & -2 & -1 & 0 & 0 & 0 & 0 & 0 \\ 0 & 0 & -1 & 0 & 0 & 4 & -2 & 0 & 0 & -1 & 0 & 0 & 0 & 0 \\ 0 & 0 & -2 & -2 & 0 & -2 & 12 & -2 & 0 & -2 & -2 & 0 & 0 & 0 \\ 0 & 0 & 0 & -2 & -2 & 0 & -2 & 12 & -2 & 0 & -2 & -2 & 0 & 0 \\ 0 & 0 & 0 & 0 & -1 & 0 & 0 & -2 & 4 & 0 & 0 & -1 & 0 & 0 \\ 0 & 0 & 0 & 0 & 0 & -1 & -2 & 0 & 0 & 6 & -2 & 0 & -1 & 0 \\ 0 & 0 & 0 & 0 & 0 & 0 & -2 & -2 & 0 & -2 & 12 & -2 & -2 & -2 \\ 0 & 0 & 0 & 0 & 0 & 0 & 0 & -2 & -1 & 0 & -2 & 6 & 0 & -1 \\ 0 & 0 & 0 & 0 & 0 & 0 & 0 & 0 & 0 & -1 & -2 & 0 & 4 & -1 \\ 0 & 0 & 0 & 0 & 0 & 0 & 0 & 0 & 0 & 0 & -2 & -1 & -1 & 4 \end{bmatrix} \tag{E2.1a}$$

And,

$$\{\varnothing\}=\begin{Bmatrix} \varnothing_1{}^{\nearrow 0} \\ \varnothing_2{}^{\nearrow 0} \\ \varnothing_3{}^{\nearrow 0} \\ \varnothing_4 \\ \varnothing_5{}^{\nearrow 0} \\ \varnothing_6{}^{\nearrow 0} \\ \varnothing_7 \\ \varnothing_8 \\ \varnothing_9{}^{\nearrow 0} \\ \varnothing_{10}{}^{\nearrow 0} \\ \varnothing_{11} \\ \varnothing_{12}{}^{\nearrow 0} \\ \varnothing_{13}{}^{\nearrow 0} \\ \varnothing_{14}{}^{\nearrow 0} \end{Bmatrix}=\begin{Bmatrix} \varnothing_4 \\ \varnothing_7 \\ \varnothing_8 \\ \varnothing_{11} \end{Bmatrix}, \text{ and } u=\begin{Bmatrix} u_1=? \\ u_2=0 \\ u_3=0 \\ u_4=? \\ u_5=0 \\ u_6=0 \\ u_7=? \\ u_8=? \\ u_9=0 \\ u_{10}=0 \\ u_{11}=? \\ u_{12}=0 \\ u_{13}=0 \\ u_{14}=0 \end{Bmatrix} \tag{E2.1}$$

The resulted constrained $[H]$ matrix after deletion the corresponding rows and columns for the boundary nodes ($\varnothing=0$) shown in Eq. (E2.1a) becomes,

$$[H]=\frac{\sqrt{3}}{6G}\begin{bmatrix} 12 & -2 & -2 & 0 \\ -2 & 12 & -2 & -2 \\ -2 & -2 & 12 & -2 \\ 0 & -2 & -2 & 12 \end{bmatrix}$$

And, the constrained $\{u\}$ vector becomes,

$$\{u\} = -\frac{2\alpha\Delta}{3}\begin{Bmatrix} 6 \\ 6 \\ 6 \\ 6 \end{Bmatrix} = -\frac{2\alpha\Delta\sqrt{3}}{4*3}\begin{Bmatrix} 6 \\ 6 \\ 6 \\ 6 \end{Bmatrix}$$

Therefore the final constrained equilibrium equations become,

$$\frac{\sqrt{3}}{6G}\begin{bmatrix} 12 & -2 & -2 & 0 \\ -2 & 12 & -2 & -2 \\ -2 & -2 & 12 & -2 \\ 0 & -2 & -2 & 12 \end{bmatrix}\begin{Bmatrix} \varnothing_4 \\ \varnothing_7 \\ \varnothing_8 \\ \varnothing_{11} \end{Bmatrix} \Rightarrow -\frac{2\sqrt{3}}{6G}\begin{bmatrix} -6 & 1 & 1 & 0 \\ 1 & -6 & 1 & 1 \\ 1 & 1 & -6 & 1 \\ 0 & 1 & 1 & -6 \end{bmatrix}\begin{Bmatrix} \varnothing_4 \\ \varnothing_7 \\ \varnothing_8 \\ \varnothing_{11} \end{Bmatrix} = -\alpha\sqrt{3}\begin{Bmatrix} 1 \\ 1 \\ 1 \\ 1 \end{Bmatrix}$$

Or,

$$\Rightarrow \begin{bmatrix} -6 & 1 & 1 & 0 \\ 1 & -6 & 1 & 1 \\ 1 & 1 & -6 & 1 \\ 0 & 1 & 1 & -6 \end{bmatrix}\begin{Bmatrix} \varnothing_4 \\ \varnothing_7 \\ \varnothing_8 \\ \varnothing_{11} \end{Bmatrix} = 3G\alpha\begin{Bmatrix} 1 \\ 1 \\ 1 \\ 1 \end{Bmatrix}$$

Example 7.3: Consider the tapered shaft, shown in Fig. 7.5, use the finite element model (three elements) to calculate the angle of twist at the end "b" and the reaction torque T at the end "a". $T = 8000$ N.m, $D = 100$ mm, $L = 400$ mm, $G = 80$ GPa.

We have,

$$[h]^e = \frac{GJ}{L}\begin{bmatrix} 1 & -1 \\ -1 & 1 \end{bmatrix}$$

Where, J: Polar moment of inertia, G is the modulus of elasticity, and L is the element length.

$J = \frac{\pi D^4}{32}$, and the polar moment of inertia for each section is given by,

$$J_1 = \frac{\pi(1.125D)^4}{32} = 1.6\frac{\pi D^4}{32}$$

$$J_2 = \frac{\pi(1.375D)^4}{32} = 3.57\frac{\pi D^4}{32}$$

$$J_3 = \frac{\pi(1.625D)^4}{32} = 6.97\frac{\pi D^4}{32}$$

And since

$$h.\theta = T$$

Or,

$$[H]\{\theta\} = \{T\}$$

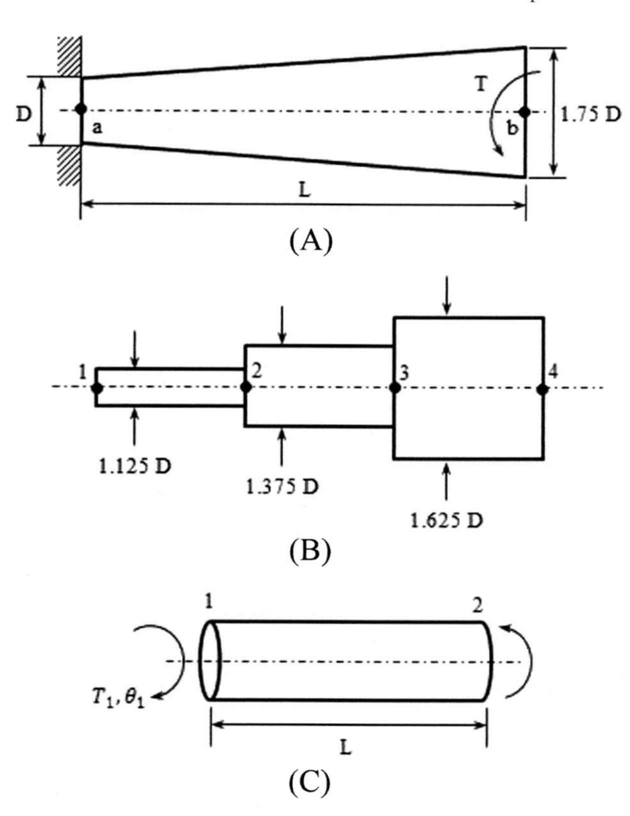

FIGURE 7.5 (A) Tapered shaft, (B) F.E. Model, (C) Shaft element under torsion.

Where the system torsion stiffness matrix, and by using the inspection method is given by,

$$[H_{sys.}] = \begin{bmatrix} h_{11}^1 & h_{12}^1 & 0 & 0 \\ h_{21}^1 & \left(h_{22}^1 + h_{11}^2\right) & h_{12}^2 & 0 \\ 0 & k_{21}^2 & \left(h_{22}^2 + h_{11}^3\right) & h_{12}^3 \\ 0 & 0 & h_{21}^3 & h_{22}^3 \end{bmatrix} = [H] \tag{E3.1}$$

And, as before,

$$\left[\tilde{h}\right] = \begin{bmatrix} [h]^1 & 0 & 0 \\ 0 & [h]^2 & 0 \\ 0 & 0 & [h]^3 \end{bmatrix}$$

Where,

$$[h]^1 = \frac{GJ_1}{l}\begin{bmatrix} 1 & -1 \\ -1 & 1 \end{bmatrix} = \frac{\pi G D^4}{32l}\begin{bmatrix} 1.6 & -1.6 \\ -1.6 & 1.6 \end{bmatrix}$$

$$[h]^2 = \frac{GJ_2}{l}\begin{bmatrix} 1 & -1 \\ -1 & 1 \end{bmatrix} = \frac{\pi GD^4}{32l}\begin{bmatrix} 3.57 & -3.57 \\ -3.57 & 3.57 \end{bmatrix}$$

$$[h]^3 = \frac{GJ_3}{l}\begin{bmatrix} 1 & -1 \\ -1 & 1 \end{bmatrix} = \frac{\pi GD^4}{32l}\begin{bmatrix} 6.97 & -6.97 \\ -6.97 & 6.97 \end{bmatrix}$$

Substitution the torsion stiffness from the elements 1, 2, and 3 in Eq. (E3.1) gives,

$$\left[\tilde{k}\right] = \frac{\pi GD^4}{32l}\begin{bmatrix} 1.6 & -1.6 & 0 & 0 & 0 & 0 \\ -1.6 & 1.6 & 0 & 0 & 0 & 0 \\ 0 & 0 & 3.57 & -3.57 & 0 & 0 \\ 0 & 0 & -3.57 & 3.57 & 0 & 0 \\ 0 & 0 & 0 & 0 & 6.97 & -6.97 \\ 0 & 0 & 0 & 0 & -6.97 & 6.97 \end{bmatrix}$$

$[H]$ Matrix can also be obtained by using the compatibility matrix as follows,

$$\begin{Bmatrix} \theta_1^1 \\ \theta_2^1 \\ \theta_1^2 \\ \theta_2^2 \\ \theta_1^3 \\ \theta_2^3 \end{Bmatrix} = \begin{bmatrix} 1 & 0 & 0 & 0 \\ 0 & 1 & 0 & 0 \\ 0 & 1 & 0 & 0 \\ 0 & 0 & 1 & 0 \\ 0 & 0 & 1 & 0 \\ 0 & 0 & 0 & 1 \end{bmatrix}\begin{Bmatrix} \theta_1 \\ \theta_2 \\ \theta_3 \\ \theta_4 \end{Bmatrix}$$

$$[c] = \begin{bmatrix} 1 & 0 & 0 & 0 \\ 0 & 1 & 0 & 0 \\ 0 & 1 & 0 & 0 \\ 0 & 0 & 1 & 0 \\ 0 & 0 & 1 & 0 \\ 0 & 0 & 0 & 1 \end{bmatrix}$$

$$[c]^t = \begin{bmatrix} 1 & 0 & 0 & 0 & 0 & 0 \\ 0 & 1 & 1 & 0 & 0 & 0 \\ 0 & 0 & 0 & 1 & 1 & 0 \\ 0 & 0 & 0 & 0 & 0 & 1 \end{bmatrix}$$

We have,

$[H] = [c]^t[\tilde{k}][c]$, therefore,

$$[H] = \frac{\pi GD^4}{32l}\begin{bmatrix} 1 & 0 & 0 & 0 & 0 & 0 \\ 0 & 1 & 1 & 0 & 0 & 0 \\ 0 & 0 & 0 & 1 & 1 & 0 \\ 0 & 0 & 0 & 0 & 0 & 1 \end{bmatrix}\begin{bmatrix} 1.6 & -1.6 & 0 & 0 & 0 & 0 \\ -1.6 & 1.6 & 0 & 0 & 0 & 0 \\ 0 & 0 & 3.57 & -3.57 & 0 & 0 \\ 0 & 0 & -3.57 & 3.57 & 0 & 0 \\ 0 & 0 & 0 & 0 & 6.97 & -6.97 \\ 0 & 0 & 0 & 0 & -6.97 & 6.97 \end{bmatrix}\begin{bmatrix} 1 & 0 & 0 & 0 \\ 0 & 1 & 0 & 0 \\ 0 & 1 & 0 & 0 \\ 0 & 0 & 1 & 0 \\ 0 & 0 & 1 & 0 \\ 0 & 0 & 0 & 1 \end{bmatrix}$$

$$[H] = \frac{\pi G D^4}{32l}\begin{bmatrix} 1.6 & -1.6 & 0 & 0 \\ -1.6 & 5.17 & -3.57 & 0 \\ 0 & -3.57 & 10.54 & -.697 \\ 0 & 0 & -6.97 & 6.97 \end{bmatrix} \tag{E3.2}$$

We have,

$$[H]\{\theta\} = \{T\}$$

And, by the substitution of the $[H]$ matrix and constructing the vectors $\{\theta\}$ and $\{T\}$ give,

$$\frac{\pi G D^4}{32l}\begin{bmatrix} 1.6 & -1.6 & 0 & 0 \\ -1.6 & 5.17 & -3.57 & 0 \\ 0 & -3.57 & 10.54 & -6.97 \\ 0 & 0 & -6.97 & 6.97 \end{bmatrix}\begin{Bmatrix} \theta_1 \\ \theta_2 \\ \theta_3 \\ \theta_4 \end{Bmatrix} = \begin{Bmatrix} 0 \\ 0 \\ 0 \\ T \end{Bmatrix}$$

By using boundary condition, $\theta_1 = 0$, reduce above equation as,

$$\frac{\pi G D^4}{32l}\begin{bmatrix} 5.17 & -3.57 & 0 \\ -3.57 & 10.54 & -6.97 \\ 0 & -6.97 & 6.97 \end{bmatrix}\begin{Bmatrix} \theta_2 \\ \theta_3 \\ \theta_4 \end{Bmatrix} = \begin{Bmatrix} 0 \\ 0 \\ T \end{Bmatrix}$$

Then,

$$5.17\theta_2 - 3.57\theta_3 = 0$$

$$-3.57\theta_2 + 10.54\theta_3 - 6.97\theta_4 = 0$$

$$\frac{\pi G D^4}{32l}(-6.97\theta_3 + 6.97\theta_4) = T \Rightarrow -6.97\theta_3 + 6.97\theta_4 = 0.00407$$

By solving the above equations,
$\theta_2 = 0.00254$ rad
$\theta_3 = 0.00368$ rad
$\theta_4 = 0.00426$ rad, the angle of twist at end b.
The torque at the end a is given by,

$$T_a = \frac{\pi G D^4}{32l}\begin{bmatrix} 1.6 & -1.6 & 0 & 0 \end{bmatrix}\begin{Bmatrix} 0 \\ 0.00254 \\ 0.00368 \\ 0.00426 \end{Bmatrix}$$

$$= 1.962 * 10^6(0 - 0.00404 + 0 + 0)$$

$$T_a \cong 7979.65 \text{ N.m}$$

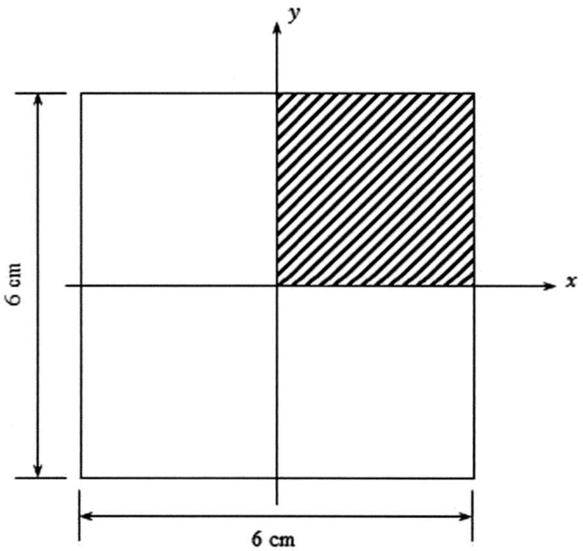

FIGURE 7.6 A rectangular shaft section.

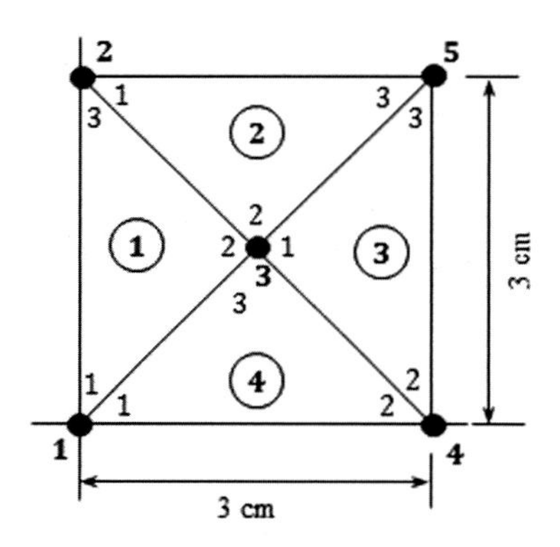

FIGURE 7.7 FE modeling of the quarter of the section shaft.

Example 7.4: Find the stresses developed in a $6*6$ cm square shaft, Fig. 7.6, when the angle of twist 3° in a length of 100 cm. Take $G = 83$ GPa.

Solution,

Idealize the region and symmetry, a quarter of the shaft is considered with the mesh shown in Fig. 7.7. The coordinate of the nodes is shown in Table 7.1. The relations between the global and local coordinates are shown in Table 7.2.

We have,

$$N_i = \frac{1}{2\Delta}\left(a_i + b_i x + c_i y\right)$$

TABLE 7.1 Coordinates of nodes for mesh.

Nodes	1	2	3	4	5
x-coordinate	0	0	1.5	3	3
y-coordinate	0	3	1.5	0	3

TABLE 7.2 Relations between the global and local coordinates.

Element		**1**	**2**	**3**	**4**
Local coordinate	1	1	2	3	1
	2	3	3	4	4
	3	2	5	5	3

$$h_{ij} = \frac{1}{4G\Delta}\left(b_i b_j + c_i c_j\right)$$

For element 1:

$$a_1 = x_2 y_1 - x_1 y_2 = 1.5 \times 0 - 0 \times 1.5 = 0$$

$$b_1 = y_2 - y_3 = 1.5 - 3 = -1.5 \text{ cm}$$

$$c_1 = x_3 - x_2 = 0 - 1.5 = -1.5 \text{ cm}$$

$$a_2 = x_3 y_2 - x_2 y_3 = 0 \times 1.5 - 1.5 \times 3 = -4.5 \text{ cm}$$

$$b_2 = y_3 - y_1 = 3 - 0 = 3 \text{ cm}$$

$$c_2 = x_1 - x_3 = 0 - 0 = 0$$

$$a_3 = x_1 y_3 - x_3 y_1 = 0 \times 3 - 0 \times 0 = 0$$

$$b_3 = y_1 - y_2 = 0 - 1.5 = -1.5 \text{ cm}$$

$$c_3 = x_2 - x_1 = 1.5 - 0 = 1.5 \text{ cm}, \text{ and, } \Delta = \frac{1}{2} \times 3 \times 1.5 = 2.25 \text{ cm}^2$$

Then,

$$[h]^1 = \frac{1}{4G\Delta}\begin{bmatrix} h_{11} & h_{12} & h_{13} \\ h_{21} & h_{22} & h_{23} \\ h_{31} & h_{32} & h_{33} \end{bmatrix} = \frac{1}{4G \times 2.25}\begin{bmatrix} 4.5 & -4.5 & 0 \\ -4.5 & 9 & -4.5 \\ 0 & -4.5 & 4.5 \end{bmatrix}$$

$$[h]^1 = \frac{1}{G}\begin{bmatrix} 0.5 & -0.5 & 0 \\ -0.5 & 1 & -0.5 \\ 0 & -0.5 & 0.5 \end{bmatrix}$$

For element 2:

$$a_1 = x_2y_1 - x_1y_2 = 1.5\times3 - 0\times1.5 = 4.5\ \text{cm}$$
$$b_1 = y_2 - y_3 = 1.5 - 3 = -1.5\ \text{cm}$$
$$c_1 = x_3 - x_2 = 3 - 1.5 = 1.5\ \text{cm}$$
$$a_2 = x_3y_2 - x_2y_3 = 3\times1.5 - 1.5\times3 = 0$$
$$b_2 = y_3 - y_1 = 3 - 3 = 0$$
$$c_2 = x_1 - x_3 = 0 - 3 = -3\ \text{cm}$$
$$a_3 = x_1y_3 - x_3y_1 = 0\times3 - 3\times3 = -9\ \text{cm}$$
$$b_3 = y_1 - y_2 = 3 - 1.5 = 1.5\ \text{cm}$$
$$c_3 = x_2 - x_1 = 1.5 - 0 = 1.5\ \text{cm}$$
$$\Delta = \frac{1}{2}\times3\times1.5 = 2.25\ \text{cm}^2$$

Then,

$$[h]^2 = \frac{1}{G}\begin{bmatrix} 0.5 & -0.5 & 0 \\ -0.5 & 1 & -0.5 \\ 0 & -0.5 & 0.5 \end{bmatrix}$$

For element 3:

$$a_1 = x_2y_1 - x_1y_2 = 3\times1.5 - 1.5\times0 = 4.5\ \text{cm}$$
$$b_1 = y_2 - y_3 = 0 - 3 = -3\ \text{cm}$$
$$c_1 = x_3 - x_2 = 3 - 3 = 0$$
$$a_2 = x_3y_2 - x_2y_3 = 3\times0 - 3\times3 = -9\ \text{cm}$$
$$b_2 = y_3 - y_1 = 3 - 1.5 = 1.5\ \text{cm}$$
$$c_2 = x_1 - x_3 = 1.5 - 3 = -1.5\ \text{cm}$$
$$a_3 = x_1y_3 - x_3y_1 = 1.5\times3 - 3\times1.5 = 0$$
$$b_3 = y_1 - y_2 = 1.5 - 0 = 1.5\ \text{cm}$$
$$c_3 = x_2 - x_1 = 3 - 1.5 = 1.5\ \text{cm}$$
$$\Delta = \frac{1}{2}\times3\times1.5 = 2.25\ \text{cm}^2$$

Then,

$$[h]^3 = \frac{1}{G}\begin{bmatrix} 1 & -0.5 & -0.5 \\ -0.5 & 0.5 & 0 \\ -0.5 & 0 & 0.5 \end{bmatrix}$$

For element 4:

$$a_1 = x_2y_1 - x_1y_2 = 3\times 0 - 0\times 0 = 0$$

$$b_1 = y_2 - y_3 = 0 - 1.5 = -1.5 \text{ cm}$$

$$c_1 = x_3 - x_2 = 1.5 - 3 = -1.5 \text{ cm}$$

$$a_2 = x_3y_2 - x_2y_3 = 1.5\times 0 - 3\times 1.5 = -4.5 \text{ cm}$$

$$b_2 = y_3 - y_1 = 1.5 - 0 = 1.5 \text{ cm}$$

$$c_2 = x_1 - x_3 = 0 - 1.5 = -1.5 \text{ cm}$$

$$a_3 = x_1y_3 - x_3y_1 = 0\times 1.5 - 1.5\times 0 = 0$$

$$b_3 = y_1 - y_2 = 0 - 0 = 0$$

$$c_3 = x_2 - x_1 = 3 - 0 = 3 \text{ cm}$$

$$\Delta = \frac{1}{2}\times 3\times 1.5 = 2.25 \text{ cm}^2$$

Then,

$$[h]^4 = \frac{1}{G}\begin{bmatrix} 0.5 & 0 & -0.5 \\ 0 & 0.5 & -0.5 \\ -0.5 & -0.5 & 1 \end{bmatrix}$$

$$[h]_{\text{sys.}} = \begin{bmatrix} {h_{11}}^1 + {h_{11}}^4 & {h_{13}}^1 & {h_{12}}^1 + {h_{13}}^4 & {h_{12}}^4 & 0 \\ {h_{31}}^1 & {h_{33}}^1 + {h_{11}}^2 & {h_{32}}^1 + {h_{12}}^2 & 0 & {h_{13}}^2 \\ {h_{21}}^1 + {h_{31}}^4 & {h_{23}}^1 + {h_{21}}^2 & \left({h_{22}}^1 + {h_{22}}^2 + {h_{11}}^3 + {h_{33}}^4\right) & {h_{12}}^3 + {h_{32}}^4 & {h_{23}}^2 + {h_{13}}^3 \\ {h_{21}}^4 & 0 & {h_{21}}^3 + {h_{23}}^4 & {h_{22}}^3 + {h_{22}}^4 & {h_{23}}^3 \\ 0 & {h_{31}}^2 & {h_{32}}^2 + {h_{31}}^3 & {h_{32}}^3 & {h_{33}}^2 + {h_{33}}^3 \end{bmatrix}$$

$$[h]_{\text{sys.}} = \frac{1}{G}\begin{bmatrix} 1 & 0 & -1 & 0 & 0 \\ 0 & 1 & -1 & 0 & 0 \\ -1 & -1 & 4 & -1 & -1 \\ 0 & 0 & -1 & 1 & 0 \\ 0 & 0 & -1 & 0 & 1 \end{bmatrix}$$

We have,

$$\{u\}^1 = -\frac{2 \propto \Delta}{3}\begin{Bmatrix} 1 \\ 1 \\ 1 \end{Bmatrix} = -\frac{2 \times 2 \times \pi \times 2.25}{3 \times 180}\begin{Bmatrix} 1 \\ 1 \\ 1 \end{Bmatrix}$$

$$\{u\}^1 = -\frac{\pi}{60}\begin{Bmatrix} 1 \\ 1 \\ 1 \end{Bmatrix}$$

$$\{u\} = -\frac{\pi}{60}\begin{Bmatrix} 2 \\ 2 \\ 4 \\ 2 \\ 2 \end{Bmatrix}$$

$$[H]\{\varnothing\} = \{u\}$$

Therefore,

$$\frac{1}{G}\begin{bmatrix} 1 & 0 & -1 & 0 & 0 \\ 0 & 1 & -1 & 0 & 0 \\ -1 & -1 & 4 & -1 & -1 \\ 0 & 0 & -1 & 1 & 0 \\ 0 & 0 & -1 & 0 & 1 \end{bmatrix}\begin{Bmatrix} \varnothing_1 \\ \varnothing_2 \\ \varnothing_3 \\ \varnothing_4 \\ \varnothing_5 \end{Bmatrix} = -\frac{\pi}{60}\begin{Bmatrix} 2 \\ 2 \\ 4 \\ 2 \\ 2 \end{Bmatrix}$$

$$\varnothing_1 = \varnothing_2 = \varnothing_4 = \varnothing_5 = 0$$

$$\frac{1}{G}\begin{bmatrix} 1 & -1 \\ -1 & 4 \end{bmatrix}\begin{Bmatrix} \varnothing_1 \\ \varnothing_3 \end{Bmatrix} = -\frac{\pi}{60}\begin{Bmatrix} 2 \\ 4 \end{Bmatrix}$$

Solving gives,

$$\varnothing_1 = -\frac{\pi G}{15}, \ \varnothing_3 = -\frac{\pi G}{30}$$

The elements shear stresses:
Element No. 1

$$\tau_{xz} = \frac{\partial \varnothing}{\partial y}$$

$$N_1 = \frac{1}{2\Delta}\left(a_1 + b_1 x + c_1 y\right) = \frac{1}{2 \times 2.25}\left(0 - 1.5x - 1.5y\right)$$

$$N_1 = -\frac{x}{3} - \frac{y}{3}$$

$$N_2 = \frac{1}{2\Delta}(a_2 + b_2x + c_2y) = \frac{1}{2 \times 2.25}(-4.5 + 3x + 0y)$$

$$N_2 = -1 + \frac{x}{1.5}$$

$$N_3 = \frac{1}{2\Delta}(a_3 + b_3x + c_3y) = \frac{1}{2 \times 2.25}(0 - 1.5x + 1.5y)$$

$$N_3 = -\frac{x}{3} + \frac{y}{3}$$

$$\varnothing(x,y) = N_1\varnothing_1 + N_2\varnothing_3 + N_3\varnothing_2 = \left(-\frac{x}{3} - \frac{y}{3}\right)\left(-\frac{\pi G}{15}\right) + \left(-1 + \frac{x}{1.5}\right)\left(-\frac{\pi G}{30}\right) + \left(-\frac{x}{3} + \frac{y}{3}\right)(0)$$

$$\varnothing(x,y) = \left(\frac{\pi G}{30}\right)\left(1 + \frac{y}{1.5}\right)$$

$$\tau_{zx} = \frac{\partial\varnothing}{\partial y} = \frac{\pi G}{45}$$

$$\tau_{zy} = \frac{\partial\varnothing}{\partial x} = 0$$

Element No. 2

$$N_1 = \frac{1}{2\Delta}(a_1 + b_1x + c_1y) = \frac{1}{2 \times 2.25}(4.5 - 1.5x + 1.5y)$$

$$N_1 = 1 - \frac{x}{3} + \frac{y}{3}$$

$$N_2 = \frac{1}{2\Delta}(a_2 + b_2x + c_2y) = \frac{1}{2 \times 2.25}(0 + 0x - 3y)$$

$$N_2 = -\frac{y}{1.5}$$

$$N_3 = \frac{1}{2\Delta}(a_3 + b_3x + c_3y) = \frac{1}{2 \times 2.25}(-9 + 1.5x + 1.5y)$$

$$N_3 = -2 + \frac{x}{3} + \frac{y}{3}$$

$$\varnothing(x,y) = N_1\varnothing_2 + N_2\varnothing_3 + N_3\varnothing_5 = \left(1 - \frac{x}{3} + \frac{y}{3}\right)(0) + \left(-\frac{y}{1.5}\right)\left(-\frac{\pi G}{30}\right) + \left(-2 + \frac{x}{3} + \frac{y}{3}\right)(0)$$

$$\varnothing(x,y) = \frac{\pi G y}{45}$$

$$\tau_{zx} = \frac{\partial \varnothing}{\partial y} = \frac{\pi G}{45}$$

$$\tau_{zy} = \frac{\partial \varnothing}{\partial x} = 0$$

Element No. 3

$$N_1 = \frac{1}{2\Delta}(a_1 + b_1 x + c_1 y) = \frac{1}{2 \times 2.25}(4.5 - 3x - 0y)$$

$$N_1 = 1 - \frac{x}{1.5}$$

$$N_2 = \frac{1}{2\Delta}(a_2 + b_2 x + c_2 y) = \frac{1}{2 \times 2.25}(-9 + 1.5x - 1.5y)$$

$$N_2 = -2 + \frac{x}{3} - \frac{y}{3}$$

$$N_3 = \frac{1}{2\Delta}(a_3 + b_3 x + c_3 y) = \frac{1}{2 \times 2.25}(0 + 1.5x + 1.5y)$$

$$N_3 = \frac{x}{3} + \frac{y}{3}$$

$$\varnothing(x,y) = N_1\varnothing_3 + N_2\varnothing_4 + N_3\varnothing_5 = \left(1 - \frac{x}{1.5}\right)\left(-\frac{\pi G}{30}\right) + \left(-2 + \frac{x}{3} - \frac{y}{3}\right)(0) + \left(\frac{x}{3} + \frac{y}{3}\right)(0)$$

$$\varnothing(x,y) = -\frac{\pi G}{30}\left(1 - \frac{x}{1.5}\right)$$

$$\tau_{zx} = \frac{\partial\varnothing}{\partial y} = 0$$

$$\tau_{zy} = \frac{\partial\varnothing}{\partial x} = \frac{\pi G}{45}$$

Element No. 4

$$N_1 = \frac{1}{2\Delta}(a_1 + b_1x + c_1y) = \frac{1}{2\times 2.25}(0 - 1.5x - 1.5y)$$

$$N_1 = -\frac{x}{3} - \frac{y}{3}$$

$$N_2 = \frac{1}{2\Delta}(a_2 + b_2x + c_2y) = \frac{1}{2\times 2.25}(-4.5 + 1.5x - 1.5y)$$

$$N_2 = -1 + \frac{x}{3} - \frac{y}{3}$$

$$N_3 = \frac{1}{2\Delta}(a_3 + b_3x + c_3y) = \frac{1}{2\times 2.25}(0 - 0x + 3y)$$

$$N_3 = \frac{y}{1.5}$$

$$\varnothing(x,y) = N_1\varnothing_1 + N_2\varnothing_4 + N_3\varnothing_3 = \left(-\frac{x}{3} - \frac{y}{3}\right)\left(-\frac{\pi G}{15}\right) + \left(-1 + \frac{x}{3} - \frac{y}{3}\right)(0) + \left(\frac{y}{1.5}\right)\left(-\frac{\pi G}{30}\right)$$

$$\varnothing(x,y) = \frac{\pi Gx}{45}$$

$$\tau_{zx} = \frac{\partial\varnothing}{\partial y} = 0$$

$$\tau_{zy} = \frac{\partial \varnothing}{\partial x} = \frac{\pi G}{45}$$

Computing Twisting moment:

$$M = 2\iint \varnothing \, dx \, dy$$

$$= \sum_{e=1}^{4} \frac{2\Delta}{3}(\varnothing_1 + \varnothing_2 + \varnothing_3) = \sum_{e=1}^{4} \frac{2 \times 2.25}{3}\left(-\frac{\pi G}{15} + 0 - \frac{\pi G}{30}\right)$$

$$M = -0.15\pi G \text{ N.cm}$$

The Exact value of the square shaft is given by (strength of materials):

$$M_t = 0.1406 \times G \times \propto (2a)^4 \quad (a = 6 \text{ cm})$$

$$= 0.1406 \times G \times \frac{2\pi}{180}(2 \times 6)^4$$

$$M_t = 0.12654\pi G \text{ N.cm}$$

The discrepancy per cent is:

$$\text{Error\%} = \frac{M - M_t}{M} \times 100\% = \frac{0.15\pi G - 0.12654\pi G}{0.15\pi G} \times 100\% = 15.64\%$$

7.2 Iso-parametric formulation of torsion problem: triangular element

For the torsion of the prismatic bar of arbitrary cross-section shape under twisting loading M_t which induces shearing stresses τ_{xz}andτ_{yz} and the angle of twist per unit length θ. We have,

$\frac{\partial^2 \varnothing}{\partial x^2} + \frac{\partial^2 \varnothing}{\partial y^2} + z = 0$ inside the domain

$\varnothing = 0$ on the boundary

Shearing stresses are given by,

$$\tau_{xz} = G\theta \frac{\partial \varnothing}{\partial y}, \quad \tau_{yz} = -G\theta \frac{\partial \varnothing}{\partial x}$$

And,

$$M_t = 2G\theta \iint_A \varnothing \, dA$$

Moreover, G is the modulus of rigidity. The stress function within a triangular element, Fig. 7.8, is assumed by,

$$\{\varnothing\} = [N(\xi, \eta)]\{\phi\}^e$$

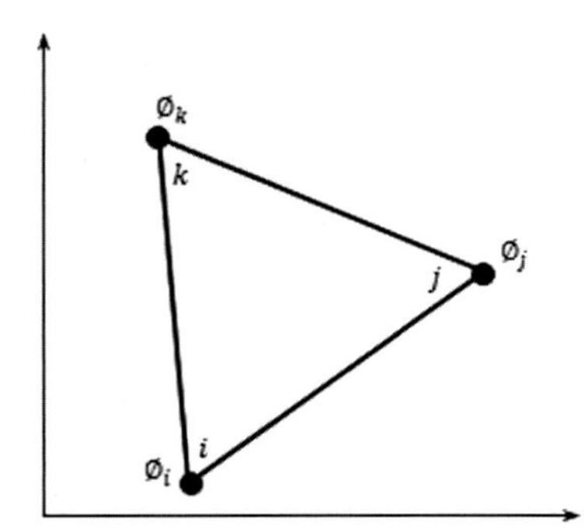

FIGURE 7.8 A typical three nodes iso-parametric element.

Where,

$[N(\xi,\eta)]^t = \begin{bmatrix} \xi & \eta & (1-\xi-\eta) \end{bmatrix} = \begin{bmatrix} N_1 & N_2 & N_3 \end{bmatrix}$. This vector contains the nodal shape functions. The derivatives will of the iso-parametric shape functions will be presented in Chapter 12, Shape Functions Determinations and Numerical Integration.

And the nodal value of torsion function $\{\phi\}^e = \begin{Bmatrix} \phi_1 \\ \phi_2 \\ \phi_3 \end{Bmatrix}$
For iso-parametric formulation

$$x(\xi,\eta) = x_1N_1 + x_2N_2 + x_3N_3$$

$$y(\xi,\eta) = y_1N_1 + y_2N_2 + y_3N_3$$

We have,

$$\begin{Bmatrix} \frac{\partial \varnothing}{\partial \xi} \\ \frac{\partial \varnothing}{\partial \eta} \end{Bmatrix} = \begin{bmatrix} \frac{\partial x}{\partial \xi} & \frac{\partial y}{\partial \xi} \\ \frac{\partial x}{\partial \eta} & \frac{\partial y}{\partial \eta} \end{bmatrix} \begin{Bmatrix} \frac{\partial \varnothing}{\partial x} \\ \frac{\partial \varnothing}{\partial y} \end{Bmatrix}$$

$$\begin{Bmatrix} \frac{\partial \varnothing}{\partial \xi} \\ \frac{\partial \varnothing}{\partial \eta} \end{Bmatrix} = [J] \begin{Bmatrix} \frac{\partial \varnothing}{\partial x} \\ \frac{\partial \varnothing}{\partial y} \end{Bmatrix}$$

Where,

$$[J] = \begin{bmatrix} x_{13} & y_{13} \\ x_{23} & y_{23} \end{bmatrix}$$

With,

$$x_{ij} = x_i - x_j$$

$$y_{ij} = y_i - y_j$$

And,

$$|J| = 2A_e$$

Where, A_e is the element area.

We have,

$$\begin{Bmatrix} \frac{\partial \varnothing}{\partial x} \\ \frac{\partial \varnothing}{\partial y} \end{Bmatrix} = [B]\{\phi\}^e = \begin{bmatrix} -\tau_{yz} & \tau_{xz} \end{bmatrix}^t = G\theta[B]\{\phi\}^e$$

Where,

$$[B] = \frac{1}{|J|}\begin{bmatrix} y_{23} & y_{31} & y_{12} \\ x_{32} & x_{13} & x_{21} \end{bmatrix} \tag{7.14}$$

The function approach which should be minimized,

$$V = G\theta^2 \iint_A \left\{ \frac{1}{2}\left(\frac{\partial \varnothing}{\partial x}\right)^2 + \left(\frac{\partial \varnothing}{\partial y}\right)^2 - 2\varnothing \right\} dA$$

Carrying out the substitution of the torsion stress function $\varnothing(x, y)$ and carrying out the integration using the iso-parametric substitution, the equilibrium equations, as stated previously will be,

$$[h]^e\{\varnothing\}^e = \{F\}^e$$

Where,

$$[h]^e = A_e[B]^t[B] \tag{7.15}$$

And,

$$\{F\}^e = \frac{2A_e}{3}\begin{Bmatrix} 1 \\ 1 \\ 1 \end{Bmatrix}$$

Carrying out the multiplication of Eq. (7.15), we will obtain the torsion element stiffness matrix as,

$$\left[h_{ij}\right]^e = \begin{bmatrix} h_{11} & h_{12} & h_{13} \\ h_{21} & h_{22} & h_{23} \\ h_{31} & h_{32} & h_{33} \end{bmatrix}$$

$$h_{ij}^e = \frac{1}{4A_e}\left(b_i b_j + a_i b_j\right)$$

Where,

$$b_i = y_j - y_k, a_i = x_j - x_k$$

$$b_j = y_k - y_i, a_j = x_k - x_i$$

$$b_k = y_i - y_j, a_k = x_i - x_j$$

$$[h]^e = \begin{bmatrix} (a_i a_i + b_i b_i) & (a_i a_j + b_i b_j) & (a_i a_k + b_i b_k) \\ & (a_j a_j + b_j b_j) & (a_j a_k + b_j b_k) \\ & & (a_k a_k + b_k b_k) \end{bmatrix} \quad (7.16)$$

Stresses

$$\{\sigma\} = \begin{Bmatrix} \tau_{xz} \\ \tau_{yz} \end{Bmatrix} = G\theta \begin{Bmatrix} \dfrac{\partial \varnothing}{\partial y} \\ -\dfrac{\partial \varnothing}{\partial x} \end{Bmatrix} = G\theta \begin{Bmatrix} \dfrac{\partial}{\partial y} \\ -\dfrac{\partial}{\partial x} \end{Bmatrix} \varnothing$$

We have,

$$\varnothing = [N]\{\varnothing\}^e$$

Or,

$$\{\sigma\} = G\theta \begin{Bmatrix} \dfrac{\partial}{\partial y} \\ -\dfrac{\partial}{\partial x} \end{Bmatrix} [N]\{\varnothing\}^e$$

$$\frac{\partial \varnothing}{\partial x} = \begin{bmatrix} \frac{\partial N_i}{\partial x} & \frac{\partial N_j}{\partial x} & \frac{\partial N_k}{\partial x} \end{bmatrix}^e = \frac{1}{2A_e}\begin{bmatrix} b_i & b_j & b_k \end{bmatrix}^e$$

$$\frac{\partial \varnothing}{\partial y} = \frac{1}{2A_e}\begin{bmatrix} a_i & a_j & a_k \end{bmatrix}^e$$

Or,

$$\{\sigma\}^e = \begin{Bmatrix} \frac{\partial \varnothing}{\partial x} \\ \frac{\partial \varnothing}{\partial y} \end{Bmatrix}^e = \begin{bmatrix} b_i & b_j & b_k \\ a_i & a_j & a_k \end{bmatrix}^e \begin{Bmatrix} \phi_i \\ \phi_j \\ \phi_k \end{Bmatrix}^e = \frac{G\theta}{2A_e}$$

Problems

P.7.1 A torsional problem in the circular shaft can be solved by the finite element method. A two-node element shown in Fig. P.7.1A can be derived by minimizing the total energy, $X = U - W$, provided that,

$U = \frac{1}{2}\int_0^l JG\left(\frac{d\theta}{dx}\right)^2 dx$ strain energy

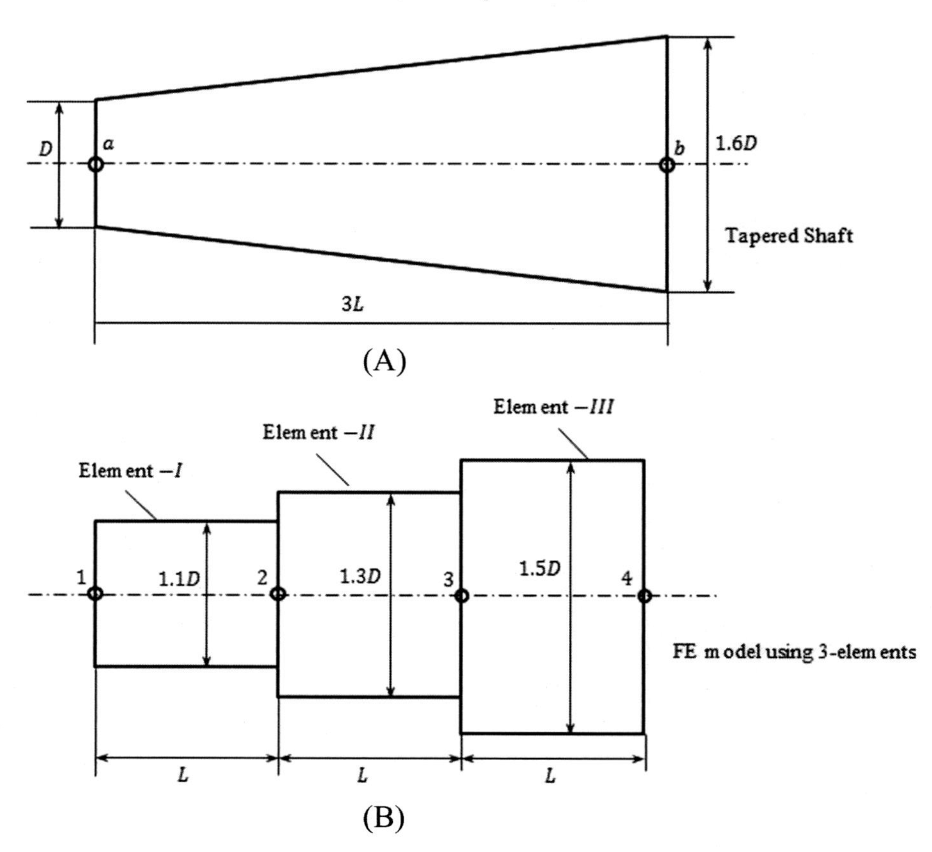

FIGURE P.7.1

$$W = T_1\theta_1 + T_2\theta_2 \text{ work done}$$

Where, T = Twisting moment, J = polar moment of area = $\frac{\pi d^4}{32}$, l = element length, G = Modulus of Rigidity of the material, d = element diameter, and θ = angle of twist.

1. Use the previous data to derive the following matrix equation,

$$\frac{GJ}{l}\begin{bmatrix} 1 & -1 \\ -1 & 1 \end{bmatrix}\begin{bmatrix} \theta_1 \\ \theta_2 \end{bmatrix} = \begin{bmatrix} T_1 \\ T_2 \end{bmatrix}$$

2. Consider the paper shaft shown in Fig. P.7.1B and C which shows the tapered shaft and the finite element mesh, where end "a" is fixed and end "b" is subjected to a torque "T". use the FE model to calculate the angle of twist at the end "b", and determine the reaction torque at node "a". $T = 10$ kN.m, $D = 100$ mm, $L = 300$ mm, $G = 8 * 10^{10}$ N/m^2.

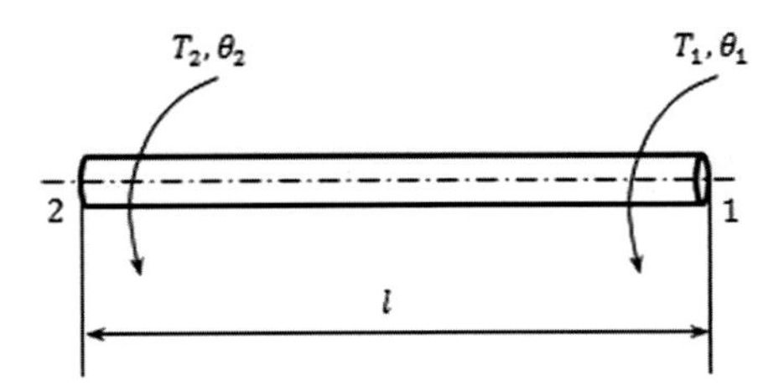

FIGURE P.7.1A

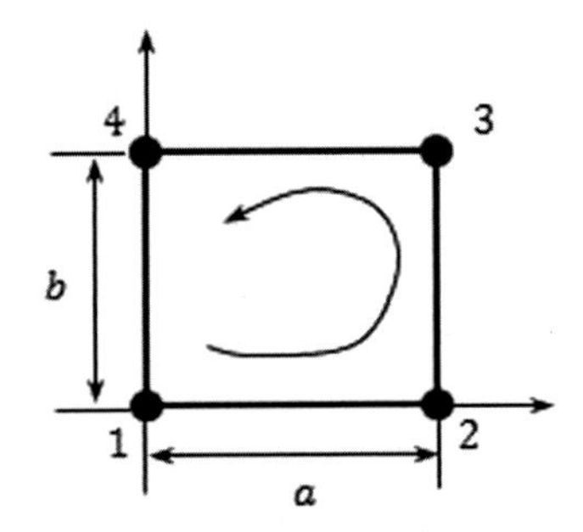

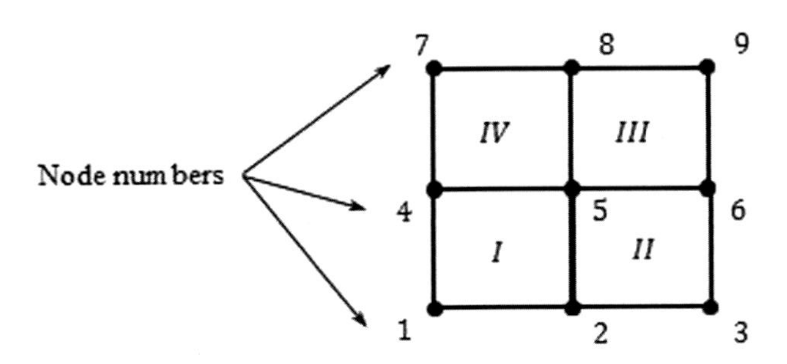

FIGURE P.7.2A

P.7.2 The torsional problem of the prismatic bar can be solved by FEM using a four-node quadrilateral element shown in Fig. P.7.2A. The dimensions and material properties are given by Fig. P.7.2B. for rectangular cross-section bar, determine in terms of θ, the following,

1. The stress function at node five
2. Shear strain and stress at nodes one and five if the applied torque is equal to 20 kN.m.

Notation

$\varnothing$: Prandtl stress function

$$\tau_{xy} = \frac{\partial \varnothing}{\partial y}, \tau_{yz} = \frac{\partial \varnothing}{\partial x}$$

u: Warping function

$$[h]^e\{u\}^e = \{F\}^e$$

$$[h]^e = \frac{1}{6G}\begin{bmatrix} (2\alpha+2\beta) & (\alpha-2\beta) & (-\alpha-\beta) & (-2\alpha+\beta) \\ & (2\alpha+2\beta) & (-2\alpha+\beta) & (-\alpha-\beta) \\ & & (2\alpha+2\beta) & (\alpha-2\beta) \\ & & & (2\alpha+2\beta) \end{bmatrix}$$

$\{F\}^e = \frac{ab}{2}\{1 \quad 1 \quad 1 \quad 1\}^t$, Nodal force vector.

$$G_{II} = 3G_I$$

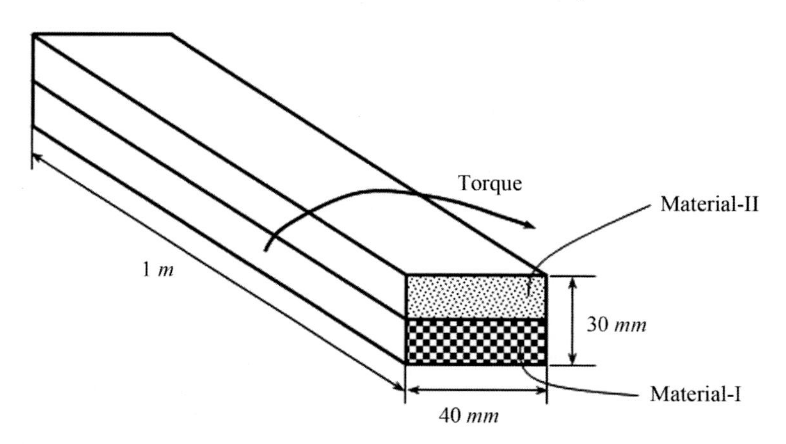

FIGURE P.7.2B

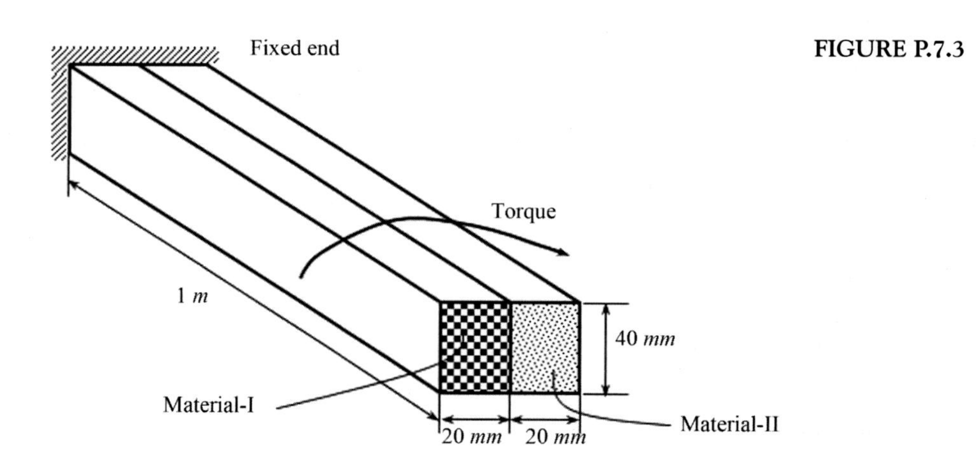

FIGURE P.7.3

P.7.3 If the beam in problem P.7.2 is positioned as shown in Fig. P.7.3, compare the angle of twist and the maximum shearing stress in the shaft section at a distance equal to 500 mm the fixed end.

P.7.4 The torsional problem of the circular shaft can be solved by the finite element method. A three-node element (Fig. P.7.4A) can be derived by minimizing the total potential energy, $X = U - W$, provided that,

$$U = \frac{1}{2}\int_0^l JG\left(\frac{d\theta}{dx}\right)^2 dx = \text{Strain energy}$$

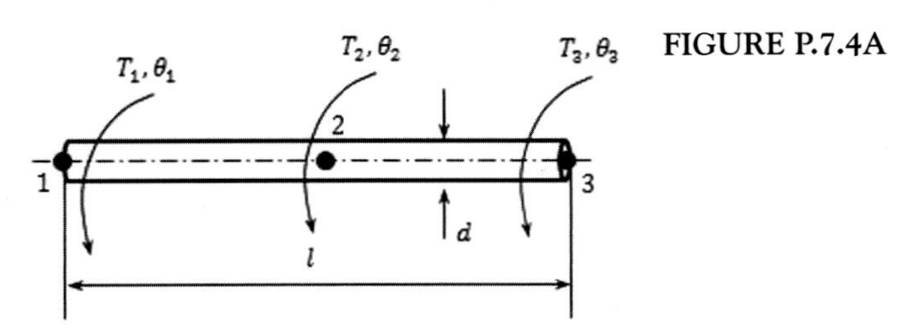

FIGURE P.7.4A

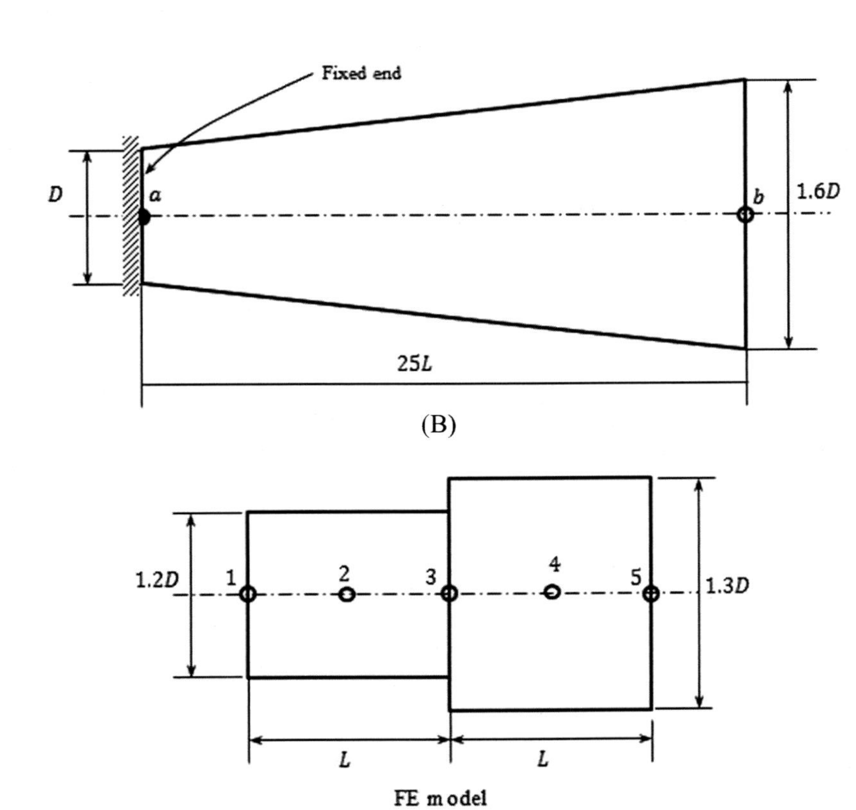

FIGURE P.7.4 Finite eelement modeling, two line elements and 5 nodes.

And,

$$W = \sum_{i=1}^{3} T_i \theta_i = \text{Work done}$$

Where, $T=$ Twisting moment, $J=$ polar moment of area $=\frac{\pi d^4}{32}$, $l=$ element length, $G=$ Modulus of Rigidity of the material, $d=$ element diameter, and $\theta=$ angle of twist.

1. Use the previous information to derive the following matrix characteristic equations for the three-node finite element.
2. Consider the tapered shaft shown in Fig. P.7.4B and C. Use the FE model shown to calculate the angle of twist at the end "b" and the reaction torque at node "a".
   $T=10$ kN.m, $D=100$ mm, $L=300$ mm, $G=8*10^{10}$ N/m^2.

P.7.5 The torsion problem of the prismatic bar can be solved by FEM using a four-node quadrilateral element, shown in Figs. P.7.5A, the dimensions and material properties are given by Fig. P.7.5B. For rectangular cross-section bar, determine in term of θ, the following:

1. The stress function at node five
2. Shear strain and stress at nodes one and five if the applied torque is equal to 10 kN.m.

Notation:
$\varnothing$: Prandtl stress function

$$\tau_{xy}=\frac{\partial\varnothing}{\partial y},\tau_{yz}=\frac{\partial\varnothing}{\partial x}$$

u: warping function

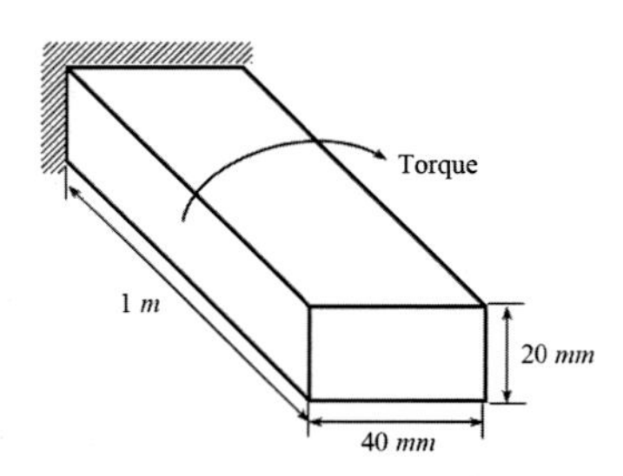

FIGURE P.7.5A Shaft under torque.

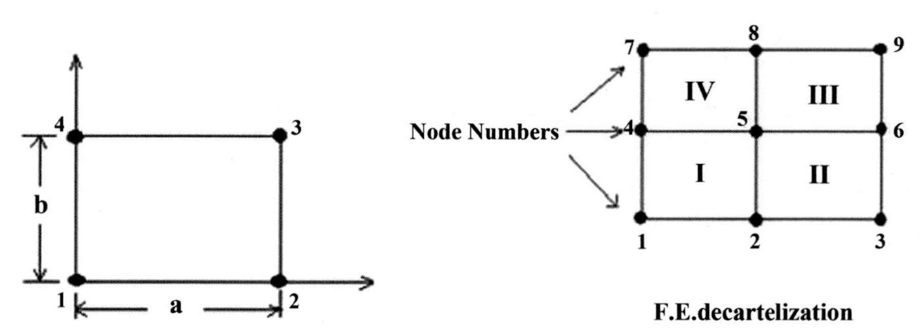

FIGURE P.7.5B Finite element discretization of the square section shafy.

$$[h]^e\{u\}^e = \{F\}^e$$

$$[h]^e = \frac{1}{6G}\begin{bmatrix} 2\alpha+2\beta & \alpha-2\beta & -\alpha-\beta & -2\alpha+\beta \\ & 2\alpha+2\beta & -2\alpha+\beta & -\alpha-\beta \\ & & 2\alpha+2\beta & \alpha-2\beta \\ \text{sum.} & & & 2\alpha+2\beta \end{bmatrix}$$

$\{F\}^e = \frac{ab}{2}\{1111\}^t$: Nodal force vector

P.7.6 For the elliptical cross-section of a shaft shown in Fig. P.7.6 with $a = 150$ mm, $b = 50$ mm

And it is made from aluminum with 1 m length and is twisted by $\theta = 3°, G = 28$ GPa and subjected to a twisting moment M_z by modeling the shaft by a triangular element compare the maximum shearing stresses obtained by using the suggested meshes with the following expressions obtained by using the torsion function derived using the theory of elasticity approach,

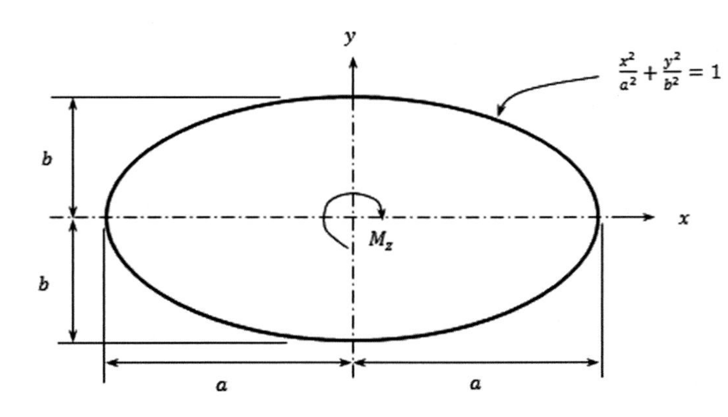

FIGURE P.7.6 Eliiptical shaft setion.

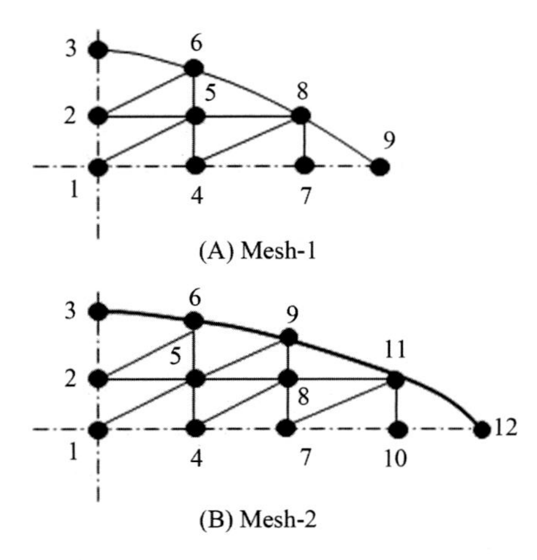

FIGURE P.7.6A AND B Finite element modeling of the elliptical shaft section (triangular elements).

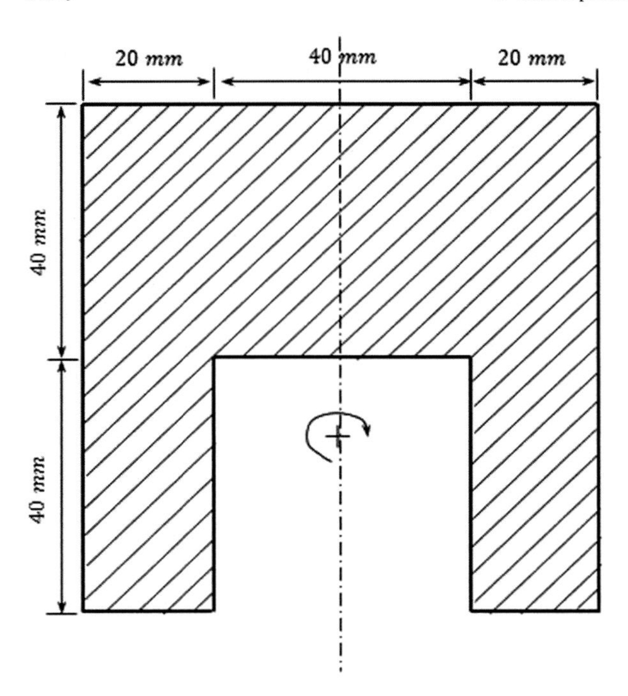

FIGURE P.7.7A Beam section under twsting load.

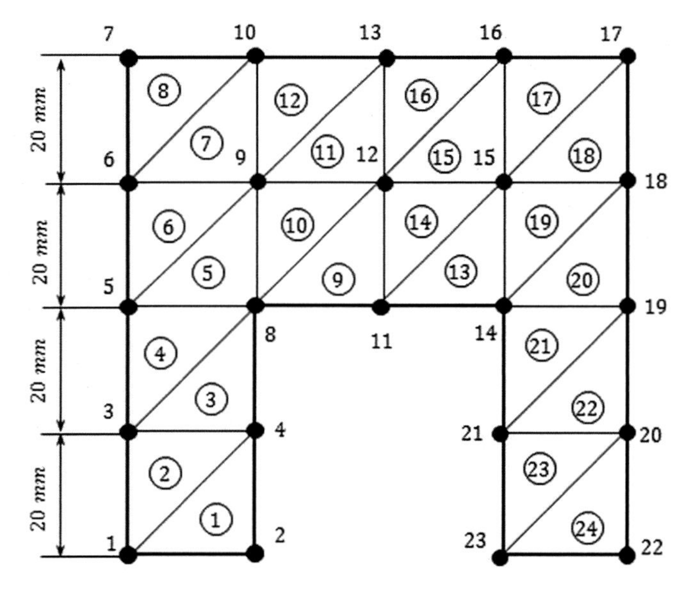

FIGURE P.7.7B F.E. Mesh for the P.7.7A.

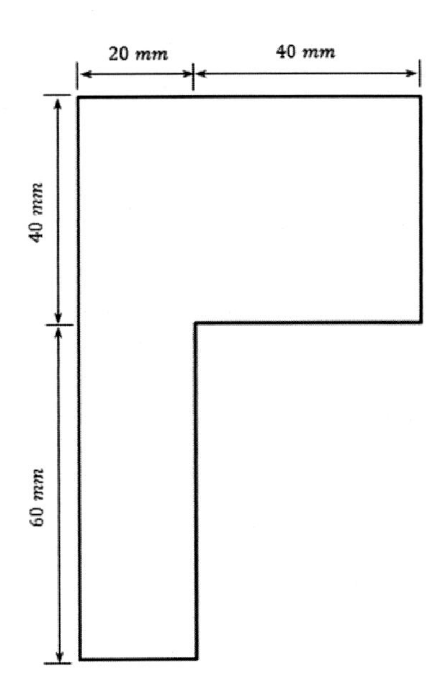

FIGURE P.7.7C Beam section.

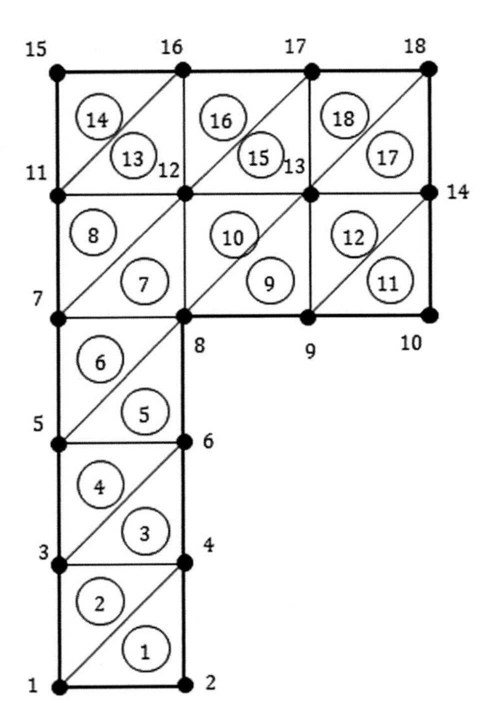

FIGURE P.7.7D F.E. Mesh for the P.7.7C.

$P_{max} = \frac{2M_z}{\pi ab^2}$ at$x = 0, y = \pm b$(ends of the minor axis)
And,

$$M_z = \frac{G\theta a^3 b^3}{(a^2 + b^2)}$$

Use the sign convections in the derived equations.
Suggested meshes (Figs. P.7.6A and B)
Note: Using the simple triangular element of three nodes, we need to approximate the curved side into line segments. The modeling needs fine mesh, that is mesh two is more approximate to the analytical solution. An iso-parametric element (Section 7.2) may be used in the modeling of such sections. The solution needs a computer program for the assembly and the solution.

P.7.7 Formulate the constrained equilibrium equations $[k]^e\{u\}^e = \{F\}^e$ for the beam sections shown in Figs. P.7.7A–P.7.7D which supports a floor slab in a building which is subjected to a twisting moment T(N.m). How do you obtain the angle of twist per unit length and each finite element's contribution to the total twisting moment? Rectangular or triangular element is recommended for the finite element modeling. You may get use the formulations of elements derived in Chapter 3, Application of Energy Methods to Plate Problems, to solve the sections' torsion problem in Figs. P.7.7A and P.7.7C.

If $T = 20$ kN.m, $G = 83$ Gpa, how do you find the angle of twist and the nodal shearing stresses?

Bibliography

[1] C.S. Desai, J.F. Abel, Introduction to the Finite Element Method, Van Nostrand, Reinhold, New York, 1972.
[2] L.R. Hermann, Elastic torsional analysis of irregular shapes, Journal of Engineering Mechanics Division 91 (EM6) (1965) 11–19.
[3] R.J. Roark, W.C. Young, Formulas for Stress and Strain, fifth ed., McGraw-Hill, New York, 1975.
[4] T.H. Richards, Energy Methods in stress Analysis, Ellis Horwood Series in Engineering Science, Ellis Horwood, 1977.
[5] K.J. Bathe, Finite Element Procedures, Prentice-Hill, Englewood Cliffs, 1996.
[6] Y.K. Cheung, M.F. Yeo, A Practical Introduction to Finite Element Analysis, Pitman, 1979.
[7] B. Irons, N. Shrive, Finite Element Primer, Ellis Horwood, 1983.
[8] D.J. Dawe, Matrix and Finite Element Displacement Analysis of Structures, Clarendon, 1984.
[9] A.P. Borsei, O.M. Sidebottom, Advanced Mechanics of Materials, fourth ed., Wiley, 1985.
[10] J.L. Tocher, B.J. Hertz, Higher order finite elements for plane stress, Journal of the Engineering Mechanics Division, Proceedings of the American Society of Civil Engineers 93 (EM4) (1967) 149–174.
[11] O.C. Zienkiewicz, B. Schimming, Torsion of non-homogeneous bars with axial symmetry, IJMSci 4 (1) (1962) 15–23.
[12] O.C. Zienkiewicz, Y.K. Cheung, Stresses in shafts, Engineering 224 (5835) (1967) 696–697.

CHAPTER

8

Axisymmetric elasticity problems

This chapter includes the finite element modelling of the axisymmetric solids subjected to an axisymmetric loading. The geometric modelling and the elasticity background required for determining the total potential energy formulation of the axisymmetric solids were presented. A simple presentation was adopted in this respect and the revolution of the three nodes triangular element is presented using the intrinsic coordinated in deriving the element stiffness matrix. The element load vector elements were presented taking into consideration the possible types of the applied forces in terms of the isoparametric triangular element. The body forces, rotating bodies, and the surface traction types of forces were included in the element load vector formulation. This chapter also contains problems for cases enjoy axisymmetric subjected to axisymmetric loading.

8.1 Geometrical description

Fig. 8.1 shows an axial-symmetric solid element formed by rotating a planar element about an axis say z-axis. The cylindrical coordinates (r,θ) and they are related to the Cartesian coordinates as follows,

$$x = r\cos\theta$$

$$x = r\sin\theta$$

$$z = z \tag{8.1}$$

The Conditions for Axisymmetric Formulation:

In order to approximate 3D-problem solid body to elastic problems the following conditions should be satisfied,

1. The structure is axisymmetric, that is the cross-section at any θ should be the same.
2. There is no loading component normal to the $r-z$ plane.
3. The load distribution in the $r-z$ plane should be the same at any θ.

Energy Methods and Finite Element Techniques
DOI: https://doi.org/10.1016/B978-0-323-88666-6.00007-1

From the above conditions,

$$\text{i. } u_\theta = 0$$

$$\text{ii. } \frac{\partial(\text{property})}{\partial\theta} = 0, \text{ that is, } \frac{\partial u_r}{\partial\theta} = 0, \frac{\partial u_z}{\partial\theta} = 0 \tag{8.2}$$

Where, u_r, u_θ, u_z are the displacement components in the r, z, and θ directions.

8.2 Three nodes triangular element

Consider the triangular section of a shell of revolution element with an axisymmetric loading as shown in Fig. 8.2. At each node, there is two d.o.f. in radial and axial directions.

FIGURE 8.1 Axisymmetric shell of revolution triangular elements.

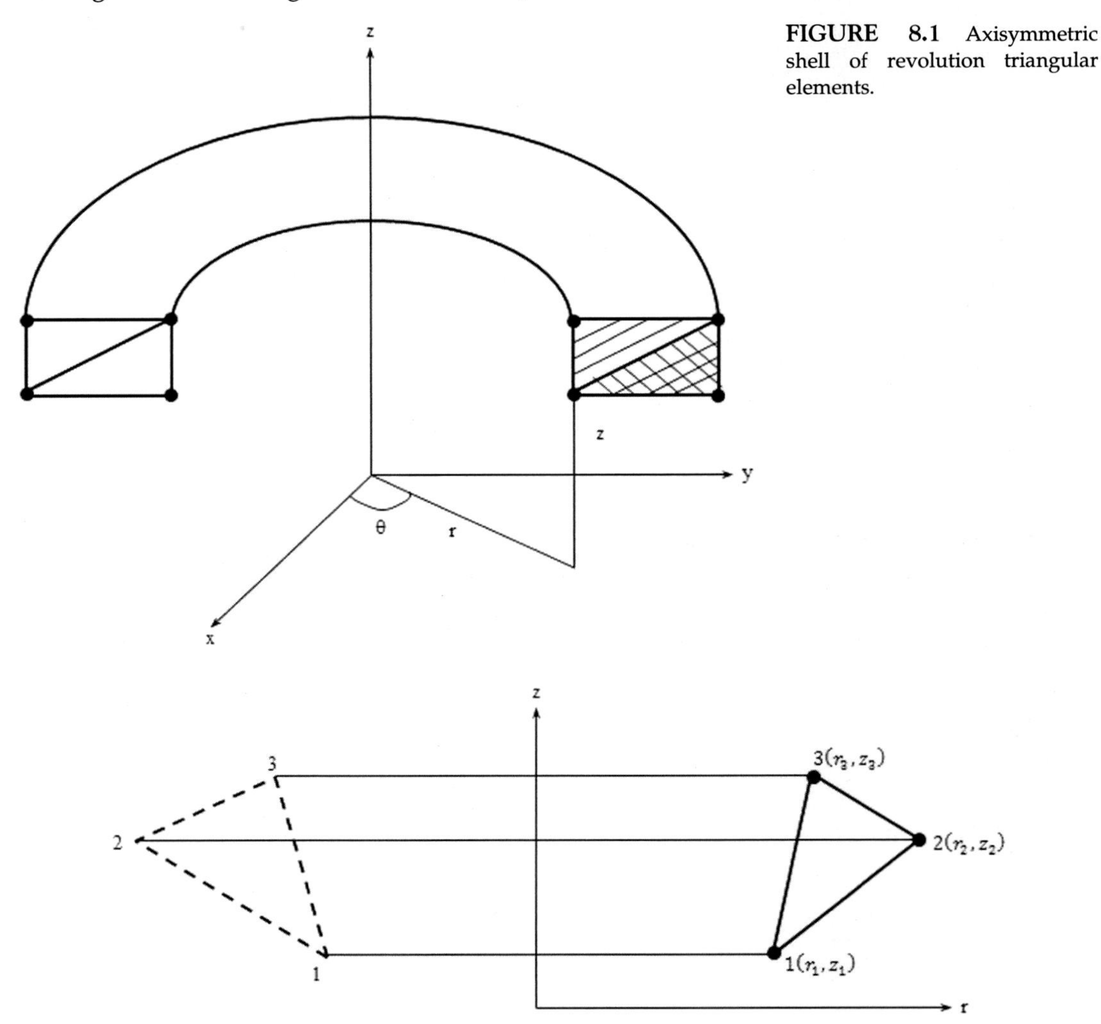

FIGURE 8.2 Triangular shell of revolution element.

The following steps explain the application of the finite element method for axisymmetric solid bodies. These are as follows,

Step-1, Definition of the nodal parameters,

The displacement vector of this type of element as follows,

$$\{\delta\} = \{ u_{r_1} \quad u_{z_1} \quad u_{r_2} \quad u_{z_2} \quad u_{r_3} \quad u_{z_3} \}$$

Where u_{r_1} is the displacement at node 1 in the r-direction, u_{z_1} Is the displacement at node 1 in the z-direction,.. . ..etc.

And the load vector is,

$$\{F\} = \{ F_{r_1} \quad F_{z_1} \quad F_{r_2} \quad F_{z_2} \quad F_{r_3} \quad F_{z_3} \} \tag{8.3}$$

Where F_{r_1} is the force at node 1 in the r-direction, F_{z_1} Is the force at node 1 in the z-direction,.. . ..etc.

Step-2, Express the displacement components at any point in terms of nodal displacements and shape functions.

Using intrinsic coordinates Fig. 8.3,

$$u_r(\xi, \eta) = \sum_{i=1}^{n} (u_r)_i N_i(\xi, \eta)$$

$$u_z(\xi, \eta) = \sum_{i=1}^{n} (u_z)_i N_i(\xi, \eta) \tag{8.4}$$

Where $N_i(\xi, \eta)$ is the shape function in terms of intrinsic coordinates.

For the three-node element,

$$u_r = \begin{bmatrix} N_1 & 0 & N_2 & 0 & N_3 & 0 \end{bmatrix} \{\delta\}$$

$$u_z = \begin{bmatrix} 0 & N_1 & 0 & N_2 & 0 & N_3 \end{bmatrix} \{\delta\} \tag{8.5}$$

Where,

$N_1 = 1 - \xi - \eta$, $N_2 = \xi$, $N_3 = \eta$, Fig. 8.3.

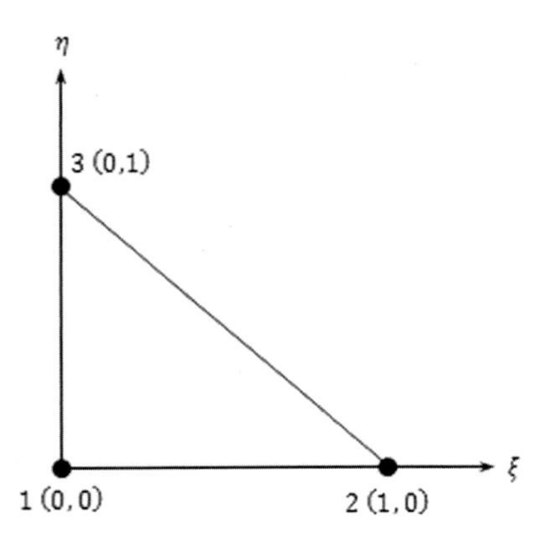

FIGURE 8.3 Triangular element using intrinsic coordinates.

Step-3, Expressing the strain components at any point in terms of nodal displacements,

$$\varepsilon_r = \frac{\partial u_r}{\partial r}$$

$$\varepsilon_z = \frac{\partial u_z}{\partial z}$$

$$\varepsilon_\theta = \frac{u_r}{r}$$

$$\gamma_{rz} = \frac{\partial u_r}{\partial z} + \frac{\partial u_z}{\partial r}$$

$$\gamma_{z\theta} = \gamma_{\theta z} = 0, \text{ this is due to symmetry.} \tag{8.6}$$

Hence,

$$\{\varepsilon_r\} = \begin{bmatrix} \frac{\partial N_1}{\partial r} & 0 & \frac{\partial N_2}{\partial r} & 0 & \frac{\partial N_3}{\partial r} & 0 \end{bmatrix} \{\delta\}$$

$$\{\varepsilon_z\} = \begin{bmatrix} 0 & \frac{\partial N_1}{\partial z} & 0 & \frac{\partial N_2}{\partial z} & 0 & \frac{\partial N_3}{\partial z} \end{bmatrix} \{\delta\}$$

$$\{\varepsilon_\theta\} = \begin{bmatrix} \frac{N_1}{r} & 0 & \frac{N_2}{r} & 0 & \frac{N_3}{r} & 0 \end{bmatrix} \{\delta\}$$

$$\{\gamma_{rz}\} = \begin{bmatrix} \frac{\partial N_1}{\partial z} & \frac{\partial N_1}{\partial r} & \frac{\partial N_2}{\partial z} & \frac{\partial N_2}{\partial r} & \frac{\partial N_3}{\partial z} & \frac{\partial N_3}{\partial r} \end{bmatrix} \{\delta\}$$

$$\{\varepsilon\} = [B]\{\delta\} \tag{8.7}$$

where $[B]$ is the strain displacement relationship.

Where

$$\{\varepsilon\} = \begin{bmatrix} \varepsilon_r & \varepsilon_z & \varepsilon_\theta & \gamma_{rz} \end{bmatrix}$$

$$[B] = \begin{bmatrix} \frac{\partial N_1}{\partial r} & 0 & \frac{\partial N_2}{\partial r} & 0 & \frac{\partial N_3}{\partial r} & 0 \\ 0 & \frac{\partial N_1}{\partial z} & 0 & \frac{\partial N_2}{\partial z} & 0 & \frac{\partial N_3}{\partial z} \\ \frac{N_1}{r} & 0 & \frac{N_2}{r} & 0 & \frac{N_3}{r} & 0 \\ \frac{\partial N_1}{\partial z} & \frac{\partial N_1}{\partial r} & \frac{\partial N_2}{\partial z} & \frac{\partial N_2}{\partial r} & \frac{\partial N_3}{\partial z} & \frac{\partial N_3}{\partial r} \end{bmatrix}$$

Chain rule. This rule is used to constructing the relationship between the intrinsic and Cartesian coordinates, and achieved as follows,

$$\frac{\partial N_i}{\partial \xi} = \frac{\partial N_i}{\partial r}\frac{\partial r}{\partial \xi} + \frac{\partial N_i}{\partial z}\frac{\partial z}{\partial \xi}$$

$$\frac{\partial N_i}{\partial \eta} = \frac{\partial N_i}{\partial r}\frac{\partial r}{\partial \eta} + \frac{\partial N_i}{\partial z}\frac{\partial z}{\partial \eta}$$

$$\begin{bmatrix} \dfrac{\partial N_i}{\partial \xi} \\ \dfrac{\partial N_i}{\partial \eta} \end{bmatrix} = [J]\begin{Bmatrix} \dfrac{\partial N_i}{\partial r} \\ \dfrac{\partial N_i}{\partial z} \end{Bmatrix} \tag{8.8}$$

where

$$\left[J\left(\frac{r,z}{\xi,\eta}\right)\right] = \begin{bmatrix} \dfrac{\partial r}{\partial \xi} & \dfrac{\partial z}{\partial \xi} \\ \dfrac{\partial r}{\partial \eta} & \dfrac{\partial z}{\partial \eta} \end{bmatrix} \text{ is the Jacobian matrix.} \tag{8.9a}$$

$$\begin{bmatrix} \dfrac{\partial N_i}{\partial r} \\ \dfrac{\partial N_i}{\partial z} \end{bmatrix} = [J]^{-1}\begin{bmatrix} \dfrac{\partial N_i}{\partial \xi} \\ \dfrac{\partial N_i}{\partial \eta} \end{bmatrix}$$

Where $[J]^{-1}$ is the inverse of the Jacobian matrix.

For the three-node triangular element, applying the isoparametric transformation,

$$r(\xi,\eta) = r_1N_1 + r_2N_2 + r_3N_3 = r_1 + \xi(r_2 - r_1) + \eta(r_3 - r_1)$$

$$z(\xi,\eta) = z_1 + \xi(z_2 - z_1) + \eta(z_3 - z_1)$$

Carrying out the differentiation according to Eq. (8.9a), the Jacobian matrix is,

$$[J] = \begin{bmatrix} (z_3 - z_1) & -(z_2 - z_1) \\ -(r_3 - r_1) & (r_2 - r_1) \end{bmatrix} \tag{8.9b}$$

where

$|J| = 2A$, A: area of the triangular element.

And

$$[J]^{-1} = \frac{1}{2A}\begin{bmatrix} (r_2 - r_1) & -(z_2 - z_1) \\ -(r_3 - r_1) & (z_3 - z_1) \end{bmatrix} \tag{8.9}$$

Step-4, Expressing the stress components at any point in terms of nodal displacements.

$$\varepsilon_r = \frac{1}{E}[\sigma_r - \nu(\sigma_z + \sigma_\theta)]$$
$$\varepsilon_z = \frac{1}{E}[\sigma_z - \nu(\sigma_r + \sigma_\theta)] \tag{8.10}$$
$$\varepsilon_\theta = \frac{1}{E}[\sigma_\theta - \nu(\sigma_r + \sigma_z)]$$
$$\gamma_{rz} = \frac{2(1+\nu)}{E}\tau_{rz}$$

or

$$\{\sigma\} = [D]\{\varepsilon\} \tag{8.11}$$

where

$$\{\sigma\} = \left\{ \sigma_r \quad \sigma_z \quad \sigma_\theta \quad \tau_{rz} \right\}$$

And $[D]$ is the elasticity matrix, and can be written for the axisymmetric problems as follows,

$$[D] = \frac{E}{(1+\nu)(1-2\nu)}\begin{bmatrix} (1-\nu) & \nu & \nu & 0 \\ & (1-\nu) & \nu & 0 \\ & & (1-\nu) & 0 \\ & & & \dfrac{(1-2\nu)}{2} \end{bmatrix}$$

Hence,

$$\{\sigma\} = [D][B]\{\delta\} \tag{8.12}$$

Step-5, Expressing the total potential energy,

$$V = U - \Omega = \frac{1}{2}\iiint_{V_{\text{element}}} \{\sigma\}^t\{\varepsilon\}d_{\text{vol.}} - \{\delta\}^t\{F\}$$

Using Eqs. (8.7) and (8.11) gives,

$$= \frac{1}{2}\iiint \{\delta\}^t[B]^t[D][B]\{\delta\}d_{\text{vol.}} - \{\delta\}^t\{F\} \tag{8.13}$$

Step-6, Minimize the total potential energy of the element shown in Fig. 8.4,

$$\frac{\partial V}{\partial\{\delta\}} = 0 \Rightarrow$$
$$[k]^e\{\delta\}^e = \{F\}^e \tag{8.14}$$

Where,

$$[k]^e = \iiint_{\text{element}} [B]^t[D][B]d_{\text{vol.}} \tag{8.15a}$$

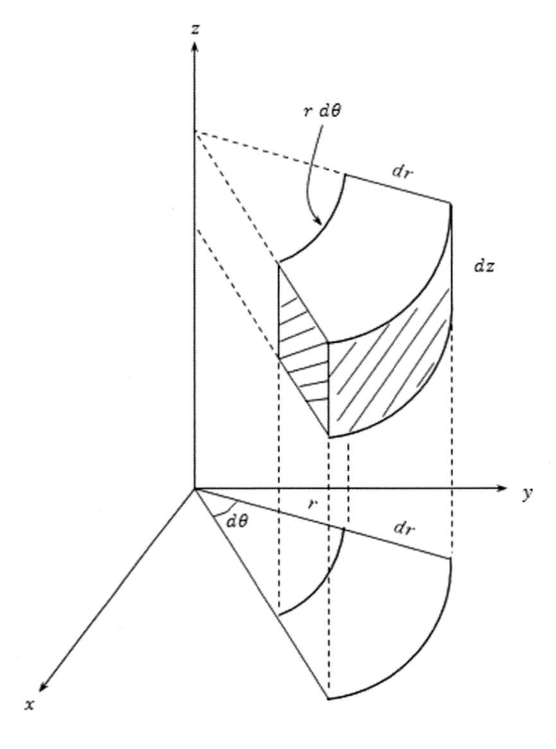

FIGURE 8.4 Sector of the solid shell of revolution element.

Is the element stiffness matrix of size (6×6), where the triangular element has six degrees of freedom. Also,

$d_{vol.} = dx\,dy\,dz$ in the $x - y - z$ system $= r\,d\theta\,dr\,dz$

We have,

$$dxdy = \left|J\left(\frac{x,y}{r,\theta}\right)\right| drd\theta \tag{8.15}$$

where

$$x = r\cos\theta$$

$$y = r\sin\theta$$

Therefore

$$\left[J\left(\frac{x,y}{r,\theta}\right)\right] = \begin{bmatrix} \frac{\partial x}{\partial r} & \frac{\partial y}{\partial r} \\ \frac{\partial x}{\partial \theta} & \frac{\partial y}{\partial \theta} \end{bmatrix} = \begin{bmatrix} \cos\theta & \sin\theta \\ -r\sin\theta & r\cos\theta \end{bmatrix}$$

$$\left|J\left(\frac{x,y}{r,\theta}\right)\right| = r\cos^2\theta + r\sin^2\theta = r$$

From Eq. (8.15a),

$$[k]^e = \iint_A [B]^t[D][B]2\pi rt\ dr\ dz$$

And

$$drdz = \left|\left[J\left(\frac{r,z}{\xi,\eta}\right)\right]\right|d\xi d\eta$$

The element stiffness matrix in terms of intrinsic coordinates is given by,

$$[k]^e = \iint_{\text{intrinsic element}} [B]^t[D][B]\left|\left[J\left(\frac{r,z}{\xi,\eta}\right)\right]\right|d\xi\ d\eta \tag{8.16}$$

Eq. (8.16) is integrated numerically using the Gaussian numerical integration technique. This will be explained in Chapter 12, Shape Functions Determinations and Numerical Integration.

8.3 Representation of the applied forces as an equivalent nodal forces

8.3.1 Body force

The body force is expressed as follows,

$$\{F\}^e = 2\pi\int \{\delta\}^t\{F\}rdA = 2\pi\int \{u_rF_r \quad u_zF_z\}r\ dA$$

From Eq. (8.4), we have,

$$\begin{Bmatrix} u_r \\ u_z \end{Bmatrix} = \begin{bmatrix} N_1 & 0 & N_2 & 0 & N_3 & 0 \\ 0 & N_1 & 0 & N_2 & 0 & N_3 \end{bmatrix}\begin{Bmatrix} u_{r_1} \\ u_{z_1} \\ u_{r_2} \\ u_{z_2} \\ u_{r_3} \\ u_{z_3} \end{Bmatrix}$$

or

$$\{u\} = [N]\{\delta\}^e$$

Therefore

$$\{F\}^e = 2\pi\int_e \{u_{r_1} \quad u_{z_1} \quad u_{r_2} \quad u_{z_2} \quad u_{r_3} \quad u_{z_3}\}\begin{Bmatrix} F_{r_1} \\ F_{z_1} \\ F_{r_2} \\ F_{z_2} \\ F_{r_3} \\ F_{z_3} \end{Bmatrix}rdA \tag{8.17}$$

Or, in compact form,

$$\{F\}^e = 2\pi\{\delta\}^t\int_e [N]^t\{F\}^e rdA$$

where

$$[N]^t = \begin{bmatrix} N_1 & 0 \\ 0 & N_1 \\ N_2 & 0 \\ 0 & N_2 \\ N_3 & 0 \\ 0 & N_3 \end{bmatrix}, \quad \{F\} = \begin{Bmatrix} F_r \\ F_z \end{Bmatrix}^e$$

And

$$r = N_1 r_1 + N_2 r_2 + N_3 r_3$$

Integrating Eq. (8.17), the equivalent nodal forces are obtained. The approximate element load vector for the body force will be as follows,

$$\{F\}^e = \begin{Bmatrix} F_{r_1} \\ F_{z_1} \\ F_{r_2} \\ F_{z_2} \\ F_{r_3} \\ F_{z_3} \end{Bmatrix} = \frac{2\pi \bar{r} A}{3} \left\{ \overline{F}_{r_1} \quad \overline{F}_{z_1} \quad \overline{F}_{r_2} \quad \overline{F}_{z_2} \quad \overline{F}_{r_3} \quad \overline{F}_{z_3} \right\}^t \tag{8.18}$$

where $\overline{F}_{r_1}, \ldots et$ indicate that the body forces are calculated at the centroids.

8.3.2 Rotating bodies

If the body rotates about z-axis, the centrifugal force per unit volume is $\rho\omega^2 r$, where ρ is the mass per unit volume and ω is the rotating speed r/s. Note that the centrifugal force is in the radial direction. Therefore

$$\{F\} = \left\{ F_r \ \ F_z \right\}^t = \left\{ \rho\omega^2 r \ \ 0 \right\}$$

And if there is gravity, the force vector,

$$\{F\} = \left\{ \rho\omega^2 r \quad -\rho g \right\}$$

8.3.3 Surface traction

Consider the two-node side 1–2 subjected to the traction as shown in Fig. 8.5, the components are T_r and T_z, the radial and tangential components.

We have,

$$r = N_1 r_1 + N_2 r_2$$

$$l_{12} = \sqrt{(r_2 - r_1)^2 + (z_2 - z_1)^2}$$

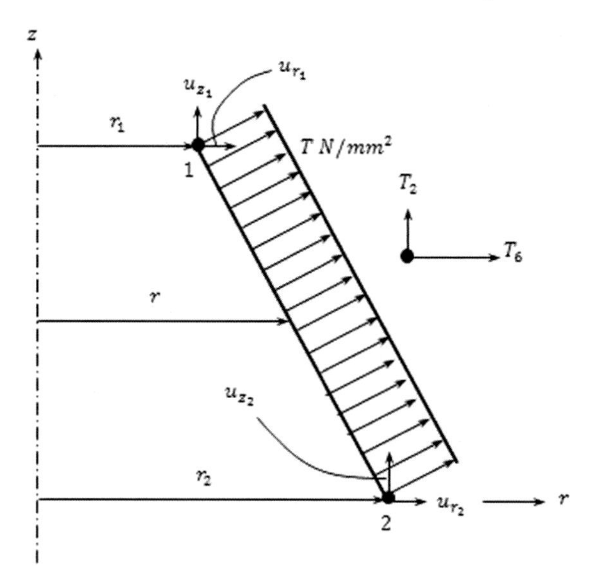

FIGURE 8.5 Surface traction.

We have, l_{12} is the side length between the circular nodes 1 and 2.

$$\{u(r,z)\} = [N]\{\delta\}$$

And

$$\{F\} = \{\delta\}^t 2\pi \int [N]^t \{T\} r dl = \{\delta\}^t \{F\}^e \tag{8.19}$$

Using Eq. (8.19) and carrying out the integration gives,

$$\{F\}^e = \begin{Bmatrix} F_{r_1} \\ F_{z_1} \\ F_{r_2} \\ F_{z_2} \end{Bmatrix} = 2\pi l_{12} \begin{Bmatrix} \left(\dfrac{2r_1 + r_2}{6}\right) & T_{r_1} \\ \left(\dfrac{2r_1 + r_2}{6}\right) & T_{z_1} \\ \left(\dfrac{r_1 + 2r_2}{6}\right) & T_{r_2} \\ \left(\dfrac{r_1 + 2r_2}{6}\right) & T_{z_2} \end{Bmatrix} \tag{8.20}$$

For cylindrical shape, put,
$r_1 = r_2$ in Eq. (8.20) gives,

$$\{F\}^e = \begin{Bmatrix} F_{r_1} \\ F_{z_1} \\ F_{r_2} \\ F_{z_2} \end{Bmatrix} = \pi r l_{12} \begin{Bmatrix} T_{r_1} \\ T_{z_1} \\ T_{r_2} \\ T_{z_2} \end{Bmatrix} \tag{8.21}$$

Where T_{r_1} is the traction in the r-direction at the circular node 1, T_{z_1} is the traction in the z-direction at the circular node 1, ... etc.

The triangular element shown in Fig. 8.6 with three nodes, the displacement field is represented by,

$$u_r(r,z) = u_{r_1}N_1 + u_{r_2}N_2 + u_{r_3}N_3$$
$$u_z(r,z) = u_{z_1}N_1 + u_{z_2}N_2 + u_{z_3}N_3 \tag{8.22}$$

It can be assumed that,

$$u_r(r,z) = \alpha_1 + \alpha_2 r + \alpha_3 z \tag{8.23}$$

Therefore

$$\begin{Bmatrix} u_{r_1} \\ u_{r_2} \\ u_{r_3} \end{Bmatrix} = \begin{bmatrix} 1 & r_1 & z_1 \\ 1 & r_2 & z_2 \\ 1 & r_3 & z_3 \end{bmatrix} \begin{Bmatrix} \alpha_1 \\ \alpha_2 \\ \alpha_3 \end{Bmatrix} \tag{8.24}$$

Solving Eq. (8.24) for α_1, α_2, and α_3 and substituting in Eq. (8.22), the shape function can be deduced where

$$N_1(r,z) + N_2(r,z) + N_3(r,z) = 1$$
$$N_1 r_1 + N_2 r_2 + N_3 r_3 = r$$
$$N_1 z_1 + N_2 z_2 + N_3 z_3 = z$$

Therefore

$$\begin{bmatrix} 1 & r_1 & z_1 \\ 1 & r_2 & z_2 \\ 1 & r_3 & z_3 \end{bmatrix} \begin{Bmatrix} N_1 \\ N_2 \\ N_3 \end{Bmatrix} = \begin{Bmatrix} 1 \\ r \\ z \end{Bmatrix} \tag{8.25}$$

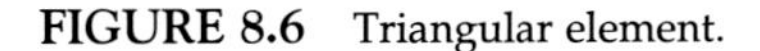

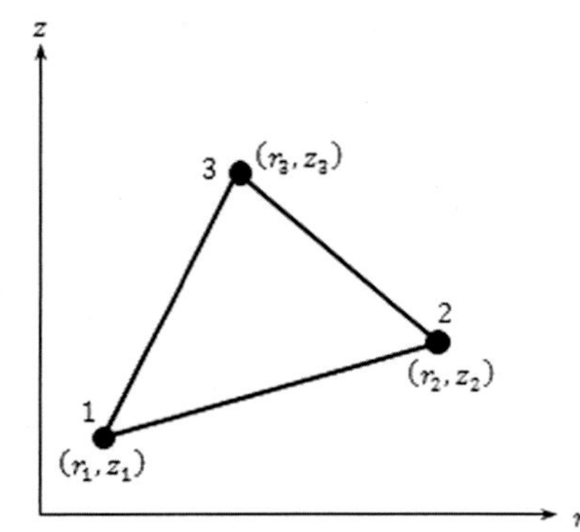

FIGURE 8.6 Triangular element.

where

$$N_1 = \frac{\begin{vmatrix} 1 & 1 & 1 \\ r & r_2 & r_3 \\ z & z_2 & z_3 \end{vmatrix}}{\begin{vmatrix} 1 & 1 & 1 \\ r_1 & r_2 & r_3 \\ z_1 & z_2 & z_3 \end{vmatrix}}, \quad N_2 = \frac{\begin{vmatrix} 1 & 1 & 1 \\ r_1 & r & r_3 \\ z_1 & z & z_3 \end{vmatrix}}{\begin{vmatrix} 1 & 1 & 1 \\ r_1 & r_2 & r_3 \\ z_1 & z_2 & z_3 \end{vmatrix}}, \quad N_3 = \frac{\begin{vmatrix} 1 & 1 & 1 \\ r_1 & r_2 & r \\ z_1 & z_2 & z \end{vmatrix}}{\begin{vmatrix} 1 & 1 & 1 \\ r_1 & r_2 & r_3 \\ z_1 & z_2 & z_3 \end{vmatrix}} \tag{8.26}$$

The area of the triangle,

$$\Delta_{123} = \frac{1}{2}\begin{vmatrix} 1 & 1 & 1 \\ r_1 & r_2 & r_3 \\ z_1 & z_2 & z_3 \end{vmatrix} = \frac{1}{2}\begin{vmatrix} 1 & r_1 & z_1 \\ 1 & r_2 & z_2 \\ 1 & r_3 & z_3 \end{vmatrix} \tag{8.27}$$

At any point $P(r, z)$ inside the triangle element, it can be deduced that,

$$N_i = \frac{1}{|A|}\left(\alpha_i + r\beta_i + z\gamma_i\right) \tag{8.28}$$

Where,

$$|A| = \begin{vmatrix} 1 & 1 & 1 \\ r_1 & r_2 & r_3 \\ z_1 & z_2 & z_3 \end{vmatrix} = 2 * \text{area of the element}$$

And

$$\alpha_1 = r_2z_3 - r_3z_2, \ \alpha_2 = r_1z_3 - r_3z_1, \ \alpha_3 = r_1z_2 - r_2z_1, \ \beta_1 = z_2 - z_3, \ \beta_2 = z_1 - z_3,$$
$$\beta_3 = z_1 - z_2, \ \gamma_1 = r_2 - r_3, \ \gamma_2 = r_1 - r_3, \ \gamma_3 = r_1 - r_2 \tag{8.29}$$

Strain matrix is given by

$$\begin{Bmatrix} \varepsilon_r \\ \varepsilon_\theta \\ \varepsilon_z \\ \gamma_{rz} \end{Bmatrix} = \begin{Bmatrix} \dfrac{\partial u_r}{\partial r} \\ \dfrac{u_r}{r} \\ \dfrac{\partial u_z}{\partial z} \\ \dfrac{\partial u_r}{\partial z} + \dfrac{\partial u_z}{\partial r} \end{Bmatrix} = \begin{bmatrix} 0 & 1 & 0 & 0 & 0 & 0 \\ 0 & 0 & 0 & 0 & 0 & 1 \\ \dfrac{1}{r} & 1 & \dfrac{z}{r} & 0 & 0 & 0 \\ 0 & 0 & 1 & 0 & 1 & 0 \end{bmatrix} \begin{Bmatrix} \alpha_1 \\ \alpha_2 \\ \alpha_3 \\ \alpha_4 \\ \alpha_5 \\ \alpha_6 \end{Bmatrix} \tag{8.30}$$

Using the value for $\alpha_1 \ldots \alpha_6$ we may rewrite Eq. (8.29) into Eq. (8.30) gives,

$$\{\varepsilon\} = \frac{1}{2A}\begin{bmatrix} \beta_1 & 0 & \beta_2 & 0 & \beta_3 & 0 \\ 0 & \gamma_1 & 0 & \gamma_2 & 0 & \gamma_3 \\ \left(\dfrac{\alpha_1}{r} + \beta_1 + \dfrac{\gamma_1 z}{r}\right) & 0 & \left(\dfrac{\alpha_2}{r} + \beta_2 + \dfrac{\gamma_2 z}{r}\right) & 0 & \left(\dfrac{\alpha_3}{r} + \beta_3 + \dfrac{\gamma_3 z}{r}\right) & 0 \end{bmatrix} \begin{Bmatrix} u_{r_1} \\ u_{z_1} \\ u_{r_2} \\ u_{z_2} \\ u_{r_3} \\ u_{z_3} \end{Bmatrix} dr \tag{8.31}$$

Rearranging Eq. (8.31) in the following simplified form,

$$\{\varepsilon\} = \begin{bmatrix} B_1 & B_2 & B_3 \end{bmatrix} \begin{Bmatrix} u_{r_1} \\ u_{z_1} \\ u_{r_2} \\ u_{z_2} \\ u_{r_3} \\ u_{z_3} \end{Bmatrix}$$

where

$$[B_i]_{i=1,3} = \begin{bmatrix} \beta_i & 0 \\ 0 & \gamma_i \\ \alpha_i + \beta_i + \dfrac{\gamma_i}{r} & 0 \\ \gamma_i & \beta_i \end{bmatrix} \tag{8.32}$$

In compact form

$$\{\varepsilon\} = [B]\{u\} \tag{8.33}$$

where the strain-displacement matrix $[B]$ is

$$[B(r,z)] = \begin{bmatrix} B_1(r,z) & B_2(r,z) & B_3(r,z) \end{bmatrix} \tag{8.34}$$

And the stress vector is given by

$$\begin{Bmatrix} \sigma_r \\ \sigma_\theta \\ \sigma_z \\ \sigma_{r\theta} \end{Bmatrix} = [D][B]\{u\} \tag{8.35}$$

where $[D]$ is the stress-strain relationship, which was given by Eq. (8.12).

The element stiffness matrix is given by

$$[k]^e = \iiint_V [B]^t[D][B]dvol$$

$$\underset{(6,6)}{[k]^e} = 2\pi \int_A \underset{(4,6)}{[B]^t} \; \underset{(4,4)}{[D]} \; \underset{(6,4)}{[B]} \; r \; dr \; dz \tag{8.36}$$

Eq. (8.36) may be integrated numerically (Chapter 12: Shape Functions Determinations and Numerical Integration)

$[B]$ is evaluated for a central point $(\bar{r}, \bar{z})$ of the element,

$$r = \bar{r} = \frac{r_1 + r_2 + r_3}{3}, \quad \bar{z} = \frac{z_1 + z_2 + z_3}{3}$$

And by define,

$$[B(\bar{r}, \bar{z})] = \boxed{B} \tag{8.37}$$

Therefore

$$[k]^e = 2\pi\bar{r}A\boxed{B}^t[D]\boxed{B} \tag{8.38}$$

Again, Eq. (8.38) is tailored to be integrated numerically.

Problems

P.8.1: for axisymmetric plane stress situations, a disk element having inner and outer radii r_1 and r_2 respectively and thickness t may be conceived. Assuming a displacement model,

$$u^e = \left[\left(1 - \frac{(r - r_1)}{(r_2 - r_1)}\right) \frac{(r - r_1)}{(r_2 - r_1)}\right] u_0^e$$

Deduce strain-displacement relations in the form,

$$\{\varepsilon\} = \begin{Bmatrix} \varepsilon_r \\ \varepsilon_\theta \end{Bmatrix} = \begin{bmatrix} B_{11} & B_{12} \\ B_{21} & B_{22} \end{bmatrix} \begin{Bmatrix} u_1 \\ u_2 \end{Bmatrix}^e = [B]\{u\}_0^e$$

Show that the element stiffness matrix is given by,

$$[k]^e = \frac{2\pi Et}{(1-\nu^2)} \int_{r_1}^{r_2} \begin{bmatrix} (B_{11}^2 + 2\nu B_{11}B_{21} + B_{21}^2) & (B_{11}B_{12} + \nu(B_{11}B_{22} + B_{12}B_{21}) + B_{21}B_{22}) \\ (B_{11}B_{12} + \nu(B_{11}B_{22} + B_{12}B_{21}) + B_{21}B_{22}) & (B_{12}^2 + 2\nu B_{12}B_{22} + B_{22}^2) \end{bmatrix} dr$$

P.8.2: For the axisymmetric bending of circular plates, disk type elements of inner and outer radii r_1 and r_2 Moreover, thickness t may be used. Does the following displacement model satisfy the continuity conditions?

$$u^e(\xi) = \left[\left(1 - \frac{3\xi^2}{l^2} + \frac{2\xi^3}{l^3}\right) \quad \left(\xi - \frac{2\xi^2}{l} + \frac{\xi^3}{l^2}\right) \quad \left(\frac{3\xi^2}{l^2} - \frac{2\xi^3}{l^3}\right) \quad \left(-\frac{\xi^2}{l} + \frac{\xi^3}{l^2}\right)\right]$$

where

$$\xi = (r - r_1) \text{ and } l = (r_2 - r_1)$$

P.8.3: The stress analysis of axisymmetrically loaded solids is a quasi-two-dimensional problem. using a coordinate system r_{oz} in a diametral plane, with oz along the z-axis, the following stress and strain vectors are relevant,

$$\{\sigma\}^t = \begin{bmatrix} \sigma_r & \sigma_\theta & \sigma_z & \tau_{rz} \end{bmatrix}$$

$$\{\varepsilon\}^t = \begin{bmatrix} \varepsilon_r & \varepsilon_\theta & \varepsilon_z & \gamma_{rz} \end{bmatrix}$$

With displacement components in the r and z directions being u and w, we have,

$$\varepsilon_r = \frac{\partial u}{\partial r}$$

$$\varepsilon_\theta = \frac{u}{r}$$

$$\varepsilon_z = \frac{\partial w}{\partial z}$$

$$\gamma_{rz} = \frac{\partial w}{\partial r} + \frac{\partial u}{\partial z}$$

The elasticity matrix is,

$$[D] = \frac{E(1-\nu)}{(1+\nu)(1-2\nu)}\begin{bmatrix} 1 & \frac{\nu}{(1-\nu)} & 0 & 0 \\ & 1 & \frac{\nu}{(1-\nu)} & 0 \\ & \text{Sym.} & 1 & 0 \\ & & & \frac{(1-2\nu)}{2(1-\nu)} \end{bmatrix}$$

If a ring-type element having a triangular cross-section is used, explain why the following displacement model is acceptable,

$$\begin{bmatrix} u \\ w \end{bmatrix} = \begin{bmatrix} N_1 & 0 & N_2 & 0 & N_3 & 0 \\ 0 & N_1 & 0 & N_2 & 0 & N_3 \end{bmatrix}\begin{Bmatrix} u_1 \\ u_2 \\ \vdots \\ u_6 \end{Bmatrix}^e$$

where

$$N_1 = \frac{1}{2A}(a_1 + b_1 z + c_1 r)$$

$$a_1 = z_2.r_3 - z_3.r_2$$

$$b_1 = r_2 - r_3$$

$$c_1 = z_3 - z_2$$

Without developing the algebraic detail, show that the stiffness matrix $[k]^e$ is (6×6).

P.8.4: The linearly distributed loading shown in Fig. P.8.4. determines the equivalent nodal loads in r and z directions for nodes 1, 2.

P.8.5: For the rotating disc shown in Fig. P.8.5a and Fig. P.8.5b, if $\omega = 2000$ rpm, $\rho = 7800\ \text{kg/m}^2$, $E = 200$ GPa, $\nu = 0.3$, use two elements of four nodes quadrilateral element by using symmetry and form the stiffness matrix and the system load vector.

Note, use the following mesh, (Fig.P.8.5b)

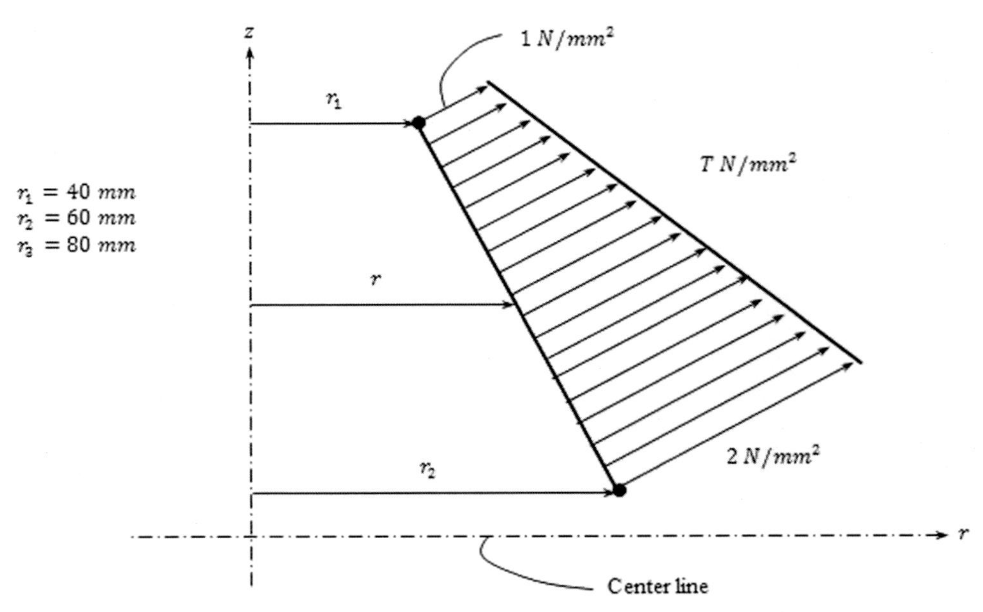

FIGURE P.8.4

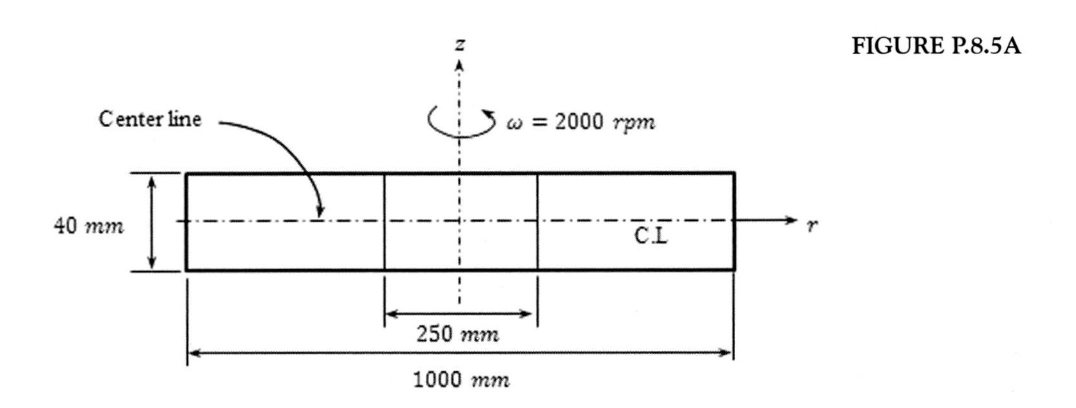

FIGURE P.8.5A

P.8.6: Obtain the nodal load vector for the uniform stress traction on the side shown in Figs. P.8.6A and P.8.6B

P.8.7: For the axisymmetric element shown in Fig. P.8.7,

1. Start from the first principle to formulate the strain-displacement relation-ship $[B]$.
2. Write down the elasticity matrix $[D]$.
3. Drive the element stiffness matrix $[k]^e$.

$$E = 70 \text{ GPa}, \quad \nu = 0.33.$$

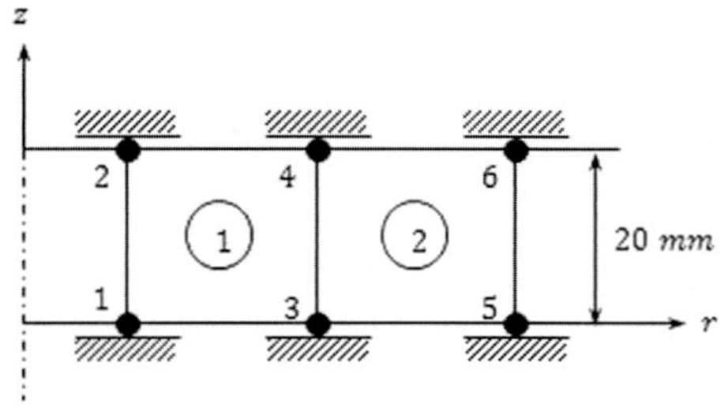

FIGURE P.8.5B

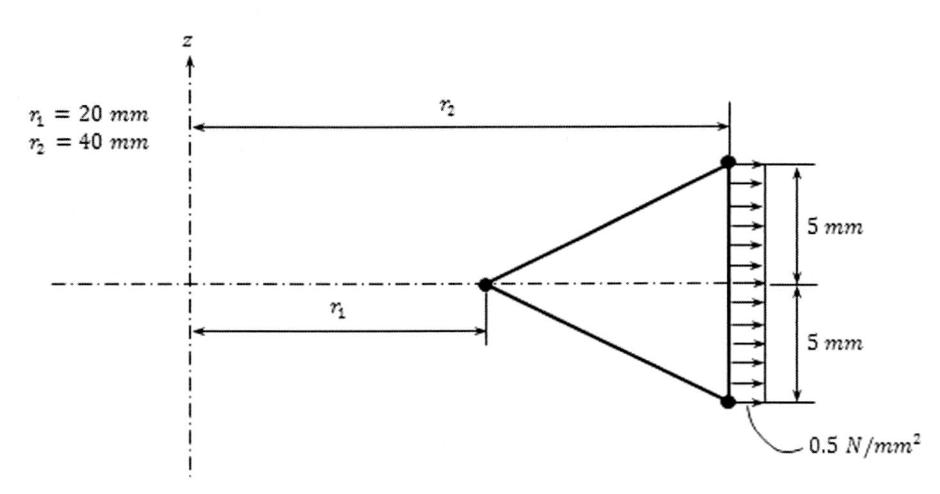

FIGURE P.8.6A

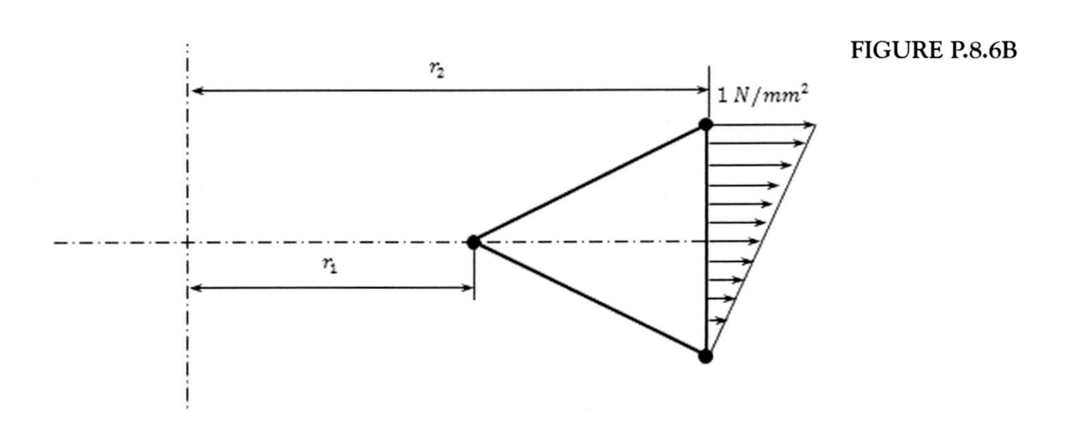

FIGURE P.8.6B

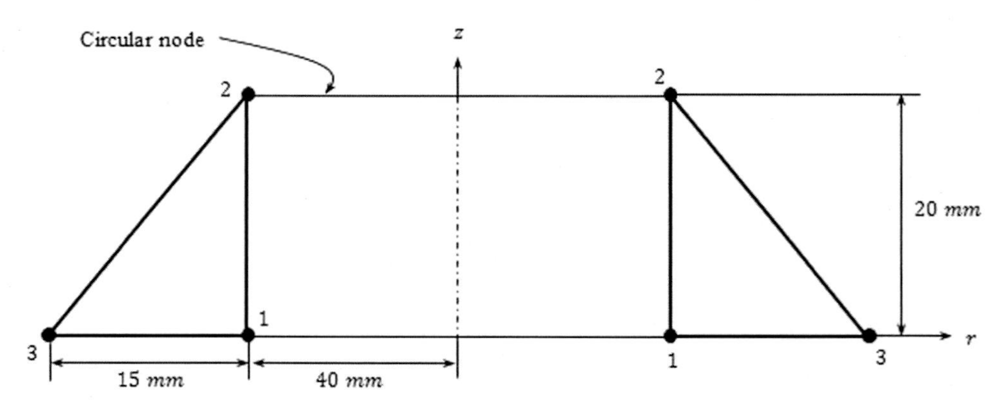

FIGURE P.8.7

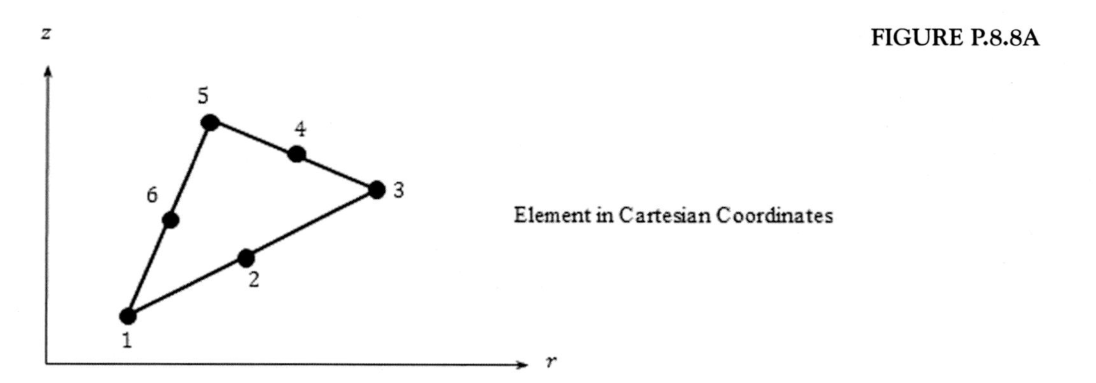

FIGURE P.8.8A

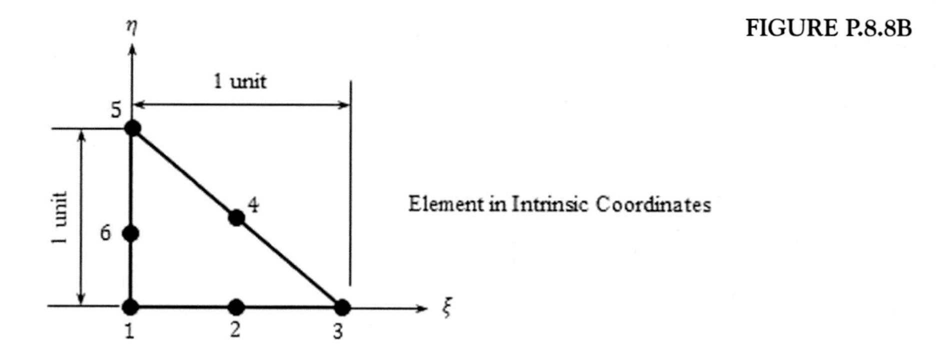

FIGURE P.8.8B

P.8.8: The six-node triangular finite element, shown in Figs. P.8.8A and P.8.8B is suggested for the stress analysis of the axisymmetric problem. The element shape functions are,

$$N_1 = L_1(2L_1 - 1)$$

$$N_2 = 4L_1L_2$$
$$N_3 = L_2(2L_2 - 1)$$
$$N_4 = 4L_2L_3$$
$$N_5 = L_3(2L_3 - 1)$$
$$N_6 = 4L_3L_1$$
$$L_1 = 1 - \eta - \xi$$
$$L_2 = \eta, \text{ and, } L_3 = \xi$$

1. using the standard algorithms for element derivation prove the element equations can be expressed as follows, $[K]\{\delta\} = \{F\}$
   where

$$\{\delta\} = \text{nodal displacement vector} = \begin{Bmatrix} (u_r)_1 \\ (u_z)_1 \\ \vdots \\ (u_r)_6 \\ (u_z)_6 \end{Bmatrix}$$

$$[K] = \iint 2\pi r[B]^t[D][B]|J|d\xi d\eta, \text{ with usual notations}$$

$$\{F\} = \text{nodal force vector} = \begin{Bmatrix} (F_r)_1 \\ (F_z)_1 \\ \vdots \\ (F_r)_6 \\ (F_z)_6 \end{Bmatrix}$$

Note that: $\{\sigma\} = [D]\{\varepsilon\}$ Where,

$$\{\varepsilon\} = \begin{Bmatrix} \varepsilon_r \\ \varepsilon_z \\ \varepsilon_\theta \\ \gamma_{rz} \end{Bmatrix} = \begin{Bmatrix} \dfrac{\partial u_r}{\partial r} \\ \dfrac{\partial u_z}{\partial z} \\ \dfrac{u_r}{r} \\ \dfrac{\partial u_r}{\partial z} + \dfrac{\partial u_z}{\partial r} \end{Bmatrix}$$

2. Explain how such integrations can be evaluated numerically in intrinsic $\xi\eta$ plane.
3. What are the essential conditions required to analyze a 3-Dim stress analysis through axisymmetric ring elements?

P.8.9: For the formulation of a plane stress axisymmetric ring , as shown in Fig. P.8.9, finite element of thickness t, inner, and outer radii r_1 and r_2 respectively, it is proposed to use the following displacement model

$$u = \frac{1}{(r_2 - r_1)}[(r_2 - r)(r - r_1)]\left\{ \begin{matrix} u_1 \\ u_2 \end{matrix} \right\}$$

Or,

$$u = [N]\{u\}_0$$

Deduce expressions for the radial and tangential strain components $\varepsilon_r, \varepsilon_\theta$ in the form

$$\{\varepsilon\} = [B]\{u\}^0$$

Accepting that the strain energy density is

$$U_0 = \frac{1}{2}\{\varepsilon\}^t[D]\{\varepsilon\}$$

And using the appropriate form of the elasticity matrix [D], develop an expression for the strain energy stored in the element in the form

$$U_e = \frac{1}{2}\{u\}^{0t}\left(2\pi\int_{r_1}^{r_2} t[\text{Matrix}]rdr\right)\{u\}^0$$

Showing that [Matrix] is 2×2 symmetric matrix.

Comment on any particular problems associated with a closing "no hole" element.

If the element is to be used in a discretization of a rotating disk (Fig. P.8.9), show how element ring nodal forces represent the inertia force $\{F\}$.

Such a rotating disk is to be analyzed by three concentric elements 1,2,3 having inner and outer radii r_1, r_2 and r_3, and r_4 respectively; $r_1 < r_2 < r_3 < r_4$. Using the notation $[K_{ij}]$

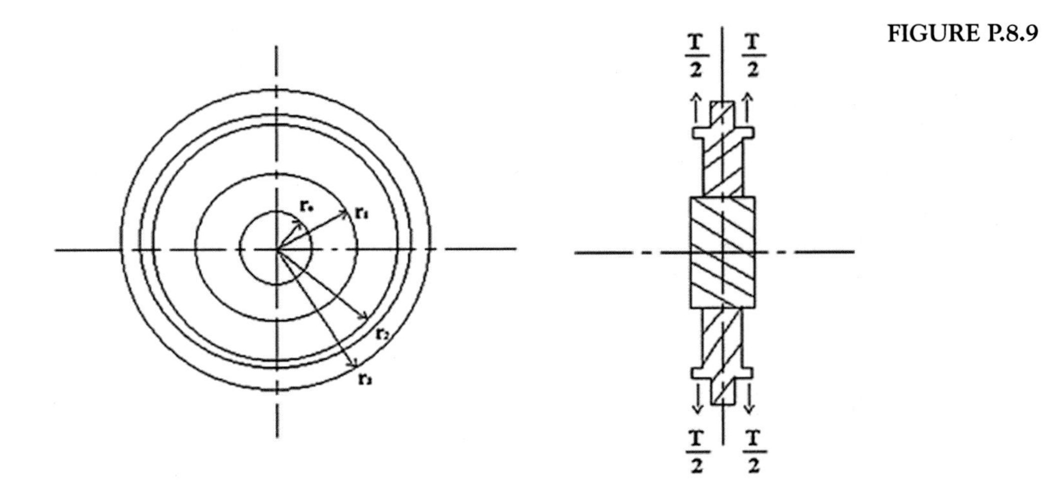

FIGURE P.8.9

for the elements of stiffness matrix for element e, $\{F_i\}$ As ring nodal forces use stationary potential energy to deduce the assembled matrix equilibrium equations for the system. With so few degrees of freedom, work directly in terms of the generalized coordinates u_1, u_2, u_3, u_4.

P.8.10: For the element shown in Fig. P.8.10, evaluate the stiffness matrices, $E = 210$ GPa, $\nu = 0.25$ for each element.

P.8.11: For an element of an axisymmetric body rotates with a constant angular velocity $\omega = 20$rpm as shown in Fig. P.8.11. Evaluate the body force matrix, the density of the element $\rho = 7800$ kg/m^3.

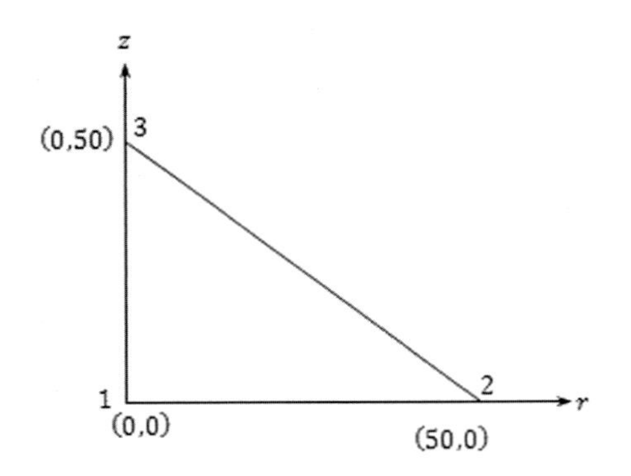

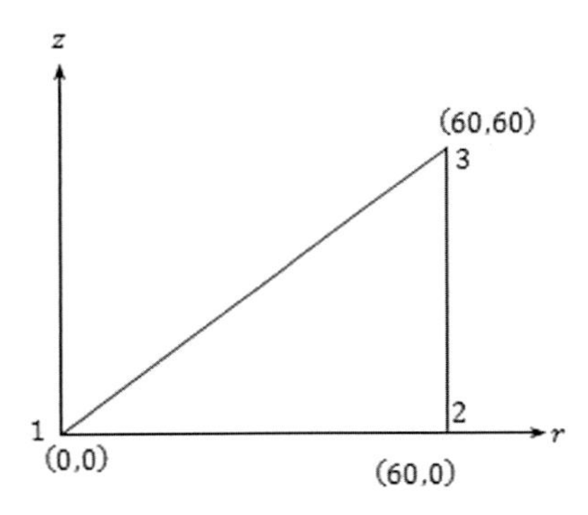

FIGURE P.8.10

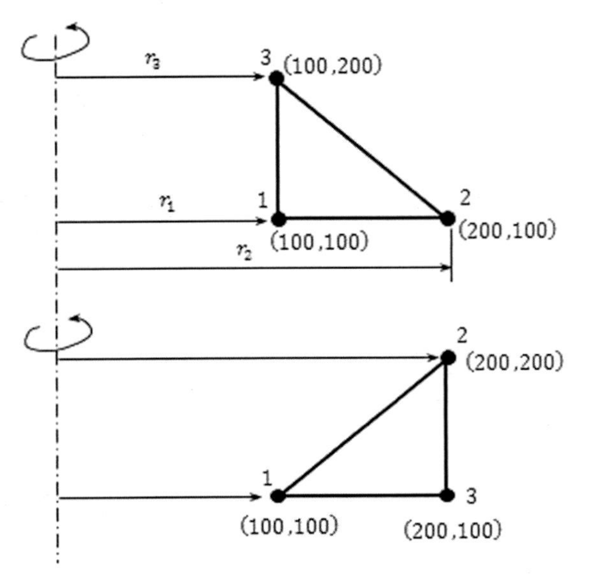

FIGURE P.8.11

Bibliography

[1] K.J. Bathe, Finite Element Procedures, Prentice-Hill, Englewood Cliffs, NJ, 1996.
[2] Y.K. Cheung, M.F. Yeo, A Practical Introduction to Finite Element Analysis, Pitman, 1979.
[3] R.D. Cook, Finite Element Modeling for Stress Analysis, Book, Wiley, 1995.
[4] J.G. Crose, Stress analysis of axisymmetric solids with asymmetric properties, AIJJA 10 (7) (1972) 866–871.
[5] C.S. Desai, J.F. Abel, Introduction to the Finite Element Method, Van Nostrand, Reinhold, New York, 1972.
[6] A. El-Zafran, Internal Report, Grandfield University (1985).
[7] R. Galagher, Finite Element Analysis Fundamentals, Prentice-Hall, Englewood Cliffs, 1975.
[8] E.J. Hearn, Mechanics of Materials, International Series on Materials Science and Technology I, Elsevier Science & Technology, 1997.
[9] K.H. Huebner, E.A. Thornton, The Finite Element Analysis for Engineers, third ed., Wiley, New York, 1995.
[10] M.J. Jweeg, Application of finite element method to study the stress distribution in shell of revolution subjected to non-axisymmetric loading, in: Proceedings of the International ASMS Conference, Signals and Systems, Cetinje, Yougoslavia 5 (1990) 3–14.
[11] M.J. Jweeg, Static bending deflections and flexural frequencies of prismatic cantilever tapered beams by Galerkin's method, A Scientific Journal, College of Engineering, University of Baghdad 4 (3) (1997) 13–24.
[12] M.J. Jweeg, A.N. Al-Timimy, B.I. Kazem, Detection and location of defects in a plate structure by vibration technique, A Scientific Journal, College of Engineering, University of Baghdad 4 (3) (1998) 31–47.
[13] N.M. Motosh, M.J. Jweeg, Study of rotating axi-symmetrical shell, Scientific Journal of M.E.C 7 (1991) 30–48.
[14] Y.R. Rashid, Three-dimensional analysis of elastic solids-II; analysis procedure, International Journal of Solids and Structures 6 (1970) 195–207.
[15] R.J. Roark, W.C. Young, Formulas for Stress and Strain, fifth ed., McGraw-Hill, New York, 1975.
[16] J. Robinson, Understanding Finite Element Stress Analysis, Robinson and Associates, Wimborne, 1973.
[17] E.L. Wilson, Structural analysis of axisymmetric solids, AIAAJ 3 (12) (1965) 2269–2274.

CHAPTER

9

Application of finite element method to three-dimensional elasticity problems

In this chapter, the formulation of three-dimensional (3D) elements will be presented following the standard step-by-step method. The stiffness matrix, in addition to the load vectors, is presented. For example, the 8-node hexahedral element is explained using the most straightforward modeling of solid bodies. In this analysis, a rigid body is assumed, and therefore there is no rotation at the nodes, we assume that there are three degrees, in x, y, and z directions of freedom at each node. The isoparametric formulation was adopted in the modeling of 3D bodies. The 8-node hexahedral isoparametric and the tetrahedral solid elements were presented in detail starting from the displacements models assumptions of u, v, and w till reaches the element stiffness matrix formula in terms of the intrinsic coordinates. Finally, selected problems were included in this chapter.

9.1 Three-dimensional elasticity relations

Consider the 3D element with dimensions dx, dy, and dz and normal and shear stresses as shown in Fig. 9.1.

From the theory of elasticity, we have, cross shears are equal for equilibrium requirements. Therefore

$$\sigma_{ij} = \sigma_{ji} \tag{9.1}$$

The strain–displacement relationship is given by,

$$\varepsilon_x = \frac{\partial u}{\partial x}, \varepsilon_y = \frac{\partial v}{\partial y}, \varepsilon_z = \frac{\partial w}{\partial z} \tag{9.2}$$

Where, u, v, and w are the displacements in x, y, and z directions.

DOI: https://doi.org/10.1016/B978-0-323-88666-6.00008-3

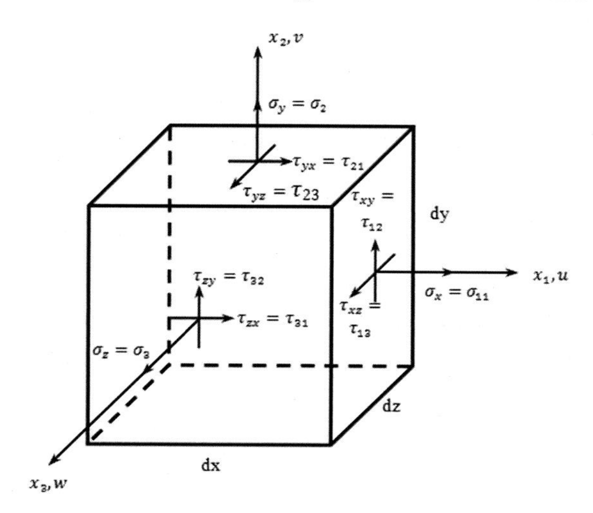

FIGURE 9.1 Three-dimensional stresses on an element.

The shear strain components are,

$$\gamma_{xy} = \frac{\partial u}{\partial y} + \frac{\partial v}{\partial x}$$

$$\gamma_{xz} = \frac{\partial u}{\partial z} + \frac{\partial w}{\partial x}$$

$$\gamma_{yz} = \frac{\partial v}{\partial z} + \frac{\partial w}{\partial y} \tag{9.3}$$

Therefore the stress and strain components can be written as follows,

$$\{\sigma\} = \begin{Bmatrix} \sigma_{11} \\ \sigma_{22} \\ \sigma_{33} \\ \sigma_{12} \\ \sigma_{13} \\ \sigma_{23} \end{Bmatrix}$$

And,

$$\{\varepsilon\} = \begin{Bmatrix} \varepsilon_{11} \\ \varepsilon_{22} \\ \varepsilon_{33} \\ \gamma_{12} \\ \gamma_{13} \\ \gamma_{23} \end{Bmatrix} \tag{9.4}$$

Where $\sigma_{11} = \sigma_{xx}$, $\sigma_{12} = \sigma_{xy}, \ldots$

And Hooke's law,

$$\{\sigma\} = [D]\{\varepsilon\} \tag{9.5}$$

Where $[D]$ is the elasticity matrix. For the particular case of homogenous isotropic material, it can be shown that,

$$[D] = \frac{E}{(1+\nu)(1-2\nu)}\begin{bmatrix} (1-\nu) & \nu & \nu & 0 & 0 & 0 \\ \nu & (1-\nu) & \nu & 0 & 0 & 0 \\ \nu & \nu & (1-\nu) & 0 & 0 & 0 \\ 0 & 0 & 0 & \frac{(1-2\nu)}{2} & 0 & 0 \\ 0 & 0 & 0 & 0 & \frac{(1-2\nu)}{2} & 0 \\ 0 & 0 & 0 & 0 & 0 & \frac{(1-2\nu)}{2} \end{bmatrix} \tag{9.6}$$

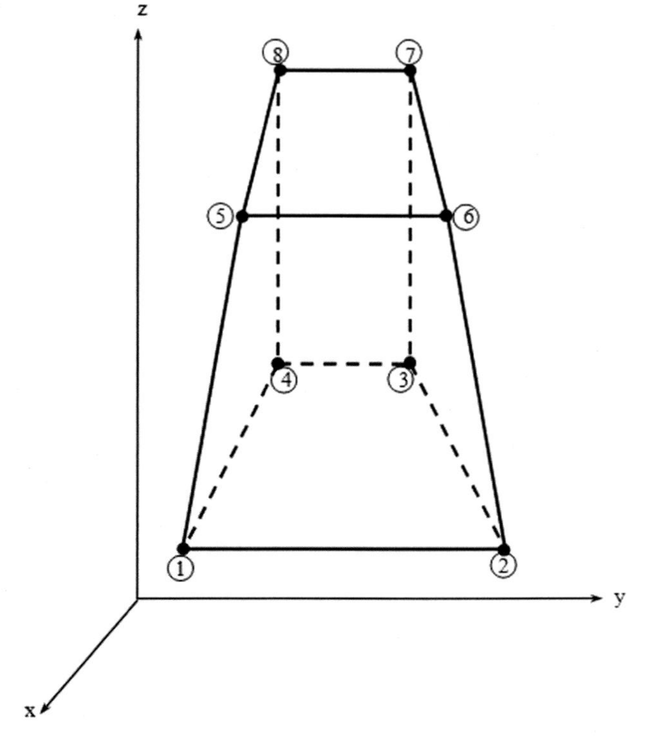

FIGURE 9.2 8-Node hexahedral element.

9.2 8-Node hexahedral element

The mathematical formulation of the 8-node hexahedral element shown in Fig. 9.2 is presented. This element is suitable for modeling the 3D solid bodies. This requires the 3D elasticity equations to be applied, and the procedure will follow the similar steps explained in the chapters five , six and seven.

9.3 Steps of formulation

The procedure of formulation of the stiffness equations is similar to those presented previously. The complexities of obtaining the stiffness and load vectors will be simplified in detail. The following steps are presented in detail. These steps are summarized as follows,

Step 1, Define the nodal parameters for the case of 3D elasticity problems, the state of displacements at any point in the domain can be defined in terms of three displacement components u, v, and w in the direction of the x, y, and z axes, respectively. For the 8-node hexahedral element using a local numbering system, the following vectors can be defined,

1. Nodal displacement vector

At each node, the displacements are u_i, v_i, and w_i, in x_i, y_i, and z_i, directions, where $i = 1, 8$ is the number of nodes of the element and can be arranged in a vector as follows,

$$\{\delta\}^t = \{u_1 \quad v_1 \quad w_1 \quad \cdots \quad \cdots \quad u_8 \quad v_8 \quad w_8\} \tag{9.7}$$

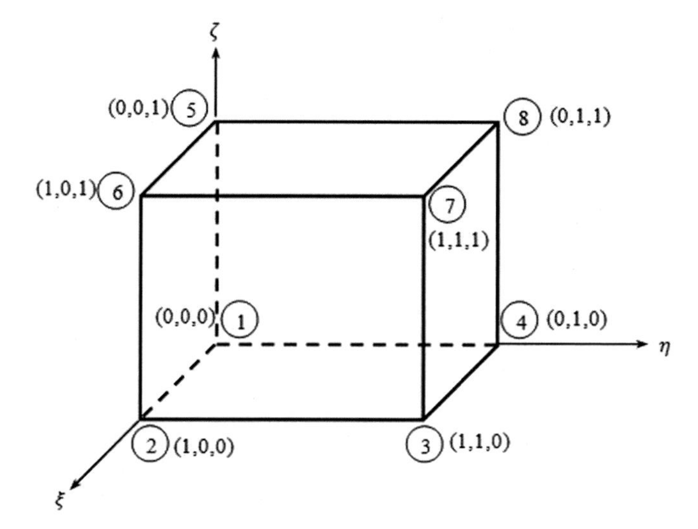

FIGURE 9.3 8-Node isoparametric hexahedral element.

2. Nodal loading vector

The loading is in x_i, y_i, and z_i, directions where $i = 1, 8$ is the number of nodes. These can be arranged in a vector containing the element load vector as follows.

$$\{F\}^t = \{F_{x_1} \quad F_{y_1} \quad F_{z_1} \quad \cdots \quad \cdots \quad F_{x_8} \quad F_{y_8} \quad F_{z_8}\} \tag{9.8}$$

3. Shape function determination

The shape function $N_i(x, y, z)$ can be found as follows. Fig. 9.3 shows a 3D element with eight nodes, and by using the natural coordinates (ξ, η, ζ) with the local numbering sequence should be adhered to be used in obtaining the system stiffness matrix concerning the global numbering. The displacement field can be assumed as follows,

$$u(x, y, z) = \sum_{i=1}^{8} u_i N_i(x, y, z)$$

$$v(x, y, z) = \sum_{i=1}^{8} v_i N_i(x, y, z)$$

$$w(x, y, z) = \sum_{i=1}^{8} w_i N_i(x, y, z) \tag{9.9}$$

To obtain the same function expressions for every 8-node hexahedral element, it is essential to transform the element into a uniform one in an intrinsic space. This may be carried out by transforming the element into a cube of unit side length in the intrinsic ξ, η, ζ space as shown in Fig. 9.3.

Step 2,

Consider one of the displacement components, say (u-component in the x-direction), it can be expressed in terms of ξ, η, ζ using an algebraic polynomial as follows,

$$u(\xi, \eta, \zeta) = \alpha_1 + \alpha_2\xi + \alpha_3\eta + \alpha_4\zeta + \alpha_5\xi\eta + \alpha_6\xi\zeta + \alpha_7\eta\zeta + \alpha_8\xi\eta\zeta$$

$$v(\xi, \eta, \zeta) = \alpha_9 + \alpha_{10}\xi + \alpha_{11}\eta + \alpha_{12}\zeta + \alpha_{13}\xi\eta + \alpha_{14}\xi\zeta + \alpha_{15}\eta\zeta + \alpha_{16}\xi\eta\zeta$$

$$w(\xi, \eta, \zeta) = \alpha_{17} + \alpha_{18}\xi + \alpha_{19}\eta + \alpha_{20}\zeta + \alpha_{21}\xi\eta + \alpha_{22}\xi\zeta + \alpha_{23}\eta\zeta + \alpha_{24}\xi\eta\zeta \tag{9.10}$$

Using the following eight conditions,

$$u = u_1, v = v_1, w = w_1, \quad \text{at,} \quad \xi = 0, \eta = 0, \zeta = 0$$

$$u = u_2, v = v_2, w = w_2, \quad \text{at,} \quad \xi = 1, \eta = 0, \zeta = 0$$

$$u = u_3, v = v_3, w = w_3, \quad \text{at,} \quad \xi = 1, \eta = 1, \zeta = 0$$

$$u = u_4, v = v_4, w = w_4, \quad \text{at,} \quad \xi = 0, \eta = 1, \zeta = 0$$

$$u = u_5, v = v_5, w = w_5, \quad \text{at}, \quad \xi = 0, \eta = 0, \zeta = 1$$

$$u = u_6, v = v_6, w = w_6, \quad \text{at}, \quad \xi = 1, \eta = 0, \zeta = 1$$

$$u = u_7, v = v_7, w = w_7, \quad \text{at}, \quad \xi = 1, \eta = 1, \zeta = 1$$

$$u = u_8, v = v_8, w = w_8, \quad \text{at}, \quad \xi = 0, \eta = 1, \zeta = 1 \tag{9.11}$$

The unknowns $\alpha_1, \alpha_2, \ldots, \alpha_8$ in Eq. (9.10) can be found as follows, by substitution of boundary conditions of Eq. (9.5) in Eq. (9.4) as follows, using the conditions of Eq. (9.11), we can write,

$$u_1 = \alpha_1,\ v_1 = \alpha_9,\ w_1 = \alpha_{17}$$

$$u_2 = \alpha_1 + \alpha_2$$

$$v_2 = \alpha_9 + \alpha_{10}$$

$$w_2 = \alpha_{17} + \alpha_{18}$$

$$u_3 = \alpha_1 + \alpha_2 + \alpha_3 + \alpha_5$$

$$v_3 = \alpha_9 + \alpha_{10} + \alpha_{11} + \alpha_{13}$$

$$w_3 = \alpha_{17} + \alpha_{18} + \alpha_{19} + \alpha_{21}$$

$$u_4 = \alpha_1 + \alpha_3$$

$$v_4 = \alpha_9 + \alpha_{11}$$

$$w_4 = \alpha_{17} + \alpha_{19}$$

$$u_5 = \alpha_1 + \alpha_4$$

$$v_5 = \alpha_9 + \alpha_{12}$$

$$w_5 = \alpha_{17} + \alpha_{20}$$

$$u_6 = \alpha_1 + \alpha_2 + \alpha_4 + \alpha_6$$

$$v_6 = \alpha_9 + \alpha_{10} + \alpha_{12} + \alpha_{14}$$

$$w_6 = \alpha_{17} + \alpha_{18} + \alpha_{20} + \alpha_{22}$$

$$u_7 = \alpha_1 + \alpha_2 + \alpha_3 + \alpha_4 + \alpha_5 + \alpha_6 + \alpha_7 + \alpha_8$$

$$v_7 = \alpha_9 + \alpha_{10} + \alpha_{11} + \alpha_{12} + \alpha_{13} + \alpha_{14} + \alpha_{15} + \alpha_{16}$$

$$w_7 = \alpha_{17} + \alpha_{18} + \alpha_{19} + \alpha_{20} + \alpha_{21} + \alpha_{22} + \alpha_{23} + \alpha_{24}$$

$$
\begin{aligned}
u_8 &= \alpha_1 + \alpha_3 + \alpha_4 + \alpha_7 \\
v_8 &= \alpha_9 + \alpha_{11} + \alpha_{12} + \alpha_{15} \\
w_8 &= \alpha_{17} + \alpha_{19} + \alpha_{20} + \alpha_{23}
\end{aligned}
\tag{9.12}
$$

Arrange Eq. (9.12) in matrix form,

$$
\begin{Bmatrix} u_1 \\ v_1 \\ w_1 \\ u_2 \\ v_2 \\ w_2 \\ u_3 \\ v_3 \\ w_3 \\ u_4 \\ v_4 \\ w_4 \\ u_5 \\ v_5 \\ w_5 \\ u_6 \\ v_6 \\ w_6 \\ u_7 \\ v_7 \\ w_7 \\ u_8 \\ v_8 \\ w_8 \end{Bmatrix}
=
\begin{bmatrix}
1&0&0&0&0&0&0&0&0&0&0&0&0&0&0&0&0&0&0&0&0&0&0&0\\
0&0&0&0&0&0&0&0&1&0&0&0&0&0&0&0&0&0&0&0&0&0&0&0\\
0&0&0&0&0&0&0&0&0&0&0&0&0&0&0&0&1&0&0&0&0&0&0&0\\
1&1&0&0&0&0&0&0&0&0&0&0&0&0&0&0&0&0&0&0&0&0&0&0\\
0&0&0&0&0&0&0&0&1&1&0&0&0&0&0&0&0&0&0&0&0&0&0&0\\
0&0&0&0&0&0&0&0&0&0&0&0&0&0&0&0&1&1&0&0&0&0&0&0\\
1&1&1&0&1&0&0&0&0&0&0&0&0&0&0&0&0&0&0&0&0&0&0&0\\
0&0&0&0&0&0&0&0&1&1&1&0&1&0&0&0&0&0&0&0&0&0&0&0\\
0&0&0&0&0&0&0&0&0&0&0&0&0&0&0&0&1&1&1&0&1&0&0&0\\
1&0&1&0&0&0&0&0&0&0&0&0&0&0&0&0&0&0&0&0&0&0&0&0\\
0&0&0&0&0&0&0&0&1&0&1&0&0&0&0&0&0&0&0&0&0&0&0&0\\
0&0&0&0&0&0&0&0&0&0&0&0&0&0&0&0&1&0&1&0&0&0&0&0\\
1&0&0&1&0&0&0&0&0&0&0&0&0&0&0&0&0&0&0&0&0&0&0&0\\
0&0&0&0&0&0&0&0&1&0&0&1&0&0&0&0&0&0&0&0&0&0&0&0\\
0&0&0&0&0&0&0&0&0&0&0&0&0&0&0&0&1&0&0&1&0&0&0&0\\
1&1&0&1&0&1&0&0&0&0&0&0&0&0&0&0&0&0&0&0&0&0&0&0\\
0&0&0&0&0&0&0&0&1&1&0&1&0&1&0&0&0&0&0&0&0&0&0&0\\
0&0&0&0&0&0&0&0&0&0&0&0&0&0&0&0&1&1&0&1&0&1&0&0\\
1&1&1&1&1&1&1&1&0&0&0&0&0&0&0&0&0&0&0&0&0&0&0&0\\
0&0&0&0&0&0&0&0&1&1&1&1&1&1&1&1&0&0&0&0&0&0&0&0\\
0&0&0&0&0&0&0&0&0&0&0&0&0&0&0&0&1&1&1&1&1&1&1&1\\
1&0&1&1&0&0&1&0&0&0&0&0&0&0&0&0&0&0&0&0&0&0&0&0\\
0&0&0&0&0&0&0&0&1&0&1&1&0&0&1&0&0&0&0&0&0&0&0&0\\
0&0&0&0&0&0&0&0&0&0&0&0&0&0&0&0&1&0&1&1&0&0&1&0
\end{bmatrix}
\begin{Bmatrix} \alpha_1 \\ \alpha_2 \\ \alpha_3 \\ \alpha_4 \\ \alpha_5 \\ \alpha_6 \\ \alpha_7 \\ \alpha_8 \\ \alpha_9 \\ \alpha_{10} \\ \alpha_{11} \\ \alpha_{12} \\ \alpha_{13} \\ \alpha_{14} \\ \alpha_{15} \\ \alpha_{16} \\ \alpha_{17} \\ \alpha_{18} \\ \alpha_{19} \\ \alpha_{20} \\ \alpha_{21} \\ \alpha_{22} \\ \alpha_{23} \\ \alpha_{24} \end{Bmatrix}
\tag{9.13}
$$

Writing Eq. (9.13) in an abbreviated form gives,

$$\{\delta\} = [A]\{\alpha\}$$

Or,

$$\{\alpha\} = [A]^{-1}\{\delta\} \tag{9.14}$$

Where,

$$[A] = \begin{bmatrix}
1 & 0 & 0 & 0 & 0 & 0 & 0 & 0 & 0 & 0 & 0 & 0 & 0 & 0 & 0 & 0 & 0 & 0 & 0 & 0 & 0 & 0 & 0 & 0 \\
0 & 0 & 0 & 0 & 0 & 0 & 0 & 0 & 1 & 0 & 0 & 0 & 0 & 0 & 0 & 0 & 0 & 0 & 0 & 0 & 0 & 0 & 0 & 0 \\
0 & 0 & 0 & 0 & 0 & 0 & 0 & 0 & 0 & 0 & 0 & 0 & 0 & 0 & 0 & 0 & 1 & 0 & 0 & 0 & 0 & 0 & 0 & 0 \\
1 & 1 & 0 & 0 & 0 & 0 & 0 & 0 & 0 & 0 & 0 & 0 & 0 & 0 & 0 & 0 & 0 & 0 & 0 & 0 & 0 & 0 & 0 & 0 \\
0 & 0 & 0 & 0 & 0 & 0 & 0 & 0 & 1 & 1 & 0 & 0 & 0 & 0 & 0 & 0 & 0 & 0 & 0 & 0 & 0 & 0 & 0 & 0 \\
0 & 0 & 0 & 0 & 0 & 0 & 0 & 0 & 0 & 0 & 0 & 0 & 0 & 0 & 0 & 0 & 1 & 1 & 0 & 0 & 0 & 0 & 0 & 0 \\
1 & 1 & 1 & 0 & 1 & 0 & 0 & 0 & 0 & 0 & 0 & 0 & 0 & 0 & 0 & 0 & 0 & 0 & 0 & 0 & 0 & 0 & 0 & 0 \\
0 & 0 & 0 & 0 & 0 & 0 & 0 & 0 & 1 & 1 & 1 & 0 & 1 & 0 & 0 & 0 & 0 & 0 & 0 & 0 & 0 & 0 & 0 & 0 \\
0 & 0 & 0 & 0 & 0 & 0 & 0 & 0 & 0 & 0 & 0 & 0 & 0 & 0 & 0 & 0 & 1 & 1 & 1 & 0 & 1 & 0 & 0 & 0 \\
1 & 0 & 1 & 0 & 0 & 0 & 0 & 0 & 0 & 0 & 0 & 0 & 0 & 0 & 0 & 0 & 0 & 0 & 0 & 0 & 0 & 0 & 0 & 0 \\
0 & 0 & 0 & 0 & 0 & 0 & 0 & 0 & 1 & 0 & 1 & 0 & 0 & 0 & 0 & 0 & 0 & 0 & 0 & 0 & 0 & 0 & 0 & 0 \\
0 & 0 & 0 & 0 & 0 & 0 & 0 & 0 & 0 & 0 & 0 & 0 & 0 & 0 & 0 & 0 & 1 & 0 & 1 & 0 & 0 & 0 & 0 & 0 \\
1 & 0 & 0 & 1 & 0 & 0 & 0 & 0 & 0 & 0 & 0 & 0 & 0 & 0 & 0 & 0 & 0 & 0 & 0 & 0 & 0 & 0 & 0 & 0 \\
0 & 0 & 0 & 0 & 0 & 0 & 0 & 0 & 1 & 0 & 0 & 1 & 0 & 0 & 0 & 0 & 0 & 0 & 0 & 0 & 0 & 0 & 0 & 0 \\
0 & 0 & 0 & 0 & 0 & 0 & 0 & 0 & 0 & 0 & 0 & 0 & 0 & 0 & 0 & 0 & 1 & 0 & 0 & 1 & 0 & 0 & 0 & 0 \\
1 & 1 & 0 & 1 & 0 & 1 & 0 & 0 & 0 & 0 & 0 & 0 & 0 & 0 & 0 & 0 & 0 & 0 & 0 & 0 & 0 & 0 & 0 & 0 \\
0 & 0 & 0 & 0 & 0 & 0 & 0 & 0 & 1 & 1 & 0 & 1 & 0 & 1 & 0 & 0 & 0 & 0 & 0 & 0 & 0 & 0 & 0 & 0 \\
0 & 0 & 0 & 0 & 0 & 0 & 0 & 0 & 0 & 0 & 0 & 0 & 0 & 0 & 0 & 0 & 1 & 1 & 0 & 1 & 0 & 1 & 0 & 0 \\
1 & 1 & 1 & 1 & 1 & 1 & 1 & 1 & 0 & 0 & 0 & 0 & 0 & 0 & 0 & 0 & 0 & 0 & 0 & 0 & 0 & 0 & 0 & 0 \\
0 & 0 & 0 & 0 & 0 & 0 & 0 & 0 & 1 & 1 & 1 & 1 & 1 & 1 & 1 & 1 & 0 & 0 & 0 & 0 & 0 & 0 & 0 & 0 \\
0 & 0 & 0 & 0 & 0 & 0 & 0 & 0 & 0 & 0 & 0 & 0 & 0 & 0 & 0 & 0 & 1 & 1 & 1 & 1 & 1 & 1 & 1 & 1 \\
1 & 0 & 1 & 1 & 0 & 0 & 1 & 0 & 0 & 0 & 0 & 0 & 0 & 0 & 0 & 0 & 0 & 0 & 0 & 0 & 0 & 0 & 0 & 0 \\
0 & 0 & 0 & 0 & 0 & 0 & 0 & 0 & 1 & 0 & 1 & 1 & 0 & 0 & 1 & 0 & 0 & 0 & 0 & 0 & 0 & 0 & 0 & 0 \\
0 & 0 & 0 & 0 & 0 & 0 & 0 & 0 & 0 & 0 & 0 & 0 & 0 & 0 & 0 & 0 & 1 & 0 & 1 & 1 & 0 & 0 & 1 & 0
\end{bmatrix}$$

Having obtained $\alpha s'$ from Eq. (9.14), Eq. (9.10) becomes,

$$\underset{\substack{\text{field displacement} \\ (8\times 1)}}{u_i(\xi,\eta,\zeta)} = \underset{\substack{\text{shape function} \\ (1\times 8)}}{\begin{bmatrix} N_1(\xi,\eta,\zeta) & \cdots & N_8(\xi,\eta,\zeta) \end{bmatrix}} \underset{\substack{\text{nodal displacement vector} \\ (8\times 1)}}{\begin{Bmatrix} u_1 \\ u_2 \\ \vdots \\ u_8 \end{Bmatrix}}$$

Similar expressions are obtained for $v_i(\xi,\eta,\zeta)$ and $w_i(\xi,\eta,\zeta)$.

$$\underset{\substack{\text{field displacement}\\(8\times 1)}}{v_i(\xi,\eta,\zeta)} = \underset{\substack{\text{shape function}\\(1\times 8)}}{\left[N_1(\xi,\eta,\zeta) \quad \cdots \quad N_8(\xi,\eta,\zeta)\right]} \underset{\substack{\text{nodal displacement vector}\\(8\times 1)}}{\begin{Bmatrix} v_1 \\ v_2 \\ \vdots \\ v_8 \end{Bmatrix}}$$

$$\underset{\substack{\text{field displacement}\\(8\times 1)}}{w_i(\xi,\eta,\zeta)} = \underset{\substack{\text{shape function}\\(1\times 8)}}{\left[N_1(\xi,\eta,\zeta) \quad \cdots \quad N_8(\xi,\eta,\zeta)\right]} \underset{\substack{\text{nodal displacement vector}\\(8\times 1)}}{\begin{Bmatrix} w_1 \\ w_2 \\ \vdots \\ w_8 \end{Bmatrix}} \tag{9.15}$$

These can be arranged in matrix form,

$$\underset{\substack{\text{field displacement vector}\\\text{of a particular position}\\(24\times 1)}}{\begin{Bmatrix} u_i \\ v_i \\ w_i \end{Bmatrix}} = \underset{(24\times 24)}{\begin{bmatrix} N_1 & 0 & 0 & N_2 & 0 & 0 & \cdots & N_8 & 0 & 0 \\ 0 & N_1 & 0 & 0 & N_2 & 0 & \cdots & 0 & N_8 & 0 \\ 0 & 0 & N_1 & 0 & 0 & N_2 & \cdots & 0 & 0 & N_8 \\ \vdots & \vdots & \vdots & \vdots & \vdots & \vdots & \vdots & \vdots & \vdots & \vdots \end{bmatrix}} \underset{(24\times 1)}{\begin{Bmatrix} u_1 \\ v_1 \\ w_1 \\ \vdots \\ u_8 \\ v_8 \\ w_8 \end{Bmatrix}} \tag{9.16}$$

Therefore we can write the expressions for the shape functions as follows,

The boundary conditions in Eq. (9.11) are substituted in Eq. (9.10), eight equations are obtained and can be arranged as follows,

The unknown parameters $\alpha_1, \alpha_2, \ldots\ldots, \alpha_8$ can be obtained using the above boundary conditions. Hence, it can be shown that,

$$N_1(\xi,\eta,\zeta) = (1-\xi)(1-\eta)(1-\zeta)$$

$$N_2(\xi,\eta,\zeta) = (\xi)(1-\eta)(1-\zeta)$$

$$N_3(\xi,\eta,\zeta) = (\xi)(\eta)(1-\zeta)$$

$$N_4(\xi,\eta,\zeta) = (1-\xi)(\eta)(1-\zeta)$$

$$N_5(\xi,\eta,\zeta) = (1-\xi)(1-\eta)(\zeta)$$

$$N_6(\xi,\eta,\zeta) = (\xi)(1-\eta)(\zeta)$$

$$N_7(\xi,\eta,\zeta) = (\xi)(\eta)(\zeta)$$

$$N_8(\xi,\eta,\zeta) = (1-\xi)(\eta)(\zeta) \tag{9.17}$$

The relationship between the Cartesian system (x, y, z) and the intrinsic system (ξ,η,ζ) can be obtained by employing the isoparametric transformation,

$$x(\xi,\eta,\zeta) = \sum_{i=1}^{8} x_i N_i(\xi,\eta,\zeta)$$

$$y(\xi,\eta,\zeta) = \sum_{i=1}^{8} y_i N_i(\xi,\eta,\zeta)$$

$$z(\xi,\eta,\zeta) = \sum_{i=1}^{8} z_i N_i(\xi,\eta,\zeta) \tag{9.18}$$

Step 3, Express the strain components at any point in terms of nodal displacements and shape functions, we have, from the theory of elasticity, the strain components in 3D problems are as follows,

$$\varepsilon_x = \frac{\partial u}{\partial x} = \sum_{i=1}^{8} u_i \frac{\partial N_i}{\partial x}$$

$$\varepsilon_y = \frac{\partial v}{\partial y} = \sum_{i=1}^{8} v_i \frac{\partial N_i}{\partial y}$$

$$\varepsilon_z = \frac{\partial w}{\partial z} = \sum_{i=1}^{8} w_i \frac{\partial N_i}{\partial z}$$

$$\gamma_{xy} = \frac{\partial u}{\partial y} + \frac{\partial v}{\partial x} = \sum_{i=1}^{8} \left(u_i \frac{\partial N_i}{\partial y} + v_i \frac{\partial N_i}{\partial x} \right)$$

$$\gamma_{yz} = \frac{\partial v}{\partial z} + \frac{\partial w}{\partial y} = \sum_{i=1}^{8} \left(v_i \frac{\partial N_i}{\partial z} + w_i \frac{\partial N_i}{\partial y} \right)$$

$$\gamma_{zx} = \frac{\partial w}{\partial z} + \frac{\partial u}{\partial x} = \sum_{i=1}^{8} \left(u_i \frac{\partial N_i}{\partial x} + w_i \frac{\partial N_i}{\partial z} \right) \tag{9.19}$$

Defining the strain tensor,

$$\{\varepsilon\}^t = \left\{ \varepsilon_x \quad \varepsilon_y \quad \varepsilon_z \quad \gamma_{xy} \quad \gamma_{yz} \quad \gamma_{zx} \right\} \tag{9.20}$$

It can be shown that,

$$\underset{(6\times1)}{\{\varepsilon\}} = \underset{(6\times24)}{[B]} \underset{(24\times1)}{\{\delta\}} \tag{9.21}$$

Where,

$$[B]^t = \left[B_1 \quad B_2 \quad \cdots \quad \cdots \quad B_6 \right]$$

$$[B] = \begin{bmatrix} \frac{\partial N_i}{\partial x} & 0 & 0 \\ 0 & \frac{\partial N_i}{\partial y} & 0 \\ 0 & 0 & \frac{\partial N_i}{\partial z} \\ \frac{\partial N_i}{\partial y} & \frac{\partial N_i}{\partial x} & 0 \\ 0 & \frac{\partial N_i}{\partial z} & \frac{\partial N_i}{\partial y} \\ \frac{\partial N_i}{\partial z} & 0 & \frac{\partial N_i}{\partial x} \end{bmatrix} : \tag{9.22}$$

Where $[B]$ is the strain–displacement matrix of size (6×24). It is a function of natural coordinates $f(\xi, \eta, \zeta)$

The following procedure can be employed in order to obtain the Cartesian derivative of the intrinsic shape functions.

Consider a function $f(\xi, \eta, \zeta)$. Using the chain rule of partial differentiation, it can be deduced that,

$$\frac{\partial f}{\partial \xi} = \frac{\partial f}{\partial x}\frac{\partial x}{\partial \xi} + \frac{\partial f}{\partial y}\frac{\partial y}{\partial \xi} + \frac{\partial f}{\partial z}\frac{\partial z}{\partial \xi}$$

$$\frac{\partial f}{\partial \eta} = \frac{\partial f}{\partial x}\frac{\partial x}{\partial \eta} + \frac{\partial f}{\partial y}\frac{\partial y}{\partial \eta} + \frac{\partial f}{\partial z}\frac{\partial z}{\partial \eta}$$

$$\frac{\partial f}{\partial \zeta} = \frac{\partial f}{\partial x}\frac{\partial x}{\partial \zeta} + \frac{\partial f}{\partial y}\frac{\partial y}{\partial \zeta} + \frac{\partial f}{\partial z}\frac{\partial z}{\partial \zeta} \tag{9.23}$$

The above equations can be expressed in matrix form as follows,

$$\begin{Bmatrix} \dfrac{\partial f}{\partial \xi} \\ \dfrac{\partial f}{\partial \eta} \\ \dfrac{\partial f}{\partial \zeta} \end{Bmatrix} = [J] \begin{bmatrix} \dfrac{\partial f}{\partial x} \\ \dfrac{\partial f}{\partial y} \\ \dfrac{\partial f}{\partial z} \end{bmatrix} \tag{9.24}$$

Where, $[J]$ is the 3D Jacobian matrix defined as follows,

$$[J] = \begin{bmatrix} \dfrac{\partial x}{\partial \xi} & \dfrac{\partial y}{\partial \xi} & \dfrac{\partial z}{\partial \xi} \\ \dfrac{\partial x}{\partial \eta} & \dfrac{\partial y}{\partial \eta} & \dfrac{\partial z}{\partial \eta} \\ \dfrac{\partial x}{\partial \zeta} & \dfrac{\partial y}{\partial \zeta} & \dfrac{\partial z}{\partial \zeta} \end{bmatrix} \tag{9.25}$$

Applying the above analysis to the intrinsic shape functions of the hexahedral element, it can be deduced that,

$$\begin{Bmatrix} \dfrac{\partial N_i}{\partial x} \\ \dfrac{\partial N_i}{\partial y} \\ \dfrac{\partial N_i}{\partial z} \end{Bmatrix} = [J]^{-1} \begin{Bmatrix} \dfrac{\partial N_i}{\partial \xi} \\ \dfrac{\partial N_i}{\partial \eta} \\ \dfrac{\partial N_i}{\partial \zeta} \end{Bmatrix}, \quad \text{where,} \quad i = (1, 2, \ldots\ldots, 8) \tag{9.26}$$

And,

$$|J| = x_{14}(y_{24}z_{34} - y_{34}z_{24}) + y_{14}(z_{24}x_{34} - z_{34}x_{24}) + z_{14}(x_{24}y_{34} - x_{34}y_{24})$$

Therefore the strain–displacement matrix $[B]$ is obtained by using Eqs. (9.25) and (9.26) as follows,

$$\begin{Bmatrix} \frac{\partial N_i}{\partial x} \\ \frac{\partial N_i}{\partial y} \\ \frac{\partial N_i}{\partial z} \end{Bmatrix} = \frac{1}{|J|}\begin{bmatrix} \left(\frac{\partial y}{\partial \eta}\frac{\partial z}{\partial \zeta} - \frac{\partial z}{\partial \eta}\frac{\partial y}{\partial \zeta}\right) & \left(\frac{\partial z}{\partial \eta}\frac{\partial x}{\partial \zeta} - \frac{\partial x}{\partial \eta}\frac{\partial z}{\partial \zeta}\right) & \left(\frac{\partial x}{\partial \eta}\frac{\partial y}{\partial \zeta} - \frac{\partial y}{\partial \eta}\frac{\partial x}{\partial \zeta}\right) \\ \left(\frac{\partial z}{\partial \xi}\frac{\partial y}{\partial \zeta} - \frac{\partial y}{\partial \xi}\frac{\partial z}{\partial \zeta}\right) & \left(\frac{\partial x}{\partial \xi}\frac{\partial z}{\partial \zeta} - \frac{\partial z}{\partial \xi}\frac{\partial x}{\partial \zeta}\right) & \left(\frac{\partial y}{\partial \xi}\frac{\partial x}{\partial \zeta} - \frac{\partial x}{\partial \xi}\frac{\partial y}{\partial \zeta}\right) \\ \left(\frac{\partial y}{\partial \xi}\frac{\partial z}{\partial \eta} - \frac{\partial z}{\partial \xi}\frac{\partial y}{\partial \eta}\right) & \left(\frac{\partial z}{\partial \xi}\frac{\partial x}{\partial \eta} - \frac{\partial x}{\partial \xi}\frac{\partial z}{\partial \eta}\right) & \left(\frac{\partial x}{\partial \xi}\frac{\partial y}{\partial \eta} - \frac{\partial y}{\partial \xi}\frac{\partial x}{\partial \eta}\right) \end{bmatrix}\begin{Bmatrix} \frac{\partial N_i}{\partial \xi} \\ \frac{\partial N_i}{\partial \eta} \\ \frac{\partial N_i}{\partial \zeta} \end{Bmatrix}$$

We have,

$$\begin{Bmatrix} \varepsilon_x \\ \varepsilon_y \\ \varepsilon_z \\ \gamma_{xy} \\ \gamma_{yz} \\ \gamma_{zx} \end{Bmatrix} = \begin{bmatrix} B(1,1) & 0 & 0 & \cdots & \cdots & \cdots & B(1,22) & 0 & 0 \\ 0 & B(2,2) & 0 & \cdots & \cdots & \cdots & 0 & B(1,23) & 0 \\ 0 & 0 & B(3,3) & \cdots & \cdots & \cdots & 0 & 0 & B(1,24) \\ B(4,1) & B(4,2) & 0 & \cdots & \cdots & \cdots & B(4,22) & B(4,23) & 0 \\ 0 & B(5,2) & B(5,3) & \cdots & \cdots & \cdots & 0 & B(5,23) & B(5,24) \\ B(6,1) & 0 & B(6,3) & \cdots & \cdots & \cdots & B(6,22) & 0 & B(6,24) \end{bmatrix}\begin{Bmatrix} u_1 \\ v_1 \\ w_1 \\ \vdots \\ u_8 \\ v_8 \\ w_8 \end{Bmatrix}$$

Or,

$$\{\varepsilon\} = [B]\{\delta\}$$

Where,

$$B(1,1) = -\frac{1}{|J|}\left(\begin{array}{c}\left(\frac{\partial y}{\partial \eta}\frac{\partial z}{\partial \zeta} - \frac{\partial z}{\partial \eta}\frac{\partial y}{\partial \zeta}\right)(1-\eta)(1-\zeta) + \left(\frac{\partial z}{\partial \eta}\frac{\partial x}{\partial \zeta} - \frac{\partial x}{\partial \eta}\frac{\partial z}{\partial \zeta}\right)(1-\xi)(1-\zeta) + \\ \left(\frac{\partial x}{\partial \eta}\frac{\partial y}{\partial \zeta} - \frac{\partial y}{\partial \eta}\frac{\partial x}{\partial \zeta}\right)(1-\xi)(1-\eta)\end{array}\right)$$

$$B(1,4) = \frac{1}{|J|}\left(\begin{array}{c}\left(\frac{\partial y}{\partial \eta}\frac{\partial z}{\partial \zeta} - \frac{\partial z}{\partial \eta}\frac{\partial y}{\partial \zeta}\right)(1-\eta)(1-\zeta) - \left(\frac{\partial z}{\partial \eta}\frac{\partial x}{\partial \zeta} - \frac{\partial x}{\partial \eta}\frac{\partial z}{\partial \zeta}\right)(\xi)(1-\zeta) - \\ \left(\frac{\partial x}{\partial \eta}\frac{\partial y}{\partial \zeta} - \frac{\partial y}{\partial \eta}\frac{\partial x}{\partial \zeta}\right)(\xi)(1-\eta)\end{array}\right)$$

$$B(1,7)=\frac{1}{|J|}\left(\left(\frac{\partial y}{\partial\eta}\frac{\partial z}{\partial\zeta}-\frac{\partial z}{\partial\eta}\frac{\partial y}{\partial\zeta}\right)(\eta)(1-\zeta)+\left(\frac{\partial z}{\partial\eta}\frac{\partial x}{\partial\zeta}-\frac{\partial x}{\partial\eta}\frac{\partial z}{\partial\zeta}\right)(\xi)(1-\zeta)-\left(\frac{\partial x}{\partial\eta}\frac{\partial y}{\partial\zeta}-\frac{\partial y}{\partial\eta}\frac{\partial x}{\partial\zeta}\right)(\xi)(\eta)\right)$$

$$B(1,10)=\frac{1}{|J|}\left(\begin{array}{c}-\left(\frac{\partial y}{\partial\eta}\frac{\partial z}{\partial\zeta}-\frac{\partial z}{\partial\eta}\frac{\partial y}{\partial\zeta}\right)(\eta)(1-\zeta)+\left(\frac{\partial z}{\partial\eta}\frac{\partial x}{\partial\zeta}-\frac{\partial x}{\partial\eta}\frac{\partial z}{\partial\zeta}\right)(1-\xi)(1-\zeta)-\\ \left(\frac{\partial x}{\partial\eta}\frac{\partial y}{\partial\zeta}-\frac{\partial y}{\partial\eta}\frac{\partial x}{\partial\zeta}\right)(1-\xi)(\eta)\end{array}\right)$$

$$B(1,13)=\frac{1}{|J|}\left(\begin{array}{c}-\left(\frac{\partial y}{\partial\eta}\frac{\partial z}{\partial\zeta}-\frac{\partial z}{\partial\eta}\frac{\partial y}{\partial\zeta}\right)(1-\eta)(\zeta)-\left(\frac{\partial z}{\partial\eta}\frac{\partial x}{\partial\zeta}-\frac{\partial x}{\partial\eta}\frac{\partial z}{\partial\zeta}\right)(1-\xi)(\zeta)+\\ \left(\frac{\partial x}{\partial\eta}\frac{\partial y}{\partial\zeta}-\frac{\partial y}{\partial\eta}\frac{\partial x}{\partial\zeta}\right)(1-\xi)(1-\eta)\end{array}\right)$$

$$B(1,16)=\frac{1}{|J|}\left(\left(\frac{\partial y}{\partial\eta}\frac{\partial z}{\partial\zeta}-\frac{\partial z}{\partial\eta}\frac{\partial y}{\partial\zeta}\right)(1-\eta)(\zeta)-\left(\frac{\partial z}{\partial\eta}\frac{\partial x}{\partial\zeta}-\frac{\partial x}{\partial\eta}\frac{\partial z}{\partial\zeta}\right)(\xi)(\zeta)+\left(\frac{\partial x}{\partial\eta}\frac{\partial y}{\partial\zeta}-\frac{\partial y}{\partial\eta}\frac{\partial x}{\partial\zeta}\right)(\xi)(1-\eta)\right)$$

$$B(1,19)=\frac{1}{|J|}\left(\left(\frac{\partial y}{\partial\eta}\frac{\partial z}{\partial\zeta}-\frac{\partial z}{\partial\eta}\frac{\partial y}{\partial\zeta}\right)(\eta)(\zeta)+\left(\frac{\partial z}{\partial\eta}\frac{\partial x}{\partial\zeta}-\frac{\partial x}{\partial\eta}\frac{\partial z}{\partial\zeta}\right)(\xi)(\zeta)+\left(\frac{\partial x}{\partial\eta}\frac{\partial y}{\partial\zeta}-\frac{\partial y}{\partial\eta}\frac{\partial x}{\partial\zeta}\right)(\xi)(\eta)\right)$$

$$B(1,22)=\frac{1}{|J|}\left(\begin{array}{c}-\left(\frac{\partial y}{\partial\eta}\frac{\partial z}{\partial\zeta}-\frac{\partial z}{\partial\eta}\frac{\partial y}{\partial\zeta}\right)(\eta)(\zeta)+\left(\frac{\partial z}{\partial\eta}\frac{\partial x}{\partial\zeta}-\frac{\partial x}{\partial\eta}\frac{\partial z}{\partial\zeta}\right)(1-\xi)(\zeta)+\\ \left(\frac{\partial x}{\partial\eta}\frac{\partial y}{\partial\zeta}-\frac{\partial y}{\partial\eta}\frac{\partial x}{\partial\zeta}\right)(1-\xi)(\eta)\end{array}\right)$$

$$\begin{aligned}B(1,2)&=B(1,3)=B(1,5)=B(1,6)=B(1,8)=B(1,9)=B(1,11)=B(1,12)\\&=B(1,14)=B(1,15)=B(1,17)=B(1,18)=B(1,20)=B(1,21)=B(1,23)=B(1,24)=0\end{aligned}$$

$$B(2,2)=-\frac{1}{|J|}\left(\begin{array}{c}\left(\frac{\partial z}{\partial\xi}\frac{\partial y}{\partial\zeta}-\frac{\partial y}{\partial\xi}\frac{\partial z}{\partial\zeta}\right)(1-\eta)(1-\zeta)+\left(\frac{\partial x}{\partial\xi}\frac{\partial z}{\partial\zeta}-\frac{\partial z}{\partial\xi}\frac{\partial x}{\partial\zeta}\right)(1-\xi)(1-\zeta)+\\ \left(\frac{\partial y}{\partial\xi}\frac{\partial x}{\partial\zeta}-\frac{\partial x}{\partial\xi}\frac{\partial y}{\partial\zeta}\right)(1-\xi)(1-\eta)\end{array}\right)$$

$$B(2,5) = \frac{1}{|J|}\begin{pmatrix} \left(\frac{\partial z}{\partial \xi}\frac{\partial y}{\partial \zeta} - \frac{\partial y}{\partial \xi}\frac{\partial z}{\partial \zeta}\right)(1-\eta)(1-\zeta) - \left(\frac{\partial x}{\partial \xi}\frac{\partial z}{\partial \zeta} - \frac{\partial z}{\partial \xi}\frac{\partial x}{\partial \zeta}\right)(\xi)(1-\zeta) - \\ \left(\frac{\partial y}{\partial \xi}\frac{\partial x}{\partial \zeta} - \frac{\partial x}{\partial \xi}\frac{\partial y}{\partial \zeta}\right)(\xi)(1-\eta) \end{pmatrix}$$

$$B(2,8) = \frac{1}{|J|}\left(\left(\frac{\partial z}{\partial \xi}\frac{\partial y}{\partial \zeta} - \frac{\partial y}{\partial \xi}\frac{\partial z}{\partial \zeta}\right)(\eta)(1-\zeta) + \left(\frac{\partial x}{\partial \xi}\frac{\partial z}{\partial \zeta} - \frac{\partial z}{\partial \xi}\frac{\partial x}{\partial \zeta}\right)(\xi)(1-\zeta) - \left(\frac{\partial y}{\partial \xi}\frac{\partial x}{\partial \zeta} - \frac{\partial x}{\partial \xi}\frac{\partial y}{\partial \zeta}\right)(\xi)(\eta)\right)$$

$$B(2,11) = \frac{1}{|J|}\begin{pmatrix} -\left(\frac{\partial z}{\partial \xi}\frac{\partial y}{\partial \zeta} - \frac{\partial y}{\partial \xi}\frac{\partial z}{\partial \zeta}\right)(\eta)(1-\zeta) + \left(\frac{\partial x}{\partial \xi}\frac{\partial z}{\partial \zeta} - \frac{\partial z}{\partial \xi}\frac{\partial x}{\partial \zeta}\right)(1-\xi)(1-\zeta) - \\ \left(\frac{\partial y}{\partial \xi}\frac{\partial x}{\partial \zeta} - \frac{\partial x}{\partial \xi}\frac{\partial y}{\partial \zeta}\right)(1-\xi)(\eta) \end{pmatrix}$$

$$B(2,14) = \frac{1}{|J|}\begin{pmatrix} -\left(\frac{\partial z}{\partial \xi}\frac{\partial y}{\partial \zeta} - \frac{\partial y}{\partial \xi}\frac{\partial z}{\partial \zeta}\right)(1-\eta)(\zeta) - \left(\frac{\partial x}{\partial \xi}\frac{\partial z}{\partial \zeta} - \frac{\partial z}{\partial \xi}\frac{\partial x}{\partial \zeta}\right)(1-\xi)(\zeta) + \\ \left(\frac{\partial y}{\partial \xi}\frac{\partial x}{\partial \zeta} - \frac{\partial x}{\partial \xi}\frac{\partial y}{\partial \zeta}\right)(1-\xi)(1-\eta) \end{pmatrix}$$

$$B(2,17) = \frac{1}{|J|}\left(\left(\frac{\partial z}{\partial \xi}\frac{\partial y}{\partial \zeta} - \frac{\partial y}{\partial \xi}\frac{\partial z}{\partial \zeta}\right)(1-\eta)(\zeta) - \left(\frac{\partial x}{\partial \xi}\frac{\partial z}{\partial \zeta} - \frac{\partial z}{\partial \xi}\frac{\partial x}{\partial \zeta}\right)(\xi)(\zeta) + \left(\frac{\partial y}{\partial \xi}\frac{\partial x}{\partial \zeta} - \frac{\partial x}{\partial \xi}\frac{\partial y}{\partial \zeta}\right)(\xi)(1-\eta)\right)$$

$$B(2,20) = \frac{1}{|J|}\left(\left(\frac{\partial z}{\partial \xi}\frac{\partial y}{\partial \zeta} - \frac{\partial y}{\partial \xi}\frac{\partial z}{\partial \zeta}\right)(\eta)(\zeta) + \left(\frac{\partial x}{\partial \xi}\frac{\partial z}{\partial \zeta} - \frac{\partial z}{\partial \xi}\frac{\partial x}{\partial \zeta}\right)(\xi)(\zeta) + \left(\frac{\partial y}{\partial \xi}\frac{\partial x}{\partial \zeta} - \frac{\partial x}{\partial \xi}\frac{\partial y}{\partial \zeta}\right)(\xi)(\eta)\right)$$

$$B(2,23) = \frac{1}{|J|}\begin{pmatrix} -\left(\frac{\partial z}{\partial \xi}\frac{\partial y}{\partial \zeta} - \frac{\partial y}{\partial \xi}\frac{\partial z}{\partial \zeta}\right)(\eta)(\zeta) + \left(\frac{\partial x}{\partial \xi}\frac{\partial z}{\partial \zeta} - \frac{\partial z}{\partial \xi}\frac{\partial x}{\partial \zeta}\right)(1-\xi)(\zeta) + \\ \left(\frac{\partial y}{\partial \xi}\frac{\partial x}{\partial \zeta} - \frac{\partial x}{\partial \xi}\frac{\partial y}{\partial \zeta}\right)(1-\xi)(\eta) \end{pmatrix}$$

$$\begin{aligned} B(2,1) &= B(2,3) = B(2,4) = B(2,6) = B(2,7) = B(2,9) = B(2,10) = B(2,12) \\ &= B(2,13) = B(2,15) = B(2,16) = B(2,18) = B(2,19) = B(2,21) = B(2,22) = B(2,24) = 0 \end{aligned}$$

$$B(3,3) = -\frac{1}{|J|}\begin{pmatrix} \left(\frac{\partial y}{\partial \xi}\frac{\partial z}{\partial \eta} - \frac{\partial z}{\partial \xi}\frac{\partial y}{\partial \eta}\right)(1-\eta)(1-\zeta) + \left(\frac{\partial z}{\partial \xi}\frac{\partial x}{\partial \eta} - \frac{\partial x}{\partial \xi}\frac{\partial z}{\partial \eta}\right)(1-\xi)(1-\zeta) + \\ \left(\frac{\partial x}{\partial \xi}\frac{\partial y}{\partial \eta} - \frac{\partial y}{\partial \xi}\frac{\partial x}{\partial \eta}\right)(1-\xi)(1-\eta) \end{pmatrix}$$

$$B(3,6) = \frac{1}{|J|}\begin{pmatrix} \left(\frac{\partial y}{\partial \xi}\frac{\partial z}{\partial \eta} - \frac{\partial z}{\partial \xi}\frac{\partial y}{\partial \eta}\right)(1-\eta)(1-\zeta) - \left(\frac{\partial z}{\partial \xi}\frac{\partial x}{\partial \eta} - \frac{\partial x}{\partial \xi}\frac{\partial z}{\partial \eta}\right)(\xi)(1-\zeta) - \\ \left(\frac{\partial x}{\partial \xi}\frac{\partial y}{\partial \eta} - \frac{\partial y}{\partial \xi}\frac{\partial x}{\partial \eta}\right)(\xi)(1-\eta) \end{pmatrix}$$

$$B(3,9) = \frac{1}{|J|}\left(\left(\frac{\partial y}{\partial \xi}\frac{\partial z}{\partial \eta} - \frac{\partial z}{\partial \xi}\frac{\partial y}{\partial \eta}\right)(\eta)(1-\zeta) + \left(\frac{\partial z}{\partial \xi}\frac{\partial x}{\partial \eta} - \frac{\partial x}{\partial \xi}\frac{\partial z}{\partial \eta}\right)(\xi)(1-\zeta) - \left(\frac{\partial x}{\partial \xi}\frac{\partial y}{\partial \eta} - \frac{\partial y}{\partial \xi}\frac{\partial x}{\partial \eta}\right)(\xi)(\eta)\right)$$

$$B(3,12) = \frac{1}{|J|}\begin{pmatrix} -\left(\frac{\partial y}{\partial \xi}\frac{\partial z}{\partial \eta} - \frac{\partial z}{\partial \xi}\frac{\partial y}{\partial \eta}\right)(\eta)(1-\zeta) + \left(\frac{\partial z}{\partial \xi}\frac{\partial x}{\partial \eta} - \frac{\partial x}{\partial \xi}\frac{\partial z}{\partial \eta}\right)(1-\xi)(1-\zeta) - \\ \left(\frac{\partial x}{\partial \xi}\frac{\partial y}{\partial \eta} - \frac{\partial y}{\partial \xi}\frac{\partial x}{\partial \eta}\right)(1-\xi)(\eta) \end{pmatrix}$$

$$B(3,15) = \frac{1}{|J|}\begin{pmatrix} -\left(\frac{\partial y}{\partial \xi}\frac{\partial z}{\partial \eta} - \frac{\partial z}{\partial \xi}\frac{\partial y}{\partial \eta}\right)(1-\eta)(\zeta) - \left(\frac{\partial z}{\partial \xi}\frac{\partial x}{\partial \eta} - \frac{\partial x}{\partial \xi}\frac{\partial z}{\partial \eta}\right)(1-\xi)(\zeta) + \\ \left(\frac{\partial x}{\partial \xi}\frac{\partial y}{\partial \eta} - \frac{\partial y}{\partial \xi}\frac{\partial x}{\partial \eta}\right)(1-\xi)(1-\eta) \end{pmatrix}$$

$$B(3,18) = \frac{1}{|J|}\left(\left(\frac{\partial y}{\partial \xi}\frac{\partial z}{\partial \eta} - \frac{\partial z}{\partial \xi}\frac{\partial y}{\partial \eta}\right)(1-\eta)(\zeta) - \left(\frac{\partial z}{\partial \xi}\frac{\partial x}{\partial \eta} - \frac{\partial x}{\partial \xi}\frac{\partial z}{\partial \eta}\right)(\xi)(\zeta) + \left(\frac{\partial x}{\partial \xi}\frac{\partial y}{\partial \eta} - \frac{\partial y}{\partial \xi}\frac{\partial x}{\partial \eta}\right)(\xi)(1-\eta)\right)$$

$$B(3,21) = \frac{1}{|J|}\left(\left(\frac{\partial y}{\partial \xi}\frac{\partial z}{\partial \eta} - \frac{\partial z}{\partial \xi}\frac{\partial y}{\partial \eta}\right)(\eta)(\zeta) + \left(\frac{\partial z}{\partial \xi}\frac{\partial x}{\partial \eta} - \frac{\partial x}{\partial \xi}\frac{\partial z}{\partial \eta}\right)(\xi)(\zeta) + \left(\frac{\partial x}{\partial \xi}\frac{\partial y}{\partial \eta} - \frac{\partial y}{\partial \xi}\frac{\partial x}{\partial \eta}\right)(\xi)(\eta)\right)$$

$$B(3,24) = \frac{1}{|J|}\begin{pmatrix} -\left(\frac{\partial y}{\partial \xi}\frac{\partial z}{\partial \eta} - \frac{\partial z}{\partial \xi}\frac{\partial y}{\partial \eta}\right)(\eta)(\zeta) + \left(\frac{\partial z}{\partial \xi}\frac{\partial x}{\partial \eta} - \frac{\partial x}{\partial \xi}\frac{\partial z}{\partial \eta}\right)(1-\xi)(\zeta) + \\ \left(\frac{\partial x}{\partial \xi}\frac{\partial y}{\partial \eta} - \frac{\partial y}{\partial \xi}\frac{\partial x}{\partial \eta}\right)(1-\xi)(\eta) \end{pmatrix}$$

$$\begin{aligned} B(3,1) &= B(3,2) = B(3,4) = B(3,5) = B(3,7) = B(3,8) = B(3,10) = B(3,11) \\ &= B(3,13) = B(3,14) = B(3,16) = B(3,17) = B(3,19) = B(3,20) = B(3,22) = B(3,23) = 0 \end{aligned}$$

$$B(4,1) = -\frac{1}{|J|}\left(\begin{array}{c}\left(\frac{\partial z}{\partial \xi}\frac{\partial y}{\partial \zeta} - \frac{\partial y}{\partial \xi}\frac{\partial z}{\partial \zeta}\right)(1-\eta)(1-\zeta) + \left(\frac{\partial x}{\partial \xi}\frac{\partial z}{\partial \zeta} - \frac{\partial z}{\partial \xi}\frac{\partial x}{\partial \zeta}\right)(1-\xi)(1-\zeta) + \\ \left(\frac{\partial y}{\partial \xi}\frac{\partial x}{\partial \zeta} - \frac{\partial x}{\partial \xi}\frac{\partial y}{\partial \zeta}\right)(1-\xi)(1-\eta)\end{array}\right)$$

$$B(4,2) = -\frac{1}{|J|}\left(\begin{array}{c}\left(\frac{\partial y}{\partial \eta}\frac{\partial z}{\partial \zeta} - \frac{\partial z}{\partial \eta}\frac{\partial y}{\partial \zeta}\right)(1-\eta)(1-\zeta) + \left(\frac{\partial z}{\partial \eta}\frac{\partial x}{\partial \zeta} - \frac{\partial x}{\partial \eta}\frac{\partial z}{\partial \zeta}\right)(1-\xi)(1-\zeta) + \\ \left(\frac{\partial x}{\partial \eta}\frac{\partial y}{\partial \zeta} - \frac{\partial y}{\partial \eta}\frac{\partial x}{\partial \zeta}\right)(1-\xi)(1-\eta)\end{array}\right)$$

$$B(4,4) = \frac{1}{|J|}\left(\begin{array}{c}\left(\frac{\partial z}{\partial \xi}\frac{\partial y}{\partial \zeta} - \frac{\partial y}{\partial \xi}\frac{\partial z}{\partial \zeta}\right)(1-\eta)(1-\zeta) - \left(\frac{\partial x}{\partial \xi}\frac{\partial z}{\partial \zeta} - \frac{\partial z}{\partial \xi}\frac{\partial x}{\partial \zeta}\right)(\xi)(1-\zeta) - \\ \left(\frac{\partial y}{\partial \xi}\frac{\partial x}{\partial \zeta} - \frac{\partial x}{\partial \xi}\frac{\partial y}{\partial \zeta}\right)(\xi)(1-\eta)\end{array}\right)$$

$$B(4,5) = \frac{1}{|J|}\left(\begin{array}{c}\left(\frac{\partial y}{\partial \eta}\frac{\partial z}{\partial \zeta} - \frac{\partial z}{\partial \eta}\frac{\partial y}{\partial \zeta}\right)(1-\eta)(1-\zeta) - \left(\frac{\partial z}{\partial \eta}\frac{\partial x}{\partial \zeta} - \frac{\partial x}{\partial \eta}\frac{\partial z}{\partial \zeta}\right)(\xi)(1-\zeta) - \\ \left(\frac{\partial x}{\partial \eta}\frac{\partial y}{\partial \zeta} - \frac{\partial y}{\partial \eta}\frac{\partial x}{\partial \zeta}\right)(\xi)(1-\eta)\end{array}\right)$$

$$B(4,7) = \frac{1}{|J|}\left(\left(\frac{\partial z}{\partial \xi}\frac{\partial y}{\partial \zeta} - \frac{\partial y}{\partial \xi}\frac{\partial z}{\partial \zeta}\right)(\eta)(1-\zeta) + \left(\frac{\partial x}{\partial \xi}\frac{\partial z}{\partial \zeta} - \frac{\partial z}{\partial \xi}\frac{\partial x}{\partial \zeta}\right)(\xi)(1-\zeta) - \left(\frac{\partial y}{\partial \xi}\frac{\partial x}{\partial \zeta} - \frac{\partial x}{\partial \xi}\frac{\partial y}{\partial \zeta}\right)(\xi)(\eta)\right)$$

$$B(4,8) = \frac{1}{|J|}\left(\left(\frac{\partial y}{\partial \eta}\frac{\partial z}{\partial \zeta} - \frac{\partial z}{\partial \eta}\frac{\partial y}{\partial \zeta}\right)(\eta)(1-\zeta) + \left(\frac{\partial z}{\partial \eta}\frac{\partial x}{\partial \zeta} - \frac{\partial x}{\partial \eta}\frac{\partial z}{\partial \zeta}\right)(\xi)(1-\zeta) - \left(\frac{\partial x}{\partial \eta}\frac{\partial y}{\partial \zeta} - \frac{\partial y}{\partial \eta}\frac{\partial x}{\partial \zeta}\right)(\xi)(\eta)\right)$$

$$B(4,10) = \frac{1}{|J|}\left(\begin{array}{c}-\left(\frac{\partial z}{\partial \xi}\frac{\partial y}{\partial \zeta} - \frac{\partial y}{\partial \xi}\frac{\partial z}{\partial \zeta}\right)(\eta)(1-\zeta) + \left(\frac{\partial x}{\partial \xi}\frac{\partial z}{\partial \zeta} - \frac{\partial z}{\partial \xi}\frac{\partial x}{\partial \zeta}\right)(1-\xi)(1-\zeta) - \\ \left(\frac{\partial y}{\partial \xi}\frac{\partial x}{\partial \zeta} - \frac{\partial x}{\partial \xi}\frac{\partial y}{\partial \zeta}\right)(1-\xi)(\eta)\end{array}\right)$$

$$B(4,11)=\frac{1}{|J|}\left(\begin{array}{c}-\left(\frac{\partial y}{\partial \eta}\frac{\partial z}{\partial \zeta}-\frac{\partial z}{\partial \eta}\frac{\partial y}{\partial \zeta}\right)(\eta)(1-\zeta)+\left(\frac{\partial z}{\partial \eta}\frac{\partial x}{\partial \zeta}-\frac{\partial x}{\partial \eta}\frac{\partial z}{\partial \zeta}\right)(1-\xi)(1-\zeta)-\\ \left(\frac{\partial x}{\partial \eta}\frac{\partial y}{\partial \zeta}-\frac{\partial y}{\partial \eta}\frac{\partial x}{\partial \zeta}\right)(1-\xi)(\eta)\end{array}\right)$$

$$B(4,13)=\frac{1}{|J|}\left(\begin{array}{c}-\left(\frac{\partial z}{\partial \xi}\frac{\partial y}{\partial \zeta}-\frac{\partial y}{\partial \xi}\frac{\partial z}{\partial \zeta}\right)(1-\eta)(\zeta)-\left(\frac{\partial x}{\partial \xi}\frac{\partial z}{\partial \zeta}-\frac{\partial z}{\partial \xi}\frac{\partial x}{\partial \zeta}\right)(1-\xi)(\zeta)+\\ \left(\frac{\partial y}{\partial \xi}\frac{\partial x}{\partial \zeta}-\frac{\partial x}{\partial \xi}\frac{\partial y}{\partial \zeta}\right)(1-\xi)(1-\eta)\end{array}\right)$$

$$B(4,14)=\frac{1}{|J|}\left(\begin{array}{c}-\left(\frac{\partial y}{\partial \eta}\frac{\partial z}{\partial \zeta}-\frac{\partial z}{\partial \eta}\frac{\partial y}{\partial \zeta}\right)(1-\eta)(\zeta)-\left(\frac{\partial z}{\partial \eta}\frac{\partial x}{\partial \zeta}-\frac{\partial x}{\partial \eta}\frac{\partial z}{\partial \zeta}\right)(1-\xi)(\zeta)+\\ \left(\frac{\partial x}{\partial \eta}\frac{\partial y}{\partial \zeta}-\frac{\partial y}{\partial \eta}\frac{\partial x}{\partial \zeta}\right)(1-\xi)(1-\eta)\end{array}\right)$$

$$B(4,16)=\frac{1}{|J|}\left(\left(\frac{\partial z}{\partial \xi}\frac{\partial y}{\partial \zeta}-\frac{\partial y}{\partial \xi}\frac{\partial z}{\partial \zeta}\right)(1-\eta)(\zeta)-\left(\frac{\partial x}{\partial \xi}\frac{\partial z}{\partial \zeta}-\frac{\partial z}{\partial \xi}\frac{\partial x}{\partial \zeta}\right)(\xi)(\zeta)+\left(\frac{\partial y}{\partial \xi}\frac{\partial x}{\partial \zeta}-\frac{\partial x}{\partial \xi}\frac{\partial y}{\partial \zeta}\right)(\xi)(1-\eta)\right)$$

$$B(4,17)=\frac{1}{|J|}\left(\left(\frac{\partial y}{\partial \eta}\frac{\partial z}{\partial \zeta}-\frac{\partial z}{\partial \eta}\frac{\partial y}{\partial \zeta}\right)(1-\eta)(\zeta)-\left(\frac{\partial z}{\partial \eta}\frac{\partial x}{\partial \zeta}-\frac{\partial x}{\partial \eta}\frac{\partial z}{\partial \zeta}\right)(\xi)(\zeta)+\left(\frac{\partial x}{\partial \eta}\frac{\partial y}{\partial \zeta}-\frac{\partial y}{\partial \eta}\frac{\partial x}{\partial \zeta}\right)(\xi)(1-\eta)\right)$$

$$B(4,19)=\frac{1}{|J|}\left(\left(\frac{\partial z}{\partial \xi}\frac{\partial y}{\partial \zeta}-\frac{\partial y}{\partial \xi}\frac{\partial z}{\partial \zeta}\right)(\eta)(\zeta)+\left(\frac{\partial x}{\partial \xi}\frac{\partial z}{\partial \zeta}-\frac{\partial z}{\partial \xi}\frac{\partial x}{\partial \zeta}\right)(\xi)(\zeta)+\left(\frac{\partial y}{\partial \xi}\frac{\partial x}{\partial \zeta}-\frac{\partial x}{\partial \xi}\frac{\partial y}{\partial \zeta}\right)(\xi)(\eta)\right)$$

$$B(4,20)=\frac{1}{|J|}\left(\left(\frac{\partial y}{\partial \eta}\frac{\partial z}{\partial \zeta}-\frac{\partial z}{\partial \eta}\frac{\partial y}{\partial \zeta}\right)(\eta)(\zeta)+\left(\frac{\partial z}{\partial \eta}\frac{\partial x}{\partial \zeta}-\frac{\partial x}{\partial \eta}\frac{\partial z}{\partial \zeta}\right)(\xi)(\zeta)+\left(\frac{\partial x}{\partial \eta}\frac{\partial y}{\partial \zeta}-\frac{\partial y}{\partial \eta}\frac{\partial x}{\partial \zeta}\right)(\xi)(\eta)\right)$$

$$B(4,22)=\frac{1}{|J|}\left(\begin{array}{c}-\left(\frac{\partial z}{\partial \xi}\frac{\partial y}{\partial \zeta}-\frac{\partial y}{\partial \xi}\frac{\partial z}{\partial \zeta}\right)(\eta)(\zeta)+\left(\frac{\partial x}{\partial \xi}\frac{\partial z}{\partial \zeta}-\frac{\partial z}{\partial \xi}\frac{\partial x}{\partial \zeta}\right)(1-\xi)(\zeta)+\\ \left(\frac{\partial y}{\partial \xi}\frac{\partial x}{\partial \zeta}-\frac{\partial x}{\partial \xi}\frac{\partial y}{\partial \zeta}\right)(1-\xi)(\eta)\end{array}\right)$$

$$B(4,23)=\frac{1}{|J|}\left(\begin{array}{c}-\left(\frac{\partial y}{\partial\eta}\frac{\partial z}{\partial\zeta}-\frac{\partial z}{\partial\eta}\frac{\partial y}{\partial\zeta}\right)(\eta)(\zeta)+\left(\frac{\partial z}{\partial\eta}\frac{\partial x}{\partial\zeta}-\frac{\partial x}{\partial\eta}\frac{\partial z}{\partial\zeta}\right)(1-\xi)(\zeta)+\\ \left(\frac{\partial x}{\partial\eta}\frac{\partial y}{\partial\zeta}-\frac{\partial y}{\partial\eta}\frac{\partial x}{\partial\zeta}\right)(1-\xi)(\eta)\end{array}\right)$$

$$B(4,3)=B(4,6)=B(4,9)=B(4,12)=B(4,15)=B(4,18)=B(4,21)=B(4,24)=0$$

$$B(5,2)=-\frac{1}{|J|}\left(\begin{array}{c}\left(\frac{\partial y}{\partial\xi}\frac{\partial z}{\partial\eta}-\frac{\partial z}{\partial\xi}\frac{\partial y}{\partial\eta}\right)(1-\eta)(1-\zeta)+\left(\frac{\partial z}{\partial\xi}\frac{\partial x}{\partial\eta}-\frac{\partial x}{\partial\xi}\frac{\partial z}{\partial\eta}\right)(1-\xi)(1-\zeta)+\\ \left(\frac{\partial x}{\partial\xi}\frac{\partial y}{\partial\eta}-\frac{\partial y}{\partial\xi}\frac{\partial x}{\partial\eta}\right)(1-\xi)(1-\eta)\end{array}\right)$$

$$B(5,3)=-\frac{1}{|J|}\left(\begin{array}{c}\left(\frac{\partial z}{\partial\xi}\frac{\partial y}{\partial\zeta}-\frac{\partial y}{\partial\xi}\frac{\partial z}{\partial\zeta}\right)(1-\eta)(1-\zeta)+\left(\frac{\partial x}{\partial\xi}\frac{\partial z}{\partial\zeta}-\frac{\partial z}{\partial\xi}\frac{\partial x}{\partial\zeta}\right)(1-\xi)(1-\zeta)+\\ \left(\frac{\partial y}{\partial\xi}\frac{\partial x}{\partial\zeta}-\frac{\partial x}{\partial\xi}\frac{\partial y}{\partial\zeta}\right)(1-\xi)(1-\eta)\end{array}\right)$$

$$B(5,5)=\frac{1}{|J|}\left(\begin{array}{c}\left(\frac{\partial y}{\partial\xi}\frac{\partial z}{\partial\eta}-\frac{\partial z}{\partial\xi}\frac{\partial y}{\partial\eta}\right)(1-\eta)(1-\zeta)-\left(\frac{\partial z}{\partial\xi}\frac{\partial x}{\partial\eta}-\frac{\partial x}{\partial\xi}\frac{\partial z}{\partial\eta}\right)(\xi)(1-\zeta)-\\ \left(\frac{\partial x}{\partial\xi}\frac{\partial y}{\partial\eta}-\frac{\partial y}{\partial\xi}\frac{\partial x}{\partial\eta}\right)(\xi)(1-\eta)\end{array}\right)$$

$$B(5,6)=\frac{1}{|J|}\left(\begin{array}{c}\left(\frac{\partial z}{\partial\xi}\frac{\partial y}{\partial\zeta}-\frac{\partial y}{\partial\xi}\frac{\partial z}{\partial\zeta}\right)(1-\eta)(1-\zeta)-\left(\frac{\partial x}{\partial\xi}\frac{\partial z}{\partial\zeta}-\frac{\partial z}{\partial\xi}\frac{\partial x}{\partial\zeta}\right)(\xi)(1-\zeta)-\\ \left(\frac{\partial y}{\partial\xi}\frac{\partial x}{\partial\zeta}-\frac{\partial x}{\partial\xi}\frac{\partial y}{\partial\zeta}\right)(\xi)(1-\eta)\end{array}\right)$$

$$B(5,8)=\frac{1}{|J|}\left(\left(\frac{\partial y}{\partial\xi}\frac{\partial z}{\partial\eta}-\frac{\partial z}{\partial\xi}\frac{\partial y}{\partial\eta}\right)(\eta)(1-\zeta)+\left(\frac{\partial z}{\partial\xi}\frac{\partial x}{\partial\eta}-\frac{\partial x}{\partial\xi}\frac{\partial z}{\partial\eta}\right)(\xi)(1-\zeta)-\left(\frac{\partial x}{\partial\xi}\frac{\partial y}{\partial\eta}-\frac{\partial y}{\partial\xi}\frac{\partial x}{\partial\eta}\right)(\xi)(\eta)\right)$$

$$B(5,9)=\frac{1}{|J|}\left(\left(\frac{\partial z}{\partial\xi}\frac{\partial y}{\partial\zeta}-\frac{\partial y}{\partial\xi}\frac{\partial z}{\partial\zeta}\right)(\eta)(1-\zeta)+\left(\frac{\partial x}{\partial\xi}\frac{\partial z}{\partial\zeta}-\frac{\partial z}{\partial\xi}\frac{\partial x}{\partial\zeta}\right)(\xi)(1-\zeta)-\left(\frac{\partial y}{\partial\xi}\frac{\partial x}{\partial\zeta}-\frac{\partial x}{\partial\xi}\frac{\partial y}{\partial\zeta}\right)(\xi)(\eta)\right)$$

$$B(5,11)=\frac{1}{|J|}\left(\begin{array}{c}-\left(\frac{\partial y}{\partial \xi}\frac{\partial z}{\partial \eta}-\frac{\partial z}{\partial \xi}\frac{\partial y}{\partial \eta}\right)(\eta)(1-\zeta)+\left(\frac{\partial z}{\partial \xi}\frac{\partial x}{\partial \eta}-\frac{\partial x}{\partial \xi}\frac{\partial z}{\partial \eta}\right)(1-\xi)(1-\zeta)-\\ \left(\frac{\partial x}{\partial \xi}\frac{\partial y}{\partial \eta}-\frac{\partial y}{\partial \xi}\frac{\partial x}{\partial \eta}\right)(1-\xi)(\eta)\end{array}\right)$$

$$B(5,12)=\frac{1}{|J|}\left(\begin{array}{c}-\left(\frac{\partial z}{\partial \xi}\frac{\partial y}{\partial \zeta}-\frac{\partial y}{\partial \xi}\frac{\partial z}{\partial \zeta}\right)(\eta)(1-\zeta)+\left(\frac{\partial x}{\partial \xi}\frac{\partial z}{\partial \zeta}-\frac{\partial z}{\partial \xi}\frac{\partial x}{\partial \zeta}\right)(1-\xi)(1-\zeta)-\\ \left(\frac{\partial y}{\partial \xi}\frac{\partial x}{\partial \zeta}-\frac{\partial x}{\partial \xi}\frac{\partial y}{\partial \zeta}\right)(1-\xi)(\eta)\end{array}\right)$$

$$B(5,14)=\frac{1}{|J|}\left(\begin{array}{c}-\left(\frac{\partial y}{\partial \xi}\frac{\partial z}{\partial \eta}-\frac{\partial z}{\partial \xi}\frac{\partial y}{\partial \eta}\right)(1-\eta)(\zeta)-\left(\frac{\partial z}{\partial \xi}\frac{\partial x}{\partial \eta}-\frac{\partial x}{\partial \xi}\frac{\partial z}{\partial \eta}\right)(1-\xi)(\zeta)+\\ \left(\frac{\partial x}{\partial \xi}\frac{\partial y}{\partial \eta}-\frac{\partial y}{\partial \xi}\frac{\partial x}{\partial \eta}\right)(1-\xi)(1-\eta)\end{array}\right)$$

$$B(5,15)=\frac{1}{|J|}\left(\begin{array}{c}-\left(\frac{\partial z}{\partial \xi}\frac{\partial y}{\partial \zeta}-\frac{\partial y}{\partial \xi}\frac{\partial z}{\partial \zeta}\right)(1-\eta)(\zeta)-\left(\frac{\partial x}{\partial \xi}\frac{\partial z}{\partial \zeta}-\frac{\partial z}{\partial \xi}\frac{\partial x}{\partial \zeta}\right)(1-\xi)(\zeta)+\\ \left(\frac{\partial y}{\partial \xi}\frac{\partial x}{\partial \zeta}-\frac{\partial x}{\partial \xi}\frac{\partial y}{\partial \zeta}\right)(1-\xi)(1-\eta)\end{array}\right)$$

$$B(5,17)=\frac{1}{|J|}\left(\left(\frac{\partial y}{\partial \xi}\frac{\partial z}{\partial \eta}-\frac{\partial z}{\partial \xi}\frac{\partial y}{\partial \eta}\right)(1-\eta)(\zeta)-\left(\frac{\partial z}{\partial \xi}\frac{\partial x}{\partial \eta}-\frac{\partial x}{\partial \xi}\frac{\partial z}{\partial \eta}\right)(\xi)(\zeta)+\left(\frac{\partial x}{\partial \xi}\frac{\partial y}{\partial \eta}-\frac{\partial y}{\partial \xi}\frac{\partial x}{\partial \eta}\right)(\xi)(1-\eta)\right)$$

$$B(5,18)=\frac{1}{|J|}\left(\left(\frac{\partial z}{\partial \xi}\frac{\partial y}{\partial \zeta}-\frac{\partial y}{\partial \xi}\frac{\partial z}{\partial \zeta}\right)(1-\eta)(\zeta)-\left(\frac{\partial x}{\partial \xi}\frac{\partial z}{\partial \zeta}-\frac{\partial z}{\partial \xi}\frac{\partial x}{\partial \zeta}\right)(\xi)(\zeta)+\left(\frac{\partial y}{\partial \xi}\frac{\partial x}{\partial \zeta}-\frac{\partial x}{\partial \xi}\frac{\partial y}{\partial \zeta}\right)(\xi)(1-\eta)\right)$$

$$B(5,20)=\frac{1}{|J|}\left(\left(\frac{\partial y}{\partial \xi}\frac{\partial z}{\partial \eta}-\frac{\partial z}{\partial \xi}\frac{\partial y}{\partial \eta}\right)(\eta)(\zeta)+\left(\frac{\partial z}{\partial \xi}\frac{\partial x}{\partial \eta}-\frac{\partial x}{\partial \xi}\frac{\partial z}{\partial \eta}\right)(\xi)(\zeta)+\left(\frac{\partial x}{\partial \xi}\frac{\partial y}{\partial \eta}-\frac{\partial y}{\partial \xi}\frac{\partial x}{\partial \eta}\right)(\xi)(\eta)\right)$$

$$B(5,21)=\frac{1}{|J|}\left(\left(\frac{\partial z}{\partial \xi}\frac{\partial y}{\partial \zeta}-\frac{\partial y}{\partial \xi}\frac{\partial z}{\partial \zeta}\right)(\eta)(\zeta)+\left(\frac{\partial x}{\partial \xi}\frac{\partial z}{\partial \zeta}-\frac{\partial z}{\partial \xi}\frac{\partial x}{\partial \zeta}\right)(\xi)(\zeta)+\left(\frac{\partial y}{\partial \xi}\frac{\partial x}{\partial \zeta}-\frac{\partial x}{\partial \xi}\frac{\partial y}{\partial \zeta}\right)(\xi)(\eta)\right)$$

$$B(5,23)=\frac{1}{|J|}\left(\begin{array}{c}-\left(\frac{\partial y}{\partial \xi}\frac{\partial z}{\partial \eta}-\frac{\partial z}{\partial \xi}\frac{\partial y}{\partial \eta}\right)(\eta)(\zeta)+\left(\frac{\partial z}{\partial \xi}\frac{\partial x}{\partial \eta}-\frac{\partial x}{\partial \xi}\frac{\partial z}{\partial \eta}\right)(1-\xi)(\zeta)+\\ \left(\frac{\partial x}{\partial \xi}\frac{\partial y}{\partial \eta}-\frac{\partial y}{\partial \xi}\frac{\partial x}{\partial \eta}\right)(1-\xi)(\eta)\end{array}\right)$$

$$B(5,24)=\frac{1}{|J|}\left(\begin{array}{c}-\left(\frac{\partial z}{\partial \xi}\frac{\partial y}{\partial \zeta}-\frac{\partial y}{\partial \xi}\frac{\partial z}{\partial \zeta}\right)(\eta)(\zeta)+\left(\frac{\partial x}{\partial \xi}\frac{\partial z}{\partial \zeta}-\frac{\partial z}{\partial \xi}\frac{\partial x}{\partial \zeta}\right)(1-\xi)(\zeta)+\\ \left(\frac{\partial y}{\partial \xi}\frac{\partial x}{\partial \zeta}-\frac{\partial x}{\partial \xi}\frac{\partial y}{\partial \zeta}\right)(1-\xi)(\eta)\end{array}\right)$$

$$B(5,1)=B(5,4)=B(5,7)=B(5,10)=B(5,13)=B(5,16)=B(5,19)=B(5,22)=0$$

$$B(6,1)=-\frac{1}{|J|}\left(\begin{array}{c}\left(\frac{\partial y}{\partial \xi}\frac{\partial z}{\partial \eta}-\frac{\partial z}{\partial \xi}\frac{\partial y}{\partial \eta}\right)(1-\eta)(1-\zeta)+\left(\frac{\partial z}{\partial \xi}\frac{\partial x}{\partial \eta}-\frac{\partial x}{\partial \xi}\frac{\partial z}{\partial \eta}\right)(1-\xi)(1-\zeta)+\\ \left(\frac{\partial x}{\partial \xi}\frac{\partial y}{\partial \eta}-\frac{\partial y}{\partial \xi}\frac{\partial x}{\partial \eta}\right)(1-\xi)(1-\eta)\end{array}\right)$$

$$B(6,3)=-\frac{1}{|J|}\left(\begin{array}{c}\left(\frac{\partial y}{\partial \eta}\frac{\partial z}{\partial \zeta}-\frac{\partial z}{\partial \eta}\frac{\partial y}{\partial \zeta}\right)(1-\eta)(1-\zeta)+\left(\frac{\partial z}{\partial \eta}\frac{\partial x}{\partial \zeta}-\frac{\partial x}{\partial \eta}\frac{\partial z}{\partial \zeta}\right)(1-\xi)(1-\zeta)+\\ \left(\frac{\partial x}{\partial \eta}\frac{\partial y}{\partial \zeta}-\frac{\partial y}{\partial \eta}\frac{\partial x}{\partial \zeta}\right)(1-\xi)(1-\eta)\end{array}\right)$$

$$B(6,4)=\frac{1}{|J|}\left(\begin{array}{c}\left(\frac{\partial y}{\partial \xi}\frac{\partial z}{\partial \eta}-\frac{\partial z}{\partial \xi}\frac{\partial y}{\partial \eta}\right)(1-\eta)(1-\zeta)-\left(\frac{\partial z}{\partial \xi}\frac{\partial x}{\partial \eta}-\frac{\partial x}{\partial \xi}\frac{\partial z}{\partial \eta}\right)(\xi)(1-\zeta)-\\ \left(\frac{\partial x}{\partial \xi}\frac{\partial y}{\partial \eta}-\frac{\partial y}{\partial \xi}\frac{\partial x}{\partial \eta}\right)(\xi)(1-\eta)\end{array}\right)$$

$$B(6,6)=\frac{1}{|J|}\left(\begin{array}{c}\left(\frac{\partial y}{\partial \eta}\frac{\partial z}{\partial \zeta}-\frac{\partial z}{\partial \eta}\frac{\partial y}{\partial \zeta}\right)(1-\eta)(1-\zeta)-\left(\frac{\partial z}{\partial \eta}\frac{\partial x}{\partial \zeta}-\frac{\partial x}{\partial \eta}\frac{\partial z}{\partial \zeta}\right)(\xi)(1-\zeta)-\\ \left(\frac{\partial x}{\partial \eta}\frac{\partial y}{\partial \zeta}-\frac{\partial y}{\partial \eta}\frac{\partial x}{\partial \zeta}\right)(\xi)(1-\eta)\end{array}\right)$$

$$B(6,7)=\frac{1}{|J|}\left(\left(\frac{\partial y}{\partial \xi}\frac{\partial z}{\partial \eta}-\frac{\partial z}{\partial \xi}\frac{\partial y}{\partial \eta}\right)(\eta)(1-\zeta)+\left(\frac{\partial z}{\partial \xi}\frac{\partial x}{\partial \eta}-\frac{\partial x}{\partial \xi}\frac{\partial z}{\partial \eta}\right)(\xi)(1-\zeta)-\left(\frac{\partial x}{\partial \xi}\frac{\partial y}{\partial \eta}-\frac{\partial y}{\partial \xi}\frac{\partial x}{\partial \eta}\right)(\xi)(\eta)\right)$$

$$B(6,9)=\frac{1}{|J|}\left(\left(\frac{\partial y}{\partial \eta}\frac{\partial z}{\partial \zeta}-\frac{\partial z}{\partial \eta}\frac{\partial y}{\partial \zeta}\right)(\eta)(1-\zeta)+\left(\frac{\partial z}{\partial \eta}\frac{\partial x}{\partial \zeta}-\frac{\partial x}{\partial \eta}\frac{\partial z}{\partial \zeta}\right)(\xi)(1-\zeta)-\left(\frac{\partial x}{\partial \eta}\frac{\partial y}{\partial \zeta}-\frac{\partial y}{\partial \eta}\frac{\partial x}{\partial \zeta}\right)(\xi)(\eta)\right)$$

$$B(6,10)=\frac{1}{|J|}\left(\begin{array}{c}-\left(\frac{\partial y}{\partial \xi}\frac{\partial z}{\partial \eta}-\frac{\partial z}{\partial \xi}\frac{\partial y}{\partial \eta}\right)(\eta)(1-\zeta)+\left(\frac{\partial z}{\partial \xi}\frac{\partial x}{\partial \eta}-\frac{\partial x}{\partial \xi}\frac{\partial z}{\partial \eta}\right)(1-\xi)(1-\zeta)-\\ \left(\frac{\partial x}{\partial \xi}\frac{\partial y}{\partial \eta}-\frac{\partial y}{\partial \xi}\frac{\partial x}{\partial \eta}\right)(1-\xi)(\eta)\end{array}\right)$$

$$B(6,12)=\frac{1}{|J|}\left(\begin{array}{c}-\left(\frac{\partial y}{\partial \eta}\frac{\partial z}{\partial \zeta}-\frac{\partial z}{\partial \eta}\frac{\partial y}{\partial \zeta}\right)(\eta)(1-\zeta)+\left(\frac{\partial z}{\partial \eta}\frac{\partial x}{\partial \zeta}-\frac{\partial x}{\partial \eta}\frac{\partial z}{\partial \zeta}\right)(1-\xi)(1-\zeta)-\\ \left(\frac{\partial x}{\partial \eta}\frac{\partial y}{\partial \zeta}-\frac{\partial y}{\partial \eta}\frac{\partial x}{\partial \zeta}\right)(1-\xi)(\eta)\end{array}\right)$$

$$B(6,13)=\frac{1}{|J|}\left(\begin{array}{c}-\left(\frac{\partial y}{\partial \xi}\frac{\partial z}{\partial \eta}-\frac{\partial z}{\partial \xi}\frac{\partial y}{\partial \eta}\right)(1-\eta)(\zeta)-\left(\frac{\partial z}{\partial \xi}\frac{\partial x}{\partial \eta}-\frac{\partial x}{\partial \xi}\frac{\partial z}{\partial \eta}\right)(1-\xi)(\zeta)+\\ \left(\frac{\partial x}{\partial \xi}\frac{\partial y}{\partial \eta}-\frac{\partial y}{\partial \xi}\frac{\partial x}{\partial \eta}\right)(1-\xi)(1-\eta)\end{array}\right)$$

$$B(6,15)=\frac{1}{|J|}\left(\begin{array}{c}-\left(\frac{\partial y}{\partial \eta}\frac{\partial z}{\partial \zeta}-\frac{\partial z}{\partial \eta}\frac{\partial y}{\partial \zeta}\right)(1-\eta)(\zeta)-\left(\frac{\partial z}{\partial \eta}\frac{\partial x}{\partial \zeta}-\frac{\partial x}{\partial \eta}\frac{\partial z}{\partial \zeta}\right)(1-\xi)(\zeta)+\\ \left(\frac{\partial x}{\partial \eta}\frac{\partial y}{\partial \zeta}-\frac{\partial y}{\partial \eta}\frac{\partial x}{\partial \zeta}\right)(1-\xi)(1-\eta)\end{array}\right)$$

$$B(6,16)=\frac{1}{|J|}\left(\left(\frac{\partial y}{\partial \xi}\frac{\partial z}{\partial \eta}-\frac{\partial z}{\partial \xi}\frac{\partial y}{\partial \eta}\right)(1-\eta)(\zeta)-\left(\frac{\partial z}{\partial \xi}\frac{\partial x}{\partial \eta}-\frac{\partial x}{\partial \xi}\frac{\partial z}{\partial \eta}\right)(\xi)(\zeta)+\left(\frac{\partial x}{\partial \xi}\frac{\partial y}{\partial \eta}-\frac{\partial y}{\partial \xi}\frac{\partial x}{\partial \eta}\right)(\xi)(1-\eta)\right)$$

$$B(6,18)=\frac{1}{|J|}\left(\left(\frac{\partial y}{\partial \eta}\frac{\partial z}{\partial \zeta}-\frac{\partial z}{\partial \eta}\frac{\partial y}{\partial \zeta}\right)(1-\eta)(\zeta)-\left(\frac{\partial z}{\partial \eta}\frac{\partial x}{\partial \zeta}-\frac{\partial x}{\partial \eta}\frac{\partial z}{\partial \zeta}\right)(\xi)(\zeta)+\left(\frac{\partial x}{\partial \eta}\frac{\partial y}{\partial \zeta}-\frac{\partial y}{\partial \eta}\frac{\partial x}{\partial \zeta}\right)(\xi)(1-\eta)\right)$$

$$B(6,19)=\frac{1}{|J|}\left(\left(\frac{\partial y}{\partial \xi}\frac{\partial z}{\partial \eta}-\frac{\partial z}{\partial \xi}\frac{\partial y}{\partial \eta}\right)(\eta)(\zeta)+\left(\frac{\partial z}{\partial \xi}\frac{\partial x}{\partial \eta}-\frac{\partial x}{\partial \xi}\frac{\partial z}{\partial \eta}\right)(\xi)(\zeta)+\left(\frac{\partial x}{\partial \xi}\frac{\partial y}{\partial \eta}-\frac{\partial y}{\partial \xi}\frac{\partial x}{\partial \eta}\right)(\xi)(\eta)\right)$$

$$B(6,21)=\frac{1}{|J|}\left(\left(\frac{\partial y}{\partial \eta}\frac{\partial z}{\partial \zeta}-\frac{\partial z}{\partial \eta}\frac{\partial y}{\partial \zeta}\right)(\eta)(\zeta)+\left(\frac{\partial z}{\partial \eta}\frac{\partial x}{\partial \zeta}-\frac{\partial x}{\partial \eta}\frac{\partial z}{\partial \zeta}\right)(\xi)(\zeta)+\left(\frac{\partial x}{\partial \eta}\frac{\partial y}{\partial \zeta}-\frac{\partial y}{\partial \eta}\frac{\partial x}{\partial \zeta}\right)(\xi)(\eta)\right)$$

$$B(6,22)=\frac{1}{|J|}\left(\begin{array}{c}-\left(\frac{\partial y}{\partial \xi}\frac{\partial z}{\partial \eta}-\frac{\partial z}{\partial \xi}\frac{\partial y}{\partial \eta}\right)(\eta)(\zeta)+\left(\frac{\partial z}{\partial \xi}\frac{\partial x}{\partial \eta}-\frac{\partial x}{\partial \xi}\frac{\partial z}{\partial \eta}\right)(1-\xi)(\zeta)+\\ \left(\frac{\partial x}{\partial \xi}\frac{\partial y}{\partial \eta}-\frac{\partial y}{\partial \xi}\frac{\partial x}{\partial \eta}\right)(1-\xi)(\eta)\end{array}\right)$$

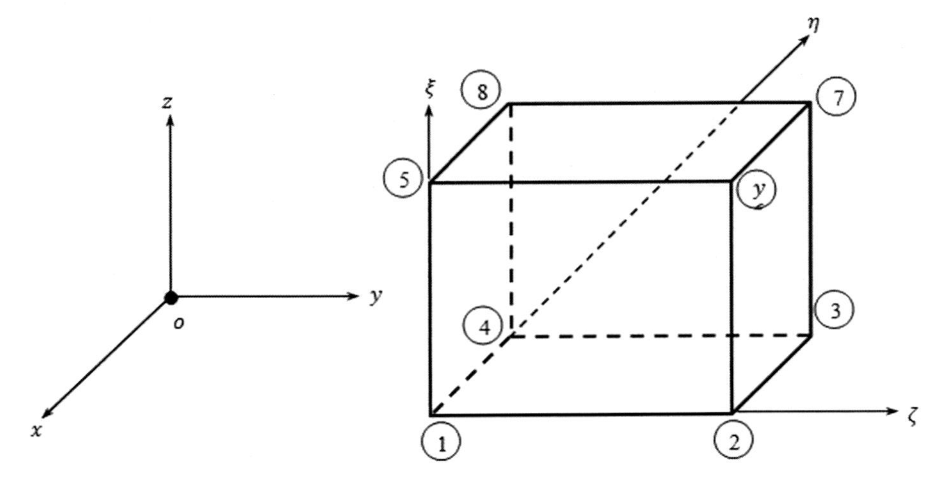

FIGURE 9.4 Parallelopiped element.

$$B(6,24) = \frac{1}{|J|}\left(\begin{array}{c} -\left(\frac{\partial y}{\partial \eta}\frac{\partial z}{\partial \zeta} - \frac{\partial z}{\partial \eta}\frac{\partial y}{\partial \zeta}\right)(\eta)(\zeta) + \left(\frac{\partial z}{\partial \eta}\frac{\partial x}{\partial \zeta} - \frac{\partial x}{\partial \eta}\frac{\partial z}{\partial \zeta}\right)(1-\xi)(\zeta) + \\ \left(\frac{\partial x}{\partial \eta}\frac{\partial y}{\partial \zeta} - \frac{\partial y}{\partial \eta}\frac{\partial x}{\partial \zeta}\right)(1-\xi)(\eta) \end{array} \right)$$

$$B(6,2) = B(6,5) = B(6,8) = B(6,11) = B(6,14) = B(6,17) = B(6,20) = B(6,23) = 0 \quad (9.27)$$

9.4 Example of parallelopiped element

Consider the parallelopiped element shown in Fig. 9.4. Since all of the opposite sides are parallel to each other, it can be deduced that,

$$x_3 - x_2 = x_4 - x_1$$

$$x_6 - x_2 = x_7 - x_3 = x_8 - x_4 = x_5 - x_1$$

Hence, it can also be deduced that,

$$x_3 = x_2 - x_1 + x_4$$

$$x_6 = x_2 - x_1 + x_5$$

$$x_7 = x_2 - 2x_1 + x_4 + x_5$$

$$x_8 = x_4 - x_1 + x_5 \quad (9.28)$$

Substituting the above equations in the following transformation equations,

$$x(\xi,\eta,\zeta)=\sum_{i=1}^{8}x_iN_i(\xi,\eta,\zeta) \tag{9.29}$$

By writing Eq. (9.29) in an extended form and taking advantage of Eq. (9.28), it can be proved that,

$$x(\xi,\eta,\zeta)=x_1+\xi(x_2-x_1)+\eta(x_4-x_1)+\zeta(x_5-x_1)$$

Similarly, it can be shown that,

$$y(\xi,\eta,\zeta)=y_1+\xi(y_2-y_1)+\eta(y_4-y_1)+\zeta(y_5-y_1)$$

$$z(\xi,\eta,\zeta)=z_1+\xi(z_2-z_1)+\eta(z_4-z_1)+\zeta(z_5-z_1) \tag{9.30}$$

Hence, by arranging Eq. (9.30) in matrix form, it can be shown that the Jacobian matrix for such an element is,

$$[J]=\begin{bmatrix}(x_2-x_1) & (y_2-y_1) & (z_2-z_1)\\ (x_4-x_1) & (y_4-y_1) & (z_4-z_1)\\ (x_5-x_1) & (y_5-y_1) & (z_5-z_1)\end{bmatrix} \tag{9.31}$$

Which is independent of ξ,η,ζ. From the rules of analytical geometry, it can be shown that,

$|J|=$ The volume of the parallelepiped element.

Step 4, Express the stress components at any point in terms of nodal displacements and shape functions. The state of stress at any point inside the domain can be defined in terms of,

$$\{\sigma\}^t=\{\sigma_x \quad \sigma_y \quad \sigma_z \quad \tau_{xy} \quad \tau_{yz} \quad \tau_{zx}\} \tag{9.32}$$

Let,

$\{\sigma_0\}=$ Vector of initial stress.

$\{\varepsilon_0\}=$ Vector of initial strain.

If the initial strains are due to thermal expansion for isotropic materials,

$$\{\varepsilon_0\}=\{\alpha T \quad \alpha T \quad \alpha T \quad 0 \quad 0 \quad 0\} \tag{9.33}$$

Where, α: Coefficient of thermal expansion. T: Temperature rise at the given point. From the theory of elasticity for a linearly elastic material, we can write,

$$\{\sigma\}=[D]\{\{\varepsilon\}-\{\varepsilon_0\}\}+\{\sigma_0\} \tag{9.34}$$

From the result of the step. 3, it can be deduced that,

$$\{\sigma\}=[D]\{\varepsilon\}-[D]\{\varepsilon_0\}+\{\sigma_0\}=[D][B]\{\delta\}-[D]\{\varepsilon_0\}+\{\sigma_0\} \tag{9.35}$$

Step 5, express the total P.E. of the element in terms of nodal displacements. It can be shown that,

$$V=U-\Omega \tag{9.36}$$

Where, $V =$ Total P.E. of the element.

$$U = \iiint_{\text{element}} \left(\int_{\{\varepsilon_0\}}^{\{\varepsilon\}} \{\sigma\}^t \{d\varepsilon\} \right) dx\, dy\, dz \tag{9.37}$$

From step 4, Eq. (9.34),

$$\{\sigma\}^t = \{\delta\}^t [B]^t [D]^t - \{\varepsilon_0\}^t [D]^t + \{\sigma_0\}^t \tag{9.38}$$

Since $[D]$ is a symmetric matrix, therefore $[D]^t = [D]$,
Hence, by using Eq. (9.38), Eq. (9.37) becomes,

$$U = \left(\begin{array}{c} \frac{1}{2}\{\delta\}^t \left(\iiint_{\text{element}} [B]^t [D][B] dx\, dy\, dz \right) \{\delta\} - \{\delta\}^t \left(\iiint_{\text{element}} [B]^t [D][\varepsilon_0] dx\, dy\, dz \right) + \\ \{\delta\}^t \left(\iiint_{\text{element}} [B]^t \{\sigma_0\} dx\, dy\, dz \right) \end{array} \right) \tag{9.39}$$

An equivalent nodal loading vector $\{P\}^e$ can be found such that,

$$V = \{\delta\}^t \{P\}^e$$

Therefore

$$\Omega = \frac{1}{2}\{\delta\}^t \left(\iiint_{\text{element}} [B]^t [D][B] dx\, dy\, dz \right) - \{\delta\}^t \{P\}^e$$

Where,

$$\{P\} = \{P\}^e + \{P\}^\varepsilon + \{P\}^\sigma$$

Where,

$$\{P\}^\varepsilon = \iiint_{\text{element}} [B]^t [D][\varepsilon_0] dx\, dy\, dz \tag{9.40}$$

$$\{P\}^\sigma = \iiint_{\text{element}} [B]^t \{\sigma_0\} dx\, dy\, dz$$

Where, $\{P\}^\varepsilon =$ is the nodal loading vector equivalent to initial and thermal strains. $\{P\}^\sigma =$ is the nodal loading vector equivalent to initial strain. $\{P\} =$ is the nodal loading vector equivalent to any other type loading.

Step 6, Apply the minimum total potential energy theorem,

$$\Omega\{\delta\} = \text{minimum} \quad \text{or} \quad \frac{\partial \Omega}{\partial \{\delta\}} = 0 \tag{9.41}$$

Which results in the following matrix equations,

$$[k]^e \{\delta\}^e = \{P\}^e \tag{9.42}$$

Where,

$$[k]^e = \iiint_{\text{element}} [B]^t[D][B]dx\, dy\, dz$$

Since $[B]$ matrix is a function of natural coordinates $f(\xi, \eta, \zeta)$ a change of differential volume from x, y, z into ξ, η, ζ.

From the rules of differentiation geometry,

$$dxdydz = |J|d\xi d\eta d\zeta \tag{9.43}$$

Hence, for the hexahedral element, a volume integral "I" can be expressed as follows,

$$I = \iiint_{\text{element}} f(\xi, \eta, \zeta)dx\, dy\, dz = \int_0^1 \int_0^1 \int_0^1 f(\xi, \eta, \zeta)|J|d\xi\, d\eta\, d\zeta \tag{9.44}$$

Or,

$$I = \sum_{r=1}^{n}\sum_{s=1}^{n}\sum_{t=1}^{n} w_r w_s w_t f(\xi, \eta, \zeta)|J| \tag{9.45}$$

The values $(\xi_r, w_r), (\eta_s, w_s), (\zeta_t, w_t)$ Can be taken from the one-dimensional modified Gauss Legendre quadrature rule. This will be explained in detail in Chapter 12, Shape Function Determinations and Numerical Integration.

The same procedure is followed to find the equivalent nodal forces and employing the numerical integration of the matrices of Eq. (9.40) for any type of loading temperature nodal loads, nodal force vector, surface traction, and body force vectors.

If we use,

$$a = \frac{(x_2 - x_1)}{2},\quad b = \frac{(y_3 - y_2)}{2},\quad c = \frac{(z_7 - z_3)}{2} \tag{9.46}$$

The Jacobian matrix is,

$$[J] = \begin{bmatrix} a & 0 & 0 \\ 0 & b & 0 \\ 0 & 0 & c \end{bmatrix}, |J| = a \times b \times c \tag{9.47}$$

Therefore

$$[k]^e = a \times b \times c \int_{-1}^{1}\int_{-1}^{1}\int_{-1}^{1} [B(\zeta, \eta, \xi)]^t[D][B(\zeta, \eta, \xi)]\zeta d\, d\eta\, d\xi \tag{9.48}$$

Using the Gaussian quadrature rule Eq. (9.45) for numerical integration give,

$$[k]^e = \sum_{r=1}^{n}\sum_{s=1}^{n}\sum_{t=1}^{n} w_r w_s w_t\, f(\zeta, \eta, \xi)[B(\zeta, \eta, \xi)]^t[D][B(\zeta, \eta, \xi)]|J| \tag{9.49}$$

9.5 Tetrahedron element

9.5.1 Element description and element stiffness $[k]^e$ determination

The element shown in Fig. 9.5 represents a tetrahedron element with nodes 1, 2, 3, and 4. At each node, there are three degrees of freedom u, v, and w with the corresponding forces F_x, F_y, and F_z, respectively. Then,

$$\{\delta\}^{e^t} = \{ u_1 \quad v_1 \quad w_1 \quad \cdots \quad \cdots \quad u_4 \quad v_4 \quad w_4 \} \tag{9.50}$$

And,

$$\{F\}^t = \{ F_{x_1} \quad F_{x_2} \quad F_{x_3} \quad \cdots \quad \cdots \quad F_{x_4} \quad F_{x_4} \quad F_{x_4} \} \tag{9.51}$$

Using Lagrange shape function, Fig. 9.6, $N_i(\xi, \eta, \zeta), i = 1, 2, 3, 4$, where,

$$\begin{aligned} N_1 &= \xi \\ N_2 &= \eta \\ N_3 &= \zeta \\ N_4 &= 1 - \xi - \eta - \zeta \end{aligned} \tag{9.52}$$

Therefore

$$\{u\} = [N]\{\delta\}$$

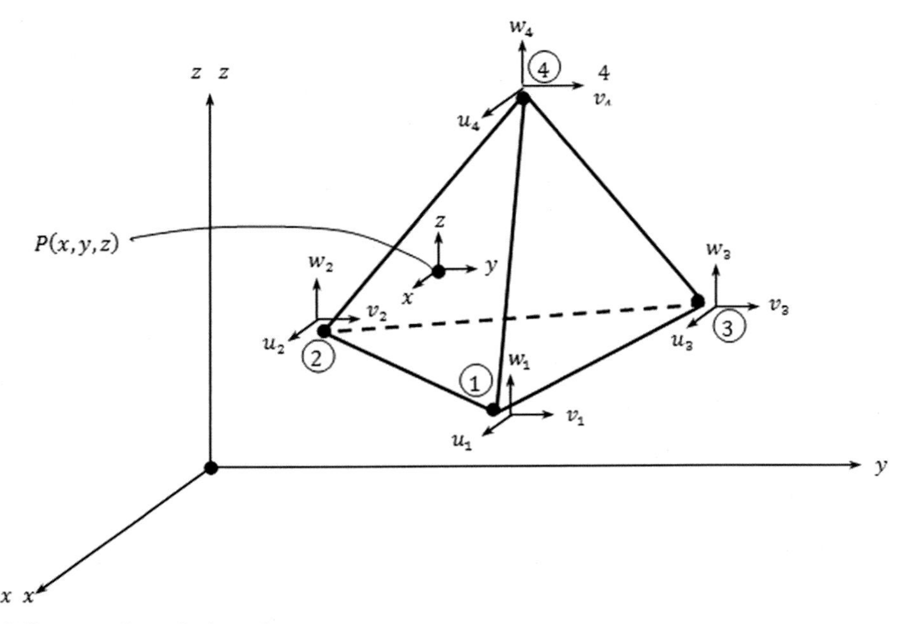

FIGURE 9.5 Typical tetrahedron element.

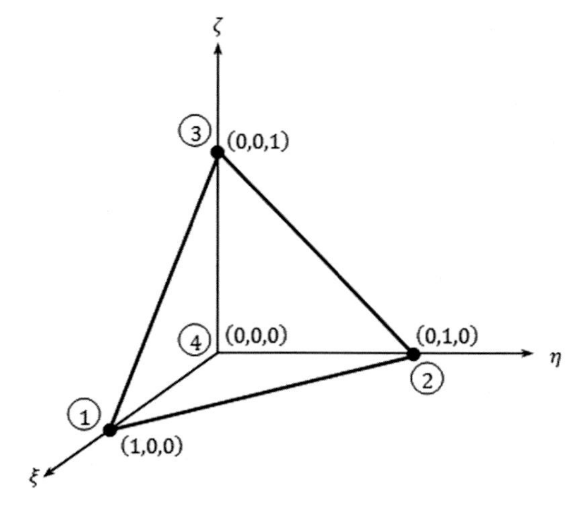

FIGURE 9.6 3D tetrahedral element shape function.

Where $\{u\}$ contains the displacements $u, v,$ and w at a typical point, and,

$$[N] = \begin{bmatrix} N_1 & 0 & 0 & N_2 & 0 & 0 & N_3 & 0 & 0 & N_4 & 0 & 0 \\ 0 & N_1 & 0 & 0 & N_2 & 0 & 0 & N_3 & 0 & 0 & N_4 & 0 \\ 0 & 0 & N_1 & 0 & 0 & N_2 & 0 & 0 & N_3 & 0 & 0 & N_4 \end{bmatrix} \tag{9.53}$$

Size of $[N] = [3 \times 12]$

Using the isoperimetric transformation to obtain the Cartesian coordinates at any point in the element as follows,

$$\begin{aligned} x &= N_1 x_1 + N_2 x_2 + N_3 x_3 + N_4 x_4 \\ y &= N_1 y_1 + N_2 y_2 + N_3 y_3 + N_4 y_4 \\ z &= N_1 z_1 + N_2 z_2 + N_3 z_3 + N_4 z_4 \end{aligned} \tag{9.54}$$

Using Eq. (9.54) and since $x_{ij} = x_i - x_j, y_{ij} = y_i - y_j$ and $z_{ij} = z_i - z_j$, gives,

$$\begin{aligned} x &= x_4 + x_{14}\xi + x_{24}\eta + x_{34}\zeta \\ y &= y_4 + y_{14}\xi + y_{24}\eta + y_{34}\zeta \\ z &= z_4 + z_{14}\xi + z_{24}\eta + z_{34}\zeta \end{aligned} \tag{9.55}$$

Using chain rule to obtain the Jacobean matrix, say, of u, we obtain,

$$\begin{Bmatrix} \frac{\partial u}{\partial \xi} \\ \frac{\partial u}{\partial \eta} \\ \frac{\partial u}{\partial \zeta} \end{Bmatrix} = [J] \begin{Bmatrix} \frac{\partial u}{\partial x} \\ \frac{\partial u}{\partial y} \\ \frac{\partial u}{\partial z} \end{Bmatrix} \tag{9.56}$$

Therefore

$$[J] = \begin{bmatrix} \frac{\partial x}{\partial \xi} & \frac{\partial y}{\partial \xi} & \frac{\partial z}{\partial \xi} \\ \frac{\partial x}{\partial \eta} & \frac{\partial y}{\partial \eta} & \frac{\partial z}{\partial \eta} \\ \frac{\partial x}{\partial \zeta} & \frac{\partial y}{\partial \zeta} & \frac{\partial z}{\partial \zeta} \end{bmatrix} = \begin{bmatrix} x_{14} & y_{14} & z_{14} \\ x_{24} & y_{24} & z_{24} \\ x_{34} & y_{34} & z_{34} \end{bmatrix} \tag{9.57}$$

Where,

$$|J| = x_{14}(y_{24}z_{34} - z_{24}y_{34}) - y_{14}(x_{24}z_{34} - z_{24}x_{34}) + z_{14}(x_{24}y_{34} - y_{24}x_{34})$$

The volume of the element,

$$V_e = \left| \int_0^1 \int_0^{1-\xi} \int_0^{1-\xi-\eta} \det[J] d\xi \, d\eta \, d\zeta \right| = \left|\det[J]\right| \int_0^1 \int_0^{1-\xi} \int_0^{1-\xi-\eta} d\xi \, d\eta \, d\zeta \tag{9.58}$$

Using the polynomial integral formula,

$$\int_0^1 \int_0^{1-\xi} \int_0^{1-\xi-\eta} \xi^m \eta^n \zeta^p \, d\xi \, d\eta \, d\zeta = \frac{m!n!p!}{(m+n+p)} \tag{9.59}$$

The volume of the element is given by,

$$V_e = \frac{1}{6}\left|\det[J]\right| \tag{9.60}$$

From Eq. (9.56),

$$\begin{Bmatrix} \frac{\partial u}{\partial x} \\ \frac{\partial u}{\partial y} \\ \frac{\partial u}{\partial z} \end{Bmatrix} = [J]^{-1} \begin{Bmatrix} \frac{\partial u}{\partial \xi} \\ \frac{\partial u}{\partial \eta} \\ \frac{\partial u}{\partial \zeta} \end{Bmatrix} \tag{9.61}$$

Where,

$$[J]^{-1} = [X] = \frac{1}{\det[J]} \begin{bmatrix} (y_{24}z_{34} - y_{34}z_{24}) & (y_{34}z_{14} - y_{14}z_{34}) & (y_{14}z_{24} - y_{24}z_{14}) \\ (z_{24}x_{34} - z_{34}x_{24}) & (z_{34}x_{14} - z_{14}x_{34}) & (z_{14}x_{24} - z_{24}x_{14}) \\ (x_{24}y_{34} - x_{34}y_{24}) & (x_{34}y_{14} - x_{14}y_{34}) & (x_{14}y_{24} - x_{24}y_{14}) \end{bmatrix} \tag{9.62}$$

Or,

$$[J]^{-1} = [X] = \begin{bmatrix} X(1,1) & X(1,2) & X(1,3) \\ X(2,1) & X(2,2) & X(2,3) \\ X(3,1) & X(2,3) & X(3,3) \end{bmatrix} \tag{9.63}$$

Therefore Eq. (9.61) becomes,

$$\begin{Bmatrix} \dfrac{\partial u}{\partial x} \\ \dfrac{\partial u}{\partial y} \\ \dfrac{\partial u}{\partial z} \end{Bmatrix} = [X] \begin{Bmatrix} \dfrac{\partial u}{\partial \xi} \\ \dfrac{\partial u}{\partial \eta} \\ \dfrac{\partial u}{\partial \zeta} \end{Bmatrix} \tag{9.64}$$

Similarly,

$$\begin{Bmatrix} \dfrac{\partial v}{\partial x} \\ \dfrac{\partial v}{\partial y} \\ \dfrac{\partial v}{\partial z} \end{Bmatrix} = [X] \begin{Bmatrix} \dfrac{\partial v}{\partial \xi} \\ \dfrac{\partial v}{\partial \eta} \\ \dfrac{\partial v}{\partial \zeta} \end{Bmatrix} \tag{9.65}$$

And,

$$\begin{Bmatrix} \dfrac{\partial w}{\partial x} \\ \dfrac{\partial w}{\partial y} \\ \dfrac{\partial w}{\partial z} \end{Bmatrix} = [X] \begin{Bmatrix} \dfrac{\partial w}{\partial \xi} \\ \dfrac{\partial w}{\partial \eta} \\ \dfrac{\partial w}{\partial \zeta} \end{Bmatrix} \tag{9.66}$$

For isoparametric formulation, we have,

$$\begin{aligned} u(\xi,\eta,\zeta) &= u_1 N_1(\xi,\eta,\zeta) + u_2 N_2(\xi,\eta,\zeta) + u_3 N_3(\xi,\eta,\zeta) + u_4 N_4(\xi,\eta,\zeta) \\ v(\xi,\eta,\zeta) &= v_1 N_1(\xi,\eta,\zeta) + v_2 N_2(\xi,\eta,\zeta) + v_3 N_3(\xi,\eta,\zeta) + v_4 N_4(\xi,\eta,\zeta) \\ w(\xi,\eta,\zeta) &= w_1 N_1(\xi,\eta,\zeta) + w_2 N_2(\xi,\eta,\zeta) + w_3 N_3(\xi,\eta,\zeta) + w_4 N_4(\xi,\eta,\zeta) \end{aligned} \tag{9.67}$$

Therefore to use, the expressions (9.62)–(9.66), it is required to derive the expressions (9.67) as needed in the calculations of strain elements. Hence,

$$\begin{aligned} \frac{\partial u}{\partial \xi} &= \frac{\partial N_1}{\partial \xi} u_1 + \frac{\partial N_2}{\partial \xi} u_2 + \frac{\partial N_3}{\partial \xi} u_3 + \frac{\partial N_4}{\partial \xi} u_4 \\ \frac{\partial u}{\partial \eta} &= \frac{\partial N_1}{\partial \eta} u_1 + \frac{\partial N_2}{\partial \eta} u_2 + \frac{\partial N_3}{\partial \eta} u_3 + \frac{\partial N_4}{\partial \eta} u_4 \\ \frac{\partial u}{\partial \zeta} &= \frac{\partial N_1}{\partial \zeta} u_1 + \frac{\partial N_2}{\partial \zeta} u_2 + \frac{\partial N_3}{\partial \zeta} u_3 + \frac{\partial N_4}{\partial \zeta} u_4 \end{aligned}$$

$$
\begin{aligned}
\frac{\partial v}{\partial \xi} &= \frac{\partial N_1}{\partial \xi}v_1 + \frac{\partial N_2}{\partial \xi}v_2 + \frac{\partial N_3}{\partial \xi}v_3 + \frac{\partial N_4}{\partial \xi}v_4 \\
\frac{\partial v}{\partial \eta} &= \frac{\partial N_1}{\partial \eta}v_1 + \frac{\partial N_2}{\partial \eta}v_2 + \frac{\partial N_3}{\partial \eta}v_3 + \frac{\partial N_4}{\partial \eta}v_4 \\
\frac{\partial v}{\partial \zeta} &= \frac{\partial N_1}{\partial \zeta}v_1 + \frac{\partial N_2}{\partial \zeta}v_2 + \frac{\partial N_3}{\partial \zeta}v_3 + \frac{\partial N_4}{\partial \zeta}v_4 \\
\frac{\partial w}{\partial \xi} &= \frac{\partial N_1}{\partial \xi}w_1 + \frac{\partial N_2}{\partial \xi}w_2 + \frac{\partial N_3}{\partial \xi}w_3 + \frac{\partial N_4}{\partial \xi}w_4 \\
\frac{\partial w}{\partial \eta} &= \frac{\partial N_1}{\partial \eta}w_1 + \frac{\partial N_2}{\partial \eta}w_2 + \frac{\partial N_3}{\partial \eta}w_3 + \frac{\partial N_4}{\partial \eta}w_4 \\
\frac{\partial w}{\partial \zeta} &= \frac{\partial N_1}{\partial \zeta}w_1 + \frac{\partial N_2}{\partial \zeta}w_2 + \frac{\partial N_3}{\partial \zeta}w_3 + \frac{\partial N_4}{\partial \zeta}w_4
\end{aligned}
\tag{9.68}
$$

We have, $\varepsilon_x = \frac{\partial u}{\partial x}$, therefore using Eq. (9.64) to obtain,

$$
\varepsilon_x = \left(\begin{array}{c}
X(1,1)\begin{bmatrix} \frac{\partial N_1}{\partial \xi} & \frac{\partial N_2}{\partial \xi} & \frac{\partial N_3}{\partial \xi} & \frac{\partial N_4}{\partial \xi} \end{bmatrix}\begin{Bmatrix} u_1 \\ u_2 \\ u_3 \\ u_4 \end{Bmatrix} + X(1,2)\begin{bmatrix} \frac{\partial N_1}{\partial \eta} & \frac{\partial N_2}{\partial \eta} & \frac{\partial N_3}{\partial \eta} & \frac{\partial N_4}{\partial \eta} \end{bmatrix}\begin{Bmatrix} u_1 \\ u_2 \\ u_3 \\ u_4 \end{Bmatrix} + \\
X(1,3)\begin{bmatrix} \frac{\partial N_1}{\partial \zeta} & \frac{\partial N_2}{\partial \zeta} & \frac{\partial N_3}{\partial \zeta} & \frac{\partial N_4}{\partial \zeta} \end{bmatrix}\begin{Bmatrix} u_1 \\ u_2 \\ u_3 \\ u_4 \end{Bmatrix}
\end{array}\right)
\tag{9.69}
$$

Similarly, for $\varepsilon_y = \frac{\partial v}{\partial y}$, also using Eq. (9.64),

$$
\varepsilon_y = \left(\begin{array}{c}
X(2,1)\begin{bmatrix} \frac{\partial N_1}{\partial \xi} & \frac{\partial N_2}{\partial \xi} & \frac{\partial N_3}{\partial \xi} & \frac{\partial N_4}{\partial \xi} \end{bmatrix}\begin{Bmatrix} v_1 \\ v_2 \\ v_3 \\ v_4 \end{Bmatrix} + X(2,2)\begin{bmatrix} \frac{\partial N_1}{\partial \eta} & \frac{\partial N_2}{\partial \eta} & \frac{\partial N_3}{\partial \eta} & \frac{\partial N_4}{\partial \eta} \end{bmatrix}\begin{Bmatrix} v_1 \\ v_2 \\ v_3 \\ v_4 \end{Bmatrix} + \\
X(2,3)\begin{bmatrix} \frac{\partial N_1}{\partial \zeta} & \frac{\partial N_2}{\partial \zeta} & \frac{\partial N_3}{\partial \zeta} & \frac{\partial N_4}{\partial \zeta} \end{bmatrix}\begin{Bmatrix} v_1 \\ v_2 \\ v_3 \\ v_4 \end{Bmatrix}
\end{array}\right)
\tag{9.70}
$$

And, $\varepsilon_z = \frac{\partial w}{\partial z}$, or,

$$\varepsilon_z = \left(\begin{array}{c} X(3,1)\left[\frac{\partial N_1}{\partial \xi} \quad \frac{\partial N_2}{\partial \xi} \quad \frac{\partial N_3}{\partial \xi} \quad \frac{\partial N_4}{\partial \xi}\right]\begin{Bmatrix} w_1 \\ w_2 \\ w_3 \\ w_4 \end{Bmatrix} + X(3,2)\left[\frac{\partial N_1}{\partial \eta} \quad \frac{\partial N_2}{\partial \eta} \quad \frac{\partial N_3}{\partial \eta} \quad \frac{\partial N_4}{\partial \eta}\right]\begin{Bmatrix} w_1 \\ w_2 \\ w_3 \\ w_4 \end{Bmatrix} + \\ X(3,3)\left[\frac{\partial N_1}{\partial \zeta} \quad \frac{\partial N_2}{\partial \zeta} \quad \frac{\partial N_3}{\partial \zeta} \quad \frac{\partial N_4}{\partial \zeta}\right]\begin{Bmatrix} w_1 \\ w_2 \\ w_3 \\ w_4 \end{Bmatrix} \end{array} \right) \tag{9.71}$$

Also,

$$\gamma_{xy} = \frac{\partial u}{\partial y} + \frac{\partial v}{\partial x}$$

From Eqs. (9.62)–(9.66), we have,

$$\frac{\partial u}{\partial y} = \left(\begin{array}{c} X(2,1)\left[\frac{\partial N_1}{\partial \xi} \quad \frac{\partial N_2}{\partial \xi} \quad \frac{\partial N_3}{\partial \xi} \quad \frac{\partial N_4}{\partial \xi}\right]\begin{Bmatrix} u_1 \\ u_2 \\ u_3 \\ u_4 \end{Bmatrix} + X(2,2)\left[\frac{\partial N_1}{\partial \eta} \quad \frac{\partial N_2}{\partial \eta} \quad \frac{\partial N_3}{\partial \eta} \quad \frac{\partial N_4}{\partial \eta}\right]\begin{Bmatrix} u_1 \\ u_2 \\ u_3 \\ u_4 \end{Bmatrix} + \\ X(2,3)\left[\frac{\partial N_1}{\partial \zeta} \quad \frac{\partial N_2}{\partial \zeta} \quad \frac{\partial N_3}{\partial \zeta} \quad \frac{\partial N_4}{\partial \zeta}\right]\begin{Bmatrix} u_1 \\ u_2 \\ u_3 \\ u_4 \end{Bmatrix} \end{array} \right)$$

$$\frac{\partial v}{\partial x} = \left(\begin{array}{c} X(1,1)\left[\frac{\partial N_1}{\partial \xi} \quad \frac{\partial N_2}{\partial \xi} \quad \frac{\partial N_3}{\partial \xi} \quad \frac{\partial N_4}{\partial \xi}\right]\begin{Bmatrix} v_1 \\ v_2 \\ v_3 \\ v_4 \end{Bmatrix} + X(1,2)\left[\frac{\partial N_1}{\partial \eta} \quad \frac{\partial N_2}{\partial \eta} \quad \frac{\partial N_3}{\partial \eta} \quad \frac{\partial N_4}{\partial \eta}\right]\begin{Bmatrix} v_1 \\ v_2 \\ v_3 \\ v_4 \end{Bmatrix} + \\ X(1,3)\left[\frac{\partial N_1}{\partial \zeta} \quad \frac{\partial N_2}{\partial \zeta} \quad \frac{\partial N_3}{\partial \zeta} \quad \frac{\partial N_4}{\partial \zeta}\right]\begin{Bmatrix} v_1 \\ v_2 \\ v_3 \\ v_4 \end{Bmatrix} \end{array} \right)$$

Then,

$$\gamma_{xy} = \left(\begin{array}{c} X(2,1)\left[\frac{\partial N_1}{\partial \xi} \quad \frac{\partial N_2}{\partial \xi} \quad \frac{\partial N_3}{\partial \xi} \quad \frac{\partial N_4}{\partial \xi}\right]\begin{Bmatrix} u_1 \\ u_2 \\ u_3 \\ u_4 \end{Bmatrix} + X(2,2)\left[\frac{\partial N_1}{\partial \eta} \quad \frac{\partial N_2}{\partial \eta} \quad \frac{\partial N_3}{\partial \eta} \quad \frac{\partial N_4}{\partial \eta}\right]\begin{Bmatrix} u_1 \\ u_2 \\ u_3 \\ u_4 \end{Bmatrix} + \\ X(2,3)\left[\frac{\partial N_1}{\partial \zeta} \quad \frac{\partial N_2}{\partial \zeta} \quad \frac{\partial N_3}{\partial \zeta} \quad \frac{\partial N_4}{\partial \zeta}\right]\begin{Bmatrix} u_1 \\ u_2 \\ u_3 \\ u_4 \end{Bmatrix} + X(1,1)\left[\frac{\partial N_1}{\partial \xi} \quad \frac{\partial N_2}{\partial \xi} \quad \frac{\partial N_3}{\partial \xi} \quad \frac{\partial N_4}{\partial \xi}\right]\begin{Bmatrix} v_1 \\ v_2 \\ v_3 \\ v_4 \end{Bmatrix} + \\ X(1,2)\left[\frac{\partial N_1}{\partial \eta} \quad \frac{\partial N_2}{\partial \eta} \quad \frac{\partial N_3}{\partial \eta} \quad \frac{\partial N_4}{\partial \eta}\right]\begin{Bmatrix} v_1 \\ v_2 \\ v_3 \\ v_4 \end{Bmatrix} + X(1,3)\left[\frac{\partial N_1}{\partial \zeta} \quad \frac{\partial N_2}{\partial \zeta} \quad \frac{\partial N_3}{\partial \zeta} \quad \frac{\partial N_4}{\partial \zeta}\right]\begin{Bmatrix} v_1 \\ v_2 \\ v_3 \\ v_4 \end{Bmatrix} \end{array} \right) \tag{9.72}$$

Similarly, for,

$$\gamma_{xz} = \frac{\partial u}{\partial z} + \frac{\partial w}{\partial x}$$

$$\frac{\partial u}{\partial z} = \left(\begin{array}{c} X(3,1)\begin{bmatrix} \frac{\partial N_1}{\partial \xi} & \frac{\partial N_2}{\partial \xi} & \frac{\partial N_3}{\partial \xi} & \frac{\partial N_4}{\partial \xi} \end{bmatrix} \begin{Bmatrix} u_1 \\ u_2 \\ u_3 \\ u_4 \end{Bmatrix} + X(3,2)\begin{bmatrix} \frac{\partial N_1}{\partial \eta} & \frac{\partial N_2}{\partial \eta} & \frac{\partial N_3}{\partial \eta} & \frac{\partial N_4}{\partial \eta} \end{bmatrix} \begin{Bmatrix} u_1 \\ u_2 \\ u_3 \\ u_4 \end{Bmatrix} + \\ X(3,3)\begin{bmatrix} \frac{\partial N_1}{\partial \zeta} & \frac{\partial N_2}{\partial \zeta} & \frac{\partial N_3}{\partial \zeta} & \frac{\partial N_4}{\partial \zeta} \end{bmatrix} \begin{Bmatrix} u_1 \\ u_2 \\ u_3 \\ u_4 \end{Bmatrix} \end{array} \right)$$

$$\frac{\partial w}{\partial x} = \left(\begin{array}{c} X(1,1)\begin{bmatrix} \frac{\partial N_1}{\partial \xi} & \frac{\partial N_2}{\partial \xi} & \frac{\partial N_3}{\partial \xi} & \frac{\partial N_4}{\partial \xi} \end{bmatrix} \begin{Bmatrix} w_1 \\ w_2 \\ w_3 \\ w_4 \end{Bmatrix} + X(1,2)\begin{bmatrix} \frac{\partial N_1}{\partial \eta} & \frac{\partial N_2}{\partial \eta} & \frac{\partial N_3}{\partial \eta} & \frac{\partial N_4}{\partial \eta} \end{bmatrix} \begin{Bmatrix} w_1 \\ w_2 \\ w_3 \\ w_4 \end{Bmatrix} + \\ X(1,3)\begin{bmatrix} \frac{\partial N_1}{\partial \zeta} & \frac{\partial N_2}{\partial \zeta} & \frac{\partial N_3}{\partial \zeta} & \frac{\partial N_4}{\partial \zeta} \end{bmatrix} \begin{Bmatrix} w_1 \\ w_2 \\ w_3 \\ w_4 \end{Bmatrix} \end{array} \right)$$

Then,

$$\gamma_{xz} = \left(\begin{array}{c} X(3,1)\begin{bmatrix} \frac{\partial N_1}{\partial \xi} & \frac{\partial N_2}{\partial \xi} & \frac{\partial N_3}{\partial \xi} & \frac{\partial N_4}{\partial \xi} \end{bmatrix} \begin{Bmatrix} u_1 \\ u_2 \\ u_3 \\ u_4 \end{Bmatrix} + X(3,2)\begin{bmatrix} \frac{\partial N_1}{\partial \eta} & \frac{\partial N_2}{\partial \eta} & \frac{\partial N_3}{\partial \eta} & \frac{\partial N_4}{\partial \eta} \end{bmatrix} \begin{Bmatrix} u_1 \\ u_2 \\ u_3 \\ u_4 \end{Bmatrix} + \\ X(3,3)\begin{bmatrix} \frac{\partial N_1}{\partial \zeta} & \frac{\partial N_2}{\partial \zeta} & \frac{\partial N_3}{\partial \zeta} & \frac{\partial N_4}{\partial \zeta} \end{bmatrix} \begin{Bmatrix} u_1 \\ u_2 \\ u_3 \\ u_4 \end{Bmatrix} + X(1,1)\begin{bmatrix} \frac{\partial N_1}{\partial \xi} & \frac{\partial N_2}{\partial \xi} & \frac{\partial N_3}{\partial \xi} & \frac{\partial N_4}{\partial \xi} \end{bmatrix} \begin{Bmatrix} w_1 \\ w_2 \\ w_3 \\ w_4 \end{Bmatrix} + \\ X(1,2)\begin{bmatrix} \frac{\partial N_1}{\partial \eta} & \frac{\partial N_2}{\partial \eta} & \frac{\partial N_3}{\partial \eta} & \frac{\partial N_4}{\partial \eta} \end{bmatrix} \begin{Bmatrix} w_1 \\ w_2 \\ w_3 \\ w_4 \end{Bmatrix} + X(1,3)\begin{bmatrix} \frac{\partial N_1}{\partial \zeta} & \frac{\partial N_2}{\partial \zeta} & \frac{\partial N_3}{\partial \zeta} & \frac{\partial N_4}{\partial \zeta} \end{bmatrix} \begin{Bmatrix} w_1 \\ w_2 \\ w_3 \\ w_4 \end{Bmatrix} \end{array} \right)$$

And,

$$\gamma_{yz} = \frac{\partial v}{\partial z} + \frac{\partial w}{\partial y}$$

$$\frac{\partial v}{\partial z}=\left(\begin{array}{c}X(3,1)\begin{bmatrix}\frac{\partial N_1}{\partial \xi} & \frac{\partial N_2}{\partial \xi} & \frac{\partial N_3}{\partial \xi} & \frac{\partial N_4}{\partial \xi}\end{bmatrix}\begin{Bmatrix}v_1\\v_2\\v_3\\v_4\end{Bmatrix}+X(3,2)\begin{bmatrix}\frac{\partial N_1}{\partial \eta} & \frac{\partial N_2}{\partial \eta} & \frac{\partial N_3}{\partial \eta} & \frac{\partial N_4}{\partial \eta}\end{bmatrix}\begin{Bmatrix}v_1\\v_2\\v_3\\v_4\end{Bmatrix}+\\X(3,3)\begin{bmatrix}\frac{\partial N_1}{\partial \zeta} & \frac{\partial N_2}{\partial \zeta} & \frac{\partial N_3}{\partial \zeta} & \frac{\partial N_4}{\partial \zeta}\end{bmatrix}\begin{Bmatrix}v_1\\v_2\\v_3\\v_4\end{Bmatrix}\end{array}\right)$$

$$\frac{\partial w}{\partial y}=\left(\begin{array}{c}X(2,1)\begin{bmatrix}\frac{\partial N_1}{\partial \xi} & \frac{\partial N_2}{\partial \xi} & \frac{\partial N_3}{\partial \xi} & \frac{\partial N_4}{\partial \xi}\end{bmatrix}\begin{Bmatrix}w_1\\w_2\\w_3\\w_4\end{Bmatrix}+X(2,2)\begin{bmatrix}\frac{\partial N_1}{\partial \eta} & \frac{\partial N_2}{\partial \eta} & \frac{\partial N_3}{\partial \eta} & \frac{\partial N_4}{\partial \eta}\end{bmatrix}\begin{Bmatrix}w_1\\w_2\\w_3\\w_4\end{Bmatrix}+\\X(2,3)\begin{bmatrix}\frac{\partial N_1}{\partial \zeta} & \frac{\partial N_2}{\partial \zeta} & \frac{\partial N_3}{\partial \zeta} & \frac{\partial N_4}{\partial \zeta}\end{bmatrix}\begin{Bmatrix}w_1\\w_2\\w_3\\w_4\end{Bmatrix}\end{array}\right)$$

Then,

$$\gamma_{xz}=\left(\begin{array}{c}X(3,1)\begin{bmatrix}\frac{\partial N_1}{\partial \xi} & \frac{\partial N_2}{\partial \xi} & \frac{\partial N_3}{\partial \xi} & \frac{\partial N_4}{\partial \xi}\end{bmatrix}\begin{Bmatrix}v_1\\v_2\\v_3\\v_4\end{Bmatrix}+X(3,2)\begin{bmatrix}\frac{\partial N_1}{\partial \eta} & \frac{\partial N_2}{\partial \eta} & \frac{\partial N_3}{\partial \eta} & \frac{\partial N_4}{\partial \eta}\end{bmatrix}\begin{Bmatrix}v_1\\v_2\\v_3\\v_4\end{Bmatrix}+\\X(3,3)\begin{bmatrix}\frac{\partial N_1}{\partial \zeta} & \frac{\partial N_2}{\partial \zeta} & \frac{\partial N_3}{\partial \zeta} & \frac{\partial N_4}{\partial \zeta}\end{bmatrix}\begin{Bmatrix}v_1\\v_2\\v_3\\v_4\end{Bmatrix}+X(2,1)\begin{bmatrix}\frac{\partial N_1}{\partial \xi} & \frac{\partial N_2}{\partial \xi} & \frac{\partial N_3}{\partial \xi} & \frac{\partial N_4}{\partial \xi}\end{bmatrix}\begin{Bmatrix}w_1\\w_2\\w_3\\w_4\end{Bmatrix}+\\X(2,2)\begin{bmatrix}\frac{\partial N_1}{\partial \eta} & \frac{\partial N_2}{\partial \eta} & \frac{\partial N_3}{\partial \eta} & \frac{\partial N_4}{\partial \eta}\end{bmatrix}\begin{Bmatrix}w_1\\w_2\\w_3\\w_4\end{Bmatrix}+X(2,3)\begin{bmatrix}\frac{\partial N_1}{\partial \zeta} & \frac{\partial N_2}{\partial \zeta} & \frac{\partial N_3}{\partial \zeta} & \frac{\partial N_4}{\partial \zeta}\end{bmatrix}\begin{Bmatrix}w_1\\w_2\\w_3\\w_4\end{Bmatrix}\end{array}\right) \tag{9.73}$$

Arranging the strain components,

$$\underset{(6\times 1)}{\{\varepsilon\}} = \underset{(6\times 12)}{[B]}\ \underset{(12\times 1)}{\{u\}}$$

The strain–displacement matrix $[B]$ is obtained by carrying out the required differentiation.

We have,

$$N_1=\xi$$

$$N_2 = \eta$$
$$N_3 = \zeta$$
$$N_4 = 1 - \xi - \eta - \zeta \tag{9.74}$$

Therefore

$$\frac{\partial N_1}{\partial \xi} = 1$$
$$\frac{\partial N_2}{\partial \eta} = 1$$
$$\frac{\partial N_3}{\partial \zeta} = 1$$
$$\frac{\partial N_4}{\partial \xi} = -1$$
$$\frac{\partial N_4}{\partial \eta} = -1$$
$$\frac{\partial N_4}{\partial \zeta} = -1$$

The other derivatives of the shape functions required in calculating the strain- displacement relationship $[B]$ concerning ξ, η, and ζ are zeros. Then,

$$\begin{Bmatrix} \varepsilon_x \\ \varepsilon_y \\ \varepsilon_z \\ \gamma_{xy} \\ \gamma_{xz} \\ \gamma_{yz} \end{Bmatrix} = \begin{bmatrix} \frac{\partial N_1}{\partial x} & 0 & 0 & \frac{\partial N_2}{\partial x} & 0 & 0 & \frac{\partial N_3}{\partial x} & 0 & 0 & \frac{\partial N_4}{\partial x} & 0 & 0 \\ 0 & \frac{\partial N_1}{\partial y} & 0 & 0 & \frac{\partial N_2}{\partial y} & 0 & 0 & \frac{\partial N_3}{\partial y} & 0 & 0 & \frac{\partial N_4}{\partial y} & 0 \\ 0 & 0 & \frac{\partial N_1}{\partial z} & 0 & 0 & \frac{\partial N_2}{\partial z} & 0 & 0 & \frac{\partial N_3}{\partial z} & 0 & 0 & \frac{\partial N_4}{\partial z} \\ \frac{\partial N_1}{\partial y} & \frac{\partial N_1}{\partial x} & 0 & \frac{\partial N_2}{\partial y} & \frac{\partial N_2}{\partial x} & 0 & \frac{\partial N_3}{\partial y} & \frac{\partial N_3}{\partial x} & 0 & \frac{\partial N_4}{\partial y} & \frac{\partial N_4}{\partial x} & 0 \\ \frac{\partial N_1}{\partial z} & 0 & \frac{\partial N_1}{\partial x} & \frac{\partial N_2}{\partial z} & 0 & \frac{\partial N_2}{\partial x} & \frac{\partial N_3}{\partial z} & 0 & \frac{\partial N_3}{\partial x} & \frac{\partial N_4}{\partial z} & 0 & \frac{\partial N_4}{\partial x} \\ 0 & \frac{\partial N_1}{\partial z} & \frac{\partial N_1}{\partial y} & 0 & \frac{\partial N_2}{\partial z} & \frac{\partial N_2}{\partial y} & 0 & \frac{\partial N_3}{\partial z} & \frac{\partial N_3}{\partial y} & 0 & \frac{\partial N_4}{\partial z} & \frac{\partial N_4}{\partial y} \end{bmatrix} \begin{Bmatrix} u_1 \\ v_1 \\ w_1 \\ u_2 \\ v_2 \\ w_2 \\ u_3 \\ v_3 \\ w_3 \\ u_4 \\ v_4 \\ w_4 \end{Bmatrix} \tag{9.75}$$

Or,

$$\{\varepsilon\} = [B]\{\delta\} \tag{9.76a}$$

If we substitute the expressions of derivatives in Eq. (9.75) into Eqs. (9.72) and (9.73) and evaluating the components of $[B]$ matrix, we deduce that,

$$[B]=\begin{bmatrix} X(1,1) & 0 & 0 & X(1,2) & 0 & 0 & X(1,3) & 0 & 0 & -X_1 & 0 & 0 \\ 0 & X(2,1) & 0 & 0 & X(2,2) & 0 & 0 & X(2,3) & 0 & 0 & -X_2 & 0 \\ 0 & 0 & X(3,1) & 0 & 0 & X(3,2) & 0 & 0 & X(3,3) & 0 & 0 & -X_3 \\ X(2,1) & X(1,1) & 0 & X(2,2) & X(1,2) & 0 & X(2,3) & X(1,3) & 0 & -X_2 & -X_1 & 0 \\ X(3,1) & 0 & X(1,1) & X(3,2) & 0 & X(1,2) & X(3,3) & 0 & X(1,3) & -X_3 & 0 & -X_1 \\ 0 & X(3,1) & X(2,1) & 0 & X(3,2) & X(2,2) & 0 & X(3,3) & X(2,3) & 0 & -X_3 & -X_2 \end{bmatrix}$$

Where,

$$X_1 = X(1,1) + X(1,2) + X(1,3)$$

$$X_2 = X(2,1) + X(2,2) + X(2,3)$$

$$X_3 = X(3,1) + X(3,2) + X(3,3)$$

Similarly, we have,

$$\begin{Bmatrix} \frac{\partial v}{\partial x} \\ \frac{\partial v}{\partial y} \\ \frac{\partial v}{\partial z} \end{Bmatrix} = [J]^{-1} \begin{Bmatrix} \frac{\partial v}{\partial \xi} \\ \frac{\partial v}{\partial \eta} \\ \frac{\partial v}{\partial \zeta} \end{Bmatrix} \tag{9.76}$$

And,

$$\begin{Bmatrix} \frac{\partial w}{\partial x} \\ \frac{\partial w}{\partial y} \\ \frac{\partial w}{\partial z} \end{Bmatrix} = [J]^{-1} \begin{Bmatrix} \frac{\partial v}{\partial \xi} \\ \frac{\partial v}{\partial \eta} \\ \frac{\partial v}{\partial \zeta} \end{Bmatrix} \tag{9.77}$$

We have,

$$\underset{(6*1)}{\{\varepsilon\}^e} = \underset{(6*12)}{[B]^e} \underset{(12*1)}{\{\delta\}^e} \tag{9.78}$$

The element stiffness matrix is given by,

$$[k]^e = \int [B]^{e^t} [D][B]^e dvol$$

Or,

$$[k]^e = V_e[B]^{e^t}[D][B]^e \tag{9.79}$$

9.5.2 Equivalent nodal forces

From potential energy, we can deduce as previously,

$$[F]^e = \int [\delta]^e\{F\}dV = [u]^t \iiint [N]^t\{F\}\det[J]d\xi\, d\eta\, d\zeta = [u]^t[F]^e \tag{9.80}$$

Using the integration rule Eq. 9.45, the equivalent nodal forces,

$$\{F\}^{eb} = \frac{v_e}{4}\{F_{x_1} \quad F_{y_1} \quad F_{z_1} \quad \cdots \quad \cdots \quad F_{x_4} \quad F_{y_4} \quad F_{z_4}\}$$

Where,

$\{F\}^{eb}$ is the body force vector, and V_e is the element volume.

For the regular traction forces, the equivalent nodal force vector is given by,

$$\{F^T\} = \int_{A_e} \{\delta\}^t\{T\}dA \tag{9.81}$$

For uniformly distributed traction T^n applied for a face,

$$\text{P.E.} = [u]^t \int_A [N]^t\{\overset{n}{T}\}dA \tag{9.82}$$

Where,

$$\{F\}^e = \frac{A_e}{3}\{T_{x_1} \quad T_{y_1} \quad T_{z_1} \quad T_{x_2} \quad T_{y_2} \quad T_{z_2} \quad T_{x_3} \quad T_{y_3} \quad T_{z_3} \quad 0 \quad 0 \quad 0\}^t \tag{9.83}$$

The equilibrium equations for a typical element is given by,

$$[k]^e\{u\}^e = \{F\}^e \tag{9.84}$$

9.5.3 System stiffness and load vector

Having obtained the element stiffness and force vectors, the assembly, of system stiffness matrix $[K]$ and the system load vector $\{F\}$ using the calculated $[k]^e$ for each element and the load vector $\{F\}^e$ for each element. The assembly follows the same construction procedure the system stiffness and load vector explained in the chapters five and six. The boundary conditions are applied in which the displacements are specified. Then, the reduced system stiffness and load vectors are obtained. The solution will give the displacements vector $\{u\}$ from which the strains and stresses are obtained as follows,

$$\{\varepsilon\} = [B]\{u\} \tag{9.85}$$

And,

$$\{\sigma\} = [D]\{\varepsilon\} \tag{9.86}$$

Then, by substation Eq. (9.75) into Eq. (9.76) gives,

$$\{\sigma\} = [D][B]\{u\} \tag{9.87}$$

Example, Formulate the matrices required to evaluate the element stiffness matrix $[k]^e$ for the element shown in Fig. 9.7.

Moreover, nodes 1, 3, 4 are fired and a load in the z-direction at node $2 = 1000$ N, $E = 200$GPa, $\nu = 0.3$, calculate the displacements at node 2.

Also, find the strain components and then calculate the stresses.

Solution,

From Eq. (9.55), we have,

$$x = 100\xi$$

$$y = 150\eta$$

$$z = 200\zeta$$

Using Eq. (9.57) gives the Jacobian matrix as follows,

$$[J] = \begin{bmatrix} 100 & 0 & 0 \\ 0 & 150 & 0 \\ 0 & 0 & 200 \end{bmatrix}$$

The determinant of Jacobian is,

$$\det[J] = 3 \times 10^6$$

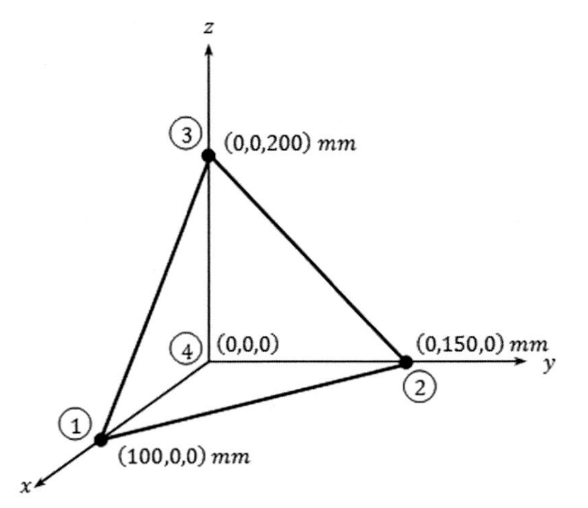

FIGURE 9.7 A tetrahedron element.

Using Eq. (9.60) gives,

$$V_e = \frac{1}{6}\left|\det[J]\right| = 5 \times 10^5 \ \text{mm}^3$$

Hence, using Eq. (9.62) gives

$$[J]^{-1} = [X] = 10^{-2} \times \begin{bmatrix} 1 & 0 & 0 \\ 0 & 0.667 & 0 \\ 0 & 0 & 0.5 \end{bmatrix}$$

And applying Eq. (9.76a) gives the strain–displacement matrix $[B]$ as follows,

$$[B] = 10^{-2} \times \begin{bmatrix} 1 & 0 & 0 & 0 & 0 & 0 & 0 & 0 & 0 & -1 & 0 & 0 \\ 0 & 0 & 0 & 0 & 0.667 & 0 & 0 & 0 & 0 & 0 & -0.667 & 0 \\ 0 & 0 & 0 & 0 & 0 & 0 & 0 & 0 & 0.5 & 0 & 0 & -0.5 \\ 0 & 1 & 0 & 0.667 & 0 & 0 & 0 & 0 & 0 & -0.667 & -1 & 0 \\ 0 & 0 & 1 & 0 & 0 & 0 & 0.5 & 0 & 0 & -0.5 & 0 & -1 \\ 0 & 0 & 0 & 0 & 0 & 0.667 & 0 & 0.5 & 0 & 0 & -0.5 & -0.667 \end{bmatrix}$$

The elasticity matrix is given by Eq. (9.6), as follows,

$$[D] = \frac{E}{(1+\nu)(1-2\nu)} \begin{bmatrix} (1-\nu) & \nu & \nu & 0 & 0 & 0 \\ \nu & (1-\nu) & \nu & 0 & 0 & 0 \\ \nu & \nu & (1-\nu) & 0 & 0 & 0 \\ 0 & 0 & 0 & \frac{(1-2\nu)}{2} & 0 & 0 \\ 0 & 0 & 0 & 0 & \frac{(1-2\nu)}{2} & 0 \\ 0 & 0 & 0 & 0 & 0 & \frac{(1-2\nu)}{2} \end{bmatrix} \tag{9.E10.1}$$

Substitute the values of modulus of elasticity and poisons ratio in Eq. (9.E1.1) as follows,

$$[D] = \frac{10^6}{26} \begin{bmatrix} 7 & 3 & 3 & 0 & 0 & 0 \\ 3 & 7 & 3 & 0 & 0 & 0 \\ 3 & 3 & 7 & 0 & 0 & 0 \\ 0 & 0 & 0 & 2 & 0 & 0 \\ 0 & 0 & 0 & 0 & 2 & 0 \\ 0 & 0 & 0 & 0 & 0 & 2 \end{bmatrix}$$

The element stiffness matrix is obtained using Eq. (9.79), as follows,

$$\underset{(12\times12)}{[k]^e} = \iiint_{V_e} \underset{(12\times6)}{[B]^t} \underset{(6\times6)}{[D]} \underset{(6\times12)}{[B]} \, dvol \quad \Rightarrow (12 \times 12)$$

If the element stiffness matrix happens to be,

$$[k]^e = 10^3 \begin{bmatrix} 1 & 2 & 0.1 & 0.4 & 1.5 & -2 & 0.1 & 2.2 & -0.2 & 1.4 & 2.8 & 1 \\ 2 & 8 & 2.1 & -0.4 & 3.5 & 4 & -4.5 & 4 & 4.6 & -2.2 & 0.3 & 2 \\ 0.1 & 2.1 & 6 & 2 & -0.4 & 2 & 0.1 & -2.4 & 4.8 & 2 & -0.6 & 2.2 \\ 0.4 & -0.4 & 2 & 3 & 2.5 & 5.5 & -0.4 & 4.4 & -2 & 6 & 1 & -3 \\ 1.5 & 3.5 & -0.4 & 2.5 & 4 & 6.8 & 8 & -0.8 & 8 & 2 & -0.4 & 5 \\ -2 & 4 & 2 & 5.5 & 6.8 & 1 & -0.3 & 4 & 6.5 & -1.8 & 2.4 & -0.8 \\ 0.1 & -4.5 & 0.1 & -0.4 & 8 & -0.3 & 8.4 & 5.6 & -0.8 & 2.2 & -5 & 4.8 \\ 2.2 & 4 & -2.4 & 4.4 & -0.8 & 4 & 5.6 & 7.9 & 6.6 & -0.4 & 5 & -0.34 \\ -0.2 & 4.6 & 4.8 & -2 & 8 & 6.5 & -0.8 & 6.6 & 2 & 4.8 & -0.3 & 6.6 \\ 1.4 & -2.2 & 2 & 6 & 2 & -1.8 & 2.2 & -0.4 & 4.8 & 10.2 & -6.6 & 3.3 \\ 2.8 & 0.3 & -0.6 & 1 & -0.4 & 2.4 & -5 & 5 & -0.3 & -6.6 & 5 & -0.3 \\ 1 & 2 & 2.2 & -3 & 5 & -0.8 & 4.8 & -0.34 & 6.6 & 3.3 & -0.3 & 12 \end{bmatrix}$$

Then,

$$[k]^e[u] = [F]$$

The equilibrium equations become,

$$10^3 \begin{bmatrix} 1 & 2 & 0.1 & 0.4 & 1.5 & -2 & 0.1 & 2.2 & -0.2 & 1.4 & 2.8 & 1 \\ 2 & 8 & 2.1 & -0.4 & 3.5 & 4 & -4.5 & 4 & 4.6 & -2.2 & 0.3 & 2 \\ 0.1 & 2.1 & 6 & 2 & -0.4 & 2 & 0.1 & -2.4 & 4.8 & 2 & -0.6 & 2.2 \\ 0.4 & -0.4 & 2 & 3 & 2.5 & 5.5 & -0.4 & 4.4 & -2 & 6 & 1 & -3 \\ 1.5 & 3.5 & -0.4 & 2.5 & 4 & 6.8 & 8 & -0.8 & 8 & 2 & -0.4 & 5 \\ -2 & 4 & 2 & 5.5 & 6.8 & 1 & -0.3 & 4 & 6.5 & -1.8 & 2.4 & -0.8 \\ 0.1 & -4.5 & 0.1 & -0.4 & 8 & -0.3 & 8.4 & 5.6 & -0.8 & 2.2 & -5 & 4.8 \\ 2.2 & 4 & -2.4 & 4.4 & -0.8 & 4 & 5.6 & 7.9 & 6.6 & -0.4 & 5 & -0.34 \\ -0.2 & 4.6 & 4.8 & -2 & 8 & 6.5 & -0.8 & 6.6 & 2 & 4.8 & -0.3 & 6.6 \\ 1.4 & -2.2 & 2 & 6 & 2 & -1.8 & 2.2 & -0.4 & 4.8 & 10.2 & -6.6 & 3.3 \\ 2.8 & 0.3 & -0.6 & 1 & -0.4 & 2.4 & -5 & 5 & -0.3 & -6.6 & 5 & -0.3 \\ 1 & 2 & 2.2 & -3 & 5 & -0.8 & 4.8 & -0.34 & 6.6 & 3.3 & -0.3 & 12 \end{bmatrix} \begin{bmatrix} 0 \\ 0 \\ 0 \\ u_2 \\ v_2 \\ w_2 \\ 0 \\ 0 \\ 0 \\ 0 \\ 0 \\ 0 \end{bmatrix} = \begin{bmatrix} ? \\ ? \\ ? \\ 0 \\ 0 \\ 1000 \\ ? \\ ? \\ ? \\ ? \\ ? \\ ? \end{bmatrix}$$

Then,

$$10^3 \begin{bmatrix} 3 & 2.5 & 5.5 \\ 2.5 & 4 & 6.8 \\ 5.5 & 6.8 & 1 \end{bmatrix} \begin{bmatrix} u_2 \\ v_2 \\ w_2 \end{bmatrix} = \begin{bmatrix} 0 \\ 0 \\ 1000 \end{bmatrix}$$

Therefore

$$u_2 = 0.07466 \text{ mm}$$

$$v_2 = 0.0993 \text{ mm}$$

$$w_2 = -0.08586 \text{ mm}$$

Problems

P.9.1 Given that the hexahedral element shown in Fig. P.9.1,

1. Derive the Jacobean matrix $[J]$ and prove that the volume of this element $= \frac{1}{2}|J|$.
2. Derive the strain–displacement relationship $[B]$.
3. Formulate the stiffness matrix and load vector using one-element modeling, if $E = 200\,\text{GPa}, \nu = 0.3, F_x = 18000\,N, F_y = 400\,N, F_z = 1000\,N$.

P.9.2 For stress analysis of the 3D problem by FEM hexahedral, pentahedral or tetrahedral families of elements can be employed. Given the 6-node pentahedral element shown in Fig. P.9.2 and Table P.9.2.

Requirements,

1. prove that the shape functions for such element are,

$$N_1 = (1 - - \eta)(1 - \xi)$$

$$N_2 = \xi(1 - \xi)$$

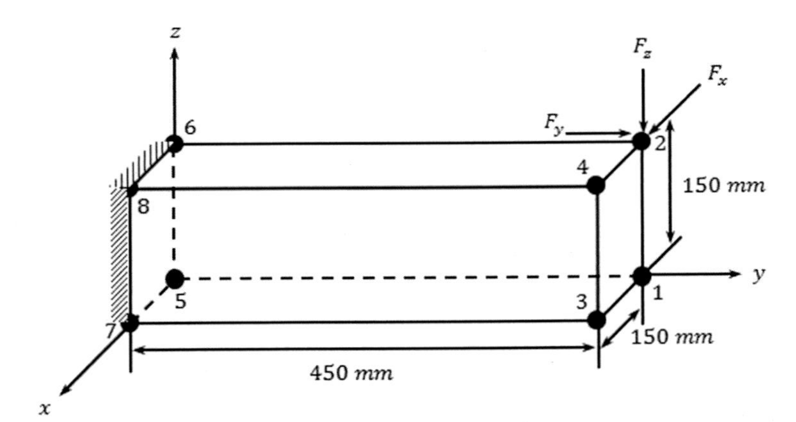

FIGURE P.9.1

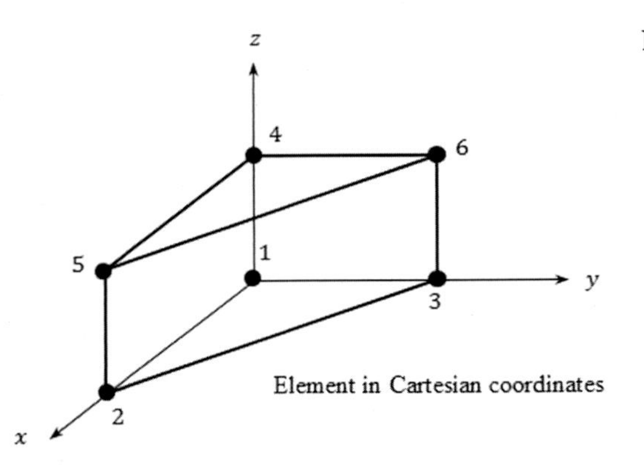

FIGURE P.9.2

TABLE P.9.2

Node	Intrinsic coordinate			Cartesian coordinate		
	ξ	ξ	η	x	y	z
1	0	0	0	0	0	0
2	1	0	0	a	0	0
3	0	1	0	0	b	0
4	0	0	1	0	0	c
5	1	0	1	a	0	c
6	0	1	1	0	b	c

$$N_3 = \eta(1-\xi)$$
$$N_2 = \xi(1-\xi)$$
$$N_3 = \eta(1-\xi)$$
$$N_4 = (1-\xi-\eta)\xi$$
$$N_5 = \xi\xi$$
$$N_6 = \eta\xi$$

2. Using isoparametric transformation proves that,

$$x = a\xi$$
$$y = b\eta$$
$$z = c\xi$$

Hence, show that the Jacobin matrix of the element can be expressed as follows,

$$J\left(\frac{x,y,z}{\xi,\eta,\xi}\right) = \begin{bmatrix} a & 0 & 0 \\ 0 & b & 0 \\ 0 & 0 & c \end{bmatrix}, [J]^{-1} = \begin{bmatrix} 1/a & 0 & 0 \\ 0 & 1/b & 0 \\ 0 & 0 & 1/c \end{bmatrix}$$

And $|J| = abc$

3. Using the standard algorithm for element stiffness matrix derivation,

1. Derive the explicit form of the $[B]$ matrix in terms of (ξ, η, ξ), (just the first six columns).

2. Show that the element stiffness matrix can be expressed as follows,

$$[k] = abc \iiint [B]^t[D][B] d\xi \, d\eta \, d\xi$$

3. Show how the stiffness matrix integral expression can be evaluated numerically. Write down block diagram for software to evaluate the $[k]$ matrix for a given element.

Hint, Total potential energy,

$$\chi = U - W = \frac{1}{2}\iiint [\sigma]^t[\varepsilon]dx\,dy\,dz - \sum_{i=1}^{6}\left(u_iF_i^x + v_iF_i^y + w_iF_i^z\right) = \min$$

With usual notations.

P.9.3 Derive the equivalent nodal forces for uniform stress traction on any side of 8-node of a solid element.

P.9.4 starting with Eq. (9.6) derive the expression for the determinant of Jacobian matrix $[J]$ for the hexahedral element shown in Fig. P.9.4. If $E = 70\,\text{GPa}, \nu = 0.28$ formulate the strain–displacement relationship $[B]$.

P.9.5 For the hexahedral element shown in Fig. 9.3, prove that the volume of this element is equal to $\frac{1}{2}|J|$ as defined in Eq. (9.25).

P.9.5.a If in Problem P.9.5,

$$u_1 = v_1 = w_1 = 0.0$$
$$u_4 = v_4 = w_4 = 0.0$$
$$u_6 = v_6 = w_6 = 0.0$$
$$u_8 = v_8 = w_8 = 0.0$$
$$u_2 = 0.005, v_2 = 0.003, w_2 = 0.01\,\text{mm}$$
$$u_3 = 0.003, v_3 = 0.002, w_3 = 0.02\,\text{mm}$$
$$u_5 = 0.008, v_5 = 0.004, w_5 = 0.015\,\text{mm}$$
$$u_7 = 0.015, v_7 = 0.006, w_7 = 0.03\,\text{mm}$$

Determine the strain components and then the stresses if $E = 200$ GPa and $\nu = 0.3$.

P.9.6 For the tetrahedron element shown in Fig. P.9.6, and using the relations developer in Section 9.2,

1. Derive the strain–displacement matrix $[B]$.
2. Derive the relations that are necessary for the calculation of the element stiffness matrix.
3. Construct a subroutine to calculate the element stiffness matrix.

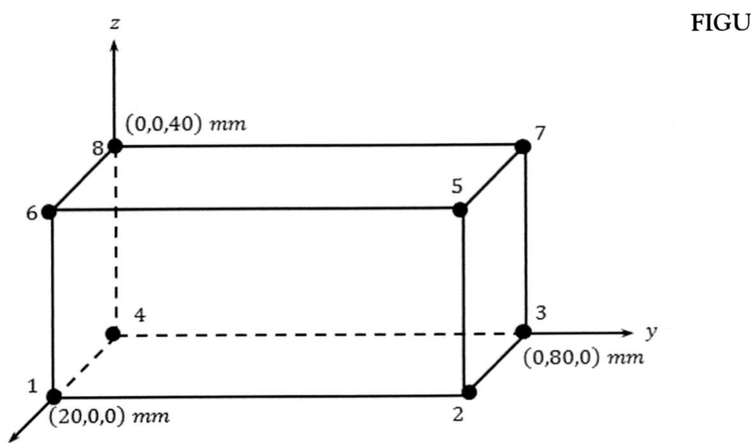

FIGURE P.9.4

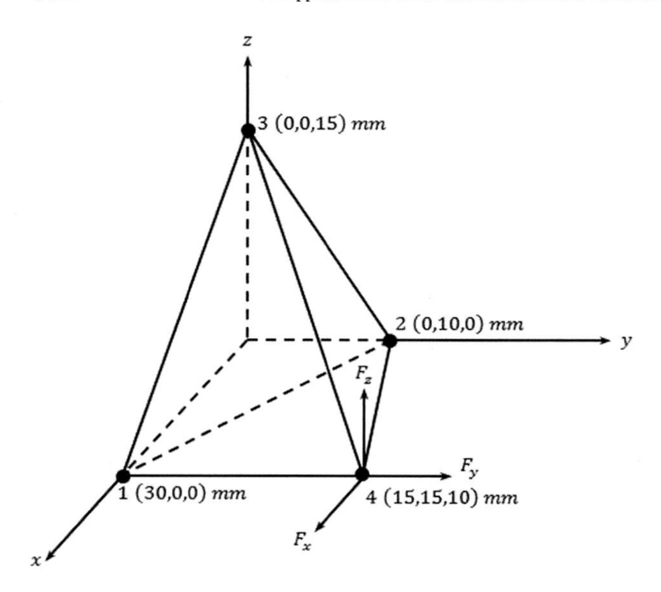

FIGURE P.9.6

4. Formulate the load vector taking into consideration the body force effects, and $F_x = 2000$ N, $F_y = 1500\ N$, and $F_z = 2250$ N.

Assume that $E = 70$ GPa, $\nu = 0.25$

P.9.6.a If in problem P.9.6, assume the nodal displacements are,

$$u_1 = v_1 = w_1 = 0.0$$

$$u_2 = 0.0,\ v_2 = 0.015,\ w_2 = 0.015$$

$$u_3 = 0.03,\ v_3 = 0.015,\ w_3 = 0.015$$

$$u_4 = 0.02,\ v_4 = 0.01,\ w_4 = 0.005$$

Formulate $[B]$ matrix and find the strain components and the stresses if $E = 70$ GPa and $\nu = 0.25$.

P.9.7 Start from the displacement field for the tetrahedron element shown in Fig. P.9.7., such that,

$$u(x, y, z) = a_1 + a_2 x + a_3 y + a_4 z$$

$$v(x, y, z) = a_5 + a_6 x + a_7 y + a_8 z$$

$$w(x, y, z) = a_9 + a_{10} x + a_{11} y + a_{12} z$$

1. Derive the strain–displacement matrix $[B]$ when the shape functions are given below,

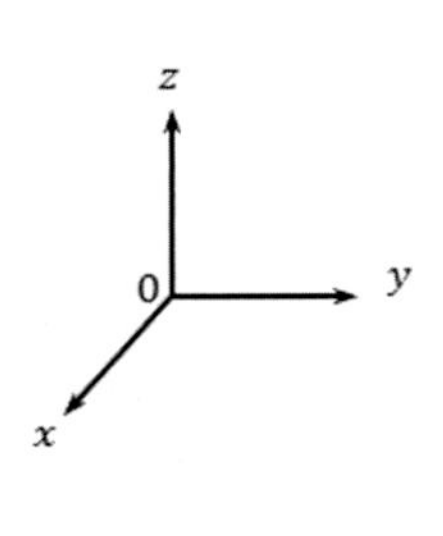

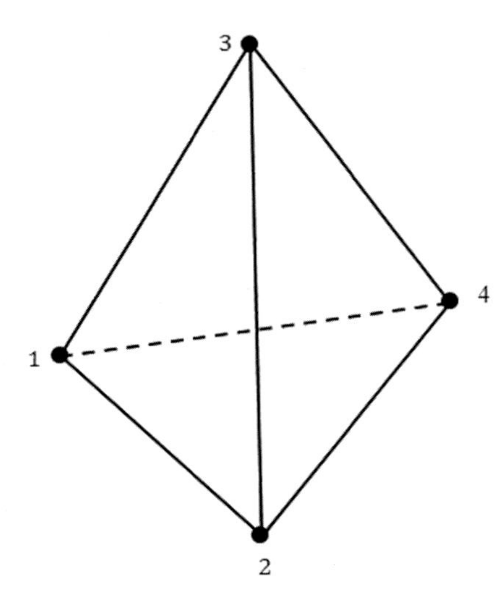

FIGURE P.9.7

$$N_1 = \frac{(\alpha_1 + \beta_1 x + \gamma_1 y + \delta_1 z)}{6V}$$

$$N_2 = \frac{(\alpha_2 + \beta_2 x + \gamma_2 y + \delta_2 z)}{6V}$$

$$N_3 = \frac{(\alpha_3 + \beta_3 x + \gamma_3 y + \delta_3 z)}{6V}$$

$$N_4 = \frac{(\alpha_4 + \beta_4 x + \gamma_4 y + \delta_4 z)}{6V}$$

Where,

$$\alpha_1 = \begin{vmatrix} x_2 & y_2 & z_2 \\ x_3 & y_3 & z_3 \\ x_4 & y_4 & z_4 \end{vmatrix}, \beta_1 = -\begin{vmatrix} 1 & y_2 & z_2 \\ 1 & y_3 & z_3 \\ 1 & y_4 & z_4 \end{vmatrix}, \gamma_1 = \begin{vmatrix} 1 & x_2 & z_2 \\ 1 & x_3 & z_3 \\ 1 & x_4 & z_4 \end{vmatrix}, \delta_1 = -\begin{vmatrix} 1 & x_2 & y_2 \\ 1 & x_3 & y_3 \\ 1 & x_4 & y_4 \end{vmatrix}$$

$$\alpha_2 = -\begin{vmatrix} x_1 & y_1 & z_1 \\ x_3 & y_3 & z_3 \\ x_4 & y_4 & z_4 \end{vmatrix}, \beta_2 = \begin{vmatrix} 1 & y_1 & z_1 \\ 1 & y_3 & z_3 \\ 1 & y_4 & z_4 \end{vmatrix}, \gamma_2 = -\begin{vmatrix} 1 & x_1 & z_1 \\ 1 & x_3 & z_3 \\ 1 & x_4 & z_4 \end{vmatrix}, \delta_2 = \begin{vmatrix} 1 & x_1 & y_1 \\ 1 & x_3 & y_3 \\ 1 & x_4 & y_4 \end{vmatrix}$$

$$\alpha_3 = \begin{vmatrix} x_1 & y_1 & z_1 \\ x_2 & y_2 & z_2 \\ x_4 & y_4 & z_4 \end{vmatrix}, \beta_3 = -\begin{vmatrix} 1 & y_1 & z_1 \\ 1 & y_2 & z_2 \\ 1 & y_4 & z_4 \end{vmatrix}, \gamma_3 = \begin{vmatrix} 1 & x_1 & z_1 \\ 1 & x_2 & z_2 \\ 1 & x_4 & z_4 \end{vmatrix}, \delta_3 = -\begin{vmatrix} 1 & x_1 & y_1 \\ 1 & x_2 & y_2 \\ 1 & x_4 & y_4 \end{vmatrix}$$

$$\alpha_4 = -\begin{vmatrix} x_1 & y_1 & z_1 \\ x_2 & y_2 & z_2 \\ x_3 & y_3 & z_3 \end{vmatrix}, \beta_4 = \begin{vmatrix} 1 & y_1 & z_1 \\ 1 & y_2 & z_2 \\ 1 & y_3 & z_3 \end{vmatrix}, \gamma_4 = -\begin{vmatrix} 1 & x_1 & z_1 \\ 1 & x_2 & z_2 \\ 1 & x_3 & z_3 \end{vmatrix}, \delta_4 = \begin{vmatrix} 1 & x_1 & y_1 \\ 1 & x_2 & y_2 \\ 1 & x_3 & y_3 \end{vmatrix}$$

2. If $E = 200\,\text{GPa}, \nu = 0.3$, determine elasticity matrix $[D]$
3. Steps of calculating the element stiffness matrix $[k]^e$.

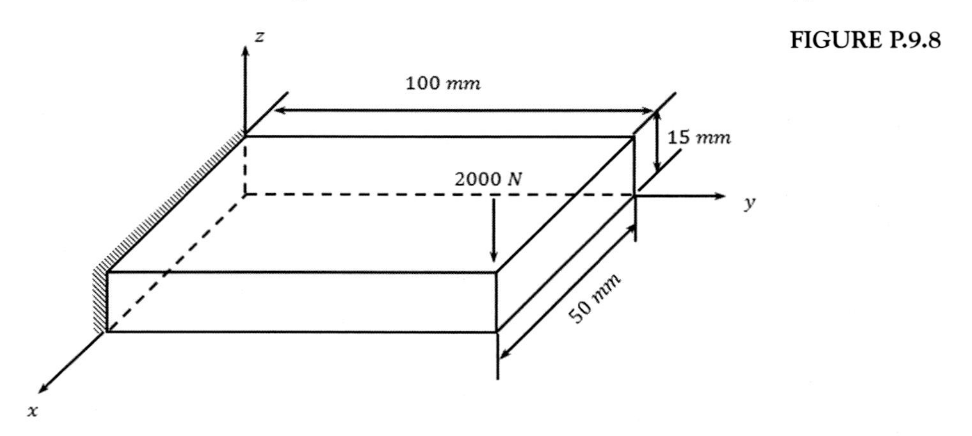

FIGURE P.9.8

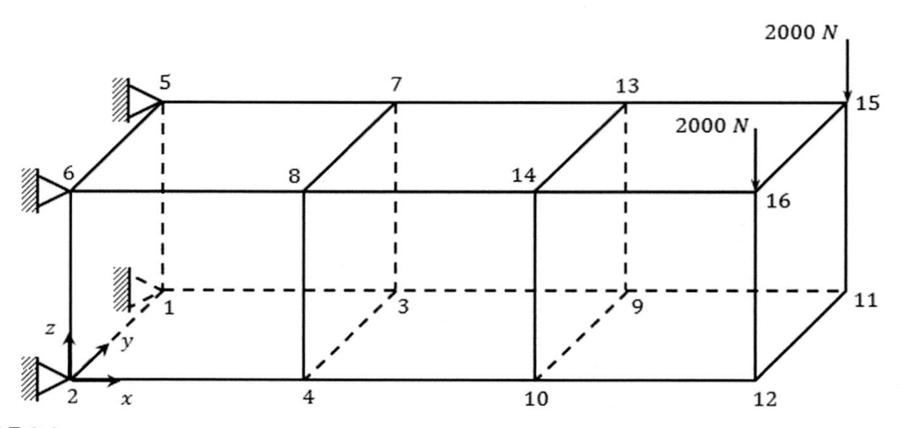

FIGURE P.9.9

P.9.7.a If the coordinate of the element shown in Fig. P.9.7 of the nodes are $1(20, 20, 40), 2(0, 0, 0), 3(0, 40, 0), 4(40, 20, 0)$ Moreover, $E = 70\text{GPa}, \nu = 0.28$, construct the necessary matrices required to determine thus element stiffness matrix.

P.9.8 For the cantilever steel beam shown in Fig. P.9.9, formulate the constrained stiffness matrix use two hexagonal elements with eight nodes. Use the $[B]$ matrix given in Eq. (9.76a) (Fig. P.9.8).

$$E = 200*10^3 N/\text{mm}^2, \nu = 0.3$$

P.9.9 The cantilever beam is shown in Fig. P.9.9,

The beam is divided into three hexahedra, and each hexahedron is then divided into five tetrahedral, required meshes and the element connectivity. Also, derive from the first principles the element stiffness matrix using the assumptions and procedure explained in Section 9.2.

Bibliography

[1] C.S. Desai, J.F. Abel, Introduction to the Finite Element Method, Van Nostrand, Reinhold, New York, 1972.
[2] Y.R. Rashid, Three-dimensional analysis of elastic solids-II; analysis procedure, International Journal of Solids and Structures 6 (1970) 195–207.
[3] R. Galagher, Finite Element Analysis Fundamentals, Prentice-Hall, Englewood Cliffs, 1975.
[4] T.H. Richards, Energy Methods in Stress Analysis, Ellis Horwood Series in Engineering Science, Ellis Horwood, Chichester, 1977.
[5] E. Hinton, D.R. Owen, An Introduction to Finite Element Computations, Pineridge Press, 1979.
[6] J.T. Oden, J.N. Reddy, Variational Methods in Theoretical Mechanics, second ed., Springer-Verlag, Berlin, 1982.
[7] J.N. Reddy, C.F. Liu, A higher–order shear deformation theory of laminated elastic shells, International Journal of engineering Science 23 (3) (1985) 319–330.
[8] C. Lanczos, The Variational Principles of Mechanics, fourth ed., Dover, New York, 1986.
[9] K.H. Huebner, E.A. Thornton, The Finite Element Analysis for Engineers, third ed., Wiley, New York, 1995.
[10] R.D. Cook, Finite Element Modeling for Stress Analysis, Book, Wiley, 1995.
[11] K.J. Bathe, Finite Element Procedures, Prentice-Hill, Englewood Cliffs, 1996.
[12] A.D. Belegundu, Finite Elements in Engineering, third ed., PH T. R. Learning, New Delhi, 2009.
[13] D.H. Norrie, G. Vries, The Finite Element Method-Fundamentals and Applications, Academic Press, London, 1973.
[14] O.C. Zienkiewicz, B.M. Irons, J. Ergatoudis, S. Ahmad, F.C. Scott, Isoparametric and Associated Element Familiesfor Two and Three Dimensional Analysis in the Finite Elemtnt Methods in Stress Analysis, Taper Press, Trondheim, 1969.
[15] M.J. Jweeg, Lecture Notes, Al-Nahrain University, 2003–2016.

CHAPTER

10

Application of finite element to the vibration problems

In this chapter, the application of the finite element method to vibration problems will be presented. The element stiffness matrix derived in the previous chapters, bar, beam, plane problems and the three-dimension applications is adopted here. The corresponding element mass matrix will be derived here depending upon Hamilton's principles. The Eigenvalue problem's construction is formulated here, and the solution gives the natural frequencies and mode shapes. The mass matrix is derived in this chapter for each element presented here, while the element stiffness matrix was taken from the previous chapters. Moreover, based on Hamilton's principles, the Eigenvalue problem is formulated. The problems are also included in this chapter.

10.1 General

The dynamical part of P.S.T.P.E. is Hamilton's principle which states that accurate motion of a system is that,

$$A = \int_{t_1}^{t_2} L\,dt = \int_{t_1}^{t_2} (T - V)dt = \text{stationary} \tag{10.1}$$

Where T is the kinetic energy and V is the total potential energy

To extend the previous work of the finite element (F.E), it is only necessary to deduce an appropriate term for the kinetic energy. The net result of the entire process is that,

$$A = L(\dot{q}_1, \dot{q}_2, \ldots, \dot{q}_n, q_1, q_2, \ldots, q_n) \tag{10.2}$$

Where, $\dot{q}$ is the velocity in the i direction.

Then $\delta A = 0$ generates n differential equations of motion. We have already discussed how to generate stiffness matrix; it only remains to determine appropriate mass matrices.

Energy Methods and Finite Element Techniques
DOI: https://doi.org/10.1016/B978-0-323-88666-6.00010-1

10.2 Application to axial vibration of a bar

1. Definition of the finite element mesh,

Assume that the tapered bar is discretized into elements and the number of nodes is $n+1$ as shown in Fig. 10.1.

2. Selection of the displacement model,
As before, we write,

$$u^e(\xi) = \underset{\uparrow - \text{ShapeFunction}}{[N(\xi)]}\{u\}_0^e \tag{10.3}$$

We will use a linear shape function as defined previously in Chapter 1, Fundamentals of Energy Methods,

$$[N(\xi)] = \left[\left(1 - \frac{\xi}{l}\right) \quad \frac{\xi}{l}\right] \tag{10.4}$$

3. Generation of discrete equations of motion,
1. To find $[k]^e$ element stiffness matrix, as derived in Chapter 5, Introduction to Finite Element Method: Bar and Beam Applications, the result is,

$$[k]^e = \frac{EA}{l}\begin{bmatrix} 1 & -1 \\ -1 & 1 \end{bmatrix} \tag{10.5}$$

With,

$$[k] = [c]^t\left[\tilde{k}\right][c]$$

2. To find $[m]^e$ element mass matrix,

The kinetic energy T^e, energy stored in a typical mass/unit length element is given by,

$$T^e = \int_0^l \frac{1}{2}\overline{m}(\dot{u})^2 d\xi \tag{10.6}$$

Where $\overline{m}$ is the mass per unit length.
We can write

$$T^e = \frac{1}{2}\int_0^l \dot{u}^t\overline{m}\,\dot{u}d\xi \tag{10.7}$$

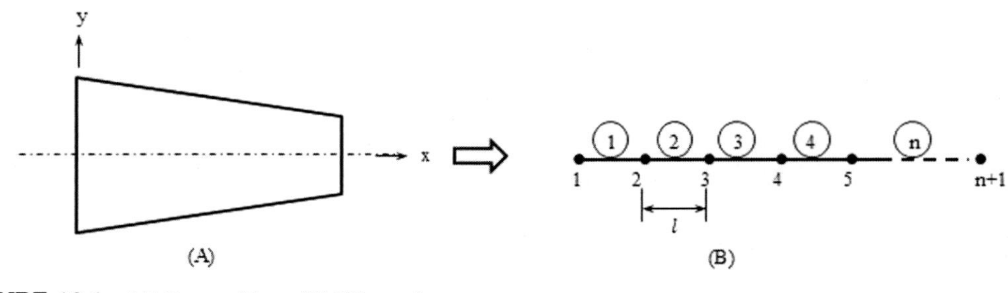

FIGURE 10.1 (A) Bar problem, (B) F.E. mesh.

Since,

$$u(\xi,t) = [N(\xi)]\{u(t)\}_0^e$$

Therefore,

$$\dot{u} = \underset{1\times 2}{[N]}\{\dot{u}\}_0^e \tag{10.8}$$

Eq. (10.7) becomes,

$$T^e = \frac{1}{2}\{\dot{u}\}_0^{e^t}\left(\int_0^l \underbrace{\underset{(2\times 1)}{[N]^t}\,\underset{(1\times 1)}{\overline{m}}\,\underset{(1\times 2)}{[N]}}_{2\times 2}\,d\xi\right)\{\dot{u}\}_0^e$$

Where,

$$[m]^e = \begin{bmatrix} \left(\int_0^l \overline{m}\left(1-\frac{\xi}{l}\right)d\xi\right) & \left(\left(\int_0^l \overline{m}\frac{\xi}{l}\left(1-\frac{\xi}{l}\right)d\xi\right)\right) \\ \left(\left(\int_0^l \overline{m}\frac{\xi}{l}\left(1-\frac{\xi}{l}\right)d\xi\right)\right) & \left(\left(\int_0^l \overline{m}\left(\frac{\xi}{l}\right)^2 d\xi\right)\right) \end{bmatrix} \tag{10.10}$$

Carrying out the integration of Eq. (10.10) gives,

$$[m]^e = \frac{\overline{m}l}{6}\begin{bmatrix} 2 & 1 \\ 1 & 2 \end{bmatrix} \tag{10.11}$$

This called the consistent mass matrix where we supposed $\overline{m} =$ constant. The element mass matrix bound to be symmetric and the diagonal element are + ve.

To find the total kinetic energy T is given by,

$$T = \sum_e T^e \tag{10.12}$$

Denoting $\{\tilde{\dot{u}}\} = [c]\{\dot{u}\}$ compatibility and,

$$[\tilde{m}] = \begin{bmatrix} [m]^1 & 0 & 0 & 0 \\ & [m]^2 & 0 & 0 \\ & \vdots & & \\ & \vdots & \cdots\cdots & [m]^n \end{bmatrix} \tag{10.13}$$

Where $[c]$ is the connection matrix between the global and local coordinates.

$$T = \frac{1}{2}\{\tilde{\dot{u}}\}^t[\overline{m}]\{\tilde{\dot{u}}\} \Rightarrow T = \frac{1}{2}\{\dot{u}\}^t\left([c]^t[\overline{m}][c]\right)\{\dot{u}\} \Rightarrow T = \frac{1}{2}\{\dot{u}\}^t[M]\{\dot{u}\} \tag{10.14}$$

Where,

$[M] = [c]^t[\overline{m}][c] =$ system mass matrix.

We notice that $[M]$'s assembly procedure is precise as the procedure for $[K]$.

10.3 Equation of motion

If the excitation force $\{P(t)\}$ is applied to the nodes, then,

$$\delta A = \delta \int_{t_1}^{t_2} \left(\frac{1}{2}\{\dot{u}\}^t[M]\{\dot{u}\} - \frac{1}{2}\{u\}^t[k]\{u\} + \{u\}^t\{P\} \right) dt = 0 \tag{10.15}$$

Consider the first term of integration in Eq. (10.15),

$$\delta \int_{t_1}^{t_2} \{\dot{u}\}^t[M]\{\dot{u}\}dt = \frac{1}{2}\left(\int_{t_1}^{t_2} \{\delta\dot{u}\}^t[M]\{\dot{u}\}dt + \int_{t_1}^{t_2} \{\dot{u}\}^t[M]\{\delta\dot{u}\}dt \right) = \int_{t_1}^{t_2} \{\delta\dot{u}\}^t[M]\{\delta\dot{u}\}dt \tag{10.16}$$

Integration by parts of Eq. (10.16) gives,

$$= \left|\{\delta\dot{u}\}^t[M]\{\dot{u}\}\right|_{t_1}^{t_2} - \int_{t_1}^{t_2} \{\delta u\}^t[M]\{\ddot{u}\}dt \tag{10.17}$$

According to Hamilton principles, the variation of u between limits $t_1$and$t_2 = 0$, we have,

$$\left|\{\delta\dot{u}\}^t[M]\{\dot{u}\}\right|_{t_1}^{t_2} = 0$$

$$\delta A = \int_{t_1}^{t_2} \left(\{\delta u\}^t(-[M]\{\ddot{u}\} - [K]\{u\} + \{P\})\right)dt = 0 \tag{10.18}$$

Since, $\{\delta u\}^t$ arbitrary, then,

$$[M]\{\ddot{u}\} + [K]\{u\} = \{P\} \tag{10.19}$$

For free vibration, $\{P\} = 0$,

$$\left([K] - \omega^2[M]\right)\{\tilde{u}\} = \{0\} \quad \text{Eigen Value Problem} \tag{10.20}$$

Example 10.1: Find the fundamental natural frequency of axial vibration for the bar shown in Fig. 10.2.

From Eqs. (10.5) and (10.11), the element stiffness matrix and the element mass matrix respectively are given by,

$$[k]^e = \frac{EA}{l}\begin{bmatrix} 1 & -1 \\ -1 & 1 \end{bmatrix}$$

$$[m]^e = \frac{\overline{m}l}{6}\begin{bmatrix} 2 & 1 \\ 1 & 2 \end{bmatrix}$$

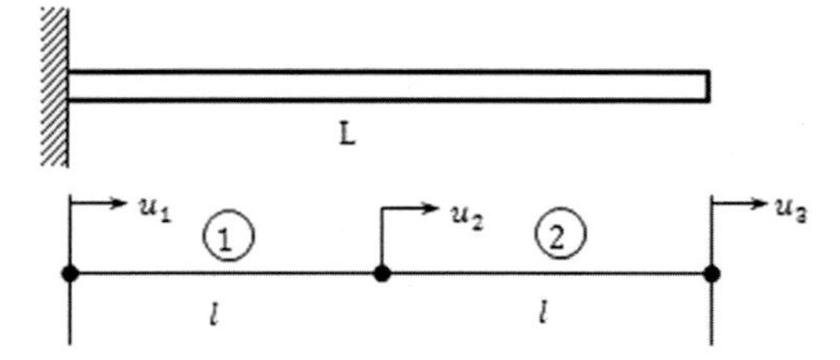

FIGURE 10.2 Bar under axial vibration.

The system mass and stiffness matrices can be assembled as discussed previously in Chapter 1, Fundamentals of Energy Methods, using two elements mesh (Fig. 10.2), are as follows,

$$[K]=\frac{EA}{l}\begin{bmatrix}1 & -1 & 0\\ -1 & (1+1) & -1\\ 0 & -1 & 1\end{bmatrix},\quad [M]=\frac{\overline{m}l}{6}\begin{bmatrix}2 & 1 & 0\\ 1 & (2+2) & 1\\ 0 & 1 & 2\end{bmatrix}$$

The E.V.P. becomes,

$$\left(\begin{bmatrix}1 & -1 & 0\\ -1 & 2 & -1\\ 0 & -1 & 1\end{bmatrix}-\frac{\omega^2\overline{m}l^2}{6EA}\begin{bmatrix}2 & 1 & 0\\ 1 & 4 & 1\\ 0 & 1 & 2\end{bmatrix}\right)\begin{Bmatrix}\tilde{u}_1\\ \tilde{u}_2\\ \tilde{u}_3\end{Bmatrix}=\{0\} \tag{E1.1}$$

Before we determine the natural frequency, we must impose the B.C which prevent R. B. as a whole structure, we have,

$$\tilde{u}_1=0$$

Hence the E.V.P. is reduced to,

$$\left(\begin{bmatrix}2 & -1\\ -1 & 1\end{bmatrix}-\lambda\begin{bmatrix}4 & 1\\ 1 & 2\end{bmatrix}\right)\begin{Bmatrix}\tilde{u}_2\\ \tilde{u}_3\end{Bmatrix}=\{0\} \tag{E1.2}$$

Where $\lambda=\frac{\omega^2\overline{m}l^2}{6EA}$

The solution requires that the determinate of,

$\left(\begin{bmatrix}2 & -1\\ -1 & 1\end{bmatrix}-\lambda\begin{bmatrix}4 & 1\\ 1 & 2\end{bmatrix}\right)=$ zero. This will give,

Solving, $\lambda_1=0.108$ and $\lambda_2=1.32$, from which the natural frequencies are calculated as follows,

$\omega_1=0.806\sqrt{\frac{EA}{\overline{m}l^2}}$ rad/sec and $\omega_2=2.815\sqrt{\frac{EA}{\overline{m}l^2}}$ rad/sec

Suppose that the R.H.S in the cantilever beam shown in Fig. 10.2 is also clamped, so that, $\tilde{u}_3=0$, too, using two elements mesh the fundamental natural frequency is,

$$\omega_1=1.732\sqrt{\frac{EA}{\overline{m}l^2}}$$

If we repeat this problem using three elements, we will find for the clamped-clamped and cantilever beam, the fundamental natural frequencies are,

$\omega_1=1.414\sqrt{\frac{EA}{\overline{m}l^2}}$ for clamped-clamped

$\omega_2=0.841\sqrt{\frac{EA}{\overline{m}l^2}}$ for clamped-clamped

And the following procedure can be used to evaluate the natural frequencies, take the clamped-clamped beam shown in Fig. 10.3.

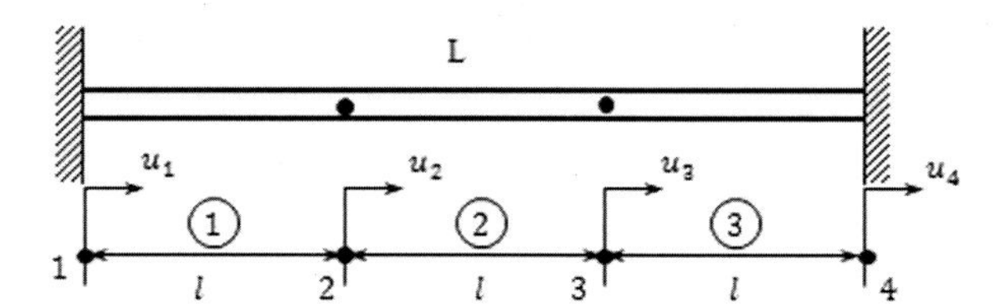

FIGURE 10.3 A clamped ended beam.

We have,

$$[K]^e = \frac{EA}{l}\begin{bmatrix} 1 & -1 & 0 & 0 \\ -1 & 2 & 1 & 0 \\ 0 & 1 & 2 & 1 \\ 0 & 0 & 1 & 1 \end{bmatrix}, [M]^e = \frac{\overline{m}l}{6}\begin{bmatrix} 2 & 1 & 0 & 0 \\ 1 & 4 & 1 & 0 \\ 0 & 1 & 4 & 1 \\ 0 & 0 & 1 & 2 \end{bmatrix} \tag{E1.3a}$$

$$\left[[K]^e - \omega^2[M]^e\right]\{u\} = \{0\}$$

Or,

$$\left[\frac{EA}{l}\begin{bmatrix} 1 & -1 & 0 & 0 \\ -1 & 2 & 1 & 0 \\ 0 & 1 & 2 & 1 \\ 0 & 0 & 1 & 1 \end{bmatrix} - \omega^2\frac{\overline{m}l}{6}\begin{bmatrix} 2 & 1 & 0 & 0 \\ 1 & 4 & 1 & 0 \\ 0 & 1 & 4 & 1 \\ 0 & 0 & 1 & 2 \end{bmatrix}\right]\begin{Bmatrix} u_1 \\ u_2 \\ u_3 \\ u_4 \end{Bmatrix} = \{0\} \tag{E1.3b}$$

1. If clamped-clamped, $u_1 = u_4 = 0$, Eq. (E1.3b) become,

$$\left(\frac{EA}{l}\begin{bmatrix} 2 & 1 \\ 1 & 2 \end{bmatrix} - \omega^2\frac{\overline{m}l}{6}\begin{bmatrix} 4 & 1 \\ 1 & 4 \end{bmatrix}\right)\begin{Bmatrix} u_2 \\ u_3 \end{Bmatrix} = \begin{Bmatrix} 0 \\ 0 \end{Bmatrix}$$

If, $\lambda = \frac{\omega^2\overline{m}l^2}{6EA}$,
Then,

$$\left|\begin{bmatrix} 2 & 1 \\ 1 & 2 \end{bmatrix} - \lambda\begin{bmatrix} 4 & 1 \\ 1 & 4 \end{bmatrix}\right| = 0$$

Or,

$$\begin{vmatrix} (2-4\lambda) & (1-\lambda) \\ (1-\lambda) & (2-4\lambda) \end{vmatrix} = 0, \lambda^2 - \frac{14}{15}\lambda + \frac{3}{15} = 0 \tag{E1.3c}$$

Then, solving Eq. (E1.3c) gives,

$$\lambda_1 = 0.333 \Rightarrow \omega_1 = 1.414\sqrt{\frac{EA}{\overline{m}l^2}}, \lambda_2 = 0.6 \Rightarrow \omega_2 = 1.897\sqrt{\frac{EA}{\overline{m}l^2}}$$

2. Cantilever Beam (Fig. 10.4), $u_1 = 0$

Eq. (E1.3b) becomes,

$$\left|\begin{bmatrix} 2 & 1 & 0 \\ 1 & 2 & 1 \\ 0 & 1 & 1 \end{bmatrix} - \lambda\begin{bmatrix} 4 & 1 & 0 \\ 1 & 4 & 1 \\ 0 & 1 & 2 \end{bmatrix}\right| = 0$$

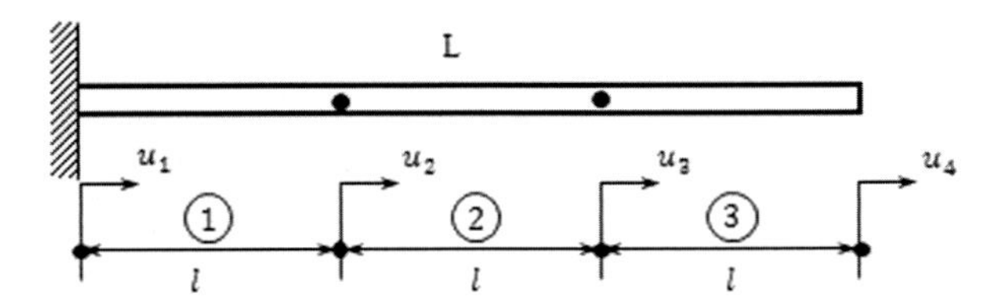

FIGURE 10.4 A cantilever beam (three elements with 4-node).

Or,

$$\begin{vmatrix} (2-4\lambda) & (1-\lambda) & 0 \\ (1-\lambda) & (2-4\lambda) & (1-\lambda) \\ 0 & (1-\lambda) & (1-2\lambda) \end{vmatrix} = 0, \lambda^3 - \frac{33}{26}\lambda^2 + \frac{12}{26}\lambda - \frac{1}{26} = 0 \tag{E1.3d}$$

Then, gives, solving Eq. (E1.3d)

$$\lambda_1 = 0.118 \Rightarrow \omega_1 = 0.841\sqrt{\frac{EA}{\overline{m}l^2}}, \lambda_2 = 0.5 \Rightarrow \omega_2 = 1.732\sqrt{\frac{EA}{\overline{m}l^2}},$$

$$\lambda_3 = 0.651 \Rightarrow \omega_3 = 1.976\sqrt{\frac{EA}{\overline{m}l^2}}$$

10.3.1 Mode shapes determination

Take the cantilever beam shown in Fig. 10.2.

To construct the mode shapes corresponding to the natural frequencies, for the cantilever beam shown in Fig. 10.2, we have, $\omega_1 = 0.108\sqrt{\frac{EA}{\overline{m}l^2}}$, and $\omega_2^2 = 0.0117\frac{EA}{\overline{m}l^2}$

Using Eq. (E1.2) and substitute the value of $\lambda_1 = \frac{\omega_1^2.\overline{m}.l^2}{6EA}$ as follows,

$$\left(\begin{bmatrix} 2 & -1 \\ -1 & 1 \end{bmatrix} - 0.108\begin{bmatrix} 4 & 1 \\ 2 & 2 \end{bmatrix}\right)\begin{Bmatrix} \tilde{u}_2 \\ \tilde{u}_3 \end{Bmatrix} = 0 \tag{E1.3}$$

The solution of Eq. (E1.3) gives the displacement vector,

$$\begin{Bmatrix} \tilde{u}_2 \\ \tilde{u}_3 \end{Bmatrix} = \begin{Bmatrix} 1 \\ 1.414 \end{Bmatrix} \text{or} \begin{Bmatrix} 0.707 \\ 1 \end{Bmatrix} \tag{E1.4}$$

Eq. (E1.4) is the fundamental mode of vibration.

Using the second natural frequency, $\omega_2 = 2.814\sqrt{\frac{EA}{\overline{m}l^2}}$, again using Eq. (E1.3) gives, $\lambda_2 = \frac{\omega_2^2.\overline{m}.l^2}{6EA}$, and for $\omega_2 = 2.814\sqrt{\frac{EA}{\overline{m}l^2}}$, or $\lambda_2 = (2.814)^2\frac{1}{6} = 1.319$

Therefore using Eq. (E1.3) as follows,

$$\left(\begin{bmatrix} 2 & -1 \\ -1 & 1 \end{bmatrix} - 1.319\begin{bmatrix} 4 & 1 \\ 2 & 2 \end{bmatrix}\right)\begin{Bmatrix} \tilde{u}_2 \\ \tilde{u}_3 \end{Bmatrix} = 0 \tag{E1.5}$$

The solution of Eq. (E1.5) gives the displacement vector as follows,

$$\begin{Bmatrix} \tilde{u}_2 \\ \tilde{u}_3 \end{Bmatrix} = \begin{Bmatrix} 1 \\ -1.414 \end{Bmatrix} \text{or} \begin{Bmatrix} -0.707 \\ 1 \end{Bmatrix} \tag{E1.6}$$

Eq. (E1.6) represents the second mode of vibration. The plot of Eqs. (E1.4) and (E1.6) are shown in Fig. 10.5A.

10.3.2 Orthogonality of mode of vibration

The mode shapes represent relative magnitudes of the displacement vector $\{\tilde{u}\}$, therefore they can be written as follows,

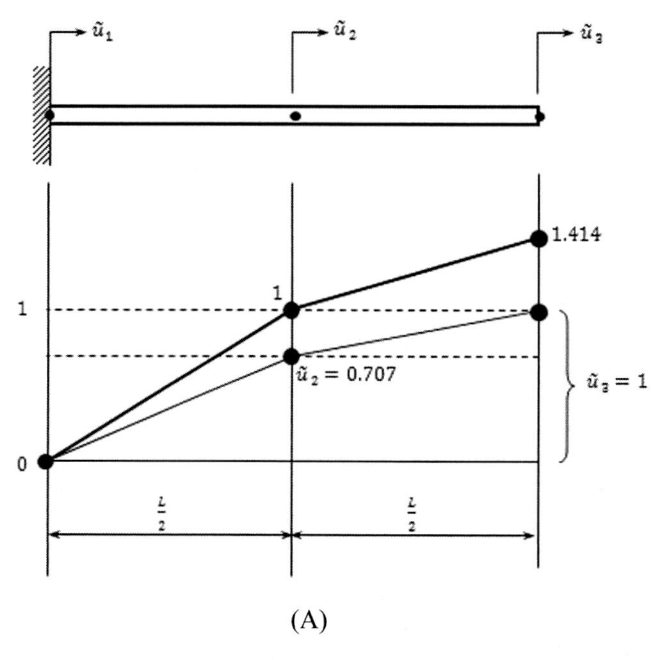

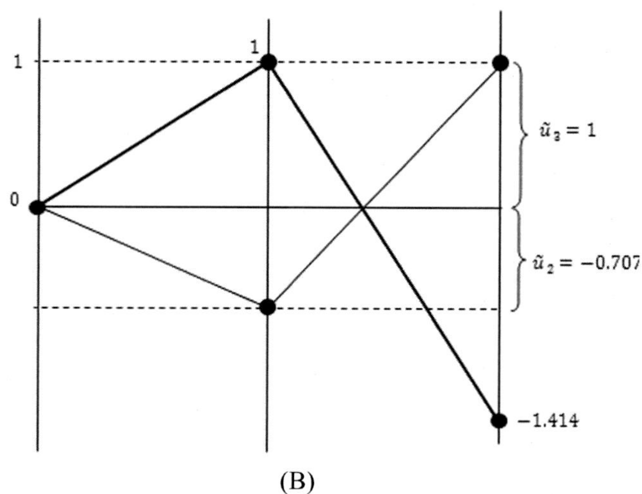

FIGURE 10.5 (A) Fundamental mode shape, (B) second mode shape.

$\alpha_i\{\tilde{u}\}_i$ where i is the corresponding mode of vibration and α_i is an arbitrary constant.

In vibration analysis, it is useful to choose α_i which makes the mode of vibration as usual to the mass matrix $[M]$. This needs,

$$\alpha_i\{\tilde{u}\}_i^t[M]\alpha_j\{\tilde{u}\}_j = \begin{cases} \text{One if } i = j \\ 0 \quad \text{if } i \neq j \end{cases} \tag{E1.7}$$

For, $\alpha_i\{\tilde{u}\}_i^t[M]\alpha_j\{\tilde{u}\}_j = 0$ is satisfied automatically, and the condition to satisfy $\alpha_i\{\tilde{u}\}_i^t[M]\alpha_j\{\tilde{u}\}_j = 1$ for $i = j$, for axial vibration in Example 10.1, we have to satisfy,

$$\alpha_i^2\{\tilde{u}\}_i^t[M]\{\tilde{u}\}_i = \alpha_i^2\{\tilde{u}\}_i^t\frac{\overline{m}.l^2}{6EA}\begin{bmatrix} 2 & 1 & 0 \\ 1 & 4 & 1 \\ 0 & 1 & 2 \end{bmatrix}\{\tilde{u}\}_i = 1 \tag{E1.8}$$

Therefore α_i^2 can be obtained from Eq. (E1.8) as follows,

$$\alpha_i^2 = \frac{6EA}{\overline{m}l^2}\cdot\frac{1}{\{\tilde{u}\}_i^t\begin{bmatrix} 2 & 1 & 0 \\ 1 & 4 & 1 \\ 0 & 1 & 2 \end{bmatrix}\{\tilde{u}\}_i}, \text{for,} \quad i = 1, 2, 3 \tag{E1.9}$$

Eq. (E1.9) gives the three arbitrary constants as follows, at,

$$\begin{Bmatrix} \tilde{u}_1 \\ \tilde{u}_2 \\ \tilde{u}_3 \end{Bmatrix} = \begin{Bmatrix} 0 \\ 1 \\ 1.414 \end{Bmatrix} \Rightarrow \{\tilde{u}\}^t\begin{bmatrix} 2 & 1 & 0 \\ 1 & 4 & 1 \\ 0 & 1 & 2 \end{bmatrix}\{\tilde{u}\} = \begin{Bmatrix} 0 & 1 & 1.414 \end{Bmatrix}\begin{bmatrix} 2 & 1 & 0 \\ 1 & 4 & 1 \\ 0 & 1 & 2 \end{bmatrix}\begin{Bmatrix} 0 \\ 1 \\ 1.414 \end{Bmatrix} = 10.826792$$

Then,

$$\alpha_1^2 = \frac{6EA}{\overline{m}l^2}\cdot\frac{1}{10.826792},$$

Then,

$$\alpha_1 = 0.744\sqrt{\frac{EA}{\overline{m}l^2}}$$

At,

$$\begin{Bmatrix} \tilde{u}_1 \\ \tilde{u}_2 \\ \tilde{u}_3 \end{Bmatrix} = \begin{Bmatrix} 0 \\ 1 \\ -1.414 \end{Bmatrix} \Rightarrow \{\tilde{u}\}^t\begin{bmatrix} 2 & 1 & 0 \\ 1 & 4 & 1 \\ 0 & 1 & 2 \end{bmatrix}\{\tilde{u}\}$$

$$= \begin{Bmatrix} 0 & 1 & -1.414 \end{Bmatrix}\begin{bmatrix} 2 & 1 & 0 \\ 1 & 4 & 1 \\ 0 & 1 & 2 \end{bmatrix}\begin{Bmatrix} 0 \\ 1 \\ -1.414 \end{Bmatrix} = 5.170792$$

Then,

$$\alpha_2^2 = \frac{6EA}{\overline{m}l^2}\cdot\frac{1}{5.170792}, \quad \text{then,} \quad \alpha_2 = 1.077\sqrt{\frac{EA}{\overline{m}l^2}}$$

Therefore the mass matrix orthogonal mode shape for the bar given in the example is,

$$1.\left(\frac{\overline{m}.l^2}{6EA}\begin{bmatrix}2 & 1 & 0\\ 1 & 4 & 1\\ 0 & 1 & 2\end{bmatrix}\right)\begin{Bmatrix}1\\ 1\\ 1\end{Bmatrix}=\frac{\overline{m}l}{2}\begin{bmatrix}1\\ 1\\ 1\end{bmatrix}$$

$$2.\left(\frac{\overline{m}.l^2}{6EA}\begin{bmatrix}2 & 1 & 0\\ 1 & 4 & 1\\ 0 & 1 & 2\end{bmatrix}\right)\begin{Bmatrix}1\\ 0\\ -1\end{Bmatrix}=\frac{\overline{m}l}{3}\begin{bmatrix}1\\ 0\\ -1\end{bmatrix}$$

$$3.\left(\frac{\overline{m}.l^2}{6EA}\begin{bmatrix}2 & 1 & 0\\ 1 & 4 & 1\\ 0 & 1 & 2\end{bmatrix}\right)\begin{Bmatrix}1\\ -1\\ 0\end{Bmatrix}=\frac{\overline{m}l}{6}\begin{bmatrix}1\\ -3\\ -1\end{bmatrix}$$

10.4 Application to transverse vibration of beams

The element stiffness matrix for the beam element is given by (Chapter 5: Introduction to Finite Element Method (F.E.M.): Bar and Beam Applications),

$$[k]^e=\frac{EI}{l^3}\begin{bmatrix}12 & 6l & -12 & 6l\\ 6l & 4l^2 & -6l & 2l^2\\ -12 & -6l & 12 & -6l\\ 6l & 2l^2 & -6l & 4l^2\end{bmatrix} \tag{10.21}$$

The procedure is precise as before. If we use Hermitian polynomial interpolation that is used for static function, the element mass matrix becomes,

$$[m]^e=\frac{\overline{m}l}{420}\begin{bmatrix}156 & 22l & 54 & -12l\\ & 4l^2 & 13l & -3l^2\\ & & 156 & -22l\\ & & & 4l^2\end{bmatrix} \tag{10.22}$$

Example 10.2: Find the fundamental natural frequency of a beam, Fig. 10.6, using two elements, E.I. = constant and m is the mass per unit length.

The system stiffness and mass matrices can be assembled by using the generalized coordinates as follows,

$$\left(\begin{bmatrix}(k_{33}^1+k_{11}^2) & (k_{34}^1+k_{12}^2) & k_{14}^2\\ & (k_{44}^1+k_{22}^2) & k_{24}^2\\ & & k_{44}^2\end{bmatrix}-\omega^2\begin{bmatrix}(m_{33}^1+m_{11}^2) & (m_{34}^1+m_{12}^2) & m_{14}^2\\ & (m_{44}^1+m_{22}^2) & m_{24}^2\\ & & m_{44}^2\end{bmatrix}\right)\begin{Bmatrix}\tilde{q}_1\\ \tilde{q}_2\\ \tilde{q}_3\end{Bmatrix}=0 \tag{E2.1}$$

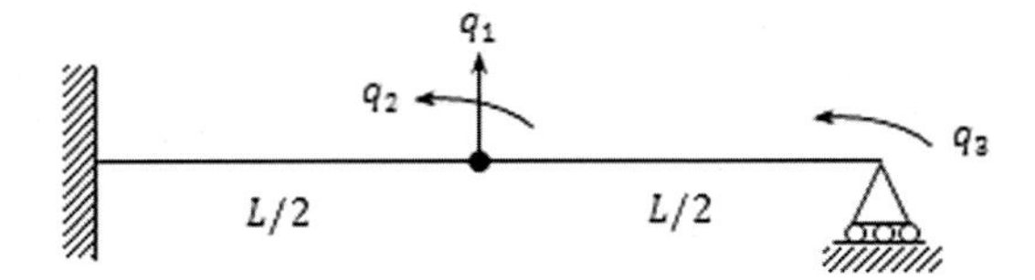

FIGURE 10.6 Beam under bending.

Choosing $l = 2$, using Eqs. (10.21) and (10.22) gives,

$$\left(\frac{EI}{l^3}\begin{bmatrix}(12+12) & (-6l+6l) & 6l \\ & (4l^2+4l^2) & 2l^2 \\ & & 4l^2\end{bmatrix} - \omega_n^2\frac{\overline{m}.l}{420}\begin{bmatrix}(156+156) & (-22l+22l) & -12l \\ & (4l^2+4l^2) & -3l^2 \\ & & 4l^2\end{bmatrix}\right)\begin{Bmatrix}\hat{q}_1 \\ \hat{q}_2 \\ \hat{q}_3\end{Bmatrix} = 0$$

Forming the Eigenvalue problem and solving gives the three natural frequencies as follows:

$$\omega_1 = 0.978\sqrt{EI\overline{m}}, \omega_2 = 3.629\sqrt{EI\overline{m}}, \omega_3 = 9.571\sqrt{EI\overline{m}}$$

Lumped mass matrix,

In this approximation technique, the element mass is distributed in the directions of the degrees of freedom involved. Therefore for the beam element lumped mass modeling, we can write,

$$[m]^e = \frac{\rho AL}{2}\begin{bmatrix}1 & 0 & 0 & 0 \\ & 0 & 0 & 0 \\ & & 1 & 0 \\ & & & 0\end{bmatrix} \tag{E2.2}$$

And the corresponding E.V.P. for the beam shown in Fig. 10.6, is,

$$\left(\frac{EI}{l^3}\begin{bmatrix}(12+12) & (-6l+6l) & 6l \\ & (4l^2+4l^2) & 2l^2 \\ & & 4l^2\end{bmatrix} - \omega_n^2\frac{\rho.A.l}{2}\begin{bmatrix}(1+1) & 0 & 0 \\ & 0 & 0 \\ & & 0\end{bmatrix}\right)\begin{Bmatrix}\hat{q}_1 \\ \hat{q}_2 \\ \hat{q}_3\end{Bmatrix} = 0 \tag{E2.3}$$

Where, $\overline{m} = \rho.A$

A solution of the Eq. (E2.3)-E.V.P., gives, let $l = 2$,

$$\omega_1 = \omega_2 = 0$$

$$\omega_3 = 0.9258\sqrt{EI\overline{m}}$$

10.5 Constant strain element

As explained in (Chapter 2: Direct Methods), the element stiffness matrix, of the element shown in Fig. 10.7, was given by,

$$[k]^e = \frac{Et}{2(1-\nu^2)}\begin{bmatrix}\frac{(1-\nu)}{2} & \frac{(1+\nu)}{2} & -1 & \frac{(1-\nu)}{2} & -\frac{(1-\nu)}{2} & -\nu \\ & \frac{(3-\nu)}{2} & -\nu & \frac{(1-\nu)}{2} & \frac{(1-\nu)}{2} & -1 \\ & & 1 & 0 & 0 & \nu \\ & & & \frac{(1-\nu)}{2} & \frac{(1-\nu)}{2} & 0 \\ & \text{Sym.} & & & \frac{(1-\nu)}{2} & 0 \\ & & & & & 1\end{bmatrix} \tag{10.23}$$

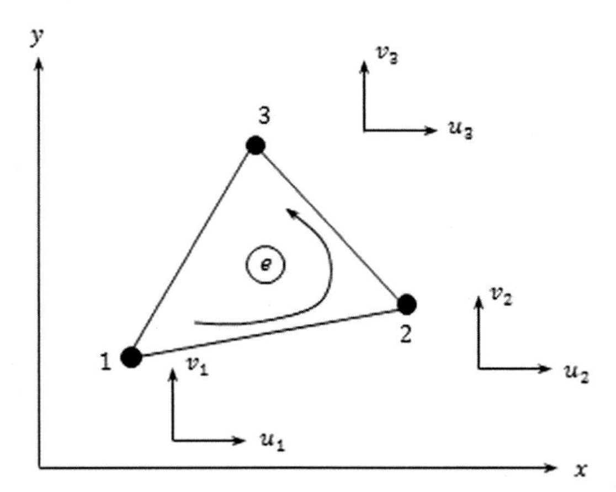

FIGURE 10.7 A typical constant strain element.

Where,

$$\{u^e\}^t = \{ u_1 \quad v_1 \quad u_2 \quad v_2 \quad u_3 \quad v_3 \}$$

And,

$$[N] = \begin{bmatrix} N_1 & 0 & N_2 & 0 & N_3 & 0 \\ 0 & N_1 & 0 & N_2 & 0 & N_3 \end{bmatrix}$$

The consistent element mass matrix is given by,

$$[m]^e = \rho t_e \int [N]^t [N] dA \tag{10.24}$$

Where, t_e: is the element thickness (mm), ρ: is the density of the element material $\left(\text{kg}/m^3\right)$.

And, since $\int_e [N]^t dA = \frac{1}{6} A_e$, $\int [N_1][N_2] dA = \frac{1}{12} A_e \ldots$, etc. carrying out the multiplication of Eq. (10.24), gives,

$$[m]^e = \frac{\rho t_e A_e}{12} \begin{bmatrix} 2 & 0 & 1 & 0 & 1 & 0 \\ & 2 & 0 & 1 & 0 & 1 \\ & & 2 & 0 & 1 & 0 \\ & & & 2 & 0 & 1 \\ & & & & 2 & 0 \\ & & & & 1 & 2 \end{bmatrix} \tag{10.25}$$

10.6 Quadrilateral elements

The quadrilateral element, for example, quads are shown in Fig. 10.8, and to be used for plane stress/strain, we have,

$$\{u^e\}^t = \{ u_1 \quad v_1 \quad u_2 \quad v_2 \quad u_3 \quad v_3 \quad u_4 \quad v_4 \}$$

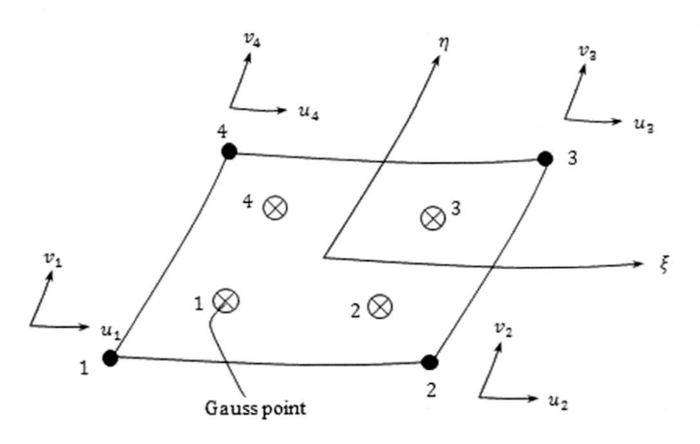

FIGURE 10.8 4 Nodes isoparametric quadrilateral element.

And,

$$[N] = \begin{bmatrix} N_1 & 0 & N_2 & 0 & N_3 & 0 & N_4 & 0 \\ 0 & N_1 & 0 & N_2 & 0 & N_3 & 0 & N_4 \end{bmatrix}$$

The consistent element mass matrix is given by,

$$[m]^e = \rho t_e \int_{-1}^{1} \int_{-1}^{1} [N]^t[N] \det J \; d\xi \; d\eta \tag{10.26}$$

This integral is evaluated numerically using the same procedure to derive the element stiffness matrix $[k]^e$ in Eq. (10.23).

The size of $[m]^e$ will be (8×8). The procedure of assembling the E.V.P. follows the standard procedure explained previously for the bar element applications. This will be derived in detail in Chapter 12, Shape Functions Determinations and Numerical Integration.

10.7 Axisymmetric triangular element

The element stiffness matrix $[k]^e$ derived in Eq. (4.16) can be used as follows,

$$[k]^e = \iint_{\text{intrinsic element}} [B]^t[D][B] \left| \left[J\left(\frac{r,z}{\xi,\eta}\right) \right] \right| d\xi \; d\eta$$

Eq. (4.16) is integrated numerically using the Gaussian numerical integration technique. The consistent element mass matrix can be derived as follows,

$$[m]^e = \int_e \rho[N]^t[N] dV = \int_e \rho[N]^t[N] 2\pi r dr \tag{10.27}$$

And since

$$r = N_1 r_1 + N_2 r_2 + N_3 r_3$$

Therefore,

$$[m]^e = 2\pi\rho\int_e (N_1r_1 + N_2r_2 + N_3r_3)[N]^t[N]dr$$

And,

$$\int_e [N_1]^3 dA = \frac{2A_e}{20}.\int_e [N_1]^2[M_2]dA = \frac{2A_e}{60},\int_e [N_1][N_2][N_3]dA = \frac{2A_e}{120}, \quad \ldots\text{etc.}$$

This gives,

$$[m]^e = \frac{\pi\rho A_e}{10}\begin{bmatrix} \frac{4}{3}r_1 + 2\bar{r} & 0 & 2\bar{r} - \frac{r_3}{3} & 0 & 2\bar{r} - \frac{r_2}{3} & 0 \\ & \frac{4}{3}r_1 + 2\bar{r} & 0 & 2\bar{r} - \frac{r_3}{3} & 0 & 2\bar{r} - \frac{r_2}{3} \\ & & \frac{4}{3}r_2 + 2\bar{r} & 0 & 2\bar{r} - \frac{r_1}{3} & 0 \\ & & & \frac{4}{3}r_2 + 2\bar{r} & 0 & 2\bar{r} - \frac{r_1}{3} \\ & & & & \frac{4}{3}r_3 + 2\bar{r} & 0 \\ & & & & & \frac{4}{3}r_3 + 2\bar{r} \end{bmatrix}$$

$$\bar{r} = \frac{r_1 + r_2 + r_3}{3} \tag{10.28}$$

10.8 Consistent element mass matrix for the 8-nodes solid element

The element stiffness matrix was derived in Section 5.3, this element is shown in Fig. 10.9, as rectangular solid and isoparametric shape.

The consistent element mass matrix is obtained as follows,

In rectangular coordinates system,

$$[m]^e = \rho\int \left[N\left(x,y,z\right)\right]^t\left[N\left(x,y,z\right)\right]dxdydz \tag{10.29}$$

And in intrinsic coordinates,

$$[m]^e = \rho\int_{-1}^{1}\int_{-1}^{1}\int_{-1}^{1} [N(\xi,\eta,\zeta)]^t[N(\xi,\eta,\zeta)]d\xi d\eta d\zeta$$

$$= \rho\sum_{i=1}^{n}\sum_{s=1}^{n}\sum_{t=1}^{n} w_r w_s w_t [N(\xi,\eta,\zeta)]^t[N(\xi,\eta,\zeta)]\left|J(\xi,\eta,\zeta)\right|$$

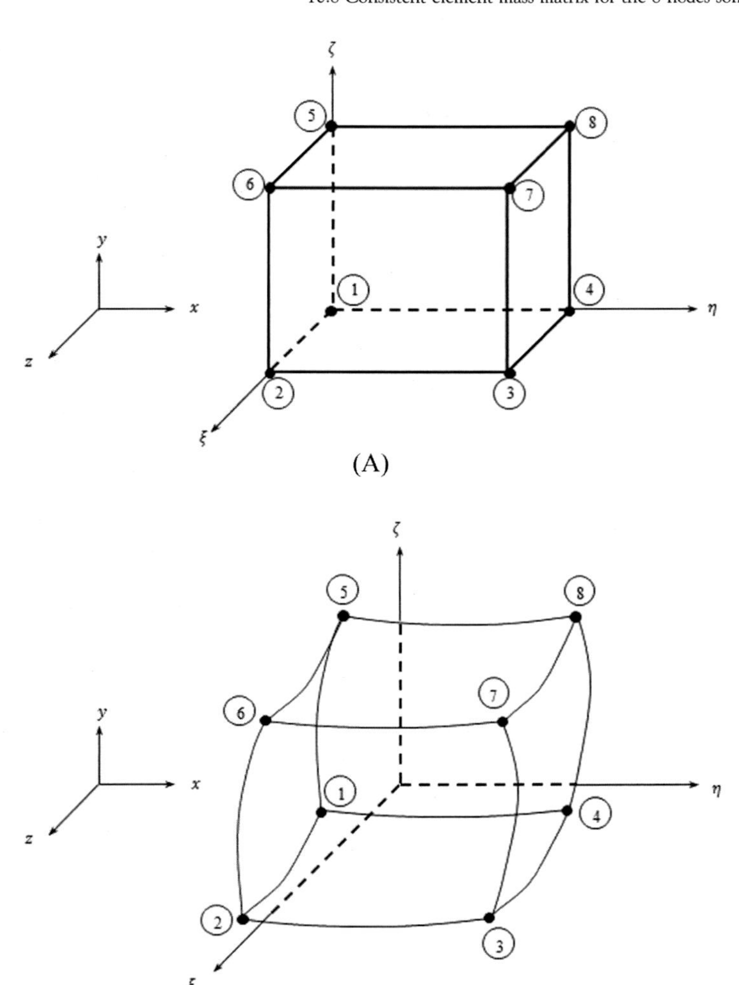

FIGURE 10.9 (A) Block element, (B) isoparametric.

And for the considered element in Section 5.3,

$$[m]^e = a \times b \times c \times \rho \int_{-1}^{1}\int_{-1}^{1}\int_{-1}^{1} [N(\xi,\eta,\zeta)]^t [N(\xi,\eta,\zeta)] d\xi d\eta d\zeta$$

Or,

$$[m]^e = a \times b \times c \times \rho \sum_{r=1}^{n}\sum_{s=1}^{n}\sum_{t=1}^{n} w_r w_s w_t [N(\xi,\eta,\zeta)]^t [N(\xi,\eta,\zeta)] \left|J(\xi,\eta,\zeta)\right| \tag{10.30}$$

Eq. (10.30) can be integrated numerically to obtain the consistent element mass matrix, and the numerical integration scheme will be presented in Chapter 12, Shape Functions Determinations and Numerical Integration.

Example 10.3: Accepting that the transverse deflection of a beam element may be approximated by

$$W(\zeta) = [N(\zeta)]\{u(t)\}_0^e \quad \text{(E3.1)}$$

Where $\{u(t)\}_0^e = [u_1 u_2 u_3 u_4]$, concerning Fig. 10.10, and $[N(\zeta)]$ is an appropriate interpolation matrix, show, without the algebraic details, how 4×4 stiffness and consistent mass matrices $[k]^e$ and $[m]^e$ maybe generated for dynamic analysis. Using Hamilton's principle

$$\delta \int_{t_1}^{t_2} (T - U)dt = 0, \quad \text{(E3.2)}$$

Show, again without algebraic detail, that the frequency equation for a free vibration Finite Element analysis of a beam is given by

$$([K] - \omega^2[M])\{q\} = \{0\} \quad \text{(E3.3)}$$

A uniform shaft carrying two rotors is supported in one rigid and one flexible bearing shown in Fig. 10.10. The shaft has mass $\overline{m}$ per unit length and bending stiffness E.I. whilst the rotors have mass and relevant moments of inertia M_1, I_1 and M_2, I_2.

Generate the assemblage stiffness and mass matrices $[M]$ and $[K]$ in readiness for frequency analysis of the system of Fig. 10.10, when the shaft is not rotating.

It will be recalled that for uniform beam element,

$$[K]^e = \frac{EI}{l^3}\begin{bmatrix} 12 & 6l & -12 & 6l \\ & 4l^2 & -6l & 2l^2 \\ & & 12 & -6l \\ \text{sym.} & & & 4l^2 \end{bmatrix}$$

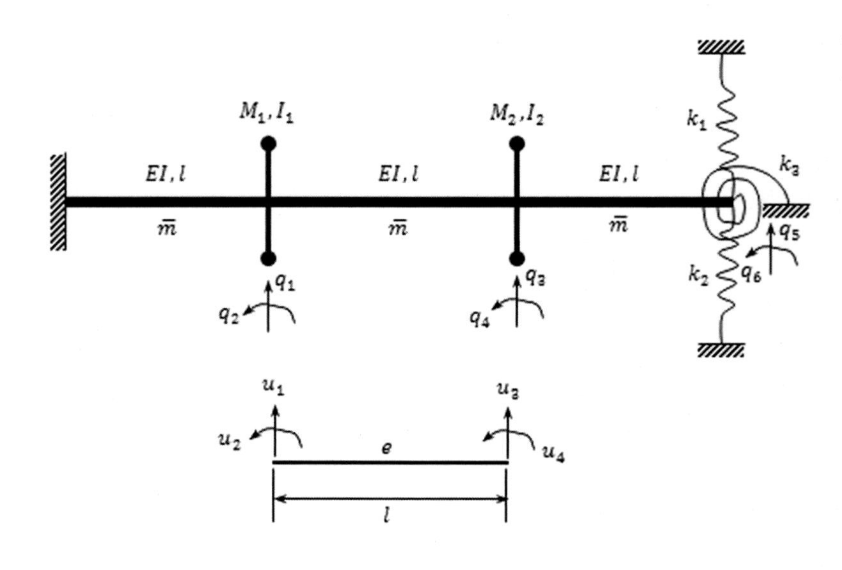

FIGURE 10.10 A shaft carrying two rotors.

$$[m]^e = \frac{\overline{m}l}{420}\begin{bmatrix} 156 & 22l & 54 & -13l \\ & 4l^2 & 13l & -3l^2 \\ & & 156 & -22l \\ \text{sym.} & & & 4l^2 \end{bmatrix} \tag{E3.4}$$

Solution,

We have from Eq. (E3.1)

$$w(\xi) = [N(\xi)]\{u(t)\}_0^e$$

Where,

$$\{u(t)\}_0^e = \begin{bmatrix} u_1 & u_2 & u_3 & u_4 \end{bmatrix}$$

To find the element stiffness matrix, $[k]$, we have to calculate the stored strain energy in the element U^e, and the total strain energy of the system is, $U = \sum U^e$

Where U^e is given by,

$$U^e = \int_0^l \frac{EI}{l}\left(\frac{d^2\xi}{d\xi^2}\right)^2 d\xi \tag{E3.5}$$

And, as before,

$$w'' = [N''(\xi)]\{u(t)\}_0^e = [B]\{u\}_0^e$$

And the bending stiffness of the beam,

$$EI = [D]$$

Therefore,

$$U^e = \frac{1}{2}\{u\}_0^{et}\left(\int_0^l [B]^t[D][B]d\xi\right)\{u\}_0^e \tag{E3.6}$$

Or,

$$U^e = \frac{1}{2}\{u\}_0^{e^t}[k]^e\{u\}_0^e \tag{E3.7a}$$

To generate mass matrix,

We should find the kinetic energy T^e stored in the element is given by,

$$T^e = \frac{1}{2}\int_0^l \overline{m}\dot{u}^2 d\xi \tag{E3.7}$$

Or,

$$T^e = \frac{1}{2}\int_0^l \dot{u}^t\overline{m}\dot{u}\ d\xi \tag{E3.8}$$

We have,

$$\dot{u} = [N]\{\dot{u}\}_0^e \tag{E3.9}$$

Put Eq. (E3.9) into Eq. (E3.8) gives,

$$T^e = \frac{1}{2}\{\dot{u}\}_0^{e^t}\left(\int_0^l [N]^t\overline{m}[N]d\xi\right)\{\dot{u}\}_0^e$$

Or,

$$T^e = \frac{1}{2}\{\dot{u}\}_0^{e^t}[m]^e\{\dot{u}\}_0^e \tag{E3.10}$$

Where $[m]^e = \left(\int_0^l [N]^t\overline{m}[N]d\xi\right)$

We have, the change of total potential energy between two-time intervals is equal to zero, that is,

$$\delta\int_{t_1}^{t_2}(T-U)dt = 0 \tag{E3.11}$$

Using Eqs. (E3.7a) and (E3.10) in Eq. (E3.11) gives,

$\delta\int_{t_1}^{t_2}\left(\frac{1}{2}\{\dot{u}\}_0^{e^t}[m]^e\{\dot{u}\}_0^e - \frac{1}{2}\{u\}_0^{e^t}[k]^e\{u\}_0^e\right)dt = 0$

We have,

$$\delta\int_{t_1}^{t_2}\left(\frac{1}{2}\{\dot{u}\}_0^{e^t}[m]^e\{\dot{u}\}_0^e\right)dt = \frac{1}{2}\left(\int_{t_1}^{t_2}\left(\{\delta\dot{u}\}_0^{e^t}[m]^e\{\dot{u}\}_0^e\right)dt + \int_{t_1}^{t_2}\left(\{\dot{u}\}_0^{e^t}[m]^e\{\delta\dot{u}\}_0^e\right)dt\right)$$

$$= \int_{t_1}^{t_2}\left(\{\delta\dot{u}\}_0^{e^t}[m]^e\{\dot{u}\}_0^e\right)dt \tag{E3.12a}$$

Integration Eq. (E3.12a) by part gives,

$$\{\delta u\}_0^{e^t}[m]^e\{\dot{u}\}_0^e\Big|_{t_1}^{t_2} - \int_{t_1}^{t_2}\left(\{\delta u\}_0^{e^t}[m]^e\{\ddot{u}\}_0^e\right)dt$$

Therefore,

$$\delta \text{ total potential energy} = 0 = -\int_{t_1}^{t_2}\left(\{\delta u\}_0^{e^t}[m]^e\{\ddot{u}\}_0^e\right)dt - \int_{t_1}^{t_2}\left(\{\delta u\}_0^{e^t}[k]^e\{u\}_0^e\right)dt \tag{E3.12}$$

Now, we can write,

$$0 = \{\delta u\}_0^{e^t}\left([m]^e\{\ddot{u}\} + [k]^e\{u\}\right) \tag{E3.13}$$

Since $\{\delta u\}_0^{e^t}$ arbitrary, Eq. (E3.13) gives,

$$[m]^e\{\ddot{u}\} + [k]^e\{u\} = 0 \tag{E3.14}$$

For the global system of equations, Eq. (E3.14) becomes,

$$[M]^e\{\ddot{u}\} + [K]^e\{u\} = 0 \tag{E3.15}$$

To solve Eq. (E3.15), let,
$\{u\} = \{\hat{u}\}\cos\omega t$, inserting this assumption in Eq. (E3.15), the E.V.P. form is obtained as follows,

$$\left([K] - \omega^2[M]\right)\{\hat{u}\} = \{0\} \tag{E3.16}$$

The inertia of the rotor E.R is given by,

$$\text{E.R} = \frac{1}{2} I\omega^2 \tag{E3.17}$$

The system stiffness matrix can be written as follows,

$$[K] = \begin{bmatrix} (k_{33}^1 + k_{11}^2) & (k_{34}^1 + k_{12}^2) & k_{13}^2 & k_{14}^2 & 0 & 0 \\ & (k_{44}^1 + k_{22}^2) & k_{23}^2 & k_{24}^2 & 0 & 0 \\ & & (k_{33}^2 + k_{11}^3) & (k_{34}^2 + k_{12}^3) & k_{13}^3 & k_{14}^3 \\ & & & (k_{44}^2 + k_{22}^3) & k_{23}^3 & k_{24}^3 \\ & & & & (k_{33}^3 + k_1 + k_2) & k_{34}^3 \\ & & & & & (k_{44}^3 + k_3) \end{bmatrix} \tag{E3.18}$$

Effect of springs and solution, the potential energy of the springs can be written as follows,

$$\text{P.E} = \frac{1}{2}\left(k_1 q_5^2 - k_2 q_5^2\right) + k_3 q_6 \tag{E3.19}$$

The contributions of the stiffness of the springs in Eq. (E3.19) are inserted in Eq. (E3.18), gives,

$$[k]^e = \frac{EI}{l^3} \begin{bmatrix} 24 & 0 & -12 & 6l & 0 & 0 \\ & 8l^2 & -6l & 2l^2 & 0 & 0 \\ & & 24 & 0 & -6l & 6l \\ & & & 8l^2 & -6l & 2l^2 \\ & & & & \left(-6l + k_1 \frac{l^3}{EI} + k_2 \frac{l^3}{EI}\right) & 0 \\ & & & & & \left(4l^2 + k_3 \frac{l^3}{EI}\right) \end{bmatrix} \tag{E3.20}$$

To construct the mass matrix, we have,
$I_1 = \frac{1}{2} M_1 \dot{q}_1^2 + \frac{1}{2} I_1 q_2^2$, $w_1 = q_2$, and,

$$I_2 = \frac{1}{2} M_2 \dot{q}_3^2 + \frac{1}{2} I_2 q_4^2, \; w_2 = q_4 \tag{E3.21}$$

Therefore the system mass matrix can be written,

$$[M] = \frac{\overline{m}l}{420} \begin{bmatrix} \left(312 + \frac{420}{\overline{m}l} M_1\right) & 0 & 54 & -13l & 0 & 0 \\ & \left(8l^2 + I_1 \frac{420}{\overline{m}l}\right) & 13l & -3l^2 & 0 & 0 \\ & & \left(M_2 \frac{420}{\overline{m}l} + 312\right) & 0 & 54 & -13l \\ & & & \left(I_2 \frac{420}{\overline{m}l} + 8l^2\right) & 13l & -13l^2 \\ & & & & 156 & -22l \\ & & & & & 4l^2 \end{bmatrix} \tag{E3.22}$$

Example 10.4: It is required to develop the mass and stiffness matrix for an element in the form of a uniform shaft in torsion, as shown in Fig. 10.11 expressing the instantaneous rotation of the shaft at x in the form

$$\theta(x,t) = a_1(t) + a_2(t)x \tag{E4.1}$$

Determine the expressions for $a_1(t)$ and $a_2(t)$ necessary to satisfy the boundary rotations u_1, u_2, x and l. then, by considering the strain and kinetic energies of the element show that,

$$[m] = \frac{I_p l}{6}\begin{bmatrix} 2 & 1 \\ 1 & 2 \end{bmatrix} \text{and}[k] = \frac{GJ}{l}\begin{bmatrix} 1 & -1 \\ -1 & 1 \end{bmatrix} \tag{E4.2}$$

Where I_p is the polar moment of inertia/unit length.

Solution,

To define the angle of a twist as a function of x, t, consider the torsion bar shown in Fig. 10.11 The angle of twist is given by Eq. (E4.1) as follows,

$$\theta(x,t) = a_1(t) + a_2(t)x$$

Or,

$$\{\theta\}^e = \begin{bmatrix} 1 & x \end{bmatrix}\begin{Bmatrix} a_1 \\ a_2 \end{Bmatrix}$$

To find the coefficients a_1, and a_2, we apply the boundary conditions as follows,

$$\theta|_{x=0} = u_1 = a_1(t)$$

$$\theta|_{x=l} = u_2 = a_1(t) + a_2(t)l \tag{E4.3}$$

Therefore,

$$u_2 = u_2 + a_2 l$$

Hence,

$$a_2 = \frac{1}{l}(u_2 - u_1)$$

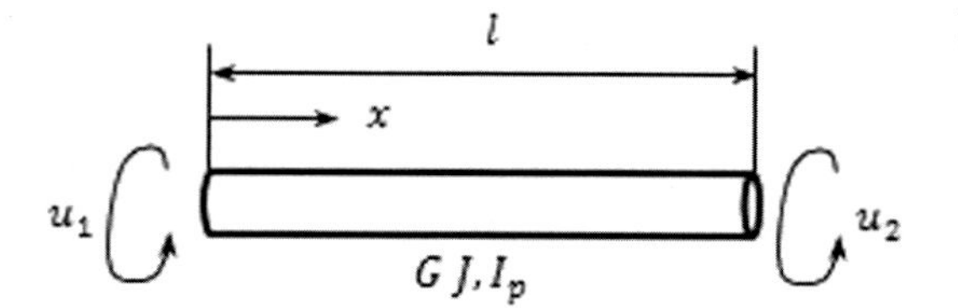

FIGURE 10.11 Torsion bar.

Therefore using the above values of a_1 and a_2, we can write,

$$\begin{Bmatrix} a_1 \\ a_2 \end{Bmatrix} = \begin{bmatrix} 1 & 0 \\ -\dfrac{1}{l} & \dfrac{1}{l} \end{bmatrix} \begin{Bmatrix} u_1 \\ u_2 \end{Bmatrix} \tag{E4.4}$$

We have,

$$\{\theta\}^e = [N(x)]\{u\}_0^e \tag{E4.5}$$

Or,

$$\{\theta(x,t)\}^e = \begin{bmatrix} 1 & x \end{bmatrix} \begin{Bmatrix} a_1 \\ a_2 \end{Bmatrix} \tag{E4.6}$$

Using Eq. (E4.4) into Eq. (E4.6) gives,

$$\{\theta\}^e = [1 \quad x] \begin{bmatrix} 1 & 0 \\ -\dfrac{1}{l} & \dfrac{1}{l} \end{bmatrix} \begin{Bmatrix} u_1 \\ u_2 \end{Bmatrix}$$

Or,

$$\{\theta\}^e = \begin{bmatrix} \left(1 - \dfrac{x}{l}\right) & \dfrac{x}{l} \end{bmatrix} \begin{Bmatrix} u_1 \\ u_2 \end{Bmatrix} \tag{E4.7}$$

The strain energy stored in the element is given by,

$$U = \frac{1}{2} GJ \int_0^l [N(x)]^2 \left(\{u\}_0^e\right)^2 dx \tag{E4.8}$$

Where

$$[N(x)] = \begin{bmatrix} \left(1 - \dfrac{x}{l}\right) & \dfrac{x}{l} \end{bmatrix} \tag{E4.9a}$$

And the kinetic energy is given by,

$$T = \frac{1}{2} \int_0^l I_p \dot{\theta}^2 dx \tag{E4.9}$$

Or,

$$[T] = \frac{1}{2} \int_0^l \{\dot{\theta}\}^{e^t} [I_p] \{\dot{\theta}\}^e dx \tag{E4.10}$$

We have,

$$\{\dot{\theta}\}^e = [N(x)]\{\dot{u}\}_0^e \tag{E4.11}$$

We have,

$$dx = du$$

Therefore Eq. (E4.14) becomes,

$$T = \frac{1}{2}\int_0^l \{\dot{u}\}_0^{et}[N(x)]^t[I_p]\{\dot{u}\}_0^e du \tag{E4.12}$$

Since,
$\{\dot{u}\}_0^{et} = \{u\}_0^{et}$ scalar, Eq. (E4.12) becomes,

$$T = \frac{1}{2}\{\dot{u}\}_0^{et}\underbrace{\int_0^l [N(x)]^t[I_p][N(x)]}_{[m]^e}\{\dot{u}\}_0^e dx \tag{E4.13}$$

Where,

$$[m]^e = \int_0^l [N(x)]^t[I_p][N(x)]dx \tag{E4.14}$$

Using the shape function $[N(x)]$ given in Eq. (E4.13a), Eq. (E4.18) becomes,

$$[m]^e = \int_0^l \frac{1}{l}\begin{bmatrix} l-x \\ x \end{bmatrix}[I_p].\frac{1}{l}\begin{bmatrix} l-x & x \end{bmatrix}dx$$

Or,
$[m]^e = \frac{1}{l^2}I_p\int_0^l \begin{bmatrix} (l-x)^2 & x(l-x) \\ x(l-x) & x^2 \end{bmatrix}dx$
Which gives,

$$[m]^e = \frac{I_p}{l^2}\begin{bmatrix} \int_0^l (l-x)^2 dx & \int_0^l x(l-x)dx \\ \int_0^l x(l-x)dx & \int_0^l x^2 dx \end{bmatrix} \tag{E4.15}$$

To achieve the integrations in Eq. (E4.19), we have,

$$\int_0^l (l-x)^2 dx = \int_0^l (l^2 - 2xl + x^2)dx = \left[l^2x - x^2l + \frac{x^3}{3}\right]_0^l$$

$$= \left[l^3 - l^3 + \frac{l^3}{3}\right] = \frac{l^3}{3}$$

And,

$$\int_0^l x(l-x)dx = \frac{l^3}{2} - \frac{l^3}{3} = \frac{1}{6}l^3$$

$$\int_0^l x^2 dx = \frac{l^3}{3}$$

Insert the above integrations in Eq. (E4.15) gives,

$$[m]^e = \frac{I_p}{l^2}\begin{bmatrix} \left(\frac{l^3}{3}\right) & \left(\frac{l^3}{6}\right) \\ \left(\frac{l^3}{6}\right) & \left(\frac{l^3}{3}\right) \end{bmatrix} = \frac{lI_p}{6}\begin{bmatrix} 2 & 1 \\ 1 & 2 \end{bmatrix} \tag{E4.16}$$

The strain energy of the bar under torsion is given by,

$$V = \frac{1}{2} GJ \int_0^l \left(\frac{\partial \theta}{\partial x}\right)^2 dx \tag{E4.17}$$

Using Eq. (E4.7)

$$\frac{\partial \theta}{\partial x} = \begin{bmatrix} -\frac{1}{l} & \frac{1}{l} \end{bmatrix} \begin{Bmatrix} u_1 \\ u_2 \end{Bmatrix} = \frac{1}{l} \begin{bmatrix} -1 & 1 \end{bmatrix} \begin{Bmatrix} u_1 \\ u_2 \end{Bmatrix} = \begin{bmatrix} N' \end{bmatrix} \{u\}_0^e \tag{E4.18}$$

Using Eq. (E4.17),

$$V = \frac{1}{2} \int_0^l \left[\frac{\partial \theta}{\partial x}\right]^t [GJ] \left[\frac{\partial \theta}{\partial x}\right] dx$$

Or,

$$V = \frac{1}{2} \int_0^l [\theta']^{e^t} [GJ][\theta']^e dx \tag{E4.19}$$

Inserting Eq. (E4.18) into Eq. (E4.19) gives,

$$V = \frac{1}{2} \int_0^l [u]^{e^t} [N'][D][N][u]^e dx$$

Or,

$$= \frac{1}{2} \{u\}_0^{e^t} \left(\int_0^l [N']^t [D][N'] dx \right) \{u\}_0^e = \frac{1}{2} \{u\}_0^{e^t} [k]^e \{u\}_0^e$$

Where,

$$[k]^e = \int_0^l [N'(x)]^t [D][N'(x)] dx \tag{E4.20}$$

Or,

$$[k]^e = \int_0^l \frac{1}{l} \begin{bmatrix} -1 \\ 1 \end{bmatrix} (GJ) \frac{1}{l} \begin{bmatrix} -1 & 1 \end{bmatrix} dx$$

$$= \frac{GJ}{l^2} \int_0^l \begin{bmatrix} 1 & -1 \\ -1 & 1 \end{bmatrix} dx = \frac{GJ}{l^2} \begin{bmatrix} \int_0^l dx & -\int_0^l dx \\ -\int_0^l dx & \int_0^l dx \end{bmatrix} \tag{E4.21}$$

Carrying out the integrations in Eq. (E4.21) gives,

$$[k]^e = \frac{GJ}{l^2}\begin{bmatrix} l & -l \\ -l & l \end{bmatrix} = \frac{GJ}{l}\begin{bmatrix} 1 & -1 \\ -1 & 1 \end{bmatrix} \tag{E4.22}$$

10.9 Consistent mass matrix for a tetrahedron element

For the tetrahedron element shown in Fig. 10.12,
The element stiffness matrix is given by Eq. (9.79) as follows,

$$[k]^e = [B]^{e^t}[D][B]^e V^e \tag{10.31}$$

With usual notations.

$[k]^e$ of size (12×12) since at each node, there is three degrees of freedom $u, v,$ and$w$ in directions $x, y,$ and$z$, respectively. The element mass matrix $[m]^e$ is given by ,

$$[m]^e = \iiint_{v^e} \rho[N]^t[N]dV \tag{10.32}$$

ρ is the element density. The shape function $[N]$ is given in Eq. (5.53),

$$[N] = \begin{bmatrix} N_1 & 0 & 0 & N_2 & 0 & 0 & N_3 & 0 & 0 & N_4 & 0 & 0 \\ 0 & N_1 & 0 & 0 & N_2 & 0 & 0 & N_3 & 0 & 0 & N_4 & 0 \\ 0 & 0 & N_1 & 0 & 0 & N_2 & 0 & 0 & N_3 & 0 & 0 & N_4 \end{bmatrix}$$

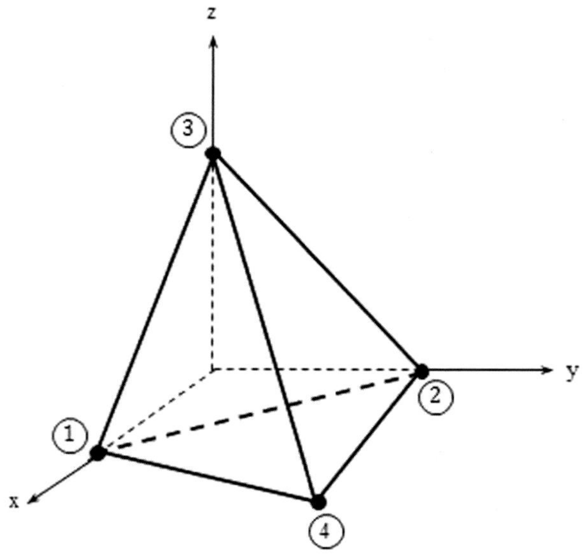

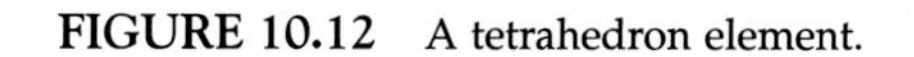
FIGURE 10.12 A tetrahedron element.

Carrying out the integrations of Eq. (10.32) numerically, the element mass matrix becomes,

$$[m]^e = \frac{\rho V^e}{20}\begin{bmatrix} 2 & 0 & 0 & 1 & 0 & 0 & 1 & 0 & 0 & 1 & 0 & 0 \\ 0 & 2 & 0 & 0 & 1 & 0 & 0 & 1 & 0 & 0 & 1 & 0 \\ 0 & 0 & 2 & 0 & 0 & 1 & 0 & 0 & 1 & 0 & 0 & 1 \\ 1 & 0 & 0 & 2 & 0 & 0 & 1 & 0 & 0 & 1 & 0 & 0 \\ 0 & 1 & 0 & 0 & 2 & 0 & 0 & 1 & 0 & 0 & 1 & 0 \\ 0 & 0 & 1 & 0 & 0 & 2 & 0 & 0 & 1 & 0 & 0 & 1 \\ 1 & 0 & 0 & 1 & 0 & 0 & 2 & 0 & 0 & 1 & 0 & 0 \\ 0 & 1 & 0 & 0 & 1 & 0 & 0 & 2 & 0 & 0 & 1 & 0 \\ 0 & 0 & 1 & 0 & 0 & 1 & 0 & 0 & 2 & 0 & 0 & 1 \\ 1 & 0 & 0 & 1 & 0 & 0 & 1 & 0 & 0 & 2 & 0 & 0 \\ 0 & 1 & 0 & 0 & 1 & 0 & 0 & 1 & 0 & 0 & 2 & 0 \\ 0 & 0 & 1 & 0 & 0 & 1 & 0 & 0 & 1 & 0 & 0 & 2 \end{bmatrix} \tag{10.33}$$

V^e: element volume.

Example 10.5: Using two elements of approximant constant reactions, but equal lengths determine an approximation to the point two modes of axial vibration of the tapered bar of Fig. 10.13.

$$E = 200 \text{ GPa}, \quad \nu = 0.3, \quad A_0 = 100 \text{ mm}^2, \quad l = 0.5 \text{ m}, \quad \rho = 7680 \text{ kg}/m^3.$$

Solution,

To solve for the axial vibration, take the finite element model shown in Fig. 10.14.

For the one-dimension modeling shown in Fig. 10.15, we have,

$$\frac{m_1}{m_2} = \frac{A_1}{A_2} = \frac{5}{4} * \frac{4}{7} = \frac{5}{7}$$

$$\overline{m}_1 = \frac{5}{7}\overline{m}_2 = \frac{\text{mass}}{\text{unitlength}}$$

$$\overline{m}_2 = \frac{7}{5}\overline{m}_1$$

If $m_1 = \overline{m}_1 = \overline{m}$, then, $\overline{m}_2 = \frac{7}{5}\overline{m}$

Element stiffness matrix $[k]^e = \frac{EA}{L}\begin{bmatrix} 1 & -1 \\ -1 & 1 \end{bmatrix}$

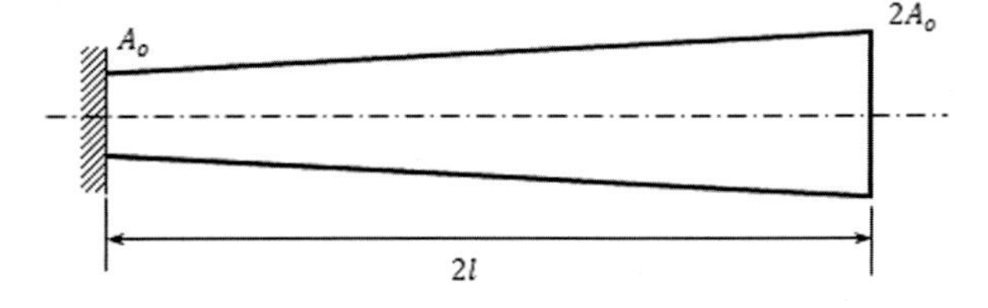

FIGURE 10.13 A tapered cantilever beam.

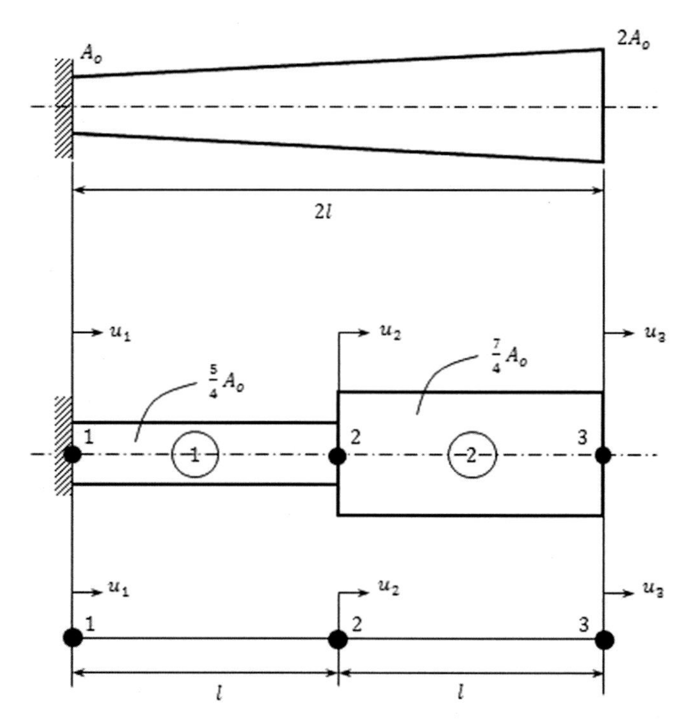

FIGURE 10.14 Finite element modeling of the tapered cantilever beam.

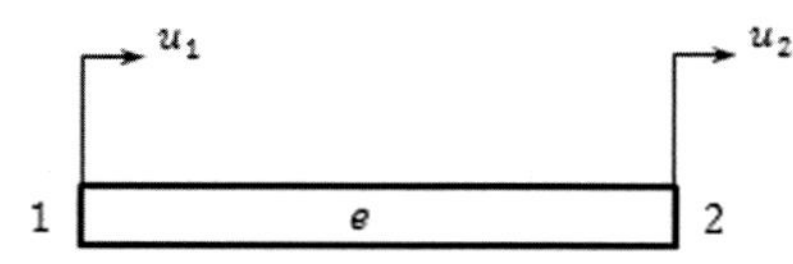

FIGURE 10.15 A typical one-dimensional element.

By inspection,

$$[k] = \frac{EA_0}{L}\begin{matrix} \\ 1 \\ 2 \\ 3 \end{matrix}\begin{matrix} \begin{matrix} 1 & & 2 & & 3 \end{matrix} \\ \begin{bmatrix} \frac{5}{4} & -\frac{5}{4} & 0 \\ -\frac{5}{4} & \left(\frac{5}{4}+\frac{7}{4}\right) & -\frac{7}{4} \\ 0 & -\frac{7}{4} & \frac{7}{4} \end{bmatrix} \end{matrix} = \frac{EA_0}{4L}\begin{bmatrix} 5 & -5 & 0 \\ -5 & 12 & -7 \\ 0 & -7 & 7 \end{bmatrix} \tag{E5.1}$$

$[m]^e = \frac{\overline{m}l}{6}\begin{bmatrix} 2 & 1 \\ 1 & 2 \end{bmatrix}, \overline{m}$ mass/unit length with a consistent rule.

Then the system mass matrix is given by,

$$[M] = \frac{\overline{m}l}{6}\begin{bmatrix} 2 & 1 & 0 \\ 1 & \left(2 + \frac{14}{5}\right) & \frac{7}{5} \\ 0 & \frac{7}{5} & \frac{14}{5} \end{bmatrix} \tag{E5.2}$$

Or,

$$[\overline{M}] = \frac{\overline{m}l}{6}\begin{bmatrix} 2 & 1 & 0 \\ 1 & \frac{24}{5} & \frac{7}{5} \\ 0 & \frac{7}{5} & \frac{14}{5} \end{bmatrix} = \frac{\overline{m}l}{30}\begin{bmatrix} 10 & 5 & 0 \\ 5 & 24 & 7 \\ 0 & 7 & 14 \end{bmatrix} \tag{E5.3}$$

We have,

$$[M]\{\ddot{u}\} + [K]\{u\} = \{0\}$$

Or,

$$\left([K] - \omega^2[M]\right)\{\hat{u}\} = \{0\}$$

Then, inserting Eqs. (E5.1) and (E5.2) gives,

$$\left(\frac{EA_0}{4l}\begin{bmatrix} 5 & -5 & 0 \\ -5 & 12 & -7 \\ 0 & -7 & 7 \end{bmatrix} - \omega^2\frac{\overline{m}l}{30}\begin{bmatrix} 10 & 5 & 0 \\ 5 & 24 & 7 \\ 0 & 7 & 14 \end{bmatrix}\right)\{\hat{u}\} = \{0\}$$

Since $u_1 = 0$ clamped end, so E.V.P. reduce to (2×2) matrix and we can directly reduce the $[M]$ & $[K]$ into (2×2) matrix.

$$\left(\begin{bmatrix} 12 & -7 \\ -7 & 7 \end{bmatrix} - \omega^2 * \frac{\overline{m}l}{30}\cdot\frac{4L}{EA_0}\begin{bmatrix} 24 & 7 \\ 7 & 14 \end{bmatrix}\right)\begin{Bmatrix} \hat{u}_2 \\ \hat{u}_3 \end{Bmatrix} = \{0\}$$

Let, $\lambda = \frac{2\omega^2\overline{m}l^2}{15EA_0}$, then,

$$\left(\begin{bmatrix} 12 & -7 \\ -7 & 7 \end{bmatrix} - \lambda\begin{bmatrix} 24 & 7 \\ 7 & 14 \end{bmatrix}\right)\{u\} = \{0\} \tag{E5.4}$$

Therefore,

$$\begin{bmatrix} 12 & -7 \\ -7 & 7 \end{bmatrix}\begin{Bmatrix} u_2 \\ u_3 \end{Bmatrix} = \lambda\begin{bmatrix} 24 & 7 \\ 7 & 14 \end{bmatrix}\begin{Bmatrix} u_2 \\ u_3 \end{Bmatrix}$$

$$\frac{1}{287}\begin{bmatrix} 24 & -7 \\ -7 & 14 \end{bmatrix}\begin{bmatrix} 12 & -7 \\ -7 & 7 \end{bmatrix}\begin{Bmatrix} u_2 \\ u_3 \end{Bmatrix} = \lambda\begin{Bmatrix} u_2 \\ u_3 \end{Bmatrix}$$

Or,

$$\frac{1}{287}\begin{bmatrix} 337 & -217 \\ 182 & 147 \end{bmatrix}\begin{Bmatrix} u_2 \\ u_3 \end{Bmatrix} = \lambda\begin{Bmatrix} u_2 \\ u_3 \end{Bmatrix}$$

$$\left(\begin{bmatrix} 1.174 & 0.756 \\ 0.634 & 0.512 \end{bmatrix} - \lambda(I)\right)\begin{Bmatrix} u_2 \\ u_3 \end{Bmatrix} = 0$$

$$\begin{bmatrix} (1.174-\lambda) & 0.756 \\ 0.634 & (0.512-\lambda) \end{bmatrix}\begin{Bmatrix} u_2 \\ u_3 \end{Bmatrix} = 0 \qquad \text{(E5.5)}$$

For a nontrivial solution, the determinant should be = 0.

$$(1.174-\lambda)(0.512-\lambda) - 0.634*0.756 = 0$$

Or,

$$\lambda^2 - 1.686\lambda + 0.6 - 0.48 = 0$$

$$\lambda^2 - 1.686\lambda + 0.12 = 0$$

Therefore,

$$\lambda_{1,2} = \frac{1.686 \pm \sqrt{(1.686)^2 - 4*0.12}}{2}$$

Or,

$$\lambda_1 = 1.611$$
$$\lambda_2 = 0.074$$

For fundamental natural frequency, we choose the smallest one.
$\lambda = 0.074$

$$0.074 = \lambda = \frac{\omega^2 \times \overline{m} \times l \times 2L}{15EA_0}$$

$$\omega^2 = \frac{15 \times EA_0 \times 0.074}{\overline{m} \times l \times 2L}$$

$$\omega^2 = \frac{EA_0}{\overline{m}l^2} \times 0.555$$

Hence,

$$\omega_1 = 0.75\sqrt{\frac{EA_0}{\overline{m}l^2}}, \omega_2 = 3.476\sqrt{\frac{EA_0}{\overline{m}l^2}}$$

Example 10.6: Formulate the eigenvalue problem defining the free vibration of the beam of Fig. 10.16 discretized into three elements as shown. EI = constant, $\overline{m}$: mass/unit length, $M = 2\overline{m}l$.

FIGURE 10.16 A beam carries a concentrated mass.

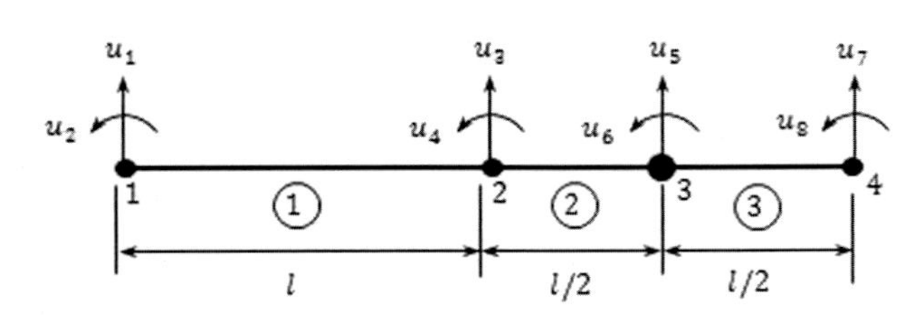

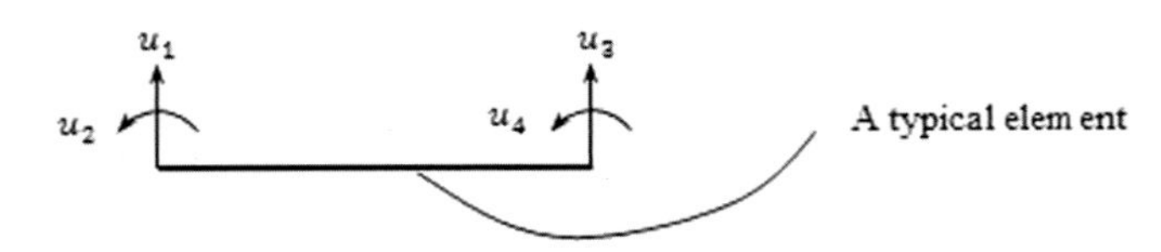

Solution,
For transverse vibration,

$$[m]^e = \frac{\overline{m}l}{420}\begin{bmatrix} 156 & 22l & 54 & -13l \\ 22l & 4l^2 & 13l & -3l^2 \\ 54 & 13l & 156 & -22l \\ -13l & -3l^2 & -22l & 4l^2 \end{bmatrix} \tag{E6.1}$$

$$[k]^e = \frac{EI}{l^3}\begin{bmatrix} 12 & 6l & -12 & 6l \\ 6l & 4l^2 & -6l & 2l^2 \\ -12 & -6l & 12 & -6l \\ 6l & 2l^2 & -6l & 4l^2 \end{bmatrix} \tag{E6.2}$$

We have,

$$\underbrace{u_1 = u_2}_{\text{Clamped}} = \underbrace{u_3 = u_7}_{\text{S.S.}} = 0 \tag{E6.3}$$

Then, there are only four d.o.f $[k]$ by inspection,

$$[K] = \frac{EI}{l^3}\begin{matrix} \\ 4 \\ 5 \\ 6 \\ 8 \end{matrix}\begin{matrix} 4 & 5 & 6 & 8 \\ \left[\begin{matrix} (k_{44}^1 + k_{22}^2) \\ k_{23}^2 \\ k_{24}^2 \\ 0 \end{matrix}\right. & \begin{matrix} k_{23}^2 \\ (k_{33}^2 + k_{11}^3) \\ (k_{34}^2 + k_{12}^3) \\ k_{14}^3 \end{matrix} & \begin{matrix} k_{24}^2 \\ (k_{34}^2 + k_{12}^3) \\ (k_{44}^2 + k_{22}^3) \\ k_{24}^3 \end{matrix} & \left.\begin{matrix} 0 \\ k_{14}^3 \\ k_{24}^3 \\ k_{44}^3 \end{matrix}\right] \end{matrix}$$

Or,

$$[K]=\frac{EI}{l^3}\begin{bmatrix} (4l^2*32+4l^2) & -\frac{6l}{16} & \frac{2l^2}{32} & 0 \\ -\frac{6l}{16} & (24*32) & 0 & \frac{6l}{16} \\ \frac{2l^2}{32} & 0 & \frac{8l^2}{32} & \frac{2l^2}{32} \\ 0 & \frac{6l}{16} & \frac{2l^2}{32} & \frac{4l^2}{32} \end{bmatrix}$$

Then,

$$[K]=\frac{EI}{4l^3}\begin{bmatrix} 20l^2 & -12l & 2l^2 & 0 \\ -12l & 96 & 0 & 12l \\ 2l^2 & 0 & 8l^2 & 2l^2 \\ 0 & 12l & 2l^2 & 4l^2 \end{bmatrix}$$

The system mass matrix is given by,

$$[M]=\frac{\overline{m}l}{1680}\begin{matrix} \\ 4 \\ 5 \\ 6 \\ 8 \end{matrix}\begin{bmatrix} \overset{4}{20l^2} & \overset{5}{26l} & \overset{6}{-3l^2} & \overset{8}{0} \\ 26l & \left(1248+\frac{1680\overline{M}}{\overline{m}l}\right) & 0 & -26l \\ -3l^2 & 0 & 8l^2 & -3l^2 \\ 0 & -26l & -3l^2 & 4l^2 \end{bmatrix}$$

Or,

$$[M]=\frac{\overline{m}l}{1680}\begin{bmatrix} 20l^2 & 26l & -3l^2 & 0 \\ 26l & 4608 & 0 & -26l \\ -3l^2 & 0 & 8l^2 & -3l^2 \\ 0 & -26l & -3l^2 & 4l^2 \end{bmatrix} \tag{E6.4}$$

$\overline{M}$ = will be added to the element (5, 5) as a concentrated load at the node.
We have,

$$[M]\{\ddot{u}\}+[K]\{u\}=\{0\}$$

$$\left([K]-\omega^2[M]\right)\{u\}=\{0\} \tag{E6.5}$$

Inserting Eqs. (E6.3) and (E6.4) into Eq. (E6.5) gives,

$$\frac{EI}{4l^3}\begin{bmatrix} 20l^2 & -12l & 2l^2 & 0 \\ -12l & 96 & 0 & 12l \\ 2l^2 & 0 & 8l^2 & 2l^2 \\ 0 & 12l & 2l^2 & 4l^2 \end{bmatrix}-\frac{\omega^2\overline{m}l}{1680}\begin{bmatrix} 20l^2 & 26l & -3l^2 & 0 \\ 26l & 4608 & 0 & -26l \\ -3l^2 & 0 & 8l^2 & -3l^2 \\ 0 & -26l & -3l^2 & 4l^2 \end{bmatrix}$$

Example 10.7: Formulate the matrix Eigenvalue problem for the system of Fig. 10.17. Use three beam elements.

Solution,

For the typical beam bending element shown in Fig. 10.18, we have,

$$[m]^e = \frac{\overline{m}l}{420}\begin{bmatrix} 156 & 22l & 54 & -13l \\ & 4l^2 & 13l & -3l^2 \\ & & 156 & -22l \\ & & & 4l^2 \end{bmatrix} \text{(element mass matrix)} \quad \text{(E7.1)}$$

$$[k]^e = \frac{\text{E.I.}}{l^3}\begin{bmatrix} 12 & 6l & -12 & -6l \\ & 4l^2 & -6l & 2l^2 \\ & & 12 & -6l \\ & & & 4l^2 \end{bmatrix} \text{(element stiffness matrix)} \quad \text{(E7.2)}$$

For $M_1,\ I_1,\ u = \hat{u}\sin\omega t$, we have,

$$\text{K.E}_1\text{of disc} = \frac{1}{2}\left(Mv^2 + I_1\overset{\dot{u}_2}{\omega^2}\right) \quad \text{(E7.3)}$$

$$\text{For } M_1, I_1 = \frac{1}{2}\left(M_1\dot{u}_1^2 + I_1\dot{u}_2^2\right) = -\frac{1}{2}\omega^2(M_1u_1 + I_1u_2) \quad \text{(E7.4)}$$

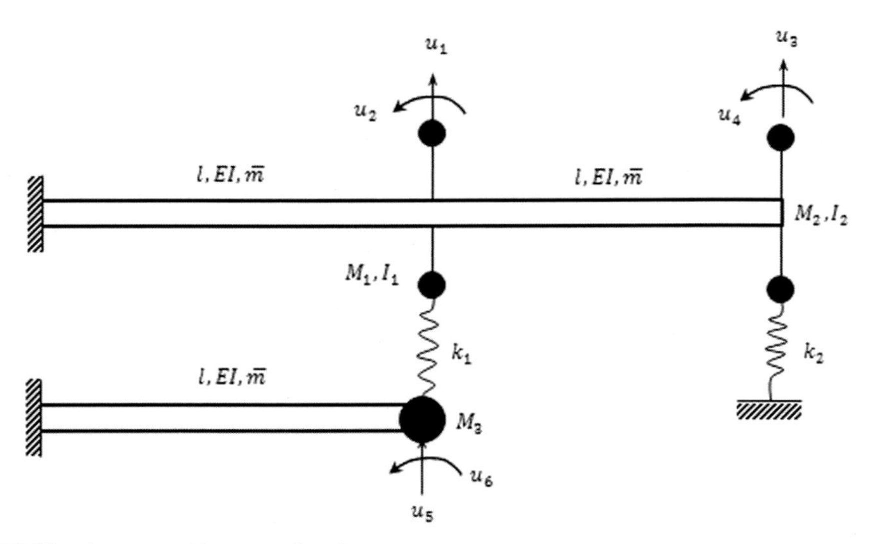

FIGURE 10.17 A system of beams and springs.

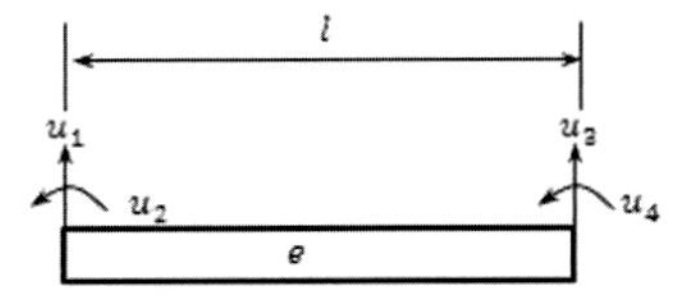

FIGURE 10.18 A typical beam element.

Similarly for M_2, I_2,

$$\text{K.E}_2 = -\frac{1}{2}\omega^2(M_2u_3 + I_2u_4) \tag{E7.5}$$

For M_3,

$$\text{K.E}_2 = -\frac{1}{2}\omega^2 M_3 u_5 \tag{E7.6}$$

Springs,
For k_2,

$$\text{P.E.} = \frac{1}{2}k_2u_4^2$$

For k_1,

$$\text{P.E.} = \frac{1}{2}k_1(u_1 - u_5)^2$$

$$= \frac{1}{2}k_1u_1^2 + \frac{1}{2}k_1u_5^2 - k_1u_1u_5$$

$$= \frac{1}{2}\left(\underbrace{k_1u_1^2}_{k_{11}} + \underbrace{k_1u_5^2}_{k_{55}} - \underbrace{2k_1u_1u_5}_{k_{15}=-k_1}\right) \tag{E7.7}$$

Stiffness matrix by inspection,

$$[K] = \begin{bmatrix} (k_{33}^1 + k_{11}^2 + k_1) & (k_{34}^1 + k_{12}^2) & k_{13}^2 & k_{14}^2 & -k_1 & 0 \\ (k_{34}^1 + k_{12}^2) & (k_{44}^1 + k_{22}^2) & k_{23}^2 & k_{24}^2 & 0 & 0 \\ k_{13}^2 & k_{23}^2 & k_{33}^2 & k_{34}^2 & 0 & 0 \\ k_{14}^2 & k_{24}^2 & k_{34}^2 & k_{44}^2 & 0 & 0 \\ -k_1 & 0 & 0 & 0 & (k_{33}^3 + k_1) & k_{34}^3 \\ 0 & 0 & 0 & 0 & k_{34}^3 & k_{44}^3 \end{bmatrix}$$

$$[K] = \begin{bmatrix} \left(24 + \frac{k_1l^3}{\text{E.I.}}\right) & 0 & -12 & 6l & -\frac{k_1l^3}{\text{E.I.}} & 0 \\ 0 & 8l^2 & -6l & 2l^2 & 0 & 0 \\ -12 & -6l & 12 & -6l & 0 & 0 \\ 6l & 2l^2 & -6l & 4l^2 & 0 & 0 \\ -\frac{k_1l^3}{\text{E.I.}} & 0 & 0 & 0 & \left(12 + \frac{k_1l^3}{\text{E.I.}}\right) & -6l \\ 0 & 0 & 0 & 0 & -6l & 4l^2 \end{bmatrix} \tag{E7.8}$$

Mass matrix by inspection, the same as the formation of stiffness matrix except M_1 will be added to m_{11},I_1 to m_{22},M_2 to m_{33},I_2 to m_{44} and M_3 to m_{55} these position of general five-coordinate.

$$
[M] = \begin{bmatrix}
\left(312 + \frac{M_1 420}{\overline{m}l}\right) & 0 & 54 & -13l & 0 & 0 \\
0 & \left(8l^2 + \frac{I_1 420}{\overline{m}l}\right) & 13l & -3l^2 & 0 & 0 \\
54 & 13l & \left(13l + \frac{M_2 420}{\overline{m}l}\right) & -22l & 0 & 0 \\
-13l & -3l^2 & -22l & \left(4l^2 + \frac{I_2 420}{\overline{m}l}\right) & 0 & 0 \\
0 & 0 & 0 & 0 & \left(156 + \frac{M_3 420}{\overline{m}l}\right) & -22l \\
0 & 0 & 0 & 0 & -22l & 4l^2
\end{bmatrix} \tag{E7.9}
$$

E.V.P.,

$$([K] - \omega^2[M])\{u\} = \{0\}$$

We can get ω by solving the E.V.P.

Problems

P.10.1 Formulate the Eigenvalue problem defining the beam's free vibration shown in Fig. P.10.1. Use three elements,

$$
[k]^e = \frac{EI}{l^3}\begin{bmatrix}
12 & 6l & -12 & 6l \\
 & 4l^2 & -6l & 2l^2 \\
 & & 12 & 6l \\
 & & & 4l^2
\end{bmatrix}
$$

$$
[m]^e = \frac{\overline{m}l}{420}\begin{bmatrix}
156 & 22l & 54 & 13l \\
22l & 4l^2 & 13l & -3l^2 \\
54 & 13l & 156 & -22l \\
13l & -3l^2 & -22l & 4l^2
\end{bmatrix}
$$

$$E = 70000\ N/\text{mm}^2$$

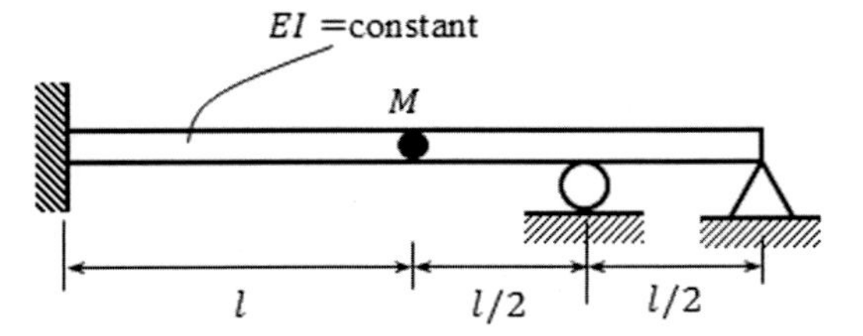

FIGURE P.10.1

$\overline{m}$: mass/unit length

$$\overline{m} = 2\ \text{kg}/m$$

$$M = 3\ \overline{m}l.$$

$$l = 1m$$

$$I = 6*10^6\text{mm}^4$$

How do you obtain the natural frequencies and the mode shapes?

P.10.2 Formulate the E.V.P. defining the free vibration of the beam shown in Fig. P.10.2. You may accept that,

$$[k]^e = \frac{EI}{l^3}\begin{bmatrix} 12 & 6l & -12 & 6l \\ & 4l^2 & -6l & 2l^2 \\ & & 12 & 6l \\ \text{Sym.} & & & 4l \end{bmatrix}$$

$$[m]^e = \frac{\overline{m}l}{420}\begin{bmatrix} 156 & 22l & 54 & 13l \\ & 4l^2 & 13l & -3l^2 \\ & & 156 & -22l \\ \text{Sym.} & & & 4l^2 \end{bmatrix}$$

EI: constant, $\overline{m}$: per unit length, $M = 2\overline{m}l$

In the right-hand end of the beam is fixed, calculate the natural frequencies and mode shape.

$$k_1 = \frac{EI}{l^3},\ k_2 = 3\frac{EI}{l^3}$$

$$l = 0.2m,\ \overline{m} = 20\text{kg}/m$$

$$I = 4*10^{-6}\text{m}^4, E = 200\ \text{Gpa}$$

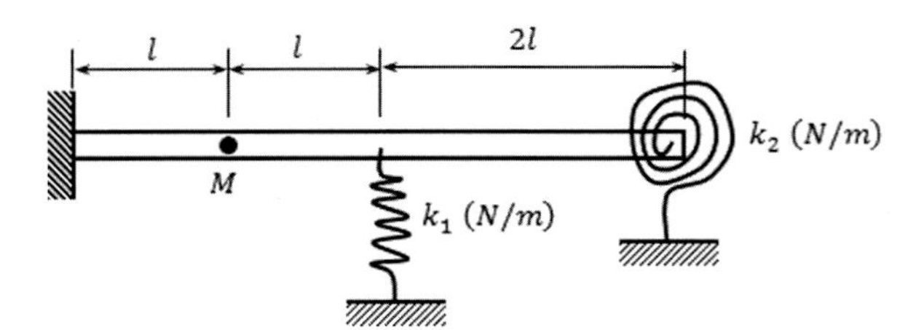

FIGURE P.10.2

P.10.3 Formulate the matrix Eigenvalue problem for the system of Fig. P.10.3. Use the three beam elements, as shown.

$$[k]^e = \frac{EI}{l^3}\begin{bmatrix} 12 & 6l & -12 & 6l \\ & 4l^2 & -6l & 2l^2 \\ & & 12 & 6l \\ \text{Sym.} & & & 4l^2 \end{bmatrix}$$

$$[m]^e = \frac{\overline{m}l}{420}\begin{bmatrix} 156 & 22l & 54 & 13l \\ & 4l^2 & 13l & -3l^2 \\ & & 156 & -22l \\ \text{Sym.} & & & 4l^2 \end{bmatrix}$$

$$k_1 = \frac{EI}{l^3}, k_2 = \frac{2EI}{l^3}$$

$$\overline{m} = 20\text{kg}/m$$

$$I = 4{*}10^{-6}\text{m}^4, E = 200\text{GPa}$$

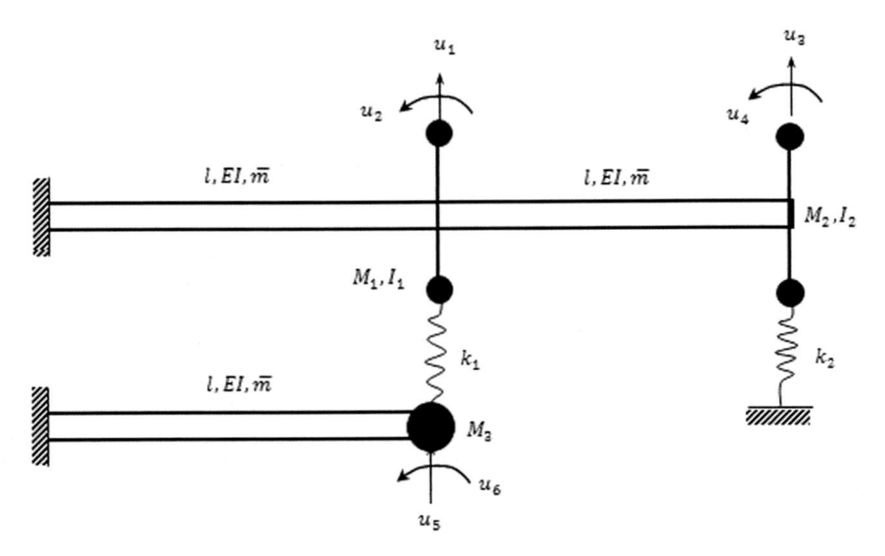

FIGURE P.10.3

P.10.4 Formulate the E.V.P. defining the free vibration of the beam shown in Fig. P.10.4. You may accept that,

$$[k]^e = \frac{EI}{l^3}\begin{bmatrix} 12 & 6l & -12 & 6l \\ & 4l^2 & -6l & 2l^2 \\ & & 12 & 6l \\ \text{Sym.} & & & 4l \end{bmatrix}$$

$$[m]^e = \frac{\overline{m}l}{420}\begin{bmatrix} 156 & 22l & 54 & 13l \\ & 4l^2 & 13l & -3l^2 \\ & & 156 & -22l \\ \text{Sym.} & & & 4l^2 \end{bmatrix}$$

$\overline{m}$: per unit length

$$M = 2\overline{m}l$$

In the right-hand end of the beam is fixed, calculate the natural frequencies and mode shape.

$$k_1 = \frac{2EI}{l^3}, k_2 = 4\frac{EI}{l^3}$$

$$l = 0.15m, \overline{m} = 10\text{kg}/m$$

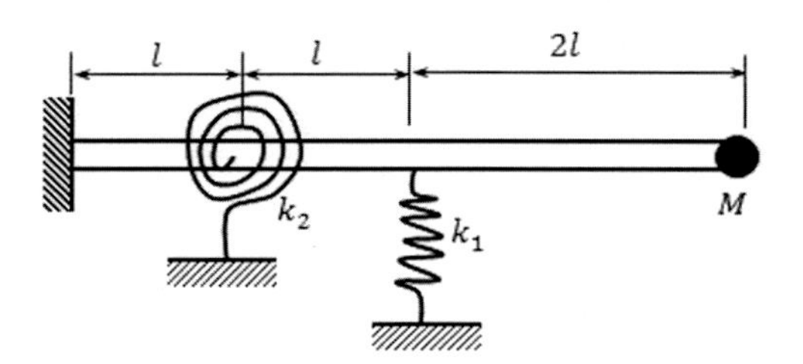

FIGURE P.10.4

P.10.5 For the analysis of in-plane vibrations of a heavy plate by the finite element technique, an element displacement model may be assumed to have formed as,

$$\{u(x,y)\} = [N(x,y)]\{u\}_0^e$$

Where, $\{u\}_0^e$ is a vector of element nodal displacement and $[N(x,y)]$ is chosen to be linear show, without developing the algebraic detail, that the element stiffness and consistent mass matrices are of order 6×6 and that a constant strain triangle element is appropriate.

Using Hamilton's principle,

$$\delta\int_{t_1}^{t_2}(T - V)dt = 0$$

Deduce, again without all the algebraic detail, that the in-plane equations of motion for a discretized two-dimensional continuum take the form,

$$[M]\{\ddot{q}\} + [K]\{q\} = \{Q\}$$

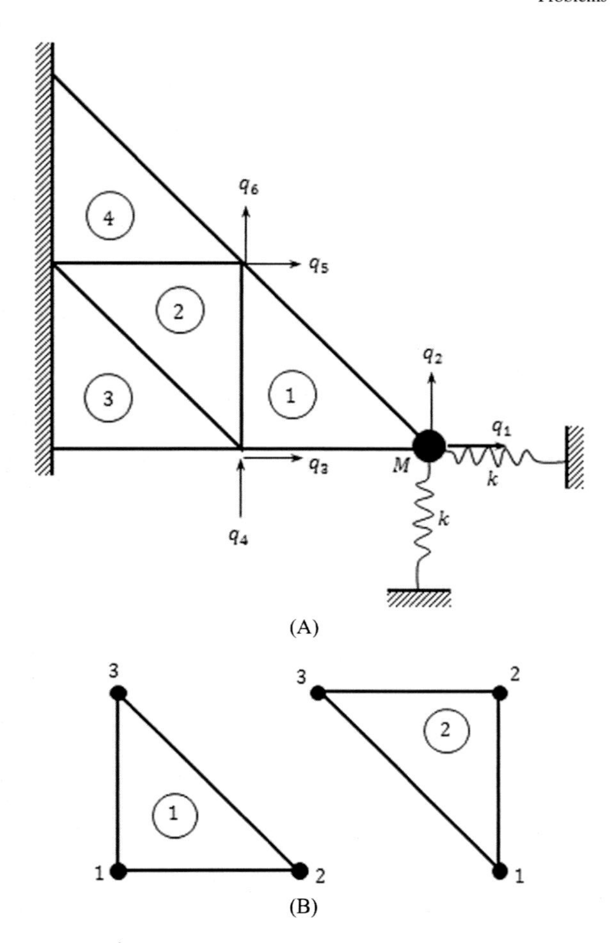

FIGURE P.10.5 (A, B) Typical element.

For some preliminary design calculations, a heavy support arrangement for an optical system has been modeled as shown in Fig. P.10.5.

By inspection, or otherwise, generate the system mass and stiffness matrices.

For, t is the thickness of the plate, and ρ is the density of the material.

Use the element stiffness matrices,

$$[k]^1 = [k]^2 = 0.55\ Et \begin{bmatrix} 1.35 & 0.65 & -1.00 & 0.35 & -0.35 & -0.30 \\ & 1.35 & -0.30 & -0.35 & -0.35 & -1.00 \\ & & 1.00 & 0 & 0 & 0.30 \\ & & & 0.35 & 0.35 & 0 \\ & \text{Sym.} & & & 0.35 & 0 \\ & & & & & 1.00 \end{bmatrix}$$

And accept that the consistent mass matrices are,

$$[m]^1 = [m]^2 = \frac{\rho t}{3}\begin{bmatrix} 0.50 & 0 & 0.25 & 0 & 0.25 & 0 \\ & 0.50 & 0 & 0.25 & 0 & 0.25 \\ & & 0.50 & 0 & 0.25 & 0 \\ & & & 0.50 & 0 & 0.25 \\ & \text{Sym.} & & & 0.50 & 0 \\ & & & & & 0.50 \end{bmatrix}$$

P.10.6 The natural frequencies of a cantilever beam's transverse vibration are increased if a distributed axial load is applied. (Turbine blades are similarly affected). Proceed as follows to estimate the effect using,

$$w(x) = \sum_{i=1}^{n} q_1.N_i(x)$$

In Fig. P.10.6, the centrifugal load dF_c has potential energy,

$$d\Omega = \left(\overline{m}(x)\omega^2 x dx\right)\frac{1}{2}\int_0^x \left(\frac{\partial w}{\partial \xi}\right)^2 d\xi$$

So,

$$\Omega = \frac{\omega^2}{2}\int_0^l \left(\overline{m}(x)x\left(\frac{\partial w}{\partial \xi}\right)^2 d\xi\right)dx$$

Use Hamilton's principle to generate the eigenvalue problem corresponding to this system. Express it in terms suitable for handling in a CAD package.

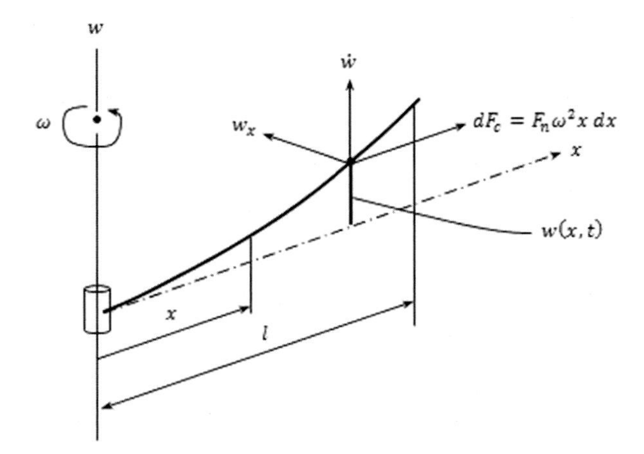

FIGURE P.10.6

Confirm that, compared to the "static" situation, each stiffness coefficient is augmented to give,

$$k_{ij} = \int_0^l EI(x)N''_i(x)N''_j(x)dx + \omega^2 \int_0^l \left(\overline{m}(x)x \int_0^x N'_i(\xi)N'_j(\xi)d\xi \right) dx$$

And each mass coefficient,

$$m_{ij} = \int_0^l \overline{m}N_i(x)N_j(x)dx + \omega^2 \int_0^l \overline{m}x^2 dx$$

P.10.7 The equation of motion for a uniform circular plate is

$$\nabla^4 w + \frac{\rho}{D}\frac{\partial^2 w}{\partial t^2} = 0$$

Where

$$\nabla^4 = \nabla^2\nabla^2 = \left(\frac{\partial^2}{\partial r^2} + \frac{1}{r}\frac{\partial}{\partial r} + \frac{1}{r^2}\frac{\partial^2}{\partial \theta^2}\right)\left(\frac{\partial^2}{\partial r^2} + \frac{1}{r}\frac{\partial}{\partial r} + \frac{1}{r^2}\frac{\partial^2}{\partial \theta^2}\right)$$

w = deflexion of plate, ρ = mass/unit surface of the plate, D = plate flexural rigidity, r, θ = polar coordinates

By arranging a harmonic solution in the time domain show that for axisymmetric modes of vibration the equation of motion reduces to

$$\frac{d^4\varnothing}{dr^4} + \frac{2}{r}\frac{d^3\varnothing}{dr^3} - \frac{1}{r^2}\frac{d^2\varnothing}{dr^2} + \frac{1}{r^3}\frac{d\varnothing}{dr} = \frac{\rho\omega^2}{D}\varnothing$$

The circular plate of Fig. P.10.7 is of six units outside radius and two units inside radius. It is clamped both along its outside edge and to a central boss along its inside edge. Obtain an approximation to the fundamental frequency by finite element method using the element stiffness and mass matrices. Divide the plate radially into four elements as shown in the diagram and show that the resulting equation can be expressed in the symmetrical form

$$\begin{bmatrix} 4\cdot 13 & -2\cdot 86 & 0.80 \\ -2.86 & 4.90 & -3.64 \\ 0.80 & -3.64 & 7.28 \end{bmatrix} \begin{bmatrix} \varnothing_1 \\ \varnothing_2 \\ \varnothing_3 \end{bmatrix} = \frac{\omega^2\rho}{D} \begin{bmatrix} 0.6 & 0 & 0 \\ 0 & 0.8 & 0 \\ 0 & 0 & 1 \end{bmatrix} \begin{bmatrix} \varnothing_1 \\ \varnothing_2 \\ \varnothing_3 \end{bmatrix}$$

Accepting that

$$\begin{bmatrix} 4\cdot 13 & -2\cdot 86 & 0.80 \\ -2.86 & 4.90 & -3.64 \\ 0.80 & -3.64 & 7.28 \end{bmatrix}^{-1} = \begin{bmatrix} .48 & .38 & .14 \\ .38 & .63 & .27 \\ .14 & .27 & .26 \end{bmatrix}$$

Determine the fundamental frequency by matrix iteration using the vector $\begin{bmatrix} 1.5 \\ 2 \\ 1 \end{bmatrix}$ as a first trial,

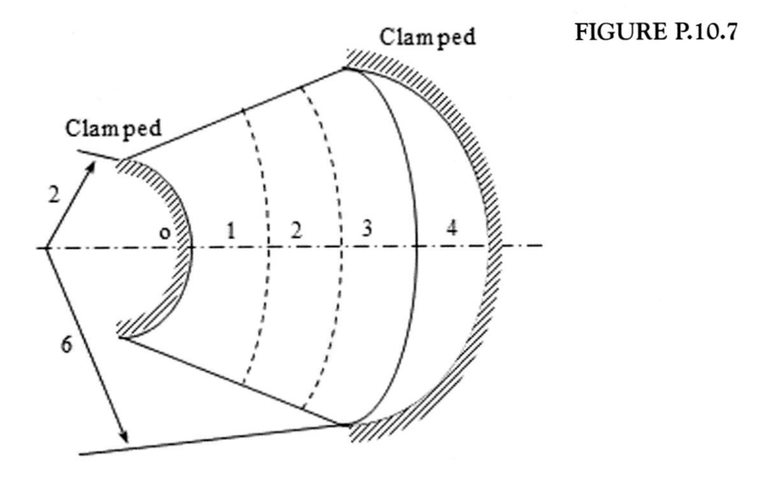

FIGURE P.10.7

P.10.8 For the cantilever beam shown in Fig. P.10.8A is subjected to a fore in direction of u_1 as shown in Fig. P.10.8B, determine the response $u_1(t), u_2(t)$ and the corresponding dynamic stresses. You may approximate the beam as one element and use the $[k]^e$and$[m]^e$

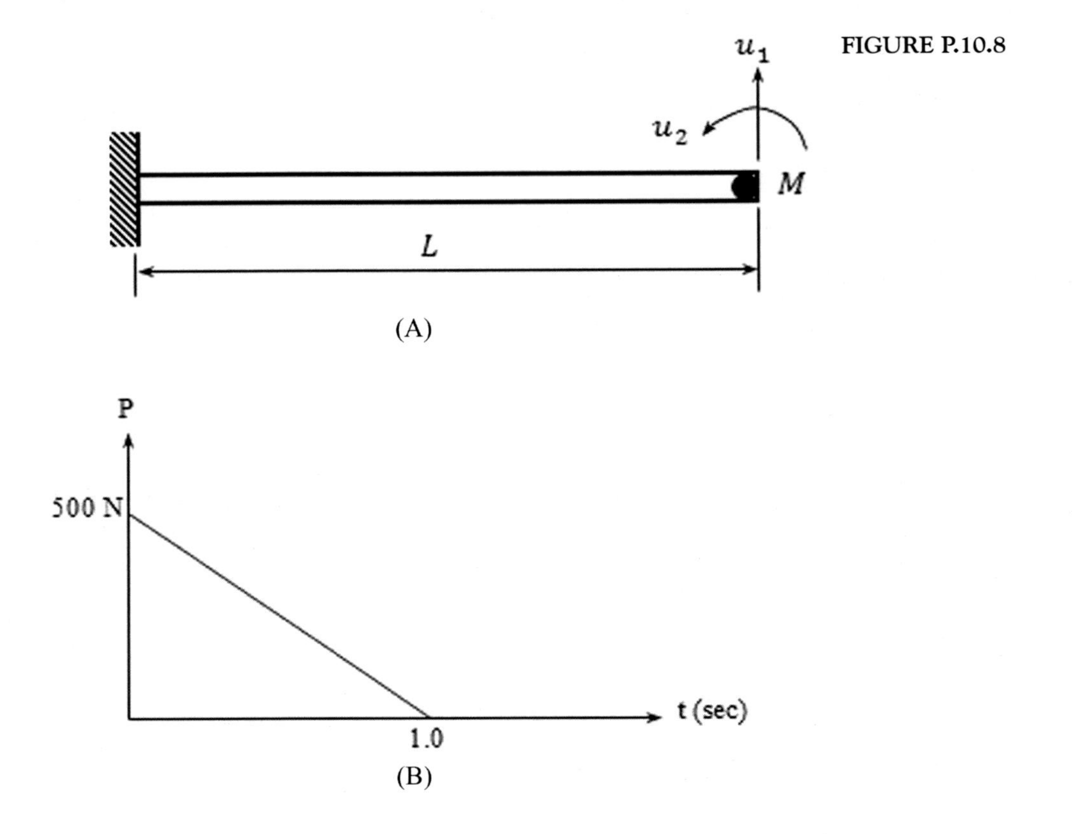

FIGURE P.10.8

for the beam element (problem P.10.1). use $E = 80\text{GPa}, \nu = 0.28, \rho = 2800\ \text{kg/m}^3, L = 0.5m$, sectional area $A = 300\text{mm}^2$ Moreover, the concentrated mass $M = 50\text{kg}$.

P.10.9 Repeat the problem P.10.8 if the loading condition shown in Fig. P.10.9 is applied in the direction of u_1.

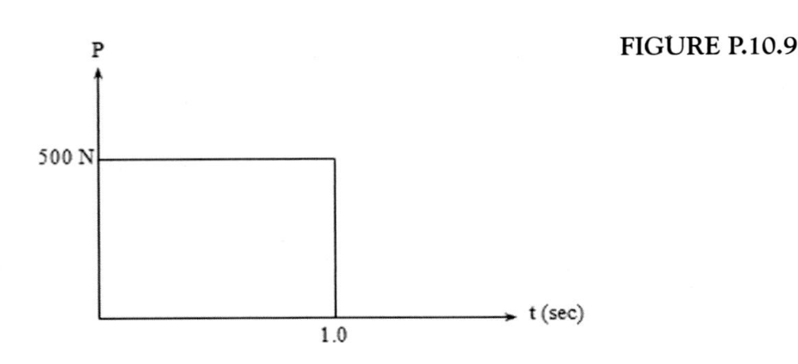

FIGURE P.10.9

P.10.10 Solve the problem P.10.8 using lumped mass matrix instead of using the consistent mass matrix.

P.10.11 The uniformly variable beam section shown in Fig. P.10.11, use two finite elements model and finds the natural frequencies and mode shapes. Use the consistent mass matrix formulation. $L = 1m, A_0 = 200\text{mm}^2, E = 200\text{GPa}, \nu = 0.3, \rho = 7800\ \text{kg}/m^3, M = 50\text{kg}$.

Also, determine the mass matrix orthogonal mode shape.

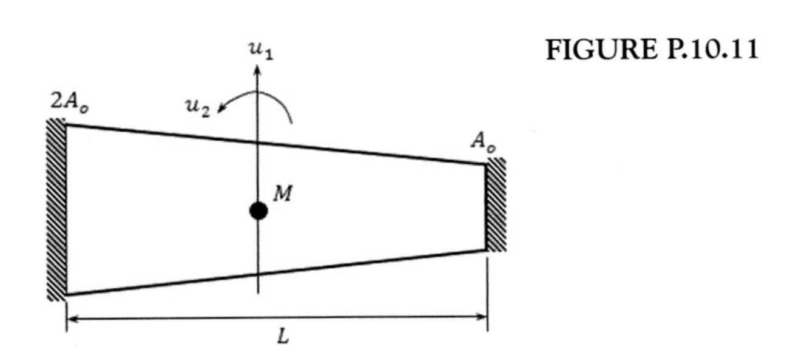

FIGURE P.10.11

P.10.12 A rigid mass M and the moment of inertia about its centroid is I_0 is attached to a cantilever beam, as shown in Fig. P.10.12. The kinetic energy of this mass is given by,

$T = \frac{1}{2}Mv^2 + \frac{1}{2}I_0\omega^2$

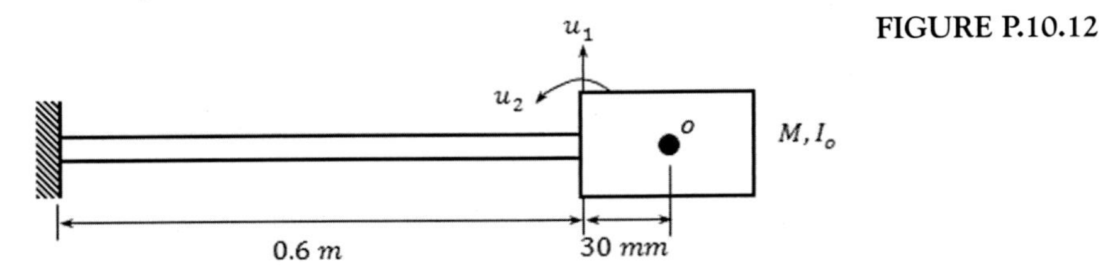

FIGURE P.10.12

Write down the contribution of M and I_0 of the mass matrix at node 1, and then the relation of v and w in terms of $\dot{u}_1$ and $\dot{u}_2$.

If $M = 50\text{kg}, I_0 = 3 \times 10^6 \text{mm}^4, L = 0.6m, E = 200\text{GPa}, \nu = 0.3, \rho = 7800\text{kg}/m^3$, determine the natural frequencies and mode shapes. Use the consistent mass matrix for the beam element mass matrix $[m]^e$ with the corresponding stiffness matrix.

P.10.13 For the beam shown in Fig. P.10.13, solve the resulted E.V.P. for the natural frequencies and the corresponding mode shapes, if $L = 250\text{mm}, \rho = 7800\ \text{kg}/m^3, E = 200\text{GPa}, \nu = 0.3,$ and $k = \frac{EI}{L^3} N/m$. The contribution of the spring stiffness k is added to node one's stiffness by writing the strain energy of the spring.

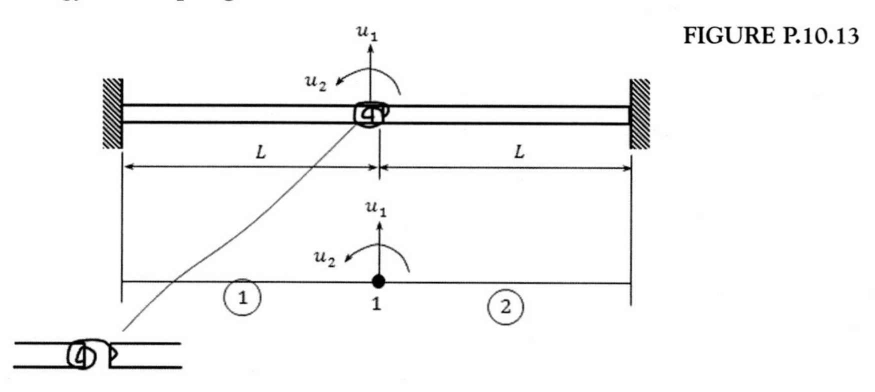

FIGURE P.10.13

P.10.14 For the plane stress element shown in Fig. P.10.14, if $E = 80\text{GPa}, \nu = 0.28,$ and $\rho = 2700\text{kg}/m^3$. The requirements are,

1. Formulate the E.V.P. using Eq. (10.23) for the element stiffness and Eq. (10.25) for the mass matrix.
2. If the nodes 1 and 3 are fixed, find the natural frequencies and mode shapes corresponding to the degrees of freedom u_3 and u_4.
3. Orthogonalize the mode shape, concerning the corresponding mass matrix.

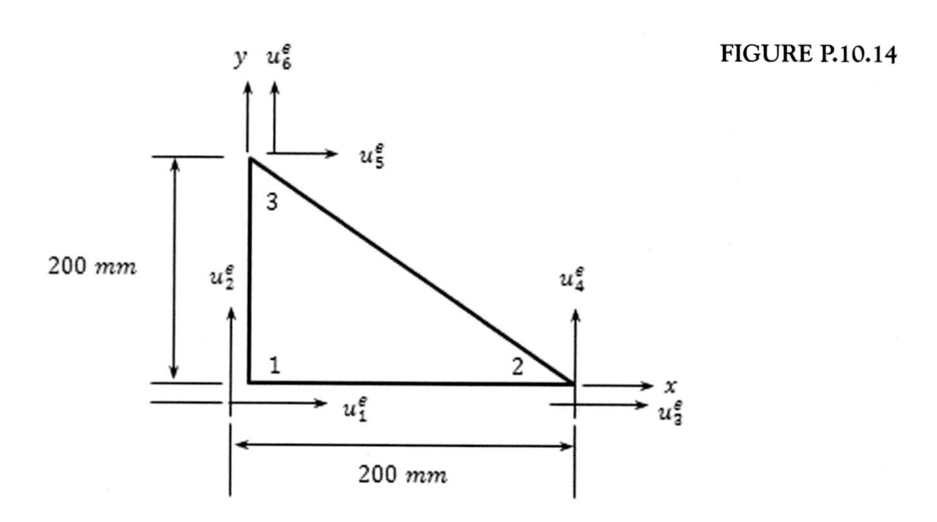

FIGURE P.10.14

P.10.15 Generate the characteristic matrix equation for the finite element discretization for the in-plane vibration of the case of Fig. P.10.15. Use the element stiffness and mass matrices obtained in Fig. P.10.14. Use the notation m'_{12}, k'_{42} etc. and assemble the system stiffness and mass matrices by inspection. Take local coordinates, as shown in elements 1 and 2.

The system shown in Fig. P.10.15A is analyzed by the finite element method using the coordinates shown on the figure.

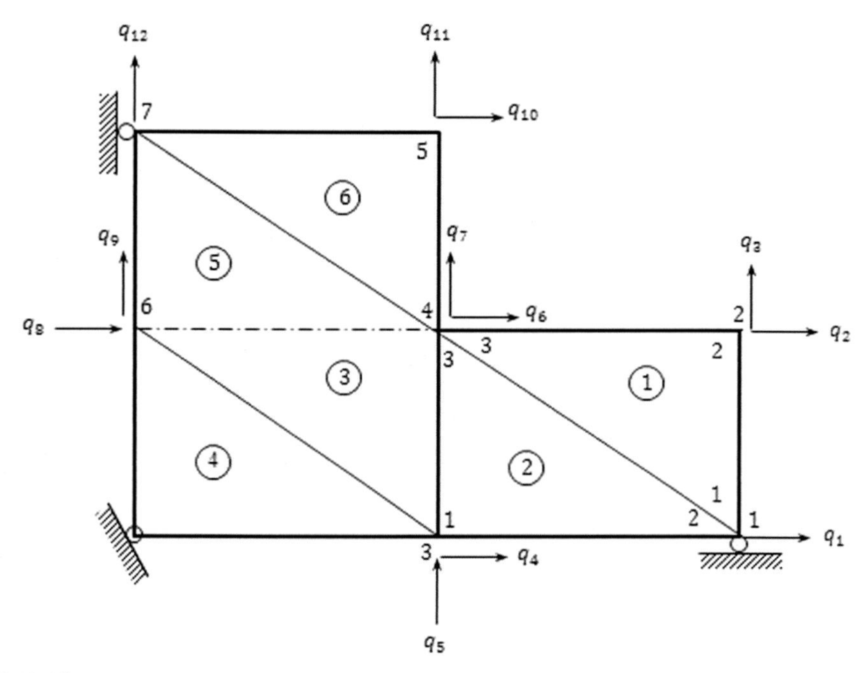

FIGURE P.10.15

The torsion tube PQP′ is rigidly fixed to all three beams and may be considered infinitely stiff in bending. Since only modes of vibration symmetric about o-o are required only five degrees of freedom are allowed.

Obtain the mass and stiffness matrices for the system if $GJ = EI/4$ and

$$I_pL = m/8.75.$$

You may accept that for the beam element shown in Fig. P.10.15A

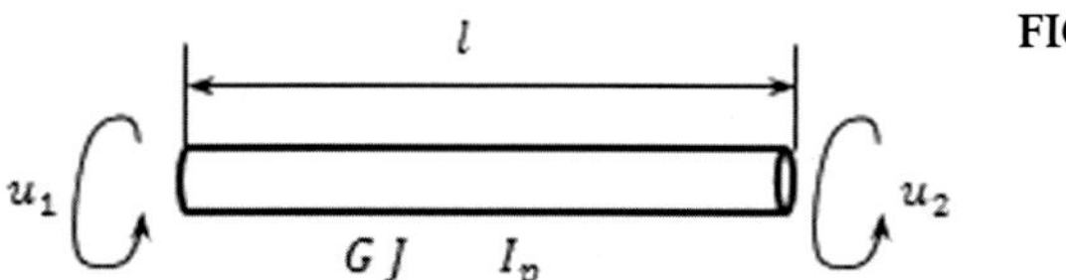

FIGURE P.10.15A

$$[m] = \frac{m}{420}\begin{bmatrix} 156l & 22l^2 & 54l & -13l^2 \\ 22l^2 & 4l^3 & 13l^2 & -3l^3 \\ 54l & 13l^2 & 156l & -22l^2 \\ -13l^2 & -3l^3 & -22l^2 & 4l^3 \end{bmatrix} \quad \text{(P15.1)}$$

$$[m] = EI\begin{bmatrix} 12/l^3 & 6/l^2 & -12/l^3 & 6/l^2 \\ -12/l^3 & 4/l & -6/l^2 & 2/l \\ -12/l^3 & -6/l^2 & 12/l^3 & -6/l^2 \\ 6/l^2 & 2/l & -6/l^2 & 4/l \end{bmatrix} \quad \text{(P15.2)}$$

Obtain the mass and stiffness matrices for the system if $GJ = \frac{EI}{4}$ and $I_pL = \frac{m}{8.75}$.
For a bar in torsion, the strain energy stored is

$$V = \frac{1}{2}GJ\int_0^l \left(\frac{\partial \theta}{\partial x}\right)^2 dx \quad \text{(P15.3)}$$

Once to obtain a value for ω. Obtain another solution for ω using a mesh interval of $h = \frac{a}{2}$ and use Richardson's Extrapolation to obtain an improved value of ω from the expression

$$\omega_{imp}^2 = \alpha_1\omega_{h1}^2 + \alpha_2\omega_{h2}^2 \quad \text{(P15.4)}$$

And the Table P.10.15 of values for α_1 and α_2:

TABLE P.10.15

h_1/h_2	α_1	α_2
2/1	− 0.333	1.333
3/2	− 0.8	1.8
3/4	− 1.285	2.285

P.10.16 If in the element shown in Fig. P.10.16A and P.10.16B, the coordinates of nodes are as follows, Node 1: (280, 0, 0), node 2: (0, 80, 0), node 3: (0, 0, 160), node 4: (100, 80, 0) and, $E = 70\text{GPa}$, $\nu = 0.3$ and $\rho = 2800\text{kg}/m^3$.

1. Construct the element mass matrix $[m]^e$.
2. If the nodes 1, 2, and 3 are fixed due to fixing the face 1 2 3 and $[k]^e$ is given by,

$$[k]^e = \frac{E}{(1+\nu)(1-2\nu)}$$

$$\begin{bmatrix}
3.55 & 2.33 & 0.88 & -1.77 & 0.88 & 3.3 & 4.55 & -2.45 & 2.25 & 2.33 & 1.6 & 2.8 \\
2.33 & 2.68 & 2.33 & 8.9 & 10.25 & 4.56 & 1.14 & -0.88 & 6.22 & 8.48 & 2.24 & 8.34 \\
0.88 & 2.33 & 12.45 & 6.66 & 2.87 & -1.56 & 16.23 & 8.6 & -9.23 & 10.12 & 8.82 & 4.22 \\
-1.77 & 8.9 & 6.66 & 3.88 & 10.23 & 16.23 & 6.43 & -0.22 & 4.89 & 18.21 & 2.66 & -0.08 \\
0.88 & 10.25 & 2.87 & 10.23 & 17.78 & 2.67 & 5.33 & -17.7 & 0.006 & 1.067 & 1.77 & 10.51 \\
3.3 & 4.56 & -1.56 & 16.23 & 2.67 & 18.54 & 2.77 & 0.44 & -9.81 & 0.086 & 2.34 & 6.32 \\
4.55 & 1.14 & 16.23 & 6.43 & 5.33 & 2.77 & 5.82 & 10.32 & 5.84 & -10.8 & 8.88 & 0.004 \\
-2.45 & -0.88 & 8.6 & -0.22 & -17.7 & 0.44 & 10.32 & 16.43 & 3.05 & 2.54 & 0.03 & 14.32 \\
2.25 & 6.22 & -9.23 & 4.89 & 0.006 & -9.81 & 5.84 & 3.05 & 8.84 & 4.34 & -4.88 & 1.25 \\
2.33 & 8.48 & 10.12 & 18.21 & 1.067 & 0.086 & -10.8 & 2.54 & 4.34 & 6.84 & 2.44 & -2.44 \\
1.6 & 2.24 & 8.82 & 2.66 & 1.77 & 2.34 & 8.88 & 0.03 & -4.88 & 2.44 & 10.08 & 15 \\
2.8 & 8.34 & 4.22 & -.08 & 10.51 & 6.32 & 0.004 & 14.32 & 1.25 & -2.44 & 15 & 11.22
\end{bmatrix}$$

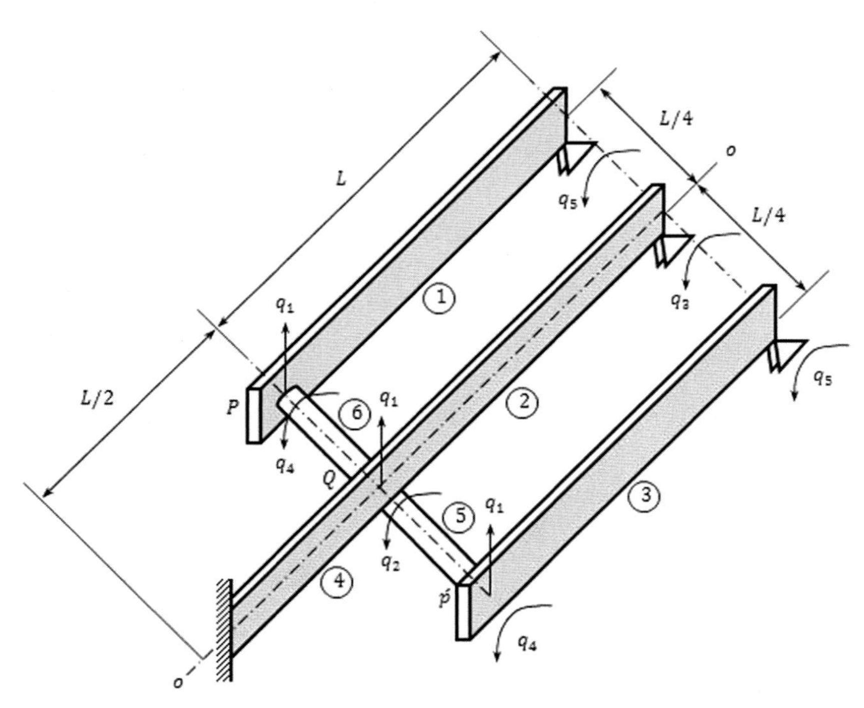

FIGURE P.10.16A

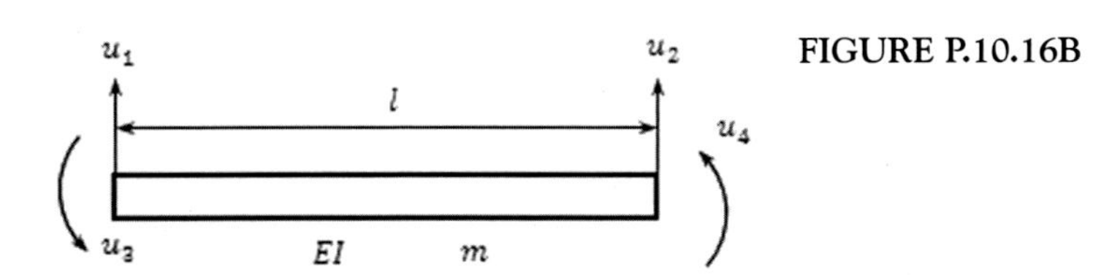

FIGURE P.10.16B

Construct the E.V.P. Explain how you achieve the E.V.P.'s solution to determine the natural frequencies and mode shapes.

P.10.17 Suppose that the axial bar shown in Example 10.6 is subjected to an axial load $F(t)$ as shown in Fig. P.10.17. Determine the

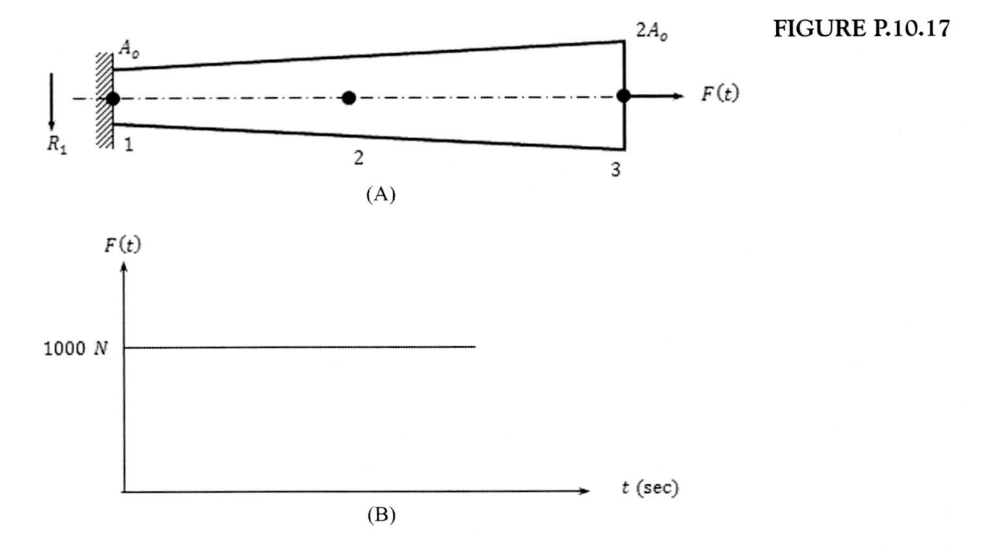

FIGURE P.10.17

P.10.18 If in the element shown in Fig. P.10.18, the coordinates of nodes are as follows, Node 1: $(280, 0, 0)$, node 2: $(0, 80, 0)$, node 3: $(0, 0, 160)$, node 4: $(100, 80, 0)$ and, $E = 70\text{GPa}$, $\nu = 0.3$ and $\rho = 2800\text{kg}/m^3$.

(c) Construct the element mass matrix $[m]^e$.

(d) If the nodes 1, 2, and 3 are fixed due to fixing the face 1 2 3 and $[k]^e$ is given by,

$$[k]^e = \frac{E}{(1+\nu)(1-2\nu)}$$

$$\begin{bmatrix} 3.55 & 2.33 & 0.88 & -1.77 & 0.88 & 3.3 & 4.55 & -2.45 & 2.25 & 2.33 & 1.6 & 2.8 \\ 2.33 & 2.68 & 2.33 & 8.9 & 10.25 & 4.56 & 1.14 & -0.88 & 6.22 & 8.48 & 2.24 & 8.34 \\ 0.88 & 2.33 & 12.45 & 6.66 & 2.87 & -1.56 & 16.23 & 8.6 & -9.23 & 10.12 & 8.82 & 4.22 \\ -1.77 & 8.9 & 6.66 & 3.88 & 10.23 & 16.23 & 6.43 & -0.22 & 4.89 & 18.21 & 2.66 & -0.08 \\ 0.88 & 10.25 & 2.87 & 10.23 & 17.78 & 2.67 & 5.33 & -17.7 & 0.006 & 1.067 & 1.77 & 10.51 \\ 3.3 & 4.56 & -1.56 & 16.23 & 2.67 & 18.54 & 2.77 & 0.44 & -9.81 & 0.086 & 2.34 & 6.32 \\ 4.55 & 1.14 & 16.23 & 6.43 & 5.33 & 2.77 & 5.82 & 10.32 & 5.84 & -10.8 & 8.88 & 0.004 \\ -2.45 & -0.88 & 8.6 & -0.22 & -17.7 & 0.44 & 10.32 & 16.43 & 3.05 & 2.54 & 0.03 & 14.32 \\ 2.25 & 6.22 & -9.23 & 4.89 & 0.006 & -9.81 & 5.84 & 3.05 & 8.84 & 4.34 & -4.88 & 1.25 \\ 2.33 & 8.48 & 10.12 & 18.21 & 1.067 & 0.086 & -10.8 & 2.54 & 4.34 & 6.84 & 2.44 & -2.44 \\ 1.6 & 2.24 & 8.82 & 2.66 & 1.77 & 2.34 & 8.88 & 0.03 & -4.88 & 2.44 & 10.08 & 15 \\ 2.8 & 8.34 & 4.22 & -.08 & 10.51 & 6.32 & 0.004 & 14.32 & 1.25 & -2.44 & 15 & 11.22 \end{bmatrix}$$

Construct the E.V.P. Explain how you achieve the E.V.P.'s solution to determine the natural frequencies and mode shapes.

(e) Use the approximation of lumped mass modeling of the element mass matrix. Compare between the modeling in (b) and (c).

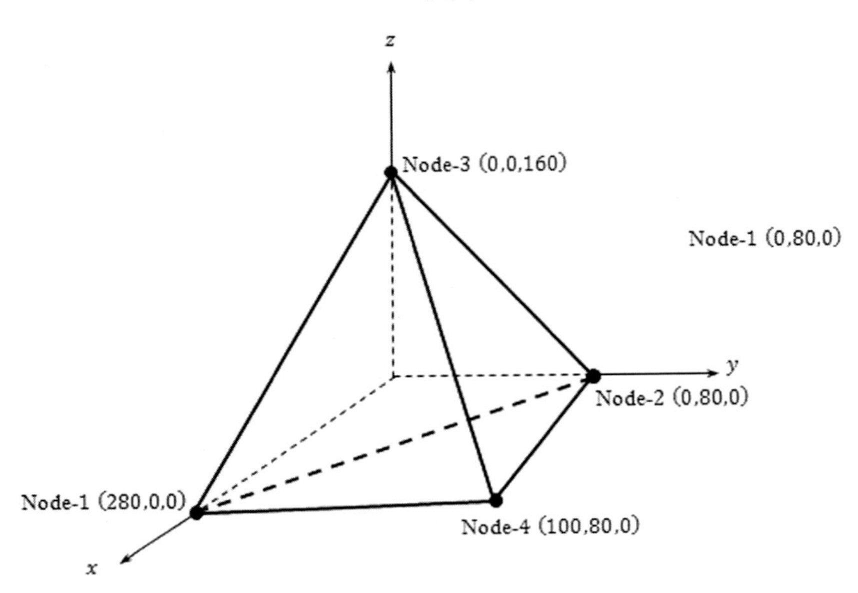

FIGURE P.10.18

Bibliography

[1] B. Irons, S. Ahmad, Techniques of Finite Elements, Wiley, New York, 1980.

[2] I. Ergatoudis, B.M. Irons, O.C. Zienkiewicz, Curved isoparametric quadrilateral elements for finite element analysis, International Journal of Solids and Structures 4 (1968) 31–42.

[3] Y. Yamada, Dynamic Analysis of Civil Engineering Structures, Recent Advances in Matrix Methods of Structural Analysis and Design, University of Alabama Press, Alabama, 1970, pp. 487–512.

[4] R.W. Clough, J. Penzien, Dynamics of Structures, McGraw-Hill, New York, 1975.

[5] M. Witts, K. Sobczyk, Dynamics response of laminated plates to random loading, Journal of Solids and Structures 16 (3) (1980).

[6] J.N. Reddy, C.F. Liu, A higher–order shear deformation theory of laminated elastic shells, International Journal of Engineering Science 23 (3) (1985) 319–330.

[7] T.J.R. Hughes, The Finite Element Method, Linear Static and Dynamic Finite Element Analysis, Prentice-Hall, Englewood Cliffs, 1987.

[8] W. Weaver, S.P. Timoshenko, D.H. Young, Vibration Problems in Engineering, fifth ed., Wiley, New York, 1990.

[9] K.J. Bathe, Finite Element Procedures, Prentice-Hill, Englewood Cliffs, 1996.

[10] M.J. Jweeg, S.K. Radi, Analysis of composite plates subjected to impact loading, in: The National Engineering Conference, Kufa, Iraq, 2000.

[11] M.J. Jweeg, M. Al-Waily, A. Abdulzahra Deli, Theoretical and numerical investigation of buckling of orthotropic hyper composite plates, International Journal of Mechanical & Mechatronics Engineering 15 (4) (2015).

[12] M.J. Jweeg, A suggested analytical solution for vibration of honeycombs sandwich combined plate structure, International Journal of Mechanical & Mechatronics Engineering 16 (2) (2016).

[13] E.N. Abbas, M.J. Jweeg, M. Al-Waily, Analytical and numerical investigations for dynamic response of composite plates under various dynamic loading with the influence of carbon multi-wall tube nano materials, International Journal of Mechanical & Mechatronics Engineering 18 (6) (2018) 1–10.

[14] M.J. Jweeg, A.A. Ahumdany, A.F.M. Jawad, Dynamic stresses and deformations investigation of the below knee prosthesis using CT-scan modeling, International Journal of Mechanical & Mechatronics Engineering 19 (1) (2019).

[15] H.I. Mansoor, M. Al-shammari, A. Al-Hamood, Theoretical analysis of the vibrations in gas turbine rotor, in: 3rd International Conference on Engineering Sciences, IOP Conference Series: Materials Science and Engineering, 2020, p. 671.
[16] M. Al-Waily, M.A. Al-Shammari, M.J. Jweeg, An analytical investigation of thermal buckling behavior of composite plates reinforced by carbon nano particles, Engineering Journal 24 (3) (2020).
[17] R.G. Anderson, B.M. Irons, O.C. Zienkiewicz, Vibration and stability of plates using finite elements, International Journal Solids and Structure 4 (1968) 1031–1055.
[18] E. Hinton, The dynamic transient analysis of axisymmetric circular plates by the finite element method, Journal of Sound and Vibation 46 (1976) 465–472.
[19] A. El-Zafarany, R.A. Cookson, Derivation of Lagrangian and Hemitian shape function for quadrilateral elements, International Journal for Numerical Methods in Engineering 23 (1986) 1939–1958.
[20] M.J. Jweeg, S.Z. Said, Effect of rotational and geometric stiffness matrices on dynamic stresses and deformations of rotating blades, Journal of the Institution of Engineers, Mechanical Engineering Division 76 (1995) 29–38.
[21] A.W. Lessa, Vibration of Plates, NASA, Washington, DC, 1969.
[22] M.J. Jweeg, W.I. Al-Azzawy, E.A. Sadiq, Free vibration of composite laminated plate subjected to cryogenic environments, in: The 6th Engineering Conference, College of Engineering, Mechanical and Nuclear Engineering III, 2009, pp, 175–191.
[23] M.J. Jweeg, A.H. Al-Hilli, M.I. Salim, Effect of support conditions on vibration characteristics in a pipe conveying fluid, in: The 2nd Regional Conference for Engineering Sciences, Al- Nahrain University, 2010, pp. 338–354.
[24] M.J. Jweeg, T.J. Ntayeesh, Active vibration control of a cantilever pipe conveying fluid using smart mterial, Innovative Systems Design and Engineering 6 (12) (2015) 53–79.
[25] M.J. Jweeg, E.Q. Hussein, K.I. Mohammed, Effects of cracks on the frequency response of a simply supported pipe conveying fluid, International Journal of Mechanical & Mechatronics Engineering 17 (5) (2017).
[26] M. Al-Waily, K.K. Resan, A.H. Al-Wazir, Z.A.A. Abud Ali, Influences of glass and carbon powder reinforcement on the vibration response and characterization of an isotropic hyper composite materials plate structure, International Journal of Mechanical & Mechatronics Engineering 17 (6) (2017).
[27] M.A. Al-Shamari, Theoretical and Experimental Investigation of a Cone-Cylinder Shell Under Transient Loading (Ph.D thesis), University of Technology, Mechanical Engineering Department, 2010.
[28] S.G. Hussein, M.A. Al-Shammari, A.M. Takhakh, M. Al-Waily, Effect of heat treatment on mechanical and vibration properties for 6061 and 2024 aluminum alloys, Journal of Mechanical Engineering Research and Developments 43 (1) (2020) 48–66.

CHAPTER

11

Steady state heat conduction

In this chapter, the temperature distribution is to be determined. The presentation will be restricted to one and two-dimensional conduction and convection heat flow. The steady-state heat conduction equations were used in the formulation of the total potential energy. The finite element equilibrium equations were derived using the minimization principles of the total potential energy of the heat conduction and convection cases. The element heat conduction, convection coefficient matrices, and the element thermal load vector were derived for bar and two-dimensional problems. Different boundary conditions were used in the heat conducted regions. Different types of case studies were included as problems.

11.1 Steady-state heat flows

Consider the bar shown in Fig. 11.1. The aim is to determine the temperature distribution for the steady-heat conduction state, where L is the bar element length, H is the heat flow entering, and $H + dH$ is the heat flow leaving in the x-direction.

Let Q is the internal heat generated per unit volume (W/m^3). usually, the heat generated by a wire carrying a current I is,

$$Q = \frac{I^2 R}{V} \tag{11.1}$$

Where R is the wire resistance, and V is the voltage.

For the element shown in Fig. 11.1, Heatflow$H$ = heatflux(q)*Area(A).

Where q is the heat lux in W/m^2

For equilibrium,

$$\text{Heatflowrateentering}(H) + \text{Heatgenerated}(Q) = \text{Heatflowleaving}(H + dH)$$

Or,

$$q.A + Q.A.dx = \left(q + \frac{\partial q}{\partial x} dx\right) A \tag{11.2}$$

Energy Methods and Finite Element Techniques
DOI: https://doi.org/10.1016/B978-0-323-88666-6.00011-3

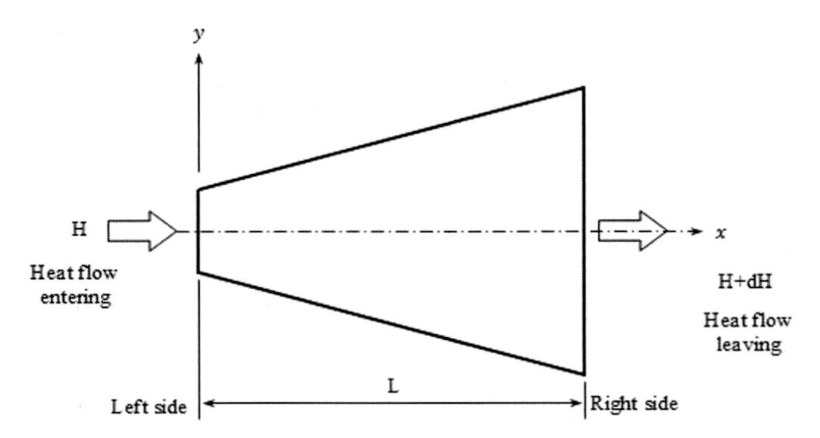

FIGURE 11.1 Steady-state heat conduction of the bar element.

Where,

$$H = q.A$$

And, $H + dH = \left(q + \frac{\partial q}{\partial x}dx\right)A$,
Substitution in Eq. (11.2) gives,

$$Q = \frac{\partial q}{\partial x} \tag{11.3a}$$

For the heat conduction in the x-direction,
$q = -k\frac{\partial T}{\partial x}$, where k is the thermal conductivity of the solid (W.m/s) and T is the temperature,
Therefore using Eq. (11.3a),

$$Q = \frac{\partial}{\partial x}\left(-k\frac{\partial T}{\partial x}\right)$$

Or,

$$Q + \frac{\partial}{\partial x}\left(k\frac{\partial T}{\partial x}\right) = 0 \tag{11.3}$$

Where Q is called source, + ve Q means heat is generated and −ve Q means heat is consumed.

11.2 Boundary conditions

The boundary conditions for the elements shown in Fig. 11.2A−C are as follows,
$T(x) = T_1$ at $x = x_1$

$$\left.\frac{dT}{dx}\right|_{x=x_2} = T'(L)\text{: Temperature gradient} \tag{11.4}$$

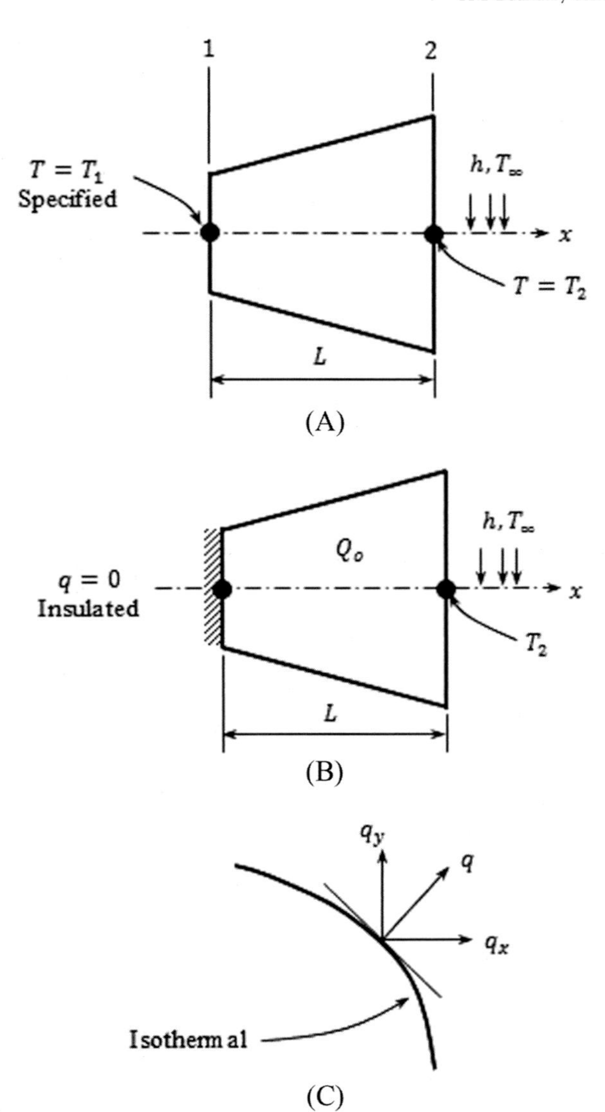

FIGURE 11.2 Boundary conditions of bar elements, (A) for the boundary conditions specified in Eq. (11.4), (B) for the boundary conditions specified in Eq. (11.5). (C) Isothermal surface.

And Eq. (11.3a),

$$\frac{d}{dx}\left(k\frac{dT}{dx}\right) + Q = 0$$

Where Q is the heat generated per unit volume.

For a perfectly insulated wall (Fig. 11.2B),

$$\left.\begin{aligned} q_{x=0} &= 0 \\ q_{x=L} &= h(T_{x=L} - T_{\infty}) \end{aligned}\right\} \tag{11.5}$$

Where h is the convection heat coefficient (W/m^2.°C).

Where

h: Convection heat flux watt coefficient (W)/m^2.°C

T_s: Surface, T_∞: fluid temperature (surroundings)

The components of heat flow (Fig. 11.2C) according to Fourier law of heat flow are as follows,

$$q_x = -k\frac{\partial T}{\partial x}, q_y = -k\frac{\partial T}{\partial y}$$

$T(x, y)$, k = Thermal conductivity coefficient = W/m.°C

$$W = 1\,\mathrm{J/s} = 1\,\mathrm{N.m/s}$$

The($-$) Sign indicates the heat is transferred in the direction of decreasing temperature.

11.3 Derivation of equilibrium equations

The heat conduction Eq. (11.3) is,

$$\frac{d}{dx}\left(k\frac{dT}{dx}\right) + Q = 0$$

Using Galerkin's method for Eq. (11.3) as an approximate solution to $T(x)$ is given by,

$$\int_0^L \left[\frac{d}{dx}\left(k\frac{dT}{dx}\right) + Q\right]\varnothing dx = \text{Residual} \tag{11.6}$$

Where Residual ≈ 0 and $\varnothing$ is a residual temperature variation, thus $\varnothing = 0$ on the boundary.

Integrating the first term by part, Eq. (11.6) becomes,

$$\varnothing.k.\frac{dT}{dx}\bigg|_0^L - \int_0^L k.\frac{d\varnothing}{dx}.\frac{dT}{dx}dx + \int_0^L \varnothing.Q.dx = 0 \tag{11.7}$$

And,

$$\varnothing.k.\frac{dT}{dx}\bigg|_0^L = \varnothing(L).k.\frac{dT}{dx}(L) - \varnothing(0).k.\frac{dT}{dx}(0) \tag{11.8a}$$

Using the boundary condition, $\varnothing(0) = 0$, and,

$$q = -k\frac{dT}{dx} = h(T_L - T_\infty)$$

Substitution the above integrations in Eq. (11.8a) into Eq. (11.7) gives,

$$-\varnothing(L).h(T_L - T_\infty) - \int_0^L k\frac{d\varnothing}{dx}\frac{dT}{dx}dx + \int_0^L \varnothing.Q.dx = 0 \tag{11.8}$$

Using the assumption,

$$T(x) = [N(x)]\{T\}^e \tag{11.9}$$

Assume that,

$$\varnothing(x) = [N(x)]\begin{Bmatrix} \varnothing_1 \\ \varnothing_2 \\ \vdots \\ \varnothing_L \end{Bmatrix} \tag{11.10}$$

Where, $\varnothing$ is the test function within the element, and,

$\{\varnothing_1 \quad \varnothing_2 \quad \cdots \quad \varnothing_L\}^t$ is the virtual displacement vector.

The shape function $[N]$ may be assumed in terms of x-coordinates system (Fig. 11.3A) or terms of non-dimensions coordinates (Fig. 11.3B), same shape functions derived previously for the bar under axial loading presented in Chapter 1, Fundamentals of Energy Methods, are as follows,

The x-coordinate system,

$$[N(x)] = \begin{bmatrix} -\frac{1}{L}(x - x_2) & \frac{1}{L}(x - x_1) \end{bmatrix}$$

Or, the intrinsic coordinate system,

$$[N(\xi)] = \begin{bmatrix} -\frac{1}{L}(1 - \xi) & \frac{1}{L}(1 + \xi) \end{bmatrix} \tag{11.11a}$$

From Eq. (11.9), we have,

$$\frac{dT}{dx} = [B]^t\{T\}^e \tag{11.11b}$$

Where $[B]$ is the temperature- temperature gradient relationship and $\{T\}^e$ is a vector that contains the nodal temperatures.

And, from Eq. (11.10), we can write,

$$\frac{d\phi}{dx} = [B]^t\{\phi\}^e$$

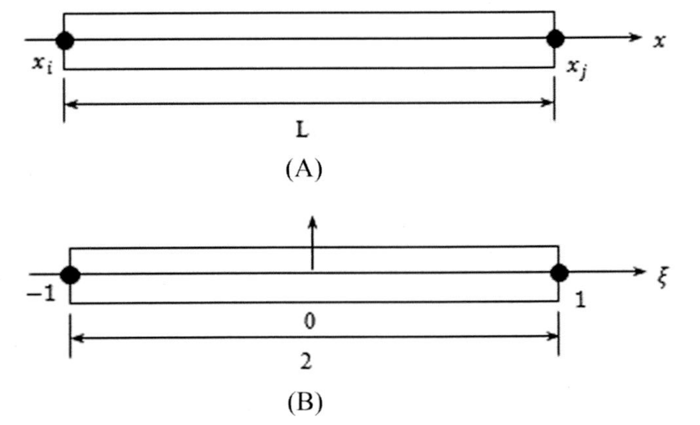

FIGURE 11.3 1D Bar Finite Element, (A) Bar element in terms of x-coordinates, (B) Bar element in terms of ξ coordinates.

Put into Eq. (11.8) gives,

$$\phi_L h(T_L - T_\infty) - \sum_e \{\varnothing\}^t \left(\frac{k_e l_e}{2} \int_{-1}^{1} [B]^t [B] d\xi\right) \{T\}^e \sum_e \{\varnothing\}^t \frac{Q_e l_e}{2} \int_{-1}^{1} [N]^t d\xi \tag{11.11c}$$

Where the subscript e stands for the element number.

Simplification of Eq. (11.11c) gives,

$$\phi_L h T_L - \phi_L h T_\infty - \{\varnothing\}^t [k]^e \{T\}^e + \{\varnothing\}^t \{F\}^e = 0$$

Where,

$$[k]^e = \left(\frac{k_e l_e}{2} \int_{-1}^{1} [B]^t [B] d\xi\right)$$

Or,

$$[k]^e = \frac{k_e}{l_e} \begin{bmatrix} 1 & -1 \\ -1 & 1 \end{bmatrix} \tag{11.11d}$$

And,

$$\{F\}^e = \frac{Q_e l_e}{2} \begin{Bmatrix} 1 \\ 1 \end{Bmatrix} \tag{11.12}$$

Where $[k]^e$ is the element heat conduction coefficient matrix, $\{F\}^e$ is the element thermal load vector, Q_e is the heat input at the current element, k_e is the thermal conductivity of the element material, and l_e is the element length.

The assembly of system heat conduction equations for $[K]$ and $\{F\}$ are obtained in a similar procedure used for the system stiffness, load vectors, and the solution of the equilibrium equations achieved using the penalty function which will be described in Appendix B.

Example 11.1: A composite wall consists of three materials, as shown in Fig. 11.4. The outer temperature is 30°C. Convection heat transfer takes place on the inner surface of the wall with $T_\infty = 600$°C and $h = 10$ W/m^2.°C, determine the temperature in the wall.

Solution,

The finite element modeling is shown in Fig. 11.5 1, 3 elements and four nodes.

The conductive element matrix for the typical element is,

$$[k]^e = \frac{k_e}{l_e} \begin{bmatrix} 1 & -1 \\ -1 & 1 \end{bmatrix}$$

Therefore,

$$[k]^1 = \frac{50}{0.2} \begin{bmatrix} 1 & -1 \\ -1 & 1 \end{bmatrix} \tag{E1.1}$$

$$[k]^2 = \frac{20}{0.1} \begin{bmatrix} 1 & -1 \\ -1 & 1 \end{bmatrix} \tag{EE1.2}$$

$$[k]^3 = \frac{70}{0.2} \begin{bmatrix} 1 & -1 \\ -1 & 1 \end{bmatrix} \tag{EE1.3}$$

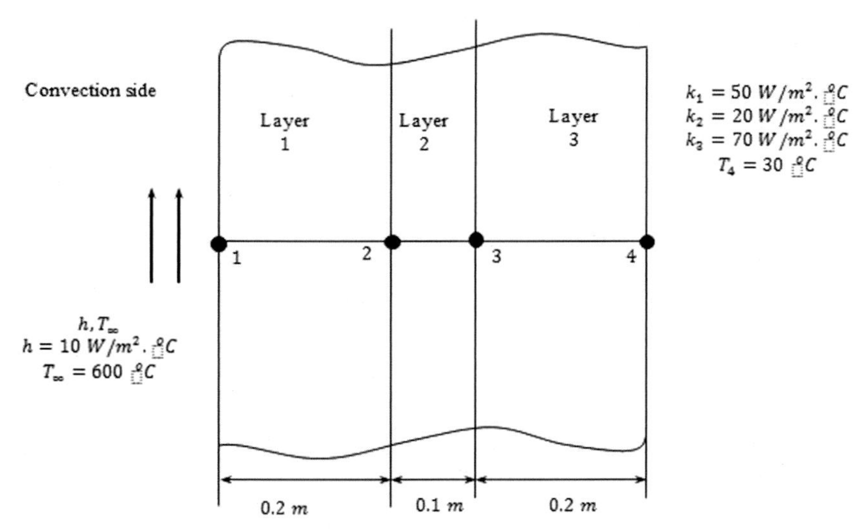

FIGURE 11.4 A composite wall.

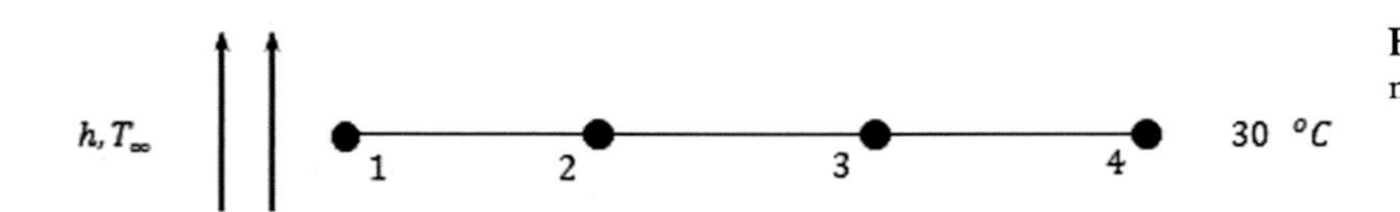

FIGURE 11.5 Finite element modeling.

The conductive system matrix can be obtained by using the standard procedure of the assembly- overlapping method- as follows,

$$[K] = \begin{bmatrix} \frac{50}{0.2} & -\frac{50}{0.2} & 0 & 0 \\ -\frac{50}{0.2} & \frac{50}{0.2} + \frac{20}{0.1} & -\frac{20}{0.1} & 0 \\ 0 & -\frac{20}{0.1} & \frac{20}{0.1} + \frac{70}{0.2} & -\frac{70}{0.2} \\ 0 & 0 & -\frac{70}{0.2} & \frac{70}{0.2} \end{bmatrix}$$

$$= \begin{bmatrix} 250 & -250 & 0 & 0 \\ -250 & 450 & -200 & 0 \\ 0 & -200 & 550 & -350 \\ 0 & 0 & -350 & 350 \end{bmatrix} \tag{EE1.4}$$

$h = 10\,\mathrm{W/m^2.°C}$ is added to the location (1, 1), since the convection occurs at node 1which gives,

$$[K] = \begin{bmatrix} 260 & -250 & 0 & 0 \\ -250 & 450 & -200 & 0 \\ 0 & -200 & 550 & -350 \\ 0 & 0 & -350 & 350 \end{bmatrix}$$

And since the heat flow vector is simply h^*T_∞. With the absence of heat generated at a node,1.

The equilibrium equation can be written as follows,

$$\begin{bmatrix} 260 & -250 & 0 & 0 \\ -250 & 450 & -200 & 0 \\ 0 & -200 & 550 & -350 \\ 0 & 0 & -350 & 350 \end{bmatrix} \begin{Bmatrix} T_1 \\ T_2 \\ T_3 \\ T_4 = 30°\mathrm{C} \end{Bmatrix} = \begin{Bmatrix} h^*T_\infty(10^*600) \\ 0 \\ 0 \\ 0 \end{Bmatrix}$$

For solving the above equation, and by using the penalty approach presented in Appendix B, which requires,

$C = \max k(i,i)^*10^4 = 550^*10^4$. while $C \times T_4$ is added to the fourth row of $\{T\}$.

This is added to the element $K(4,4)$ which yields to,

$$\begin{bmatrix} 260 & -250 & 0 & 0 \\ -250 & 450 & -200 & 0 \\ 0 & -200 & 550 & -350 \\ 0 & 0 & -350 & 350 \end{bmatrix} \begin{Bmatrix} T_1 \\ T_2 \\ T_3 \\ 30 \end{Bmatrix} = \begin{Bmatrix} 6000 \\ 0 \\ 0 \\ 550^*10^4{}^*30 \end{Bmatrix}$$

The standard procedure of solving these equations is by deleting the fourth row and fourth column and modifying the load vector according to Appendix B

$$F_3 = 0 + 30 = 30$$

$$\begin{bmatrix} 260 & -1 & 0 \\ -1 & 450 & -1 \\ 0 & -1 & 550 \end{bmatrix} \begin{Bmatrix} T_1 \\ T_2 \\ T_3 \end{Bmatrix} = \begin{Bmatrix} 6000 \\ 0 \\ 30 \end{Bmatrix}$$

Solving give,

$$T_1 = 90.42°\mathrm{C},\ T_2 = 70.04°\mathrm{C},\ T_3 = 44.56°\mathrm{C}$$

11.4 Application of three nodes constant strain triangular element

The temperature at any point in the field in terms of natural coordinates shown in Fig. 11.6 is given by,

$$T(x,y) = L_1T_1 + L_2T_2 + L_3T_3 = \begin{bmatrix} L_1 & L_2 & L_3 \end{bmatrix} \begin{Bmatrix} T_1 \\ T_2 \\ T_3 \end{Bmatrix}$$

$$T(x,y) = [N(x,y)]\{T\}_0^e \tag{11.13}$$

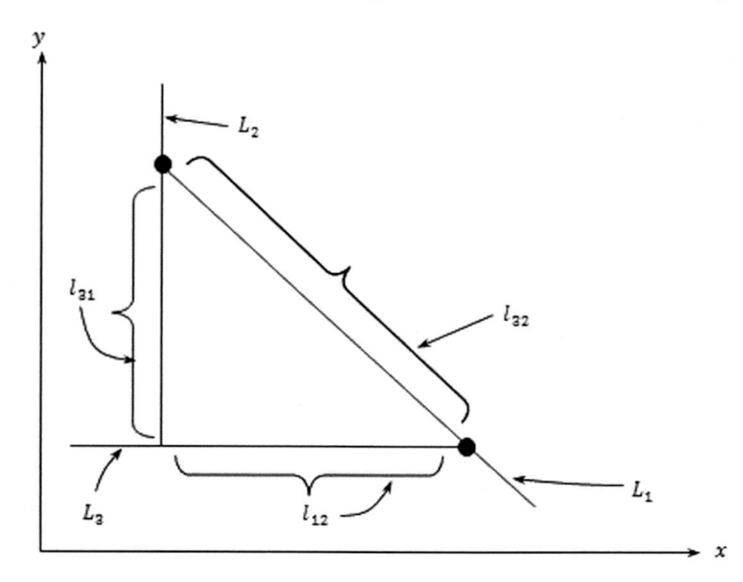

FIGURE 11.6 A typical three nodes triangular element.

And the global coordinates can be defined by the same interpolation function is as follows,

$$\begin{Bmatrix} x \\ y \end{Bmatrix} = [N(L_1, L_2, L_3)]\{T\}_0^e \tag{11.14}$$

L1, L2, and L3 are the shape functions derived using natural coordinates system.
Using Eq. (11.14), we have,

$$q_{cd,x} = -k\frac{\partial T}{\partial x} = -k\frac{\partial [N]}{\partial x}\{T\}_{\text{side}} \tag{11.15}$$

Where,

$q_{cd,x}$: Heat conduction flow in x-direction per unit area.
And,

$$[N] = \begin{bmatrix} L_1 & L_2 & L_3 \end{bmatrix}$$

Instead of using the shape functions in L_1, L_2, and L_3, we can use them in terms of rectangular coordinates as presented and derived in Chapter 6, Two Dimensional Problems: Application of Plane Strain and Stress. Fig. 11.7 is repeated here to be used in the derivative of the heat conduction matrix $[k_{\text{cond}}]$. for the two-dimensional region.

$$\text{In fact}[N_1] = \underset{\uparrow - \text{scalarfunction}}{N_1} \begin{bmatrix} 1 & 0 \\ 0 & 1 \end{bmatrix}. \tag{11.16a}$$

Where,

$$N_1 = \frac{1}{2\Delta}(a_1 + b_1 x + c_1 y) \tag{11.16b}$$

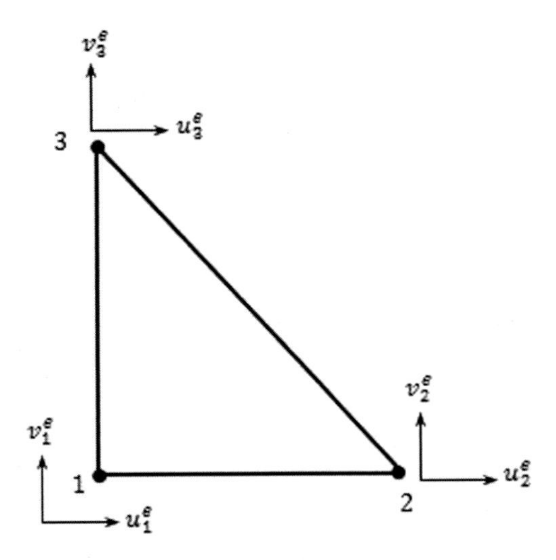

FIGURE 11.7 A typical finite element three nodes and 6 degrees of freedom.

Where,

$$a_1 = x_2y_3 - x_3y_2, b_1 = y_2 - y_3, c_1 = x_3 - x_2$$

N_2 (x,y) and N_3 (x,y) are found by cyclic permutations.
Therefore we have,

$$\frac{\partial[N]}{\partial x} = \frac{1}{|J|}\left[\,(y_2 - y_3) \quad (y_3 - y_1) \quad (y_1 - y_3)\,\right] \tag{11.16c}$$

Using the same notations used in the derivation of the element stiffness matrix for the plane problems explained in Chapter 2, Direct Methods, we can write,

$$= \frac{1}{|J|}\left[y_{23} \quad y_{31} \quad y_{12}\right]$$

Where $|J|$ is the Jacobian determinant.
And,

$$\frac{\partial[N]}{\partial y} = \frac{1}{|J|}\left[(x_2 - x_3) \quad (x_1 - x_3) \quad (x_2 - x_1)\right] \tag{11.16d}$$

$$= \frac{1}{|J|}\left[x_{23} \quad x_{13} \quad x_{21}\right]$$

Where,

$$|J| = \begin{vmatrix} 1 & 1 & 1 \\ x_1 & x_2 & x_3 \\ y_1 & y_2 & y_3 \end{vmatrix} = 2A_e \tag{11.16}$$

Where,
A_e is the element area.

11.4.1 Determination of heat conduction matrices $[k]_{\text{cond}}$

If $k_x = k_y = k$, the heat conduction matrix on x-direction is given by,

$$[k]_{\text{cond},x} = -k\int_A \frac{\partial [N]^t}{\partial x}\frac{\partial [N]}{\partial x} dx\, dy$$

$$= -k\frac{\partial [N]^t}{\partial x}\frac{\partial N}{\partial x}\int_A dx\, dy \tag{11.17}$$

Carrying out the differentiation of the shape functions as achieved in Eqs. (11.16a) and (11.16b) and the integration requirements give,

$$[k]_{\text{cond},x} = \frac{k}{2|J|}\begin{bmatrix} x_{32}x_{32} & x_{32}x_{13} & x_{32}x_{21} \\ x_{13}x_{32} & x_{13}x_{13} & x_{13}x_{21} \\ x_{21}x_{32} & x_{21}x_{13} & x_{21}x_{21} \end{bmatrix} \tag{11.18}$$

Similarly, the heat conduction matrix in the y-direction is given by,

$$[k]_{\text{cond},y} = -k\int_A \frac{\partial [N]^t}{\partial y}\frac{\partial [N]}{y} dx\, dy$$

$$= \frac{k}{2|J|}\begin{bmatrix} y_{23}y_{23} & y_{23}y_{31} & y_{23}y_{12} \\ y_{31}y_{23} & y_{31}y_{31} & y_{31}y_{12} \\ y_{23}y_{12} & y_{12}y_{31} & y_{12}y_{12} \end{bmatrix} \tag{11.19}$$

Where,

$$x_{32} = x_3 - x_2 \ldots \text{etc.}$$

And,

$$y_{32} = y_3 - y_2 \ldots \text{etc.}$$

11.4.2 Determination of boundary convection matrix

The convection matrix is given by,

$$[k]_{cv} = h\int_l [N]^t[N]dl$$

$$= h\int_l \begin{bmatrix} L_1L_1 & L_1L_2 & L_1L_3 \\ L_2L_1 & L_2L_2 & L_2L_3 \\ L_3L_1 & L_3L_2 & L_3L_3 \end{bmatrix} dl \tag{11.20}$$

We may achieve the integration by using,

$$\int_l L^\alpha L^\beta dl = \frac{\alpha_i \beta_i}{(\alpha + \beta + 1)!} l$$

The integration is carried out on each side of the element.

For side, $L_1 = 0$, Eq. (11.20) gives,

$$[k]_{cv,l_{32}} = h\int_{l_{32}} \begin{bmatrix} 0 & 0 & 0 \\ 0 & L_2L_2 & L_2L_3 \\ 0 & L_3L_2 & L_3L_3 \end{bmatrix} dl \tag{11.21}$$

The integration of the terms in Eq. (11.21) gives,
$\int_{l_{32}} L_2L_2 dl = \frac{2!}{(2+1)!} l_{32} = \frac{1}{3} l_{32}$...etc., this gives,

$$[k]_{cv,32} = \frac{hl_{32}}{6}\begin{bmatrix} 0 & 0 & 0 \\ 0 & 2 & 1 \\ 0 & 1 & 2 \end{bmatrix} \tag{11.22a}$$

Similarly, for the other sides,

$$[k]_{cv,l_{31}} = \frac{hl_{31}}{6}\begin{bmatrix} 2 & 0 & 1 \\ 0 & 0 & 0 \\ 1 & 0 & 2 \end{bmatrix} \tag{11.22b}$$

$$[k]_{cv,l_{12}} = \frac{hl_{12}}{6}\begin{bmatrix} 2 & 1 & 0 \\ 1 & 2 & 0 \\ 0 & 0 & 0 \end{bmatrix} \tag{11.22c}$$

11.4.3 Determination of boundary convection vector $\{Q\}_{cv}$

The boundary convection vector is given by,

$$\{Q\}_{cv} = h\int_l [N]^t T_\infty dl \tag{11.22}$$

This vector is calculated for the element boundaries as follows,

$$\{Q\}_{cv} = hT_\infty \int_l \begin{Bmatrix} L_1 \\ L_2 \\ L_3 \end{Bmatrix} dl$$

For side l_{32},

$$\{Q\}_{cv,l_{32}} = hT_\infty \int_{l_{32}} \begin{Bmatrix} 0 \\ L_2 \\ L_3 \end{Bmatrix} dl$$

$$= \frac{hT_\infty l_{32}}{2}\begin{Bmatrix} 0 \\ 1 \\ 1 \end{Bmatrix} \tag{11.23a}$$

For side l_{21},

$$\{Q\}_{cv,l_{21}} = \frac{hT_\infty l_{21}}{2}\begin{Bmatrix}1\\1\\0\end{Bmatrix} \tag{11.23b}$$

For side l_{13},

$$\{Q\}_{cv,l_{13}} = \frac{hT_\infty l_{13}}{2}\begin{Bmatrix}1\\0\\1\end{Bmatrix} \tag{11.23c}$$

11.4.4 Determination of applied heat vector $\{Q\}_b$

The applied boundary heat vector $\{Q\}_b$ is given by,

$$\{Q\}_b = k\int_s [N]^t \frac{\partial T}{\partial n} ds \tag{11.23d}$$

Where n is normal to the surface.

And, in case of thermal insulation, $Q = 0$.

From energy principles, the applied boundary heat flux, q_b, must equal the negative of the region conduction heat flow vector and can be written as follows,

$q_b = -q_{cd} = -k\frac{\partial T}{\partial n}$, in n-direction normal to s.

Put into Eq. (11.23c) gives,

$$\{Q\}_b = \int_s [N]^t q_b ds \tag{11.23e}$$

This is analogue to Eq. (11.22) by replacing hT_∞ by q_b, for example, if the boundary flux q_b is constant over the side l_{32} of the triangle shown in Fig. 11.6, the heat flow vector is given by,

$$\{Q\}_b = q_b\frac{l_{32}}{2}\begin{Bmatrix}0\\1\\1\end{Bmatrix} \tag{11.23}$$

And in case of insulation, $\{Q\}_b = \{0\}$.

Example 11.2: For the region shown in Fig. 11.8 (2 elements), formulate the equilibrium equations.

$$h = 1.0\ \mathrm{W/cm^2.^\circ C}$$

$$k = 20\ \mathrm{W/cm^2.^\circ C}$$

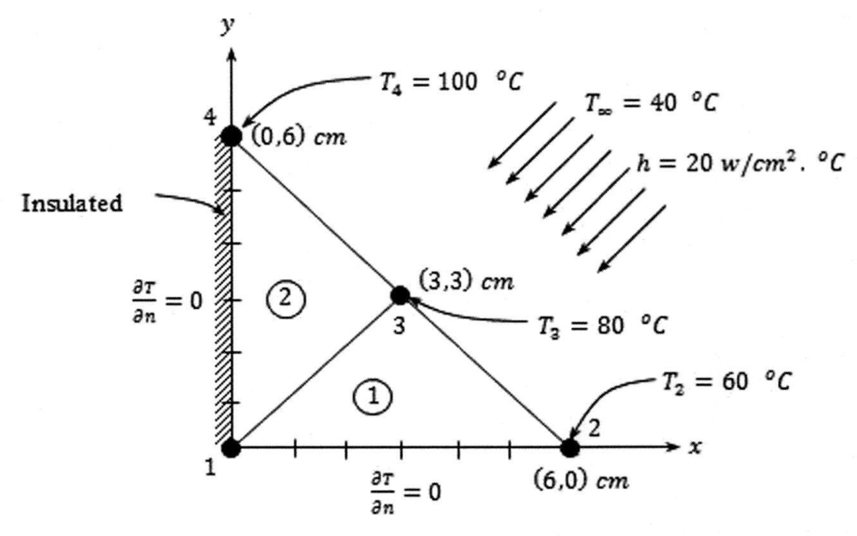

FIGURE 11.8 A triangular region for 2D heat transfer problem.

$$T_\infty = 40\ \mathrm{C}^\circ.\ ^\circ\mathrm{C}$$

Side 1–4 is perfectly insulated that is, $Q_1 = Q_4 = 0$

Steps 1 & 2 of the solution

1. Isolate the elements and assign the local and global numbering, as shown in Fig. 11.9.
2. Determine the x and y-coordinates, lengths of sides, $|J|$ Jacobin, and the coefficients in matrices in Eqs. (11.18) and (11.19), Table 11.1.

The Jacobian matrix is given by Eq. (11.16) and is calculated as follows:

$$|J| = (0-3)(0-3) - (6-3)(0-3)$$

$= 9 + 9 = 18$: used for both elements (Fig. 11.10).

For element No. 1

We have,

$$l_{12} = 6\,\mathrm{cm},\ l_{23} = 4.24\,\mathrm{cm},\ l_{31} = 4.24\,\mathrm{cm},\ l_{14} = 6\,\mathrm{cm}$$

$$y_{23} = y_2 - y_3 = 0 - 3 = -3\,\mathrm{cm}$$

$$y_{31} = y_3 - y_1 = 3 - 0 = 3\,\mathrm{cm}$$

$$y_{12} = y_1 - y_2 = 0 - 0 = 0\,\mathrm{cm}$$

$$x_{32} = x_3 - x_2 = 3 - 6 = 3\,\mathrm{cm}$$

$$x_{13} = x_1 - x_3 = 0 - 3 = -3\,\mathrm{cm}$$

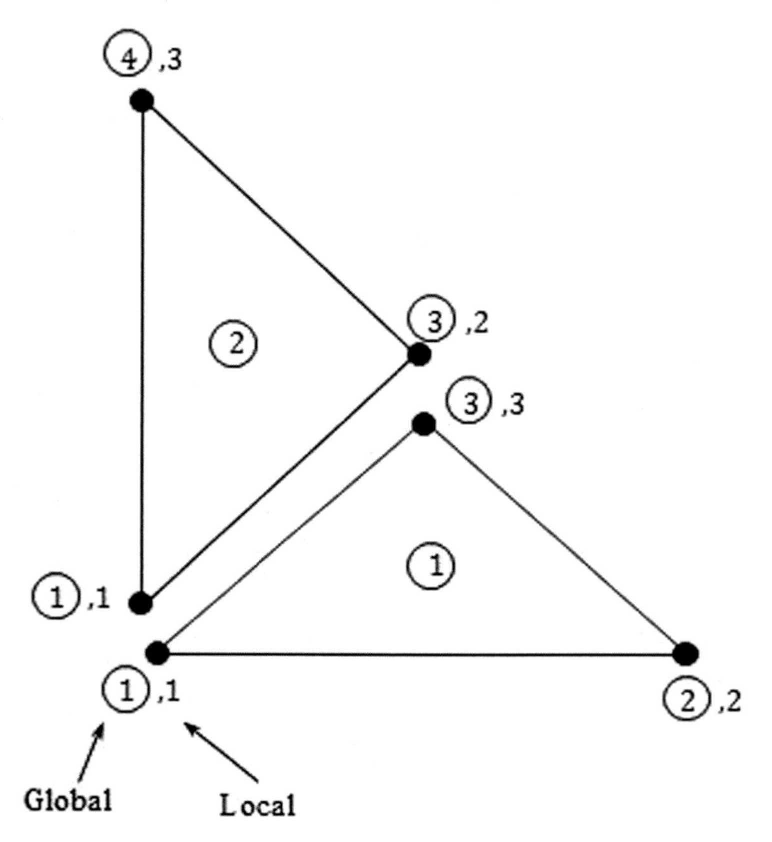

FIGURE 11.9 A finite element model, two elements and four nodes.

TABLE 11.1 x and y coordinate of the nodes.

Node	1	2	3	4
x (cm)	0	6	3	0
y (cm)	0	0	3	6

$$x_{21} = x_2 - x_1 = 6 - 0 = 6\,\text{cm}$$

11.4.4.1 Heat conduction matrices determination

The heat conduction matrices are obtained by using Eqs. (11.18) and (11.19) as follows,

$$[k]_{cd,x} = \frac{20}{2*18}\begin{bmatrix} 9 & -9 & 0 \\ -9 & 9 & 0 \\ 0 & 0 & 0 \end{bmatrix} = \begin{bmatrix} 9 & -9 & 0 \\ -9 & 9 & 0 \\ 0 & 0 & 0 \end{bmatrix} \tag{E2.1}$$

And,

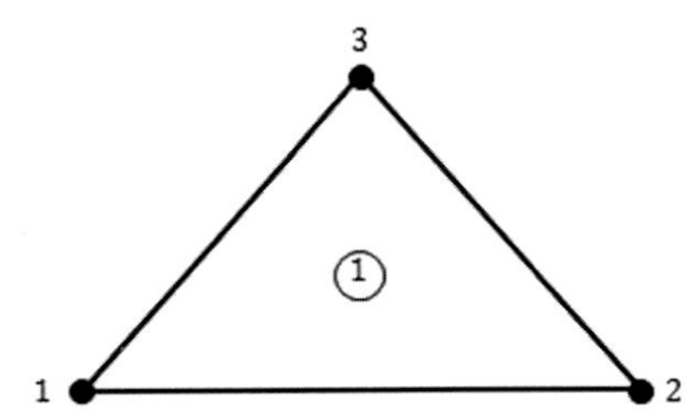

FIGURE 11.10 The element No. 1 of the triangular region.

$$[k]_{cd,y} = \frac{20}{2*18}\begin{bmatrix} 9 & 9 & -18 \\ 9 & 9 & -18 \\ -18 & -18 & 36 \end{bmatrix} = \begin{bmatrix} 9 & 9 & -18 \\ 9 & 9 & -18 \\ -18 & -18 & 36 \end{bmatrix} \tag{E2.2}$$

11.4.4.2 Convection matrix determination

The convection of the element 1 is only on side 2–3, that is, $L_1 = 0$. Using Eq. (11.20) and regarding Fig. 11.8, we have,

$$[k]_{cv} = \frac{hl_{23}}{6}\begin{bmatrix} 0 & 0 & 0 \\ 0 & 2 & 1 \\ 0 & 1 & 2 \end{bmatrix} = \frac{1*4.24}{6}\begin{bmatrix} 0 & 0 & 0 \\ 0 & 2 & 1 \\ 0 & 1 & 2 \end{bmatrix} = \begin{bmatrix} 0 & 0 & 0 \\ 0 & 1.47 & 0.736 \\ 0 & 0.736 & 1.47 \end{bmatrix} \tag{E2.3}$$

The convection load vector will be on the side l_{23} only, and can be written using Eq. (11.23b) as follows,

$$\{Q\}_{cv,l_{23}} = \frac{1*40*4.24}{2}\begin{bmatrix} 0 \\ 1 \\ 1 \end{bmatrix} = \begin{Bmatrix} 0 \\ 84.8 \\ 84.8 \end{Bmatrix} \tag{E2.4}$$

The total conduction + convection matrices for element No. 1 as follows,

$$[K] = [k]_{cd,x} + [k]_{cd,y} + [k]_{cv} \tag{E2.5}$$

Using Eqs. (E2.1)–(E2.3) into Eq. (E2.5) gives,

$$[K] = \begin{bmatrix} 9 & -9 & 0 \\ -9 & 9 & 0 \\ 0 & 0 & 0 \end{bmatrix}_{cd,x} + \begin{bmatrix} 9 & 9 & -18 \\ 9 & 9 & -18 \\ -18 & -18 & 36 \end{bmatrix}_{cd,y} + \begin{bmatrix} 0 & 0 & 0 \\ 0 & 1.47 & 0.736 \\ 0 & 0.736 & 1.47 \end{bmatrix}_{cv}$$

Or,

$$= \begin{bmatrix} 18 & 0 & -18 \\ 0 & 19.47 & -17.264 \\ -18 & -17.264 & 37.47 \end{bmatrix} \tag{E2.6}$$

The equilibrium equations for element No. 1,

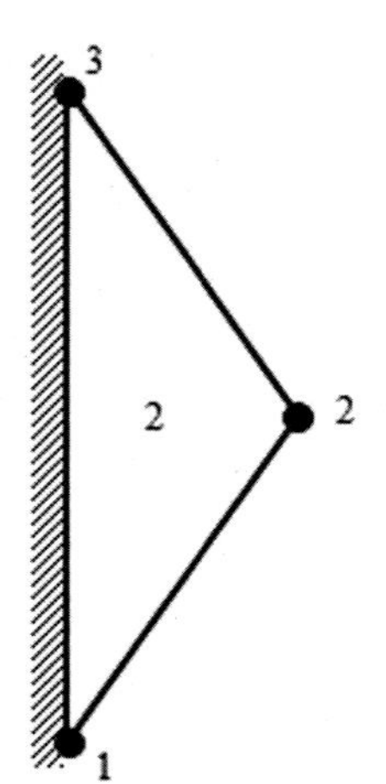

FIGURE 11.11 The element No. 2 of the triangular region with an isolated region 1–3.

$$\begin{bmatrix} 18 & 0 & -18 \\ 0 & 19.47 & -17.264 \\ -18 & -17.264 & 37.47 \end{bmatrix} \begin{Bmatrix} T_1 \\ T_2 \\ T_3 \end{Bmatrix} = \begin{Bmatrix} 0 \\ 84.8 \\ 84.8 \end{Bmatrix} \tag{E2.7}$$

For element No. 2, Fig. 11.11
We have,

1. As before, $|J| = 18$,

$$l_{12} = 6 \text{ cm}, \ l_{23} = 4.24 \text{ cm}, \ l_{31} = 6\text{cm}$$

$$y_{23} = 3 - 6 = -3 \text{ cm}$$

$$y_{31} = 6 - 0 = 6 \text{ cm}$$

$$y_{12} = 0 - 3 = -3 \text{ cm}$$

$$x_{32} = 0 - 3 = -3 \text{ cm}$$

$$x_{13} = 0 - 6 = -6 \text{ cm}$$

$$x_{21} = 3 - 0 = 3 \text{ cm}$$

Using Eqs. (11.18) and (11.19) gives,

$$[k]_{cd,x} = \frac{20}{2*10} \begin{bmatrix} 9 & -18 & 9 \\ -18 & 36 & -18 \\ 9 & -18 & 9 \end{bmatrix} = \begin{bmatrix} 9 & -18 & 9 \\ -18 & 36 & -18 \\ 9 & -18 & 9 \end{bmatrix} \tag{E2.8}$$

$$[k]_{cd,y} = \frac{20}{2*10} \begin{bmatrix} 9 & 18 & -9 \\ 18 & 36 & -18 \\ -9 & -18 & 9 \end{bmatrix} = \begin{bmatrix} 9 & 18 & -9 \\ 18 & 36 & -18 \\ -9 & -18 & 9 \end{bmatrix} \tag{E2.9}$$

For the convection matrix, the only side l_{23} is considered, and by using Eq. (11.20) gives,

$$[k]_{cv,2} = \frac{1*l_{23}^{\nearrow=4.24}}{6}\begin{bmatrix}0 & 0 & 0\\ 0 & 2 & 1\\ 0 & 1 & 2\end{bmatrix} = \begin{bmatrix}0 & 0 & 0\\ 0 & 1.47 & 0.736\\ 0 & 0.736 & 1.47\end{bmatrix} \quad \text{(E2.10)}$$

Determination of convection force vector $\{Q\}_{cv}$, Eq. (11.23b)

$$\{Q\}_{cv,l_{23}} = \frac{hT_\infty l_{23}}{2}\begin{Bmatrix}0\\1\\1\end{Bmatrix} = \frac{1*40*4.24}{2}\begin{Bmatrix}0\\1\\1\end{Bmatrix} = \begin{Bmatrix}0\\88.4\\88.4\end{Bmatrix} = \{Q\}_{cv,l_{34}} \quad \text{(E2.11)}$$

The total conductive + convection matrices for element no. 2 as follows,

$$[K] = \begin{bmatrix}9 & -18 & 9\\ -18 & 36 & -18\\ 9 & -18 & 9\end{bmatrix} + \begin{bmatrix}9 & 18 & -9\\ 18 & 36 & -18\\ -9 & -18 & 9\end{bmatrix} + \begin{bmatrix}0 & 0 & 0\\ 0 & 1.47 & 0.736\\ 0 & 0.736 & 1.47\end{bmatrix}$$

$$= \begin{bmatrix}18 & 0 & 0\\ 0 & 73.47 & -35.264\\ 0 & -35.264 & 19.47\end{bmatrix} \quad \text{(E2.12)}$$

Therefore the equilibrium equations for element No. 2,

$$\begin{bmatrix}18 & 0 & 0\\ 0 & 73.47 & -35.264\\ 0 & -35.264 & 19.47\end{bmatrix}\begin{Bmatrix}T_1\\T_3\\T_4\end{Bmatrix} = \begin{Bmatrix}0\\88.4\\88.4\end{Bmatrix} \quad \text{(E2.13)}$$

1. Assembly, the Assembly is achieved for elements No: 1 and No: 2 according to their contributions at the nodes as follows,

$$\begin{bmatrix} \overset{1}{k_{11}^1 + k_{11}^2} & \overset{2}{k_{12}^2} & \overset{3}{k_{13}^1 + k_{12}^2} & \overset{4}{k_{13}^2}\\ k_{21}^1 & k_{22}^1 & k_{23}^1 & 0\\ k_{31}^1 + k_{12}^2 & k_{32}^1 & k_{33}^1 + k_{22}^2 & k_{23}^2\\ k_{31}^2 & 0 & k_{32}^2 & k_{33}^2\end{bmatrix}\begin{Bmatrix}T_1\\T_2\\T_3\\T_4\end{Bmatrix} = \begin{Bmatrix}0\\88.4\\88.4+88.4\\88.4\end{Bmatrix}\begin{matrix}1\\2\\3\\4\end{matrix} \quad \text{(E2.14)}$$

11.4.4.3 Determination of heat flow

Note that at the boundary either Q or T is known or both. If it is assumed that Q_1 is unknown, and $Q_2 = 0$, this requires that the temperature is known at node 1, assume $T_1 = 100°C$, and by substitution the values of the heat conduction matrix (E2.14) from the heat conduction matrices for the elements 1 and 2 calculated in Eqs. (E2.6) and (E2.12) gives,

$$\begin{bmatrix} \overset{1}{36} & \overset{2}{0} & \overset{3}{-18} & \overset{4}{0}\\ 0 & 19.47 & -17.264 & 0\\ -18 & -17.264 & 110.94 & -35.264\\ 0 & 0 & -35.264 & 19.47\end{bmatrix}\begin{Bmatrix}T_1 = 100\\T_2\\T_3\\T_4\end{Bmatrix} = \begin{Bmatrix}0\\88.4\\88.4+88.4\\88.4\end{Bmatrix} \quad \text{(E2.15)}$$

The solution requires that the equilibrium equations are modified as follows,

$$\begin{bmatrix} \overset{1}{36} & \overset{2}{0} & \overset{3}{-18} & \overset{4}{0} \\ 0 & 19.47 & -17.264 & 0 \\ -18 & -17.264 & 110.94 & -35.264 \\ 0 & 0 & -35.264 & 19.47 \end{bmatrix} \begin{Bmatrix} T_1 \\ T_2 \\ T_3 \\ T_4 \end{Bmatrix} = \begin{Bmatrix} 0 \\ 88.4 \\ 88.4+88.4 \\ 88.4 \end{Bmatrix}$$

Since, we have specified the temperature at node number 1, $C = \max k(i,i)*10^4 = 110.94*10^4$, while $C \times T_1$ is added to the first row of $\{F_{th}\}$.

It is added to the element $K(1,1)$ which yields to,

$$\begin{bmatrix} \overset{1}{36+100.94*10^4} & \overset{2}{0} & \overset{3}{-18} & \overset{4}{0} \\ 0 & 19.47 & -17.264 & 0 \\ -18 & -17.264 & 110.94 & -35.264 \\ 0 & 0 & -35.264 & 19.47 \end{bmatrix} \begin{Bmatrix} 100 \\ T_2 \\ T_3 \\ T_4 \end{Bmatrix} = \begin{Bmatrix} 100.94*10^4*100 \\ 88.4 \\ 176.8 \\ 88.4 \end{Bmatrix} \quad \text{(E2.16)}$$

According to the penalty principles, deleting the first row of Eq. (E2.16) and modifying the load vector of Eq. (E2.16) which results in,

$$\begin{bmatrix} \overset{1}{19.47} & \overset{2}{-17.264} & \overset{3}{} \overset{4}{0} \\ -17.264 & 110.94 & -35.264 \\ 0 & -35.264 & 19.47 \end{bmatrix} \begin{Bmatrix} T_2 \\ T_3 \\ T_4 \end{Bmatrix} = \begin{Bmatrix} 84.8 \\ 1976.8 \\ 84.8 \end{Bmatrix} \quad \text{(E2.17)}$$

Solving Eq. (E2.17), gives,

$$T_2 = 66.341^\circ\text{C},\ T_3 = 69.697^\circ\text{C},\ T_4 = 130.775^\circ\text{C}$$

For solving Eq. (E2.16), we may use Dirichlet boundary conditions, that is, enforcing $T_1 = 100^\circ$C and deleting the corresponding first row with replacing $k(1,1) = 1$ this will give,

$$\begin{bmatrix} 1 & 0 & 0 & 0 \\ 0 & 19.47 & -17.264 & 0 \\ -18 & -17.264 & 110.94 & -35.264 \\ 0 & 0 & -35.264 & 19.47 \end{bmatrix} \begin{Bmatrix} T_1 \\ T_2 \\ T_3 \\ T_4 \end{Bmatrix} = \begin{Bmatrix} 100 \\ T_2 \\ T_3 \\ T_4 \end{Bmatrix}$$

This will result in,

$$\begin{bmatrix} \overset{2}{19.47} & \overset{3}{-17.264} & \overset{4}{0} \\ -17.264 & 110.94 & -35.264 \\ 0 & -35.264 & 19.47 \end{bmatrix} \begin{Bmatrix} T_2 \\ T_3 \\ T_4 \end{Bmatrix} = \begin{Bmatrix} 84.8 \\ 1976.8 \\ 84.8 \end{Bmatrix}$$

Which is identical to Eq. (E2.17).

The heat flow per unit area of each element can be obtained from Eq. (11.15) as follows,

$$q_{cd,x} = -k\frac{\partial T}{\partial x} = -k\frac{\partial [N]}{\partial x}\{T_i\} = -\frac{k}{|J|}\begin{bmatrix} y_{23} & y_{31} & y_{12} \end{bmatrix}\begin{Bmatrix} T_1 \\ T_2 \\ T_3 \end{Bmatrix}$$

$$q_{cd,y} = -\frac{k}{|J|}\begin{bmatrix} x_{32} & x_{13} & x_{21} \end{bmatrix}\begin{Bmatrix} T_1 \\ T_2 \\ T_3 \end{Bmatrix}$$

For element No. 1,

$$q_{cd,x} = -\frac{20}{18}\begin{bmatrix} 3 & -3 & 6 \end{bmatrix}\begin{Bmatrix} T_1 \\ T_3 \\ T_4 \end{Bmatrix} \tag{E2.18}$$

$$q_{cd,y} = -\frac{20}{18}\begin{bmatrix} -3 & 3 & 0 \end{bmatrix}\begin{Bmatrix} T_1 \\ T_3 \\ T_4 \end{Bmatrix} \tag{E2.19}$$

For element No. 2,

$$q_{cd,x} = -\frac{20}{18}\begin{bmatrix} -3 & -6 & 3 \end{bmatrix}\begin{Bmatrix} T_1 \\ T_3 \\ T_4 \end{Bmatrix} \tag{E2.20}$$

$$q_{cd,y} = -\frac{20}{18}\begin{bmatrix} -3 & 6 & -3 \end{bmatrix}\begin{Bmatrix} T_1 \\ T_3 \\ T_4 \end{Bmatrix} \tag{E2.21}$$

11.5 4-Node quadrilateral element

To illustrate the application of the F.E.M. to heat conduction problem, we are going to present 4-node quadrilateral element isoparametric type shown in Fig. 11.12. This will follow the same procedure of the standard Formulation of the finite element technique.

The governing differential equation for steady-state heat conduction in anisotropic media without internal heat source is the Laplace equation,

$$\nabla^2 T = 0 \tag{11.24}$$

Where T is the temperature at a point where,

$$\nabla^2 = \frac{\partial^2}{\partial x^2} + \frac{\partial^2}{\partial y^2}$$

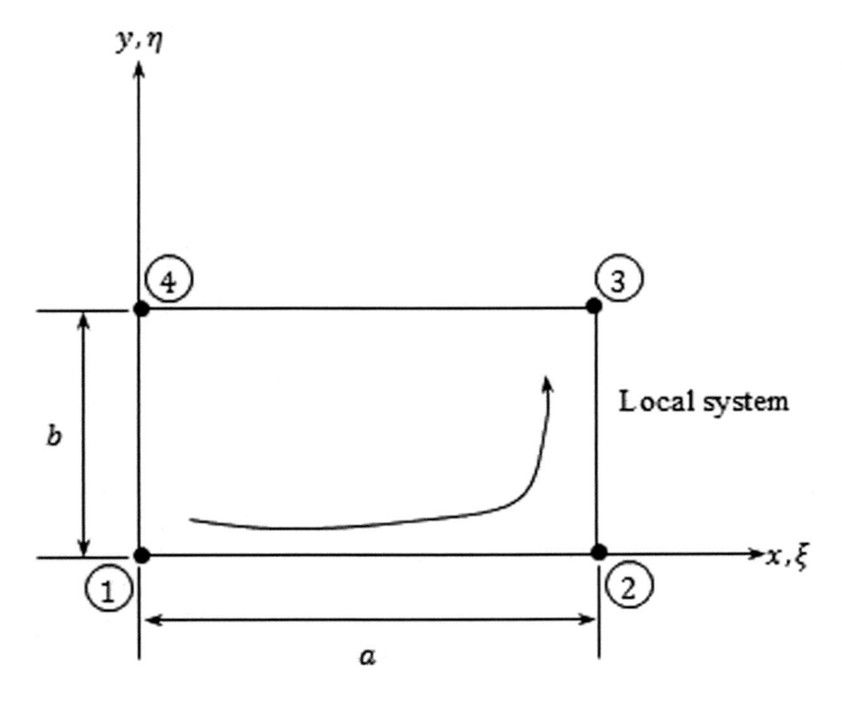

FIGURE 11.12 The 4-node quadrilateral element.

The following equation governs the Steady-state heat conduction problem problems,

$$\frac{\partial}{\partial x}\left(k_x\frac{\partial T}{\partial x}\right)+\frac{\partial}{\partial y}\left(k_y\frac{\partial T}{\partial y}\right)+Q=0 \tag{11.25}$$

where $q_x=-k_x\frac{\partial T}{\partial x}$, and $q_y=-k_y\frac{\partial T}{\partial y}$

Or,

The total potential energy χ

$$\chi=\frac{1}{2}\iint_D\left[k_x\left(\frac{\partial T}{\partial x}\right)^2+k_y\left(\frac{\partial T}{\partial y}\right)^2\right]dxdy-\iint_D Q\,T\,dx\,dy=\text{extremum} \tag{11.26}$$

Steps,

The element is shown in Fig. 11.12 has four nodes and 4 degrees of freedom T_1, T_2, T_3, and T_4. This type of element is widely used in modeling two-dimensional region. The procedure of using this element in heat applications is demonstrated as follows,

Step-1, Assume a solution in terms of nodal parameters and shape functions,

$$T(x,y)=T_1N_1(x,y)+T_2N_2(x,y)+T_3N_3(x,y)+T_4N_4(x,y) \tag{11.27}$$

Where,

$$N_1=\left(1-\frac{x}{a}\right)\left(1-\frac{y}{b}\right)=(1-\xi)(1-\eta)$$

$$N_2 = \frac{x}{a}\left(1 - \frac{y}{b}\right) = \xi(1 - \eta)$$

$$N_3 = \frac{x\,y}{a\,b} = \xi\eta$$

$$N_4 = \left(1 - \frac{x}{a}\right)\frac{y}{b} = (1 - \xi)\eta \tag{11.28a}$$

$$\xi = \frac{x}{a}, \eta = \frac{y}{b}$$

Where ξ and η are the intrinsic coordinates.

Step-2: Determination of element matrices.

To evaluate the element matrices, substitute the assumed solution in the variation statement,

$$\chi = \chi_1 + \chi_2$$

$$\chi_1 = \frac{1}{2}\int_0^b\int_0^a \left[k_x\left(\frac{\partial T}{\partial x}\right)^2 + k_y\left(\frac{\partial T}{\partial y}\right)^2\right]dx\,dy$$

$$\chi_2 = -\int_0^b\int_0^a QT(x,y)dx\,dy \tag{11.28}$$

The integrand of χ_1,

$$\chi_1 = \begin{Bmatrix} \frac{\partial T}{\partial x} & \frac{\partial T}{\partial y} \end{Bmatrix} \begin{Bmatrix} k_x\frac{\partial T}{\partial x} \\ k_y\frac{\partial T}{\partial y} \end{Bmatrix} = \begin{Bmatrix} \frac{\partial T}{\partial x} & \frac{\partial T}{\partial y} \end{Bmatrix} \begin{bmatrix} k_x & 0 \\ 0 & k_y \end{bmatrix} \begin{Bmatrix} \frac{\partial T}{\partial x} \\ \frac{\partial T}{\partial y} \end{Bmatrix} = \{g\}^t[D]\{g\} \tag{11.29}$$

We have,

$$\{g\} = [B]\{T\},$$

Where $[B]$ is the temperature—temperature gradient relationship and $\{T\}$ is the vector containing the nodal temperatures. They are written as follows,

$$[B] = \begin{bmatrix} \frac{\partial N_1}{\partial x} & \frac{\partial N_2}{\partial x} & \frac{\partial N_3}{\partial x} & \frac{\partial N_4}{\partial x} \\ \frac{\partial N_1}{\partial y} & \frac{\partial N_2}{\partial y} & \frac{\partial N_3}{\partial y} & \frac{\partial N_4}{\partial y} \end{bmatrix}$$

$$\{T\} = \begin{Bmatrix} T_1 & T_2 & T_3 & T_4 \end{Bmatrix}$$

Hence,

$$\{g\}^t[D]\{g\} = \{T\}^t[B]^t[D][B]\{T\} \tag{11.30}$$

And,

$$\chi_1 = \frac{1}{2}\{T\}^t\left[\int_0^b\int_0^a [B]^t[D][B]dx\,dy\right]\{T\} \tag{11.31}$$

Similarly,

$$\chi_2 = -\{T\}^t\left[\int_0^b\int_0^a Q[N]dx\,dy\right] \tag{11.32}$$

And,

$$\chi = \chi_1 + \chi_2 = \frac{1}{2}\{T\}^t\left[\int_0^b\int_0^a [B]^t[D][B]dx\,dy\right]\{T\} - \{T\}^t\left[\int_0^b\int_0^a Q[N]dx\,dy\right] \tag{11.33}$$

Step-4, Extremise χ,
The minimization of total potential energy is given by,

$$\frac{\partial\chi}{\partial T} = 0$$

Using Eq. (11.33) in the above equilibrium requirements gives,

$$\frac{\partial\chi}{\partial T} = \left[\int_0^b\int_0^a [B]^t[D][B]dx\,dy\right]\{T\} - \int_0^b\int_0^a Q[N]dx\,dy = \{0\} \tag{11.34}$$

We can express Eq. (11.34) in the finite element standard form as follows,

$$[k]^e\{T\} = \{F\}^e$$

Where,

$$[k]^e = \int_0^b\int_0^a [B]^t[D][B]dx\,dy \tag{11.35}$$

$$\{F\}^e = \int_0^b\int_0^a Q[N]dx\,dy \tag{11.36}$$

The integration of the Eqs. (11.35) and (11.36) is performed numerically. Note that the size of the matrix of the $[k]^e = (4\times 4)$ and the size of $\{F\}^e = (4\times 1)$.

However, the final form of the stiffness matrix if $k_x = k_y = k$ is:

$$[k]^e = \frac{k}{6}\begin{bmatrix} (2\alpha+2\beta) & (\alpha-2\beta) & (-\alpha-\beta) & (2\alpha+\beta) \\ & (2\alpha+2\beta) & (-2\alpha+\beta) & (-\alpha-\beta) \\ & \text{sym.} & (2\alpha+2\beta) & (\alpha-2\beta) \\ & & & (2\alpha+2\beta) \end{bmatrix} \tag{11.37}$$

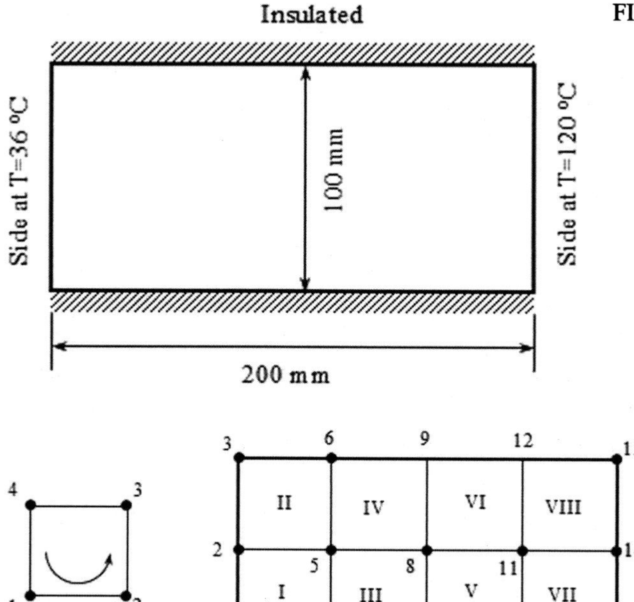

FIGURE 11.13 2D heat conduction region.

FIGURE 11.14 (A) The finite element mash 4-element of 15 nodes, (B) The typical element.

Where,

$$\alpha = \frac{a}{b}, \beta = \frac{b}{a}$$

The Eq. (11.37) matrix is simila r to that used in the boundary value problem presented in Chapter 3, Application of Energy Methods to Plate Problems, for the torsion problem modeling.

Example 11.3: Formulate the conduction matrices and how do you obtain the steady-state temperature distribution. Consider heat conduction in the domain shown in Fig. 11.13.

For such a problem, the heat generated is unknown at any node where the temperature is specified, discretization, consider the following simple finite element mesh, four rectangular elements connected with nine nodes as shown in Fig. 11.14.

Step-2, Formulation of equations for each finite element in the selected mesh.

For, $a = 1, \ b = 1$, using Eq. (11.37), we have,

$$[k]^e = \frac{k}{6}\begin{bmatrix} 4 & -1 & -2 & -1 \\ & 4 & -1 & -2 \\ & \text{sym.} & 4 & -1 \\ & & & 4 \end{bmatrix} \tag{E3.1}$$

Note,
In this case $[k]^{\mathrm{I}} = [k]^{\mathrm{II}} = \ldots\ldots\ldots = [k]^{\mathrm{VIII}} = [k]^e$
The equilibrium equations for each element is given by,

$$[k]^e\{T\}^e = \{F\}^e \tag{E3.2}$$

Applying the above matrix formulation for the given elements in the mesh, the following can be deduced,
Element-I,

$$[k]^e \begin{Bmatrix} T_1 \\ T_4 \\ T_5 \\ T_2 \end{Bmatrix} = \begin{Bmatrix} Q_1 \\ Q_4 \\ Q_5 \\ Q_2 \end{Bmatrix}_{\mathrm{I}} \tag{E3.3}$$

Element-II,

$$[k]^e \begin{Bmatrix} T_2 \\ T_5 \\ T_6 \\ T_3 \end{Bmatrix} = \begin{Bmatrix} Q_2 \\ Q_5 \\ Q_6 \\ Q_3 \end{Bmatrix}_{\mathrm{II}} \tag{E3.4}$$

Element-III,

$$[k]^e \begin{Bmatrix} T_4 \\ T_7 \\ T_8 \\ T_5 \end{Bmatrix} = \begin{Bmatrix} Q_4 \\ Q_7 \\ Q_8 \\ Q_5 \end{Bmatrix}_{\mathrm{III}} \tag{E3.5}$$

Element-IV,

$$[k]^e \begin{Bmatrix} T_5 \\ T_8 \\ T_9 \\ T_6 \end{Bmatrix} = \begin{Bmatrix} Q_5 \\ Q_8 \\ Q_9 \\ Q_6 \end{Bmatrix}_{\mathrm{IV}} \tag{E3.6}$$

Element-V,

$$[k]^e \begin{Bmatrix} T_7 \\ T_{10} \\ T_{11} \\ T_8 \end{Bmatrix} = \begin{Bmatrix} Q_7 \\ Q_{10} \\ Q_{11} \\ Q_8 \end{Bmatrix}_{\mathrm{V}} \tag{E3.7}$$

Element-VI,

$$[k]^e \begin{Bmatrix} T_8 \\ T_{11} \\ T_{12} \\ T_9 \end{Bmatrix} = \begin{Bmatrix} Q_8 \\ Q_{11} \\ Q_{12} \\ Q_9 \end{Bmatrix}_{\mathrm{VI}} \tag{E3.8}$$

Element-VII,

$$[k]^e \begin{Bmatrix} T_{10} \\ T_{13} \\ T_{14} \\ T_{11} \end{Bmatrix} = \begin{Bmatrix} Q_{10} \\ Q_{13} \\ Q_{14} \\ Q_{11} \end{Bmatrix}_{\text{VII}} \tag{E3.9}$$

Element-VIII,

$$[k]^e \begin{Bmatrix} T_{11} \\ T_{14} \\ T_{15} \\ T_{12} \end{Bmatrix} = \begin{Bmatrix} Q_{11} \\ Q_{14} \\ Q_{15} \\ Q_{12} \end{Bmatrix}_{\text{VIII}} \tag{E3.10}$$

Step-3, Assembly of the equations for the whole domain,

$$[k]_{\text{sys.}} = \frac{k}{6}\begin{bmatrix}
4 & -1 & 0 & -1 & -2 & 0 & 0 & 0 & 0 & 0 & 0 & 0 & 0 & 0 & 0 \\
 & 8 & -1 & -2 & -2 & -2 & 0 & 0 & 0 & 0 & 0 & 0 & 0 & 0 & 0 \\
 & & 4 & 0 & -2 & -1 & 0 & 0 & 0 & 0 & 0 & 0 & 0 & 0 & 0 \\
 & & & 8 & -2 & 0 & -1 & -2 & 0 & 0 & 0 & 0 & 0 & 0 & 0 \\
 & & & & 16 & -2 & -2 & -2 & -2 & 0 & 0 & 0 & 0 & 0 & 0 \\
 & & & & & 8 & 0 & -2 & -1 & 0 & 0 & 0 & 0 & 0 & 0 \\
 & & & & & & 8 & -2 & 0 & -1 & -2 & 0 & 0 & 0 & 0 \\
 & & & & & & & 16 & -2 & -2 & -2 & -2 & 0 & 0 & 0 \\
 & & & & & & & & 8 & 0 & -2 & -1 & 0 & 0 & 0 \\
 & & & & & & & & & 8 & -2 & 0 & -1 & -2 & 0 \\
 & & & & & & & & & & 16 & -2 & -2 & -2 & -2 \\
 & & & & & & & & & & & 8 & 0 & -2 & -1 \\
 & & & & & & & & & & & & 4 & -1 & 0 \\
 & & & & & & & & & & & & & 8 & -1 \\
 & & & & & & & & & & & & & & 4
\end{bmatrix} \tag{E3.11}$$

The thermal load vector $\{Q\}$ and the temperature vector $\{T\}$ is given as follows,

$$\{Q\} = \begin{Bmatrix} Q_1 \\ Q_2 \\ Q_3 \\ Q_4 \\ Q_5 \\ Q_6 \\ Q_7 \\ Q_8 \\ Q_9 \\ Q_{10} \\ Q_{11} \\ Q_{12} \\ Q_{13} \\ Q_{14} \\ Q_{15} \end{Bmatrix}, \{T\} = \begin{Bmatrix} T_1 \\ T_2 \\ T_3 \\ T_4 \\ T_5 \\ T_6 \\ T_7 \\ T_8 \\ T_9 \\ T_{10} \\ T_{11} \\ T_{12} \\ T_{13} \\ T_{14} \\ T_{15} \end{Bmatrix} \tag{E3.12}$$

Then,

$$[k]_{sys.}\begin{Bmatrix} T_1 \\ T_2 \\ T_3 \\ T_4 \\ T_5 \\ T_6 \\ T_7 \\ T_8 \\ T_9 \\ T_{10} \\ T_{11} \\ T_{12} \\ T_{13} \\ T_{14} \\ T_{15} \end{Bmatrix} = \begin{Bmatrix} Q_1 \\ Q_2 \\ Q_3 \\ Q_4 \\ Q_5 \\ Q_6 \\ Q_7 \\ Q_8 \\ Q_9 \\ Q_{10} \\ Q_{11} \\ Q_{12} \\ Q_{13} \\ Q_{14} \\ Q_{15} \end{Bmatrix}$$

Step-4, Boundary condition verification: From Fig. 11.13, we can write,

$$T_1 = T_2 = T_3 = 36\ ^{\circ}\mathrm{C}$$

$$T_{13} = T_{14} = T_{15} = 120\ ^{\circ}\mathrm{C}$$

$$Q_4 = Q_5 = Q_6 = Q_7 = Q_8 = Q_9 = Q_{10} = Q_{11} = Q_{12} = 0 \qquad \text{(E3.13)}$$

Using Eq. (E3.13) for the $\{Q\}$ vector in Eq. (E3.12), the reduced $\{Q\}$ vector will result in the following equilibrium equations,

$$\frac{k}{6}\begin{bmatrix} 4 & -1 & 0 & -1 & -2 & 0 & 0 & 0 & 0 & 0 & 0 & 0 & 0 & 0 & 0 \\ & 8 & -1 & -2 & -2 & -2 & 0 & 0 & 0 & 0 & 0 & 0 & 0 & 0 & 0 \\ & & 4 & 0 & -2 & -1 & 0 & 0 & 0 & 0 & 0 & 0 & 0 & 0 & 0 \\ & & & 8 & -2 & 0 & -1 & -2 & 0 & 0 & 0 & 0 & 0 & 0 & 0 \\ & & & & 16 & -2 & -2 & -2 & -2 & 0 & 0 & 0 & 0 & 0 & 0 \\ & & & & & 8 & 0 & -2 & -1 & 0 & 0 & 0 & 0 & 0 & 0 \\ & & & & & & 8 & -2 & 0 & -1 & -2 & 0 & 0 & 0 & 0 \\ & & & & & & & 16 & -2 & -2 & -2 & -2 & 0 & 0 & 0 \\ & & & & & & & & 8 & 0 & -2 & -1 & 0 & 0 & 0 \\ & & & & & & & & & 8 & -2 & 0 & -1 & -2 & 0 \\ & & & & & & & & & & 16 & -2 & -2 & -2 & -2 \\ & & & & & & & & & & & 8 & 0 & -2 & -1 \\ & & & & & & & & & & & & 4 & -1 & 0 \\ & & & & & & & & & & & & & 8 & -1 \\ & & & & & & & & & & & & & & 4 \end{bmatrix}\begin{Bmatrix} 36 \\ 36 \\ 36 \\ T_4 \\ T_5 \\ T_6 \\ T_7 \\ T_8 \\ T_9 \\ T_{10} \\ T_{11} \\ T_{12} \\ 120 \\ 120 \\ 120 \end{Bmatrix} = \begin{Bmatrix} Q_1 \\ Q_2 \\ Q_3 \\ 0 \\ 0 \\ 0 \\ 0 \\ 0 \\ 0 \\ 0 \\ 0 \\ 0 \\ Q_{13} \\ Q_{14} \\ Q_{15} \end{Bmatrix}$$

Where $k = 24$ w/cm.°C

Then,

$$T_4 = T_5 = T_6 = 57^{\circ}\mathrm{C}$$

$$T_7 = T_8 = T_9 = 78^{\circ}\mathrm{C}$$

$$T_{16} = T_{11} = T_{12} = 99^{\circ}\mathrm{C}$$

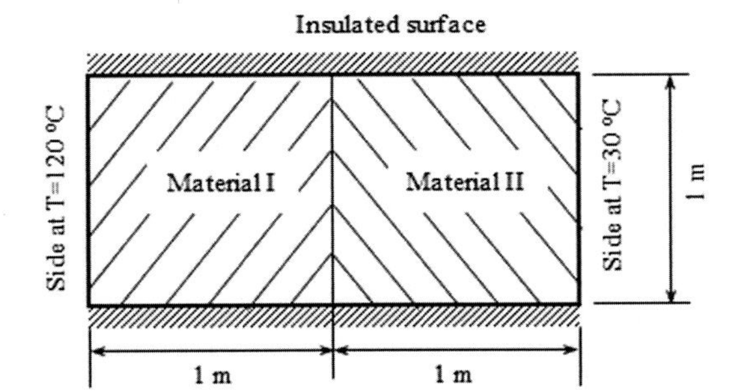

FIGURE 11.15 A heat conduction region.

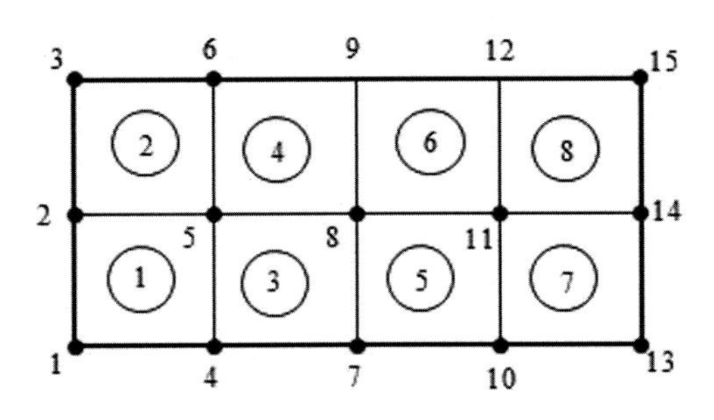

FIGURE 11.16 The finite element mesh, 15 nodes, 8 elements.

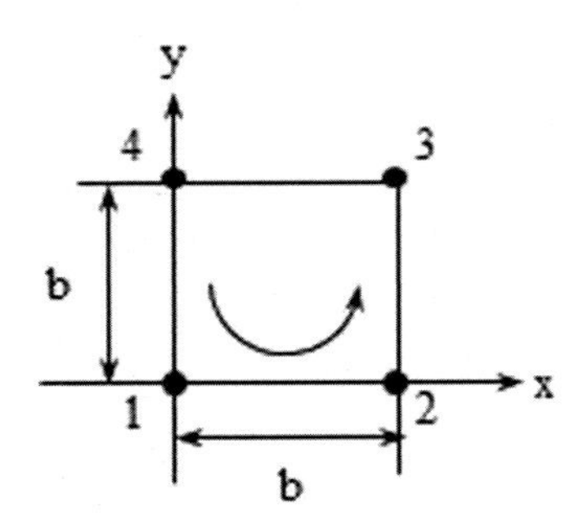

FIGURE 11.17 The typical finite element local node numbering.

And,

$$Q_1 = Q_3 = -252 \text{ w/cm}$$

$$Q_2 = -504 \text{ w/cm}$$

$$Q_{13} = Q_{15} = 252 \text{ w/cm}, \ Q_{14} = 504 \text{ w/cm}$$

Example 11.4: If the region shown in Fig. 11.15, as is modeled in Fig. 11.16, and is required to solve the steady-state heat conduction in this two-dimensional region, find the temperature distribution in this region.

For material-I, $k_x = k_y = 30\,\text{W/m/°C}$
For material-II, $k_x = k_y = 42\,\text{W/m/°C}$
Use the 4-node rectangular element, as shown in Fig. 11.17,

Solution,

Assume $T(x,y) = a_1 + a_2x + a_3y + a_4xy$ and in terms of nodal values of temperature and the shape function as follows,

$$T(x,y) = T_1N_1(x,y) + T_2N_2(x,y) + T_3N_3(x,y) + T_4N_4(x,y) \quad \text{(E4.1)}$$

Where,

$$N_1(x,y) = \frac{(x-a)(y-b)}{(ab)}$$

$$N_2(x,y) = \frac{x(y-b)}{(ab)}$$

$$N_3(x,y) = \frac{xy}{(ab)}$$

$$N_4(x,y) = \frac{(x-a)}{(ab)} \quad \text{(E4.2)}$$

We have,

$[K]^e\{T\}^e = \{Q\}^e$ where,

$$\{T\}^e = \{T_1 \quad T_2 \quad T_3 \quad T_4\}^t$$

And,

$$[k]^e = \frac{k}{6}\begin{bmatrix} (2\alpha+2\beta) & (\alpha-2\beta) & (-\alpha-\beta) & (2\alpha+\beta) \\ & (2\alpha+2\beta) & (-2\alpha+\beta) & (-\alpha-\beta) \\ & \text{sym.} & (2\alpha+2\beta) & (\alpha-2\beta) \\ & & & (2\alpha+2\beta) \end{bmatrix} \quad \text{(E4.3)}$$

For,

$$k_x = k_y = k,\ k_{\text{I}} = 30\frac{\text{W}}{\text{m.°C}}, k_{\text{II}} = 42\frac{\text{W}}{\text{m.°C}} \quad \text{(E4.4)}$$

$$\alpha = \frac{a}{b}, \beta = \frac{b}{a}$$

In our problem $\frac{a}{b} = 1$. The boundary conditions,

$$T_1 = T_2 = T_3 = 120\ °\text{C},$$

$$T_{13} = T_{14} = T_{15} = 30\ °\text{C} \quad \text{(E4.5)}$$

$$[K_I]_{\text{sys.}} = [K]^1 = [K]^2 = [K]^3 = [K]^4 = \frac{k_I}{6}\begin{bmatrix} 4 & -1 & -2 & -1 \\ & 4 & -1 & -2 \\ & & 4 & -1 \\ \text{Sym.} & & & 4 \end{bmatrix}$$

$$[K_I]_{sys.} = \begin{bmatrix} 20 & -5 & -10 & -5 \\ & 20 & -5 & -10 \\ & & 20 & -5 \\ \text{Sym.} & & & 20 \end{bmatrix} \tag{E4.6}$$

And,

$$[K_{II}]_{sys.} = [K]^5 = [K]^6 = [K]^7 = [K]^8 = \frac{k_{II}}{6} \begin{bmatrix} 4 & -1 & -2 & -1 \\ & 4 & -1 & -2 \\ & & 4 & -1 \\ \text{Sym.} & & & 4 \end{bmatrix}$$

$$[K_{II}]_{sys.} = \begin{bmatrix} 28 & -7 & -14 & -7 \\ & 28 & -7 & -14 \\ & & 28 & -7 \\ \text{Sym.} & & & 28 \end{bmatrix} \tag{E4.7}$$

For,

$$\{T\}^1 = \begin{Bmatrix} T_1 \\ T_4 \\ T_5 \\ T_2 \end{Bmatrix}, \{T\}^2 = \begin{Bmatrix} T_2 \\ T_5 \\ T_6 \\ T_3 \end{Bmatrix}, \{T\}^3 = \begin{Bmatrix} T_4 \\ T_7 \\ T_8 \\ T_5 \end{Bmatrix}, \{T\}^4 = \begin{Bmatrix} T_5 \\ T_8 \\ T_9 \\ T_6 \end{Bmatrix}$$

$$\{T\}^5 = \begin{Bmatrix} T_7 \\ T_{10} \\ T_{11} \\ T_8 \end{Bmatrix}, \{T\}^6 = \begin{Bmatrix} T_8 \\ T_{11} \\ T_{12} \\ T_9 \end{Bmatrix}, \{T\}^7 = \begin{Bmatrix} T_{10} \\ T_{13} \\ T_{14} \\ T_{11} \end{Bmatrix}, \{T\}^8 = \begin{Bmatrix} T_{11} \\ T_{14} \\ T_{15} \\ T_{12} \end{Bmatrix}$$

And,

$$\{Q\}^1 = \begin{Bmatrix} Q_1 \\ Q_4 \\ Q_5 \\ Q_2 \end{Bmatrix}, \{Q\}^2 = \begin{Bmatrix} Q_2 \\ Q_5 \\ Q_6 \\ Q_3 \end{Bmatrix}, \{Q\}^3 = \begin{Bmatrix} Q_4 \\ Q_7 \\ Q_8 \\ Q_5 \end{Bmatrix}, \{Q\}^4 = \begin{Bmatrix} Q_5 \\ Q_8 \\ Q_9 \\ Q_6 \end{Bmatrix}$$

$$\{Q\}^5 = \begin{Bmatrix} Q_7 \\ Q_{10} \\ Q_{11} \\ Q_8 \end{Bmatrix}, \{Q\}^6 = \begin{Bmatrix} Q_8 \\ Q_{11} \\ Q_{12} \\ Q_9 \end{Bmatrix}, \{Q\}^7 = \begin{Bmatrix} Q_{10} \\ Q_{13} \\ Q_{14} \\ Q_{11} \end{Bmatrix}, \{Q\}^8 = \begin{Bmatrix} Q_{11} \\ Q_{14} \\ Q_{15} \\ Q_{12} \end{Bmatrix} \tag{E4.8}$$

The system stiffness matrix will be of the size (15×15),

$$[K]_{Sys.} = \begin{bmatrix}
k_{11}^1 & k_{14}^1 & 0 & k_{12}^1 & k_{13}^1 & 0 & 0 & 0 & 0 & 0 & 0 & 0 & 0 & 0 & 0 \\
 & (k_{44}^1 + k_{11}^2) & k_{14}^2 & k_{24}^1 & (k_{34}^1 + k_{12}^2) & k_{13}^2 & 0 & 0 & 0 & 0 & 0 & 0 & 0 & 0 & 0 \\
 & & k_{44}^2 & 0 & k_{24}^2 & k_{34}^2 & 0 & 0 & 0 & 0 & 0 & 0 & 0 & 0 & 0 \\
 & & & (k_{22}^1 + k_{11}^3) & (k_{23}^1 + k_{14}^3) & 0 & k_{12}^3 & k_{13}^3 & 0 & 0 & 0 & 0 & 0 & 0 & 0 \\
 & & & & (k_{33}^1 + k_{22}^2 + k_{44}^3 + k_{11}^4) & (k_{23}^2 + k_{14}^4) & k_{24}^3 & (k_{34}^3 + k_{12}^4) & k_{13}^4 & 0 & 0 & 0 & 0 & 0 & 0 \\
 & & & & & (k_{33}^2 + k_{44}^4) & 0 & k_{24}^4 & k_{34}^4 & 0 & 0 & 0 & 0 & 0 & 0 \\
 & & & & & & (k_{22}^3 + k_{11}^5) & (k_{23}^3 + k_{14}^5) & 0 & k_{12}^5 & k_{13}^5 & 0 & 0 & 0 & 0 \\
 & & & & & & & (k_{33}^3 + k_{22}^4 + k_{44}^5 + k_{11}^6) & (k_{23}^4 + k_{14}^6) & k_{24}^5 & (k_{34}^5 + k_{12}^6) & k_{13}^6 & 0 & 0 & 0 \\
 & & & & & & & & (k_{33}^4 + k_{44}^6) & 0 & k_{24}^6 & k_{34}^6 & 0 & 0 & 0 \\
 & & & & & & & & & (k_{22}^5 + k_{11}^7) & (k_{23}^5 + k_{14}^7) & 0 & k_{12}^7 & k_{13}^7 & 0 \\
 & & & & & & & & & & (k_{33}^5 + k_{22}^6 + k_{44}^7 + k_{11}^8) & (k_{23}^6 + k_{14}^6) & k_{24}^7 & (k_{34}^7 + k_{12}^8) & k_{13}^8 \\
 & & & & & & & & & & & (k_{33}^6 + k_{44}^6) & 0 & k_{24}^8 & k_{34}^8 \\
 & & & & & & & & & & & & k_{22}^7 & k_{23}^7 & 0 \\
 & & & & & & & & & & & & & (k_{33}^7 + k_{22}^8) & k_{23}^8 \\
 & & & & & & & & & & & & & & k_{33}^8
\end{bmatrix} \tag{E4.9}$$

and,

$$\{T\} = \begin{Bmatrix} T_1 \\ T_2 \\ T_3 \\ T_4 \\ T_5 \\ T_6 \\ T_7 \\ T_8 \\ T_9 \\ T_{10} \\ T_{11} \\ T_{12} \\ T_{13} \\ T_{14} \\ T_{15} \end{Bmatrix}, \{Q\} = \begin{Bmatrix} Q_1 \\ Q_2 \\ Q_3 \\ Q_4 \\ Q_5 \\ Q_6 \\ Q_7 \\ Q_8 \\ Q_9 \\ Q_{10} \\ Q_{11} \\ Q_{12} \\ Q_{13} \\ Q_{14} \\ Q_{15} \end{Bmatrix}$$

The equilibrium equations will be as follows,

$$[K]_{\text{sys.}}\begin{Bmatrix} T_1 \\ T_2 \\ T_3 \\ T_4 \\ T_5 \\ T_6 \\ T_7 \\ T_8 \\ T_9 \\ T_{10} \\ T_{11} \\ T_{12} \\ T_{13} \\ T_{14} \\ T_{15} \end{Bmatrix} = \begin{Bmatrix} Q_1 \\ Q_2 \\ Q_3 \\ Q_4 \\ Q_5 \\ Q_6 \\ Q_7 \\ Q_8 \\ Q_9 \\ Q_{10} \\ Q_{11} \\ Q_{12} \\ Q_{13} \\ Q_{14} \\ Q_{15} \end{Bmatrix}$$

Application of boundary conditions,

$$T_1 = T_2 = T_3 = 120\ ^\circ\text{C}, T_{13} = T_{14} = T_{15} = 30\ ^\circ\text{C}$$

$$Q_4 = Q_5 = Q_6 = Q_7 = Q_8 = Q_9 = Q_{10} = Q_{11} = Q_{12} = 0 \quad \text{(E4.10)}$$

$$\begin{bmatrix} 20 & -5 & 0 & -5 & -10 & 0 & 0 & 0 & 0 & 0 & 0 & 0 & 0 & 0 & 0 \\ & 40 & -5 & -10 & -10 & -10 & 0 & 0 & 0 & 0 & 0 & 0 & 0 & 0 & 0 \\ & & 20 & 0 & -10 & -5 & 0 & 0 & 0 & 0 & 0 & 0 & 0 & 0 & 0 \\ & & & 40 & -10 & 0 & -5 & -10 & 0 & 0 & 0 & 0 & 0 & 0 & 0 \\ & & & & 80 & -10 & -10 & -10 & -10 & 0 & 0 & 0 & 0 & 0 & 0 \\ & & & & & 40 & 0 & -10 & -5 & 0 & 0 & 0 & 0 & 0 & 0 \\ & & & & & & 48 & -12 & 0 & -7 & -14 & 0 & 0 & 0 & 0 \\ & & & & & & & 96 & -12 & -14 & -14 & -14 & 0 & 0 & 0 \\ & & & & & & & & 48 & 0 & -14 & -7 & 0 & 0 & 0 \\ & & & & & & & & & 56 & -14 & 0 & -7 & -14 & 0 \\ & & & & & & & & & & 122 & -14 & -14 & -14 & -14 \\ & & & & & & & & & & & 56 & 0 & -14 & -7 \\ & & & & & & & & & & & & 28 & -7 & 0 \\ & & & & & & & & & & & & & 56 & -7 \\ & & & & & & & & & & & & & & 28 \end{bmatrix}\begin{Bmatrix} 120 \\ 120 \\ 120 \\ T_4 \\ T_5 \\ T_6 \\ T_7 \\ T_8 \\ T_9 \\ T_{10} \\ T_{11} \\ T_{12} \\ 30 \\ 30 \\ 30 \end{Bmatrix} = \begin{Bmatrix} Q_1 \\ Q_2 \\ Q_3 \\ 0 \\ 0 \\ 0 \\ 0 \\ 0 \\ 0 \\ 0 \\ 0 \\ 0 \\ Q_{13} \\ Q_{14} \\ Q_{15} \end{Bmatrix} \quad \text{(E4.11)}$$

Then,

$$T_4 = T_5 = T_6 = 93.75\ ^\circ\text{C}$$

$$T_7 = T_8 = T_9 = 67.5\ ^\circ\text{C}$$

$$T_{10} = T_{11} = T_{12} = 48.75\ ^\circ\text{C}$$

$$Q_1 = Q_3 = 393.75\ \text{W/m}$$

$$Q_2 = 787.5\ \text{W/m}$$

$$Q_{13} = Q_{15} = -393.75\ \text{W/m}$$

$$Q_{14} = -787.5\ \text{W/m} \quad \text{(E4.12)}$$

Example 11.5: For the regions shown in Fig. 11.18A and B, determine the temperature at point *P*.

Solution,
(a) Assuming,

$$T(x, y) = T_1N_1 + T_2N_2 + T_3N_3 \tag{E5.1}$$

$$2\Delta = |J| = \begin{vmatrix} 1 & 1 & 1 \\ x_1 & x_2 & x_3 \\ y_1 & y_2 & y_3 \end{vmatrix}$$

For the triangular region,

$$x_1 = 1, y_1 = 2$$

$$x_2 = 3, y_2 = 1$$

$$x_3 = 2.5, y_3 = 4$$

$$2\Delta = \begin{vmatrix} 1 & 1 & 1 \\ 1 & 3 & 2.5 \\ 2 & 1 & 4 \end{vmatrix} = 5.5 \tag{E5.2}$$

$$N_1 = \frac{1}{2\Delta}(a_1 + b_1x + c_1y)$$

Where,

$$a_1 = x_2.y_3 - x_3.y_2 = 9.5$$

$$b_1 = y_2 - y_3 = -3$$

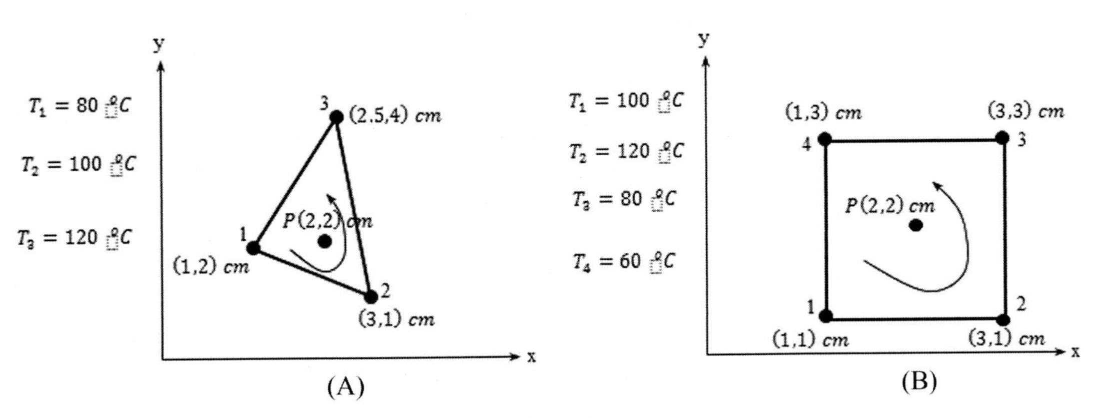

FIGURE 11.18 (A) Triangular region, (B) Rectangular region.

$$c_1 = x_3 - x_2 = -0.5$$

Then,

$$N_1 = \frac{1}{5.5}(9.5 - 3x - 0.5y) \quad \text{(E5.3)}$$

And,

$$N_2 = \frac{1}{2\Delta}(a_2 + b_2x + c_2y)$$

For,

$$a_2 = x_3.y_1 - x_1.y_3 = 1$$

$$b_2 = y_3 - y_1 = 2$$

$$c_2 = x_1 - x_3 = -1.5$$

Then,

$$N_2 = \frac{1}{5.5}(1 + 2x - 1.5y) \quad \text{(E5.4)}$$

And,

$$N_3 = \frac{1}{2\Delta}(a_3 + b_3x + c_3y)$$

For,

$$a_3 = x_1.y_2 - x_2.y_1 = -5$$

$$b_3 = y_1 - y_2 = 1$$

$$c_3 = x_2 - x_1 = 2$$

Then,

$$N_3 = \frac{1}{5.5}(-5 + x + 2y) \quad \text{(E5.5)}$$

At, $P(2,2)$, $N_1 = \frac{2.5}{5.5}, N_2 = \frac{2}{5.5}, N_3 = \frac{1}{5.5}$

$$T(2,2) = 80\frac{2.5}{5.5} + 100\frac{2}{5.5} + 120\frac{1}{5.5} = 94.545^\circ\text{C} \quad \text{(E5.6)}$$

(b) For the rectangular region, assuming (Fig. 11.19),

$$T(x,y) = T_1N_1 + T_2N_2 + T_3N_3 + T_4N_4 \quad \text{(E5.7)}$$

For,

$$x' = 1, y' = 1$$

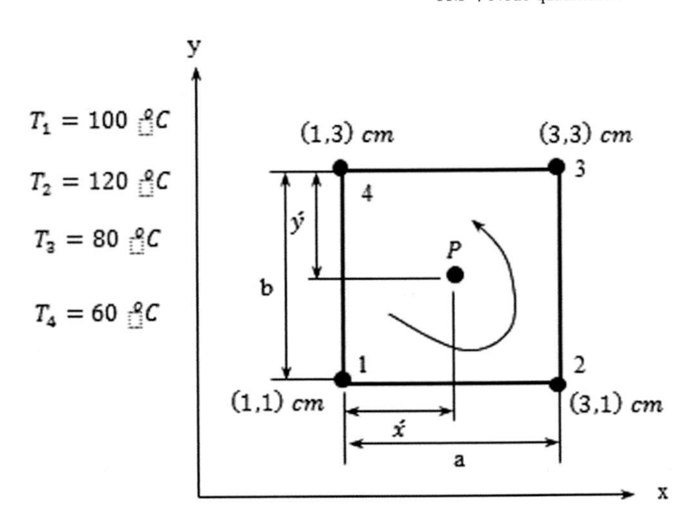

FIGURE 11.19 Nine nodes iosparametric elements.

$$a = b = 2$$

Then,

$$N_1 = \left(1 - \frac{x'}{a}\right)\left(1 - \frac{y'}{b}\right) = 0.25$$

$$N_2 = \frac{x'}{a}\left(1 - \frac{y'}{b}\right) = 0.25 \qquad \text{(E5.8)}$$

$$N_3 = \frac{x'}{a} - \frac{y'}{b} = 0.25$$

$$N_1 = \left(1 - \frac{x'}{a}\right)\frac{y'}{b} = 0.25$$

At, $P(2,2)$,

$$T(2,2) = 100{*}0.25 + 120{*}0.25 + 80{*}0.25 + 60{*}0.25 = 90^\circ\text{C} \qquad \text{(E5.9)}$$

Example 11.6: The coordinates of the nodes of a triangular element of thickness 2 mm as shown in Fig. 11.20. convection takes place from all three edges of the element. If $Q = 200\ \text{W/cm}^3$, $k = 100\ \text{W/cm.}^\circ\text{C}$, $h = 150\ \text{W/cm}^2.^\circ\text{C}$, and $T_\infty = 30^\circ\text{C}$. Determine,

1. $[K]_{\text{cond},x}, [K]_{\text{cond},y}$
2. $[K]_{\text{conv}}$ For the three sides of the triangle.
3. Boundary convection vectors.
4. Applied heat vector $\{Q\}_b$.

Solution,

1. To defining $[k]$, we have,

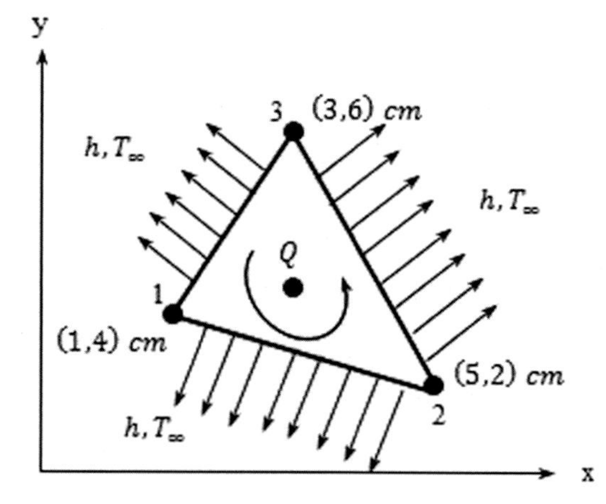

FIGURE 11.20 A triangle of thickness 2 mm.

$$|J| = \begin{vmatrix} 1 & 1 & 1 \\ x_1 & x_2 & x_3 \\ y_1 & y_2 & y_3 \end{vmatrix} = \begin{vmatrix} 1 & 1 & 1 \\ 1 & 5 & 3 \\ 4 & 2 & 6 \end{vmatrix} = 12 \quad \text{(E6.1)}$$

For,

$$x_1 = 1, x_2 = 5, x_3 = 3$$
$$y_1 = 4, y_2 = 2, y_3 = 6$$
$$x_{21} = x_2 - x_1 = 4$$
$$x_{13} = x_1 - x_3 = -2$$
$$x_{32} = x_3 - x_2 = -2$$
$$y_{12} = y_1 - y_2 = 2$$
$$y_{31} = y_3 - y_1 = 2$$
$$y_{23} = y_2 - y_3 = -4 \quad \text{(E6.2)}$$

$$[k]_{\text{cond}_x} = \frac{k}{2|J|}\begin{bmatrix} x_{32}x_{32} & x_{32}x_{13} & x_{32}x_{21} \\ x_{13}x_{32} & x_{13}x_{13} & x_{13}x_{21} \\ x_{21}x_{32} & x_{21}x_{13} & x_{21}x_{21} \end{bmatrix} = \frac{100}{2*12}\begin{bmatrix} (-2*-2) & (-2*-2) & (-2*4) \\ (-2*-2) & (-2*-2) & (-2*4) \\ (4*-2) & (4*-2) & (4*4) \end{bmatrix}$$

Then,

$$[k]_{\text{cond}_x} = \frac{25}{6}\begin{bmatrix} 4 & 4 & -8 \\ 4 & 4 & -8 \\ -8 & -8 & 16 \end{bmatrix} \quad \text{(E6.3)}$$

$$[k]_{\text{cond}_y} = \frac{k}{2|J|}\begin{bmatrix} y_{23}y_{23} & y_{23}y_{31} & y_{23}y_{12} \\ y_{31}y_{23} & y_{31}y_{31} & y_{31}y_{12} \\ y_{32}y_{12} & y_{12}y_{31} & y_{12}y_{12} \end{bmatrix} = \frac{25}{6}\begin{bmatrix} 16 & -8 & -8 \\ -8 & 4 & 4 \\ -8 & 4 & 4 \end{bmatrix}$$

2. To defining

$$l_{21} = \sqrt{(x_2 - x_1)^2 + (y_2 - y_1)^2} = \sqrt{(5-1)^2 + (2-4)^2}$$

$$l_{21} = 2\sqrt{5}$$

$$l_{32} = 2\sqrt{5}$$

$$l_{13} = 2\sqrt{2}$$

$$[k]_{\text{conv},l_{21}} = \frac{hl_{21}}{6}\begin{bmatrix} 2 & 1 & 0 \\ 1 & 2 & 0 \\ 0 & 0 & 0 \end{bmatrix} = \frac{150*2\sqrt{5}}{6}\begin{bmatrix} 2 & 1 & 0 \\ 1 & 2 & 0 \\ 0 & 0 & 0 \end{bmatrix} = 50\sqrt{5}\begin{bmatrix} 2 & 1 & 0 \\ 1 & 2 & 0 \\ 0 & 0 & 0 \end{bmatrix} \quad \text{(E6.4)}$$

$$[k]_{\text{conv},l_{32}} = \frac{hl_{32}}{6}\begin{bmatrix} 0 & 0 & 0 \\ 0 & 2 & 1 \\ 0 & 1 & 2 \end{bmatrix} = 50\sqrt{5}\begin{bmatrix} 0 & 0 & 0 \\ 0 & 2 & 1 \\ 0 & 1 & 2 \end{bmatrix} \quad \text{(E6.5)}$$

$$[k]_{\text{conv},l_{13}} = \frac{hl_{13}}{6}\begin{bmatrix} 2 & 0 & 1 \\ 0 & 0 & 0 \\ 1 & 0 & 2 \end{bmatrix} = 50\sqrt{2}\begin{bmatrix} 2 & 0 & 1 \\ 0 & 0 & 0 \\ 1 & 0 & 2 \end{bmatrix} \quad \text{(E6.6)}$$

3. The applied heat vector can be calculated as follows,

$$Q_{\text{conv},l_{21}} = \frac{hT_\infty l_{21}}{2}\begin{Bmatrix} 1 \\ 1 \\ 0 \end{Bmatrix} = \frac{150*30*2\sqrt{5}}{2}\begin{Bmatrix} 1 \\ 1 \\ 0 \end{Bmatrix} = 4500\sqrt{5}\begin{Bmatrix} 1 \\ 1 \\ 0 \end{Bmatrix} \quad \text{(E6.7)}$$

$$Q_{\text{conv},l_{32}} = \frac{hT_\infty l_{32}}{2}\begin{Bmatrix} 0 \\ 1 \\ 1 \end{Bmatrix} = 4500\sqrt{5}\begin{Bmatrix} 0 \\ 1 \\ 1 \end{Bmatrix} \quad \text{(E6.8)}$$

$$Q_{\text{conv},l_{13}} = \frac{hT_\infty l_{13}}{2}\begin{Bmatrix} 1 \\ 0 \\ 1 \end{Bmatrix} = 4500\sqrt{2}\begin{Bmatrix} 1 \\ 0 \\ 1 \end{Bmatrix} \quad \text{(E6.9)}$$

4. The applied heat vector $\{Q\}_b$ is obtained as follows,

$$A_e = \frac{|J|}{2} = \frac{12}{2} = 6\,\text{cm}^2$$

$$\{F\}_{\text{th}} = \frac{QA_e}{3}\begin{Bmatrix} 1 \\ 1 \\ 1 \end{Bmatrix} = \frac{200*6}{3}\begin{Bmatrix} 1 \\ 1 \\ 1 \end{Bmatrix} = 400\begin{Bmatrix} 1 \\ 1 \\ 1 \end{Bmatrix} \quad \text{(E6.10)}$$

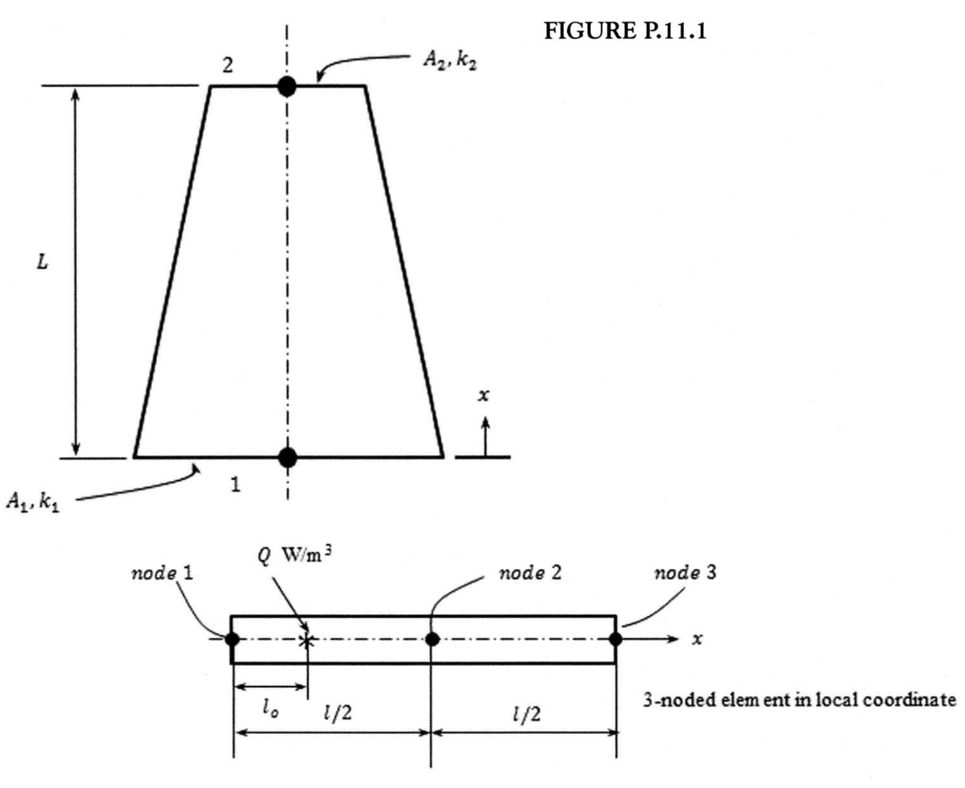

FIGURE P.11.1

FIGURE P.11.2

Problems

P.11.1 The 2-node heat conduction element, shown in Fig. P.11.1 has been used in 1-dim analysis. Derive the element characteristic equations, if the value of the thermal conductivity "k" and the cross-sectional area "A" vary linearly as,

$$k(x) = k_1 N_1 + k_2 N_2$$

$$A(x) = A_1 N_1 + A_2 N_2$$

Where A_1 and A_2, k_1 and k_2 are the values of the area "A" and "k" at node 1 and 2 respectively.

P.11.2 The 3-node heat condition element shown in Fig. P.11.2 has been used in one-dimensional analysis. Derive the element characteristic equations, if the value of the thermal conductivity "k" vary as

$$k = k_1 N_1 + k_2 N_2 + k_3 N_3$$

Where, k_1, k_2, k_3 are the values of k at nodes 1, 2, and 3, respectively. N_1, N_2, N_3 are the shape functions. Q is the local point heat source at $x = l_0$, and there is no heat transfer around the circumferential of the element. Hint, the variational statement is,

$$\chi = \frac{1}{2}\iiint k_x\left(\frac{\partial T}{\partial x}\right)^2 dx\,dy\,dz + \frac{1}{2}\iiint_{\text{surface}} h\left(T^2 - 2T_\infty T\right)ds - \iiint QT\,dx\,dy\,dz$$

(with usual notations)

P.11.3 heating cable at point $P(x = 5, y = 4\text{ cm})$ has been placed in a medium, as shown in Fig. P.11.3. The medium has conductivities of $k_x = 10\text{ W/cm}^\circ\text{k}$ and $k_y = 15\text{ W/cm}^\circ\text{k}$. The upper and lower surface are subjected to -20°C and 20°C respectively.

1. Determine the temperature at the points A, B, C, D and P using 1-element, of 4-node. Hint: for 4-node rectangular element,

$$[k]^e = \begin{bmatrix} (2\alpha + 2\beta) & (\alpha - 2\beta) & (-\alpha - \beta) & (-2\alpha + \beta) \\ & (2\alpha + 2\beta) & (-2\alpha + \beta) & (-\alpha - \beta) \\ & & (2\alpha + 2\beta) & (\alpha - 2\beta) \\ & & & (2\alpha + 2\beta) \end{bmatrix} + \begin{bmatrix} 0 & 0 & 0 & 0 \\ 0 & 0 & 0 & 0 \\ 0 & 0 & 2 & 1 \\ 0 & 0 & 1 & 2 \end{bmatrix} * \left(\frac{ha}{6}\right)$$

$$[Q]^e = \begin{bmatrix} 1 \\ 1 \\ 1 \\ 1 \end{bmatrix} \frac{abQ}{4} + \frac{h_\infty a T_\infty}{2} \begin{bmatrix} 0 \\ 0 \\ 1 \\ 1 \end{bmatrix}$$

$$[k]^e[T] = [Q]^e$$

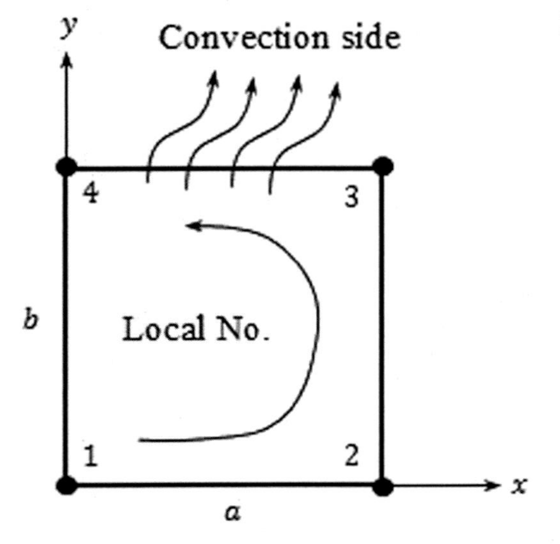

FIGURE. P.11.3A

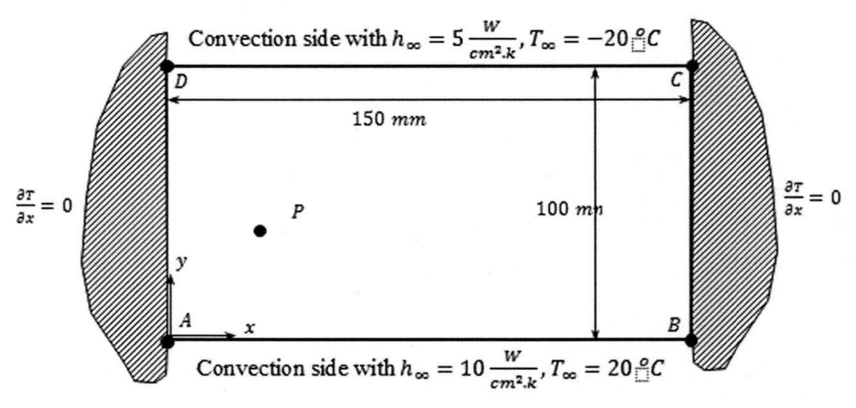

FIGURE P.11.3

$\alpha = \frac{ak_x}{b}, \beta = \frac{ak_y}{b}$, with usual notations, Q is heat generated per unit volume and $h = 1.0\ \text{W/cm}^2.°\text{C}$. (Fig. P.11.3A)

2. Heat flow $\left(\frac{\partial T}{\partial x}, \frac{\partial T}{\partial y}\right)$ At point P (Fig. P.11.3)

P.11.4 The 2-Dim heat transfer problem is one of the B.V.P. which can be solved by F.E. M such a problem is governed by the following variational statements,
by F.E.M such a problem is governed by the following variational statements,

$\chi = \iint_{\Omega_{\text{element}}} \frac{1}{2}\left[k_x\left(\frac{\partial T}{\partial x}\right)^2 + k_y\left(\frac{\partial T}{\partial y}\right)^2\right] dx\, dy - \iint Q\, T\, dx\, dy + \int_s \frac{1}{2} h\left(T^2 - T_\infty T\right) ds$ (with usual notations)

1. If the problem is to be solved by the 3-node triangular element, shown in Fig. P.11.4A. show that the matrix equation is,

$$[k]^e[T]^e = [Q]^e$$

Where,

$$[k]^e = [k]_{\text{cond}} + [k]_{\text{conv}} = \text{element characteristic matrix.}$$

$$[k]_{\text{cond}} = \frac{1}{8A_e}\left(k_y\beta_j\beta_i + k_y\gamma_i\gamma_j\right) i = 1, 2, 3,\ j = 1, 2, 3$$

$$[Q]^e = [Q]_{\text{cond}} + [Q]_{\text{conv}}$$

$$[Q]_{\text{cond}} = \overline{Q}\begin{bmatrix} \alpha_1 + \beta_1\overline{x} + \gamma_1\overline{y} \\ \alpha_2 + \beta_2\overline{x} + \gamma_2\overline{y} \\ \alpha_3 + \beta_3\overline{x} + \gamma_3\overline{y} \end{bmatrix}$$

Q = heat generation/unit volume = constant

$$\overline{x} = \frac{x_1 + x_2 + x_3}{3}, \overline{y} = \frac{y_1 + y_2 + y_3}{3}$$

$A_e =$ area of the element

FIGURE P.11.4A

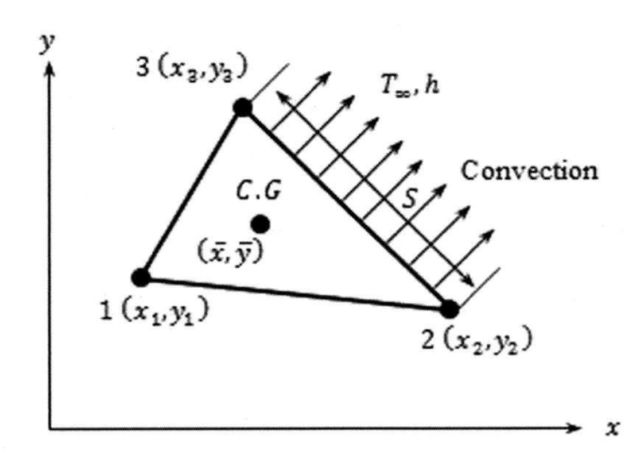

FIGURE P.11.4B

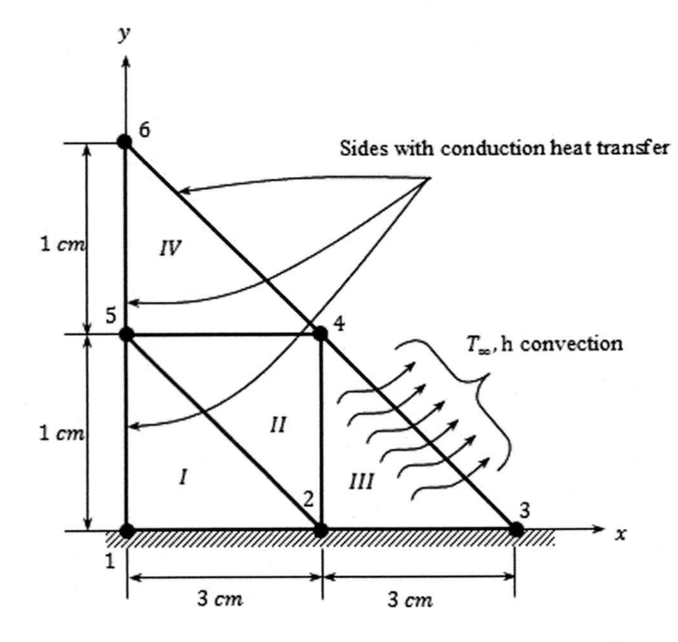

$$\alpha_1 = x_2y_3 - x_3y_2,\ \alpha_2 = x_3y_1 - x_1y_3,\ \alpha_3 = x_1y_2 - x_2y_1$$

$$\beta_1 = y_2 - y_3,\ \beta_2 = y_3 - y_1,\ \beta_3 = y_1 - y_2$$

$$\gamma_1 = x_3 - x_2,\ \gamma_2 = x_1 - x_3,\ \gamma_3 = x_2 - x_1$$

2. Write down the explicit form of [*k*] convection and [*Q*] convection.
3. In Fig. P.11.4B, four elements are insulated in the *x*-axis, solve the temperature of nodes 1,2 and 3.

   If k = thermal conductivity = 20 W/cm

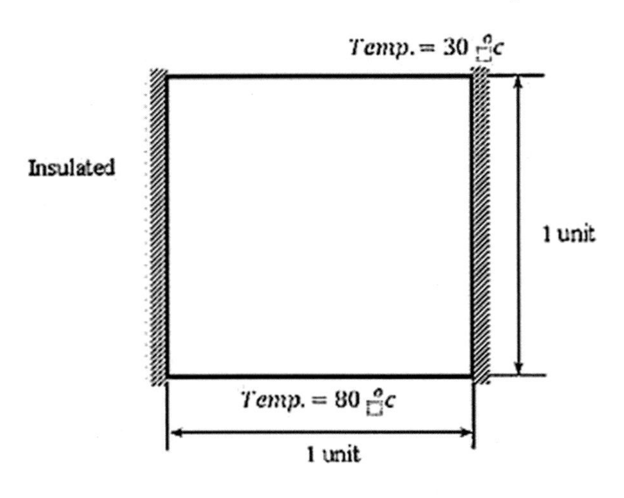

FIGURE P.11.5

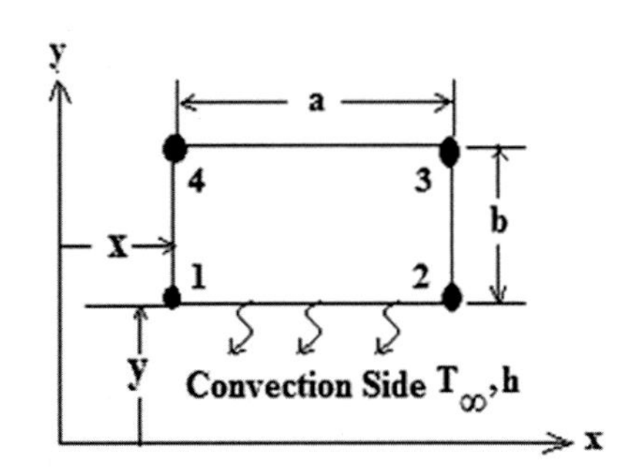

FIGURE P.11.6A

$$h = 1.0/\text{cm}^2.^\circ\text{C}$$
$$T_\infty = 40^\circ\text{C}$$
$$T_6 = 300^\circ\text{C}$$
$$T_4 = 200^\circ\text{C}, \;\; T_5 = 250^\circ\text{C}$$

P.11.5 Find the temperature distribution for the domain shown in Fig. P.11.5 by only considering heat conduction.

P.11.6 The variation statement for 2-dimensional heat transfer problem, with usual notations, is:

$$x = \frac{1}{2}\iint_{\text{element}}\left[K_x\left(\frac{\partial T}{\partial x}\right)^2 + K_y\left(\frac{\partial T}{\partial y}\right)^2\right]d_xd_y - \iint QTd_xd_y \qquad + h\int_{\Gamma\text{conv}}(T^2 - 2T_\infty{*}T)d\Gamma_c$$

Show that the matrix equations for the 4-node rectangular element with side 1–2 in convection shown in Fig. P.11.6A, can be expressed as follows,

FIGURE P.11.6B

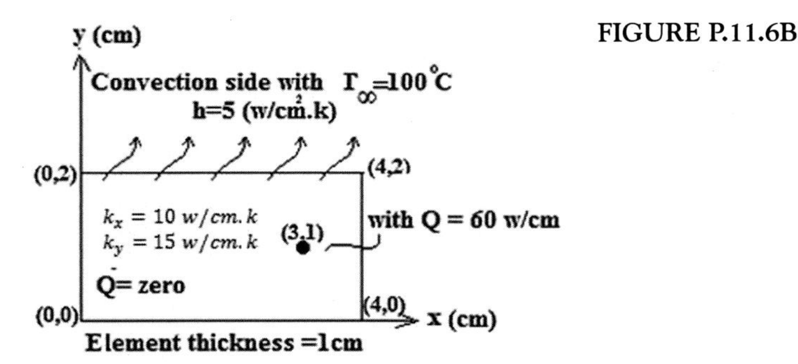

$$([K_{\text{cond}}] + [K_{\text{conv}}])[T] = \{Q_b\} + \{Q_{\text{conv}}\}$$

$$[K_{\text{cond}}] = \text{Conduction element matrix} = \frac{1}{6}\begin{bmatrix} 2\alpha+2\beta & \alpha-2\beta & -\alpha-\beta & -2\alpha+\beta \\ & 2\alpha+2\beta & -2\alpha+\beta & -\alpha-\beta \\ & \text{sum.} & 2\alpha+2\beta & \alpha-2\beta \\ & & & 2\alpha+2\beta \end{bmatrix}$$

$$\alpha = \frac{(K_y a)}{b}, \beta = \frac{(K_x b)}{a}$$

$$[K_{\text{conv}}] = \text{Convection element matrix} = \frac{ha}{6}\begin{bmatrix} 2 & 1 & 0 & 0 \\ 1 & 2 & 0 & 0 \\ 0 & 0 & 0 & 0 \\ 0 & 0 & 0 & 0 \end{bmatrix}$$

$$\{Q_b\} = \int_0^b \int_0^a Q[N] dx\, dy$$

$$[K_{\text{conv}}] = \int_0^a hT_\infty [\text{N}] d\Gamma_{12} = \text{Convection nodal force vector.}$$

$$\{T\}^e = \{T_1 T_2 T_3 T_4\}\ldots\text{Nodal temperature vector for the element, and}$$

$$[N]^e = \{N_1 N_2 N_3 N_4\}\ldots\text{Shape function vector}$$

A line source with $Q = 60$ Watt/cm is located at point (3, 1) in the element shown in Fig. P.11.6B.

Determine the matrix characteristic equations for the element.

P.11.7 The 6-node triangular finite element, shown in Fig. P.11.7, is suggested for the heat transfer of axisymmetric problems. The element shape functions are,

$$N_1 = L_1(2L_1 - 1)$$

$$N_2 = L_1 L_2$$

$$N_3 = L_2(2L_2 - 1)$$

$$N_4 = 4L_2 L_3$$

$$N_5 = L_3(2L_3 - 1)$$

$$N_6 = 4L_3 L_1$$

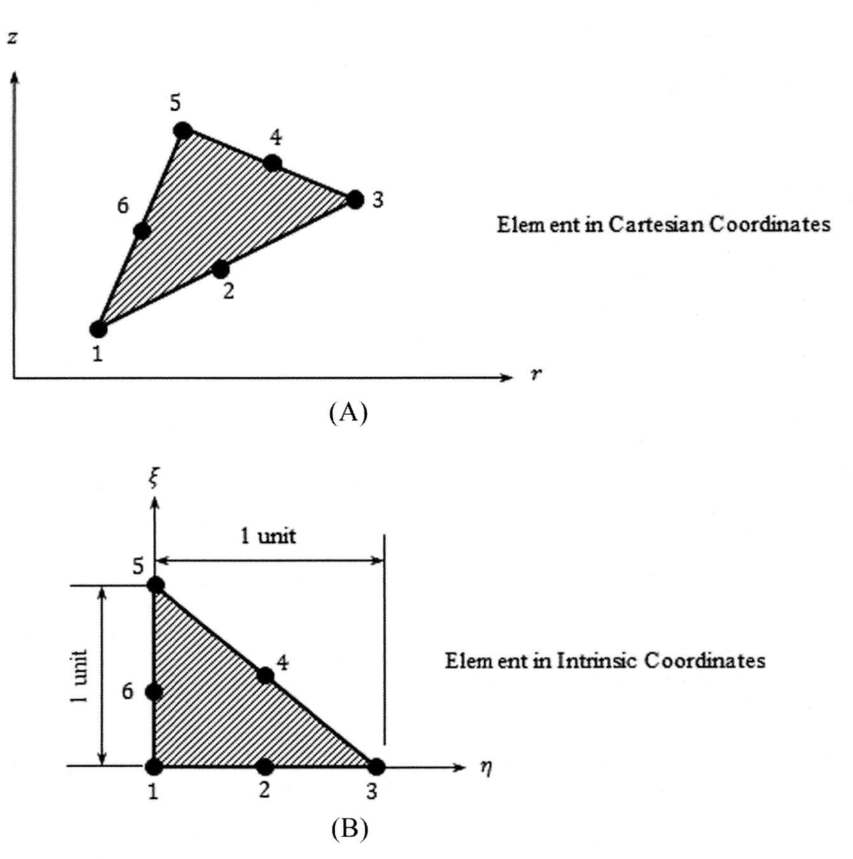

FIGURE P.11.7

$$L_1 = 1 - \eta - \xi,\ L_2 = \eta,\ L_3 = \xi$$

1. What are the necessary conditions required to analyze 3-Dim heat transfer problem through axis matrix ring elements?
2. Using the standard algorithms for element derivation prove that the element equations can be expressed as follows,

$$\left([k]_{\text{conv}} + [k]_{\text{cond}}\right)[T] = [Q]_{\text{cond}} + [Q]_{\text{conv}}$$

Where,

$$[k]_{\text{cond}} = \iint_{\Omega} [B]^t[D][B]2\pi r^2 dr\, dz$$

$$[k]_{\text{conv}} = 2\pi h \iint_{S_{\text{conv}}} [N][N]^t r^2 dS$$

$$[Q]_{\text{conv}} = 2\pi hT_{\infty}\iint_{S_{\text{conv}}}[N]r^2dS$$

$$[Q]_{\text{cond}} = \iint_{\Omega}\boxed{Q}[N]d\Omega$$

$$\boxed{Q} = \text{heat generated/unit volume}$$

$$[B] = \begin{bmatrix} \left(\dfrac{\partial N_i}{\partial r}\right) \\ \left(\dfrac{\partial N_i}{\partial z}\right) \end{bmatrix} i = 1,\ldots,6$$

And,

$$[D] = \begin{bmatrix} k_r & 0 \\ 0 & k_z \end{bmatrix}$$

Hint, variational statement,

$$\chi = \iint_{\text{volume}}\left[k_r.r\left(\frac{\partial T}{\partial r}\right)^2 + T_r\left(\frac{\partial T}{\partial z}\right)^2 - 2QrT\right]dvol + \frac{1}{2}\iint_{S_{\text{conv}}}h.r(T^2 - T_{\infty}T)dS = \text{minimum}$$

Show also how $[K_{\text{conv}}]$ and $[K_{\text{cond}}]$ Can be evaluated numerically by using a modified Gaussian quadrature $\zeta - \eta$ plane.

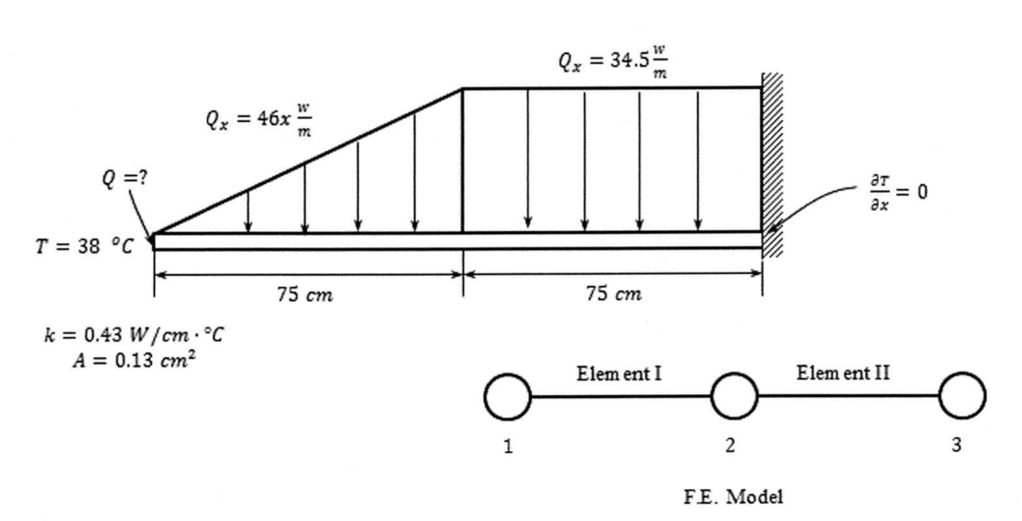

FIGURE P.11.8A

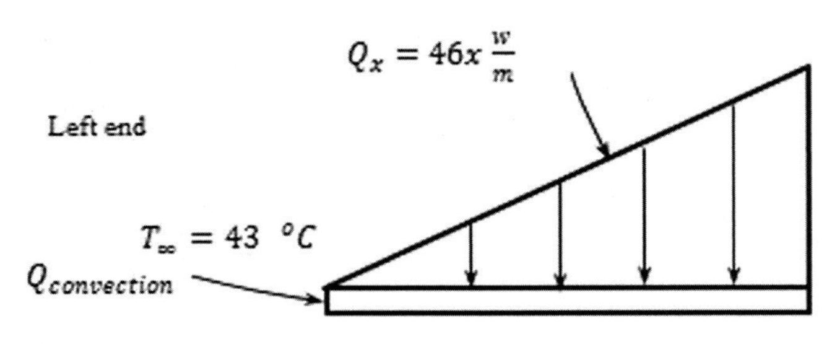

FIGURE P.11.8B .

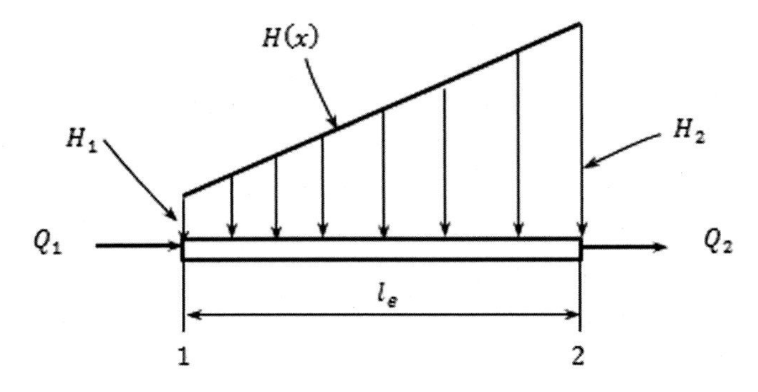

FIGURE P.11.8C

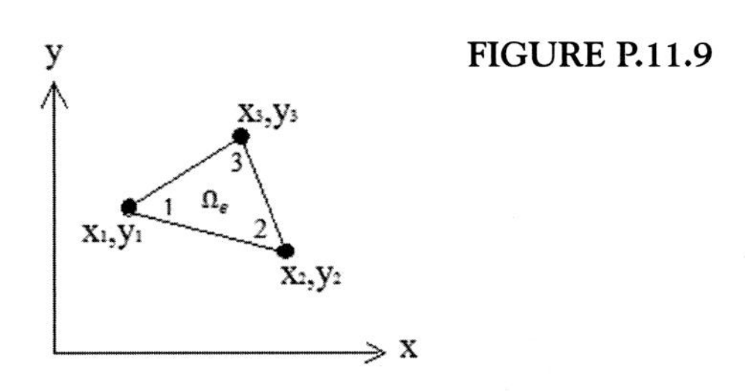

FIGURE P.11.9

P.11.8 The bar shown in Fig. P.11.8A. is subjected to a linearly varying and constant distributed heat source. The right end is against an insulated boundary; the left end is held at a temperature of 100°C.

1. Determine the temperature distribution in the bar using two finite elements with 2-nodes. Also, find the heat flow of the left end.
2. if the left end has heat transfer by convection with temperature and coefficient as defined in Fig. P.11.8B. determine the heat flow and temperature distribution.

Hint: the finite element equation for the 2-node element is (Fig. P.11.8C)

$$\frac{AK}{le}\begin{bmatrix} 1 & -1 \\ -1 & 1 \end{bmatrix}\begin{bmatrix} T_1 \\ T_2 \end{bmatrix} = \begin{bmatrix} Q_1 \\ Q_2 \end{bmatrix} + \frac{le}{6}\begin{bmatrix} 2 & 1 \\ 1 & 2 \end{bmatrix}\begin{bmatrix} H_1 \\ H_2 \end{bmatrix}$$

TABLE P.11.9 Weighting factors and function arguments used in Gauss-Legender formulas.

Points	Weighting factors	Function arguments	Truncation error
2	1.000000000	− 0.577350269	f^4
	1.000000000	0.577350269	
3	0.555555556	− 0.774596669	f^6
	0.888888889	0.0	
	0.555555556	0.774596669	
4	0.347854845	− 0.861136312	f^8
	0.652145155	− 0.339981044	
	0.652145155	0.339981044	
	0.347854845	0.861136312	
5	0.236926885	± 0.906179846	f^{10}
	0.478628670	± 0.538469310	
	0.568888889	0.0	
6	0.171324492	± 0.932469514	f^{12}
	0.360761573	0.661209386	
	0.467913935	± 0.238619186	

P.11.9 The 2-Dim potential problem (i.e., 2-Dim heat conduction and torsion of prismatic bar) is one of the boundary value problems that F.E.M. (Table P.11.9).

Suppose that the problem is to be solved employing 3-node triangular element shown in Fig. P.11.9. The shape function is:

$$N_1 = \frac{1}{2A}(a_1 + b_1x + c_1y), i = 1, 2, 3$$

Where,

A: the area of the element

$$a_1 = x_2y_3 - x_3y_2 \quad b_1 = y_2 - y_3 \quad c_1 = x_3 - x_2$$
$$a_2 = x_1y_3 - x_3y_1 \quad b_2 = y_1 - y_3 \quad c_2 = x_3 - x_1$$
$$a_3 = x_1y_2 - x_2y_1 \quad b_3 = y_1 - y_2 \quad c_3 = x_2 - x_1$$

Show that the matrix equation of the element can be expressed as follows:

$$[k]^e\{T\}^e = \{F\}^e$$

Where:

$$[k]^e = \frac{C}{2A}(b_ib_y + c_ic_j) \quad i = 1.2.3 \text{ and } j = 1.2.3$$

$c = k$ ... thermal conductivity ... for 2-Dim heat transfer
$c = 1/G$ ... for torsion of prismatic bar.
G = the modulus of rigidity.

$$\{T\}^e = \begin{Bmatrix} T_1 \\ T_2 \\ T_3 \end{Bmatrix}, \text{ with usual notations}$$

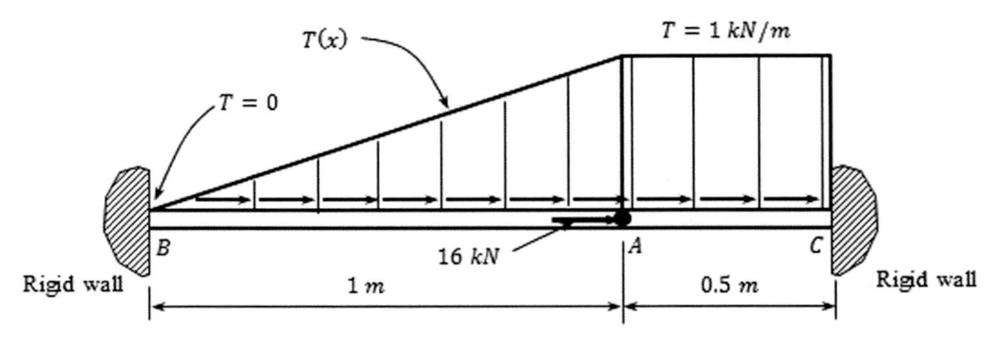

FIGURE P.11.10A Rigid foundations.

$$[F]^e = \iint_{\Omega_e} Q[N]dxdy\ldots\text{for 2-Dim heat conduction}$$

$$[F]^e = \iint_{\Omega_e} [N]dxdy\ldots\text{for torsion of prismatic bar}$$

$$[N] = \begin{bmatrix} N_1 \\ N_2 \\ N_3 \end{bmatrix}\ldots\text{Shape functions vector}$$

$Q\ldots$ is the heat generation/unit volume
Hints: Variational statements.

$$x = \frac{1}{2}\iint_{\Omega_{\text{element}}}\left[K_x\left(\frac{\partial T}{\partial x}\right)^2 + K_y\left(\frac{\partial T}{\partial y}\right)^2\right]d_x d_y - \iint_{\Omega_e} QTd_x d_y = \text{extremism for 2-}D\text{ heat conduction}$$

or:

$$x = \frac{1}{2G}\iint_{\Omega_e}\left(\frac{\partial w}{\partial x}\right)^2 + \left(\frac{\partial w}{\partial y}\right)^2 dx\,dy - 2\iint_{\Omega_e} wdx\,dy\ldots\text{for torsion of prismatic bars with usual notations.}$$

P.11.10 A steel rod is attached to rigid walls at each end. the rod is subjected to distributed loading $T(x)$ Furthermore, the concentrated force of 16 kN, as shown below. Use a suitable number of 2-noded elements to calculate,

1. The displacement at point (A).
2. The stress distribution in zone "A.B."
3. Prove that the F.E.M. results satisfy the equilibrium conditions.
4. If the rod is attached to elastic foundation at ends B and C with elastic stiffness $k_B = k_C = 3.5$ N.m, determine the reactions and displacements at the ends.

Hint, Finite element equation for 2-node bar element is (Figs. P.11.10A and P.11.10B),
$\frac{AE}{l}\begin{bmatrix} 1 & -1 \\ -1 & 1 \end{bmatrix}\begin{bmatrix} u_1 \\ u_2 \end{bmatrix} = \begin{bmatrix} F_1 \\ F_2 \end{bmatrix} + \frac{l}{6}\begin{bmatrix} 2 & 1 \\ 1 & 2 \end{bmatrix}\begin{bmatrix} T_1 \\ T_2 \end{bmatrix}$, with usual notations.

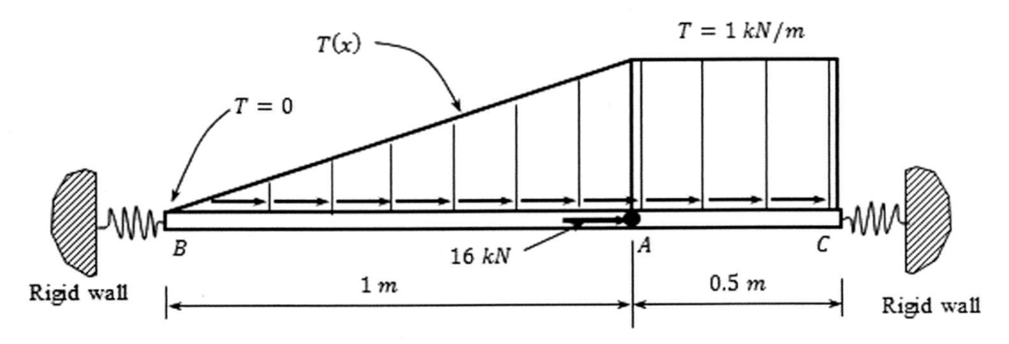

FIGURE P.11.10B Elastic foundation.

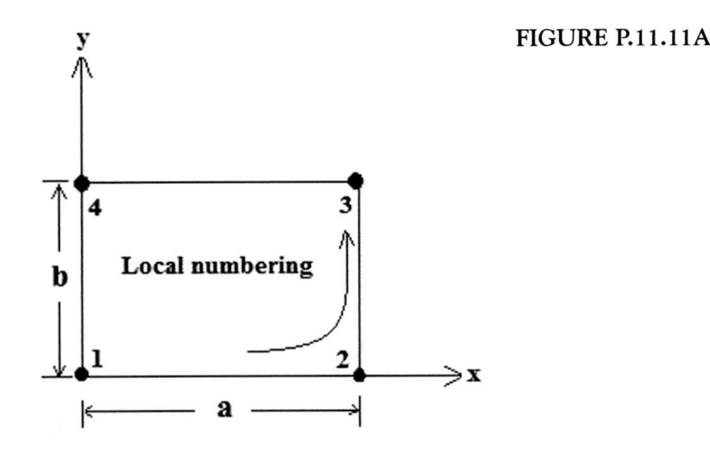

FIGURE P.11.11A

P.11.11 The two dimensional, steady-state, heat conduction problem is one of the boundary- value problems which can be solved using F.E.M. Such a problem is governed by the following variation statement

$X = \frac{1}{2}\iint_I \left[K_x\left(\frac{\partial T}{\partial x}\right)^2 + K_y\left(\frac{\partial T}{\partial y}\right)^2\right]dxdy - \iint QTdx\,dy$ = Extremism symbols having their usual meaning. Requirements,

If the problem is to be solved by means of four-nodes rectangular element shown in Fig. P.11.11A.

Prove that the field function $T(x,y)$ can be expressed in terms of nodal values and shape functions as follows:

$$T(x,y) = \sum_{i=j}^{\infty} T_1 N_1(x,y)$$

Where:

$$N_1(x,y) = \frac{(x-a)(y-b)}{ab},\quad N_2(x,y) = \frac{-x(y-b)}{ab}$$

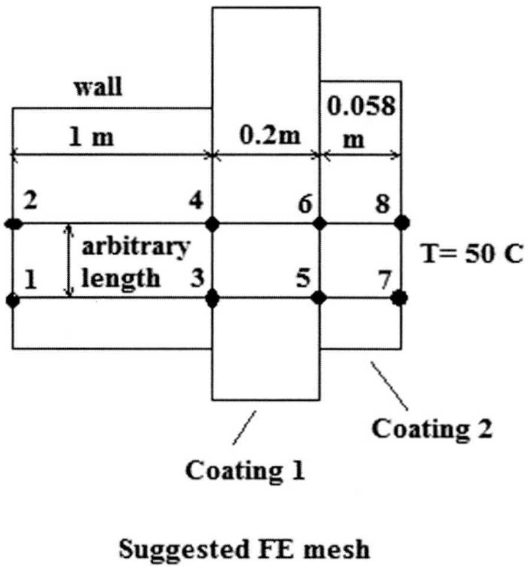

FIGURE P.11.11B

wall
1 m
0.2m
0.058 m
2 4 6 8
arbitrary length
1 3 5 7
T= 50 C
Coating 2
Coating 1

Suggested FE mesh

$$N_3(x,y) = \frac{xy}{ab}, \quad N_4(x,y) = \frac{-y(x-a)}{ab}$$

2i) Show that the matrix equations of the element can be expressed as follows:

$$[K]^e[T]^e = [Q]^e$$

$$[T]^t = [T_1 T_2 T_3 T_4]$$

$$[K]^e \begin{bmatrix} 2\alpha + 2\beta & \alpha - 2\beta & -\alpha - \beta & -2\alpha + \beta \\ \alpha - 2\beta & 2\alpha + 2\beta & -2\alpha + \beta & -\alpha - \beta \\ -\alpha - \beta & -2\alpha + \beta & 2\alpha + 2\beta & \alpha - 2\beta \\ -2\alpha + \beta & -\alpha - \beta & \alpha - 2\beta & 2\alpha + 2\beta \end{bmatrix}$$

$$\alpha = \frac{aK_x}{b}, \text{ and } \beta = \frac{bK_y}{a}$$

Note: "It is not recommended to go through all the details of integrating the $[K]^e$ terms"

Use the above element to find the temperature at the coating surface for the wall shown in Fig. 11.14B (Fig. P.11.11B).

Thermal conductivities $(K_x = K_y = K)$

Wall 150 J/m.s.°C

Coating 1 12 J/m.s.°C

Coating 2 30 J/m.s.°C

P.11.12 Using the quadrilateral element presented in Chapter 7, Torsion Problem, Formulate the stiffness matrix using the suggested mesh in Fig. P.11.12. The equations that give the total heat flow through the wall. Take the average of the values at the inner and outer surfaces, and then write down the equation that gives the conduction shape factor S from $q_{\text{total}} = k.S.°C$, where, $\Delta T_{\text{total}} = (500 - 100)°C = 400°C$. Write down the equilibrium equations.

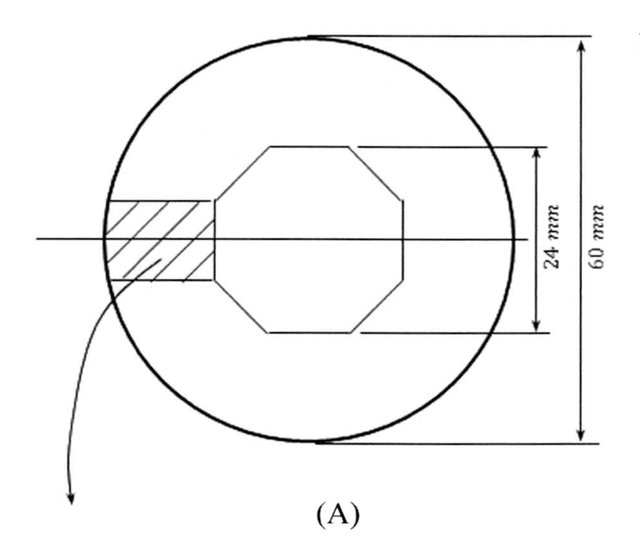

(A)

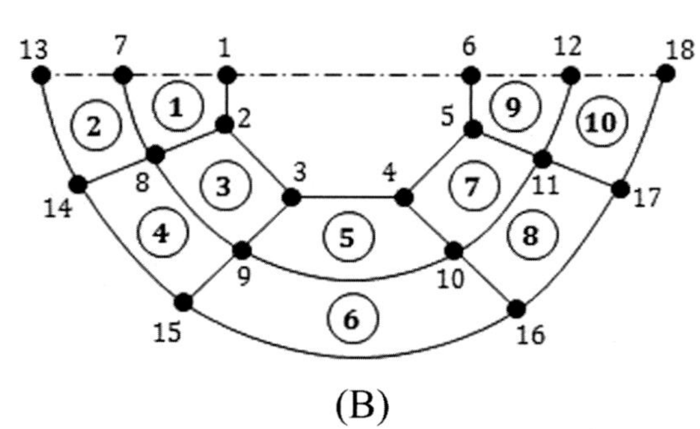

(B)

FIGURE P.11.12 . F.E. Mesh.

300 mm

$k = 0.7\ W/m^2.\ ^\circ C$

28 °C

$h = 40\ W/m^2.\ ^\circ C$

$T_\infty = -15\ ^\circ C$

FIGURE P.11.13

Apply the boundary conditions and write the reduced equilibrium equations. You may use a quarter mesh instead of the half shown in Fig. P.11.12.

Comment on the procedure of solving the equations to obtain the distribution of wall temperature.

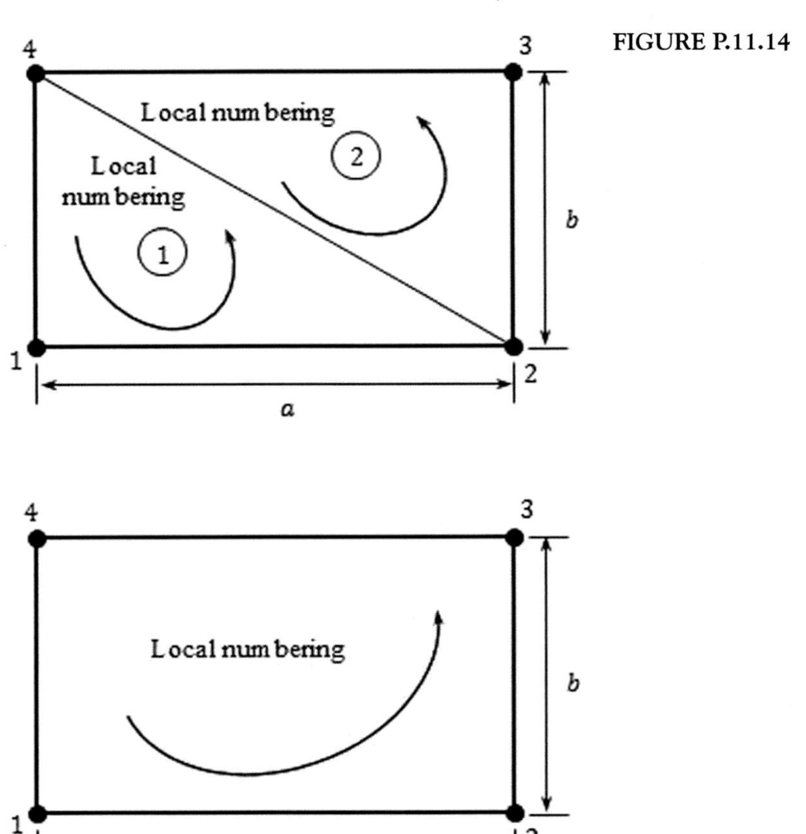

FIGURE P.11.14

P.11.13 For the brick wall shown in Fig. P.11.13, if the length of the brick = 300 mm, and the coefficient of heat conduction $k = 0.7\,W/m^2.°C$, and the brick is subjected to the conditions shown in the Fig. 11.14. The heat transfer associated with the outside surface is $h = 40\ W/m^2.°C$. Determine the steady-state temperature distribution within the wall and also the heat flux through the wall. Use the two-element model and assume one-dimensional heat flow.

P.11.14 (a) Consider a rectangular element of sides a and b and thickness t idealized as two triangular elements and one rectangular element as shown in Fig. P.11.14, respectively.

Derive the assembled conduction matrix $[k]_1$ for the rectangle.

(b) Compare the results of (a) with the conduction matrix of a rectangular element given by,

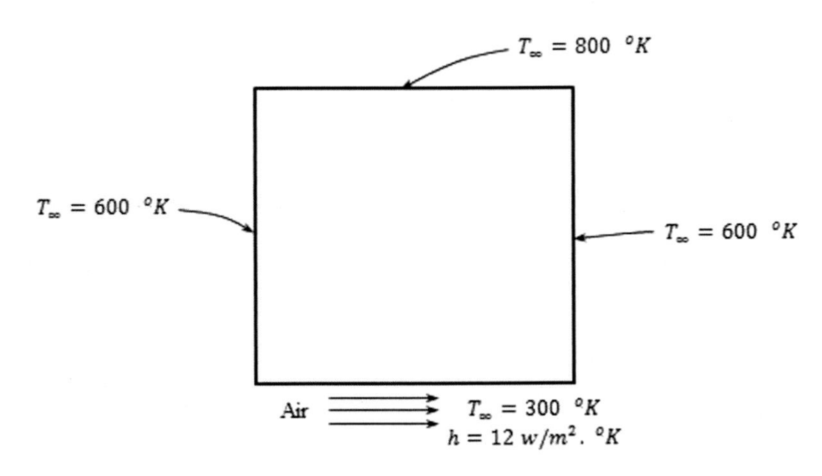

FIGURE P.11.15

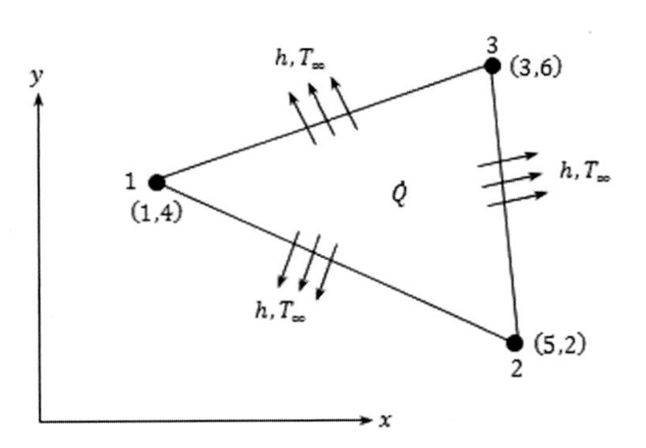

FIGURE P.11.16

$$[k]_1\text{rect.} = \frac{k+b}{4a}\begin{bmatrix} 1 & 1 & -1 & -1 \\ 1 & 1 & -1 & -1 \\ -1 & -1 & 1 & 1 \\ -1 & -1 & 1 & 1 \end{bmatrix}$$

P.11.15 An industrial furnace is supported on along column of fireclay brick (1×1 m) on a side Fig. P.11.15. During steady-state operation, installation is such that three surfaces of the column are maintained at 600°K while the remaining surface is exposed to an airstream for which $T_\infty = 300$°K and $h = 12\,\text{W/m}^2.$°k. Formulate the equilibrium equation required to obtain the column's temperature distribution and the heat rate to the airstream per unit length at the column. $k = 1\,\text{W/m}^2.$°k.

P.11.16 The coordinate of the nodes of a triangular element of thickness $= 2$ mm as shown in Fig. P.11.16 convection takes place form all three edges of the element. If $\dot{Q} = 200\ \text{W/cm}^3.$°C, $k = 100\ \text{W/m.}$°C, $h = 150\ \text{W/cm}^2.$°C and $T_\infty = 30$°C, determine the following matrices,

1. $[k]_1^e$, and $[k]_2^e$.

2. Vectors $\{F_{\text{th}}\}_1^e$, $\{F_{\text{th}}\}_2^e$.

Bibliography

[1] W.M. Rohsenow, H.Y. Choi, 2nd Printing Heat, Mass, and Momentum Transfer, Prentice-Hall, Englewood Cliffs, 1963.
[2] R.J. Melosh, Basis for derivation of matrices for the direct stiffness method, Journal of the American Institute of Aeronautics and Astronautics I 7 (1963) 1631–1637.
[3] J. Robinson, Understanding Finite Element Stress Analysis, Robinson and Associates, Wimborne, 1973.
[4] G.R. Cower, The shear coefficients in Timoshenko's beam theory, Journal of Applied Mechanics 33 (1966) 335–340.
[5] O.C. Ziewnkiewcz, R.L. Taylor, J.M. Too, Reduced integration technique in general analysis of plates and shells, International Journal of Numerical Engineering 3 (1971) 275–290.
[6] P. Pederson, Some properties of linear strain triangles and optimal finite element models, International Journal for Numerical Methods in Engineering 7 (1973) 415–430.
[7] L.J. Segerlind, Applied Finite Element Analysis, Wiley, New York, 1976.
[8] H.C. Huang, Finite Element Analysis for Heat Transfer, Theory and Software, Springer–Verlag, London, 1994.
[9] J.N. Reddy, Energy Principles and Variation Methods in Applied Mechanics, John Wiley and Sons, 2002.
[10] L.G. Tham, Y.K. Cheung, Numerical solution of heat conduction problems by parabolic time-space element, International Journal for Numerical Methods in Engineering 18 (1982) 467–474.
[11] W.R. Abdul-Majeed, M.J. Jweeg, A.N. Jameel, Restrained edges effect on the dynamics of thermoplastic plates under different end conditions, Al-Khwarizmi Engineering Journal 8 (2) (2012) 1–11.
[12] M.J. Jweeg, M. Al-Waily, A.K. Muhammad, K.K. Resan, Effects of temperature on the characterisation of a new design for a non-articulated prosthetic foot, in: IOP Conference Series: Materials Science and Engineering, 2nd International Conference on Engineering Sciences 433, 2018.

Index

Note: Page numbers followed by "f" refer to figures.